Phytochemicals in Soybeans

Functional Foods and Nutraceuticals

Series Editor

John Shi, Ph.D.
Guelph Food Research Center, Canada

Phytochemicals in Soybeans

Bioactivity and Health Benefits

Edited by
Yang Li and Baokun Qi

CRC Press
Taylor & Francis Group
Boca Raton London New York

CRC Press is an imprint of the
Taylor & Francis Group, an **informa** business

First edition published 2022
by CRC Press
6000 Broken Sound Parkway NW, Suite 300, Boca Raton, FL 33487-2742

and by CRC Press
4 Park Square, Milton Park, Abingdon, Oxon, OX14 4RN

© 2022 Taylor & Francis Group, LLC

CRC Press is an imprint of Taylor & Francis Group, LLC

Library of Congress Cataloging-in-PublicationData

Names: Li, Yang (Professor and doctoral supervisor of college of food science), editor. | Qi, Baokun, editor.
Title: Phytochemicals in soybeans bioactivity and health benefits / edited by Yang Li, Baokun Qi.
Description: First edition. | Boca Raton : CRC Press, 2022. | Includes bibliographical references and index.
Identifiers: LCCN 2021033272 | ISBN 9780367466619 (hardback) | ISBN 9781032169972 (paperback) | ISBN 9781003030294 (ebook)
Subjects: LCSH: Functional foods. | Phytochemicals. | Soybean--Health aspects. | Nutrition.
Classification: LCC QP144.F85 P4833 2022 | DDC 613.2--dc23
LC record available at https://lccn.loc.gov/2021033272

ISBN: 978-0-367-46661-9 (hbk)
ISBN: 978-1-032-16997-2 (pbk)
ISBN: 978-1-003-03029-4 (ebk)

DOI: 10.1201/9781003030294

Typeset in Times
by Deanta Global Publishing Services, Chennai, India

Contents

Series Editor's Preface

Chronic diseases such as cancer and cardiovascular diseases have the highest mortality of any illness around the world. Oxidative stress induced by reactive oxygen species (ROS) is considered to play an important role in major diseases. Mental stress, a high-energy diet, and food and environmental contamination may be the top three main contributory factors. Soybean products, which contain plenty of bioactive phytochemicals such as isoflavones, saponins, phytic acids, phytosterols, trypsin inhibitors, and peptides, can improve human health and nutrition. Extensive research has also implicated soybean phytochemicals as functioning in cholesterol reduction, cardiovascular disease prevention, diabetic symptoms prevention, bone loss prevention, and cancer prevention.

The book, titled *Phytochemicals in Soybeans: Bioactivity and Health Benefits*, covers all aspects of the bioactive health-promoting components of soybeans and soybean products and their health benefits. Full of up-to-date information, this new book joins the series on "Functional Foods and Nutraceuticals" from CRC Press. The book introduces readers to the latest information on bioactive health-promoting components of soybean foods and their health-promoting benefits, and also includes some emerging technologies for soybean processing and new products. The book can furnish a better understanding of some information on traditional ethnic functional foods and their dietary guides and applications. The information from the book will give readers an opportunity to meet special common concepts from around the world and promote soybean products as functional foods with advanced processing technology; it is also useful to meet the interests of academic and research groups. There is a great deal of interest now in the areas of nutraceuticals and functional foods on traditional ethnic health-promoting ingredients and foods for beneficial human health. The book can also serve as scientific reading material for college and university students majoring in food science, nutrition, pharmaceutical science, and botanical science.

The "Functional Foods and Nutraceuticals" series is appropriate for academic use; it will be a good scientific reference for food science and technology, nutrition science, and pharmaceutical science faculty and students. The series can also serve as a reference for food science professionals in either government or industry who are pursuing functional food, food ingredient developments, and R&D in food companies. Readers will obtain current and sound scientific knowledge and information relating to functional food products and new developments. It is our hope that the scientific community will appreciate our efforts in the preparation and promotion of this series and its impact on the advancing frontiers of functional foods and nutraceuticals.

John Shi

Series Editor, "Functional Foods and Nutraceuticals", CRC Press
Guelph Research and Development Centre
Agriculture and Agri-Food Canada and University of Guelph
Guelph, ON

Editors' Preface

Soybeans can be made into various traditional soy foods such as tofu, miso, soy drinks, breakfast cereals, energy bars, and soy cakes. Soybean represents an excellent source of high-quality protein with a low content in saturated fat. It also contains a great amount of dietary fiber and its isoflavone content makes it singular among other legumes. Extensive research has been carried out on the benefits of soybean, its characterization, and its positive health effects. Growing evidence has shown that consumption of soybeans may reduce the risk of osteoporosis, have a beneficial role in chronic renal disease, lower plasma cholesterol, exhibit antiatherosclerotic activity, and decrease the risk of coronary heart disease.

Soy foods are suggested to provide protective effects to reduce cancer development in the breast, intestine, liver, bladder, prostate, skin, and stomach. They do not contain cholesterol that is related to arteriosclerosis and heart attack. There are quite a number of bioactive minor components in soybean foods that can improve human health and nutrition. Soybean lecithin as a crucial ingredient in human neurons can renew cells, lower blood fat, and clear superoxide to protect humans from cardiovascular disease and senile dementia. Soybean saponin as a natural antioxidant can scavenge free radicals and mediate human immune systems to prevent cancers. Several types of polypeptides have been isolated from soybeans and are associated with specific effects on the human body. Dietary fibers from soybeans and by-products are a very useful supplement for senior people. All the above reasons make bioactive health-promoting components from soybeans and their products an excellent food source for human health and nutrition today.

Soybean foods have generated a lot of interest recently as a result of evidence that populations consuming large amounts of soybeans have a lower risk of some chronic diseases, most notably heart disease and cancer. Most of the studies have been focused on soybean protein as a possible source of prevention against cardiovascular disease. This positive effect may be due to a decrease in serum cholesterol concentrations. In addition, there are many studies on isoflavones, non-nutritive substances, associated with the prevention and treatment of different chronic diseases. Moreover, some studies have shown the health properties of soy dietary fiber. Therefore, it would be interesting to consider the replacement of animal-based foods with soybean foods in order to obtain some nutritional benefits.

Phytochemicals in Soybeans: Bioactivity and Health Benefits is a detailed description of the bioactive health-promoting components of soybeans, soybean products, and health benefits. The functional bioactive health-promoting components in soybeans and their products such as active protein and peptides, phospholipids, isoflavones, oligosaccharides, lecithin, polypeptides, dietary fibers, etc. have aroused increasing attention from consumers and the food industry as vital beneficial substances for human nutrition and health.

This book contains 22 chapters that cover the most recent information associated with soybean products and their health benefits. Topics of the book chapters include the role of soybeans in human nutrition and health, its composition and physicochemical properties, the action mechanism of its physiologic function, processing engineering technology, food safety, and quality control. The book covers the fundamental knowledge and up-to-date information of chemical characteristics of health-promoting components of soybean, and its impacts on human health, as well as its industrial prospects.

We thank all of the contributing authors for their cooperation in preparing the book chapters. We also thank Dr. Stephen Zollo, Dr. John Shi, and Dr. Laura Piedrahita for their encouragement and help. We are confident that the book can serve as an excellent reference for those interested in the science and technology of bioactive components from soybean products as health-promoting foods.

About the Editors

Yang Li is a professor and doctoral supervisor at the College of Food Science, Northeast Agricultural University, Harbin, China, and a young top-notch recruit of the Thousand Talents Program. He is a post-doctoral researcher at the National Soybean Engineering Technology Research Center, director of the Food Branch of the China Grain and Oil Society and the Natural Product Engineering Society of Heilongjiang Province, winner of the Heilongjiang Youth May Fourth Prize, member of the China Youth Science and Technology Association, vice president of the Heilongjiang Youth Technology Association, member of the Institute of Food Technologists, and reviewer of many prestigious periodicals and magazines at home and abroad. He was awarded the honorary titles of Distinguished Scholar, "academic backbone", Young Talent, Youth May 4th Medal honoree, and Pioneer of Scientific Research by Northeastern Agricultural University. Since 2004, he has engaged in research work on cereals, oil, and vegetable protein. He has published more than 200 papers, including SCI retrieval papers; 28 papers as first author/corresponding author; corresponding author of an ESI paper titled "Effects of ultrasound on the structure and physical properties of black bean protein isolates", which was cited multiple times at the top 1% of excellent papers of the same publication year in the academic field of agricultural sciences; 110 EI retrieval papers, including 45 as first author/corresponding author; 138 core retrieval papers, including 56 as first author/corresponding author; teaching reform of the 13; and eight books, including one monograph (alone) and three as chief editor. He has obtained 48 authorized patents, including 43 national invention patents (the first inventor of 16, the second inventor of 24), one application for an international patent, and 11 patents for the invention of technology transfer.

Baokun Qi is an associate professor at the College of Food Science, Northeast Agricultural University, Harbin, China. Since 2013, he has presided over seven national projects, including for the National Natural Science Foundation of China, and participated in eight national projects and six provincial projects. He has published 34 papers as the first and corresponding author, including ten SCI retrieval papers and 16 EI retrieval papers, and participated in writing three teaching materials as chief editor. He has obtained 43 authorized patents, including 11 patents for the invention of technology transfer. He has received seven awards including the second prize of Science and Technology Progress in Heilongjiang Province, the first prize of the Inventive Achievement Award of the China Invention Association, the first prize of Heilongjiang Agricultural Science and Technology, the third prize of the Shen Nong China Agricultural Science and Technology Award, and the first prize of Technology Invention of Chinese Institute of Food Science and Technology (ranked second); he has also obtained two result identifications. In November 2015, Heilongjiang Assy Food Technology Co., Ltd. was founded during the period of his doctoral studies. The company focused on providing technical consulting, technical services, transformation of results, food production and sales, etc. In 2016, he was hired as the entrepreneurial supervisor of Northeast Agricultural University. In 2018, he won the Gold Medal of the 4th Heilongjiang Province Internet + College Entrepreneurship Competition and the National Silver Award (ranked first).

Contributors

Md Minhajul Abedin
Institute of Bioresources and Sustainable Development
Regional Centre Sikkim
Tadong, India

Precious E. Agboinghale
Department of Biochemistry
Afe Babalola University
Ado-Ekiti, Nigeria

Akhunzada Bilawal
College of Food Science
Northeast Agricultural University
Harbin, China

Jane Mara Block
Department of Food Science and Technology
Federal University of Santa Catarina
Florianópolis, Brazil

Oriol Comas-Basté
Department of Nutrition, Food Sciences, and Gastronomy
Food and Nutrition Torribera Campus
University of Barcelona
Santa Coloma de Gramenet, Spain

Adriano Costa de Camargo
Nutrition and Food Technology Institute
University of Chile
Santiago, Chile

Runu Chakraborty
Department of Food Technology and Biochemical Engineering
Jadavpur University
Kolkata, India

Rounak Chourasia
Institute of Bioresources and Sustainable Development
Regional Centre Sikkim
Tadong, India

Jian-Yong Chua
Department of Food Science and Technology
National University of Singapore
Singapore

Joana Costa
REQUIMTE-LAQV
Faculdade de Farmácia
Universidade do Porto
Porto, Portugal

Jelena Cvejić
Faculty of Medicine
University of Novi Sad
Novi Sad, Serbia

Renan Danielski
Department of Biochemistry
Memorial University of Newfoundland
St. John's, NL

Philip Davy
School of Environmental and Life Sciences
The University of Newcastle
Ourimbah, Australia

Sriloy Dey
Department of Food Science
University of Arkansas
Fayetteville, AR

Sakamon Devahastin
Advanced Food Processing Research Laboratory
Department of Food Engineering
King Mongkut's University of Technology Thonburi
Bangkok, Thailand

Abhishek Dutta
Department of Chemical Engineering
İzmir Institute of Technology
İzmir, Turkey

Xuejing Fan
College of Food Science
Northeast Agricultural University
Harbin, China

Zhen Feng
College of Food Science
Northeast Agricultural University
Harbin, China

Alejandra García-Alonso
Department of Nutrition and Food Science
Complutense University of Madrid
Madrid, Spain

Navam Hettiarachchy
Department of Food Science
and
Institute of Food Science and Engineering
University of Arkansas
Fayetteville, AR

Darry L. Holliday
Food Science Program
University of Holly Cross
New Orleans, LA

Hao Hu
College of Food Science and Technology
Huazhong Agricultural University
Wuhan, China

Elza Iouko Ida
Department of Food Science & Technology
The State University of Londrina
Paraná, Brazil

Ignasius Radix A.P. Jati
Department of Food Technology
Widya Mandala Surabaya Catholic University
Surabaya, Indonesia

Zhanmei Jiang
College of Food Science
Northeast Agricultural University
Harbin, China

Hyun Jo
Department of Applied Biosciences
Kyungpook National University
Daegu, Republic of Korea

Su Jin Jung
Clinical Trial Center for Functional Food
Jeonbuk National University Hospital
Jeonju, Republic of Korea

Ibrahim Khalifa
Food Technology Department
Benha University
Moshtohor, Egypt

Milica Atanacković Krstonošić
Faculty of Medicine
University of Novi Sad
Novi Sad, Serbia

M. Luz Latorre-Moratalla
Department of Nutrition, Food Sciences, and
Gastronomy
Food and Nutrition Torribera Campus
University of Barcelona
Santa Coloma de Gramenet, Spain

Jeong-Dong Lee
Department of Applied Biosciences
Kyungpook National University
Daegu, Republic of Korea

Jiaxin Li
College of Food Science and Engineering
Jilin Agricultural University
Changchun, China

Yang Li
College of Food Science
Northeast Agricultural University
Harbin, China

Hanifah Nuryani Lioe
Department of Food Science and Technology
IPB University
Bogor, Indonesia

Gefei Liu
College of Food Science
Northeast Agricultural University
Harbin, China

He Liu
College of Food Science and Engineering
Bohai University
Jinzhou, China

Shao-Quan Liu
Department of Food Science and Technology
National University of Singapore
Singapore

Isabel Mafra
REQUIMTE-LAQV
Faculdade de Farmácia
Universidade do Porto
Porto, Portugal

Inmaculada Mateos-Aparicio
Department of Nutrition and Food Science
Complutense University of Madrid
Madrid, Spain

Mira Mikulić
Faculty of Medicine
University of Novi Sad
Novi Sad, Serbia

Runni Mukherjee
Department of Food Technology and
Biochemical Engineering
Jadavpur University
Kolkata, India

Soma Mukherjee
Food Science Program
University of Holly Cross
New Orleans, LA

Nelly C. Muñoz-Esparza
Department of Nutrition, Food Sciences, and
Gastronomy
Food and Nutrition Torribera Campus
University of Barcelona
Santa Coloma de Gramenet, Spain

Chalida Niamnuy
Department of Chemical Engineering
Kasetsart University
Bangkok, Thailand

Itaciara Larroza Nunes
Department of Food Science and Technology
Federal University of Santa Catarina
Florianópolis, Brazil

Oluwaseyi A. Ogunrinola
School of Nutrition Sciences
University of Ottawa
Ottawa, ON

Ikenna C. Ohanenye
School of Nutrition Sciences
University of Ottawa
Ottawa, ON

Innocent U. Okagu
Department of Biochemistry
University of Nigeria
Nsukka, Nigeria

Loreni Chiring Phukon
Institute of Bioresources and Sustainable
Development
Regional Centre Sikkim
Tadong, India

Baokun Qi
College of Food Science
Northeast Agricultural University
Harbin, China

Yali Qiao
College of Food Science
Northeast Agricultural University
Harbin, China

Amit Kumar Rai
Institute of Bioresources and Sustainable
Development
Regional Centre Sikkim
Tadong, India

Srinivasan J. Rayaprolu
Liberty Pharma
Liberty, MO

Araceli Redondo-Cuenca
Department of Nutrition and Food Science
Complutense University of Madrid
Madrid, Spain

Dinabandhu Sahoo
Department of Botany
University of Delhi
New Delhi, India

Darija Sazdanić
Faculty of Medicine
University of Novi Sad
Novi Sad, Serbia

Swati Sharma
Institute of Bioresources and Sustainable
Development
Regional Centre Sikkim
Tadong, India

Dong Hwa Shin
Shin Dong Hwa Food Research Institute
Jeonbuk National University
Jeonju, Republic of Korea

Hong Song
College of Food Science and Engineering
Bohai University
Jinzhou, China

Jong Tae Song
Department of Applied Biosciences
Kyungpook National University
Daegu, Republic of Korea

Syada Nizer Sultana
Department of Applied Biosciences
Kyungpook National University
Daegu, Republic of Korea

Mohammed Sharif Swallah
College of Food Science and Engineering
Jilin Agricultural University
Changchun, China

Ahmed Taha
Department of Food Science
Faculty of Agriculture (Saba Basha)
Alexandria University
Alexandria, Egypt

and

State Research Institute
Center for Physical Sciences and Technology
Vilnius, Lithuania

Badrut Tamam
Department of Nutrition
Polytechnic of Health
Denpasar, Indonesia

Gerson Lopes Teixeira
Department of Food Science and Technology
Federal University of Santa Catarina
Florianópolis, Brazil

María Dolores Tenorio-Sanz
Department of Nutrition and Food Science
Complutense University of Madrid
Madrid, Spain

Natalia Toro-Funes
Universidad Internacional de Valencia
Valencia, Spain

Chibuike C. Udenigwe
School of Nutrition Sciences

and

Department of Chemistry and Biomolecular
Sciences
University of Ottawa
Ottawa, ON

M. Teresa Veciana-Nogués
Department of Nutrition, Food Sciences, and
Gastronomy
Food and Nutrition Torribera Campus
University of Barcelona
Santa Coloma de Gramenet, Spain

M. Carmen Vidal-Carou
Department of Nutrition, Food Sciences, and
Gastronomy
Food and Nutrition Torribera Campus
University of Barcelona
Santa Coloma de Gramenet, Spain

Caterina Villa
REQUIMTE-LAQV
Faculdade de Farmácia
Universidade do Porto
Porto, Portugal

María José Villanueva-Suarez
Department of Nutrition and Food Science
Complutense University of Madrid
Madrid, Spain

Weng Chan Vong
Department of Food Science and Technology
National University of Singapore
Singapore

Quan V. Vuong
School of Environmental and Life Sciences
The University of Newcastle
Ourimbah, Australia

Sainan Wang
College of Food Science and Engineering
Jilin Agricultural University
Changchun, China

Shengnan Wang
College of Food Science and Engineering
Bohai University
Jinzhou, China

Lina Yang
College of Food Science and Engineering
Bohai University
Jinzhou, China

Hansong Yu
College of Food Science and Engineering
Jilin Agricultural University

and

Division of Soybean Processing
Soybean Research & Development Centre
Chinese Agricultural Research System
Changchun, China

Shuang Zhai
China (HeiLongJiang) Intellectual Property
Protection Center
Harbin, China

Dayu Zhou
College of Food Science and Engineering
Bohai University
Jinzhou, China

1 Health Perspectives on Soy Isoflavones

*Mira Mikulić, Milica Atanacković Krstonošić,
Darija Sazdanić, and Jelena Cvejić*

CONTENTS

1.1 INTRODUCTION

Isoflavones represent a subclass of chemical compounds with estrogen-like activities, classified as phytoestrogens. They are usually found in different plant species of the *Fabaceae* family, being the most abundant in soybeans, red clover, alfalfa, kidney beans, chickpeas, or kudzu (Mortensen et al., 2009; Mazur et al., 1998). Soybeans (*Glycine max* [L.], *Fabaceae*) are generally considered the major source of isoflavones in the human diet (Messina et al., 2017).

DOI: 10.1201/9781003030294-1

Epidemiological investigations, especially in soy-consuming Asian countries, have provided much data relating to the interconnection between soy intake and lower incidence of certain chronic and hormone-dependent diseases (Messina, 2016; Applegate et al., 2018; Somekawa et al., 2001). The potential health benefits attributed specifically to soy isoflavones encompass cancer prevention and treatment, menopausal symptoms alleviation, as well as lowering the risk of osteoporosis, cardiovascular diseases, and some other chronic diseases (Messina, 2016; Xiao et al., 2017). The consumer interest in soy-based food products, as well as isoflavones-containing supplements, has increased significantly in recent decades. The reason is the higher awareness of a wide population related to the purported health benefits of isoflavones (Messina et al., 2017).

Isoflavones belong to a class of compounds called phytoalexins, which are synthesized and accumulated in plants during stress conditions, thus having a protective role (Boue et al., 2009). Furthermore, isoflavones are generated in plants by the same phenylpropanoid biosynthetic pathway as flavonoids. These compounds are synthesized from phenylalanine and four molecules of malonyl CoA through several reactions leading to a final formation of 3-phenyl-4H-1-benzopyran-4-one (3-phenylchromen-4-one IUPAC name) (Figure 1.1) (Barnes, 2010). Although nonsteroidal, such compounds bear remarkable similarity in structure to 17-β-estradiol and therefore are considered naturally occurring plant estrogens or phytoestrogens (Pilsakova et al., 2010).

Soy isoflavones include three groups of compounds in four chemical forms: Aglycones genistein (5,7,4'-trihydroxyisoflavone), daidzein (7,4'-dihydroxyisoflavone) and glycitein (7,4'-dihydroxy-6-methoxyisoflavone); the corresponding 7-O-β-glucosides (genistin, daidzin, glycitin), malonyl glycosides (6''-O-malonylgenistin, 6''-O-malonyldaidzin, 6''-O-malonylglycitin), and acetyl glycosides (6''-O-acetylgenistin, 6''-O-acetyldaidzin, 6''-O-acetylglycitin) (Lee et al., 2004) (Figure 1.2).

Total isoflavone content in soybeans varies greatly, and authors from different regions reported values from 0.4 up to 9.5 mg/g of soybean seed, with usual average values ranging from 1 to 4 mg/g (Chiari et al., 2004; Cvejić et al., 2009; Lee et al., 2003; Wiseman et al., 2002). Regarding the isoflavone composition, from 12 different isoflavones generally present in soybeans, malonyl daidzin and malonyl genistin are the compounds usually found in the highest concentration (Bursać et al., 2017; Cvejić et al., 2011). Malonyl forms are known to be dominant in raw and unprocessed seeds. Actually, it is reported that the sum of ß-glucosides and malonyl glucosides represents 90% or more of the total isoflavones (Bursać et al., 2017; Lee et al., 2010). Hence, the least present forms in soybeans are aglycons or acetyl glucosides. Among three isoflavone types (compounds derived from the same aglycone) present in soy, daidzeins, or genistens are the most abundant, while glyciteins are generally the least present isoflavone type (on average 5–15% of total isoflavones) (Cvejić et al., 2011; Tepavčević et al., 2010).

It is known that various factors like genotype, environmental factors, location, and crop season can influence this characteristic (Hoeck et al., 2000; Tsai et al., 2007; Riedl et al., 2007). Namely, genetic factors are recognized as the most important ones, and it is indicated that phytoestrogen content and composition in soy could be inheritable traits (Cvejić et al., 2011; Miladinović et al., 2019), implying that breeding of cultivars with desirable isoflavone concentration is possible.

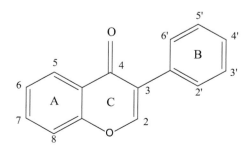

FIGURE 1.1   General structure of isoflavones.

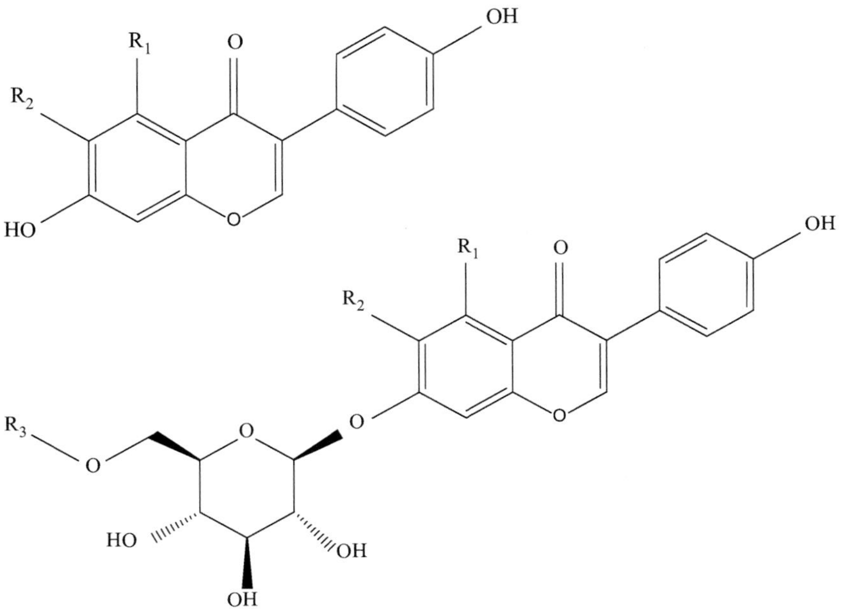

FIGURE 1.2 Chemical structures of soy isoflavones: Aglycones, 7-O-β-glycosides, acetyl glycosides, and malonyl glycosides.

	Isoflavone	R_1	R_2	R_3
Aglycones	Daidzein	H	H	-
	Genistein	OH	H	-
	Glycitein	H	OCH₃	-
7-O-β-glycosides	Daidzin	H	H	H
	Genistin	OH	H	H
	Glycitin	H	OCH₃	H
Acetyl glycosides	6''-O-acetyldaidzin	H	H	COCH₃
	6''-O-acetylgenistin	OH	H	COCH₃
	6''-O-acetylglycitin	H	OCH₃	COCH₃
Malonyl glycosides	6''-O-malonyldaidzin	H	H	COCH₂COOH
	6''-O-malonylgenistin	OH	H	COCH₂COOH
	6''-O-malonylglycitin	H	OCH₃	COCH₂COOH

1.2 MECHANISM OF ACTION

1.2.1 ESTROGENIC AND ANTI-ESTROGENIC ACTIVITY

The occurrence of plant-derived compounds that are able to exert estrogenic and anti-estrogenic activities has been known since the 1940s when it was observed that animals fed with isoflavone-rich subterranean clover developed symptoms of diverse reproductive disorders (Bennetts et al., 1946).

Even though isoflavones are non-steroidal compounds, due to their molecular structure and weight similar to endogenous estrogen (17ß-estradiol) (Figure 1.3), they are able to bind to estrogen receptors (ER) and therefore are referred to as estrogen-like molecules or diphenolic non-steroidal estrogens (Pilsakova et al., 2010).

When discussing the structural similarity of isoflavones (e.g. genistein) to 17ß-estradiol, one of the usual presentations of genistein used for comparison is orientation A in Figure 1.3. Barnes (2010) suggested that the proper way of genistein orientation is actually B (Figure 1.3), according to

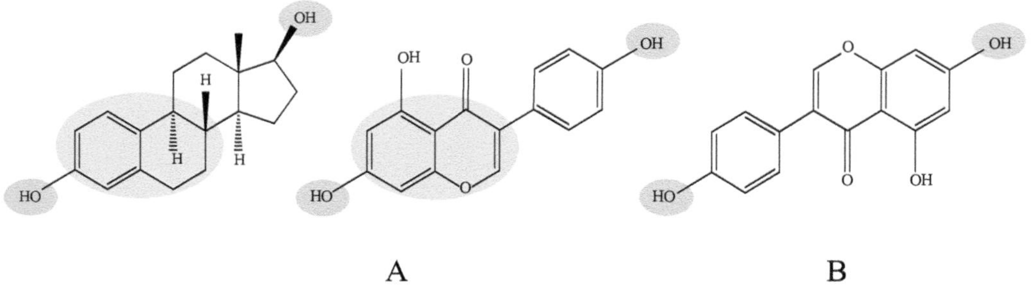

A   B

17ß-estradiol   genistein (presented in two different orientations – A, B)

FIGURE 1.3   Structural similarity of estradiol and genistein.

results obtained by crystallographic study (Pike et al., 1999). A significant characteristic of isofla-vones is the presence of a phenolic ring in the structure, which principally represents a prerequisite for binding to the estrogen receptors (Setchell, 1998).

Estrogen receptors are intranuclear binding proteins and members of the nuclear steroid receptor superfamily, which act as ligand-dependent transcription factors. There are two estrogen receptor subtypes, alpha and beta (ERα and ERβ). Estrogen receptor subtypes are highly homologous, with the overall homology of the proteins' ligand binding sites being 55% (Kuiper et al., 1996). However, a divergence of a few amino acids of the receptors induces different affinities toward various ligands (Kuiper et al., 1997).

Upon binding to a ligand, ER undergoes conformational changes and dimerizes, allowing the receptor-ligand complex to attach to specific DNA regulatory sequences, the estrogen response elements (ERE) in the promoter region of target genes. Through this action, ER modulates the rate of transcription of various specific target genes (Kuiper et al., 1997; Zhao et al., 2008; Welboren et al., 2009).

In addition, ERα can bind indirectly to the DNA in the nonclassical pathway through binding to other transcription factors, such as specificity protein 1 (Sp1) or nuclear factor kappa b (NF-κB) (Welboren et al., 2009).

Isoflavones can bind to both estrogen receptor subtypes, alpha (ERα) and beta (ERβ), with signif-icantly higher binding and transactivation affinity toward ERβ isoform (Kuiper et al., 1996; Kuiper et al., 1998; Hwang et al., 2006). Moreover, the structural features of isoflavone molecules, such as the particular position and number of hydroxyl groups, seem to govern the binding affinity rate. Particularly, the elimination of one hydroxyl group leads to the decreased binding affinity of daid-zein compared to genistein (Kuiper et al., 1998).

Isoflavones exert dual biological function depending on coexisting estrogen concentrations since isoflavones compete with endogenous estrogen for binding to ER. At the physiologic levels of estrogen, binding of less potent isoflavones to ER represses full estrogen activity and therefore they act as antagonists. Furthermore, such an inhibitory effect on estrogen action is more promi-nent with ERβ (Hwang et al., 2006). In contrast, in a state of lower levels of endogenous estrogen, isoflavones exhibit estrogenic activity acting as ER agonists (Hwang et al., 2006; Pilsakova et al., 2010).

Hence, different tissue distribution and relative binding affinities of estrogen receptor iso-forms, as well as estrogen-level-dependent action of isoflavones, determine specific biological activities of soy phytoestrogens. In this regard, isoflavones are considered selective estrogen receptor modulators that are expected to have tissue-selective beneficial effects (Setchell, 2017). Therefore, in postmenopausal women, soy isoflavones may display estrogenic activity and reduce the frequency of hot flushes (Xiao et al., 2017; Setchell, 2017). Isoflavones may affect bone remodeling in osteoporosis through the modulation of target genes transcription via estrogen receptors in osteoblasts (Xiao et al., 2017). Isoflavones are considered effective in skin health and quality maintenance due to the high expression of ERβ in the skin tissue. Consequently, an expression of target genes such as collagen and elastin may be increased (Setchell, 2017). Moreover, given that ERα activation stimulates and ERβ activation inhibits cell proliferation in the breast tissue, isoflavones may impact the proliferation rate of some breast cancer types (Křížová et al., 2019).

In addition to direct binding to estrogen receptors, isoflavones can modulate the levels of reproductive hormones. They inhibit several important enzymes involved in estrogen biosynthesis. Firstly, soy isoflavones may regulate testosterone biosynthesis in part due to inhibition of 3β-hydroxysteroid dehydrogenase, which catalyzes one of the reactions that convert the steroid cholesterol into the sex hormone testosterone (Hu et al., 2010; Le Bail et al., 2000). Moreover, soy isoflavones genistein and daidzein might be able to inhibit aromatase activity (by enzyme activity and gene expression), the enzyme responsible for the conversion of androstenedione and testosterone to estrone and 17ß-estradiol, respectively (Amaral et al., 2017; Lephart, 2015). Furthermore, soy isoflavones are able to reduce the conversion of testosterone to more potent androgen 5α-dihydrotestosterone by the inhibition of 5α-reductase (Bae et al., 2012).

1.2.2 ANTIOXIDANT ACTIVITY

Free radicals and reactive oxygen species (ROS) such as superoxide anion, hydroxyl radical, and hydrogen peroxide are continuously produced during normal oxygen metabolism in humans. The excessive ROS production or diminished cellular antioxidant capacity induces oxidative stress and subsequent oxidative damage to membrane lipids, proteins, and nucleic acids (Park et al., 2011; Yu et al., 2016a; Xiao et al., 2017). Maintaining good health and prevention of various diseases caused by oxidative stress could be achieved by taking food containing natural antioxidants. The results of *in vitro* and *in vivo* studies have indicated that isoflavones exert strong antioxidant activity (Mirahmadi et al., 2017; Meng et al., 2017; Lee & Park, 2018). There are several suggested mechanisms of antioxidant actions of soy isoflavones.

In vitro studies have shown that genistein, daidzein, and equol are effective in scavenging free radicals, superoxide anion radicals, hydroxyl radicals, and hydrogen peroxide. Isoflavone hydroxyl groups 7-OH and 4'-OH are responsible for their oxygen radical scavenging ability, and the B ring of the structure seems to be the most active site (Figure 1.1 and 1.2). Moreover, the antioxidant capacity of the phytoestrogens tested decreased in the following order: Equol > genistein > daidzein (Kladna et al., 2016). The absence of a 5-OH hydroxyl group in the structure of genistein appears to cause its lower scavenging potential. Moreover, equol exhibits higher antioxidant potency likely due to the loss of the 2,3-double bond and the 4-oxo group (Rimbach et al., 2003). Increased reactive oxygen species in diabetic mice is abolished by genistein due to its significant scavenging capacity (Valsecchi et al., 2011). In addition, soy isoflavones exert antioxidant activity through the attenuation of gene expression of ROS-generating enzymes, such as NOX2 (Meng et al., 2017).

Malondialdehyde (MDA) is considered a biomarker of oxidative damage to lipids since it is produced by lipid peroxydation. Elevated MDA levels indicate an increased lipid peroxidation which is subsequently able to induce cellular dysfunction and tissue injury (Islam et al., 2020). Soy isoflavones are effective in reducing elevated MDA levels (Morvaridzadeh et al., 2020; Mirahmadi et al., 2017).

Furthermore, isoflavones possess the ability to restore the impaired activity of antioxidant enzymes including catalase, superoxide dismutase, and glutation peroxidase (Valsecchi et al., 2011; Yoon & Park, 2014; Mirahmadi et al., 2017; Meng et al., 2017). Soybean isoflavones might increase the antioxidant capacities through superoxide dismutase up-regulation via nuclear factor erythroid 2-related factor 2 (Nrf2) (Li & Zhang, 2017). Soy isoflavones may induce transcription of the transcriptional coactivator peroxisome proliferator-activated receptor γ coactivator 1 alpha (PGC1-α), subsequently increasing superoxide dismutase and glutation peroxidase levels (Pang et al., 2020). Moreover, daidzein displays antioxidant activity in HK-2 kidney proximal tubular cell line by up-regulating the expression of manganese superoxide dismutase via FOXO3 transcription factor (Lee & Park, 2018).

Additionally, soy isoflavones induce protein expression of hem oxygenase-1 (HO-1) and NAD(P) H: Quinone oxidoreductase 1 (NQO1) via the Nrf2/ARE pathway or PPARγ. Such proteins are phase II antioxidant enzymes, representing important components of the cellular defense mechanism against oxidative stress (Siow et al., 2007; Park et al., 2011; Zhai et al., 2013; Zhang et al., 2013a). These antioxidant activities of isoflavones could be important in cardiovascular protection and may have therapeutic potential for neurodegenerative diseases related to oxidative stress. (Siow et al., 2007; Park et al., 2011; Zhang et al., 2013a)

1.2.3   ANTI-INFLAMMATORY ACTIVITY

Inflammation is a biological reaction of tissues to pathogens, irritants, or injury. Namely, the benefit of inflammation is life-saving and the inflammation is considered a protective process when it is localized, specific, rapid, and well controlled. However, uncontrolled inflammation contributes to the development of various diseases, including diabetes, atherosclerosis, cancer, etc. (Wellen & Hotamisligil, 2005; Libby, 2012; Mantovani et al., 2008; Yu et al., 2016a).

The anti-inflammatory properties of isoflavones have been recently reported and even though there are still mechanisms underlying anti-inflammatory effects that are not elucidated, some of them have been revealed (Yu et al., 2016a).

Isoflavones are able to inhibit typical pro-inflammatory enzymes involved in the generation of relevant inflammatory mediators. Genistein inhibits phospholipase A2, a key enzyme in the inflammatory reactions that trigger the release of arachidonic acid from membrane phospholipids (Dharmappa et al., 2010). After release, free arachidonic acid is afterward metabolized by cyclooxygenase (COX) or lipoxygenase (LOX) and the pro-inflammatory lipid mediators are generated (Mukherjee et al., 1994). An inducible isoform of cyclooxygenase enzymes cyclooxygenase 2 (COX2) catalyzes prostaglandin synthesis in inflammatory cells as a response to endotoxins or pro-inflammatory cytokines (Seibert et al., 1994). Genistein has the ability to suppress COX2 expression at the transcriptional level resulting in reduced prostaglandin levels (Jeong et al., 2014; Mirahmadi et al., 2017; Takaoka et al., 2018). Furthermore, lipoxygenase catalyzes the production of leukotrienes from released arachidonic acid (Lewis et al., 1990). Genistein and daidzein exhibit noncompetitive lipoxygenase inhibition with genistein being an almost twice more potent inhibitor than daidzein (Mahesha et al., 2007; Vicas et al., 2011). A catechol group in ring A, together with a reduction of ring C in an isoflavone structure are important features for the inhibitory activity of isoflavones (Mascayano et al., 2013).

Moreover, isoflavones inhibit nitric oxide (NO) overproduction by inducible nitric oxide synthase (iNOS) without affecting endothelial (eNOS) and neuronal (nNOS) nitric oxide synthases. Soy isoflavones attenuate excessive NO production by inhibition of both the activity and expression of iNOS (Sheu et al., 2001; Jeong et al., 2014; Mirahmadi et al., 2017).

Reduced production of pro-inflammatory cytokines and chemokines is a main mechanism by which isoflavones exert anti-inflammatory effects (Yu et al., 2016a). The expression of various inflammatory cytokines is reduced by soy isoflavones (Takaoka et al., 2018; Li et al., 2014; Jeong et al., 2014). Since the transcription factor NF-κβ has a substantial role in inflammation, isoflavones might exert their anti-inflammatory effects by suppressing the phosphorylation of NF-κβ inhibitor and by decreasing the translocation of free NF-κβ into the nucleus (Li et al., 2014; Takaoka et al., 2018). Moreover, the anti-inflammatory properties of isoflavones might be attributed to the interruption of the ROS/Akt/NF-κβ signaling pathway (Li et al., 2014). Genistein also inhibits the activation of a toll-like receptor which leads to downstream inhibition of NF-κβ activation and thus consequently abrogates overproduction of inflammatory mediators and cytokines (Jeong et al., 2014). Moreover, daidzein suppresses enhanced expression of cytokines and chemokines via activation of peroxisome proliferator-activated nuclear receptors α and γ (PPAR-α/γ) (Sakamoto et al., 2016).

1.2.4 OTHER MECHANISMS

Administration of soy protein with genistein induced a decrease in fasting blood glucose and a concomitant increase in plasma insulin levels in diabetic rats, suggesting the protective effect on β-cell function in diabetes (Lee, 2006). Indeed, soy isoflavones improve islet cell survival and pro-liferation, thus preserving islet mass in addition to facilitating insulin production (Yang et al., 2011; Fu et al., 2012; El-Kordy & Alshahrani, 2015; Yousefi et al., 2017). Moreover, genistein directly acts on pancreatic β-cells exerting insulinotropic activity. The glucose-stimulated insulin secretion is augmented through the stimulation of the cyclic adenosine monophosphate (cAMP) / protein kinase A (PKA) signaling pathway (Ohno et al., 1993; Liu et al., 2006). Soy isoflavones might also improve insulin sensitivity and increase glucose uptake into skeletal muscles (Cederroth et al., 2008). Genistein directly stimulates glucose uptake by increasing the translocation of glucose transporter 4 to the plasma membrane (Ha et al., 2012).

The hypocholesterolemic effects of isoflavones (Perumal et al., 2019) could be partly due to the modulation of cholesterol metabolism-related hormones, including steroid and thyroid hormones (Rababah et al., 2015). Elevated hepatic triiodothyronine concentrations induced by isoflavones might potentiate the degradation of cholesterol to bile acids and further excretion (Šošić-Jurjević et al., 2019). Moreover, these phytochemicals up-regulate the expression of hepatic LDL-receptors, which might contribute to LDL-cholesterol clearance (Tang et al., 2015). Soy isoflavones signifi-cantly improve serum triglyceride levels via activation of PPARα receptors as well as via a PPARα-independent pathway (Mezei et al., 2006). Since oxidized LDL initiates the chronic inflammatory process involved in the pathogenesis of atherosclerosis, the ability of soy isoflavones to provide protection against oxidative modification of LDL represents an important role in atherosclerosis prevention (Cano et al., 2010; Nagarajan, 2010; Yamakoshi et al., 2000). Moreover, soy isoflavones may reduce the levels of electronegative LDL subfraction, which has pro-inflammatory properties and shows oxidative changes (Damasceno et al., 2007).

Soy isoflavones are attracting heightened attention due to their antihypertensive effects (Liu et al., 2012). The ability of soy isoflavones to modulate the vascular tone is one of the mechanisms involved in blood pressure control (Martin et al., 2008). The vascular tone is determined by a delicate balance between vasoconstrictor and vasodilator influences. Soy isoflavones favor vascular tone reduction through several mechanisms. Soy isoflavones increase nitric oxide (NO) production in endothelial cells, which subsequently causes smooth muscle cells' relaxation and concomitant vasodilation (Sun et al., 2015; Hall et al., 2008). They exert genomic stimulation of eNOS (endothe-lial nitric oxide synthase) expression, resulting in enhanced NO production and alleviated hyper-tension (Si & Liu, 2008; Mahn et al., 2005; Ohkura et al., 2015; Jourkesh et al., 2017). Moreover, non-genomic eNOS activation is another mechanism through which soy isoflavones may stimulate NO production (Yang et al., 2010; Joy et al., 2006). Isoflavones may attenuate nitric oxide inactiva-tion by reactive oxygen species due to their antioxidant activity. Therefore, they increase nitric oxide half-life and bioactivity (Park et al., 2005; Vera et al., 2005; Carr & Frei, 2000). Furthermore, since prostacyclin (prostaglandin I2, PGI2) is well-known as one of the important endothelial-derived relaxing factors (Loh et al., 2018), soy isoflavones may exert vasodilating effects by stimulating prostacyclin production in endothelial cells (García-Martínez et al., 2003; Hermenegildo et al., 2005). Isoflavones might also affect the levels of endothelial-derived constrictor factors, such as a decrease of endothelin-1, endothelin-2, or endothelin-converting enzyme-1 (Squadrito et al., 2002; Ambra et al., 2006). In addition, endothelial-independent mechanisms are also involved in the vas-cular tone modulation of soy isoflavones (Lee & Man, 2003). The inhibitory effect on calcium channels contributes to endothelial-independent vasorelaxation induced by soy isoflavones (Figtree et al., 2000; Li et al., 2004). Soy isoflavones activate calcium-activated potassium channels in the cell membrane of the vascular smooth muscle, which also leads to endothelial-independent relaxation (Nevala et al., 2001).

Anticancer activities of isoflavones in various cancers are a result of multiple mechanisms. Uncontrolled proliferation of cancer cells is inhibited by soy isoflavones through the modulation of expression and function of different cell cycle regulators (cyclins, cyclin-dependent kinases, and cyclin-dependent kinase inhibitors), tumor suppressors, and growth factors (Lee et al., 2012; Seo et al., 2011; Choi & Kim, 2008). Moreover, these phytochemicals induce apoptosis by down-regulation of the antiapoptotic and up-regulation of proapoptotic genes (Chen et al., 2015a; Prietsch et al., 2014). Epigenetic changes play an important regulatory role in the expression of various genes associated with neoplastic processes. Isoflavones enhance the transcription of several tumor suppressor genes through epigenetic mechanisms (Bosviel et al., 2012; Adjakly et al., 2011; Vardi et al., 2010). Soy isoflavones can exhibit anti-angiogenic activity by inhibiting central mediators of angiogenesis, such as vascular endothelial growth factor (Yazdani et al., 2016). Inhibition of cancer invasion and metastatic spread of primary tumors by isoflavones is associated with the down-regulation of matrix metalloproteases, involved in the degradation of extracellular matrix and cancer cells invasion (Hussain et al., 2012; Magee et al., 2014; Yazdani et al., 2016). It is also demonstrated that soy isoflavones could be potent agents for the enhancement of the efficacy of radio therapy. The purported mechanism includes sensitizing tumor cells to radiation, as well as protection of normal tissue by decreasing inflammatory response induced by radiation (Hillman, 2018).

1.3 METABOLISM AND BIOAVAILABILITY

Isoflavone metabolism has been extensively studied, and it has a very important role in revealing the true health potential of these compounds in the human organism. The short presentation of metabolism is given in Figure 1.4.

It is well known that conjugated forms of isoflavones, abundant in raw soybean, cannot be absorbed intact in the gastrointestinal tract of healthy adults. Thus, upon consumption, phase I

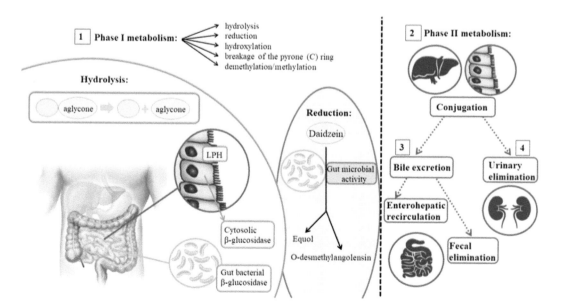

FIGURE 1.4   Isoflavone metabolism. *LPH - lactate-phlorizin-hydrolase.

metabolism takes place, and first includes hydrolysis of glucoside isoflavone forms – the step essential for their absorption (Setchell et al., 2002a). Deglycosylation is quite efficient and takes place along the gastrointestinal tract by the activity of the brush border membrane (by the mucosal lactate-phlorizin-hydrolases – LPH) and the gut bacterial ß-glucosidase activity. Additionally, glucosides can be transported into the gut epithelium and hydrolyzed by cytosolic ß-glucosidases (Németh et al., 2003; Day et al., 1998; Day et al., 2000). Aglycones are absorbed by passive diffusion due to their lipophilicity (Setchell et al., 2002a).

Plasma and urine isoflavone concentrations peak 1–2 h and again 4–6 h after soy intake, suggesting that isoflavones are first partially absorbed in the small intestine, as the major site of hydrolysis (Rowland et al., 2003), and then later in the large intestine, where the gut microbiota has the crucial role (Franke et al., 2004; Franke et al., 2014).

The most abundant metabolites are further formed by reduction, but also reactions of oxidation, hydroxylation, methylation/demethylation, and the breakage of the pyrone ring can occur leading to the formation of different isoflavone metabolites (Heinonen et al., 2002; Heinonen et al., 2003). Metabolites formed by reduction are mostly the result of microbial activity in the gut and they include equol, dihydrodaidzein, and O-desmethylangolensin as the main daidzein metabolites, and dihydrogenistein, 5-OH-equol, and 6'-OH-O-desmethylangolensin as genistein metabolites (Heinonen et al., 2002; Kelly et al., 1993; Barnes, 2010). Glycitein, the least investigated soy isoflavone, at first showed to be relatively stable (Heinonen et al., 2003), since there was only a minor degree of bacterial demethoxylation to daidzein detected in the colon. Nevertheless, several reduced glycitein metabolites were detected by Heinonen et al. (2003) and Simons et al. (2005).

Phase II metabolism of isoflavones refers to reactions of conjugation, mainly in the liver or in the gut mucosa, that contribute to faster urinary excretion of produced water-soluble metabolites (de Cremoux et al., 2010). Since there are two conjugation positions on genistein and daidzein (7 and 4'), monoglucuronides, monosulfates, diglucoronides, or disulfates and mixed conjugates can occur (Shelnutt et al., 2002). Isoflavone conjugates can be excreted via bile and hydrolyzed again in the colon by the microbial enzymes, so aglycones can be reabsorbed and undergo enterohepatic recirculation (Barnes, 2010). Isoflavones are mainly eliminated by urine in conjugated forms, and a small percentage by feces (Xu et al., 1995).

Conflicting results concerning several pharmacokinetic parameters for daidzein and genistein were published. Chang and Choue (2013) found that daidzein plasma's half-life was longer than that of genistein after consumption of soy products with different aglycon/glucoside ratios. Also, Setchell et al. (2001) determined that the half-life of elimination for daidzein is 9.4 h, for genistein 6.8 h, and for glycitein 8.9 h. Other authors reported that genistein has a longer half-life compared to daidzein (Izumi et al., 2000; Watanabe et al., 1998). Setchell et al. (2001) determined a high volume of distribution for daidzein (236 L) and genistein (161 L), and the highest for glycitein (415 L), showing extensive tissue distribution of these compounds.

Inconsistent results were obtained also on the bioavailability of different isoflavones. The bioavailability of genistein was reported to be much greater than that of daidzein, while the bioavailability of both isoflavones was higher when consumed as β-glucosides compared to aglycones. A study by Rüfer et al. (2008) confirmed that when given in pure form, daidzin has a higher bioavailability than its aglycon daidzein. Zubik and Meydani (2003) found that the bioavailability of genistein and daidzein is similar when aglycone and corresponding glycoside forms are ingested. Conversely, Chang and Choue (2013) showed that higher isoflavone concentrations in plasma are obtained when soy-based food rich in aglycones is consumed. Glycitein is very well absorbed and its bioavailability is similar to that of daidzein (Shinkaruk et al., 2012).

1.3.1 Bioactive Metabolites

The most studied isoflavone metabolites are equol (7-hydroxy-3-(4'-hydroxyphenyl)-chroman) and O-DMA (O-desmethylangolensin) (Figure 1.5), both active metabolites of daidzein formed by

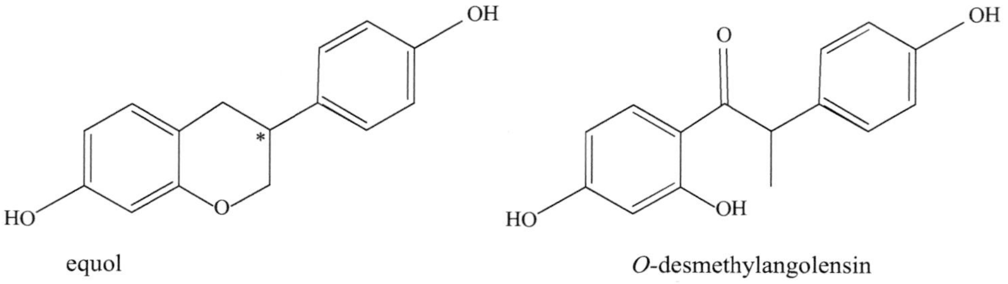

FIGURE 1.5   Structures of daidzein metabolites – equol and *O*-Desmethylangolensin (*O*-DMA).

microbiota. Although it is considered that genistein is the most biologically active soy isoflavone, because it has the highest *in vitro* affinity for estrogen receptors (Kuiper et al., 1998), it should be emphasized that actually the most potent metabolite, equol, is the product of daidzein biotransformation (Yuan et al., 2007).

It is noticed that the majority of the population is capable of *O*-DMA production (80–90%), while equol is produced only by 30–40% of the population (this percentage is higher in the Asian than in the Western population, and also in vegetarians) (Atkinson et al., 2005; Cassidy et al., 1994; Kelly et al., 1993). Thus, it is presumed that metabolite producers could have more benefits from soy isoflavone consumption than non-producers (Setchell et al., 2002b). Observational studies imply that there is no connection between equol-producer and *O*-DMA-producer phenotypes, and thus that different bacteria are responsible for gut metabolism (Frankenfeld et al., 2004a). It is believed that gut microflora composition, genetic factors (phenotype), and diet contribute to interindividual differences concerning isoflavone metabolism (Lampe, 2009). Previously, it was thought that equol production phenotype is a stable characteristic, but a few recent findings point out that some individuals may be capable of changing their equol production status over time, primarily by modification of dietary habits (Franke et al., 2014; Mayo et al., 2019).

Equol is a nonsteroidal estrogenic compound and exclusive product of gut bacterial metabolism (Setchell et al., 2002b). This compound has been in the focus of many studies in the recent period (Davinelli et al., 2017; Shor et al., 2012; Ahuja et al., 2017) because it is an interesting example of a metabolite with higher estrogenic potential than its parent compound (Setchell et al., 2002b). Since equol is a chiral compound (it has one chiral carbon atom), it has been shown that only S enantiomer has estrogenic activity – it has the ability to bind to both types of estrogen receptors, although it has a higher affinity for ERß. The affinity for ERß of S-equol is 20% and only 1% for R enantiomer if compared with 17ß-estradiol. It is interesting that S-equol is the only enantiomer found in human blood and urine samples, proving also that intestinal bacteria are enantiospecific in synthesizing only this biologically active form (Setchell et al., 2005). It was shown that equol production also depends on the consumed daidzein form. If only aglycones are ingested they are absorbed passively in the small intestine, but if daidzin or other glucoside forms are consumed they cannot be absorbed before hydrolysis, and thus they stay in the gut long enough to undergo bacterial metabolism and the production of equol (Zubik & Meydani, 2003).

Setchell et al. (2002b) suggested that equol could be an important marker for the assessment of isoflavone *in vivo* activity. There are numerous studies examining the mechanism of action of equol (Yuan et al., 2007) and an association of individual capability of equol production with the prevention of certain chronic diseases (Lampe, 2009). Besides previously mentioned estrogenic activity, antiandrogen properties (Lund et al., 2004) and high antioxidant potential (Choi & Kim, 2014; Jackman et al., 2007) are attributed to this compound. Recently, researchers were focused on determining the clinical effectiveness of equol supplementation on different conditions and diseases, for instance on menopausal symptoms (Davinelli et al., 2017; Aso et al., 2012), blood pressure, vascular function (Liu et al., 2016a), skin aging (Oyama et al., 2012), and bone resorption (Tousen et al., 2011).

Statements about O-DMA biological activity are conflicting since it is in some publications referred to as a very active metabolite (de Cremoux et al., 2010), and in others rather inactive (Lopes et al., 2017). O-DMA, with its cleaved ring, is chemically less similar to 17ß-estradiol than daidzein or equol (Frankenfeld, 2011). In some *in vitro* experiments, it was indicated that this compound has a certain potential in the treatment of breast cancer and hepatocellular carcinoma (Choi & Kim, 2013; Choi et al., 2013). Also *in vitro* antioxidant activity was established, and a beneficial effect on atherosclerosis was suggested (Hodgson et al., 1996). Interesting findings are published by Frankenfeld et al. (2006), that O-DMA producers have higher bone mineral density compared to non-O-DMA producers. Also, a higher percentage of mammographic density, a marker of breast cancer risk, was observed in O-DMA producers than in non-producers (Frankenfeld et al., 2004).

These findings are pointing out the importance of recognizing isoflavone-metabolizing phenotypes. However, the biological effects of soy isoflavone metabolites are yet to be confirmed and human studies are still needed in order to reveal their true potential.

1.3.2 Intestinal Microbiota Influence

Gut microbiota has a crucial role in different phases of isoflavone metabolism. It is known that isoflavones are not even absorbed without intestinal microbiota (Bowey et al., 2003). Certain bacteria species are able to hydrolyze isoflavones by the action of the β-glucosidase to improve bioavailability, such as *Lactobacillus casei*, *Lactobacillus rhamnosus*, *Lactobacillus plantarum*, *Lactobacillus fermentum*, *Lactobacillus acidophilus*, *Lactobacillus bulgaricus*, *Bifidobacterium breve*, and *Bifidobacterium bifidum* (Marazza et al., 2012; Rekha & Vijayalakshmi, 2011; Hati et al., 2015).

Moreover, intestinal microflora has an important role in the transformation of daidzein, producing a few different metabolites such as equol, dihydrodaidzein, and O-DMA, as previously mentioned. In recent years, a significant number of human bacteria strains capable of forming equol have been isolated, most of them belonging to the *Coriobacteriaceae* family (Clavel et al., 2014). Among equol-producing species, there are *Adlercreutzia equolifaciens* (Maruo et al., 2008), *Slackia isoflavoniconvertens* (Matthies et al., 2009), *Slackia equolifaciens* (Jin et al., 2010), some *Eggerthella* sp. (Yokoyama & Suzuki, 2008; Yu et al., 2008), *Bifidobacterium breve* ATCC 15700T, *Bifidobacterium longum* BB536 (Elghali et al., 2012), and certain *Lactobacillus* sp. (Kwon et al., 2018), as well as several others (Mayo et al., 2019). Also, it was noticed that some equol-producing bacteria (e.g. *Slackia isoflavoniconvertens*) can also transform genistein to 5-OH-equol, a metabolite with high antioxidant properties (Matthies et al., 2012). Concerning O-DMA production, it was found that some *Clostridium* strains and *Eubacterium ramulus* can produce this metabolite by cleaving the C-ring (Li et al., 2015a; Yokoyama et al., 2010). Gaya et al. (2018) showed that the strain *Enterococcus faecium* INIA P553 is capable of production of O-DMA and 6-OH-O-DMA with high efficiency and that it could be used as a probiotic.

Bearing in mind the influence of microbiota on isoflavone metabolism, it was a logical step to try to improve the bioavailability of these compounds using probiotics in combination with isoflavone supplementation, especially in cases of disrupted microflora, e.g. after antibiotic treatment. Benvenuti and Setnikar (2011) presented improvement of genistein pharmacokinetic parameters after treatment with a supplement containing soy isoflavones (genistin 30 mg + daidzin 30 mg) plus calcium and vitamin D3 in combination with *Lactobacillus sporogenes* compared to treatment without probiotic culture. The lack of positive results for daidzein was explained by the possible degradation of this compound by *L. sporogenes* to equol.

Additionally, certain bacteria strains could be used in the production of functional food with higher levels of isoflavone aglycones and active metabolites, improving digestibility and bioavailability. Hence, Peirotén et al. (2020) showed that the production of a fermented soy drink rich in soy aglycones, O-DMA, 6'-OH-O-DMA, dehydrodaidzein, and tetrahydrodaidzein is possible using certain *Bifidobacterium* strains and lactic acid bacteria.

Therefore, the general idea in the future is to use well-characterized metabolite-producing bacteria strains as probiotics, or in the process of production of enriched functional food in order to increase the bioavailability and the health effects of soy isoflavones. Also, it is generally known that certain natural polyphenolics can alter the composition of the gut microbiome, but in perspective, investigations should reveal the nature of this influence of soy isoflavones on specific intestinal bacteria (Cassidy & Minihane, 2017).

There are still controversies concerning isoflavone metabolism since there are many individual factors that should be taken into account such as dietary habits, pathophysiological condition, gut microflora status, enzyme polymorphism, or phenotype (equol and O-DMA producers or non-producers) (de Cremoux et al., 2010). Also, the conflicting results of different isoflavone clinical studies could be explained with these individual differences and numerous factors that play roles in the final bioavailability of certain isoflavones.

1.4 HEALTH EFFECTS

1.4.1 MENOPAUSAL SYMPTOMS

Menopause is a natural process characterized by the decline of estrogen which produces numerous effects including vasomotor symptoms like hot flashes and night sweats, vaginal dryness, etc. Due to their estrogen-like activity and binding for ERß receptors, isoflavones are recognized as beneficial for the alleviation of menopausal symptoms (Ko, 2014). In Table 1.1, data from several meta-analyses concerning the effects of isoflavone intake on menopausal symptoms are summarized. Most of the studies observed improvement in the number of hot flashes after isoflavone supplementation. However, a meta-analysis performed by Chen et al. (2015b) failed to show a beneficial effect on other menopausal symptoms expressed as a Kupperman index. Also, the importance of duration of intervention as well as the dosage of isoflavones is pointed out (Taku et al., 2012; Li et al., 2015b).

Oral intake of isoflavones was also found effective in the improvement of vaginal dryness score (Franco et al., 2016). Additionally, a study by Lima et al. (2013) found that vaginal application of isoflavone gel (4%, 1 g/day) during 12 weeks led to significant improvement of vaginal atrophy and cell maturation values in postmenopausal women. The effects were similar to those obtained with the application of conjugated equine estrogen cream.

Additionally, published clinical trials measure and report vasomotor symptoms in different ways, which aggravates comparison and synthesis of data. Therefore, in order to elucidate isoflavone effects on menopausal symptoms, it is desirable to introduce standardized outcome measures in further research.

1.4.2 CANCER

The newest data of the American Cancer Society estimated for 2020 show that breast and prostate cancer are leading types of cancer cases diagnosed in the USA, and the second cause of cancer mortality in women and men respectively, after lung cancer in both sexes (Siegel et al., 2020). Indeed, cancer is one of the major health problems, and its prevention is the crucial part of combatting this disease. Therefore, isoflavones have been extensively studied as cancer-preventive compounds since their beneficial effects were observed in Asians, who generally have a lower incidence of sex-hormone-dependent cancers like breast and prostate cancer compared to the Western population (Wu et al., 2015).

In Table 1.2 some of the recent meta-analyses dealing with isoflavone effects on different types of cancer are presented.

Zhao et al. (2019) found that there is a significant correlation between high intake of soy foods and breast cancer risk reduction, which was not the case with moderate consumption. Moreover, breast cancer recurrence was lowered when more than 10 mg of isoflavones per day was consumed (Nechuta et al., 2012). Kucuk (2017) confirmed that soy foods can prevent breast cancer and that

TABLE 1.1

Review and Meta-Analysis Studies of Isoflavone Effects on Menopausal Symptoms

Reference	Study Type / No. of Subjects	Intervention	Outcome Measures	Conclusion
Bolanos et al., 2010	Meta-analysis 19 placebo-controlled clinical trials 1,419 participants	Genistein 54 mg/day Daidzein 60 mg/day Soy supplement 42-134.4 mg/day Soy extract 33.3-76 mg/day MFT 12 weeks	Number of hot flashes Score of vasomotor symptoms	High heterogenicity of the results Minimal heterogenicity found only for genistein and daidzein intervention Results combined based on the type of supplement used showed significant tendency in favor of soy
Taku et al., 2012	Systematic review 19 placebo-controlled clinical trials 1,992 participants Meta-analysis 17 placebo-controlled clinical trials	Isoflavone supplements Isoflavone dosage 30-135 mg/day MFT 6 weeks	Hot flash frequency (13 trials) Hot flash severity (10 trials) Hot flash frequency (13 trials) Hot flash severity (9 trials)	10 trials reported statistically significant effects for hot flash frequency 2 trials reported statistically significant effects for hot flash severity Ingestion of 30 to 80 mg/day of isoflavones for 6 weeks to 12 months significantly reduced hot flash frequency Supplements containing more than 18.8 mg of genistein were more than twice as potent at reducing hot flash frequency than lower genistein supplements Daily ingestion of an average of 62.8 mg/day of isoflavones for 12 weeks to 12 months significantly reduced hot flash severity by 26.2%
Li et al., 2015b	Model-based meta-analysis 17 placebo-controlled clinical trials 1,710 participants	Soy tablet Soy capsule Soy powder Muffins with isoflavones Soy extract Soy beverage Isolated soy protein Isoflavone dosage 30-200 mg/day MFT 4 weeks	Effects on hot flashes	Maximal percentage change of hot flash reduction by soy isoflavones was 25.2% (57% of estradiol effect) Time interval of 13.4 weeks was needed for soy isoflavones to achieve half of their maximal effects (estradiol required 3.09 weeks) Treatment intervals of 12 weeks are too short for soy isoflavones Soy isoflavones have slight and slow effects in attenuating menopausal hot flashes compared with estradiol

(Continued)

TABLE 1.1 (CONTINUED)

Review and Meta-Analysis Studies of Isoflavone Effects on Menopausal Symptoms

Reference	Study Type No. of Subjects	Intervention	Outcome Measures	Conclusion
Chen et al., 2015b	Meta-analysis 15 placebo-controlled clinical trials 1,753 participants	Isoflavone dosage 25–100 mg/day Equol 5 mg/day (one study)* MFT 3 months	Kupperman index (hot flushes [vasomotor], paresthesia, insomnia, nervousness, melancholia, vertigo, weakness, arthralgia or myalgia, headache, palpitations, and formication) changes (7 trials) Daily hot flush frequency (10 trials) Likelihood of side-effects (5 studies)	No significant treatment effect of phytoestrogens (isoflavones) on Kupperman index compared to placebo Phytoestrogens result in a significantly greater reduction in hot flash frequency compared to placebo No significant difference in side effects between groups
Franco et al., 2016	Meta-analysis 12 placebo-controlled clinical trials	Dietary soy isoflavones 33.3–90 mg/day Supplements and extracts of soy isoflavones 50–118 mg isoflavone/day 6–12 mg isoflavone/day transdermal 10 mg equol/day 63 mg daidzein/day 70 mg genistin and daidzin/day MFT 6 weeks	Number of hot flashes in 24 h (10 trials) Vaginal dryness (2 trials)	Dietary and supplemental soy isoflavones resulted in improvement in daily number of hot flashes and vaginal dryness score

Note

*One study included in the meta-analysis was an intervention with *Trifolium pratense* (40 mg); MFT = minimal follow-up time.

TABLE 1.2

Review and Meta-Analysis Studies of Isoflavone Effects on Cancer

Reference	Study Type No. of Subjects	Intervention	Outcome Measures	Conclusion
Zhao et al., 2019	Meta-analysis 16 prospective cohort studies 11,169 breast cancers cases 648,913 participants Broad demographic range Follow up 2–14.1 years	High and/or moderate vs low intake of isoflavone High and/or moderate vs low intake of soy food	Breast cancer risk	No significant association between the consumption of isoflavones and moderate consumption of soy foods and the risk of breast cancer Significant correlation between high intake of soy foods and reduced risk of breast cancer
Nechuta et al., 2012	In-depth analysis 4 prospective cohort studies 9,514 US and Chinese breast cancer survivors Mean follow up 7.4 years	Postdiagnosis soy food consumption Isoflavone intake Chinese women (45.9 ± 38.3 mg/day) US women (3.2 ±9.8 mg/day)	Breast cancer outcome	Consumption of 10 mg or more of isoflavones per day was associated with a statistically significant reduced risk of breast cancer recurrence and a nonsignificant reduced risk of all-cause and breast cancer–specific mortality
Zhang et al., 2017	Multiethnic cohort Breast Cancer Family Registry 6,235 women with breast cancer Mean follow up 9.4 years	Dietary soy isoflavone intake < 0.34 mg 0.34–0.67 0.67–1.5 ≥ 1.5 mg/day	All-cause mortality	Women with the highest intake of dietary isoflavone intake (≥ 1.5 mg/day) had a 21% decrease in all-cause mortality compared with women with the lowest intake This effect was limited to women who had tumors that were negative for hormone receptors and did not receive hormone therapy Genistein was the major isoflavone consumed
Applegate et al., 2018	Meta-analysis 30 studies 15 case-control studies 8 cohort studies 7 nested case-control studies 266,699 participants 21,612 prostate cancer reported Minimum follow up 4 years	Total soy food intake Unfermented and fermented soy food Isoflavone intake Circulating isoflavone levels	Risk of prostate cancer (PC) Risk of advanced prostate cancer	Total soy food, genistein, and daidzein intake inversely associated with the risk of PC Only unfermented soy food was associated with decreased PC risk Circulating levels of genistein and daidzein were not associated with the risk of PC No significant reduction in the risk of advanced PC

(Continued)

TABLE 1.2 (CONTINUED)

Review and Meta-Analysis Studies of Isoflavone Effects on Cancer

Reference	Study Type No. of Subjects	Intervention	Outcome Measures	Conclusion
Yu et al., 2016b	Systematic review and meta-analysis 17 studies 13 case-control studies 4 prospective cohorts	Soy food consumption (tofu, soybean, soy milk, miso soup) (< 1 per month up to ≥ 8 times per week) Isoflavone consumption (from 1 mg up to 60 mg/day)	Colorectal cancer risk	Soy isoflavone consumption reduced colorectal cancer risk by 23% Significant protective effect was observed with soy foods/products in Asian populations
You et al., 2018	Meta-analysis 12 studies 6 case-control studies 6 cohort studies 596,553 participants	Dietary isoflavone intake Low 0.01–20 mg/day High 1.1–75.5 mg/day	Gastric cancer risk	No significant association between dietary isoflavones intake and risk of gastric cancer Subgroup analysis showed similar results
Tse & Eslick, 2016	Meta-analysis 40 studies 22 case-control studies 18 cohort studies 633,476 participants 13,639 gastric cancer cases	Fermented soy Miso ≥ 2/week to ≥ 5/week Soy 5/week to ≥ 1/day and ≥ 25 g/day up to 254 g/day Tofu ≥ 1/day up to 3/week Isoflavone > 1.1 mg/day to 128 mg/day	Gastrointestinal cancer risk	Inverse association between isoflavone intake and gastrointestinal cancer Effects of soy much weaker Correlation significant in colorectal cancer and among Asian population
Liu et al., 2016b	Meta-analysis 23 randomized controlled trials 2 perimenopausal women 21 postmenopausal women 2,305 participants	9 studies of soy-based isoflavone supplements 3 studies of red clover–based isoflavone supplements 1 study of *Pueraria mirifica*–based isoflavone supplements 7 studies of soy foods 1 study of soy protein powder 2 studies of synthetic isoflavone Isoflavone dosage 5–154 mg/day Duration 12–156 weeks	Effects on endometrial thickness (risk factor of endometrial cancer in peri- and postmenopausal women)	Stratified analysis showed that a daily dose of more than 54 mg (in aglycone equivalents) could decrease the endometrial thickness by 0.26 mm Decrease in the endometrial thickness was observed in North American studies, but increase was observed in Asian studies Effects vary in different populations and depend on administered dose

they can also be beneficial to women with this diagnosis. On the other hand, Qiu and Jiang (2019) concluded that pre-diagnosis isoflavone intake has a very small effect on the overall survival of breast cancer patients.

Interestingly, in a meta-analysis by Applegate et al. (2018), only the consumption of nonfermented soy food was associated with reduced prostate cancer risk, while total soy food and individual isoflavones showed to be inversely related to this risk. Conversely, high genistein and daidzein intake was correlated with decreased risk of prostate cancer by He et al. (2015). Ajdžanović et al. (2019) explained various mechanisms of isoflavone action against prostate cancer and pointed out their promising antimetastatic properties. Conversely, elevated risk of advanced prostate cancer was connected with dietary isoflavone intake by Reger et al. (2018).

Extensive research by Liu et al. (2016b) indicated that isoflavone doses higher than 54 mg/day could be beneficial in the prevention of endometrial cancer. Also, an inverse relationship between isoflavone intake and ovarian cancer risk in women in southern China was observed (Lee et al., 2014). Soy isoflavone consumption seems to have a protective effect against colorectal cancer (Yu et al., 2016b). Nachvak et al. (2019) in their review and meta-analysis associated higher intake of soy with decreased risk of mortality from gastric colorectal and lung cancer and found that an intake increase of 10 mg of isoflavones per day may reduce mortality from cancer by 7%.

Although isoflavones have indisputable potential in the prevention of different types of cancers, researchers are still very cautious when isoflavone recommendations to post-diagnosed patients are considered (Zhang et al., 2016).

1.4.3 CARDIOVASCULAR DISEASE

It is recognized that soy and isoflavone consumption could play an important role in the development of cardiovascular disease (CVD). Although the Food and Drug Administration (FDA) approved a food labeling health claim for soy protein as protective against coronary heart disease (FDA, 1999), later reports indicated modest effects of soy food on cardiovascular health and cholesterol levels (Sacks et al., 2006; Blanco Mejia et al., 2019).

Soybean intake of 8.5 to 17 times per week was associated with a lower incidence of cardiometabolic syndrome (CMS) and central obesity in Korean Women. The observed inverse association was less pronounced in groups with intake > 17 times/week of soy foods. However, when tofu consumption was excluded from the analysis, a decrease in CMS and central obesity was observed in both groups, indicating that the processing method of soy foods plays an important role in achieving desired effects (Jun et al., 2020).

On the other hand, three large-cohort prospective studies including male and female health professionals from the United States showed that higher isoflavone and tofu intake was associated with a moderately lower risk for the development of CVD. The effects of tofu consumption were primarily observed for younger women before menopause and postmenopausal women without hormone replacement therapy (Ma et al., 2020). It is suggested that these effects could be the consequence of masking the effects of isoflavones with hormone replacement therapy or a higher expression of vascular ER in younger women (Gavin et al., 2009). No significant correlation between soymilk and lower risk of CVD was noticed, which could be due to the presence of other constituents in soy milk (emulsifiers, sugar, etc.). A study conducted in the UK compared the effects of six months' daily intake of soy protein (15 g) or soy protein (15 g) with isoflavones (66 mg) on early menopausal women with cardiovascular disease risk markers. Reduction in systolic blood pressure and fasting glucose and insulin levels were observed in treatment with isoflavones compared to treatment with soy protein only. There were no effects on lipid profile or diastolic blood pressure in either group (Sathyapalan et al., 2018). Two meta-analyses were performed in order to evaluate the effects of soy isoflavones on blood pressure (Liu et al., 2012) and plasma lipoprotein (a) concentration (Simental-Mendia et al., 2018). Soy isoflavone consumption in a range of 65–153 mg per day together with soy protein during 1–12 months lowered blood pressure in hypertensive subjects. On the other hand, no significant effects of soy isoflavone treatment on plasma lipoprotein (a) were observed. A recently published umbrella review of the effects of soy and isoflavone consumption found the effect on CVD as one of the most noticeable. The review identified a significant beneficial association between isoflavones and endothelial function, systolic and diastolic blood pressure, while soy but not isoflavones demonstrated benefits concerning CVD, stroke, or coronary heart disease (Li et al., 2020).

1.4.4 Osteoporosis

Osteoporosis is one of the major health issues affecting postmenopausal women with decreased production of estrogen (Seeman, 2004). It is associated with an increased rate of bone remodeling (or bone turnover), decreased bone mineral density (BMD), increased risk of fracture, as well as changes in bone turnover markers (urine deoxypyridinoline, osteocalcin, etc.) (Taku et al., 2011; Eastel & Szulc, 2017).

A meta-analysis including 19 studies concerning the effects of isoflavone supplements on osteoporosis revealed that soy isoflavones increased BMD by 54% and decreased the bone resorption marker deoxypyridinoline by 23% compared to the baseline. Higher mean difference changes were observed for postmenopausal women at an isoflavone dose above 75 mg per day (Wei et al., 2012).

Recently, a systematic review and meta-analysis of a total of 52 clinical trials on the effect isoflavones on BMD and bone turnover markers was published. Isoflavones induced a significant increase of BMD of the lumbar spine, hip, and femoral neck. These effects were observed in all places for normal-weight subjects while in overweight/obese subjects beneficial effects were noticed only for the femoral neck. Effective doses were $\geq$ 90 mg/day for the lumbar spine and hip and < 90 mg/day for the femoral neck. Also, it was shown that intervention of more than one year is necessary for obtaining significant beneficial effects (Akhlaghi et al., 2019). Concerning markers of bone turnover, osteoprotegerin, pyridinoline, and C-telopeptide were affected by isoflavones, while no changes were observed for osteocalcin and alkaline phosphatase. Three markers were reduced in groups with body mass index $\geq$ 25 kg/m2 and doses below 90 mg/day. Generally, high heterogenicity between studies was observed.

Additionally, the effects of soy isoflavone, calcium, and soy isoflavone combined with calcium on BMD were evaluated over six months in perimenopausal women. It was observed that isoflavone combined with calcium had the most pronounced positive effect on BMD compared to isoflavones or calcium alone. Also, only the combination of isoflavone and calcium led to a decrease in FSH and LH levels in perimenopausal women (Zhang et al., 2019).

1.4.5 Cognitive Functions

Isoflavones are considered potential substances for the improvement of cognitive functions due to their ability to cross the blood–brain barrier and their affinity for ERβ located in the hippocampus and prefrontal cortex (González et al., 2007).

A recently published meta-analysis of 15 randomized controlled trials performed by Cui et al. (2020a) confirmed that soy isoflavones may improve cognitive functions especially in the domain of memory. The effect was recognized in menopausal women, but also in postmenopausal women and men. High isoflavone intake (> 100 mg/day) showed slightly better results, but the qualitative content of consumed isoflavones (daidzein, genistein) should be also considered. Additionally, the duration of study did not significantly influence the outcome, but this could be due to the overall short duration of the analyzed studies (the longest study included in the analysis lasted two years). Three studies from different groups of authors (White et al., 2000; Hogervorst et al., 2008; Xu et al., 2015a) observed that intake of tofu was inversely associated with cognitive functions, while Hogervorst et al. (2008) additionally related high intake of tempeh with better memory. In a later study, Hogervorst et al. (2011) reported that the negative effect of tofu is not significant in older adults (mean age 80), while the positive relationship between tempeh and memory remained. These results were explained with lifestyle changes (less tofu and more tempeh consumption) due to the dissemination of the Hogervorst et al.

(2008) study. It has been suggested that tempeh could be more favorable than tofu due to higher levels of daidzein and genistein as well as higher folate content (Wang & Murphy, 1994).

Further research should be more focused on isoflavones' potential benefits on cognitive functions and dementia in women as well as men. Also, analysis of large samples during a longer duration period is needed.

1.4.6 Diabetes

Diabetes mellitus type 2 (DMT2) is a complex metabolic disorder including insulin resistance and relative insulin deficiency leading to hyperglycemia (Zhang et al., 2013b). It has already been suggested that isoflavones could have antidiabetic potential through several mechanisms, including lowering insulin resistance and inflammation, reducing adipose tissue mass, and improving lipid parameters in subjects diagnosed with this disease (Duru et al., 2018).

Zhang et al. (2013b) confirmed through meta-analysis that soy isoflavone supplementation could be beneficial for body weight reduction, glucose, and insulin control in the plasma of non-Asian postmenopausal women. Additionally, shorter supplementation (less than six months) with a lower dose (less than 100 mg) could reduce body weight, especially in normal-weight women (BMI less than 30). Longer supplementation (> 6 months) with a lower dose of isoflavones was found effective for the reduction of blood glucose. The effects of soy isoflavone supplementation on glucose homeostasis in menopausal women were evaluated through another meta-analysis performed by Fang et al. (2016). Beneficial effects were observed for fasting blood glucose levels, insulin resistance, as well as the HOMA-IR of participants. Although high heterogenicity among results of isoflavone supplementation was noticed, it was suggested that these results were more consistent for subgroups that used genistein alone or applied a low dosage of isoflavones (< 70 mg per day).

Also, a performed meta-analysis of eight studies with 19 reports showed that soy products and soy constituents (isoflavones and proteins) may be associated with a lower risk of DMT2 development. The effect is more pronounced in Asians and females, probably due to the traditionally high intake of soy products by the Asian population as well as the receptor-mediated effect of isoflavones that modulates insulin action. Besides isoflavones, soy protein has been recognized as the main protective factor for the development of DMT2 (Li et al., 2018).

A recent study conducted on 8,296 subjects confirmed a sex-specific association between soy protein, isoflavone, daidzein, and genistein intake and lower DMT2 incidence. The beneficial effect was observed only for female participants, of which 83.5% were in menopause. It is hypothesized that the phytoestrogenic effect of isoflavones is especially important for menopausal women because they are more susceptible to the development of DMT2 due to the increased risk of elevation of fasting levels of glucose. On the other hand, it could be that the high intake of isoflavones in men could lead to lower androgen activity which is also associated with the development of DMT2 (Woo et al., 2021). Similarly, Konishi et al. (2019) observed that high soy intake was inversely associated with the risk of DMT2 in Japanese women. The study was performed on a total of 13,521 participants (5,883 men and 7,638 women) during ten years of follow-up. Further research on specific roles and the mechanism of action of isoflavones on DMT2 development is necessary in order to completely elucidate this phenomenon.

1.4.7 Polycystic Ovary Syndrome

Polycystic ovary syndrome (PCOS) is one of the most common disorders in women of reproductive age with prevalence from 8.7% up to 17.8% depending on the applied diagnostic criteria (March et al., 2010). Complications related to this condition are the consequence of disturbances in reproductive, endocrine, and metabolic functions. Some of the main symptoms include amenorrhea, hirsutism, hyperandrogenism, infertility, as well as increased risk of developing insulin resistance, type 2 diabetes mellitus, and cardiovascular disease (Norman et al., 2007).

Results concerning isoflavones' potential benefits in women with PCOS so far have been somewhat inconsistent.

A pilot study published in 2008 analyzed the effects of six months of genistein supplementation (36 mg/day) on a group of 12 obese, hyperinsulinemic, and dyslipidemic women diagnosed with PCOS. Significant improvement was observed in total cholesterol and triglycerides level, while no changes were detected in anthropometric features, hormonal status, menstrual cycle, or glycoinsulinemic metabolism. A placebo group was not included in the study (Romualdi et al., 2008).

Two studies conducted during 2016 and 2018 considered the influence of soy isoflavone or soy diet on biological makers of women with PCOS. A study performed by Jamilian and Asemi (2016) showed that consumption of 50 mg/day of soy isoflavones during 12 weeks improved markers of insulin resistance, hormonal status, triglycerides, and biomarkers of oxidative stress compared with the placebo group. However, the intervention did not affect weight, BMI, other lipid profiles, or inflammatory factors. On the other hand, a study from 2018 analyzed the influence of intake of a soy diet containing 75 mg of total phytoestrogens but also fiber, magnesium, potassium, and polyunsaturated fatty acids. After eight weeks, BMI, glycemic control, total testosterone, triglycerides, VLDL-cholesterol, and malondialdehyde significantly decreased, while a significant increase of NO and total glutathione compared to the control was noticed (Karamali et al., 2018). It was suggested that a soy diet rich with high unsaturated fat and fiber could be responsible for these beneficial effects observed in patients with PCOS.

A study conducted on 146 subjects with PCOS showed that intake of 18 mg of genistein twice daily led to the reduction of LDL cholesterol as well as luteinizing hormone, triglyceride, dehydroepiandrosterone sulfate, and testosterone compared to the control group (Khani et al., 2011). No changes in HDL cholesterol and follicle-stimulating hormone were observed. Additionally, a meta-analysis of four studies showed that soy isoflavones decreased total testosterone levels but had no significant effect on follicle-stimulating hormones (Zilaee et al., 2020). However, it was pointed out that variations in the duration of intervention, dosage, and number of participants are obstacles to drawing a conclusion.

A recent publication by Haudum et al. (2020) pointed out that three consecutive days of isoflavone consumption (soy drink twice per day, approximately 25 mg of isoflavones per drink) resulted in improvement in fasting glucose and insulin sensitivity in PCOS patients compared to the control group. Also, beneficial effects after intervention were observed in predicted stool metagenomic pathways and microbial alpha diversity, especially among equol producers.

1.4.8 Skin

Skin aging is a complex process that involves different factors – environmental, chronologic, genetic, and hormonal (Shu & Maibach, 2011). It is considered that a decrease in estrogen levels as a consequence of menopause or other causes has a direct effect on the skin, leading to skin dryness and thinning, a decrease in collagen and elastic fibers, as well as sebum content (Rzepecki et al., 2019).

Soy isoflavones are recognized as a promising alternative for hormonal therapy in dealing with signs of aging due to estrogen deficiency without undesirable side effects (Yingngam & Rungseevijitprapa, 2012). A pilot study from 2009 on 30 menopausal women showed that treatment with 100 mg/day of isoflavone-rich extract during six months lead to an increase in epithelial thickness and the number of collagen and elastic fibers. Also, an increase in dermal blood vessels was noticed (Accorsi-Neto et al., 2009). Three randomized, double-blind, 17-beta-estradiol controlled clinical trials were conducted to evaluate the efficiency of 4% genistein gel on skin morphological parameters (Moraes et al., 2009), hyaluronic acid concentration (Patriarca et al., 2013), and the facial skin collagen (Silva et al., 2017) of postmenopausal women. Obtained data suggested that isoflavone gel had a positive effect on epidermal thickness, the number of dermal papillae, fibroblasts, and blood vessels (Moraes et al., 2009), increased hyaluronic concentration (Patriarca et al., 2013), and collagen type I and III production (Silva et al., 2017). Although all three studies observed beneficial effects of genistein, results indicated that estradiol treatment was more efficient for all investigated parameters.

On the other hand, isoflavone treatment did not significantly improve skin sebum content, skin blood circulation, or skin sagging after 12 weeks of application on postmenopausal women. However, significantly improved skin thickness/echogenicity was observed (Farwick et al., 2014).

1.4.9 OTHER EFFECTS

Most published studies on isoflavone health effects are focused on the female population, due to isoflavones' estrogen-like properties. However, some additional effects of isoflavones are recognized also in men. A study performed by Yuan et al. (2019) performed on 1,319 Chinese men observed a negative correlation between semen genistein and equol and sperm concentration and count. On the other hand, semen daidzein was positively associated with sperm motility. Another study examined the effects of dietary isoflavone intake of Japanese men (total 1,335 participants) on depressive symptoms. It was concluded that intake of higher levels of isoflavones leads to lowering depressive symptoms and could contribute to the preservation of mental health. It is speculated that this mechanism involves similar patterns recognized in women, i.e. binding to selective estrogen receptor β (Bodo & Rissman, 2006) and interacting with the serotonergic and dopaminergic systems (Cui et al., 2020b). Additionally, inverse relationships between intake of total soy products, tofu, tofu products, fermented soybeans, boiled soybeans, miso soup, and isoflavones and depressive symptoms during pregnancy were noticed in Japanese pregnant women. No such effects were noticed for intake of soymilk during pregnancy (Miyake et al., 2018).

Beneficial effects soy isoflavones were tested on patients with poorly controlled asthma. Patients received 100 mg of isoflavone supplement split into two doses during 24 weeks. No improvement in lung function or clinical outcomes compared to a placebo was observed in asthmatic patients with the high PAI-1–producing genotype (Smith et al., 2015). However, in a later study, the same intervention led to a reduced number of severe asthma exacerbations in asthmatic patients with the high PAI-1–producing genotype. It is suggested that PAI-1 polymorphisms could be genetic biomarkers for the treatment of asthma with isoflavones (Cho et al., 2019).

A study performed on a total of 1,076 Japanese adults suggested that high daily isoflavone intake from food (26.74–83.06 mg/1,000 kcal/day) is positively correlated with optimal sleep duration and quality. The observed relationship is not affected by age, sex, BMI, or lifestyle habits (Cui et al., 2015).

Soy isoflavones have been recently recognized for their beneficial effects on gastrointestinal health, including motility, secretion, morphology, and barrier function (Al-Nakkash & Kubinski, 2020). Also, they could be useful in the treatment of irritable bowel disease through inhibition of inflammation as well as regulation of intestinal flora (Wu et al., 2020). A study by Głąbska et al. (2017) showed that higher daidzein and total isoflavone dietary intake is associated with a lower incidence of abdominal pain in patients with ulcerative colitis in remission. However, higher glycitein intake was correlated with higher constipation incidence. Another study conducted on 56 patients with ulcerative colitis in remission showed that a high intake of dietary isoflavones and daidzein may contribute to the absence of fecal mucus. On the other hand, it was suggested that higher daidzein could be responsible for higher fecal pus, but further research is necessary in order to clarify these effects (Skolmowska et al., 2019). Jalili et al. (2015) showed that supplementation with 20 mg of isoflavones twice per day (10 mg of daidzein, 8.5 mg genistein, 1.5 mg glycitein) for six weeks improved the quality of life of patients with irritable bowel syndrome (IBS), but had no effect on the severity of symptoms compared to a placebo. The same group of authors noticed reduced plasma inflammatory markers and fecal protease activity in female IBS patients after six weeks of the same isoflavone supplementation compared to placebo. The effect was more pronounced when isoflavone supplements were combined with vitamin D. No significant effects of supplementation on antioxidant status were observed (Jalili et al., 2019).

1.5 SOURCES AND EXPOSURE

1.5.1 Food

Soybeans and different soy-based products are the richest sources of phytoestrogens in the human diet (Cvejić et al., 2012). Bearing in mind that the number of vegans and vegetarians is rising, the consumption of different soy-based food is consistently increasing all over the world. Thus, many soy products have been developed and marketed as functional foods (Callou et al., 2010).

Soy food has been consumed in Asian countries for centuries, so there is a great variety of different products. Generally, nonfermented (fresh or roasted soybeans, soybean sprouts, soymilk, tofu, toasted soy protein flour) and fermented soy (miso, natto, soy sauce, soy paste, tempeh) foods can be distinguished (Zaheer & Akhtar, 2017) (Figure 1.6). Traditionally, in Asia, fermented products are popular and frequent in the diet, so China is well known for soybean sauce, stinky tofu (a form of fermented tofu), douchi (fermented black beans); Japan for miso and natto; Korea for cheonggukjang and doenjang (types of soy paste); and Indonesia for tempeh (Kwon et al., 2010; Jayachandran & Xu, 2019). In Western countries, most soy-based foods are produced by using soy flour, or by adding soy protein isolate in different products for the improvement of appearance or protein content (Ko, 2014; Barnes, 2010). The new class of so-called "second generation soy food" has been developed, including soy ice cream, soy yogurt, soy hot dogs, soy bread, etc.

The United States Department of Agriculture (USDA) has provided a database for the isoflavone content of selected food, which actually represents a compilation of internationally published scientific papers on the concentrations of isoflavones in food and is updated when new studies are published. The latest update was released in 2015 (USDA, 2015). Isoflavone content of some of the selected soy products from the database is presented in Table 1.3.

It can be noticed that isoflavone content varies greatly in soy products because of the difference in the raw material used and also due to different processing techniques applied during production. Wang and Murphy (1996) established that manufacturing steps like soaking, heat treatment, coagulation, or alkaline extraction cause substantial loss of isoflavones. Although isoflavone aglycones are relatively heat stable, baking or cooking can change the composition of the glucoside forms (Xu et al., 2002). Heat treatments that include moist heat convert malonyl glucosides to ß-glucosides (deacetylation), while dry heat (like frying in oil) induces limited conversion (decarboxylation) of

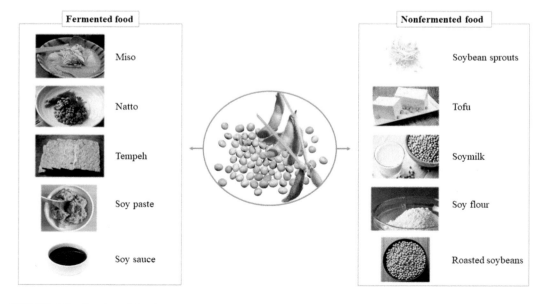

FIGURE 1.6 Soy-based food.

TABLE 1.3

Isoflavone Content in Selected Soy Foods

Food Description	Mean Content of Total Isoflavones (mg/100 g)
Soybeans, mature seeds, dry roasted (includes soy nuts)	148.50
Soy flour, different processing	150.94–178.10
Soy protein concentrate, aqueous washed	94.65
Soy protein concentrate, produced by alcohol extraction	11.49
Soy protein isolate	91.05
Instant beverage, soy, powder, not reconstituted	109.51
Soy drink	7.85
Soy protein drink	81.65
Soy yogurt	33.17
Soymilk, different types	0.70–10.73
Miso	41.45
Natto	82.29
Tempeh, different processing	35.64–72.80
Tofu, different types	9.39–83.20
Soy sauce, different types	0.10–1.18
Tempeh burger	29.00
Soy hot dog, frozen, unprepared	1.00
Soy cheese, different types	6.02–25.72

Source: USDA, 2015.

malonyl glucosides to acetyl forms (Uzzan & Labuza, 2004). Also, fermentation, usually obtained with various microorganisms such as *Bacillus subtilis* and *Aspergillus*, is known to cause conversion of glucoside forms to corresponding aglycones (Kuo et al., 2006; Chun et al., 2007; Barnes, 2010). For example, Chung et al. (2011) determined that, during the production of tofu (soy curd) and soy paste (cheonggukjang), malonyl glucosides present in soy seed were significantly decreased and hydrolyzed to aglycones by *B. subtilis*, with the result that aglycones represented 45% of total isoflavones in tofu and even 85% in soy paste. Loss of significant amounts of isoflavones during tofu making was probably due to heating, leaching in water, filtering, and coagulation (whey exclusion) (Chung et al., 2011). The loss of total isoflavones during processing can be up to 65% from the initial value of the raw material (Jackson et al., 2002). Also, during the production of soy protein concentrate, alcohol wash can be applied, and soy molasses is formed as a by-product with high isoflavone content, while the final product has very low isoflavone concentration (Table 1.3) (Uzzan & Labuza, 2004; Waggle & Bryan, 2003). Fermentation can improve physicochemical and sensory quality, as well as digestibility and nutritional value of soy food (Xu et al., 2015b). Therefore, recent studies are directed toward the assessment of the health benefits of fermented soy products. Jayachandran and Xu (2019) pointed out that the use of fermented soy foods has several advantages, showing that this type of product possesses antidiabetic, antioxidative, anti-inflammatory, antihypertensive, and anticancer potential.

Population exposure to isoflavones varies greatly in different world regions. Reported isoflavone intake among Japanese women ranged from 12 to 118.9 mg/day (47.2 ± 23.6 mg/day on average), and the main sources were tofu, natto, and miso (Arai et al., 2000; Wakai et al., 1999). The estimated daily intake of total isoflavones in women in Hong Kong ranged from 0.1 to 28.5 mg/day (7.8 ± 5.6 mg on average), where tofu was also the main dietary source of isoflavones (Chan et al., 2007). For the Korean population, the estimated isoflavone intake was around 23 mg/day

(Surh et al., 2006). Comparing the major soy-consuming countries, Messina et al. (2006) confirmed that isoflavone intake was higher in adults in Japan (25–50 mg/day) than in Singapore and Hong Kong, while in China those values varied greatly according to region. On the other hand, the estimated isoflavone intake is much lower in Europe (< 5 mg/day), although it is continuously increasing (Campos, 2020). Generally, the average isoflavone intake of the general population in Western countries is considered to be 0.1–3.3 mg/day (Mortensen et al., 2009; Zamora-Ros et al., 2012). Higher isoflavone intake – from 3 to 12 mg/day – can be observed in Western vegetarians and soy-consumers, and in vegans can even be 75 mg/day (Esch et al., 2016; Brouns, 2002; Mortensen et al., 2009). A study performed in France also showed that isoflavone intake higher than 45 mg/day is associated with a vegan diet or with health-aware intensive soy consumers (Lee et al., 2019). The so-called "Japanese paradox", referring to long life expectancy and low mortality from coronary heart disease in spite of higher total cholesterol, has contributed to the rising world popularity of soy foods, as they are a very important part of the diet in Japan (Nakamura & Ueshima, 2014). It is evident that there is a high variability of isoflavone concentration and profiles in soy products within the same category, therefore, estimation of isoflavone exposure of certain population groups based on questionnaires and official databases on isoflavone content in food gives us only a rough picture of the state (Mortensen et al., 2009; Esch et al., 2016).

Phytoestrogens can also be found in food as contaminants – for example, in cow milk, originating usually from cows fed with red clover (Hoikkala et al., 2007). This aspect is important since they present possible endocrine disruptors. It is indicated that, actually, equol content in milk could be used as a marker of animal exposure to isoflavones (Bláhová et al., 2016).

1.5.2 Dietary Supplements

Dietary supplements containing soy isoflavones are widely present on the market and they are mainly recommended as a natural alternative to hormone replacement therapy in alleviating menopausal symptoms. Moreover, they appear to be beneficial to women undergoing infertility treatments (Gaskins & Chavarro, 2018). The increased interest in the usage of soy-based dietary supplements was also caused by concerns about the potential side effects of hormone therapy (Cagnacci & Venier, 2019).

The total isoflavone content in soy-based dietary supplements can vary among producers, and even among series of the same producer (César et al., 2006). Stürz et al. (2008) determined that the total isoflavone content varied from 12.8 to 69.7 mg/sample in 11 dietary supplements containing soy extract. Nurmi et al. (2002) obtained lower values compared to declared isoflavone content and determined a range of 0.14–24.5 mg of isoflavones per tablet or capsule. Andres et al. (2015) found that soy-based supplements contained primarily glycoside isoflavone forms – genistin and daidzin – and that total isoflavone content varied from 32 to 80 mg per capsule in six analyzed products.

Generally, the analysis of dietary supplements based on soy by different authors showed that the determined content of total isoflavones often does not comply with data given by the manufacturer. Data on the individual isoflavone content are not provided on the label, although these products can have very different isoflavone profiles due to various processing techniques or primary raw material used (Boniglia et al., 2009; Stürz et al., 2008; Nurmi et al., 2002). Isoflavone composition variations may also have an impact on the therapeutic efficacy of these products (Kuiper et al., 1998). It was also pointed out that supplement labeling is generally unclear regarding the way of expressing isoflavone concentration – as aglycone equivalents or the sum of all isoflavone conjugates, which makes the control of their content complicated (Stürz et al., 2008; Nurmi et al., 2002; Almeida et al., 2015). Additionally, different information was given by manufacturers about recommended daily intake, and when calculated according to determined isoflavone content in supplements it can vary from 5.4 to 186 mg per day (Prabhakaran et al., 2006; Boniglia et al., 2009).

The field of food supplements is still not clearly regulated (Almeida et al., 2015), so the observed problems concerning the labeling and the standardization of content and composition of active ingredients should be emphasized (Andres et al., 2015) in order to have clear insight into the quality and potential health benefit of certain products on the market.

1.5.3 INFANT FORMULAS

Soy protein–based infant formulas are generally recommended in cases when breastfeeding is not possible and when there is a certain medical condition in the infant, such as galactosemia, lactose intolerance, or allergy to cow milk protein-based formula. In European countries, they are rarely used (Morandi et al., 2005), while in the USA they comprise about 20% of formulas on the market (Bhatia & Greer, 2008). These formulas draw the attention of researchers because of their high isoflavone content, thus raising the question about the safety of early life exposure to phytoestrogen compounds (Setchell et al., 1998).

In a recent publication, the determined total isoflavone content in aglycone equivalents in seven soy-based infant formulas from the Brazilian market varied from 1.12 to 5.20 mg/100 g (Fonseca et al., 2014). In some previous studies even higher concentrations were reported – up to 26.7 mg/100 g (Murphy et al., 1997; Setchell et al., 1997; Genovese & Lajolo, 2002; Irvine et al., 1998), and even 42.7 mg/100 g in one Italian study (Morandi et al., 2005). In the USDA database, reported values for isoflavones in infant formulas go up to 28 mg/100 g (USDA, 2015). When daily isoflavone intake for infants fed exclusively with soy-based formula is calculated, values can vary from 0.8 to 12 mg of isoflavones per kg of body weight, depending on age, which can lead to several times higher exposure compared to adults consuming high amounts of soy food (Murphy et al., 1997; Setchell et al., 1997; Morandi et al., 2005; Fonseca et al., 2014).

On the other hand, there are still controversies concerning the metabolism of isoflavones in infants. There are indications that infants are capable of digesting isoflavones as efficiently as adults (Irvine et al., 1998). Also, it is noted that isoflavone bioavailability may be higher in children than in adults (Halm et al., 2007), which could explain certain mentioned health benefits of early life exposure to isoflavones (Lee et al., 2009). Conversely, some findings suggest that, due to undeveloped microbiota, infants younger than four months cannot absorb these compounds (Setchell et al., 2002a).

Results of some animal studies raised the concern about the possible influence of early life exposure to phytoestrogens on reproductive organs (Wang et al., 2013; Wang et al., 2014), but soy-based infant formulas have been used for decades, and the best indicator of their safety is that there is still no clear proof that the present isoflavones cause any reproductive or other hormone-related problems in later life (Fonseca et al., 2014).

1.6 SAFETY ASPECTS

1.6.1 ADVERSE EFFECTS

The estrogen-like properties of isoflavones, although generally considered beneficial, have also raised concern for decades about the possible adverse effects in humans (Messina, 2010a). It is indicated that risks are mainly dependent on the individual's health status, and can even be connected with ethnicity (Campos & Costa, 2012). However, most of the data implying that isoflavones can have a negative influence on the reproductive system are obtained in animal studies (Jefferson et al., 2007; Jefferson et al., 2005), while clinical proofs about their unwanted effects are scarce (Messina, 2016).

Still, there are certain examples of adverse effects connected with isoflavone intake that demanded attention and additional explanations. There are studies reporting feminizing effects on men (Martinez & Lewi, 2008; Goodin et al., 2007; Chavarro et al., 2008), but these results were debated and it was indicated that these effects cannot be obtained in doses even much higher than usually consumed by Asian males (Messina, 2010b).

Furthermore, the ability of isoflavones to interact with thyroid function and hormones, demonstrating thyroid deficiency by thyroid peroxidase inhibition, raised some concerns (Divi et al., 1997; Messina, 2008). In one clinical study, it was reported that the risk of hypothyroidism was increased with 16 mg of isoflavone intake (Sathyapalan et al., 2011). Additionally, in a case report published by Nakamura et al. (2017), hypothyroidism caused by a soy isoflavon–based drink was observed in a patient with chronic lymphocytic thyroiditis. It was generally concluded that individuals with subclinical hypothyroidism, iodine deficiency, or thyroid disfunction should be careful with isoflavone supplementation, but these findings should be further investigated (Messina, 2010a; Hüser et al., 2018).

Meta-analysis of the phytoestrogen side effects in clinical trials showed that only a moderately elevated rate of gastrointestinal problems was observed (Tempfer et al., 2009). Messina (2010a) pointed out that the only reason for removing soy from a diet could be a proven allergic reaction. It is often discussed whether certain population groups, such as breast cancer patients or children, are at higher risk of exerting adverse isoflavone hormonal effects, but these claims are not supported by the clinical and epidemiologic data (Messina, 2016). Conversely, it is even suggested that in breast cancer patients, higher soy isoflavone consumption can result in a significant reduction in tumor recurrence and better survival (Messina, 2014; Chi et al., 2013). Moreover, there are indications that early life exposure to isoflavones can act preventively for breast cancer risk in adults, but the safety of high isoflavone intake by infants is still controversial. Testa et al. (2018) recently assessed the risks of soy-based infant formula usage and concluded that there is no clear proof of negative effects of isoflavones on development, although in children with congenital hypothyroidism additional monitoring is needed. Also, it is pointed out that fetal and infant risk of isoflavone exposure by vegan mothers is still to be determined. The US National Center for Complementary and Integrative Health (NCCIH) has confirmed the safety of soy food consumption in women cured of or with risk for breast cancer, although this statement did not include isoflavone dietary supplements (NCCIH, 2020). The European Food Safety Authority also concluded that isoflavones do not negatively affect the breast, thyroid, or uterus of healthy postmenopausal women (EFSA, 2015). Still, caution with high doses is needed in women with breast cancer risk (Touillaud et al., 2019).

1.6.2 RECOMMENDATIONS

There is no general consensus about the most health-beneficial daily dose of isoflavones, but certain conclusions about optimal dosage range can be drawn from numerous clinical and epidemiological studies and traditional consumption in Asia.

Clinical studies have been performed using different sources of isoflavones – from pure compounds to different soy foods and dietary supplements – which contain various amounts of isoflavones as previously stated, so there are often difficulties in connecting the obtained health effect with the applied isoflavone dose. Namely, in most previously mentioned clinical studies with positive health outcomes, isoflavone doses from 30 to 80 mg or in some cases > 100 mg/day were found to be effective. Rizzo and Baroni (2018) confirmed that 60–100 mg of isoflavones daily is needed for obtaining certain health effects. These values also correspond to those consumed by populations in countries like Japan or China where soy food is a part of everyday diet. Watanabe and Uehara (2019) concluded that, although isoflavone intake in Japan can go up to 200 mg/day, the recommended dosage may be around 100 mg/day. In Western countries, these doses are achieved only by vegans or vegetarians, or by isoflavone supplement users.

1.7 CONCLUDING REMARKS

The plethora of beneficial health effects of soy isoflavones results from their estrogen-like molecular structure and consequential binding to estrogen receptors. Moreover, antioxidant and anti-inflammatory as well as different tissue-specific mechanisms are also involved in the prevention and treatment of various diseases. Although many clinical studies provide data on the positive effects of isoflavones on menopausal symptoms, cancer, osteoporosis, and cardiovascular disease, the overall results are not convincing. Therefore, when dealing with this topic, different aspects must be considered. First of all, the biological activity of equol as the main microbial metabolite of daidzein should be elucidated. Also, the variable equol-producing capacity of individuals implies that the gut microbiota status can significantly affect the final health outcome of isoflavone intake. Moreover, data obtained in human trials are heterogeneous and have certain limitations. They can underestimate the potential of soy isoflavones due to the relatively short duration, non-standardized outcome measures, and inter-individual variability among tested subjects at baseline. Results of long-term epidemiological studies are influenced by variable approaches in the assessment of dietary isoflavone intake, as well as a lack of precise data on real isoflavone content and composition in specific foods. Also, concerning isoflavone supplementation, poor quality control leads to great variation in isoflavone amounts, so series of products should always be analyzed before application.

Future studies should provide insight into the types of soy products or food that are most favorable as isoflavone sources. Individual differences observed in subjects should be taken into consideration when analyzing outcomes, and a desirable gut microbiota profile could be revealed. Also, new aspects of isoflavone use are to be further investigated.

Finally, although isoflavones have been extensively studied in the last decades, definite conclusions on their health benefits and safety are still to be drawn.

ACKNOWLEDGMENT

This work is a part of a project financed by the Ministry of Science and Technological Development, Republic of Serbia, no. TR31022.

REFERENCES

Accorsi-Neto, A., Haidar, M., Simões, R., Simões, M., Soares-Jr, J., Baracat, E. (2009). Effects of isoflavones on the skin of postmenopausal women: a pilot study. *Clinics*, 64(6): 505–510.

Adjakly, M., Bosviel, R., Rabiau, N., Boiteux, J.P., Bignon, Y.J., Guy, L., Bernard-Gallon, D. (2011). DNA methylation and soy phytoestrogens: quantitative study in DU-145 and PC-3 human prostate cancer cell lines. *Epigenomics*, 3(6): 795–803.

Ahuja, V., Miura, K., Vishnu, A., Fujiyoshi, A., Evans, R., Zaid, M., Miyagawa, N., Hisamatsu, T., Kadota, A., Okamura, T., Ueshima, H., Sekikawa, A. (2017). Significant inverse association of equol-producer status with coronary artery calcification but not dietary isoflavones in healthy Japanese men. *British Journal of Nutrition*, 117: 260–266.

Ajdžanović, V., Filipović, B., Miljić, D., Mijatović, S., Maksimović-Ivanić, D., Miler, M., Živanović, J., Milošević, V. (2019). Prostate cancer metastasis and soy isoflavones: a dogfight over a bone. *EXCLI Journal*, 18: 106–126.

Akhlaghi, M., Nasab, M.G., Riasatian, M., Sadeghi, F. (2019). Soy isoflavones prevent bone resorption and loss, a systematic review and meta-analysis of randomized controlled trials. *Critical Reviews in Food Science and Nutrition*, 10: 1–15.

Almeida, I.M.C., Rodrigues, F., Sarmento, B., Alves, R.C., Oliveira, M.B.P.P. (2015). Isoflavones in food supplements: chemical profile, label accordance and permeability study in Caco-2 cells. *Food & Function*, 6: 938–946.

Al-Nakkash, L., Kubinski, A. (2020). Soy isoflavones and gastrointestinal health. *Current Nutrition Reports*, 9: 193–201. doi: 10.1007/s13668-020-00314-4.

Amaral, C., Toloi, M.R.T., Vasconcelos, L.D., Fonseca, M.J.V., Correia-da-Silva, G., Teixeira, N. (2017). The role of soybean extracts and isoflavones in hormone-dependent breast cancer: aromatase activity and biological effects. *Food & Function*, 8(9): 3064–3074.

Ambra, R., Rimbach, G., de Pascual Teresa, S., Fuchs, D., Wenzel, U., Daniel, H., Virgili, F. (2006). Genistein affects the expression of genes involved in blood pressure regulation and angiogenesis in primary human endothelial cells. *Nutrition, Metabolism and Cardiovascular Diseases*, 16(1): 35–43.

Andres, S., Hansen, U., Niemann, B., Palavinskas, R., Lampen, A. (2015). Determination of the isoflavone composition and estrogenic activity of commercial dietary supplements based on soy or red clover. *Food & Function*, 6(6): 2017–2025.

Applegate, C.C., Rowles, J.L., Ranard, K.M., Jeon, S., Erdman, J.W. (2018). Soy consumption and the risk of prostate cancer: an updated systematic review and meta-analysis. *Nutrients*, 10(1): 40–65.

Arai, Y., Watanabe, S., Kimira, M., Shimoi, K., Mochizuki, R., Kinae, N. (2000). Dietary Intakes of flavonols, flavones and isoflavones by Japanese women and the inverse correlation between quercetin intake and plasma LDL cholesterol concentration. *The Journal of Nutrition*, 130: 2243–2250.

Aso, T., Uchiyama, S., Matsumura, Y., Taguchi, M., Nozaki, M., Takamatsu, K., Ishizuka, B., Kubota, T., Mizunuma, H., Ohta, H. (2012). A natural S-equol supplement alleviates hot flushes and other menopausal symptoms in equol nonproducing postmenopausal Japanese women. *Journal of Womens Health (Larchmt)*, 21(1): 92–100.

Atkinson, C., Frankenfeld, C.L., Lampe, J.W. (2005). Gut bacterial metabolism of the soy isoflavone daidzein: exploring the relevance to human health. *Experimental Biology and Medicine*, 230: 155–170.

Bae, M., Woo, M., Kusuma, I.W., Arung, E.T., Yang, C.H., Kim, Y.U. (2012). Inhibitory effects of isoflavonoids on rat prostate testosterone 5α-reductase. *Journal of Acupuncture and Meridian Studies*, 5(6): 319–322.

Barnes, S. (2010). The biochemistry, chemistry and physiology of the isoflavones in soybeans and their food products. *Lymphatic Research and Biology*, 8(1): 89–98.

Bennetts, H.W., Underwood, E.J., Shier, F.L. (1946). A specific breeding problem of sheep on subterranean clover pastures in Western Australia. *Australian Veterinary Journal*, 22: 2–12.

Benvenuti, C., Setnikar, I. (2011). Effect of *Lactobacillus sporogenes* on oral isoflavones bioavailability: single dose pharmacokinetic study in menopausal women. *Arzneimittelforschung*, 61(11): 605–609.

Bhatia, J., Greer, F. (2008). Use of soy protein-based formulas in infant feeding. *Pediatrics*, 121: 1062–1068.

Bláhová, L., Kohoutek, J., Procházková, T., Prudíková, M., Bláha, L. (2016). Phytoestrogens in milk: overestimations caused by contamination of the hydrolytic enzyme used during sample extraction. *Journal of Dairy Science*, 99: 6973–6982.

Blanco Mejia, S., Messina, M., Li, S.S., Viguiliouk, E., Chiavaroli, E., Khan, T.A., Srichaikul, K., Mirrahimi, A., Sievenpiper, J.L., Kris-Etherton, P., Jenkins, D.J.A. (2019). A meta-analysis of 46 studies identified by the FDA demonstrates that soy protein decreases circulating LDL and total cholesterol concentrations in adults. *The Journal of Nutrition*, 149: 968–981.

Bodo, C., Rissman, E.F. (2006). New roles for estrogen receptor beta in behavior and neuroendocrinology. *Frontiers in Neuroendocrinology*, 27: 217–232.

Bolanos, R., Del Castillo, A., Francia, J. (2010). Soy isoflavones versus placebo in the treatment of climacteric vasomotor symptoms: systematic review and meta-analysis. *Menopause*, 17(3): 660–666.

Boniglia, C., Carratu, B., Gargiulo, R., Giammarioli, S., Mosca, M., Sanzini, E. (2009). Content of phytoestrogens in soy-based dietary supplements. *Food Chemistry*, 115: 1389–1392.

Bosviel, R., Dumollard, E., Déchelotte, P., Bignon, Y.J., Bernard-Gallon, D. (2012). Can soy phytoestrogens decrease DNA methylation in BRCA1 and BRCA2 oncosuppressor genes in breast cancer? *Omics: A Journal of Integrative Biology*, 16(5): 235–244.

Boue, S.M., Cleveland, T.E., Carter-Wientjes, C., Shih, B.Y., Bhatnagar, D., McLachlan, J.M., Burow, M.E. (2009). Phytoalexin-enriched functional foods. *Journal of Agricultural and Food Chemistry*, 57(7): 2614–2622.

Bowey, E., Adlercreutz, H., Rowland, I. (2003). Metabolisms of isoflavones and lignans by the gut microflora — a study in germfree and human flora associated rats. *Food and Chemical Toxicology*, 41(5): 631–636.

Brouns, F. (2002). Soya isoflavones: a new and promising ingredient for the health foods sector. *Food Research International*, 35: 187–193.

Bursać, M., Atanacković-Krstonošić, M., Miladinović, J., Malenčić, Đ., Gvozdenović, Lj., Cvejić-Hogervorst, J. (2017). Isoflavone composition, total phenolic content and antioxidant capacity of soybeans with colored seed coat. *Natural Product Communications*, 12(4): 527–532.

Cagnacci, A., Venier, M. (2019). The controversial history of hormone replacement therapy. *Medicina*, 55(9): 602.

Callou, K.R.D.A., Sadigov, S., Lajolo, F.M., Genovese, M.I. (2010). Isoflavones and antioxidant capacity of commercial soy-based beverages: effect of storage. *Journal of Agricultural and Food Chemistry*, 58: 4284–4291.

Campos, M.G. (2020). Soy isoflavones. In Xiao, J., et al. (Eds.), *Handbook of dietary phytochemicals*. Springer, Singapore. doi: 101007/978-981-13-1745-3_8-1.

Campos, M.G., Costa, M.L. (2012). Possible risks in Caucasians by consumption of isoflavones extracts based. In Eissa, A.A. (Ed.), *Structure and function of food engineering*. InTechOpen, London. pp. 205–214. doi: 10.5772/47837.

Cano, A., García-Pérez, M.Á., Tarín, J.J. (2010). Isoflavones and cardiovascular disease. *Maturitas*, 67(3): 219–226.

Carr, A., Frei, B. (2000). The role of natural antioxidants in preserving the biological activity of endothelium-derived nitric oxide. *Free Radical Biology and Medicine*, 28(12): 1806–1814.

Cassidy, A., Bingham, S., Setchell, K.D.R. (1994). Biological effects of a diet of soy protein rich in isoflavones on the menstrual cycle of premenopausal women. *The American Journal of Clinical Nutrition*, 60: 333–340.

Cassidy, A., Minihane, A.-M. (2017). The role of metabolism (and the microbiome) in defining the clinical efficacy of dietary flavonoids. *The American Journal of Clinical Nutrition*, 105: 10–22.

Cederroth, C.R., Vinciguerra, M., Gjinovci, A., Kühne, F., Klein, M., Cederroth, M., Caille, D., Suter, M., Neumann, D., James, R.W., Doerge, D.R., Wallimann, T., Meda, P., Foti, M., Rohner-Jeanrenaud, F., Vassalli, J.D., Nef, S. (2008). Dietary phytoestrogens activate AMP-activated protein kinase with improvement in lipid and glucose metabolism. *Diabetes*, 57(5): 1176–1185.

César, I.C., Braga, F.C., Soares, C.D.V., Nunan, E.A., Pianetti, G.A., Condessa, F.A., Barbosa, T.A.F., Campos, L.M.M. (2006). Development and validation of a RP-HPLC method for quantification of isoflavone aglycones in hydrolyzed soy dry extracts. *Journal of Chromatography B*, 836: 74–78.

Chan, S.G., Ho, S.C., Kreiger, N., Darlington, G., So, K.F., Chong, P.Y.Y. (2007). Dietary sources and determinants of soy isoflavone intake among midlife Chinese women in Hong Kong. *The Journal of Nutrition*, 137: 2451–2455.

Chang, Y., Chou, R. (2013). Plasma pharmacokinetics and urinary excretion of isoflavones after ingestion of soy products with different aglycone/glucoside ratios in South Korean women. *Nutrition Research and Practice*, 7(5): 393–399.

Chavarro, J.E., Toth, T.L., Sadio, S.M., Hauser, R. (2008). Soy food and isoflavone intake in relation to semen quality parameters among men from an infertility clinic. *Human Reproduction*, 23: 2584–2590.

Chen, J., Duan, Y., Zhang, X., Ye, Y., Ge, B., Chen, J. (2015a). Genistein induces apoptosis by the inactivation of the IGF-1R/p-Akt signaling pathway in MCF-7 human breast cancer cells. *Food & Function*, 6(3): 995–1000.

Chen, M-N., Lin, C-C., Liu, C-F. (2015b). Efficacy of phytoestrogens for menopausal symptoms: a meta-analysis and systematic review. *Climacteric*, 18: 260–269.

Chi, F., Wu, R., Zeng, Y.-C., Xing, R., Liu, J., Xu, Z.-G. (2013). Post-diagnosis soy food intake and breast cancer survival: a meta-analysis of cohort studies. *Asian Pacific Journal of Cancer Prevention*, 14: 2407–2412.

Chiari, L., Piovesan, N.D., Naoe, L.K., José, I.C., Viana, J.M.S., Moreira, M.A., de Barros, E.G. (2004). Genetic parameters relating isoflavone and protein content in soybean seeds. *Euphytica*, 138: 55–60.

Cho, S.H., Jo, A., Casale, T., Jeong, S.J., Hong, S.J., Cho, J.K., Holbrook, J.T., Kumar, R., Smith, L.J. (2019). Soy isoflavones reduce asthma exacerbation in asthmatic patients with high PAI-1–producing genotypes. *The Journal of Allergy and Clinical Immunology*, 144 (1): 109–117.

Choi, E.J., Kim, G.H. (2008). Daidzein causes cell cycle arrest at the G1 and G2/M phases in human breast cancer MCF-7 and MDA-MB-453 cells. *Phytomedicine*, 15(9): 683–690.

Choi, E.J., Kim, G.H. (2013). O-desmethylangolensin inhibits the proliferation of human breast cancer MCF-7 cells by inducing apoptosis and promoting cell cycle arrest. *Oncology Letters*, 6(6): 1784–1788.

Choi, E.J., Kim, G.H. (2014). The antioxidant activity of daidzein metabolites, O-desmethylangolensin and equol, in HepG2 cells. *Molecular Medicine Reports*, 9: 328–332.

Choi, E.J., Lee, J.-I., Kim, G.-H. (2013). Anticancer effects of O-desmethylangolensin are mediated through cell cycle arrest at the G2/M phase and mitochondrial-dependent apoptosis in Hep3B human hepatocellular carcinoma cells. *International Journal of Molecular Medicine*, 31: 726–730.

Chun, J., Kim, G.M., Lee, K.W., Choi, I.D., Kwon, G.H., Park, J.Y., Jeong, S.J., Kim, J.S., Kim, J.H. (2007). Conversion of isoflavone glucosides to aglycones in soymilk by fermentation with lactic acid bacteria. *Journal of Food Science*, 72: M39–44.

Chung, I.-M., Seo, S.-H. Ahn, J.-K., Kim, S-H. (2011). Effect of processing, fermentation, and aging treatment to content and profile of phenolic compounds in soybean seed, soy curd and soy paste. *Food Chemistry*, 127: 960–967.

Clavel, T., Lepage, P., Charrier, C. (2014). The family *Coriobacteriaceae*. In Rosenberg, E., DeLong, E.F., Lory, S., Stackebrandt, E., Thompson, F. (Eds.), *The prokaryotes actinobacteria*. Berlin/Heidelberg, Germany: Springer, pp. 201–238.

Cui, C., Birru, R.L., Snitz, B.E., Ihara, M., Kakuta, C., Lopresti, B.J., Aizenstein, H.J., Lopez, O.L., Mathis, C.A., Miyamoto, Y., Kuller, L.H., Sekikawa, A. (2020a). Effects of soy isoflavones on cognitive function: a systematic review and meta-analysis of randomized controlled trials. *Nutrition Reviews*, 78(2): 134–144.

Cui, Y., Huang, C., Momma, H., Niu, K., Nagatomi, R. (2020b). Daily dietary isoflavone intake in relation to lowered risk of depressive symptoms among men. *Journal of Affective Disorders*, 261: 121–125.

Cui, Y., Niu, K., Huang, C., Momma, H., Guan, L., Kobayashi, Y., Guo, H., Chujo, M., Otomo, A., Nagatomi, R. (2015). Relationship between daily isoflavone intake and sleep in Japanese adults: a cross-sectional study. *Nutrition Journal*, 14: 127.

Cvejić, J., Bursać, M., Atanacković, M. (2012). Phytoestrogens: "Estrogene-like" phytochemicals. In Atta-ur-Rahman (Ed.), *Studies in natural products chemistry, bioactive natural products*. Elsevier, Amsterdam. Chapter 1, 38, pp. 1–35.

Cvejić, J., Malenčić, Đ., Tepavčević, V., Poša, M., Miladinović, J. (2009). Determination of phytoestrogen composition in soybean cultivars in Serbia. *Natural Product Communications*, 4: 1–6.

Cvejić, J., Tepavčević, V., Bursać, M., Miladinović, J., Malenčić, Đ. (2011). Isoflavone composition in F1 soybean progenies. *Food Research International*, 44: 2698–2702.

Damasceno, N.R.T., Apolinário, E., Flauzino, F.D., Fernandes, I., Abdalla, D.S.P. (2007). Soy isoflavones reduce electronegative low-density lipoprotein (LDL−) and anti-LDL− autoantibodies in experimental atherosclerosis. *European Journal of Nutrition*, 46(3): 125–132.

Davinelli, S., Scapagnini, G., Marzatico, F., Nobile, V., Ferrara, N., Corbi, G. (2017). Influence of equol and resveratrol supplementation on health-related quality of life in menopausal women: a randomized, placebo-controlled study. *Maturitas*, 96: 77–83.

Day, A.J., Cañada, F.J., Díaz, J.C., Kroon, P.A., Mclauchlan, R., Faulds, C.B., Plumb, G.W., Morgan, M.R.A., Wiliamson, G. (2000). Dietary flavonoid and isoflavone glycosides are hydrolysed by the lactate site of lactate phlorizin hydrolase. *FEBS Letters*, 468(2): 166–170.

Day, A.J., DuPont, M.S., Ridley, S., Rhodes, M., Rhodes, M.J., Morgan, M.R., Williamson, G. (1998). Deglycosylation of flavonoid and isoflavonoid glycosides by human small intestine and liver beta-glucosidase activity. *FEBS Letters*, 436: 71–75.

de Cremoux, P., This, P., Leclercq, G., Jacquot, Y. (2010). Controversies concerning the use of phytoestrogens in menopause management: bioavailability and metabolism. *Maturitas*, 65: 334–339.

Dharmappa, K.K., Mohamed, R., Shivaprasad, H.V., Vishwanath, B.S. (2010). Genistein, a potent inhibitor of secretory phospholipase A 2: a new insight in down regulation of inflammation. *Inflammopharmacology*, 18(1): 25–31.

Divi, R.L., Chang, H.C., Doerge, D.R. (1997). Anti-thyroid isoflavones from soybean: isolation, characterization, and mechanisms of action. *Biochemical Pharmacology*, 54: 1087–1096.

Duru, K.C., Kovaleva, E.G., Danilova, I.G., van der Bijl, P., Belousova, A.V. (2018). The potential beneficial role of isoflavones in type 2 diabetes mellitus. *Nutrition Research*, 59: 1–15.

Eastell, R., Szulc, P. (2017). Use of bone turnover markers in postmenopausal osteoporosis. *The Lancet Diabetes & Endocrinology*, 5(11): 908–923.

EFSA (European Food Safety Authority) Panel on Food Additives and Nutrient Sources added to Food (ANS). (2015). Risk assessment for peri- and post-menopausal women taking food supplements containing isolated isoflavones. *EFSA Journal*, 13(10): 4246.

Elghali, S., Mustafa, S., Amid, M., Yaizd, M., Ismail, A., Abas, F. (2012). Bioconversion of daidzein to equol by *Bifidobacterium breve* 15700 and *Bifidobacterium longum* BB536. *Journal of Functional Foods*, 4: 736–745.

El-Kordy, E.A., Alshahrani, A.M. (2015). Effect of genistein, a natural soy isoflavone, on pancreatic β-cells of streptozotocin-induced diabetic rats: histological and immunohistochemical study. *Journal of Microscopy and Ultrastructure*, 3(3): 108–119.

Esch, H.L., Kleider, C., Scheffler, A., Lehmann, L. (2016). Isoflavones: toxicological aspects and efficacy. In Gupta, R.C. (Ed.), *Nutraceuticals: efficacy, safety and toxicity*. Academic Press, London. pp. 465–487.

Fang, K., Dong, H., Wang, D., Gong, J., Huang, W., Lu, F. (2016). Soy isoflavones and glucose metabolism in menopausal women: a systematic review and meta-analysis of randomized controlled trials. *Molecular Nutrition & Food Research*, 60(7): 1602–1614.

Farwick, M., Köhler, T., Schild, J., Mentel, M., Maczkiewitz, U., Pagani, V., Bonfigli, A., Rigano, L., Bureik, D., Gauglitz, G.G. (2014). Pentacyclic triterpenes from terminalia arjuna show multiple benefits on aged and dry skin. *Skin Pharmacology and Physiology*, 27: 71–81.

Figtree, G.A., Griffiths, H., Lu, Y.Q., Webb, C.M., MacLeod, K., Collins, P. (2000). Plant-derived estrogens relax coronary arteries in vitro by a calcium antagonistic mechanism. *Journal of the American College of Cardiology*, 35(7): 1977–1985.

Fonseca, N.D., Villar, M.P.M., Donangelo, C.M., Perrone, D. (2014). Isoflavones and soyasaponins in soy infant formulas in Brazil: profile and estimated consumption. *Food Chemistry*, 143: 492–498.

Food and Drug Administration (FDA). (1999). Food labeling: health claims; soy protein and coronary heart disease. Final rule. *Federal Register*, 64(206): 57700–57733. http://www.fda.gov/

Franco, O.H., Chowdhury, R., Troup, J., Voortman, T., Kunutsor, S., Kavousi, M., Oliver-Williams, C., Muka, T. (2016). Use of plant-based therapies and menopausal symptoms a systematic review and meta-analysis. *JAMA*, 315(23): 2554–2563.

Franke, A.A., Custer, L.J., Hundahl, S.A. (2004). Determinants for urinary and plasma isoflavones in humans after soy intake. *Nutrition and Cancer*, 50(2): 141–154.

Franke, A.A., Lai, J.F., Halm, B.M. (2014). Absorption, distribution, metabolism, and excretion of isoflavonoids after soy intake. *Archives of Biochemistry and Biophysics*, 559: 24–28.

Frankenfeld, C.L. (2011). O-desmethylangolensin: the importance of equol's lesser known cousin to human health. *Advances in Nutrition*, 2: 317–324.

Frankenfeld, C.L., Atkinson, C., Thomas, W.K., Goode, E.L., Gonzalez, A., Jokela, T., Wähälä, K., Schwartz, S.M., Li, S.S., Lampe, J.W. (2004a). Familial correlations, segregation analysis, and nongenetic correlates of soy isoflavone-metabolizing phenotypes. *Experimental Biology and Medicine (Maywood)*, 229: 902–913.

Frankenfeld, C.L., McTiernan, A., Aiello, E.J., Thomas, W.K., LaCroix, K., Schramm, J., Schwartz, S.M., Holt, V.L., Lampe, J.W. (2004). Mammographic density in relation to daidzein-metabolizing phenotypes in overweight, postmenopausal women. *Cancer Epidemiology, Biomarkers & Prevention*, 13: 1156–1162.

Frankenfeld, C.L., McTiernan, A., Thomas, W.K. (2006). Postmenopausal bone mineral density in relation to soy isoflavone-metabolizing phenotypes. *Maturitas*, 53(3): 315–324.

Fu, Z., Gilbert, E.R., Pfeiffer, L., Zhang, Y., Fu, Y., Liu, D. (2012). Genistein ameliorates hyperglycemia in a mouse model of nongenetic type 2 diabetes. *Applied Physiology, Nutrition, and Metabolism*, 37(3): 480–488.

García-Martínez, M.C., Hermenegildo, C., Tarín, J.J., Cano, A. (2003). Phytoestrogens increase the capacity of serum to stimulate prostacyclin release in human endothelial cells. *Acta Obstetricia et Gynecologica Scandinavica*, 82(8): 705–710.

Gaskins, A.J., Chavarro, J.E. (2018). Diet and fertility: a review. *American Journal of Obstetrics & Gynecology*, 218(4): 379–389.

Gavin, K.M., Seals, D.R., Silver, A.E., Moreau, K.L. (2009). Vascular endothelial estrogen receptor alpha is modulated by estrogen status and related to endothelial function and endothelial nitric oxide synthase in healthy women. *The Journal of Clinical Endocrinology and Metabolism*, 94: 3513–3520.

Gaya, P., Peirotén, Á., Álvarez, I., Medina, M., José, M.L. (2018). Production of the bioactive isoflavone O-desmethylangolensin by *Enterococcus faecium* INIA P553 with high efficiency. *Journal of Functional Foods*, 40: 180–186.

Genovese, M.I., Lajolo, F.M. (2002). Isoflavones in soy based foods consumed in Brazil: levels, distribution and estimated intake. *Journal of Agricultural and Food Chemistry*, 50: 5987–5993.

Głąbska, D., Guzek, D., Grudzińska, D., Lech, G. (2017). Influence of dietary isoflavone intake on gastrointestinal symptoms in ulcerative colitis individuals in remission. *World Journal of Gastroenterology*, 23: 5356–5363.

González, M., Cabrera-Socorro, A., Perez-Garcia, C.G., Fraser, J.D., Lopez, F.J., Alonso, R., Meyer, G. (2007). Distribution patterns of estrogen receptor a and b in the human cortex and hippocampus during development and adulthood. *Journal of Comparative Neurology*, 503: 790–802.

Goodin, S., Shen, F., Shih, W.J., Dave, N., Kane, M.P., Medina, P., Lambert, G.H., Aisner, J., Gallo, M., DiPaola, R.S. (2007). Clinical and biological activity of soy protein powder supplementation in healthy male volunteers. *Cancer Epidemiology, Biomarkers & Prevention*, 16: 829–833.

Ha, B.G., Nagaoka, M., Yonezawa, T., Tanabe, R., Woo, J.T., Kato, H., Chung, U.I., Yagasaki, K. (2012). Regulatory mechanism for the stimulatory action of genistein on glucose uptake in vitro and in vivo. *The Journal of Nutritional Biochemistry*, 23(5): 501–509.

Hall, W.L., Formanuik, N.L., Harnpanich, D., Cheung, M., Talbot, D., Chowienczyk, P.J., Sanders, T.A. (2008). A meal enriched with soy isoflavones increases nitric oxide-mediated vasodilation in healthy postmenopausal women. *The Journal of Nutrition*, 138(7): 1288–1292.

Halm, B.M., Ashburn, L.A., Franke, A.A. (2007). Isoflavones from soya foods are more bioavailable in children than adults. *British Journal of Nutrition*, 98: 998–1005.

Hati, S., Vij, S., Singh, B.P., Manda, S. (2015). β-Glucosidase activity and bioconversion of isoflavones during fermentation of soymilk. *Journal of the Science of Food and Agriculture*, 95: 216–220.

Haudum, C., Lindheim, L., Ascani, A., Trummer, C., Horvath, A., Münzker, J., Obermayer-Pietsch, B. (2020). Impact of short-term isoflavone intervention in polycystic ovary syndrome (PCOS) patients on microbiota composition and metagenomics. *Nutrients*, 12: 622.

He, J., Wang, S., Zhou, M., Yu, W., Zhang, Y., He, X. (2015). Phytoestrogens and risk of prostate cancer: a meta-analysis of observational studies. *World Journal of Surgical Oncology*, 13: 231.

Heinonen, S-M., Hoikkala, A., Wähälä, K., Adlercreutz, H. (2003). Metabolism of the soy isoflavones daidzein, genistein and glycitein in human subjects. Identification of new metabolites having an intact isoflavonoid skeleton. *Journal of Steroid Biochemistry & Molecular Biology*, 87: 285–299.

Heinonen, S-M., Wähälä, K., Adlercreutz, H. (2002). Metabolism of isoflavones in human subjects. *Phytochemistry Reviews*, 1: 175–182.

Hermenegildo, C., Oviedo, P.J., García-Pérez, M.A., Tarín, J.J., Cano, A. (2005). Effects of phytoestrogens genistein and daidzein on prostacyclin production by human endothelial cells. *Journal of Pharmacology and Experimental Therapeutics*, 315(2): 722–728.

Hillman, G.G. (2018). Soy isoflavones protect normal tissues while enhancing radiation responses. *Seminars in Radiation Oncology*, 29: 62–71.

Hodgson, J.M., Croft, K.D., Puddey, I.B., Mori, T.A., Beilin, L.J. (1996). Soybean isoflavonoids and their metabolic products inhibit *in vitro* lipoprotein oxidation in serum. *Journal of Nutritional Biochemistry*, 7: 664–669.

Hoeck, J.A., Fehr, W.R., Murphy, P.A., Welke, G.A. (2000). Influence of genotype and environment on isoflavone contents of soybean. *Crop Science*, 4: 48–51.

Hogervorst, E., Mursjid, F., Priandini, D., Setyawan, H., Ismael, R.I., Bandelow, S., Rahardjo, T.B. (2011). Borobudur revisited: soy consumption may be associated with better recall in younger, but not in older, rural Indonesian elderly. *Brain Research*, 1379: 206–212.

Hogervorst, E., Sadjimim, T., Yesufu, A., Kreager, P., Rahardjo, T.B. High tofu intake is associated with worse memory in elderly Indonesian men and women. (2008). *Dementia and Geriatric Cognitive Disorders*, 26(1): 50–57.

Hoikkala, A., Mustonen, E., Saastamoinen, I., Jokela, T., Taponen, J., Saloniemi, H., Wähälä, K. (2007). High levels of equol in organic skimmed Finnish cow milk. *Molecular Nutrition & Food Research*, 51: 782–786.

Hu, G.X., Zhao, B.H., Chu, Y.H., Zhou, H.Y., Akingbemi, B.T., Zheng, Z.Q., Ge, R.S. (2010). Effects of genistein and equol on human and rat testicular 3β-hydroxysteroid dehydrogenase and 17β-hydroxysteroid dehydrogenase 3 activities. *Asian Journal of Andrology*, 12(4): 519–526.

Hüser, S., Guth, S., Joost, H.G., Soukup, S.T., Köhrle, J., Kreienbrock, L., Diel, P., Lachenmeier, D.W., Eisenbrand, G., Vollmer, G., Nöthlings, U., Marko, D., Mally, A., Grune, T., Lehmann, L., Steinberg, P., Kulling, S.E. (2018). Effects of isoflavones on breast tissue and the thyroid hormone system in humans: a comprehensive safety evaluation. *Archives of Toxicology*, 92: 2703–2748.

Hussain, A., Harish, G., Prabhu, S.A., Mohsin, J., Khan, M.A., Rizvi, T.A., Sharma, C. (2012). Inhibitory effect of genistein on the invasive potential of human cervical cancer cells via modulation of matrix metalloproteinase-9 and tissue inhibitiors of matrix metalloproteinase-1 expression. *Cancer Epidemiology*, 36(6): e387–e393.

Hwang, C.S., Kwak, H.S., Lim, H.J., Lee, S.H., Kang, Y.S., Choe, T.B., Hur, H.G., Han, K.O. (2006). Isoflavone metabolites and their in vitro dual functions: they can act as an estrogenic agonist or antagonist depending on the estrogen concentration. *The Journal of Steroid Biochemistry and Molecular Biology*, 101(4–5): 246–253.

Irvine, C.H.G., Shand, N., Fitzpatrick, M., Alexander, S.L. (1998). Daily intake and urinary excretion of genistein and daidzeína by infants fed soy-or dairy-based infant formulas. *The American Journal Clinical Nutrition*, 68: 1462S–1465S.

Islam, A., Islam, M.S., Uddin, M.N., Hasan, M.M.I., Akanda, M.R. (2020). The potential health benefits of the isoflavone glycoside genistin. *Archives of Pharmacal Research*, 43: 395–408.

Izumi, T., Piskula, M.K., Osawa, S., Obata, A., Tobe, K., Saito, M., Kataoka, S., Kubota, Y., Kikuchi, M. (2000). Soy isoflavone aglycones are absorbed faster and in higher amounts than their glucosides in humans. *The Journal of Nutrition*, 130: 1695–1699.

Jackman, K.A., Woodman, O.L., Chrissobolis, S., Sobey, C.G. (2007). Vasorelaxant and antioxidant activity of the isoflavone metabolite equol in carotid and cerebral arteries. *Brain Research*, 1141: 99–107.

Jackson, C.J.C., Dini, J.P., Lavandier, C., Rupasinghe, H.P.V., Faulkner, H., Poysa, V., Buzzell, D., DeGrandis, S. (2002). Effects of processing on the content and composition of isoflavones during manufacturing of soy beverage and tofu. *Process Biochemistry*, 37: 1117–1123.

Jalili, M., Vahedi, H., Janani, L., Poustchi, H., Malekzadeh, R., Hekmatdoost, A. (2015). Soy isoflavones supplementation for patients with irritable bowel syndrome: a randomized double blind clinical trial. *Middle East Journal of Digestive Diseases*, 7(3): 170–176.

Jalili, M., Vahedi, H., Poustchi, H., Hekmatdoost, A. (2019). Soy isoflavones and cholecalciferol reduce inflammation, and gut permeability, without any effect on antioxidant capacity in irritable bowel syndrome: a randomized clinical trial. *Clinical Nutrition ESPEN*, 34: 50–54.

Jamilian, M., Asemi, Z. The effects of soy isoflavones on metabolic status of patients with polycystic ovary syndrome. (2016). *The Journal of Clinical Endocrinology & Metabolism*, 101: 3386–3394.

Jayachandran, M., Xu, B. (2019). An insight into the health benefits of fermented soy products. *Food Chemistry*, 271: 362–371.

Jefferson, W.N., Padilla-Banks, E., Newbold, R.R. (2005). Adverse effects on female development and reproduction in CD-1 mice following neonatal exposure to the phytoestrogen genistein at environmentally relevant doses. *Biology of Reproduction*, 73: 798–806.

Jefferson, W.N., Padilla-Banks, E., Newbold, R.R. (2007). Disruption of the female reproductive system by the phytoestrogen genistein. *Reproductive Toxicology*, 23: 308–316.

Jeong, J.W., Lee, H.H., Han, M.H., Kim, G.Y., Kim, W.J., Choi, Y.H. (2014). Anti-inflammatory effects of genistein via suppression of the toll-like receptor 4-mediated signaling pathway in lipopolysaccharide-stimulated BV2 microglia. *Chemico-Biological Interactions*, 212: 30–39.

Jin, J.S., Kitahara, M., Sakamoto, M., Hattori, M., Benno, Y. (2010). *Slackia equolifaciens* sp. nov., a human intestinal bacterium capable of producing equol. *International Journal of Systematic and Evolutionary Microbiology*, 60: 1721–1724.

Jourkesh, M., Choobineh, S., Soori, R., Ravasi, A.A. (2017). Effect of combined endurance-resistance training and soy extract supplementation on expression of eNOS gene in ovariectomized rats. Archives of medical sciences. *Atherosclerotic Diseases*, 2: e76–e81.

Joy, S., Siow, R.C., Rowlands, D.J., Becker, M., Wyatt, A.W., Aaronson, P.I., Coen, C.W., Kallo, I., Jacob, R., Mann, G.E. (2006). The isoflavone equol mediates rapid vascular relaxation Ca^{2+}-independent activation of endothelial nitric-oxide synthase/Hsp90 involving ERK1/2 and Akt phosphorylation in human endothelial cell. *Journal of Biological Chemistry*, 281(37): 27335–27345.

Jun, S-H., Shin, W-K., Kim, Y. (2020). Association of soybean food intake and cardiometabolic syndrome in Korean Women: Korea national health and nutrition examination survey (2007 to 2011). *Diabetes & Metabolism Journal*, 44:143–157.

Karamali, M., Kashanian, M., Alaeinasab, S., Asemi, Z. (2018). The effect of dietary soy intake on weight loss, glycaemic control, lipid profiles and biomarkers of inflammation and oxidative stress in women with polycystic ovary syndrome: a randomised clinical trial. *Journal of Human Nutrition and Dietetics*, 31(4): 533–543. doi: 10.1111/jhn.12545.

Kelly, G.E., Nelson, C., Waring, M.A., Joannou, G.E., Reeder, A.Y. (1993). Metabolites of dietary (soya) isoflavones in human urine. *Clinica Chimica Acta*, 223: 9–22.

Khani, B., Mehrabian, F., Khalesi, E., Eshraghi, A. (2011). Effect of soy phytoestrogen on metabolic and hormonal disturbance of women with polycystic ovary syndrome. *Journal of Research in Medical Sciences*, 16(3): 297–302.

Kładna, A., Berczyński, P., Kruk, I., Piechowska, T., Aboul-Enein, H.Y. (2016). Studies on the antioxidant properties of some phytoestrogens. *Luminescence*, 31(6): 1201–1206.

Ko, K.P. (2014). Isoflavones: chemistry, analysis, functions and effects on health and cancer. *Asian Pacific Journal of Cancer Prevention*, 15(17): 7001–7010.

Konishi, K., Wada, K., Yamakawa, M., Goto, Y., Mizuta, F., Koda, S., Uji, T., Tsuji, M., Nagata, C. (2019). Dietary soy intake is inversely associated with risk of type 2 diabetes in Japanese women but not in men. *The Journal of Nutrition*, 149: 1208–1214.

Křížová, L., Dadáková, K., Kašparovská, J., Kašparovský, T. (2019). Isoflavones. *Molecules*, 24(6): 1076–1104.

Kucuk, O. (2017). Soy foods, isoflavones, and breast cancer. *Cancer*, 123(11): 1901–1903.

Kuiper, G.G., Carlsson, B.O., Grandien, K.A.J., Enmark, E., Häggblad, J., Nilsson, S., Gustafsson, J.A. (1997). Comparison of the ligand binding specificity and transcript tissue distribution of estrogen receptors α and β. *Endocrinology*, 138(3): 863–870.

Kuiper, G.G., Enmark, E., Pelto-Huikko, M., Nilsson, S., Gustafsson, J.A. (1996). Cloning of a novel receptor expressed in rat prostate and ovary. *Proceedings of the National Academy of Sciences*, 93(12): 5925–5930.

Kuiper, G.G., Lemmen, J.G., Carlsson, B.O., Corton, J.C., Safe, S.H., van Der Saag, P.T., van Der Burg, B., Gustafsson, J.A. (1998). Interaction of estrogenic chemicals and phytoestrogens with estrogen receptor β. *Endocrinology*, 139(10): 4252–4263.

Kuo, L.C., Cheng, W.Y., Wu, R.Y., Huang, C.J., Lee, K.T. (2006). Hydrolysis of black soybean isoflavone glycosides by *Bacillus subtilis* natto. *Applied Microbiology and Biotechnology*, 73: 314–320.

Kwon, D.Y., Daily, J.W., Kim, H.J., Park, S. (2010). Antidiabetic effects of fermented soybean products on type 2 diabetes. *Nutrition Research*, 3: 1–13.

Kwon, J.E., Lim, J., Kim, I., Kim, D., Kang, S.C. (2018). Isolation and identification of new bacterial stains producing equol from *Pueraria lobata* extract fermentation. *PLoS ONE*, 13: e0192490.

Lampe, J.W. (2009). Is equol the key to the efficacy of soy foods? *The American Journal of Clinical Nutrition*, 89(suppl): 1664S–1667S.

Le Bail, J.C., Champavier, Y., Chulia, A.J., Habrioux, G. (2000). Effects of phytoestrogens on aromatase, 3β and 17β-hydroxysteroid dehydrogenase activities and human breast cancer cells. *Life Sciences*, 66(14): 1281–1291.

Lee, A., Beaubernard, L., Lamothe, V., Bennetau-Pelissero, C. (2019). New evaluation of isoflavone exposure in the French population. *Nutrients*, 11: 2308.

Lee, A.H., Su, D., Pasalich, M., Tang, L., Binns, C.W., Qiu, L. (2014). Soy and isoflavone intake associated with reduced risk of ovarian cancer in southern Chinese women. *Nutrition Research*, 34: 302–307.

Lee, J., Ju, J., Park, S., Hong, S.J., Yoon, S. (2012). Inhibition of IGF-1 signaling by genistein: modulation of E-cadherin expression and downregulation of β-catenin signaling in hormone refractory PC-3 prostate cancer cells. *Nutrition and Cancer*, 64(1): 153–162.

Lee, J., Park, S.H. (2018). Daidzein has anti-oxidant activity in normal human kidney tubular HK-2 cells via FOXO3/SOD2 pathway. *Journal of Food and Nutrition Research*, 6(9): 557–560.

Lee, J., Renita, M., Fioritto, R.J., St. Martin, S.K., Schwartz, S.J., Vodovotz, Y. (2004). Isoflavone characterization and antioxidant activity of Ohio soybeans. *Journal of Agricultural and Food Chemistry*, 52(9): 2647–2651.

Lee, J.-S. (2006). Effects of soy protein and genistein on blood glucose, antioxidant enzyme activities, and lipid profile in streptozotocin-induced diabetic rats. *Life Sciences*, 79(16): 1578–1584.

Lee, M.Y., Man, R.Y. (2003). The phytoestrogen genistein enhances endothelium-independent relaxation in the porcine coronary artery. *European Journal of Pharmacology*, 481(2–3): 227–232.

Lee, S.A., Shu, X.-O., Li, H., Yang, G., Cai, H., Wen, W., Ji, B.-T., Gao, J., Gao, Y.-T., Zheng, W. (2009). Adolescent and adult soy food intake and breast cancer risk: results from the Shanghai Women's Health Study. *The American Journal Clinical Nutrition*, 89(6): 1920–1926.

Lee, S.J., Ahn, J.K., Kim, S.H., Kim, J.T., Han, S.J., Jung, M.Y., Chung, I.M. (2003). Variations in isoflavones of soybean cultivars with location and storage duration. *Journal of Agricultural and Food Chemistry*, 51: 3382–3389.

Lee, S.J., Seguin, P., Kim, J.J., Moon, H.I., Ro, H.M., Kim, E.H., Seo, S.H., Kang, E.Y., Ahn, J.K., Chung, I.M. (2010). Isoflavones in Korean soybeans differing in seed coat and cotyledon color. *Journal of Food Composition and Analysis*, 23: 160–165.

Lephart, E.D. (2015). Modulation of aromatase by phytoestrogens. *Enzyme Research*, ID 594656. doi: 10.1155/2015/594656

Lewis, R.A., Austen, K.F., Soberman, R.J. (1990). Leukotrienes and other products of the 5-lipoxygenase pathway: biochemistry and relation to pathobiology in human diseases. *New England Journal of Medicine*, 323(10): 645–655.

Li, H.F., Wang, L.D., Qu, S.Y. (2004). Phytoestrogen genistein decreases contractile response of aortic artery *in vitro* and arterial blood pressure *in vivo*. *Acta Pharmacologica Sinica*, 25(3): 313–318.

Li, J., Li, J., Yue, Y., Hu, Y., Cheng, W., Liu, R., Pan, X., Zhang, P. (2014). Genistein suppresses tumor necrosis factor α-induced inflammation via modulating reactive oxygen species/Akt/nuclear factor κB and adenosine monophosphate-activated protein kinase signal pathways in human synoviocyte MH7A cells. *Drug Design, Development and Therapy*, 8: 315–323.

Li, L., Lv, Y., Xu, L., Zheng, Q. (2015b). Quantitative efficacy of soy isoflavones on menopausal hot flashes. *British Journal of Clinical Pharmacology*, 79(4): 593–604.

Li, M., Li, H., Zhang, X.L., Chen, B.-H., Hao, Q.H., Wang, S.Y. (2015a). Enhanced biosynthesis of O-desmethylangolensin from daidzein by a novel oxygen-tolerant cock intestinal bacterium in the presence of atmospheric oxygen. *Journal of Applied Microbiology*, 118: 619–628.

Li, N., Wu, X., Zhuang, W., Xia, L., Chen, Y., Zhao, R., Yi, M., Wan, Q., Du, L., Zhou, Y. (2020). Soy and isoflavone consumption and multiple health outcomes: umbrella review of systematic reviews and meta-analyses of observational studies and randomized trials in humans. *Molecular Nutrition & Food Research*, 64: 1900751.

Li, W., Ruan, W., Peng, Y., Wang, D. (2018). Soy and the risk of type 2 diabetes mellitus: a systematic review and meta-analysis of observational studies. *Diabetes Research and Clinical Practice*, 137: 190–199.

Li, Y., Zhang, H. (2017). Soybean isoflavones ameliorate ischemic cardiomyopathy by activating Nrf2-mediated antioxidant responses. *Food & Function*, 8(8): 2935–2944.

Libby, P. (2012). Inflammation in atherosclerosis. *Arteriosclerosis, Thrombosis, and Vascular Biology*, 32(9): 2045–2051.

Lima, S.M., Yamada, S.S., Reis, B.F., Postigo, S., Galvão da Silva, M.A., Aoki, T. (2013). Effective treatment of vaginal atrophy with isoflavone vaginal gel. *Maturitas*, 74(3):252–258.

Liu, D., Zhen, W., Yang, Z., Carter, J.D., Si, H., Reynolds, K.A. (2006). Genistein acutely stimulates insulin secretion in pancreatic β-cells through a cAMP-dependent protein kinase pathway. *Diabetes*, 55(4): 1043–1050.

Liu, J., Yuan, F., Gao, J., Shan, B., Ren, Y., Wang, H., Gao, Y. (2016b). Oral isoflavone supplementation on endometrial thickness: a meta-analysis of randomized placebo-controlled trials. *Oncotarget*, 7(14): 17369–17379.

Liu, X.X., Li, S.H., Chen, J.Z., Sun, K., Wang, X.J., Wang, X.G., Hui, R.T. (2012). Effect of soy isoflavones on blood pressure: a meta-analysis of randomized controlled trials. *Nutrition, Metabolism & Cardiovascular Diseases*, 28: 463–470.

Liu, Z.M., Ho, S.C., Chen, Y.M., Xie, Y.J., Huang, Z.G., Ling, W.H. (2016a). Research protocol: effect of natural S-equol on blood pressure and vascular function – a six-month randomized controlled trial among equol non-producers of postmenopausal women with prehypertension or untreated stage 1 hypertension. *BMC Complementary and Alternative Medicine*, 16: 89.

Loh, Y.C., Tan, C.S., Ch'ng, Y.S., Yeap, Z.Q., Ng, C.H., Yam, M.F. (2018). Overview of the microenvironment of vasculature in vascular tone regulation. *International Journal of Molecular Sciences*, 19(1): 120–135.

Lopes, D.B., de Queirós, L.D., de Ávila, A.R.A., Monteiro, N.E.S., Macedo, G.A. (2017). The importance of microbial and enzymatic bioconversions of isoflavones in bioactive compounds. In Grumezescu, A.M., Holban, A.M. (Eds.), *Food bioconversion. Handbook of food bioengineering*, pp. 55–93 doi: 10.1016/B978-0-12-811413-1.00002-4.

Lund, T.D., Munson, D.J., Haldy, M.E., Setchell, K.D.R., Lephart, E., Handa, R.J. (2004). Equol is a novel anti-androgen that inhibits prostate growth and hormone feedback. *Biology and Reproduction*, 70(4): 1188–1195.

Ma, L., Liu, G., Ding, M., Zong, G., Hu, F.B., Willett, W.C., Rimm, E.B., Manson, J.E., Sun, Q. (2020). Isoflavone intake and the risk of coronary heart disease in US men and women results from 3 prospective cohort studies. *Circulation*, 141: 1127–1137.

Magee, P.J., Allsopp, P., Samaletdin, A., Rowland, I.R. (2014). Daidzein, R-(+) equol and S-(−) equol inhibit the invasion of MDA-MB-231 breast cancer cells potentially via the down-regulation of matrix metalloproteinase-2. *European Journal of Nutrition*, 53(1): 345–350.

Mahesha, H.G., Singh, S.A., Rao, A.A. (2007). Inhibition of lipoxygenase by soy isoflavones: evidence of isoflavones as redox inhibitors. *Archives of Biochemistry and Biophysics*, 461(2): 176–185.

Mahn, K., Borrás, C., Knock, G.A., Taylor, P., Khan, I.Y., Sugden, D., Poston, L., Ward, J.P., Sharpe, R.M., Vina, J., Aaronson, P.I., Mann., G.E. (2005). Dietary soy isoflavone-induced increases in antioxidant and eNOS gene expression lead to improved endothelial function and reduced blood pressure in vivo. *The FASEB journal*, 19(12): 1755–1757.

Mantovani, A., Allavena, P., Sica, A., Balkwill, F. (2008). Cancer-related inflammation. *Nature*, 454(7203): 436–444.

Marazza, J.A., Nazareno, M.A., de Giori, G.S., Garro, M.S. (2012). Enhancement of the antioxidant capacity of soymilk by fermentation with *Lactobacillus rhamnosus*. *Journal of Functional Foods*, 4(3): 594–601.

March, W.A., Moore, V.M., Willson, K.J., Phillips, D.I.W., Norman, R.J., Davies, M.J. (2010). The prevalence of polycystic ovary syndrome in a community sample assessed under contrasting diagnostic criteria. *Human Reproduction*, 25(2): 544–551.

Martin, D., Song, J., Mark, C., Eyster, K. (2008). Understanding the cardiovascular actions of soy isoflavones: potential novel targets for antihypertensive drug development. *Cardiovascular & Haematological Disorders-Drug Targets*, 8(4): 297–312.

Martinez, J., Lewi, J.E. (2008). An unusual case of gynecomastia associated with soy product consumption. *Endocrine Practice*, 14: 415–418.

Maruo, T., Sakamoto, M., Ito, C., Toda, T., Benno, Y. (2008). *Adlercreutzia equolifaciens* gen. nov., sp. nov., an equol-producing bacterium isolated from human faeces, and emended description of the genus *Eggerthella*. *International Journal of Systematic and Evolutionary Microbiology*, 58(5): 1221–1227.

Mascayano, C., Espinosa, V., Sepúlveda-Boza, S., Hoobler, E.K., Perry, S. (2013). In vitro study of isoflavones and isoflavans as potent inhibitors of human 12-and 15-lipoxygenases. *Chemical Biology & Drug Design*, 82(3): 317–325.

Matthies, A., Blaut, M., Braune, A. (2009). Isolation of a human intestinal bacterium capable of daidzein and genistein conversion. *Applied and Environmental Microbiology*, 75(6): 1740–1744.

Matthies, A., Loh, G., Blaut, M., Braune, A. (2012). Daidzein and genistein are converted to equol and 5-hydroxy-equol by human intestinal *Slackia isoflavoniconvertens* in gnotobiotic rats. *The Journal of Nutrition*, 142: 40–66.

Mayo, B., Vázquez, L., Flórez, A.B. (2019). Equol: a bacterial metabolite from the daidzein isoflavone and its presumed beneficial health effects. *Nutrients*, 11(9): 2231.

Mazur, W.M., Duke, J.A., Wähälä, K., Rasku, S., Adlercreutz, H. (1998). Isoflavonoids and lignans in legumes: nutritional and health aspects in humans. *The Journal of Nutritional Biochemistry*, 9(4): 193–200.

Meng, H., Fu, G., Shen, J., Shen, K., Xu, Z., Wang, Y., Jin, B., Pan, H. (2017). Ameliorative effect of daidzein on cisplatin-induced nephrotoxicity in mice via modulation of inflammation, oxidative stress, and cell death. Oxidative Medicine and Cellular Longevity, 2017, ID 3140680.

Messina, M. (2008). Investigating the optimal soy protein andisoflavone intakes for women: a perspective. *Women's Health*, 4(4): 337–356.

Messina, M. (2010a). Insights gained from 20 years of soy research. *The Journal of Nutrition*, 140: 2289S–2295S.

Messina, M. (2010b). Soybean isoflavone exposure does not have feminizing effects on men: a critical examination of the clinical evidence. *Fertility and Sterility*, 93: 2095–2104.

Messina, M. (2014). Soy foods, isoflavones, and the health of postmenopausal women. *The American Journal of Clinical Nutrition*, 100(suppl): 423S–430S.

Messina, M. (2016). Soy and health update: evaluation of the clinical and epidemiologic literature. *Nutrients*, 8(12): 754–796.

Messina, M., Nagata, C., Wu, A.H. (2006). Estimated Asian adult soy protein and isoflavone intakes. *Nutrition and Cancer*, 55: 1–12.

Messina, M., Rogero, M.M., Fisberg, M., Waitzberg, D. (2017). Health impact of childhood and adolescent soy consumption. *Nutrition Reviews*, 75(7): 500–515.

Mezei, O., Li, Y., Mullen, E., Ross-Viola, J.S., Shay, N.F. (2006). Dietary isoflavone supplementation modulates lipid metabolism via PPARα-dependent and-independent mechanisms. *Physiological Genomics*, 26(1): 8–14.

Miladinović, J., Đorđević, V., Beleševic-Tubić, S., Petrović, K., Ćeran, M., Cvejić, J., Bursać, M., Miladinović, D. (2019). Increase of isoflavones in the aglycone form in soybeans by targeted crossings of cultivated breeding material. *Scientific Reports*, 9:10341.

Mirahmadi, S.-M.-S., Shahmohammadi, A., Rousta, A.-M., Azadi, M.-R., Fahanik-Babaei, J., Baluchnejadmojarad, T., Roghani, M. (2017). Soy isoflavone genistein attenuates lipopolysaccharide-induced cognitive impairments in the rat via exerting anti-oxidative and anti-inflammatory effects. *Cytokine*, 104: 151–159.

Miyake, Y., Tanaka, K., Okubo, H., Sasaki, S., Furukawa, S., Arakawa, M. (2018). Soy isoflavone intake and prevalence of depressive symptoms during pregnancy in Japan: baseline data from the Kyushu Okinawa Maternal and Child Health Study. *European Journal of Nutrition*, 57(2): 441–450.

Moraes, A.B., Haidar, M.A., Junior, M.S., Simoes, M.J., Baracat, E.C., Patriarca, M.T. (2009). The effects of topical isoflavones on postmenopausal skin: double-blind and randomized clinical trial of efficacy. *European Journal of Obstetrics & Gynecology and Reproductive Biology*, 146: 188–192.

Morandi, S., D'Agostina, A., Ferrario, F., Arnoldi, A. (2005). Isoflavone content of Italian soy food products and daily intakes of some specific classes of consumers. *European Food Research and Technology*, 221:84–91.

Mortensen, A., Kulling, S.E., Schwartz, H., Rowland, I., Ruefer, C.E., Rimbach, G., Cassidy, A., Magee, P., Millar, J., Hall, W.L., Birkved, F.K., Sorensen, I.K., Sontag, G. (2009). Analytical and compositional aspects of isoflavones in food and their biological effects. *Molecular Nutrition & Food Research*, 53 Supplement 2: S266–S309.

Morvaridzadeha, M., Nachvaka, S.M., Agahb, S., Sepidarkishc, M., Dehghanid, F., Rahimloue, M., Pizarrof, A.B., Heshmatia, J. (2020). Effect of soy products and isoflavones on oxidative stress parameters: a systematic review and meta-analysis of randomized controlled trials. *Food Research International*, 137: 109578.

Mukherjee, A.B., Miele, L., Pattabiraman, N. (1994). Phospholipase A2 enzymes: regulation and physiological role. *Biochemical Pharmacology*, 48(1): 1–10.

Murphy, P.A., Song, T., Buseman, G., Barua, K. (1997). Isoflavones in soy-based infant formulas. *Journal of Agricultural and Food Chemistry*, 45: 4635–4638.

Nachvak, S.M., Moradi, S., Anjom-Shoae, J., Rahmani, J., Nasiri, M., Maleki, V., Sadeghi, O. (2019). Soy, soy isoflavones, and protein intake in relation to mortality from all causes, cancers, and cardiovascular diseases: a systematic review and dose-response meta-analysis of prospective cohort studies. *Journal of the Academy of Nutrition and Dietetics*, 119(9): 1483–1500.

Nagarajan, S. (2010). Mechanisms of anti-atherosclerotic functions of soy-based diets. *The Journal of Nutritional Biochemistry*, 21(4): 255–260.

Nakamura, Y., Ohsawa, I., Goto, Y., Tsuji, M., Oguchi, T., Sato, N., Kiuchi, Y., Fukumura, M., Inagaki, M., Gotoh, H. (2017). Soy isoflavones inducing overt hypothyroidism in a patient with chronic lymphocytic thyroiditis: a case report. *Journal of Medical Case Reports*, 11: 253.

Nakamura, Y., Ueshima, H. (2014). Japanese diet: an explanation for Japanese paradox. *Journal of Food Nutrition and Dietetics*, 1(1): 7.

National Center for Comparative and Integrative Health - NCCIH (2020). https://nccih.nih.gov/health/soy/ataglance.htm#mind

Nechuta, S.J., Caan, B.J., Chen, W.Y., Lu, W., Chen, Z., Kwan, M.L., Flatt, S.W., Zheng, Y., Zheng, W., Pierce, J.P., Shu, X.O. (2012). Soy food intake after diagnosis of breast cancer and survival: an in-depth analysis of combined evidence from cohort studies of US and Chinese women. *The American Journal of Clinical Nutrition*, 96: 123–132.

Németh, K., Plumb, G.W., Berrin, J-G., Juge, N., Jackob, R., Naim, H.Y., Wiliamson, G., Swallow, D.M., Kroon, P.A. (2003). Deglycosylation by small intestinal epithelial cell β-glucosidases is a critical step in the absorption and metabolism of dietary flavonoid glycosides in humans. *European Journal of Nutrition*, 42: 29–42.

Nevala, R., Paukku, K., Korpela, R., Vapaatalo, H. (2001). Calcium-sensitive potassium channel inhibitors antagonize genistein-and daidzein-induced arterial relaxation *in vitro*. *Life Sciences*, 69(12): 1407–1417.

Norman, R.J., Dewailly, D., Legro, R.S., Hickey, T.E. (2007). Polycystic ovary syndrome. *Lancet*, 370: 685–697.

Nurmi, T., Mazur, W., Heinonen, S., Kokkonen, J., Adlercreutz, H. (2002). Isoflavone content of the soy based supplements. *Journal of Pharmaceutical and Biomedical Analysis*, 28: 1–11.

Ohkura, Y., Obayashi, S., Yamada, K., Yamada, M., Kubota, T. (2015). S-equol partially restored endothelial nitric oxide production in isoflavone-deficient ovariectomized rats. *Journal of Cardiovascular Pharmacology*, 65(5): 500–507.

Ohno, T., Kato, N., Ishii, C., Shimizu, M., Ito, Y., Tomono, S., Kawazu, S. (1993). Genistein augments cyclic adenosine 3′5′-monophosphate (cAMP) accumulation and insulin release in MIN6 cells. *Endocrine Research*, 19(4): 273–285.

Oyama, A., Ueno, T., Uchiyama, S., Aihara, T., Miyake, A., Kondo, S., Matsunaga, K. (2012). The effects of natural S-equol supplementation on skin aging in postmenopausal women: a pilot randomized placebo-controlled trial. *Menopause*, 19(2): 202–210.

Pang, D., Yang, C., Luo, Q., Li, C., Liu, W., Li, L., Zou, Y., Feng, B., Chen, Z., Huang, C. (2020). Soy isoflavones improve the oxidative stress induced hypothalamic inflammation and apoptosis in high fat diet-induced obese male mice through PGC1-alpha pathway. *Aging (Albany NY)*, 12(9): 8710–8727.

Park, E., Shin, J.I., Park, O.J., Kang, M.H. (2005). Soy isoflavone supplementation alleviates oxidative stress and improves systolic blood pressure in male spontaneously hypertensive rats. *Journal of Nutritional Science and Vitaminology*, 51(4): 254–259.

Park, J.S., Jung, J.S., Jeong, Y.H., Hyun, J.W., Le, T.K.V., Kim, D.H., Choi, E.C., Kim, H.S. (2011). Antioxidant mechanism of isoflavone metabolites in hydrogen peroxide-stimulated rat primary astrocytes: critical role of hemeoxygenase-1 and NQO1 expression. *Journal of Neurochemistry*, 119(5): 909–919.

Patriarca, M.T., Barbosa de Moraes, A.R., Nader, H.B., Petri, V., Martins, J.R., Gomes, R.C., Soares Jr, J.M. (2013). Hyaluronic acid concentration in postmenopausal facial skin after topical estradiol and genistein treatment: a double-blind, randomized clinical trial of efficacy. *Menopause*, 20(3): 336–341.

Peirotén, Á., Gaya, P., Álvarez, I., Landete, J.M. (2020). Production of *O*-desmethylangolensin, tetrahydro-daidzein, 6′-hydroxy-*O*-desmethylangolensin and 2-(4-hydroxyphenyl)-propionic acid in fermented soy beverage by lactic acid bacteria and *Bifidobacterium* strains. *Food Chemistry*, 318: 126521.

Perumal, D.K., Adhimoolam, M., Ivan, E.A., Rajamohammed, M.A. (2019). Effects of soy isoflavone genistein on lipid profile and hepatic steatosis in high-fat-fed Wistar rats. *National Journal of Physiology, Pharmacy and Pharmacology*, 9(9): 856–861.

Pike, A.C., Brzozowski, A.M., Hubbard, R.E., Bonn, T., Thorsell, A.G., Engström, O., Ljunggren, J., Gustafsson, J.A., Carlquist, M. (1999). Structure of the ligand-binding domain of oestrogen receptor beta in the presence of a partial agonist and a full antagonist. *The EMBO Journal*, 18: 4608–4618.

Pilsakova, L., Riecanský, I., Jagla, F. (2010). The physiological actions of isoflavone phytoestrogens. *Physiological Research*, 59(5): 651–664.

Prabhakaran, M.P., Hui, L.S., Perera, C.O. (2006). Evaluation of the composition and concentration of iso-flavones in soy based supplements, health products and infant formulas. *Food Research International*, 39: 730–738.

Prietsch, R.F., Monte, L.D., Da Silva, F.A., Beira, F.T., Del Pino, F.A.B., Campos, V.F., Collares, T., Pinto, L.S., Spanevello, R.M., Gamaro, G.D., Braganhol, E. (2014). Genistein induces apoptosis and autophagy in human breast MCF-7 cells by modulating the expression of proapoptotic factors and oxidative stress enzymes. *Molecular and Cellular Biochemistry*, 390(1–2): 235–242.

Qiu, S., Jiang, C. (2019). Soy and isoflavones consumption and breast cancer survival and recurrence: a sys-tematic review and meta-analysis. *European Journal of Nutrition*, 58: 3079–3090.

Rababah, T.M., Awaisheh, S.S., Al-Tamimi, H.J., Brewer, S. (2015). The hypocholesterolemic and hormone modulation effects of isoflavones alone or co-fermented with probiotic bacteria in hypercholesterolemic rats model. *International Journal of Food Sciences and Nutrition*, 66(5): 546–552.

Reger, M.K., Zollinger, T.W., Liu, Z., Jones, J.F., Zhang, J. (2018). Dietary intake of isoflavones and coumes-trol and the risk of prostate cancer in the prostate, lung, colorectal and ovarian cancer screening trial. *International Journal of Cancer*, 142: 719–728.

Rekha, C.R., Vijayalakshmi, G. (2011). Isoflavone phytoestrogens in soymilk fermented with β-glucosidase pro-ducing probiotic lactic acid bacteria. *International Journal of Food Sciences and Nutrition*, 62(2): 111–120.

Riedl, K.M., Lee, J.H., Renita, M., Martin, S.K., Schwartz, S.J., Vodovotz, Y. (2007). Isoflavone profiles, phe-nol content, and antioxidant activity of soybean seeds as influenced by cultivar and growing location in Ohio. *Journal of the Science of Food and Agriculture*, 87: 1197–1206.

Rimbach, G., De Pascual-Teresa, S., Ewins, B.A., Matsugo, S., Uchida, Y., Minihane, A.M., Turner, R., Vafeiadou, K., Weinberg, P.D. (2003). Antioxidant and free radical scavenging activity of isoflavone metabolites. *Xenobiotica*, 33(9): 913–925.

Rizzo, G., Baroni, L. (2018). Soy, soyfoods and their role in vegetarian diets. *Nutrients*, 10: 43.

Romualdi, D., Costantini, B., Campagna, G., Lanzone, A., Guido, M. (2008). Is there a role for soy isoflavones in the therapeutic approach to polycystic ovary syndrome? Results from a pilot study. *Fertility and Sterility*, 90(5), 1826–1833.

Rowland, I., Faughnan, M., Hoey, L., Wahala, K., Williamson, G., Cassidy, A. (2003). Bioavailability of phyto-oestrogens. *British Journal of Nutrition*, 89 (Suppl. 1): S45–S58.

Rüfer, C.E., Bub, A., Möseneder, J., Winterhalter, P., Stürtz, M., Kulling, S.E. (2008). Pharmacokinetics of the soybean isoflavone daidzein in its aglycone and glucoside form: a randomized, double-blind, crossover study. *The American Journal of Clinical Nutrition*, 87: 1314–1323.

Rzepecki, A.K., Murase, J.E., Juran, R., Fabi, S.G., McLellan, B.N. (2019). Estrogen-deficient skin: the role of topical therapy. *International Journal of Women's Dermatology*, 5: 85–90.

Sacks, F.M., Lichtenstein, A., Van Horn, L., Harris, W., Kris-Etherton, P., Winston, M., American Heart Association Nutrition Committee. (2006). Soy protein, isoflavones, and cardiovascular health: an American Heart Association Science Advisory for professionals from the Nutrition Committee. *Circulation*, 113(7): 1034–1044.

Sakamoto, Y., Kanatsu, J., Toh, M., Naka, A., Kondo, K., Iida, K. (2016). The dietary isoflavone daidzein reduces expression of pro-inflammatory genes through PPARα/γ and JNK pathways in adipocyte and macrophage co-cultures. *PLoS One*, 11(2): e0149676–e0149689.

Sathyapalan, T., Aye, M., Rigby, A.S., Thatcher, N.J., Dargham, S.R., Kilpatrick, E.S., Atkin, S.L. (2018). Soy isoflavones improve cardiovascular disease risk markers in women during the early menopause. *Nutrition, Metabolism and Cardiovascular Diseases*, 28(7): 691–697.

Sathyapalan, T., Manuchehri, A.M., Thatcher, N.J., Rigby, A.S., Chapman, T., Kilpatrick, E.S., Atkin, S.L. (2011). The effect of soy phytoestrogen supplementation on thyroid status and cardiovascular risk mark-ers in patients with subclinical hypothyroidism: a randomized, double-blind, crossover study. *The Journal of Clinical Endocrinology & Metabolism*, 96: 1442–1449.

Seeman, E. (2004). Estrogen, androgen, and the pathogenesis of bone fragility in -women and men. *Current Osteoporosis Reports*, 2(3): 90–96.

Seibert, K., Zhang, Y., Leahy, K., Hauser, S., Masferrer, J., Perkins, W., Lee, L., Isakson, P. (1994). Pharmacological and biochemical demonstration of the role of cyclooxygenase 2 in inflammation and pain. *Proceedings of the National Academy of Sciences*, 91(25): 12013–12017.

Seo, Y.J., Kim, B.S., Chun, S.Y., Park, Y.K., Kang, K.S., Kwon, T.G. (2011). Apoptotic effects of genistein, biochanin-A and apigenin on LNCaP and PC-3 cells by p21 through transcriptional inhibition of polo-like kinase-1. *Journal of Korean Medical Science*, 26(11): 1489–1494.

Setchell, K.D. (1998). Phytoestrogens: the biochemistry, physiology, and implications for human health of soy isoflavones. *The American Journal of Clinical Nutrition*, 68(6) Supplement: 1333S–1346S.

Setchell, K.D. (2017). The history and basic science development of soy isoflavones. *Menopause*, 24(12): 1338–1350.

Setchell, K.D., Brown, N.M., Zimmer-Nechemias, L., Brashear, W.T., Wolfe, B.E., Kirschner, A.S., Heubi, J.E. (2002a). Evidence for lack of absorption of soy isoflavone glycosides in humans, supporting the crucial role of intestinal metabolism for bioavailability. *American Journal of Clinical Nutrition*, 76: 447–453.

Setchell, K.D.R., Brown, N.M., Desai, P., Zimmer-Nechemias, L., Wolfe, B.E., Brashear, W.T., Kirschner, A.S., Cassidy, A., Heubi, J.E. (2001). Bioavailability of pure isoflavones in healthy humans and analysis of commercial soy isoflavone supplements. *The Journal of Nutrition*, 131 Supplement: 1362S–1375S.

Setchell, K.D.R., Brown, N.M., Lydeking-Olsen, E. (2002b). The clinical importance of the metabolite equol – A clue to the effectiveness of soy and its isoflavones. *The Journal of Nutrition*, 132: 3577–3584.

Setchell, K.D.R., Clerici, C., Lephart, E.D., Cole, S.J., Heenan, C., Castellani, D., Wolfe, B.E., Nechemias-Zimmer, L., Brown, N.M., Lund, T.D., Handa, R.J., Heubi, J.E. (2005). S-Equol, a potent ligand for estrogen receptor ß, is the exclusive enantiomeric form of the soy isoflavone metabolite produced by human intestinal bacterial flora. *The American Journal of Clinical Nutrition* 81: 1072–1079.

Setchell, K.D.R., Zimmer-Nechemias, L., Cai, J., Heubi, J.E. (1997). Exposure of infants to phytoestrogens from soy-based infant formula. *Lancet*, 350: 23–27.

Setchell, K.D.R., Zimmer-Nechemias, L., Cai, J., Heubi, J.E. (1998). Isoflavone content of infant formulas and the metabolic fate of these phytoestrogens in early life. *The American Journal of Clinical Nutrition*, 68(suppl): 1453S–1461S.

Shelnutt, S.R., Cimino, C.O., Wiggins, P.A., Ronis, M.J.J., Badger, T.M. (2002). Pharmacokinetics of the glucuronide and sulfate conjugates of genistein and daidzein in men and women after consumption of a soy beverage. *American Journal of Clinical Nutrition*, 76: 588–594.

Sheu, F., Lai, H.H., Yen, G.C. (2001). Suppression effect of soy isoflavones on nitric oxide production in RAW 264.7 macrophages. *Journal of Agricultural and Food Chemistry*, 49(4): 1767–1772.

Shinkaruk, S., Durand, M., Lamothe, V., Carpaye, A., Martinet, A., Chantre, P., Vergne, S., Nogues, X., Moore, N., Bennetau-Pelissero, C. (2012). Bioavailability of glycitein relatively to other soy isoflavones in healthy young Caucasian men. *Food Chemistry*, 135: 1104–1111.

Shor, D., Sathyapalan, T., Atkin, S.L., Thatcher, N.J. (2012). Does equol production determine soy endocrine effects? *European Journal of Nutrition*, 51: 389–398.

Shu, Y.Y., Maibach, H.I. (2011). Estrogen and skin. *American Journal of Clinical Dermatology*, 12(5): 297–311.

Si, H., Liu, D. (2008). Genistein, a soy phytoestrogen, upregulates the expression of human endothelial nitric oxide synthase and lowers blood pressure in spontaneously hypertensive rats. *The Journal of Nutrition*, 138(2): 297–304.

Siegel, R.L., Miller, K.D., Jemal, A. (2020). Cancer statistics, 2020. *CA: A Cancer Journal for Clinicians*, 70: 7–30.

Silva, L.A., Carbonel, A.A., de Moraes, A.R.B., Simões, R.S., Sasso, G.R.D.S., Goes, L., Nunes, W., Simoes, M.J., Patriarca, M.T. (2017). Collagen concentration on the facial skin of postmenopausal women after topical treatment with estradiol and genistein: a randomized double-blind controlled trial. *Gynecological Endocrinology*, 33(11): 845–848.

Simental-Mendía, L.E., Gotto, A.M. Jr., Atkin, S.L., Banach, M., Pirro, M., Sahebkar, A. (2018). Effect of soy isoflavone supplementation on plasma lipoprotein(a) concentrations: a meta-analysis. *Journal of Clinical Lipidology*, 12(1): 16–24.

Simons, A.L., Renouf, M., Hendrich, S., Murphy, P.A. (2005). Metabolism of glycitein (7,4′-dihydroxy-6-methoxy-isoflavone) by human gut microflora. *Journal of Agricultural and Food Chemistry*, 53: 8519–8525.

Siow, R.C., Li, F.Y., Rowlands, D.J., de Winter, P., Mann, G.E. (2007). Cardiovascular targets for estrogens and phytoestrogens: transcriptional regulation of nitric oxide synthase and antioxidant defense genes. *Free Radical Biology and Medicine*, 42(7): 909–925.

Skolmowska, D., Głąbska, D., Guzek, D., Lech, G. (2019). Association between dietary isoflavone intake and ulcerative colitis symptoms in Polish Caucasian individuals. *Nutrients*, 11(8): 1936.

Smith, L.J., Kalhan, R., Wise, R.A., Sugar, E.A., Lima, J.J., Irvin, C.G., Dozor, A.J, Holbrook, J.T., American Lung Association Asthma Clinical Research Centers. (2015). Effect of a soy isoflavone supplement on lung function and clinical outcomes in patients with poorly controlled asthma a randomized clinical trial. *JAMA*, 313(20): 2033–2043.

Somekawa, Y., Chiguchi, M., Ishibashi, T., Aso, T. (2001). Soy intake related to menopausal symptoms, serum lipids, and bone mineral density in postmenopausal Japanese women. *Obstetrics & Gynecology*, 97(1): 109–115.

Šošić-Jurjević, B., Lütjohann, D., Renko, K., Filipović, B., Radulović, N., Ajdžanović, V., Trifunović, S., Nestorović, N., Živanović, J., Stojanoski, M.M., Köhrle, J., Milošević, V. (2019). The isoflavones genistein and daidzein increase hepatic concentration of thyroid hormones and affect cholesterol metabolism in middle-aged male rats. *The Journal of Steroid Biochemistry and Molecular Biology*, 190: 1–10.

Squadrito, F., Altavilla, D., Morabito, N., Crisafulli, A., D'Anna, R., Corrado, F., Ruggeri, P., Campo, G.M., Calapai, G., Caputi, A.P., Squadrito, G. (2002). The effect of the phytoestrogen genistein on plasma nitric oxide concentrations, endothelin-1 levels and endothelium dependent vasodilation in postmenopausal women. *Atherosclerosis*, 163(2): 339–347.

Stürtz, M., Lander, V., Schmid, W., Winterhalter, P. (2008). Quantitative determination of isoflavones in soy based nutritional supplements by high-performance liquid chromatography. *Journal of Consumer Protection and Food Safety*, 3: 127–136.

Sun, L., Hou, Y., Zhao, T., Zhou, S., Wang, X., Zhang, L., Yu, G. (2015). A combination of genistein and magnesium enhances the vasodilatory effect via an eNOS pathway and BK_{Ca} current amplification. *Canadian Journal of Physiology and Pharmacology*, 93(4): 215–221.

Surh, J., Kim, M.-J., Koh, E., Kim, Y.-K.L., Kwon, H. (2006). Estimated intakes of isoflavones and coumestrol in Korean population. *International Journal of Food Sciences and Nutrition*, 57(5–6): 325–344.

Takaoka, O., Mori, T., Ito, F., Okimura, H., Kataoka, H., Tanaka, Y., Koshiba, A., Kusuki, I., Shigehiro, S., Amami, T., Kitawaki, J. (2018). Daidzein-rich isoflavone aglycones inhibit cell growth and inflammation in endometriosis. *The Journal of Steroid Biochemistry and Molecular Biology*, 181: 125–132.

Taku, K., Melby, M.K., Kronenberg, F., Kurzer, M.S., Messina, M. (2012). Extracted or synthesized soybean isoflavones reduce menopausal hot flash frequency and severity: systematic review and meta-analysis of randomized controlled trials. *Menopause*, 19(7): 776–790.

Taku, K., Melbyb, M.K., Nishic, N., Omorid, T., Kurzere, M.S. (2011). Soy isoflavones for osteoporosis: an evidence-based approach. *Maturitas*, 70: 333–338.

Tang, C., Zhang, K., Zhao, Q., Zhang, J. (2015). Effects of dietary genistein on plasma and liver lipids, hepatic gene expression, and plasma metabolic profiles of hamsters with diet-induced hyperlipidemia. *Journal of Agricultural and Food Chemistry*, 63(36): 7929–7936.

Tempfer, C.B., Froese, G., Heinze, G., Bentz, E.K., Hefler, L.A., Huber, J.C. (2009). Side effects of phytoestrogens: a meta-analysis of randomized trials. *The American Journal of Medicine*, 122: 939–946.

Tepavčević, V., Atanacković, M., Miladinović, J., Malenčić, Dj, Popović, J., Cvejić, J. (2010). Isoflavone composition, total polyphenolic content and antioxidant activity in soybeans of different origin. *Journal of Medicinal Food*, 13(3): 1–8.

Testa, I., Salvatori, C., Di Cara, G., Latini, A., Frati, F., Troiani, S., Principi, N., Esposito, S. (2018). Soy-based infant formula: are phyto-oestrogens still in doubt? *Frontiers in Nutrition*, 5(110): 1–8.

Touillaud, M., Gelot, A., Mesrine, S., Bennetau-Pelissero, C., Clavel-Chapelon, F., Arveux, P., Bonnet, F., Gunter, M., Boutron-Ruault, M.-C., Fournier, A. (2019). Use of dietary supplements containing soy isoflavones and breast cancer risk among women aged >50 y: a prospective study. *The American Journal of Clinical Nutrition*, 109(3), 597–605.

Tousen, Y., Ezaki, J., Fujii, Y., Ueno, T., Nishimuta, M., Ishimi, Y. (2011). Natural S-equol decreases bone resorption in postmenopausal, non-equol-producing Japanese women: a pilot randomized, placebo-controlled trial. *Menopause*, 18(5): 563–574.

Tsai, H.S., Huang, L.J., Lai, Y.H., Chang, J.C., Lee, R.S., Chiou, R.Y.Y. (2007). Solvent effects on extraction and HPLC analysis of soybean isoflavones and variations of isoflavone compositions as affected by crop season. *Journal of Agricultural and Food Chemistry*, 55: 7712–7715.

Tse, G., Eslick, G.D. (2016). Soy and isoflavone consumption and risk of gastrointestinal cancer: a systematic review and meta-analysis. *European Journal of Nutrition*, 55: 63–73.

U.S. Department of Agriculture, Agricultural Research Service. (2015). USDA database for the isoflavone content of selected foods, Release 2.1. Nutrient Data Laboratory, http://www.ars.usda.gov/nutrientdata/isoflav

Uzzan, M., Labuza, T.P. (2004). Critical issues in R&D of soy isoflavone – enriched foods and dietary supplements. *Journal of Food Science*, 69(3): CRH77-CRH86.

Valsecchi, A.E., Franchi, S., Panerai, A.E., Rossi, A., Sacerdote, P., Colleoni, M. (2011). The soy isoflavone genistein reverses oxidative and inflammatory state, neuropathic pain, neurotrophic and vasculature deficits in diabetes mouse model. *European Journal of Pharmacology*, 650(2–3): 694–702.

Vardi, A., Bosviel, R., Rabiau, N., Adjakly, M., Satih, S., Dechelotte, P., Boiteux, J.P., Fontana, L., Bignon, Y.J., Guy, L., Bernard-Gallon, D.J. (2010). Soy phytoestrogens modify DNA methylation of GSTP1, RASSF1A, EPH2 and BRCA1 promoter in prostate cancer cells. *In Vivo*, 24(4): 393–400.

Vera, R., Galisteo, M., Villar, I.C., Sanchez, M., Zarzuelo, A., Pérez-Vizcaíno, F., Duarte, J. (2005). Soy isoflavones improve endothelial function in spontaneously hypertensive rats in an estrogen-independent manner: role of nitric-oxide synthase, superoxide, and cyclooxygenase metabolites. *Journal of Pharmacology and Experimental Therapeutics*, 314(3): 1300–1309.

Vicas, S.I., Chedea, V.S., Socaciu, C. (2011). Inhibitory effects of isoflavones on soybean lipoxygenase-1 activity. *Journal of Food Biochemistry*, 35(2): 613–627.

Waggle, D.H., Bryan, B.A. (2003). Protein Technologies International Inc., assignee. Recovery of isoflavones from soy molasses. U.S. Patent 2,003,0129,263.

Wakai, K., EGAMI, I., Kato, K., Kawamura, T., Tamakoshi, A., Lin, Y., Nakayama, T., Wada, M., Ohno, Y. (1999). Dietary intake and sources of isoflavones among Japanese. *Nutrition and Cancer*, 33(2): 139–145.

Wang, H.-J., Murphy, P.A. (1994). Isoflavone content in commercial soybean foods. *Journal of Agricultural and Food Chemistry*, 42: 1666–1673.

Wang, H.-J., Murphy, P.A. (1996). Mass balance study of isoflavones during soybean processing. *Journal of Agricultural and Food Chemistry*, 44: 2377–2383.

Wang, W., Sun, Y., Liu, J., Li, Y., Li, H., Xiao, S., Weng, S., Zhang, W. (2014). Soy isoflavones administered to rats from weaning until sexual maturity affect ovarian follicle development by inducing apoptosis. *Food and Chemical Toxicology*, 72: 51–60.

Wang, W., Zhang, W., Liu, J., Sun, Y., Li, Y., Li, H., Xiao, S., Shen, X. (2013). Metabolomic changes in follicular fluid induced by soy isoflavones administered to rats from weaning until sexual maturity. *Toxicology and Applied Pharmacology*, 269(3): 280–289.

Watanabe, S., Uehara, M. (2019). Health effects of soy and isoflavones. In Singh, R.B., Watson, R.R., Takahashi, T. (Eds.), *The role of functional food security in global health*, pp. 379–394. doi: 10.1016/B978-0-12-813148-0.00022-0.

Watanabe, S., Yamaguchi, M., Sobue, T., Takahashi, T., Miura, T., Arai, Y., Mazur, W., Wähälä, K., Adlercreutz, H. (1998). Pharmacokinetics of soybean isoflavones in plasma, urine and feces of men after ingestion of 60 g baked soybean powder (kinako). *The Journal of Nutrition* 128: 1710–1715.

Wei, P., Liu, M., Chen, Y., Chen, D-C. (2012). Systematic review of soy isoflavone supplements on osteoporosis in women. *Asian Pacific Journal of Tropical Medicine*, 5(3): 243–248.

Welboren, W.J., Sweep, F.C., Span, P.N., Stunnenberg, H.G. (2009). Genomic actions of estrogen receptor: what are the targets and how are they regulated?. *Endocrine-Related Cancer*, 16(4): 1073–1089.

Wellen, K.E., Hotamisligil, G.S. (2005). Inflammation, stress, and diabetes. *The Journal of Clinical Investigation*, 115(5): 1111–1119.

White, L.R., Petrovitch, H., Ross, G.W., Masaki, K., Hardman, J., Nelson, J., Davis, D., Markesbery, W. (2000). Brain aging and midlife tofu consumption. *Journal of the American College of Nutrition*, 19(2): 242–255.

Wiseman, H.C., Clarke, D.B., Barnes, K.A., Bowey, E. (2002). Isoflavone aglycone and glucoconjugate of high- and low-soy UK foods used in nutritional studies. *Journal of Agricultural and Food Chemistry*, 50: 1404–1410.

Woo, H.W., Kim, M.K., Lee, Y-H., Shin, D.H., Shin, M-H., Choi, B.Y. (2020). Sex-specific associations of habitual intake of soy protein and isoflavones with risk of type 2 diabetes. *Clinical Nutrition*, 40(1): 127–136.

Wu, Y.C, Zheng, D., Sun, J.J., Zou, Z.K., Ma, Z.L. (2015). Meta-analysis of studies on breast cancer risk and diet in Chinese women. *International Journal of Clinical and Experimental Medicine*, 8: 73–85.

Wu, Z-Y., Sang, L-X., Chang, B. Isoflavones and inflammatory bowel disease. (2020). *World Journal of Clinical Cases*, 8(11): 2081–2091.

Xiao, Y., Zhang, S., Tong, H., Shi, S. (2017). Comprehensive evaluation of the role of soy and isoflavone supplementation in humans and animals over the past two decades. *Phytotherapy Research*, 32(3): 384–394.

Xu, L., Du, B., Xu, B. (2015b). A systematic, comparative study on the beneficial health components and antioxidant activities of commercially fermented soy products marketed in China. *Food Chemistry*, 174: 202–213.

Xu, X., Harris, K.S., Wang, H.J., Murphy, P.A., Hendrich, S. (1995). Bioavailability of soybean isoflavones depends upon gut microflora in women. *The Journal of Nutrition*, 125: 2307–2315.

Xu, X., Xiao, S., Rahardjo, T.B., Hogervorst, E. (2015a). Tofu intake is associated with poor cognitive performance among community-dwelling elderly in China. *Journal of Alzheimer's Disease*, 43(2): 669–675.

Xu, Z., Wu, Q., Godber, S. (2002). Stabilities of daidzin, genistin and generation of derivates during heating. *Journal of Agricultural and Food Chemistry*, 50:7402–7406.

Yamakoshi, J., Piskula, M.K., Izumi, T., Tobe, K., Saito, M., Kataoka, S., Obata, A., Kikuchi, M. (2000). Isoflavone aglycone–rich extract without soy protein attenuates atherosclerosis development in cholesterol-fed rabbits. *The Journal of Nutrition*, 130(8): 1887–1893.

Yang, W., Wang, S., Li, L., Liang, Z., Wang, L. (2011). Genistein reduces hyperglycemia and islet cell loss in a high-dosage manner in rats with alloxan-induced pancreatic damage. *Pancreas*, 40(3): 396–402.

Yang, Y., Nie, W., Yuan, J., Zhang, B., Wang, Z., Wu, Z., Guo, Y. (2010). Genistein activates endothelial nitric oxide synthase in broiler pulmonary arterial endothelial cells by an Akt-dependent mechanism. *Experimental & Molecular Medicine*, 42(11): 768–776.

Yazdani, Y., Rad, M.R.S., Taghipour, M., Chenari, N., Ghaderi, A., Razmkhah, M. (2016). Genistein suppression of matrix metalloproteinase 2 (MMP-2) and vascular endothelial growth factor (VEGF) expression in mesenchymal stem cell like cells isolated from high and low grade gliomas. *Asian Pacific Journal of Cancer Prevention*, 17(12): 5303–5307.

Yingngam, B., Rungseevijitprapa, W. (2012). Molecular and clinical role of phytoestrogens as anti-skin-ageing agents: a critical overview. *Phytopharmacology*, 3(2): 227–244.

Yokoyama, S., Suzuki, T. (2008). Isolation and characterization of a novel equol-producing bacterium from human feces. *Bioscience, Biotechnology and Biochemistry*, 72: 2660–2666.

Yokoyama, S.-I., Niwa, T., Osawa, T., Suzuki, T. (2010). Characterization of an O-desmethylangolensin-producing bacterium isolated from human feces. *Archives of Microbiology*, 192: 15–22.

Yoon, G., Park, S. (2014). Antioxidant action of soy isoflavones on oxidative stress and antioxidant enzyme activities in exercised rats. *Nutrition Research and Practice*, 8(6): 618–624.

You, J., Sun, Y., Bo, Y., Zhu, Y., Duan, D., Cui, H., Lu, Q. (2018). The association between dietary isoflavones intake and gastric cancer risk: a meta-analysis of epidemiological studies. *BMC Public Health*, 18: 510.

Yousefi, H., Karimi, P., Alihemmati, A., Alipour, M.R., Habibi, P., Ahmadiasl, N. (2017). Therapeutic potential of genistein in ovariectomy-induced pancreatic injury in diabetic rats: the regulation of MAPK pathway and apoptosis. *Iranian Journal of Basic Medical Sciences*, 20(9): 1009–1015.

Yu, J., Bi, X., Yu, B., Chen, D. (2016a). Isoflavones: anti-inflammatory benefit and possible caveats. *Nutrients*, 8(6): 361–347.

Yu, Y., Jing, X., Li, H., Zhao, X., Wang, D. (2016b). Soy isoflavone consumption and colorectal cancer risk: a systematic review and meta-analysis. *Scientific Reports*, 6: 25939.

Yu, Z.T., Yao, W., Zhu, W.Y. (2008). Isolation and identification of equol-producing bacterial strains from cultures of pig faeces. *FEMS Microbiology Letters*, 282: 73–80.

Yuan, G., Liu, Y., Liu, G., Wei, L., Wen, Y., Huang, S., Guo, Y., Zou, F., Cheng, J. (2019). Associations between semen phytoestrogens concentrations and semen quality in Chinese men. *Environment International*, 129: 136–144.

Yuan, J-P., Wang, J-H., Liu, X. (2007). Metabolism of dietary soy isoflavones to equol by human intestinal microflora – implications for health. *Molecular Nutrition & Food Research*, 51: 765–781.

Zaheer, K., Akhtar, H.M. (2017). An updated review of dietary isoflavones: nutrition, processing, bioavailability and impacts on human health. *Critical Reviews in Food Science and Nutrition*, 57(6): 1280–1293.

Zamora-Ros, R., Knaze, V., Luján-Barroso, L., Kuhnle, G.G.C., Mulligan, A.A., Touillaud, M. et al. (2012). Dietary intakes and food sources of phytoestrogens in the European Prospective Investigation into Cancer and Nutrition (EPIC) 24-hour dietary recall cohort. *European Journal of Clinical Nutrition*, 66: 932–941.

Zhai, X., Lin, M., Zhang, F., Hu, Y., Xu, X., Li, Y., Liu, K., Ma, X., Tian, X., Yao, J. (2013). Dietary flavonoid genistein induces Nrf2 and phase II detoxification gene expression via ERKs and PKC pathways and protects against oxidative stress in Caco-2 cells. *Molecular Nutrition & Food Research*, 57(2): 249–259.

Zhang, F.F., Haslam, D.E., Terry, M.B., Knight, J.A., Andrulis, I.L., Daly, M.B., Buys, S.S., John, E.M. (2017). Dietary isoflavone intake and all-cause mortality in breast cancer survivors: the Breast Cancer Family Registry. *Cancer*, 123(11): 2070–2079.

Zhang, H.Y., Cui, J., Zhang, Y., Wang, Z.L., Chong, T., Wang, Z.-M. (2016). Isoflavones and prostate cancer: a review of some critical issues. *Chinese Medical Journal*, 129: 341–347.

Zhang, T., Wang, F., Xu, H.X., Yi, L., Qin, Y., Chang, H., Mi, M.T., Zhang, Q.Y. (2013a). Activation of nuclear factor erythroid 2-related factor 2 and PPARγ plays a role in the genistein-mediated attenuation of oxidative stress-induced endothelial cell injury. *British Journal of Nutrition*, 109(2): 223–235.

Zhang, X., Liu, Y., Xu, Q., Zhang, Y., Liu, L., Li, H., Li, F., Liu, Z., Yang, X., Yu, X., Kong, A. (2019). The effect of soy isoflavone combined with calcium on bone mineral density in perimenopausal Chinese women: a 6-month randomized double-blind placebo-controlled study. *International Journal of Food Sciences and Nutrition*, 71(4): 473–481.

Zhang, Y.B., Chen, W.H., Guo, J.J., Fu, Z.H., Yi, C., Zhang, M., Na, X.L. (2013b). Soy isoflavone supplementation could reduce body weight and improve glucose metabolism in non-Asian postmenopausal women - a meta-analysis. *Nutrition*, 29: 8–14.

Zhao, C., Dahlman-Wright, K., Gustafsson, J.Å. (2008). Estrogen receptor β: an overview and update. *Nuclear Receptor Signaling*, 6(1): e003–e013.

Zhao, T-T., Jin, F., Li, J-G., Xu, Y-Y., Dong, H-T., Liu, Q., Xing, P., Zhu, G-L., Xu, H., Miao, Z-F. (2019). Dietary isoflavones or isoflavone-rich food intake and breast cancer risk: a meta-analysis of prospective cohort studies. *Clinical Nutrition*, 38: 136–145.

Zilaee, M., Mansoori, A., Seyed Ahmad, H., Marjan Mohaghegh, S., Asadi, M., Hormoznejad, R. (2020). The effects of soy isoflavones on total testosterone and follicle-stimulating hormone levels in women with polycystic ovary syndrome: a systematic review and meta-analysis. *The European Journal of Contraception & Reproductive Health Care*, 25(4): 305–310.

Zubik, L., Meydani, M. (2003). Bioavailability of soybean isoflavones from aglycone and glucoside forms in American women. *The American Journal of Clinical Nutrition*, 77: 1459–1465.

2 Soybean Isoflavone Profile
A New Quality Index in Food Application and Health

Elza Iouko Ida and Adriano Costa de Camargo

CONTENTS

2.1 INTRODUCTION

Soybeans (*Glycine max* [L.] Merr.) are rich sources of protein and oil and substantial amounts of carbohydrates, dietary fiber, vitamins, and minerals. Furthermore, they also present phytochemicals that are associated with the prevention and/or potential use as an adjuvant in the treatment of many chronic diseases. Isoflavones are classified as phytoestrogens and belong to a subclass of flavonoids with limited distribution in nature. However, these molecules are found in significant amounts in soybean and derived products (Golbitz & Jordan, 2006). Flavonoids (e.g. anthocyanidins, flavanones, flavonols, flavanonols, flavans, flavanols, flavones, and isoflavones) are present in different tissues from various plant families (Shahidi & Ambigaipalan, 2015), while isoflavones are found in just a few botanical families. Soybean is, therefore, the primary source of isoflavones. The aglycones (e.g. daidzein, genistein, and glycitein) are synthesized by the phenylpropanoid pathways being mainly stored in the vacuole such as malonyl glucosides and β-glucosides (Graham, 1991; Kudou et al., 1991).

Soybean isoflavones occur in four groups according to their chemical structure, namely aglycones or free forms (daidzein, genistein, and glycitein); glucosides form or β-glucosides (daidzin, genistin, and glycitin); and two glucosides conjugate forms, 6″-O-malonyl-glucosides or malonyl glucosides (malonyl daidzin, malonyl genistin, and malonyl glycitin) and 6″-O-acetylglucosides or acetyl glucosides (acetyl dadzin, acetyl genistin, and acetyl glycitin), with a total of 12 different forms (Figure 2.1). β-glucoside forms have a glucose molecule bound to position seven of the benzene rings while the conjugated forms are esterified at carbon six of the glucose molecule (Liu, 2004). However, during this decade, Yanaka et al. (2012) postulated that soybean contains 15 species of isoflavones, including isoflavones 3 succinyl glucosides which have been detected in natto (a fermented soybean product), but they have been little investigated.

Aglycones are the most bioactive forms of isoflavones and have greater absorption capacity in the intestine than that of their respective conjugates. Therefore, it is desirable to obtain soybean

DOI: 10.1201/9781003030294-2

45

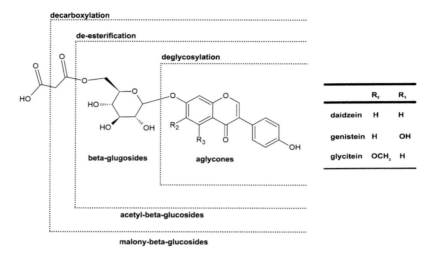

FIGURE 2.1 Isoflavones of soybeans and possible interconversion via deglycosylation, de-esterification, and decarboxylation, thus generating their respective aglycones, β-glucosides, acetyl-β-glucosides, or malonyl-β-glucosides.

products containing aglycones, which can be achieved by different processes that induce the conversion of glucosides isoflavones to aglycones (Izumi et al., 2000). In plants, the isoflavones are synthesized and have antioxidant properties (Patel et al., 2001) that during normal metabolism protect cells against free-radical oxidative damage. The anticancer activity of genistein and daidzein has also been attributed to the antagonistic and agonistic activities of estrogen receptors (Birt et al., 2001). Soybean isoflavones (genistein, daidzein, and glycitein) are, therefore, bioactive compounds of non-steroidal and phenolic nature with mildly estrogenic properties and often referred to as phytoestrogen (Zaheer & Akhtar, 2017). However, the estrogenic action of soybean is beyond the mandate of this chapter and readers may find this information in Vitale et al. (2013).

The positive effects of soybean isoflavones have been studied in pathologies such as breast, colon, and prostate cancers (Messina et al., 1994, 2006; Dong et al., 2013; Mahmoud et al., 2014), cardiovascular diseases (Zhang et al., 2012), osteoporosis (Wei et al., 2012), menopausal symptoms (Messina, 1998, 2014; Li et al., 2014); obesity and diabetes (Choi et al., 2008; Most et al., 2013), and cognitive functions in the aging process (Henderson et al., 2012). However, according to Zaheer and Akhtar (2017), there are some clinical trials that have been found to be inconsistent and there is an urgent need to address these issues respecting the international guidelines.

Most biological activities of isoflavones are attributed to their aglycone forms (Chen et al., 2012). According to Dong et al. (2013), these biomolecules have also been shown to be beneficial to human health due to their antioxidant, antihypertensive, and potential anticarcinogenic effects. The anticarcinogenic effect of isoflavones has been explained by the antioxidant capacity of genistein, which is an estrogen receptor (ER)-selective binding phytoestrogen, with a greater affinity for ERβ. In addition, it inhibits tyrosine kinases and DNA topoisomerases I and II (Kuriyama et al., 2013).

In order to be bioaccessible, soybean isoflavones must be hydrolyzed in their corresponding aglycones. Accordingly, to better understand their bioavailability and consequent bioefficacy, it is necessary to focus on the form in which they exist in products after harvest and upon processing (Larkin et al., 2008). The conversion of conjugated isoflavones to their respective aglycones depends on several factors. The conversion process can occur by the action of the endogenous β-glucosidase enzyme during the development and/or germination of the seed (Ismail & Hayes, 2005; Ribeiro et al., 2007; Paucar-Menacho et al., 2010; Yoshiara et al., 2012), soaking (Góes-Favoni et al., 2010; De Lima & Ida, 2014; Falcão et al., 2018), or during processing to obtain derived products (Baú &

Ida, 2015). This enzyme can also be produced by suitable microorganisms used to obtain fermented soybean products which render a higher content of aglycones, such as milk fermented with kefir (Baú et al., 2015); soy flour fermented with *Aspergillus oryzae* CCT 4359 (Silva et al., 2011) or with *Aspergillus oryzae* IOC 399/1998 or *Monasccusn purpureus* NRRL 1992 (Handa et al., 2014); sufu; Cheng et al., 2011; Cai et al., 2016); douchi (Wang et al., 2007); tempeh (Nakajima et al., 2005; Haron et al., 2009; Ferreira et al., 2011; Bavia et al., 2012; Borges et al., 2016; Kuligowski et al., 2017); and natto (Hassim et al., 2015).

The content and profile of isoflavones is an important characteristic of the soybean seeds and is greatly influenced by many factors, such as geographical and environmental conditions as well as genotype/variety (Eldridge & Kwolek, 1983; Wang & Murphy, 1994b; Carrão-Panizzi et al., 1999; Tsukamoto et al., 1995; Hoeck et al., 2000; Lee et al., 2003; Laurenz et al., 2017; Teekachunhatean et al., 2013; Kim et al., 2014), maturity (Seguin et al., 2004; Ribeiro et al., 2007; Britz et al., 2011; Zhang et al., 2014), and others. The profile and isoflavone content of soybeans BRS 284, BRS 257, and BRS 216 show differences, with 64% to 80% predominance of malonyl glucosides, followed by 17% to 33% β-glucosides and 1.1% to 2.6% aglycones, and acetyl glucosides were not detected (Quinhone Jr. & Ida, 2015; Andrade et al., 2016; Silva et al., 2020).

The contents and composition of isoflavones of soybeans also differ according to their components (e.g. seed coat, cotyledon, axis). The concentration of isoflavones is higher in the hypocotyl, followed by the cotyledon and seed coat (Eldridge & Kwolek, 1983; Ribeiro et al., 2006; Silva et al., 2012; Yoshiara et al., 2018b; Quinhone Jr. & Ida, 2014; Kim et al., 2007). According to Yoshiara et al. (2018b), the average total isoflavone content in the hypocotyls was 6.5 times higher than in cotyledons. In hypocotyls, the isoflavones profile consisted of β-glucosides, malonyl glucosides, and aglycones and indicated that 98.36% were conjugated isoflavones and 1.64% aglycones. In cotyledons, the profile consisted of β-glucosides, malonyl glucosides, and aglycones and indicated that 99.3% were conjugated isoflavones and 0.63% aglycones. The isoflavone content in the soybean components is also influenced by variety, field conditions, and methods of quantification (Xu & Chang, 2008). In general, after extraction using different solvents, high-performance liquid chromatography (HPLC) is used to quantify these different forms of isoflavones and has been expressed in mg per 100 g; μg g^{-1}; mg g^{-1}; or μmol g^{-1}. More recently, hyphenated techniques, such as liquid chromatography coupled with tandem mass spectrometry (LC–MS^n), have also been employed, thus increasing the reliability of the results (de Camargo et al., 2019a).

The different conditions employed during soybean germination influence the isoflavone profile of each sprout component (Zhu et al., 2005; Ribeiro et al., 2006; Lee et al., 2007; Phommalth et al., 2008a, 2008b; Yuan et al., 2009; Quinhone Jr. & Ida, 2014; Yoshiara et al., 2018b) or of the germinated soybean as a whole (Kim et al., 2006; Shi et al., 2010; Paucar-Menacho et al., 2010; Huang et al., 2014; Quinhone Jr. & Ida, 2015; Silva et al., 2020). Depending on the conditions of the germination process, it is possible to obtain soybean sprouts or soybean germinated with high content of isoflavone aglycones (Paucar-Menacho et al., 2010; Quinhone Jr. & Ida, 2014; Yoshiara et al., 2018b; Silva et al., 2020).

Processing conditions and the type applied in soybeans also affect the profile and isoflavone content in soybean products (Xu et al., 2002; Lee & Lee, 2009; Andrade et al., 2016; Muliterno et al., 2017; Czaikoski et al., 2018; Guimarães et al., 2020). In unfermented products, the distribution of isoflavones glucosides (genistin and daidzin) was higher than in fermented soybean products, which contain higher aglycone isoflavones content (Wang & Murphy, 1994a).

There is an increased interest from academia, the food industry, and consumers as well as from health professionals in better understanding the contents and distribution of specific soybean isoflavones and derived products. However, the variability in the level of isoflavones in soybean and products thereof makes it difficult to carry out clinical and/or bioavailability studies (Larkin et al., 2008). The identities and quantities of soybean isoflavones are crucial in biotechnology and in conventional breeding programs (Azam et al., 2020) and for obtaining functional products or ingredients with a high content of isoflavones in the aglycone form (Bustamante-Rangel et al.,

2018). Considering the aspects mentioned, the following sections will discuss the various factors that influence the isoflavone profile of soybean seeds and their components. A special focus will be made on their identities and quantities. Finally, the new quality index in food application and health will be addressed and discussed.

2.2 ISOFLAVONE PROFILE IN SOYBEAN SEEDS

The content and profile of isoflavones is an important characteristic of the seed and is greatly influenced by many variables, such as environmental factors (climate, soil, temperature), seed variety, seed weight, seed development, genetics, growing locations and seasons, harvest date, genotype (Eldridge & Kwolek, 1983; Graham, 1991; Kudou et al., 1991; Wang & Murphy, 1994b; Tsukamoto et al., 1995; Carrão-Panizzi et al., 1999; Hoeck et al., 2000; Lee et al., 2003; Lee et al., 2008a; Teekachunhatean et al., 2013; Kim et al., 2014; Laurenz et al., 2017), and maturity (Seguin et al., 2004; Ribeiro et al., 2007; Britz et al., 2011; Zhang et al., 2014). It has also been shown that the isoflavone content is a characteristic transmitted from parent genotypes to their hybrids (Cvejić et al., 2011; Azam et al., 2020). Therefore, understanding the specific content and profile of isoflavones in soybean seeds may help to predict the potential benefits of these constituents for consumer health, thus allowing breeding programs to select seeds and/or develop new varieties. Finally, considering the nature and content of these constituents is also important to choose the most appropriate raw materials for the food, nutraceutical/supplements, and pharmaceutical industries.

The total isoflavone content of soybean varieties planted in four different locations varied between 116 and 309 mg g^{-1}, whereas for the same variety and planting location, the content varied between 46 and 195 mg g^{-1}. According to the authors (Eldridge & Kwolek, 1983), the isoflavones were concentrated in the hypocotyl while presenting a low content in the hulls. Other environmental factors (climate, soil, and temperature), varieties, and different seed weights (small, medium, and large) of soybean also influenced the profile and content of isoflavones (Lee et al., 2008a). The isoflavone profile of soybean may also be related to the development of seeds. The content of genistin and malonyl genistin has been found to increase at the end of the seed development stage, while the content of daidzin and malonyl daidzin increases throughout the period of seed development (Kudou et al., 1991).

The content and distribution of 12 different forms of soybean isoflavones of eight American and three Japanese soybean varieties were influenced by genetics, cultivation year, place of growth, and climatic conditions. Crop year seemed to have a greater influence on isoflavones content than location. The total isoflavone contents of the seven American and three Japanese varieties ranged from 2.05 to 4.21 mg g^{-1} and 2.04 to 2.34 mg g^{-1}, respectively. Acetylated isoflavones were detected in small amounts or were not detected because these isoflavones are usually generated upon heat treatment. The predominant isoflavones were malonylgenistin, genistin, malonyldaidzin, and daidzin (Wang & Murphy, 1994b). However, Carrão-Panizzi et al. (1999) observed that factors such as genetics, environment, and different planting locations significantly influenced the total isoflavones content of Brazilian soybean seeds, whose variation was 2.18 mg g^{-1} for the cultivar IAS 5 and 0.27 mg g^{-1} for the FT-Estrela cultivar. The isoflavone content decreased significantly when the seeds were grown at high temperatures during development and at different locations and planting dates. The isoflavone content (daidzin + genistin + malonyldaidzin + malonylgenistin) of the Higomusume soybean variety, sown in May, was 0.19 mg g^{-1}, and of the Lee variety, sown in July, was 3.51 mg g^{-1} (Tsukamoto et al., 1995).

The total and individual isoflavone contents of soybean seeds were significantly influenced by interactions genotype, genotype × year, genotype × location, and genotype × year × location (Hoeck et al., 2000). The total isoflavone content in μg g^{-1} varied from 145–209 (daidzin), 64–81(glycitin), 180–277 (genistin), 1,002–1,835 (6"-O-malonyldaidzin), 146–196 (6"-O-malonylglycitin), 948–1,604 (6"-O-malonylgenistin), 1–17(daizein), 2–3 (glycitein), and 13–14 (genistein). The predominance among the different isoflavones was, therefore, for the malonyl group. Investigating the genotype, year, location, and their interactions, Lee et al. (2003) observed that the genotype ("Geomjeong1") factor had a higher total isoflavone content (5,413.7 ug g^{-1}) and the environment and genetics were the factors that most influenced the total isoflavone content. According to Laurenz et al. (2017), agronomic treatments and temporal variability between the plating year had a greater influence on the levels of isoflavones genistein (1.03–5.24 g kg^{-1}) and daidzein (2.07–11.02 g kg^{-1}) of soybean seeds compared to that of the location.

The isoflavone content in soybean seeds grown in Thailand was dependent on multiple genetic and environmental factors (Teekachunhatean et al., 2013). They observed that the β-glucosides (daidzin and genistin) were the predominant forms of isoflavones extracted in soybeans instead of the malonyl glucosides. The corresponding aglycones (daidzein and genistein) were detected in negligible amounts or not detected. Glycine has not been determined due to its low concentration.

The content and profile of 12 isoflavones from 44 soybean varieties of *Glycine max* (L.) Merrill from China, Japan, and Korea were influenced by the genetic characteristics of the plants (variety, weight, and color of the seeds), tolerance to disease, and resistance to insects (Kim et al., 2014). After employing chemometric techniques, they found that the isoflavone content varied little between the varieties. The content of daidzin, genistin, and glycitin correlated with the respective content of malonyl glucosides, and the content of daidzein correlated with the content of genistein. The CS02554 variety showed the highest level of total isoflavones (1.71 mg g^{-1}), including glucosides (0.82 mg g^{-1}), malonyl glucosides (0.78 mg g^{-1}), acetyl glucosides (0.026 mg g^{-1}), and aglycones (0.06 mg g^{-1}). This variety, due to specific metabolites compositions, was recommended for breeding programs to develop new varieties of soybean.

The profile and isoflavone content of 20 soybean cultivars from different maturity groups and grown in two locations in Montreal, Canada, were significantly affected according to the environment, cultivar, place, and year of planting (Seguin et al., 2004). Depending on the environment, early maturity groups showed total isoflavone contents that varied from 0.36 to 2.24 mg g^{-1}. The different maturity groups of 18 soybean cultivars grown under the same conditions showed that the total isoflavone content varied from 0.71–1.74 mg g^{-1} for early cultivars; 0.62–1.28 mg g^{-1} for semi-early cultivars; and 0.80–1.28 mg g^{-1} for medium maturity group (Ribeiro et al., 2007). The highest total isoflavone content was observed in the BRS 212 cultivar, belonging to the early maturity group, and the lowest was for the BRS 232 cultivar from the semi-early maturity group. The predominant isoflavone forms were malonylgenistin (0.24–0.81 mg g^{-1}) and malonyldaidzin (0.15–0.38 mg g^{-1}) in all cultivars, corresponding to 67%, while β-glucosides and aglycones forms corresponded to 31% and 2%, respectively. The malonylglycitin was detected in only three cultivars and the acetylated form was not detected.

Soybeans from 15 lines of four maturity groups, grown in three locations in Maryland, with a complete and normal season represented by a hot and dry year presented a variation in the total isoflavone content between 4.7 and 8.7 µmol g^{-1} of seeds on a dry basis (Britz et al., 2011). When these soybeans were planted in warmer regions, the total isoflavone content decreased by 50% compared to in the colder regions. The contents of total isoflavones, genistein, and daidzein were not influenced according to the latter maturity lines at any location or planting date combination, while the glycitein content was more variable. The isoflavone profile of 40 soybean cultivars from different groups planted in northern and southern China and in different years was significantly influenced by the mentioned factors. The total isoflavone content ranged from 551.15 to 7,584.07 ug g^{-1}. This investigation aimed to select soybean cultivars with characteristics of desirable isoflavone content (Zhang et al., 2014). The total isoflavone content of 20 F1 soybean progenies had a significant, high correlation with the parents and corresponding F1 progenies. The content of total isoflavones in soybean varied from 1.56 to 3.66 mg g^{-1} in dry soybean. It was suggested that the isoflavone content may be a characteristic derived from the parental genotype for their F1 progenies and, therefore, the improvement of the genotypes may be directed towards obtaining soybeans with better characteristics in the isoflavone content (Cvejić et al., 2011).

The profile and content of soybean isoflavones from 1,168 accessions of soybeans collected from different ecoregions of China in three locations over two years indicated significant differences between accession genotypes, growth year, accession types, and ecoregions of origin. The total isoflavone content in soybean accessions ranged from 0.74 mg g^{-1} to 5.25 mg g^{-1}. The metabolite profiling combined with chemometric methods was a powerful tool for assessing soybean seed quality. According to the authors (Azam et al., 2020), soybean accessions with high total isoflavone concentrations can be used as raw materials for the food and pharmaceutical industries and may serve as novel germplasm for the improvement of soybean isoflavone quantity and composition.

2.3   ISOFLAVONES IN SOYBEAN SEEDS AND THEIR COMPONENTS

Soybean seeds present different characteristics in the profile and content of isoflavones which, as mentioned before, depend on the variety and other factors. BRS 284, BRS 257, and BRS 216 soybeans seeds showed differences in composition, with 64% to 80% predominance of malonyl glucosides, followed by 17% to 33% β-glucosides and 1.1% to 2.6% aglycones, and acetyl glucosides were not detected (Quinhone Jr. & Ida, 2015; Andrade et al., 2016; Silva et al., 2020). However, vegetable-type soybean or edamame contains 85% malonyl glucosides, 15% β-glucosides and aglycones, and acetyl glucosides were not detected (Czaikoski et al., 2018).

According to Liu (1997), the cotyledons contribute to 90% of soybean seeds, followed by the hull (8%) and hypocotyl axis (2%), which is supported by the similar composition reported by other authors in commercial soybeans (Yuan et al., 2009; Quinhone Jr. & Ida, 2014). BRS 213 soybeans have 10.0% seed coat, 86.8% cotyledons, and 3.2% radicles (Ribeiro et al., 2006). Furthermore, one should bear in mind that the composition of the components may also vary according to the seed size (large, medium, and small) (Phommalth et al., 2008a). In summary, it is possible to state that the contribution of cotyledons ranges from 87–90%, the hull contributes to 8–10%, and the hypocotyl axis to 2–3%.

The content and composition of isoflavones of soybean seeds differ according to their components (e.g. seed coat, cotyledon, and axis). The content of the isoflavones in two varieties of soybeans on an equal weight basis is higher in the hypocotyl (14.0–17.5 mg g^{-1}), followed by cotyledon (1.5–3.9 mg g^{-1}) and seed coat (0.1–0.2 mg g^{-1}) (Eldridge & Kwolek, 1983).

The mature soybeans and hypocotyls present different content and composition of the nine forms of isoflavone glucosides. The total isoflavone content in the hypocotyl (containing plumula and radicle) was 5.5 to 6 times higher than in cotyledons and absent in the hull. Glycitin isoflavones occurred only in the hypocotyl (Kudou et al., 1991).

The profile and content of isoflavones vary according to the components and varieties of soybeans. In BRS 213 soybean, the total isoflavone content in the radicle (10.71 mg g^{-1}) was 25.2 times higher than in cotyledons (0.42 mg g^{-1}). Of the total isoflavones (daidzein + genistein + genistin + daidzin) from soybean (20.54 mg g^{-1}), 48.2% was in the radicles (daidzein + genistein) and 51.8% in the cotyledons (genistin + daidzin). In contrast, the seed coat contained considerably lower amounts of isoflavones (0.01 mg g^{-1}) (Ribeiro et al., 2006). Likewise, in four soybean cultivars (BRS 184, BRS 216, BRS 257, and BRS 267), the average total isoflavone content in the hypocotyls (20.30 mg g^{-1}) was 9.6 times higher than in cotyledons (2.10 mg g^{-1}), while the presence of isoflavones was not detected in the seed coat, and in soybeans, aglycones isoflavones were absent or in undetected amounts (Silva et al., 2012).

BRS 257 soybean was also investigated by Yoshiara et al. (2018a), however, they found that the average total isoflavone content in the hypocotyls (23.30 mg g^{-1}) was 6.5 times higher than in cotyledons (3.56 mg g^{-1}). In hypocotyls, the isoflavone profile consisting of β-glucosides (6.56 mg g^{-1}), malonyl glucosides (16.35 mg g^{-1}), and aglycones (0.38 mg g^{-1}) indicated that 98.36% were conjugated isoflavones and 1.64% were aglycones. Whereas in cotyledons, the profile consisting of β-glucosides (0.66 mg g^{-1}), malonyl glucosides (2.87 mg g^{-1}), and aglycones (0.02 mg g^{-1}) indicated that 99.3% were conjugated isoflavones and 0.63% aglycones. In BRS 284 soybeans, although the isoflavone content was expressed in μmol g^{-1}, Quinhone Jr. and Ida (2014) observed that the total isoflavone content in the embryonic axis (53.8 μmol g^{-1}) was 7.1 times higher than in cotyledons (7.56 μmol g^{-1}). In this soybean, cotyledons and embryonic axis showed similar proportions of β-glucosides, malonyl glucosides, and aglycones isoflavones. In relation to total isoflavones content, the soybean, cotyledons, and embryonic axis showed 26.0%, 27.2%, and 29.4% β-glucosides; 72.9%, 71.4%, and 70.0% malonyl glucosides; and 1.2%, 1.4%, and 0.6% aglycones, respectively. Acetyl glucosides were not detected in soybean.

The components of nine soybean varieties showed different contents and profiles of the 12 forms of isoflavones (Kim et al., 2007). The average content of total isoflavone was 2.88 mg g^{-1} in the embryo; 0.57 mg g^{-1} in the whole seed; 0.32 mg g^{-1} in the cotyledon; and 0.033 mg g^{-1} in the seed coat. The highest total isoflavone content of the soybean components occurred in the embryo (5.70 mg g^{-1}) in Geomjeongkong 2; the whole seed (1.32 mg g^{-1}) in Geomjeongolkong, followed by the cotyledon (0.95 mg g^{-1}) in Heugcheongkong, and the seed coat (0.05 mg g^{-1}) in Keunolkong. In general, the embryos had the highest isoflavone content, while the seed coats had the lowest content. The predominance in the isoflavone content of the three components of soybean occurred in the order of malonyl glucosides > glucosides > acetyl glucosides > aglycones, with malonyl glucosides contributing 66% to 79.0% of the total isoflavone content.

The profile and isoflavone content of the components of black soybean (black seed husks) are influenced largely by variety, cultivation conditions, and quantification methods (Xu & Chang, 2008). These black soybeans are widely utilized as a healthy food and plant material in China, Taiwan, Korea, and Japan. The content of total isoflavones in black soybean, according to components of grain, was on the order of whole black soybean (0.96 mg g^{-1}), dehulled soybean (0.89 mg g^{-1}), and seed coat (0.048 mg g^{-1}). Dehulled black soybeans possessed 93.10% of total isoflavones, while the seed coats possessed 4.97% of total isoflavones. This is especially important because, in peanuts, which are also legume seeds, skin removal has a significant negative impact on the content of phenolic compounds of the seed (de Camargo et al., 2017). Therefore, one should bear in mind that each legume seed may behave differently in terms of the distribution of phenolic compounds.

Black soybeans showed almost all 12 forms of isoflavones since only glycitein was not detected. The highest proportion of isoflavones, in whole black soybeans and dehulled black soybeans, was 66.5% for 6''-O-malonyl-β-glucosides, followed by 24.5% for 7-O-β-glucosides, while 6''-O-acetyl-β-glucosides and aglycones isoflavones occurred in small proportions. Whole black soybean and dehulled black soybean exhibited similar isoflavones contents in β-glucosides, malonyl glucosides, acetyl glucosides, and aglycones (glycitein not detected), while in the seed coat these contents were low.

Berger et al. (2008) observed that the content and composition of isoflavones in hypocotyls seem to be less influenced by the environment than in cotyledons, and in the hypocotyls contain a specific individual isoflavone, that is, glycitein and its conjugated forms. In different varieties of soybeans and with the same weight of the components, Phommalth et al. (2008a) also observed the order of increasing total isoflavone content from seed coat to cotyledon and axis. In cotyledons, they observed that the highest and lowest total isoflavone content was for the Aga3 (0.48 mg g^{-1}) and Pungsannamulkong (0.20 mg g^{-1}) varieties. In the axis, the highest and lowest total isoflavone content was for the Taekwangkong (0.06 mg g^{-1}) and Hwangkeumkong (0.02 mg g^{-1}) varieties.

Hypocotyls and cotyledons of commercial soybeans presented different content and profile of daidzein, genistein, glycitein, and their malonyl-, acetyl-, and nonconjugated β-glucosides (Yuan et al., 2009). The hypocotyl had a content of total isoflavones 7.8 times higher than in cotyledons. In hypocotyls, the most abundant isoflavones corresponded to daidzein and its conjugated glucosides (59.6%), followed by the series (aglycones and conjugates) of glycitein (26.6%) and genistein (13.8%). In contrast, the most abundant isoflavones in the cotyledons corresponded to genistein and its conjugated glucosides at 61.9%, followed by the daidzein and their conjugated forms (at 38.1%). Furthermore, the only form of glycitein was detected in the cotyledon. The contribution of different forms of isoflavones in the hypocotyls and cotyledons was quite similar and followed the decreasing order of malonyl glucosides (69.1% to 69.4%) > β-glucosides (25.4% to 27.1%) > aglycones (5.2% to 3.8%) while acetyl glucosides were not detected.

2.4 ISOFLAVONE BIOTRANSFORMATION UPON GERMINATION

Soybean germination is a simple and low-cost process that has been used to obtain soybean sprouts (Figure 2.2) with better sensory quality, nutritive value, and functional characteristics. The germination of soybean can promote a significant increase in the content of some nutrients, such as vitamins, phytosterols, tocopherols, and isoflavones. The cotyledon and embryonic axis (roots, hypocotyls, and radicles) are the components of soybean sprouts, also known as germinated soybeans. Depending on the conditions of the germination, changes may occur, thus affecting the isoflavone profile and total isoflavone content. This process has different outcomes in each component of the sprouts or when one considers the entire sprout. Soybean sprouts can, therefore, exhibit a specific isoflavone composition which increases their usefulness as a functional food and/or ingredient as well as in the manufacture of nutraceutical products and food supplements.

Hypocotyls with 0.5, 2.5, and 6.5 mm lengths of two germinated soybean varieties (Hutcheson and Caviness) showed an increase in the profile and content of isoflavones (Zhu et al., 2005). Germination at 40° C was carried out after soybean soaking for 12 h. The maximum content of total isoflavones (2.49 and 2.78 mg g^{-1}), genistein and daidzein (1.50 and 1.52 mg g^{-1}), and their β-glucosides conjugates (0.67 and 0.90 mg g^{-1}) of the germinated soybean varieties (Hutcheson and Caviness) occurred when the length of the hypocotyl reached 0.5 mm and 2.5 mm, respectively. In the respective lengths of the hypocotyls, there was a high increase in the content of malonyl genistin (1.305 mg g^{-1} and 1.308 mg g^{-1}) and malonyl daidzin (0.476 mg g^{-1} and 0.677 mg g^{-1}) for Hutcheson and Caviness varieties, respectively. After these stages of germination, the content of these isoflavones decreased. The genistein and daidzein content was higher soon after the soaking of the soybeans. The glycitein content and its β-glucosides conjugates were maintained during germination. Under controlled conditions of soybean germination, it was possible to increase the isoflavone content.

The radicles and cotyledons of BRS 213 soybeans germinated for 72 h at 25° C under light conditions showed changes in the profile and content of isoflavones (Ribeiro et al., 2006). Through regression analysis, it was found that the total isoflavone content in the radicles decreased 6.3 times,

FIGURE 2.2 Soybeans subjected to germination.

with a decrease in the content of malonyl glucosides (malonyl daidzin, malonyl glycitin, and malonyl genistin), β-glucosides (daidzin and glycitin), and aglycone (genistein) and an increase in the content of aglycone (daidzein). In cotyledons, the total isoflavones and malonyl genistin and malonyl daidzin content increased, while the content of β-glucosides (daidzin and genistin) did not change. In addition, cotyledons did not detect β-glucoside (glycitin), aglycones (glycitein and daidzein), malonyl glycitin, and acetyl glucosides isoflavones. Changes in the total isoflavones content in the radicles and cotyledons were associated with the activity of β-glucosidase and possibly with the physiological metabolism of seeds.

Cotyledons, hypocotyls, and roots of 17 soybean varieties germinated for five days under dark and light conditions presented different compositions and isoflavones content (Lee et al., 2007). The average total isoflavones content in cotyledons was 2.16 mg g^{-1} (green sprout) and 2.54 mg g^{-1} (yellow sprout); in hypocotyls, it was 1.17 mg g^{-1} (green sprout) and 1.13 mg g^{-1} (yellow sprout); and in roots, it was 2.40 mg g^{-1} (green sprout) and 2.85 mg g^{-1} (yellow sprout). The total isoflavones content between the three components was different, with the roots showing the highest content, followed by cotyledons and hypocotyls. The content of daidzin in the green (0.77 mg g^{-1}) and yellow (0.89 mg g^{-1}) sprouts increased more than four times in relation to the non-germinated seeds (0.18 mg g^{-1}). The yellow sprouts contained a higher content of genistin (1.12 mg g^{-1}), while the green sprouts (0.15 mg g^{-1}) and yellow sprouts (0.15 mg g^{-1}) had a higher glycitin content than the non-germinated seeds. Regarding the total isoflavones content, the cotyledons of green and yellow sprouts were constituted by 67% genistin, 28% daidzin, and 4% glycitin. The content of malonyl glucosides was higher in cotyledons, while the content of glucosides was higher in hypocotyls and roots.

The various parts of the sprouts of two germinated soybean cultivars, one an Aga3 cultivar of small size and high content of total isoflavones (7.24 mg g^{-1}) and the other a Pungsannamulkong cultivar of wide use and low content of total isoflavones (1.23 mg g^{-1}), showed variation in isoflavones content according to genotype and germination period (Phommalth et al., 2008a). Germination for ten days was carried out in a controlled germination chamber, with temperature at 20° C, 80% humidity, and exposure to light for 12 h a day. The sprouts (cotyledon + hypocotyl + root) of soybeans germinated for seven days showed higher content of total isoflavones for both cultivars (Aga3 with 10.78 mg g^{-1} and Pungsannamulkong with 3.55 mg g^{-1}). The shoots of samples collected on day ten showed a reduction of 12.80% and 8.52% in the total isoflavone content for Aga3 and Pungsannamulkong, respectively. In the cotyledons of the Aga3 soybean sprouts, the total isoflavone content increased significantly, while in the Pungsannamulkong soybean sprouts it decreased. However, in the root of the soybean sprouts of both cultivars, an increase in the total isoflavone content was observed, while in the hypocotyls it decreased.

The cotyledons and hypocotyls of soybeans germinated for 108 h and 120 h under natural light and uncontrolled conditions showed alterations in the profile and content of daidzein, genistein, glycitein, and their malonyl, acetyl, and nonconjugated β-glucosides during the night and according to the time germination (Yuan et al., 2009). In germinated soybean, the content of β-glucosides decreased while the content of malonyl glucosides and aglycones increased, with predominance in the content of malonyl glucosides. In the hypocotyls, after 108 h of germination, low content of β-glucosides and aglycones was detected, while the content of malonyl glucosides was the most abundant form. In the hypocotyls, the aglycone contents markedly increased and were the most prominent form after 120 h of germination, while a small amount of malonyl glucosides and β-glucosides was detected. In cotyledons after 108 h of germination, malonyl glucosides were the most abundant forms, followed by β-glucosides. In contrast, only a small amount of aglycones was detected. In the cotyledons after 120 h of germination, the aglycones content also increased and malonyl genistin was the main isoflavone, followed by genistein, β-glucosides genistin, malonyl daidzin, daidzein, and β-glucosides daidzin. In soybean germination at night, there was an increase in the content of aglycones in the cotyledons and hypocotyls, while the content of malonyl glucosides decreased, although the mechanism of these changes due to the circadian rhythm has not been fully clarified.

The different components of BRS 284 soybeans germinated up to 168 h showed changes in the profile and isoflavones content according to the germination time, and mathematical models related to the process were developed by the authors (Quinhone Jr. & Ida, 2014). The cotyledons of the sprouts showed a higher content of total isoflavones, whereas only aglycones were detected in the hypocotyls and radicles. In sprouts cotyledons with 168 h of germination, the content of β-glucosides (daidzin and genistin) and malonyl glucosides (malonyl daidzin and malonyl genistin) increased, while the content of glycitin, malonyl glycin, and aglycones was not detected. In hypocotyls and radicles, the content of β-glucosides (daidzin and genistin) and malonyl glucosides (malonyl genistin and malonyl glycin) decreased up to 96 h and became constant up to 168 h of germination. However, in hypocotyls and radicles in 168 h of germination, the content of aglycones (genistein and daidzein) increased. The authors suggested that to obtain soybean sprouts bearing a higher level of total isoflavones, soybean germination should be carried out up to 168 h and cotyledons would be the best source. The cotyledons did not show any presence of aglycones. Therefore, in order to obtain soybean sprouts with a higher content of isoflavones aglycones, the radicles and hypocotyls must be included.

The different components of soybeans germinated for 144 h in the presence of light, showed a different profile and isoflavone content and increased antioxidant potential (Yoshiara et al., 2018a). The components of the germinated soybean had the following total isoflavone content: Cotyledon (1.64 mg g^{-1}), radicle (0.41 mg g^{-1}), hypocotyl (0.43 mg g^{-1}), and epicotyl (2.11 mg g^{-1}). In germinated soybean, the percentage of conjugated isoflavones was 8.9% in cotyledon, 0% in radicle or hypocotyl, and 83.8% in epicotyl. They highlighted that the percentage of aglycones isoflavones was 91% in cotyledon, 100% in radicle or hypocotyl, and 16.1% in epicotyl. The germination for 144 h was ideal to obtain radicles and hypocotyls with 100% isoflavone aglycones. These components were considered ideal for studies aimed at isolation or application as functional ingredients.

As with each component of soybean sprouts, the profile and content of isoflavones were also investigated in whole sprouts. Soybean sprouts of seven cultivars germinated for 120 h at 25° C in dark conditions (produced yellow soybean sprouts) and in green and yellow boxes (produced green soybean sprouts), showed changes in the profile and isoflavones content (Kim et al., 2006). Green Sowonkong soybean sprouts (2.79 mg g^{-1}) had a higher total isoflavones content than yellow Pureunkong soybean sprouts (0.75 mg g^{-1}). In green and yellow soybean sprouts, a high content was observed in the three forms of aglycones (daidzein, genistein, and glycitein) with a lower amount of glycitein. In the sprouts of three cultivars, glycitein was not detected. In soybean sprouts, the malonyl glucosides (daidzin, glycitin, and genistin) were the predominant forms of isoflavones. The production of soybean sprouts under colored light sources was feasible and had a higher total isoflavones content than non-germinated soybean.

BRS 258 soybean sprouts germinated in different time and temperature conditions showed changes in the profile and isoflavones content according to proposed mathematical models (Paucar-Menacho et al., 2010). Depending on the treatments, the total content of isoflavones and aglycones of soybean sprouts varied from 2.33 to 2.94 mg g^{-1} and 0.04 to 0.64 mg g^{-1}, respectively. The model allowed us to estimate that the content of aglycones (daidzein and genistein) increased by 154% in soybean sprouts germinated for 63 h at 30° C. This increase possibly occurred due to the hydrolysis of the glucosides isoflavones during the soaking of the grains and germination process, whose increase in genistein content was 306%. Isoflavones acetyl glucosides and glycitein were not detected. The model can also be used to estimate isoflavones transformations during germination that can be useful for human health and nutrition.

The germination of soybeans for seven days, with or without light exposure, changed the content of isoflavones (Shi et al., 2010). In contrast, samples germinated in the absence of light maintained a yellow color, while those exposed to light gradually changed to a green color due to the photosynthesis of chlorophyll. The malonyl daidzin (43.8%) and malonyl genistin (34.8%) were the predominant forms of isoflavones in soybean during the early stage of germination. During the germination, malonyls (daidzin and genistin) were converted to their glucosides (daidzin and genistin) and aglycones (daidzein and genistein) forms. Acetyl glucoside forms of isoflavones (e.g. daidzin, genistin, and glycitin) were not investigated since they are generally present only in fairly low concentrations in soybeans and soybean sprouts. The germination of soybeans to obtain sprouts is a beneficial process for health.

The germination time influenced the increase in the content and changed the composition of iso-flavones (Huang et al., 2014). Soybeans germinated for three days showed a higher content of total isoflavones (4.68 mg g^{-1}) and aglycones genistein and daidzein (3.99 mg g^{-1}) with an increase of 3.0 and 2.5 times, respectively, in comparison with the non-germinated counterparts. After four days of germination, the total isoflavones and aglycones content decreased 80% compared to the third day, while the β-glucosides content decreased 76%. The soybean thus germinated can be a good raw material that contains substances beneficial to health.

BRS 284 soybean sprouts germinated at different times for 168 h with natural light showed high yield (632%) and changes in the profile and content of the different forms of isoflavones (Quinhone Jr. & Ida, 2015). The germination time had a quadratic effect on the content of daidzin, genistin, and genistein; a linear effect for the malonyl daidzin content; and was constant for malonyl glycitin content. The content of glycitin and glycitein was not detected in soybean sprouts. After 168 h of germination, the total content of aglycones in soybean sprouts represented 2.5% of the total iso-flavones and was 2.9-fold higher than in non-germinated soybean which had only 1.8% genistein.

The soaking time, irrigation frequency, and germination time of the BRS 216 soybean influenced the length, yield, and content of isoflavones and vitamin C of the sprouts (Silva et al., 2020). In the proposed mathematical model, only the germination time had a significant negative linear effect on the content of malonyl glucosides, aglycones, and total isoflavones. According to the mathematical model, the best soybean germination condition was 6 h of soaking, three days of germination, and irrigation every 8 h. Under these conditions, soybean sprouts showed higher yield (166%), preserved chemical composition, and adequate sprout length (8.4 cm), in addition to exhibiting a higher con-tent of β-glucosides (1.51 µmol g^{-1}), malonyl glucosides (10.43 µmol g^{-1}), aglycones (0.44 µmol g^{-1}), total isoflavones (12.38 µmol g^{-1}), and vitamin C (2.38 mg g^{-1}) than non-germinated soybeans. Since no heat treatment was employed, acetyl glucosides were not detected. Therefore, under suit-able germination conditions, it was possible to produce soybean sprouts showing better physical characteristics, improved yield, and higher content of total isoflavones, isoflavones in the aglycone form, and vitamin C.

2.5   ISOFLAVONES IN SOYBEAN PRODUCTS

The profile and composition of isoflavones in products soybean depend on several factors, includ-ing food processing. According to Wang and Murphy (1994a), the concentration and distribution of the 12 forms of isoflavones in 29 commercial soybean foods were influenced by the soybean variety, processing method, and type and quantity of the soybean ingredient used. In traditional unfermented soybean foods, the total isoflavones content was 1.62 mg g^{-1} for roasted soybean and 1.00–1.18 mg g^{-1} for instant soybean beverage powder, and these contents were 2–3 times higher than in fermented soybeans with a content of 0.62 mg g^{-1} for tempeh, 0.59 mg g^{-1} for bean paste, 0.29 mg g^{-1} for miso, and 0.39 mg g^{-1} for fermented bean curd. As for tofu, an unfermented soybean product, the total isoflavone content was 0.33 mg g^{-1}. The distribution of isoflavones glucosides (genistin and daidzin) in unfermented products was higher than in fermented soybean products that had low isoflavones glucosides content. In fermented soybean products, aglycones isoflavones were the main forms of isoflavones, probably due to hydrolysis during fermentation by different microor-ganisms, which produces β-glucosidases. In hydrolysis, some malonyl glucosides isoflavones might be changed into glucosides, and then into aglycones.

The roasted soybean produced by the frying of the hydrated soybean showed significant amounts of acetyl glucoside isoflavones and a lower content of malonyl glucosides than unprocessed soybean, which is explained by the heat treatment. Second-generation soybean foods (soybean hot dog, tempeh burger, tofu yogurt, soybean parmesan, cheddar cheese, mozzarella cheese, flat noodle, and others) contained only 6% to 20% of the total isoflavones content of soybean and varied according to the quantity and type of soybean ingredient used. Hendrich et al. (1994) postulated that the content and profile of isoflavones in a wide variety of soybean products would be useful for future studies in human and animal nutrition. Since suggested anticarcinogenic doses for soybean isoflavones ranged from 1.5 to 2.0 mg (kg body weight)$^{-1}$ day^{-1} there are several options for these foods for human consumption.

The isoflavone profile of several soybean products consumed in Brazil was investigated in order to estimate the daily isoflavone intake of these products (Genovese & Lajolo, 2002). The total isoflavones content varied from 2 to 100 mg/100 g (wet basis, expressed as aglycones), with higher content for textured soybean protein. The soybean beverages content ranged from 12 to 83 mg/L. Soybean sauce, miso, and tofu had isoflavone contents of 5.7 mg/L, 20 mg/100 g, and 7 mg/100 g, respectively. The β-glucosides were the predominant form of the isoflavones in the products, except for fermented products (e.g. miso, shoyu) and "diet shake" in which the aglycones were present in the highest proportions. Therefore, it was estimated that the daily intake of isoflavone from soybean-based products for infants fed with soybean-based formulas was 1.6 to 6.6 mg/kg of body weight. According to Devi et al. (2009), the isoflavones contents (daidzein and genistein) of four soybean cultivars varied between 0.52 and 0.98 mg g^{-1} and, out of 26 soybean products, the isoflavones content decreased in the following order: Soybean sprouts (0.60–0.79 mg g^{-1}), soybean flour (0.50–0.77 mg g^{-1}), soymilk (0.11–0.15 mg g^{-1}), soybean meals (0.09–0.14 mg g^{-1}), and soybean sauce (0.032–0.054 mg g^{-1}). This section will address the isoflavone profile of industrial soybean products by focusing on unfermented and fermented products.

2.5.1 Isoflavones in Unfermented Products

According to Grun et al. (2001), the tofu heating time (10, 20, 30, and 40 min) and temperature (80, 90, and 100° C) changed the profile of genistein, daidzein, genistin, daidzin, and their acetyl- and malonyl- β-glucosides. The total isoflavone content of tofu decreased significantly and possibly was due to the leaching of isoflavones into the water. In tofu treatments, the content of isoflavones of the genistein series was little affected, while the content of the daidzein series decreased significantly which resulted in a reduction in the total isoflavones content. Changes in the profile of the daidzein series possibly occurred due to the little decarboxylation of the malonyl glucosides to the acetyl glucosides, but considerable de-esterification of the malonyl- and acetyl- glucosides to the β-glucosides (Figure 2.1). The decreases in the content of the daidzein series were strongly dependent on the temperature, which suggested a possible thermal degradation of daidzein in addition to losses leaching.

The profile and content of the 12 forms of soybean isoflavones (including aglycones and glucosides) were affected during the processing/preparation of soymilk and tofu (Jackson et al., 2002; Kao et al., 2004). According to Jackson et al. (2002), the mean recoveries of isoflavones from soybean beverages and tofu were 54% and 36%, respectively. A representative picture of soymilk is shown in Figure 2.3. During processing, the detectable levels of the aglycones, glucosides, and acetyl glucosides groups increased, while the corresponding malonyl glucosides decreased. However, Kao et al. (2004) observed that an increase in the temperature and soybean soaking time also increased the content of the aglycones (daidzein, glycitein, and genistein), while the other nine forms of isoflavones decreased due to conversion to aglycones by cleavage of the ester linkages or glucosides group. The content of glucosides (daidzin, glycitin, and genistin) and acetyl genistin increased upon the thermal processing of soymilk, while the content of malonyl glucosides (daidzin, glycitin, and genistin) decreased. Furthermore, the content of aglycones remained constant. The isoflavones acetyl daidzin and acetyl glycitin were not detected. Tofu produced with 0.3% calcium sulfate (as a protein coagulant) had a higher content of total isoflavones (2.27 mg g^{-1}).

FIGURE 2.3   Soymilk, the most popular beverage produced from soybean grains.

The application of heat treatments significantly influenced the isoflavone profile of raw soymilk (Huang et al., 2006). The dry solid contents were higher for daidzein (0.31 mg g^{-1}), followed by glycitein (0.09 mg g^{-1}) and genistein (0.03 mg g^{-1}). The aglycones isoflavones were generated by the action of native soybean β-glucosidase during the soybean soaking, a pre-treatment that is conducted prior to soymilk manufacture. Heat treatment of raw soymilk, depending on the temperature and exposure time, caused an increase or decrease in the genistein content. In the initial phase of the treatment of raw soymilk at 95° and 121° C, the conversion of genistin to genistein occurred, and the prolonged treatment reduced the content of genistein due to chemical reactions, such as Maillard-type reactions and auto-degradation.

Soybean soaking is a very important process used during the preparation of several soybean products (e.g. soymilk, tofu, tempeh, and others). The soaking also affects the profile of isoflavones in the final products due to biotransformation (De Lima & Ida, 2014; Falcão et al., 2018). In fact, the hydrothermal treatment of soybeans has been found to favor the activity of endogenous β-glucosidases which leads to the conversion of isoflavones β-glucosides into aglycones (Góes-Favoni et al., 2010). Tofu, which is represented in Figure 2.4, can be consumed as such or may be used to produce sufu, a popular fermented tofu product in China.

During soybean soaking, the content of aglycones isoflavones formed showed no difference from the average content of cooked tempeh (1,922–2,968 μg kg^{-1}) or tofu (1,667–2,782 μg kg^{-1}) but was strongly dependent on the variety of soybean used (Mo et al., 2013). Soybeans reached a maximum content of aglycones (1.22 μmol g^{-1}) and a minimum of isoflavones β-glucosides and low activity of β-glucosidases when soaked for 6 h at 55° C, while isoflavones malonyl glucosides were unstable (de Lima & Ida, 2014). According to Falcão et al. (2018) the pre-treatment of soybean by ultrasound (5 min at 55° C and intensity of 19.5 Wcm^{-2}) prior to the soaking converted the conjugated isoflavones

FIGURE 2.4   Tofu, which is represented in the picture, can be consumed as such or may be used to produce sufu, a popular fermented tofu product in China.

into their corresponding aglycones and maintained a high activity of endogenous β-glucosidases. Furthermore, this pre-treatment was found advantageous because it reduced the soaking time (2 h) of the soybean as evaluated by the moisture content and hardness.

De Lima et al. (2019) used first-order kinetic models to investigate the interconversion and loss of isoflavones (malonyl glucosides, β-glucosides, and aglycones) during soybean soaking at various temperatures (25°, 40°, 55°, and 70° C). Soaking at 25° C did not influence the content of isoflavones of the daidzein, glycitein, and genistein series. In contrast, a higher hydrolysis rate constant, which corresponded to the conversion of β-glucosides to their corresponding aglycones due to the action of endogenous β-glucosidase, was observed at 55° C. In addition, a decrease in the hydrolysis rate was noted at 70° C and all isoflavone forms of the genistein and daidzein series were thermally stable. Mathematical modeling of the kinetics during soybean soaking provided better insights into the mechanisms of interconversion and losses of the isoflavones.

Soymilk obtained after hot grinding extracted a higher content of isoflavones (glucosides and aglycones) than cold grinding. However, the direct (6.16 μmol g^{-1}) or indirect (6.21 μmol g^{-1}) heating with ultra-high temperature (UHT) of the extract did not influence the isoflavones content (Prabhakaran et al., 2006). High hydrostatic pressure in soaked soybean at the beginning of the treatment also did not influence the content of isoflavones. Nevertheless, under pressure combined with heat treatment at 75° C, the isoflavone profile shifted from malonyl glucosides toward β-glucosides and the concentration of isoflavone in soymilk increased from 4.32 to 6.06 μmol g^{-1} (Jung et al., 2008). In the thermal processing of soymilk, both at pH 7 and pH 9, there was significant interconversion in the content of isoflavones, with a decrease in malonyl glucosides and a simultaneous increase in glucosides. In the extracts, there are interactions of isoflavones with proteins and that possibly influenced the precision and evaluation of these effects. There were significant losses in the total isoflavones content when the soymilk extraction was assisted by enzymes (Nufer et al., 2009). However, these authors did not address the contents of aglycones.

From modifications of some stages of processing of soymilk and use of the central design of composites, it was possible to obtain soymilk with a high content of isoflavones aglycones (Baú & Ida, 2015). In the condition of temperature (50° C) and time (2.7 h) of optimized incubation of soymilk, it was possible to predict by the mathematical model the obtaining of soymilk with the maximum conversion of isoflavones β-glucosides into aglycones (2.92 μmol g^{-1}). Aglycone content was 16.2 times greater than the control of soymilk due to the possible action of endogenous β-glucosidases from soybean. After heat treatment at 97° C for 25 min, the content of aglycones (daidzein and genistein) of soymilk remained constant, while the content of glycitein and β-glucosides increased and malonyl glucosides decreased. At this temperature, the ester linkages of malonyl glucosides can be cleaved, giving rise to the corresponding β-glucosides (Figure 2.1). This reduction in the content of malonyl glucosides can also be attributed to its hydrolysis to form aglycones or β-glucosides without the involvement of β-glucosidases. In contrast, at this temperature, the decarboxylation of malonyl glucosides did not occur for the formation of their respective acetyl derivatives. This process can be useful and of great interest to consumers or the food industry who wish to make other soybean products with a high content of isoflavones aglycones and possibly higher biological activities.

The different grinding conditions (room temperature, cold and hot) and heating methods (traditional stove cooking, 1-phase UHT, and 2-phase UHT) used to obtain soymilk (Prosoy and black soybean) influenced the profile and isoflavones content (Zhang et al., 2015). The total isoflavones content varied after cold, room temperature, and hot grinding and was 3,917, 5,013, and 5,949 nmol g^{-1} for Prosoy and 4,073, 3,966, and 4,284 nmol g^{-1} for black soybeans, respectively. In general, and after grinding, the content of different forms of isoflavones decreased (e.g. β-glucosides, acetyl glycitin, malonyl genistin, and malonyl daidzin), but malonyl glycitin increased. The aglycones isoflavones, except those after hot grinding, also exhibited increasing patterns. Throughout the process, interconversion, degradation, leaching, and heat-induced release of the content of each isoflavone were observed.

Hemicellulase-assisted extraction (HAE) and ultrasound-assisted extraction (UAE) during the manufacture of soymilk were found to increase the activity of β-glucosidases by 1.3- and 1.5-fold, respectively. Likewise, the content of aglycones was respectively 1.7- and 2.4-fold higher compared to those of the control (Silva et al., 2019). UAE produced an equimolar conversion from conjugated isoflavones to their respective aglycones. Both extracts showed higher antiradical activity towards ABTS radical cation and peroxyl radical, compared to those of the control. The aglycone isoflavones/conjugated isoflavones (aISO/cISO) ratio correlated with the antiradical activity. Therefore, it has been suggested for soybean processors to use this new quality index as an indicator of the antioxidant potential or oxidative stability of the product, which will be fully explained in the next section.

Thermal treatments, which are commonly used to produce soybean products in the industry, can affect the isoflavone profile. It was first proven that isoflavones in the malonyl group are the main and predominant isoflavones in soybean seeds, with 66% malonyl genistin and malonyl daidzin. Malonyl isoflavones are thermally unstable and at 80° C their content decreases significantly and there is an increase in the content of isoflavones glucosides and acetyl glucosides, with the exception of the 6''-O-acetyl genistin isoflavone (Kudou et al., 1991). Glucoside isoflavones (daidzin, glycitin, and genistin), purified from defatted soybean flour and heated above 135° C for 3 min, produced maximum isoflavone contents of acetyl daidzin and acetyl genistin, daidzein, glycitein, and genistein (Xu et al., 2002). Defatted soybean flour was described by Handa et al. (2014) to contain 78% malonyl glucosides, 13.5% β-glucosides, 4.5% acetyl glucosides, and 3.8% aglycones. Glycitein and acetyl glycitin were not detected. The presence of acetyl glucosides in this soybean flour was possibly due to the heat treatment given to the soybean to extract the oil. Changes in the profile of isoflavones also occurred after heat treatments of raw soybeans using oven-drying, roasting, or explosive puffing (Lee & Lee, 2009). Drying the soaked for 12 h raw soybeans in an oven at 100° C for 120 min, decreased the content of malonyl isoflavones and increased β-glucosides and aglycones. Regression analysis showed that malonyl genistin had higher slopes of decreases than malonyl daidzin in oven-dried soybeans. Roasting of raw soybeans at 200° C (7, 14, and 20 min) and using explosive puffing (490, 588, and 686 kPa) decreased the content of malonyl glucosides and increased the content of acetyl glucosides and β-glucosides isoflavones. The dry heat processing led to the formation of acetyl glucosides from malonyl glucosides via decarboxylation (Figure 2.1). Roasting caused a higher loss in total isoflavones content than oven-drying and explosive puffing, which may be due to the higher temperature of treatment. In whole soybean flour, heat treatment also changes the profile and isoflavones content (Andrade et al., 2016). Mathematical models were established to estimate the process parameters in obtaining whole soybean flour with improved isoflavone conversions. In the thermal treatment (200° C for 20 min) in the oven, there was a 2.5-fold reduction in malonyl glucosides content and an increase of 1.2-fold and 3.5-fold in β-glucosides and aglycone contents, respectively. The conversion of malonyl glucosides occurred by descaboxylation reaction.

Okara is the residue obtained after processing soymilk and has high moisture content and high nutritive value. The temperature in okara's drying kinetics (Page model) using an oven dryer with hot air circulation (50°, 60°, and 70° C) influenced the reduction, conversion, and degradation of isoflavones (Muliterno et al., 2017). At 70° C there was a significant reduction in the total isoflavones content, possibly due to thermal degradation. In the initial drying stage at 50° C, the isoflavones daidzin and genistin were converted into aglycones (daidzein and genistein), probably due to the residual action of β-glucosidase. However, after 6 h of drying at 50° C, there was a decrease in the content of genistein and daidzein. The applied kinetic model was useful to select the best drying condition to obtain dry okara with adequate content of aglycones isoflavones. Besides that, the high moisture content of okara is a limiting problem for its use as an ingredient. The okara dried in a forced-air oven (FAD at 40°, 50°, 60°, and 70° C), microwaves (MWD), and lyophilization (FRD) showed different transformations in the profile and isoflavones content (Guimarães et al., 2020). The total isoflavones content of the dried okara varied between 1.84 and 2.07 μmol g^{-1}. The dry okara with FAD at 70° C showed the highest content of aglycones isoflavones (1.87 μmol g^{-1}) and represented 98% of the total isoflavones content, while the FRD dried okara had the lowest aglycones content (1.50 μmol g^{-1}) and represented 72% of the total isoflavones content.

Vegetable-type soybean or edamame in canned form with the addition of zinc in the formulation changed the content of different forms of isoflavones in relation to fresh grains (Czaikoski et al., 2018). These canned soybeans showed a 50% increase in the content of isoflavone glucosides and a 50% reduction in the content of isoflavone malonyl glucosides. In the process, the conversion by decarboxylation of malonyl isoflavones to β-glucosides probably occurred and the thermal treatment was not sufficient to form acetyl glucosides.

2.5.2 Isoflavone Profile as Affected by Fermentation

Soybean fermented foods contain high content of aglycones, while non-fermented foods have mostly the conjugated forms (Liu, 1997). The different types of fermentation applied to soybean and derived products may result in the bioconversion of conjugated isoflavones into their corresponding aglycones. Soybean fermentation also decreases antinutritional constituents and improves the nutritive and digestibility quality of the final product. Fermented soybean foods have been recommended for the prevention and treatment of chronic diseases (Murakami et al., 1984).

According to Chun et al. (2007), soymilk fermented with lactic acid bacteria species showed a change in the profile and concentration of isoflavones. The best microorganism used was *L. paraplantarum* KM, which in 6 h of fermentation converted the glucosides genistin (100%), daidzin (90%), and glycitin (61%) into their aglycones. In the conversion, there was an increase of seven and six times in the content of genistein and daidzein, respectively. For soymilk fermentation, Hati et al. (2014) used the *Lactobacillus rhamnosus* C6 strain and also observed the biotransformation of isoflavones β-glucosides (daidzin and genistin) into aglycones (daidzein and genistein). This strain produced greater activity of β-glucosidases and in addition, three strains (*Lactobacillus rhamnosus*

C6, *L. rhamnosus* C2, and *Lactobacillus casei* NCDC297) were recommended to be selected and used in the development of functional soybean foods enriched with aglycones isoflavones, such as soy yogurt, soy cheese, soy drinks, and soy dahi.

Soymilk fermented with commercial kefir culture produced β-glucosidases and resulted in 100% bioconversion of glycitin and daidzin and 89% of genistin into the corresponding aglycones. Soymilk fermented had 1.67 µmol g^{-1} of daidzein, 0.28 µmol g^{-1} of glycitein and 1.67 µmol g^{-1} of genistein (Baú et al., 2015). Kefir-fermented soymilk promoted a two-fold increase in the content of isoflavones aglycones and nine-fold in the content of total phenolics. These compounds, after simulation in the digestive system *in vitro*, increased significantly (Fernandes et al., 2017). The soymilk fermented with kefir was recommended as a good source of isoflavones in the aglycone form and total phenolics.

Fungus-fermented soybean flour also promotes the transformation of isoflavones glucosides into aglycones. Whole soybean flour fermented with *Aspergillus oryzae* CCT 4359 at 30° C for 48 h promoted significant biotransformation of isoflavones glucosides into aglycones (Silva et al., 2011). This fermented flour presented 75.5% of aglycones, showing the following isoflavones profile: Total glucosides (0.14 mg g^{-1}) with daidzin (0.02 mg g^{-1}), glycitin (0.02 mg g^{-1}), and genistin (0.09 mg g^{-1}); total aglycones (0.44 mg g^{-1}) with daidzein (0.13 mg g^{-1}), glycitein (0.03 mg g^{-1}), and genistein (0.27 mg g^{-1}); and total isoflavones (0.59 mg g^{-1}). Fermented flour may be employed as an ingredient to produce food formulations bearing improved functional properties.

Defatted soybean flour fermented in a solid state with *Aspergillus oryzae* IOC 399/1998 or *Monascus purpureus* NRRL1992 and using the central rotational composite design produced soybean flours fermented with a high isoflavones content (Handa et al., 2014). Both fermentations produced β-glucosidases, however, the *Monascus purpureus* NRRL1992 was more efficient in converting isoflavones β-glucosides into aglycones. Flours fermented with *Monascus purpureus* NRRL1992 presented 48.3% of aglycones and the following isoflavones profile and content: Daidzin (0.29 mg g^{-1}), genistin (211.17 mg g^{-1}), and glycitin (27.47 mg g^{-1}); malonyl daidzin (162.95 mg g^{-1}), malonyl genistin (374.00 mg g^{-1}), and malonyl glycitin (not detected); acetyl daidzin (not detected), acetyl genistin (54.07 mg g^{-1}), and acetyl glycitin (not detected); daidzein (88.65 mg g^{-1}), genistein (693.51 mg g^{-1}), and glycitein (not detected); and total isoflavones (1617.07 mg g^{-1}). Under the same solid-state fermentation, *Aspergillus oryzae* IOC 399/1998 produced 10.7 times more β-glucosidase than *Monascus purpureus* NRRL1992. However, defatted soybean flour fermentation with *Monascus purpureus* NRRL1992 produced a more efficient bioconversion of isoflavones glucosides to aglycones. According to Handa et al. (2019), solid-state fermentation parameters of defatted soybean flour by *Monascus purpureus* or *Aspergillus oryzae* influenced the biotransformation of isoflavones into aglycones, the total phenolic content, and the antioxidant capacity. The definition of fermentation parameters is, therefore, essential to increase the content of bioactive compounds in fermented products owing to their promise for applications in the food industry.

Sufu is a popular fermented tofu product in China (Li-Jun et al., 2004). Sufu's isoflavones profile was significantly changed upon fermentation. In red sufu, aglycones levels increased, while corresponding glucosides levels decreased and corresponded to 99.7% and 0.3% of the total isoflavones, respectively. In a pure starter fermentation culture with *Actinomucor elegans* As3.227 (solid fungus) at 28° C, 48 h corresponded to the fastest period to induce the isoflavone conversion. Changes in the composition of isoflavones were related to the activity of β-glucosidases during the fermentation of sufu, which was also influenced by the NaCl amount used. During the preparation of sufu using *Mucor flavus* (*M. flavus*) at low temperature, changes in isoflavones content and production of β-glucosidase occurred (Cheng et al., 2011). The content of aglycones has been found to increase, while the content of glucosides isoflavones correspondents decreased (Figure 2.5). The production and accumulation of β-glucosidases by *M. flavus* contributed to the transformation of glucosides isoflavones into aglycones, which were influenced by NaCl supplementation.

Douchi is also a popular fermented soybean food in China. During the processing of soybean to obtain the douchi (i.e. pre-treatment with *Aspergillus oryzae* 3.951 and addition of NaCl)

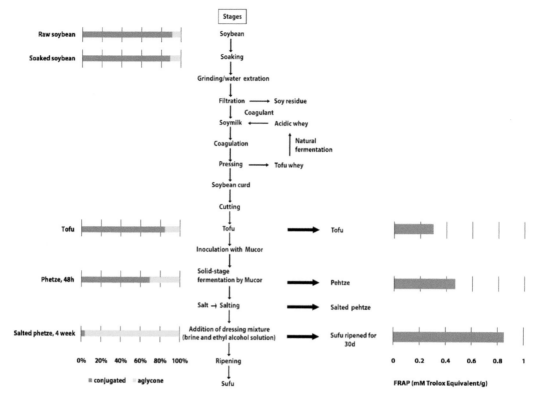

FIGURE 2.5   Representative isoflavone profile change and ferric reducing antioxidant power (FRAP) during the manufacture of sufu. Isoflavone profile adapted from Cheng et al. (2011) and FRAP adapted from Cai et al. (2016).

biotransformation of isoflavones as well as changes in the activity of endogenous β-glucosidases were observed (Wang et al., 2007). In the process, 61% of isoflavones were lost during pre- and post-fermentation when 10% NaCl was employed. There was an increase in the content of agly-cones and a decrease in the content of β-glucosides, malonyl glucosides, and acetyl glucosides. These transformations were related to β-glucosidases activity, which was affected by NaCl supplementation.

Tempeh, a fermented soybean food (usually with *Rhizopus spp.*) highly appreciated for its attrac-tive flavor, texture, and superior digestibility, is a traditional fermented soybean food from Indonesia (Nout & Kiers, 2005). Aglycones isoflavones (daidzein and genistein) are the main isoflavones in tempeh and are responsible for antioxidative capacity (Murakami et al., 1984). The isoflavones composition of the tempehs was changed during the fermentation (0–48 h) of yellow (Tamahomare) and black (Sakushuu-kuro) soybeans and defatted yellow soybean granules (Nakajima et al., 2005). After 24 h of fermentation of yellow and black soybean tempehs, the content of the three types of isoflavones aglycones increased and varied from 0.16 to 0.17 mg g^{-1}, while the content of the nine types of isoflavones glucosides varied from 0.86 to 0.87 mg g^{-1} and the average total isoflavones content was 1.03 mg g^{-1}. The granular germ of defatted-yellow soybean tempeh showed an exces-sively high content of isoflavones (aglycones and glucosides) of approximately 8.80 mg g^{-1}, 44% of which was aglycones isoflavones. According to Haron et al. (2009) the total isoflavones (daidzein and genistein) content of Malaysian commercial tempeh was 0.26 mg daidzein and 0.28 mg genis-tein while fried tempeh contained 0.35 mg daidzein and 0.31 mg genistein per g (fresh weight basis).

Changes in the isoflavone profile of tempeh were also observed by Ferreira et al. (2011) at different stages of the processing and refrigeration of two soybean cultivars: BR 36 with low (0.14; 0.74; and 0.00 mg g^{-1} dry weight) and IAS 5 with high (0.46; 2.30; and 0.07 mg g^{-1} dry weight) content of glucosides, malonyl glucosides, and aglycones isoflavones. After 24 h of fermentation with *Rhizopus microsporus* var. *oligosporus*, the content of glucosides isoflavones decreased and aglycones increased significantly due to the cooking of the grains as well as the action of β-glucosidases during fermentation. After storage for up to 24 h under refrigeration, the isoflavone profile remained constant. The tempehs made with both soybeans had an average content of glucosides, malonyl glucosides, and aglycones isoflavones of 0.73, 0.11, and 0.12 mg g^{-1} and 0.17, 0.38, and 0.36 mg g^{-1}, respectively. Tempehs with higher aglycones content have been recommended as a functional food due to the benefits to human health.

Tempehs prepared with different soybean cultivars (BRS 216, BRS 232, BRS 257, and BRS 267) had a high content of aglycone isoflavones (Bavia et al., 2012). The tempehs were obtained from the soaking of the soybeans, cooking of the cotyledons for 30 min, and fermentation with the fungus *R.oligosporus* for 26 h at 31° C. The profile and isoflavones content of soybeans and tempehs showed significant differences between cultivars. The content of aglycones isoflavones increased by about 50% in tempehs when compared to soybeans grains. The different soaking conditions (6, 12, and 18 h) of the grains, cooking time of the cotyledons (15, 30, and 45 min), and fermentation (18, 24, and 30 h) with *Rhizopus oligosporus* at 37° C during tempeh production changed the profile and distribution of the different isoflavone forms (Borges et al., 2016). According to the application of the central composite design, it was possible to verify that the highest bioconversion of β-glucosides isoflavones into aglycones occurred when the grains were soaked for 6 h, cotyledons were cooked for 15 min, and subjected to 18 h fermentation.

Soybean tempeh fermented with *Rhizopus oligosporus* NRRL 5905 showed high contents of aglycones (daidzein and genistein) due to bioconversion of isoflavones glucosides (Kuligowski et al., 2017). After five days of fermentation, the total isoflavone content (daidzin + daidzein + genistin + genistein) decreased 2–3.5 times with respect to unfermented soybean. Depending on the strain used, the genistein content was 8–10 times higher than unfermented soybeans.

Natto is a Japanese soybean product produced by fermentation with *Bacillus subtilis natto*. Natto (*B. natto*) fermented black soybean varieties detam 2 using *B. natto* strain IFO 3335 increased the content of aglycones isoflavones which is beneficial for health. After fermentation, both genistein (0.10 mg g^{-1} in defatted natto) and daidzein (0.09 mg g^{-1} in defatted natto) concentration increased up to eight times as much as those in the raw soybean that contained 0.013 and 0.011 mg g^{-1} dry weight (Hasim et al., 2015).

According to Yanaka et al. (2012), soybean isoflavones are composed of 15 species, including the three succinyl glucosides. The total isoflavones content of 14 soybean foods and eight supplements, including three succinyl glucosides, ranged from 45 to 735 µg g^{-1} and from 1.304 to 90.224 µg g^{-1}, respectively. High levels of succinyl glucosides were detected in natto, which varied from 30 to 80 µg g^{-1} and corresponded to 4.1–10.9% of the total isoflavones content. However, in soybean flour, the content of succinyl glucosides was 59 µg g^{-1} and corresponded to 4.6% of the total isoflavone content. In natto (or cheonggukjang), Park et al. (2010) found two new isoflavone metabolites (succinyl-β-daidzin and succinyl-β-daidzin). Evaluating pure standards (daidzin, genistin, daidzein, and genistein) and mixtures of isoflavones extracted from roasted soybeans containing *Bacillus* with daidzin and genistin produced succinyl-β-daidzin and succinyl-β-genistin, respectively, while the pure aglycone standards (daidzein and genistein) did not produce succinyl derivatives. In order to expand the use of soybean molasses, vinegar was obtained through alcoholic and acetic fermentations and showed changes in the isoflavones content (Miranda et al., 2020). After the fermentations, the contents of total phenolics and isoflavones decreased. The isoflavones profile of vinegar was constituted only by the three forms of isoflavones: Genistin, glycitin, and daidzein.

2.6 ANTIOXIDANT PROPERTIES AND THE NEW QUALITY INDEX IN FOOD APPLICATION AND HEALTH

The ability of phenolic compounds in scavenging free radicals stems from single electron transfer (SET) or hydrogen atom transfer (HAT). The antioxidant potential for DPPH radical and ABTS radical cation, as well as against peroxyl radicals (ORAC method) is usually measured as a first screening on the possible effects in food and/or biological systems. The effect against other reactive oxygen species (e.g. hydroxyl and superoxide radical) may also be measured (Xiao et al., 2015), but these methods are less commonly used than the former ones. Higher ORAC and FRAP values have been found to translate to a higher reduction in the activation of nuclear factor kappa B (NF-κB) in RAW 264.7 macrophages, a process mediated by oxidative stress (de Camargo et al., 2019a). Falcão et al. (2019) also demonstrated that DPPH and ABTS results may partially anticipate the reduction in the activation of NF-κB using the same cell model. Therefore, both studies indicated that colorimetric methods could still be used considering their simplicity and widespread application.

The antioxidant activity of phenolic compounds may also be explained by their reducing power and/or metal-chelating activity towards ferric and ferrous ions, respectively. Therefore, the ferric reducing antioxidant power towards Fe^{3+} and the ability of natural antioxidants to chelate metals such as Fe^{2+} also needs to be contemplated. In fact, according to Braughler et al. (1986), the ratio of Fe^{3+}/Fe^{2+} is important for quick initiation of lipid oxidation via through the Fenton reaction. They also suggested that their optimum ratios are between 1:1 to 7:1. Accordingly, decreasing the amounts of Fe^{3+} in the system may be useful. Nevertheless, during the Fenton Reaction, Fe^{2+} can be again oxidized to the ferric form, and the latter is one more time reduced to Fe^{2+}. Therefore, clearly, the reaction takes place over and over following their cyclic nature. It has been postulated (de Camargo et al., 2018) that an "ideal antioxidant" should not only be a good reducing agent but must also exhibit chelating capacity. The methods used as indicators of antioxidant capacity have been extensively revised by Shahidi and Zhong (2015).

Since 1994 acrylamide has been classified as probably carcinogenic to humans (Group 2A) according to the International Agency for Research on Cancer (IARC, 2019). A study by Cheng et al. (2015) supported the effectiveness of isoflavones (e.g. daidzin, genistin, daidzein, and genistein) and other flavonoids in decreasing the formation of acrylamide. The rate of inhibition of acrylamide formation affected by flavonoids correlated well with the change of antioxidant capacity (TEAC) measured by DPPH, ABTS, or FRAP assay. This highlights an important food application of soybean isoflavones as well as the relevance of the mentioned assays in anticipating potential benefits in a food model system.

According to the World Health Organization, cardiovascular diseases account for most global deaths from non-communicable disease (WHO, 2018). The positive effects of soybean consumption and the risk of coronary heart disease in women, considering equol, a bioactive metabolite of soybean daidzein (Yuan et al., 2007), were reported by Zhang et al. (2012). Low-density lipoprotein-cholesterol (LDL-c) levels have been used as predictors of death from cardiovascular diseases (CVD) (Pekkanen et al., 1990). A meta-analysis of controlled human trials by Weggemans and Trautwein (2003) concluded that consumption of soybean-associated isoflavones is not related to changes in LDL-c. The development of atheromatous plaques stems from the uptake of oxidized LDL-c, via scavenger receptors, thus leading to cholesterol accumulation and foam cell formation. Therefore, oxidized LDL-c is the actual entity involved in the early event of atherosclerosis (de Camargo et al., 2019b).

The antiradical activity of genistin and daidzin towards ABTS radical cation was reported by Ruiz-Larrea et al. (1997). These authors demonstrated that not only genistein but also its conjugated form (genistin) were able to delay the oxidation of LDL-c, which may be explained by its reductive properties against Cu^{2+} (used to promote oxidation) as well as by scavenging of alkoxyl and peroxyl radicals generated in the medium. Furthermore, the efficacies of the tested molecules were of genistein > genistin > biochanin A, which demonstrated that both aglycone and conjugated forms of genistein exhibit antioxidant properties which are higher than those of the methylated form (biochanin A), a characteristic also supported by using the ABTS assay.

Monascus-fermented soybeans contain natural antioxidants able to scavenge hydroxyl radicals (Lee et al., 2008b). Peroxyl and hydroxyl radicals induce DNA strand breakage, and DNA-damage signaling/repair are crucial pathways to the etiology of human cancers (Khanna & Jackson, 2001). *Lactobacillus plantarum* B1–6 fermented soybean whey has been shown to better decrease in DNA damage than unfermented soybean whey (Xiao et al., 2015). These results were anticipated by the higher antiradical activity of fermented samples towards hydroxyl radicals, which were also used to induce DNA strand breakage. According to a meta-analysis of epidemiologic studies by Yang et al. (2011), soybean intake is associated with lower lung cancer risk. Soybean consumption was also associated with reduced risk of breast cancer incidence, recurrence, and mortality (Fritz et al., 2013).

The soluble fraction of soybean meal subjected to fermentation by *Bacillus amyloliquefaciens* SWJS22 exhibited higher scavenging activity towards DPPH and peroxyl radicals as well as reducing power. According to these *in vitro* results, the authors tested the sample obtained after 54 h of fermentation in a D-galactose-induced aging mice model. In summary, biomarkers of antioxidant capacity in the serum and in the liver (e.g. superoxide dismutase, catalase, glutathione peroxidase, total antioxidative capacity, and malondialdehyde) were improved in the animals treated with lyophilized fermented soybean meal supernatant at a dose of 500 mg/kg body weight (Yang et al., 2019). Almost all conjugated isoflavones were converted into aglycones during the manufacture of white sufu (Cai et al., 2016), which is supported by previous studies (Cheng et al., 2011; Li-Jun et al., 2004). Ferric reducing antioxidant power (FRAP) and the scavenging towards ABTS radical cation increased by 2.80 and 1.60 times, respectively. Furthermore, the ABTS radical cation scavenging activity was significantly correlated to aglycone contents.

A correlational analysis is commonly used to investigate the concentration-dependent antioxidant action. Results by our research team (Silva et al., 2019) and from the literature (Cai et al., 2016) demonstrated that, by simply employing the Pearson correlation, one would find a negative correlation between some or even all conjugated isoflavones malonyl glucosides (malonyl daidzin, malonyl glycitin, and malonyl genistin) and β-glucosides (daidzin, glycitin, and genistin) and the antiradical activity towards ABTS radical cation and/or peroxyl radical. However, a negative correlation implies that these molecules present a pro-oxidant role, which is not true since the antiradical activity of conjugated isoflavones is supported by the literature (Ruiz-Larrea et al., 1997). Lending support to these latter authors, the ability of individual soybean isoflavones (including conjugates) in counteracting LDL-c oxidation *in vitro* was tested by Lee et al. (2005). The authors concluded that, although to a different extent, all molecules inhibited LDL-c oxidation to a certain extent. Therefore, when molecular conversion is taking place, it is likely inappropriate to discuss the results based on the "negative correlation" between malonyl glucosides and/or β-glucosides and the antioxidant capacity.

Considering the aforementioned and aiming to simplify the situation for both academia and industry, we have proposed an index (Silva et al., 2019) to anticipate the antioxidant potential of soybean products (Table 2.1). The ratio was calculated considering total aglycones and total conjugates (aISO/cISO) as well as considering individual isoflavones, namely daidzein, glycitein, and genistein which were calculated similarly (e.g. aDAI/cDAI, aGLY/cGLY, and aGEN/cGEN). According to this index, a high positive correlation existed between the aISO/cISO ratio and both radicals, namely ABTS radical cation (r = 0.8846, $p < 0.05$) and peroxyl radical (r = 0.8471, $p < 0.05$). In addition, all individual ratios showed a significant positive correlation with the mentioned assays. Therefore, in the present chapter, we have collected literature data and employed the proposed index to better understand if it could be used in different food processing.

Gamma irradiation can be used to decrease and/or completely eliminate pathogenic microorganisms as well as in order to achieve insect disinfestation. However, the treatment may be detrimental to antioxidants such as tocopherol, anthocyanins, and vitamin C. In contrast, gamma-radiation has been found to induce enhancement of the antioxidant capacity of soybean (Variyar et al., 2004). The authors reported a decrease in the content of glycosidic conjugates (e.g. diadzin, glycitin, and genistin) and an

TABLE 2.1

Study Case on the Application of ISOF Index Values During the Manufacturing of Unfermented and Fermented Soybean Products*

Control			Soybean Product			Antiradical activity	Control	Product	Ref.
aISO	cISO	aISO/cISO	aISO	cISO	aISO/cISO				
Soymilk			Soymilk obtained upon hemicellulase-assisted extraction						
0.26	2.19	0.12	0.44	0.63	0.26	ORAC (mM TE/g)	78.9	107.6	A
Soymilk			Soymilk obtained upon ultrasound-assisted extraction (kHz and 50 W/cm²)						
0.26	2.19	0.12	0.63	1.83	0.34	ORAC (mM TE/g)	78.9	102.8	A
BRS 257 soybeans			BRS 257 soybeans subjected to ultrasound (20 kHz and 24 W/cm²)						
0.22	3.24	0.07	0.32	2.64	0.12	ABTS (µM TE/g)	120	275.3	B
JS 335 soybeans			Gamma-irradiated soybeans (5.0 kGy)						
6.34	145.7	0.04	8.05	77.58	0.10	DPPH (inhibition %)	24.5	33.9	C
Soybeans (time zero)			Germinated soybeans (day 4)						
1.61	1.81	0.89	3.20	0.52	6.15	DPPH (µM TE/g)	0.01	0.13	D
Soybeans			Tempeh						
27.82	30.77	0.90	62.9	7.26	8.66	DPPH (IC₅₀ mg/mL)	10**	2.67*	E
Black soybeans fermented by *Bacillus natto* (time zero)			Black soybeans fermented by *Bacillus natto* (48 h)						
292	522	0.56	424	316	1.34	DPPH (inhibition %)	4.52	22.8	F

Sources: A, Silva et al. (2019); B, Falcao et al. (2019); C, Variyar et al. (2004); D, Huang et al. (2014); E, Ahmad et al. (2015); F, Hu et al. (2010).

Notes

*Except for Silva et al. (2019), the ISOF index values were all calculated for the present book chapter. Abbreviations: aISO = aglycone isoflavones; cISO = conjugated isoflavones; TE = Trolox equivalent.

**Concentration required to inhibit 50% of DPPH radical. The higher the IC_{50}, the lower the antiradical activity.

increase in their respective aglycones as a function of the dose employed (up to 5 kGy). We have calculated the aISO/cISO ratio and found that our proposed index correlates well (0.9614, $p < 0.01$) with the DPPH radical scavenging activity. Furthermore, the conversion rate, as evaluated by the ratio aglycone molecule/conjugated form was in the order of genistein > daidzein > glycitein. This trend agrees with the antioxidant capacity of the mentioned aglycones towards DPPH (Chen et al., 2005). More importantly, the study by Variyar et al. (2004) is a typical example that, especially in cases of food processing, the total isoflavone content is not the best biomarker since, despite the higher antiradical activity, the samples showed lower total isoflavone content than that of their non-irradiated counterpart.

Ultrasound is also employed in food preservation. Enzyme and microorganism inactivation are among the possible applications. Many studies report the use of ultrasound-assisted extraction to increase the phenolic yield (Lai et al., 2013). However, this is not the only process taking place when this procedure is used in soybeans and their industrial products. In fact, the conversion of conjugated isoflavones to their respective aglycones has also been contemplated (Falcão et al., 2019; Silva et al., 2019). Calculating the aISO/cISO ratio using data published by Falcao et al. (2019), it is possible to note that a higher aISO/cISO ratio in samples subjected to ultrasound at 24 W/cm^2 not only showed higher antiradical activity against DPPH radical and ABTS radical cation, but also presented a higher ability in inhibiting the activation of NF-κB in RAW 264.7 macrophages. NF-κB activation, which is involved in inflammatory responses, is known to be mediated by oxidative stress. Hemicellulose- and ultrasound-extraction of soymilk (Silva et al., 2019) changes the isoflavone profile (Figure 2.6A) as well as the aISO/cISO ratio (Figure 2.6B).

Germination has been found to increase the nutritional value of the feedstock. Furthermore, it can reduce the content of undesirable substances such as flatulence-inducing components (Kim et al., 2013; Martín-Cabrejas et al., 2008; Paucar-Menacho et al., 2010). The contents of tocopherols, which also present antioxidant capacity, have also been found to be positively affected by germination (Lee et al., 2007; Paucar-Menacho et al., 2010; Shi et al., 2010; Yuan et al., 2009). The structural pattern of isoflavones changes upon germination. Yoshiara et al. (2018b) highlighted that, despite the lower total isoflavone content, germinated soybean components including the cotyledons, the best source of aglycones, showed higher antiradical activity than their non-germinated counterparts.

The kinetic changes were investigated by Huang et al. (2014). Germination decreased the concentration of conjugated isoflavones with a parallel increase of aglycones (Figure 2.7A). This behavior translated to a higher antiradical activity as evaluated by the DPPH method (up to 13 times higher). The aISO/cISO ratio increased (Figure 2.7B) and correlated well (r = 0.9139, $p < 0.05$) with the antiradical activity. Changes in DPPH radical scavenging upon cheonggukjang soybean fermentation using *Bacillus subtilis* CS90 were also reported by Cho et al. (2011). The level of DPPH radical scavenging activity increased from 53.6% to 93.9%. Although isoflavones remain the main flavonoids, it is important to highlight that catechin gallate and epicatechin gallate showed decreased concentrations during each fermentation time (0 to 60 h). In contrast, a parallel increase in the amounts of catechin, epicatechin, and gallic acid was observed. In addition to exhibiting a higher DPPH radical scavenging, fermented soybeans also have shown greater reducing power, as evaluated by FRAP assay (Han et al., 2015).

Tempeh, which is produced by fermentation, has been found to possess greater antiradical activity towards DPPH and metal chelation than that of its feedstock, namely soybeans. The aDAI/cDAI ratio increased from 2.7 (soybean) to 56.4 (tempeh) while the respective aGEN/cGEN ratio raised from 0.45 to 3.66. Furthermore, the aISO/cISO ratio increased from 0.9 to 8.7 (Ahmad et al., 2015). A similar trend was observed by Hu et al. (2010). Furthermore, according to data processed for the present manuscript, the aGEN/cGEN (r = 0.8232, $p < 0.01$) and aDAI/cDAI (r = 0.8544, $p < 0.01$) ratio positively and significantly correlated with DPPH. Likewise, the aISO/cISO ratio correlated well with the mentioned assay (r = 0.8455, $p < 0.05$). According to Chen et al. (2005), the ability of aglycone isoflavones to scavenge DPPH was in the order of genistein > daidzein > glycitein. Interestingly, their respective hydroxylated forms (8-OH-genistein, 8-OH-daidzein, and 8-OH-glycitein) showed 21.4, 36.4, and 24.3 times higher antiradical activity.

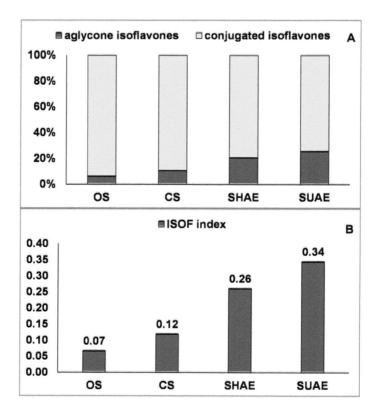

FIGURE 2.6 Concentration of different forms of isoflavones in soymilk as affected by hemicellulase-assisted extraction and ultrasound-assisted extraction (A). The ISOF index is calculated by the aglycone isoflavones/conjugated isoflavones (aISO/cISO) ratio, considering total aglycones and total conjugates (B). OS, CS, SHAE, and SUAE stand for original soybeans, control soymilk, soymilk obtained upon hemicellulase-assisted extraction, and soymilk ultrasound-assisted extraction, respectively. Adapted from Silva et al. (2019).

In addition to the reducing power, phenolics from unfermented and fermented black soybean have shown to chelating ability against Fe^{2+} (Juan & Chou, 2010) which, as mentioned before, could be classified as "ideal antioxidant" due to their potential to delay and/or stop the Fenton reaction that also generates reactive oxygen species. As expected, the metal chelation capacity of fermented black soybeans was higher than that of their unfermented counterparts. The chelating capacity of *Monascus* fermented soybeans on Fe^{2+} is supported by an earlier study by Lee et al. (2008).

In summary, by collecting data that is not limited to our research group, this contribution demonstrates that our proposed index could be successfully employed to anticipate better antioxidant characteristics during different physical, chemical, and/or biotechnological treatments. Therefore, future research in academia and industry may employ the ISOF ratio during the manufacture of soybean products as a new quality index. It is important to highlight that, for now, only when paired raw and processed samples are produced and compared it would be possible to obtain reliable ISOF index values. In such a case, a higher index may translate into better outcomes in terms of food application and potential health benefits.

2.7 SUMMARY

Soybean isoflavones, a subclass of flavonoids, can prevent and/or decrease the generation of heat-generated food toxicants that may jeopardize human health. Additionally, by decreasing the risk

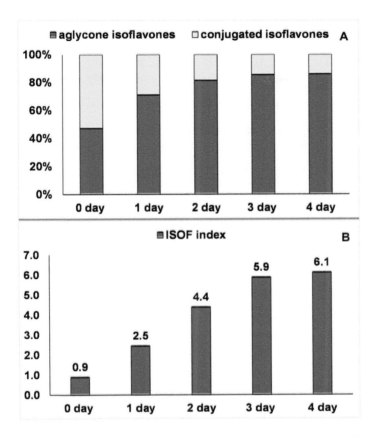

FIGURE 2.7   Kinetic changes of individual isoflavone content (mg/g) in soybean during germination (A). The ISOF index is calculated by the aglycone isoflavones/conjugated isoflavones (aISO/cISO) ratio, considering total aglycones and total conjugates (B). OS, CS, SHAE, and SUAE stand for original soybeans, control soymilk, soymilk obtained upon hemicellulase-assisted extraction, and soymilk ultrasound-assisted extraction, respectively. Adapted from Huang et al. (2014).

factors for the development of non-communicable diseases, soybean isoflavones have been associated with a myriad of benefits to human health. These positive effects on food and health have been attributed, at least in part, to the antioxidant properties of soybean isoflavones, which are higher in those present in the aglycone form. The isoflavone profile of soybean seeds and their components (e.g. seed coat, cotyledon, and axis) as well as their biotransformation upon germination and fermentation are addressed in this chapter. The interconversion of isoflavones upon physical treatments is also contemplated. The application of the new quality index, namely the ISOF index, in terms of its ability to anticipate a higher antioxidant capacity and potentially improve outcomes in food application and health, was introduced and explored. Literature data supports the use of the new ISOF index during different physical, chemical, and/or biotechnological treatments.

REFERENCES

Ahmad, A., Ramasamy, K., Majeed, A.A., Mani, V. (2015). Enhancement of beta-secretase inhibition and antioxidant activities of tempeh, a fermented soybean cake through enrichment of bioactive aglycones. *Pharmaceutical Biology*, 3: 758–766.

Andrade, J.C., Mandarino, J.M., Kurozawa, L.E., Ida, E.I. (2016). The effect of thermal treatment of whole soybean flour on the conversion of isoflavones and inactivation of trypsin inhibitors. *Food Chemistry*, 194: 1095–1101.

Azam, M., Zhang, S., Abdelghany, A.M., Shaibu, A.S., Feng, Y., Li, Y., Tian, Y., Hong, H., Li, B., Sun, J. (2020). Seed isoflavone profiling of 1168 soybean accessions from major growing ecoregions in China. *Food Research International*, 130: 108957.

Baú, T.R., Garcia, S., Ida, E.I. (2015). Changes in soymilk during fermentation with kefir culture: Oligosaccharides hydrolysis and isoflavone aglycone production. *International Journal of Food Science and Nutrition*, 66: 845–850.

Baú, T.R., Ida, E.I. (2015). Soymilk processing with higher isoflavone aglycone content. *Food Chemistry*, 183: 161–168.

Bavia, A.C.F., Silva, C.E. da, Ferreira, M.P., Santos Leite, R., Mandarino, J.M.G., CarrãoPanizzi, M.C. (2012). Chemical composition of tempeh from soybean cultivars specially developed for human consumption. *Ciência e Tecnologia de Alimentos*, 32: 613–620.

Berger, M., Rasolohery, C.A., Cazalis, R., Dayde, J. (2008). Isoflavone accumulation kinetics in soybean seed cotyledons and hypocotyls: Distinct pathways and genetic controls. *Crop Science*, 48: 700–708.

Birt, D.F., Hendrich, S., Wang, W. (2001). Dietary agents in cancer prevention: Flavonoids and isoflavonoids. *Pharmacology & Therapeutics*, 90: 157–177.

Borges, C.W.C., Carrão-Panizzi, M.C., Mandarino, J.M.G., Silva, J.B., Benedetti, S., Ida, E.I. (2016). Contents and bioconversion of β-glycoside isoflavones to aglycones in the processing conditions of soybean tempeh. *Pesquisa Agropecuária Brasileira*, 51: 271–279.

Braughler, J.M., Duncan, L.A., Chase, R.L. (1986). The involvement of iron in lipid peroxidation. Importance of ferric to ferrous ratios in initiation. *Journal of Biological Chemistry*, 261: 10282–10289.

Britz, S.J., Schomburg, C.J., Kenworthy, W.J. (2011). Isoflavones in seeds of field-grown soybean:variation among genetic lines and environmental effects. *Journal of the American Oil Chemists Society*, 88: 827–832.

Bustamante-Rangel, M., Delgado-Zamarreno, M.M., Perez-Martın, L., Rodrıguez-Gonzalo, E., Domınguez-Alvarez, J. (2018). Analysis of isoflavones in foods. *Comprehensive Reviews in Food Science and Food Safety*, 17: 391–411.

Cai, R-C., Li, L., Yang, M., Cheung, H.Y., Fu, L. (2016). Changes in bioactive compounds and their relationship to antioxidant activity in white sufu during manufacturing. *International Journal of Food Science & Technology*, 52(7): 1721–1730.

Carrão-Panizzi, M.C., Beleia, A.D.P., Kitamura, K., Oliveira, M.C.N. (1999). Effects of genetics and environment of isoflavone content of soybean from different region of Brazil. *Pesquisa Agropecuária Brasileira*, 34(10): 1787–1795.

Chen, K-I., Erth, M-H., Su, N-W., Liu, E-H, Chou, C-C., Chenh, K-C. (2012). Soyfoods and soybean products: From traditional use to modern applications. *Applied Microbiology and Biotechnology*, 96: 9–22.

Chen, Y.C., Sugiyama, Y., Abe, N., Kuruto-Niwa, R., Nozawa, R., Hirota, A. (2005). DPPH radical-scavenging compounds from dou-chi, a soybean fermented food. *Bioscience Biotechnology and Biochemistry*, 69: 999–1006.

Cheng, J., Chen, X., Zhao, S., Zhang, Y. (2015). Antioxidant-capacity-based models for the prediction of acrylamide reduction by flavonoids. *Food Chemistry*, 168: 90–99.

Cheng, Y-Q., Zhu Y-P., Hu, Q., Li, L.-T., Saito, M., Zhang, S.X., Yin L-J. (2011). Transformation of isoflavones during Sufu. (A traditional Chinese fermented soybean curd) production by fermentation with *Mucor flavus* at low temperature. *International Journal of Food Properties*, 14(3): 629–639.

Cho, K.M., Lee, J.H., Yun, H.D., Ahn, B.Y., Kim, H., Seo, W.T. (2011). Changes of phytochemical constituents (isoflavones, flavanols, and phenolic acids) during cheonggukjang soybeans fermentation using potential probiotics *Bacillus subtilis* CS90. *Journal of Food Composition and Analysis*, 24: 402–410.

Choi, M.S., Jung, U.J., Yeo, J., Kim, M.J., Lee, M.K. (2008). Genistein and daidzein prevent diabetes onset by elevating insulin level and altering hepatic gluconeogenic and lipogenic enzyme activities in non-obese diabetic (NOD) mice. *Diabetes Metabolism Research and Reviews*, 24: 74–81.

Chun, J., Kim, G., Lee, K., Choi, I.-D., Kwon, G-H., Park, J.-Y., Jeong, S-J., Kim, J.-S., Kim, J.H. (2007). Conversion of isoflavone glucosides to aglycones in soymilk by fermentation with lactic acid bacteria. *Journal of Food Science*, 72(2): M39–M44.

Cvejić, J., Tepavčević, V., Bursać, M., Miladinović, J., Maleňić, D. (2011). Isoflavone composition in F1 soybean progenies. *Food Research International*, 44: 2698–2702.

Czaikoski, K., Leite, R.S., Mandarino, J.M.G., Carrão-Panizzi, M.C., Da Silva, J.B., Ida, E.I. (2018). Physicochemical characteristics of canned vegetable-type soybean processed with zinc at different pasteurization times. *Pesquisa Agropecuária Brasileira*, 53: 840–848.

de Camargo, A.C., Biasoto, A.C.T., Schwember, A.R., Granato, D., Rasera, G.B., Franchin, M., Rosalen, P.L., Alencar, S.M., Shahidi, F. (2019a). Should we ban total phenolics and antioxidant screening methods?

The link between antioxidant potential and activation of NF-κB using phenolic compounds from grape by-products. *Food Chemistry*, 290: 229–238.

de Camargo, C.A., Favero, T.B., Morzelle, C.M., Franchin, M., Alvarez-Parrilla, E., de la Rosa, A.L., Geraldi, V.M., Maróstica Júnior, R.M., Shahidi, F., Schwember, R.A. (2019b). Is chickpea a potential substitute for soybean? Phenolic bioactives and potential Health benefits. *International Journal of Molecular Sciences*, 20: 2644.

de Camargo, A.C., Regitano-d'Arce, M.A.B., Shahidi, F. (2017). Phenolic profile of peanut by-products: Antioxidant potential and inhibition of alpha-glucosidase and lipase activities. *Journal of the American Oil Chemists' Society*, 94: 959–971.

de Camargo, A.C., Schwember, A.R., Parada, R., Garcia, S., Maróstica, M.R. Franchin, M., Regitano-d'Arce, M.A.B., Shahidi, F. (2018). Opinion on the hurdles and potential health benefits in value-added use of plant food processing by-products as sources of phenolic compounds. *International Journal of Molecular Sciences*, 19: 3498.

De Lima, F.S., Handa, C.L., Fernandes, M.S., Rodrigues, D., Kurozawa, L.E., Ida, E.I. (2019). Kinetic modeling of the conversion and losses of isoflavones during soybean soaking. *Journal of Food Engineering*, 26: 171–177.

De Lima, F.S., Ida, E.I. (2014). Optimisation of soybean hydrothermal treatment for the conversion of b-glucoside isoflavones to aglycones. *LWT - Food Science and Technology*, 56(2): 232–239.

Devi, M.K.A., Gondi, M., Sakthivelu, G., Giridhar, P., Rajasekaran, T., Ravishankar, G.A. (2009). Functional attributes of soybean seeds and products, with reference to isoflavone content and antioxidant activity. *Food Chemistry*, 114: 771–776.

Dong, X., Xu, W., Sikes, R.A., Wu, C. (2013). Combination of low dose of genistein and daidzein has synergistic preventive effects on isogenic human prostate cancer cells when compared with individual soy isoflavone. *Food Chemistry*, 141: 1923–1933.

Eldridge, A.C., Kwolek, W.F. (1983). Soybean isoflavones: Effect of environment and variety on composition. *Journal of Agricultural and Food Chemistry*, 31: 394–396.

Falcão, H.G., Handa, C.L., Silva, M.B.R., De Camargo, A.C., Shahidi, F. Kurozawa, L.E., Ida, E.I. (2018). Soybean ultrasound pre-treatment prior to soaking affects β-glucosidase activity, isoflavone profile and soaking time. *Food Chemistry*, 269: 404–412.

Falcão, H.G., Silva, M.B.R., de Camargo, A.C., Shahidi, F., Franchin, M., Rosalen, P.L., Alencar, S.M., Kurozawa, L.E., Ida, E.I. (2019). Optimizing the potential bioactivity of isoflavones from soybeans via ultrasound pretreatment: Antioxidant potential and NF-κB activation. *Journal of Food Biochemistry*, 43: e13018: 1–11.

Fernandes, M.d.S., Lima, F.S., Rodrigues, D., Handa, C., Guelfi, M., Garcia, S., Ida, E.I. (2017). Evaluation of the isoflavone and total phenolic contents of kefir-fermented soymilk storage and after the in vitro digestive system simulation. *Food Chemistry*, 229: 373–380.

Ferreira, M.P., da Silva, M.P., de Oliveira, M.C.B., Mandarino, J.M., da Silva, J.B., Ida, E.I., Carrão-Panizzi, M.C. (2011). Changes in the isoflavone profile and in the chemical composition of tempeh during processing and refrigeration. *Pesquisa Agropecuária Brasileira*, 46(11): 1555–1561.

Fritz, H., Seely, D., Flower, G., Skidmore, B., Fernandes, R., Vadeboncoeur, S., Kennedy, D., Cooley, K., Wong, R., Sagar, S., Sabri, E., Fergusson, D. (2013). Soy, red clover, and isoflavones and breast cancer: A systematic review. *PLoS One*, 8: e81968.

Genovese, I.M., Lajolo, M.F. (2002). Isoflavones in soy-based foods consumed in Brazil: Levels, distribution, and estimated intake. *Journal of Agricultural and Food Chemistry*, 50: 5987–5993.

Goes-Favoni, S.P., Carrao-Panizzi, M.C., Beleia, A. (2010). Changes of isoflavone in soybean cotyledons soaked in different volumes of water. *Food Chemistry*, 19: 1605–1612.

Golbitz, P., Jordan, J. (2006). Soyfoods: Market and products. In: *Soy applications in food*. Edited by Riaz, M.N. CRC Press Taylor & Francis Group, Boca Raton, FL.

Graham, T.L. (1991). Flavonoid and isoflavonoid distribution in developing soybean seedling tissues and in seed and root exudates. *Plant Physiology*, 95: 594–603.

Grun, I.U., Adhikari, K., Li, C., Li, Y., Lin, B., Zhang, J., Fernando, L.N. (2001). Changes in the profile of genistein, daidzein, and their conjugates during thermal processing of tofu. *Journal of Agricultural and Food Chemistry*, 49: 2839–2843.

Guimarães, R.M., Ida, E.I., Falcão, H.G., De Resende, T.A.M., Silva, J.S. Alves, C.C.F, da Silva, M.A.P., Egea, M.B. (2020). Evaluating technological quality of okara flours obtained by different drying processes. *LWT-Food Science and Technology*, 123: 109062.

Han, S.S., Hur, S.J., Lee, S.K. (2015). A comparison of antioxidative and anti-inflammatory activities of sword beans and soybeans fermented with Bacillus subtilis. *Food & Function*, 6: 2736–2748.

Handa, C.L., Couto, U.R., Vicensoti, A.H., Georgetti, S.R., Ida, E.I. (2014). Optimisation of soy flour fermentation parameters to produce βglucosidase for bioconversion into aglycones. *Food Chemistry*, 152: 56–65.

Handa, C.L., De Lima, F.S., Guelf, M.F.G., Fernandes, M. da S., Georgetti, S.R., Ida, E.I. (2019). Parameters of the fermentation of soybean flour by *Monascus purpureus* or *Aspergillus oryzae* on the production of bioactive compounds and antioxidant activity. *Food Chemistry*, 271: 274–283.

Haron, H., Ismail, A., Azlan, A., Shahar, S., Peng, L.S. (2009). Daidzein and genestein contents in tempeh and selected soy products. *Food Chemistry*, 115: 1350–1356.

Hasim, H., Astuti, P., Falah, S., Faridah, D.N. (2015). *Bacillus subtilis* natto fermentation to improve aglycone isoflavones content of black soybean varieties detam 2. *International Food Research Journal*, 22: 2558–2564.

Hati, S., Vij, S., Singh, B.P., Mandal, S. (2014). β-Glucosidase activity and bioconversion of isoflavones during fermentation of soymilk Subrota. *Journal of the Science of Food and Agriculture*, 95: 216–220.

Henderson, V.W., St-John, J.A., Hodis, H.N., Kono, N., McCleary, C.A., Franke, A.A., Mack, W.J. (2012). Long-term soy isoflavone supplementation and cognition in women - a randomized, controlled trial. *Neurology*, 78: 1841–1848.

Hendrich, S., Lee, K-W., Xu, X., Wang, H-J. Murphy, P.A. (1994). Defining food components as new nutrients. *Journal of Nutrition*, 124 Supplement: 1789S–1792S.

Hoeck, J.A., Fehr, W.R., Murphy, P.A., Welke, G.A. (2000). Influence of genotype and environment on isoflavone contents of soybean. *Crop Science*, 40: 48–51.

Hu, Y.J., Ge, C.R., Yuan, W., Zhu, R.J., Zhang, W.J., Du, L.J., Xue, J. (2010). Characterization of fermented black soybean natto inoculated with Bacillus natto during fermentation. *Journal of the Science of Food and Agriculture*, 90: 1194–1202.

Huang, H., Liang, H., Kwok, K-C. (2006). Effect of thermal processing on genistein, daidzein and glycitein content in soymilk. *Journal of the Science of Food and Agriculture*, 86: 1110–1114.

Huang, X., Cai, W., Xu, B. (2014). Kinetic changes of nutrients and antioxidant capacities of germinated soybean (*Glycine max* L.) and mung bean (*Vigna radiata* L.) with germination time. *Food Chemistry*, 143: 268–276.

IARC. (2019). Monograph on the evaluation of carcinogenic risk to humans, World Health Organization, Report of the Advisory Group to Recommend Priorities for the IARC Monographs during 2020–2024. https://monographs.iarc.fr/wp-content/uploads/2019/10/IARCMonographs-AGReport-Priorities_2020 -2024.pdf (accessed on 14 August 2020).

Ismail, B., Hayes, K. (2005). β-Glycosidase activity toward different glycosidic forms of isoflavones. *Journal of Agricultural and Food Chemistry*, 53: 4918–4924.

Izumi, T., Piskula, M., Osawa, S., Obata, A., Tobe, K., Saito, M., Kataoka, S., Kubota, Y., Kikuchi, M. (2000). Soy isoflavone aglycones are absorbed faster and in higher amounts than their glucosides in humans. *Journal of Nutrition*, 130: 1695–1699.

Jackson, C.J.C., Dini, J.P., Lavandier, C., Rupasinghe, H.P.V., Faulkner, H., Poysa, V., Buzzell, D., DeGrandis, S. (2002). Effects of processing on the content and composition of isoflavones during manufacturing of soy beverage and tofu. *Process Biochemistry*, 37: 1117–1123.

Juan, M.Y., Chou, C.C. (2010). Enhancement of antioxidant activity, total phenolic and flavonoid content of black soybeans by solid state fermentation with *Bacillus subtilis* BCRC 14715. *Food Microbiology*, 27: 586–591.

Jung, S., Murphy, P.A., Sala, I. (2008). Isoflavone profiles of soymilk as affected by high-pressure treatments of soymilk and soybeans. *Food Chemistry*, 111: 592–598.

Kao, T.H., Lu, Y.F., Hsieh, H.C., Chen, B.H. (2004). Stability of isoflavone glucosides during processing of soymilk and tofu. *Food Research International*, 37: 891–900.

Khanna, K.K., Jackson, S.P. (2001). DNA double-strand breaks: Signaling, repair and the cancer connection. *Nature Genetics*, 27: 247–254.

Kim, E.H., Kim, S.H., Chung, J.I., Chi, H.Y., Kim, J.A., Chung, I.M. (2006). Analysis of phenolic compounds and isoflavones in soybean seeds (*Glycine max* (L.) Merill) and sprouts grown under different conditions. *European Food Research and Technology*, 222: 201–208.

Kim, J.A., S.B. Hong, W.S. Jung, C.Y. Yu, K.H. Ma, J.G. Gwag, I.M. Chung. (2007). Comparison of isoflavones composition in seed, embryo, cotyledon and seed coat of cooked-with-rice and vegetable soybean (*Glycine max* L.) varieties. *Food Chemistry*, 102: 738–744.

Kim, J.K., Kim, E.H., Park, I., Yu, B.R., Lim, J.D., Lee, Y.S., Lee, J.H., Kim, S.H., Chung, I.M. (2014). Isoflavones profiling of soybean [*Glycine max* (L.) Merrill] germplasms and their correlations with metabolic pathways. *Food Chemistry*, 153: 258–264.

Kim, S.-L., Lee, J.-E., Kwon, Y.-U., Kim, W.-H., Jung, G.-H., Kim, D.-W., Lee, C.-K., Lee, Y.-Y., Kim, M.-J., Kim, Y.-H., Hwang, T.-Y., Chung, I.-M. (2013). Introduction and nutritional evaluation of germinated soy germ. *Food Chemistry*, 136: 491–500.

Kudou, S., Fleury, Y., Welti, D., Magnolato, D., Uchida, T., Kitamura, K., Okubo, K. (1991). Malonyl isoflavone glycosides in soybean seeds (*Glycine max* Merrill). *Journal Agricultural and Biological Chemistry*, 55: 2227–2233.

Kuligowski, M., Pawlowska, K., Jasińska-Kuligowska, I., Nowak, J. (2017). Isoflavone composition, polyphenols content and antioxidative activity of soybean seeds during tempeh fermentation. *Journal CyTA-Journal of Food*, 15(1): 27–33.

Kuriyama, I., Yoshida, H., Mizushina, Y., Takahashi, Y. (2013). Inhibitory effect of isoflavones from processed soybeans on human DNA topoisomerase II activity. *Journal Plants Biochemistry and Physiology*, 1: 106.

Larkin, T., Price, W.E., Astheimer, L. (2008). The key importance of soy isoflavone bioavailability to understanding health benefits. *Critical Reviews in Food Science and Technology*, 48: 538–552.

Lai, J.X., Xin, C., Zhao, Y., Feng, B., He, C.F., Dong, Y.M., Fang, Y., Wei, S.M. (2013). Optimization of ultrasonic assisted extraction of antioxidants from black soybean (*Glycine max* var) sprouts using response surface methodology. *Molecules*, 18: 1101–1110.

Laurenz, R., Tumbalam, P., Naeve, S., Thelen, K.D. (2017). Determination of isoflavone (genistein and daidzein) concentration of soybean seed as affected by environment and management inputs. *Journal of the Science of Food and Agricultural*, 97: 3342–3347.

Lee, C.H., Yang, L., Xu, J.Z., Yeung, S.Y.V., Huang, Y., Chen, Z.Y. (2005). Relative antioxidant activity of soybean isoflavones and their glycosides. *Food Chemistry*, 90: 735–741.

Lee, S, Lee, J. (2009). Effects of oven-drying, roasting, and explosive puffing process on isoflavone distributions in soybeans. *Food Chemistry*, 112: 316–320.

Lee, S.J., Ahn, J.K., Khanh, T.D., Chun, S.C., Kim, S.L., Ro, H.M., Song, H.K., Chung, I.M. (2007). Comparison of isoflavone concentrations in soybean (*Glycine max* (L.) Merrill) sprouts grown under two different light conditions. *Journal of Agricultural and Food Chemistry*, 55: 9415–9421.

Lee, S.J., Ahn, J.K., Kim, S.H., Kim, J.T., Hahn, S.J., Jung, M.Y., Chung, I.M. (2003). Variation in isoflavone of soybean cultivars with location and storage duration. *Journal of Agricultural and Food Chemistry*, 51: 3382–3389.

Lee, S.J., Kim, J.J., Moon, H.I., Ahn, J.K., Chun, S.C., Jung, W.S., Lee, O.K., Chung, I.M. (2008). Analysis of isoflavones and phenolic compounds in Korean soybean [*Glycine max* (L.) Merrill] seeds of different seed weights. *Journal of Agricultural and Food Chemistry*, 56: 2751–2758.

Lee, Y.L., Yang, J.H., Mau, J.L. (2008). Antioxidant properties of water extracts from *Monascus* fermented soybeans. *Food Chemistry*, 106(3): 1128–1137.

Li, L., Lv, Y., Xu, L., Zheng, Q. (2014). Quantitative efficacy of soy isoflavones on menopausal hot flashes. *Brazilian Journal of Clinical Pharmacology*, 79(4): 593–604.

Li-Jun, Y., Li-Te, L., Zai-Gui, L., Tatsumi, E., Saito, M. (2004). Changes in isoflavone contents and composition of sufu (fermented tofu) during manufacturing. *Food Chemistry*, 87: 587–592.

Liu, K. (1997). *Soybeans: Chemistry, technology and utilization*. Chapman & Hall, New York.

Liu, K. (2004). *Soybeans as functional foods and ingredients*. Edited by KeShun Liu. AOCS Press, Champaign, IL.

Mahmoud, A.M., Yang, W., Bosland, M.C. (2014). Soy isoflavones and prostate cancer: A review of molecular mechanisms. *The Journal of Steroid Biochemistry and Molecular Biology*, 140: 116–132.

Martín-Cabrejas, M.A., Díaz, M.F., Aguilera, Y., Benítez, V., Mollá, E., Esteban, R.M. (2008). Influence of germination on the soluble carbohydrates and dietary fibre fractions in non-conventional legumes. *Food Chemistry*, 107: 1045–1052.

Messina, M. (1998). Soy foods: An alternative to hormone replacement therapy. *Vegetarian Nutrition and Health Letter*, 1(5).

Messina, M. (2014). Soy Foods, isoflavones, and the health of postmenopausal women. *The American Journal of Clinical Nutrition*, 100 Supplement 1: 423S–430S.

Messina, M., Kucuk, O., Lampe, J.W. (2006). An overview of the health effects of isoflavones with an emphasis on prostate cancer risk and prostate-specific antigen levels. *Journal American Oil Association Chemistry International*, 89: 1121–1134.

Messina, M.J., Persky, V., Setchell, K.D.R., Barnes, S. (1994). Soy intake and cancer risk: A review of the in-vitro and in-vivo data. *Nutrition and Cancer*, 21: 113–131.

Miranda, L.C.R., Gomes, R.J., Mandarino, J.M.G., Ida, E.I., Spinosa, W.A. (2020). Acetic acid fermentation of soybean molasses and characterisation of the produced vinegar. *Food Technology and Biotechnology*, 58(1): 84–90.

Mo, H., Kariluoto, S., Piironen, V., Zhu, Y., Sanders, M.G., Vincken, J-P., Wolkers-Rooijackers, J., Rob Nout, M.J. (2013). Effect of soybean processing on content and bioaccessibility of folate, vitamin B12 and isoflavones in tofu and tempe. *Food Chemistry*, 141(3): 2418–2325.

Most, J., Goossens, G.H., Jocken, J.W., Blaak, E.E. (2013). Short-term supplementation with specific combination of dietary polyphenols increases energy expenditure and alters substrate metabolism in overweight subjects. *International Journal of Obesity (London)*, 38(5): 698–706.

Muliterno, M.M., Rodrigues, D., De Lima, F.S., Ida, E.I., Kurozawa, L.E. (2017). Conversion/degradation of isoflavones and color alterations during the drying of okara. *LWT-Food Science and Technology*, 75: 512–519.

Murakami, H., Asakawa, T., Teao, J., Matsushita, S. (1984). Antioxidative stability of tempeh and liberation of isoflavones by fermentation. *Agricultural and Biological Chemistry*, 48(12): 2971–2975.

Nakajima, N., Nozaki, N., Ishihara, K., Ishikawa, A., Tsuji, H. (2005). Analysis of isoflavone content in tempeh, a fermented soybean, and preparation of a new isoflavone-enriched tempeh. *Journal of Bioscience and Bioengineering*, 100: 685–687.

Nkhata, S.G., Ayua, E., Kamau, E.H., Shingiro, J.B. (2018). Fermentation and germination improve nutritional value of cereals and legumes through activation of endogenous enzymes. *Food Science and Nutrition*, 6(8): 2446–2458.

Nout, M.J.R., Kiers, J.L. (2005). A review: Tempe fermentation, innovation and functionality: Update into the third millenium. *Journal of Applied Microbiology*, 98: 789–805.

Nufer, K., Ismail, B., Hayes, K.D. (2009). The effects of processing and extraction conditions on content, profile, and stability of isoflavones in a soymilk system. *Journal of Agricultural and Food Chemistry*, 57: 1213–1218.

Park, U.C., Jeong, K.M., Park, M.H., Yeu, J., Park, M.S., Kim, J.M., Ahn, S.M., Chang, P.S., Lee, J. (2010). Formation of succinyl genistin and succinyl daidzin by *Bacillus* species. *Journal Food Science*, 75: C128–C133.

Patel, R.P., Boersma, B.J., Crawford, J.H., Hogg, N., Kirk, M., Kalyanaraman, B., Parks, D.A., Barnes, S., Darley-Usmar, V. (2001). Antioxidant mechanisms of isoflavones in lipid systems: Paradoxical effects of peroxyl radical scavenging. *Free Radical Biology and Medicine*, 31: 1570–1581.

Paucar-Menacho, L.M., Berhow, M.A., Mandarino, J.M.G., Chang, Y.K., Mejia, E.G.D. (2010). Effect of time and temperature on bioactive compounds in germinated Brazilian soybean cultivar BRS 258. *Food Research International*, 43: 1856–1865.

Pekkanen, J., Linn, S., Heiss, G., Suchindran, C.M., Leon, A., Rifkind, B.M., Tyroler, H.A. (1990). Ten-year mortality from cardiovascular disease in relation to cholesterol level among men with and without pre-existing cardiovascular disease. *New England Journal of Medicine*, 322: 1700–1707.

Phommalth, S., Jeong, Y.S., Kim, Y.H., Dhakal, K.H., Hwang, Y.H. (2008b). Effects of light treatment on isoflavone content of germinated soybean seeds. *Journal of Agricultural and Food Chemistry*, 56: 10123–10128.

Phommalth, S., Jeong, Y.-S., Kim, Y.-H., Hwang, Y.-H. (2008a). Isoflavone composition within each structural part of soybean seeds and sprouts. *Journal Crop Science and Biotechnology*, 11: 57–62.

Prabhakaran, M.P., Hui, L.S., Perera, C.O. (2006). Evaluation of the composition and concentration of isoflavones in soy-based supplements, health products and infant formulas. *Food Research International*, 39: 730–738.

Quinhone Jr, A., Ida, E.I. (2014). Isoflavones of the soybean components and the effect of germination time in the cotyledons and embryonic axis. *Journal of Agricultural and Food Chemistry*, 62(33): 8452–8459.

Quinhone Jr, A., Ida, E.I. (2015). Profile of the contents of different forms of soybean isoflavones and the effect of germination time on these compounds and the physical parameters in soybean sprouts. *Food Chemistry*, 166: 173–178.

Ribeiro, M.L.L., Mandarino, J.M.G., Carrao-Panizzi, M.C., Oliveira, M.C.N., Campo, C.B.H., Nepomuceno, A.L., Ida, E.I. (2006). β-Glucosidase activity and isoflavone content in germinated soy bean radicles and cotyledons. *Journal of Food Biochemistry*, 30: 453–465.

Ribeiro, M.L.L., Mandarino, J.M.G., Carrao-Panizzi, M.C., Oliveira, M.C.N., Campo, C.B.H., Nepomuceno, A.L., Ida, E.I. (2007). Isoflavone content and β-glucosidase activity in soybean cultivars of different maturity groups. *Journal of Food Composition and Analysis*, 20: 19–24.

Ruiz-Larrea, M.B., Mohan, A.R., Paganga, G., Miller, N.J., Bolwell, G.P., Rice-Evans, C.A. (1997). Antioxidant activity of phytoestrogenic isoflavones. *Free Radical Research*, 26: 63–70.

Seguin, P., Zheng, W., Smith, D.L., Deng, W. (2004). Isoflavone content of soybean cultivars grown in eastern Canada. *Journal of the Science of Food and Agriculture*, 84: 1327–1332.

Shahidi, F., Ambigaipalan, P. (2015). Phenolics and polyphenolics in foods, beverages and spices: Antioxidant activity and health effects – A review. *Journal of Functional Foods*, 18: 820–897.

Shahidi, F., Zhong, Y. (2015). Measurement of antioxidant activity. *Journal of Functional Foods*, 18(Part B): 757–781.

Shi, H., Nam, P., Ma, A. (2010). Comprehensive profiling of isoflavones, phytosterols, tocopherols, minerals, crude protein, lipid, and sugar during soybean (*Glycine max*) germination. *Journal of Agricultural and Food Chemistry*, 58: 4970–4976.

Silva, C.E., Carrao-Panizzi, M.C., Mandarino, J.M.G., Leite, R.S., Mônaco, A.P.A. (2012). Isoflavone contents of whole soybeans and their components, obtained from different cultivars (*Glycine max* (L.) Merrill). *Brazilian Journal of Food Technology*, 15: 150–156.

Silva, L.H., Celeghini, R.M.S., Chang, Y.K. (2011). Effect of the fermentation of whole soybean flour on the conversion of isoflavones from glycosides to aglycones. *Food Chemistry*, 128: 640–644.

Silva, M.B.R., Falcão, H.G., Kurozawa, L.E., Prudencio, S.H., de Camargo, A.C., Ida, E.I. (2019). Ultrasound- and hemicellulase-assisted extraction increase β-glucosidase activity, the content of isoflavone agly- cones and antioxidant potential of soymilk. *Journal of Food Bioactives*, 6: 140–147.

Silva, M.B.R., Leite, R.S., Oliveira, M.A., Ida, E.I. (2020). Germination conditions influence the physi- cal characteristics, chemical composition, isoflavones, and vitamin C of soybean sprouts. *Pesquisa Agropecuária Brasileira*, 55: e01409.

Teekachunhatean, S., Hanprasertpong, N., Teekachunhatean, T. (2013). Factors affecting isoflavone content in soybean seeds grown in Thailand. *International Journal of Agronomy*, v: 1–11.

Thompson, M.J. (2010). *Isoflavones: Biosynthesis, occurrence and health effects*. Nova Publishers, New York.

Tsukamoto, C., Shimada, S., Igita, K., Kudou, S., Kokubun, M., Okubo, K., Kitamura, K. (1995). Factors affecting isoflavone content in soybean seeds: Changes in isoflavones, saponins, and composition of fatty acids at different temperatures during seed development. *Journal of Agricultural and Food Chemistry*, 43: 1184–1192.

Variyar, P.S., Limaye, A., Sharma, A. (2004). Radiation-induced enhancement of antioxidant contents of soy- bean (*Glycine max* Merrill). *Journal of Agricultural and Food Chemistry*, 52: 3385–3388.

Vitale, D.C., Piazza, C., Melilli, B., Drago, F., Salomone, S. (2013). Isoflavones: Estrogenic activity, biological effect and bioavailability. *European Journal of Drug Metabolism and Pharmacokinetics*, 38: 15–25.

Wang, H., Murphy, P.A. (1994b). Isoflavone composition of American and Japanese soybeans in Iowa: Effects of variety, crop year, and location. *Journal of Agricultural and Food Chemistry*, 42: 1674–1677.

Wang, H., Murphy, P.A. (1994a). Isoflavone content in commercial soybean foods. *Journal of Agricultural and Food Chemistry*, 42: 1666–1673.

Wang, L.J., Yin, L.J., Li, D., Zou, L., Saito, M., Tatsumi, E., Li, L.T. (2007). Influences of processing and NaCl supplementation on isoflavone contents and composition during douchi manufacturing. *Food Chemistry*, 101: 1247–1253.

Weggemans, R.M., Trautwein, E.A. (2003). Relation between soy-associated isoflavones and LDL and HDL cholesterol concentrations in humans: A meta-analysis. *European Journal of Clinical Nutrition*, 57: 940–946.

Wei, P., Liu, M., Chen, Y., Chen, D.C. (2012). Systematic review of soy isoflavone supplements on osteoporosis in women. *Asian Pacific Journal of Tropical Medicine*, 5: 243–248.

World Health Organization. (2018). Noncommunicable diseases. Available online: https://www.who.int/news -room/fact-sheets/detail/noncommunicable-diseases#:~:text=Cardiovascular%20diseases%20account %20for%20most,and%20diabetes%20(1.6%20million). (accessed on 14 August 2020).

Xiao, Y., Wang, L.X., Rui, X., Li, W., Chen, X.H., Jiang, M., Dong, M.S. (2015). Enhancement of the antioxi- dant capacity of soy whey by fermentation with *Lactobacillus plantarum* B1-6. *Journal of Functional Foods*, 12: 33–44.

Xu, B.J., Chang, S.K.C. (2008). Antioxidant capacity of seed coat, dehulled bean, and whole black soybeans in relation to their distributions of total phenolics, phenolic acids, anthocyanins, and isoflavones. *Journal of Agricultural and Food Chemistry*, 56: 8365–8373.

Xu, Z., Wu, Q., Godber, J.S. (2002). Stabilities of daidzin, glycitin, genistin, and generation of derivatives dur- ing heating. *Journal of Agricultural and Food Chemistry*, 50(25): 7402–7406.

Yanaka, K., Takebayashi, J., Matsumoto, T., Ishimi, Y. (2012). Determination of 15 isoflavone isomers in soy foods and supplements by high-performance liquid chromatography. *Journal of Agricultural and Food Chemistry*, 60: 4012–4016.

Yang, J., Wu, X.B., Chen, H.L., Sun-Waterhouse, D., Zhong, H.B., Cui, C. (2019). A value-added approach to improve the nutritional quality of soybean meal byproduct: Enhancing its antioxidant activity through fermentation by *Bacillus amyloliquefaciens* SWJS22. *Food Chemistry*, 272: 396–403.

Yang, W.S., Va, P., Wong, M.Y., Zhang, H.L., Xiang, Y.B. (2011). Soy intake is associated with lower lung cancer risk: Results from a meta-analysis of epidemiologic studies. *American Journal of Clinical Nutrition*, 94: 1575–1583.

Yoshiara, L.Y., Madeira, T., de Camargo, A.C., Shahidi, F., Ida, E. (2018a). Multistep optimization of β-glucosidase extraction from germinated soybeans (*Glycine max* L. Merril) and recovery of isoflavone aglycones. *Foods*, 7(110): 1–13.

Yoshiara, L.Y., Madeira, T.B., Ribeiro, M.L.L., Mandarino, J.M.G., Carrao-Panizzi, M.C., Ida, E.I. (2012). β-Glucosidase activity of soybean (*Glycine max*) embryonic axis germinated in the presence or absence of light. *Journal of Food Biochemistry*, 36: 699–705.

Yoshiara, L.Y, Mandarino, J.M.G, Carrão-Panizzi, M.C., Madeira, T.B, da Silva, J.B., de Camargo, A.C, Shahidi, F., Ida, E.I. (2018b). Germination changes the isoflavone profile and increases the antioxidant potential of soybean. *Journal of Food Bioactives*, 3: 144–150.

Yuan, J.-P., Liu, Y.-B., Peng, J., Wang, J.-H., Liu, X. (2009). Changes of isoflavone profile in the hypocotyls and cotyledons of soybeans during dry heating and germination. *Journal Agricultural and Food Chemistry*, 57: 9002–9010.

Yuan, J.P., Wang, J.H., Liu, X. (2007). Metabolism of dietary soy isoflavones to equol by human intestinal microflora: Implications for health. *Molecular Nutrition and Food Research*, 51: 765–781.

Zaheer, K., Akhtar, M.H. (2017). An updated review of dietary isoflavones: Nutrition, processing, bioavailability and impacts on human health. *Critical Reviews in Food Science and Nutrition*, 57(6): 1280–1293.

Zhang, J., Ge, Y., Han, F., Li, B., Yan, S., Sun, J., Wang, L. (2014). Isoflavone content of soybean cultivars from maturity group 0 to VI grown in Northern and Southern China. *Journal American Oil Chemistry Society*, 91: 1019–1028.

Zhang, X., Gao, Y.T., Yang, G., Li, H., Cai, Q., Xiang, Y.B., Ji, B.T., Franke, A.A., Zheng, W., Shu, X.O. (2012). Urinary isoflavonoids and risk of coronary heart disease. *International Journal Epidemiology*, 41: 1367–1375.

Zhang, Y., Sam, K.C., Chang, S.K.C., Liu, Z. (2015). Isoflavone profile in soymilk as affected by soybean variety, grinding, and heat-processing methods. *Journal of Food Science*, 80: C983–C988.

Zhu, D., Hettiarachchy, N.S., Horax, R., Chen, P. (2005). Isoflavone contents in germinated soybean seeds. *Plant Foods for Human Nutrition*, 60: 147–151.

3 Bioactive Peptides from Soybeans and Derived Products

Hanifah Nuryani Lioe and Badrut Tamam

CONTENTS

3.1 INTRODUCTION

Bioactive peptides have attracted many researchers in the past decades. The peptides have 2 to 30 amino acid residues with molecular weights less than 6,000 Da. Among the reported bioactive peptides, those from soybeans and derived products are the most frequently investigated. They have many bioactivities, such as antihypertensive, antioxidant, antimicrobial, antidiabetic, anticancer, antithrombotic, and immunomodulatory effects. The soy proteins mainly consisting of glycinin and β-conglycinin are the main sources of bioactive peptides (Gibbs et al., 2004).

The bioactive peptides have been known to be present in soybeans, soymilk, fermented soybeans (tempeh, miso, soy sauce, tofuyo, douchi, and so on), soy protein hydrolysates (especially those produced from enzymatic hydrolysis using proteases), defatted soy protein hydrolysates, and soy-based products such as infant formula with various bioactivities. Therefore, the utilization of soybeans to provide functional food rich in bioactive peptides becomes challenging future research. Moreover, the use of soybean byproducts, that is, defatted soybeans, has become an increased interest due to the economical reasons to produce bioactive peptides naturally.

Other than glycinin and β-conglycinin as the main precursors for providing bioactive peptides from soybeans, trypsin inhibitors (Kunitz and Bowman-Birk), lectins, lipoxygenase, allergen P34, proteins that bind to sucrose, α-amylase, urease, oleosins, and several other small amounts of protein including enzymes (Herman, 2014; Gomes et al., 2014) might also be important in producing bioactive peptides. However, different conditions of protein hydrolysis or soybean fermentation have revealed different bioactive peptides (Gibbs et al., 2004; Tamam et al., 2019).

3.2 BIOACTIVE PEPTIDES FROM SOYBEANS AND THEIR PRODUCTS

Bioactive peptides are defined as protein fragments with amino acid sequences that have biological activities, such as antioxidant, antihypertensive, antithrombotic, anti-adipogenic, antimicrobial, anti-inflammatory, and immunomodulatory effects (Castro & Sato, 2015). They can be obtained from vegetable and animal protein sources. From vegetable protein sources, generally, bioactive peptides come from grains such as wheat, rice, corn, oats, and rye, as well as from legumes such as soybeans, peas, and chickpeas. Among these plant protein sources, soybeans are the most studied material as a source of bioactive peptides (Ortiz-Martinez et al., 2014). Several bioactive peptides from soybeans have been isolated or have been studied for their potential functional properties and physiological activities such as antihypertensive, hypocholesterolemia, antiobesity, antimicrobial, antioxidant, antidiabetic, and anticancer effects (Capriotti et al., 2015; Coscueta et al., 2016; Nakahara et al., 2010; Sato et al., 2018; Tamam et al., 2019). Antihypertensive peptides are the most studied peptides from soybeans and their products.

Bioactive peptides are not active if in the parent protein sequence. The peptide will be active if it is removed from its parent protein through enzymatic hydrolysis (during fermentation or by adding proteases). However, the bioavailability of the bioactive peptides present in food is still a challenge. The digestive stability of simple peptides with two to three amino acid residues, as an example, has been exhibited during *in vitro* testing using pepsin, chymotrypsin, and trypsin (Kuba et al., 2003). The bioavailability is important to be tested through *in vivo* testing using animal experiments before the peptides are used for a clinical study using humans (Gallego et al., 2018).

Gibbs et al. (2004) reported producing bioactive peptides from natto (using *Bacillus subtilis*) in Petri dishes and tempeh (using *Rhizopus oligosporus*) in Petri dishes. The peptides produced have ACE inhibitor, antithrombotic, surface tension, and antioxidant activity. All peptides produced are derived from glycinin. β-Conglycinin is more stable against proteolytic enzyme attack even with many enzymes (Gibbs et al., 2004).

The bioactivity of a bioactive peptide can be determined by the presence of amino acid residues in its sequence (Castro & Sato, 2015). The relationship of the presence of certain amino acids with their functional properties can be seen in Table 3.1. It can be inferred that most bioactive peptides have hydrophobic amino acid residues in their sequences (Table 3.1). Dashper et al. (2007) also stated that in general, antimicrobial peptides are hydrophobic. Chen et al. (1998)

TABLE 3.1

Relationship between the Presence of Amino Acid Residues and the Bioactivities of Peptides

No	Amino Acid Presence and/or Location	Bioactivity	References
1	Aromatic residues on C terminal (Phe, Trp, Tyr)	Antihypertensive	Wijesekara et al. (2011); Castro & Sato (2015)
2	Aliphatic residues on C terminal (Gly, Ala, Val, Pro, Leu, Ile, Met)	Antihypertensive	Wijesekara et al. (2011); Castro & Sato (2015)
3	Val or Ile on N terminal	Antihypertensive	Wijesekara et al. (2011); Castro & Sato (2015)
4	Hydrophobic residues (Gly, Ala, Val, Pro, Leu, Ile, Met) and (Phe, Trp, Tyr)	Antihypertensive	Haque & Chand (2008); Sanjukta & Rai (2016)
5	Positive charge residues (Lys, Arg, His)	Antihypertensive	Haque & Chand (2008); Sanjukta & Rai (2016))
6	Pro on C terminal	Antihypertensive	Haque & Chand (2008); Sanjukta & Rai (2016)
7	Metal reducing amino acid residues (Phe, Tyr, Met, Lys, Cys)	Antioxidant	Carrasco-Castilla et al. (2012)
8	Amino acid residues with phenol, indol, imidazole functional groups (Trp, Tyr, Phe, His)	Antioxidant	Duan et al. (2014)
9	Having high hydrophobicity (but rich in Gly and Leu)	Antimicrobial	Amadou et al. (2013)
10	Having residual amino acids with cation (Lys, Arg, His)	Antimicrobial	Amadou et al. (2013); Castro & Sato (2015)

stated that the activity of antioxidant peptides is influenced by the hydrophobicity of these peptides. Furthermore, Duan et al. (2014) explained that aromatic (hydrophobic) amino acids are able to act as proton donors to radicals that lose their electrons and are able to pick up these radicals efficiently. Wijesekara et al. (2011) explained that the presence of peptides with aromatic or aliphatic (hydrophobic) groups was strongly associated with ACE inhibitory activity. Hydrophobic peptides can increase the interaction with non-polar amino acid residues in the active site of the ACE enzyme.

Bioactive peptides with antihypertensive activity have a different presence of amino acid residues. Antihypertensive peptides are related to the presence of aromatic and aliphatic amino acids such as Pro, Phe, and Tyr on C terminal and Val or Ile on N terminal (Wijesekara et al., 2011). The presence of hydrophobic amino acids (Tyr, Phe, Trp, Ala, Ile, Val, Met) or positively charged amino acids (Arg and Lys), as well as Pro at the C terminal position of the ACE-inhibiting peptide, shows better affinity with ACE (Haque & Chand, 2008; Sanjukta & Rai, 2016). Several peptide sequences found in soybeans and derived products that have been reported to have antihypertensive activity can be seen in Table 3.2.

The presence of Trp, Tyr, Met, Lys, and Cys residues has the ability to reduce Fe^{3+} ion into Fe^{2+} ion and bind Fe^{2+} and Cu^{2+} ions (Carrasco-Castilla et al., 2012). Aromatic amino acids such as Trp, Tyr, Phe, and His which have phenol, indole, and imidazole groups can act as proton donors to electron-deficient radicals and efficiently capture radical compounds, trap lipids and decompose imidazole rings (Sarmadi & Ismail, 2010). Some of the peptide sequences that have been reported to have antioxidant activity can be seen in Table 3.3.

Soybean fermentation seems to produce peptides with antioxidant properties such as in natto (Iwai et al., 2002) and douchi (Wang et al., 2008), as well as fermented soybean broth (Yang et al., 2000) and a *Bacillus subtilis*–fermented soybean (Sanjukta et al., 2015). Douchi extract could exhibit a free radical scavenging activity (by *in vitro* test) and an increase in superoxide dismutase activities in the liver and kidney (by *in vivo* animal test). Natto fermented by *B. subtilis* has shown a reduction in lipid peroxidation in the liver and aorta by *in vivo* animal testing (Iwai et al., 2002). The hydrolysis of soybean proteins through fermentation could provide more antioxidant activity, such as in sufu, a fermented tofu, a traditional Chinese soybean product fermented by *Aspergillus oryzae*, which showed a higher DPPH radical scavenging activity compared to unfermented tofu (Huang et al., 2011).

Generally, antimicrobial peptides have similarities in their amino acid chains, which are composed of cationic and hydrophobic amino acid residues. The presence of the Arg residue in the peptide sequence also plays an important role in antimicrobial activity, which can increase interactions with bacterial cell walls due to its cationic nature (Amadou et al., 2013; Castro & Sato, 2015). Bacterial cells are rich in negatively charged phospholipids facing the outside environment, making it easier to interact with peptides, most of which are positively charged (Matsuzaki, 1999). The hydrophobicity of peptides containing Gly and Leu was reported as a potential antimicrobial molecule (Amadou et al., 2013). Some of the peptide sequences reported to have antimicrobial properties can be seen in Table 3.4.

Exploration of bioactive peptides from fermented soybeans is always challenging, due to the various types of fermented soybean products found in many countries, particularly in Asian countries. Investigation of bioactive peptides in tempeh, a *Rhizopus oligosporus*–fermented soybean product by a 48-hour solid-state fermentation (Sitanggang et al., 2020; Tamam et al., 2019), becomes interesting due to the potential use of tempeh as a functional food, known worldwide. A number of bioactive peptides found in tempeh through the fractionation by ultrafiltration membranes (to obtain fractions less than 3,000 Da or less than 5,000 Da) and with or without isolation by chromatography and HPLC, is presented in Table 3.5. The peptides have antihypertensive, antioxidant, antidiabetic, anticancer, immunomodulatory, antimicrobial, and antithrombotic bioactivities (Table 3.5).

TABLE 3.2

Antihypertensive Peptides from Soybeans and Derived Products: Sequences and Their Properties*

No	Peptide Sequence	Sample	Reference	Peptide Length (Residues)	MW (Da)	pI	Net Charge	Hydrophobicity (Kcal/mol)
1	TIIPLPV	Soybeans and soymilk	Capriotti et al. (2015)	7	751.48	5.36	0	4.48
2	YVVFK	e-Soy protein hydrolysate	Kodera & Nio (2006)	5	654.37	9.48	1	7.36
3	YLAGNQ	Soybeans and soymilk	Capriotti et al. (2015)	6	664.32	5.37	0	9.21
4	IPPGVPYWT	Soybeans and soymilk	Capriotti et al. (2015)	9	1,028.53	5.45	0	5.34
5	ASYDTKF	Soybeans and soymilk	Capriotti et al. (2015)	7	830.39	6.52	0	13.13
6	DTKF	Soybeans and soymilk	Capriotti et al. (2015)	4	509.25	6.55	0	12.88
7	NWGPLV	Soybeans and soymilk	Capriotti et al. (2015)	6	684.35	5.41	0	6.24
8	PNNKPFQ	Soybeans and soymilk	Capriotti et al. (2015)	7	843.42	9.84	1	11.74
9	DQTPRVF	Soybeans and soymilk	Capriotti et al. (2015)	7	861.43	6.56	0	12.34
10	VPP	Fermented soybeans (miso)	Inoue et al. (2009)	3	311.18	5.69	0	7.72
11	IPP	Fermented soybeans (miso)	Inoue et al. (2009)	3	325.20	5.62	0	7.06
12	GY	Fermented soy seasoning	Nakahara et al. (2010)	2	238.09	5.5	0	8.34
13	AF	Fermented soy seasoning	Nakahara et al. (2010)	2	236.11	5.53	0	6.69
14	VP	Fermented soy seasoning	Nakahara et al. (2010)	2	214.13	5.69	0	7.58
15	AI	Fermented soy seasoning	Nakahara et al. (2010)	2	202.13	5.6	0	7.28
16	VG	Fermented soy seasoning	Nakahara et al. (2010)	2	174.10	5.64	0	8.59
17	LIVTG	Soy concentrate	Vallabha & Tiku (2014)	5	503.15	5.58	0	6.47
18	IA	Soy protein	Castro & Sato (2015)	2	202.13	5.58	0	7.28
19	YLAGNG	Soy protein	Castro & Sato (2015)	6	593.28	5.45	0	9.59
20	FFL	Soy protein	Castro & Sato (2015)	3	425.23	5.46	0	3.23
21	IYLL	Soy protein	Castro & Sato (2015)	4	520.33	5.55	0	3.57
22	AW	Fermented soy seasoning	Nakahara et al. (2010)	2	275.12	5.55	0	6.31
23	AY	Fermented soy seasoning	Nakahara et al. (2010)	2	252.11	5.5	0	7.69
24	GW	Fermented soy seasoning	Nakahara et al. (2010)	2	261.11	5.55	0	6.96
25	SY	Fermented soy seasoning	Nakahara et al. (2010)	2	268.10	5.37	0	7.65
26	PGTAVFK	Soybean	McClean et al. (2014)	7	718.4	9.80	1	10.57
27	LVGQGS	Fermented soybeans	Rho et al. (2009)	6	559.29	5.5	0	9.72
28	HHL	Fermented soybeans	Shin et al. (2001)	3	405.21	8.04	0	11.31
29	ND	Fermented soymilk	Tsai et al. (2006)	2	247.08	3.02	-1	12.39

(Continued)

TABLE 3.2 (CONTINUED)

Antihypertensive Peptides from Soybeans and Derived Products: Sequences and Their Properties*

No	Peptide Sequence	Sample	Reference	Peptide Length (Residues)	MW (Da)	pI	Net Charge	Hydrophobicity (Kcal/mol)
30	LVQGS	Fermented soybeans	Rho et al. (2009)	5	502.27	5.5	0	8.57
31	IFL	Fermented soybeans (tofuyo)	Kuba et al. (2003)	3	391.24	5.57	0	3.82
32	WL	Fermented soybeans (tofuyo)	Kuba et al. (2003)	2	317.17	5.69	0	4.56
33	APAMR	Soy infant formula	Puchalska et al. (2014)	5	544.27	10.73	1	10.18
34	EAPRY	Soy infant formula	Puchalska et al. (2014)	5	634.31	6.58	0	13.27
35	IPSEVLS	Soy infant formula	Puchalska et al. (2014)	7	743.41	3.20	-1	9.76
36	KHFLA	Soy infant formula	Puchalska et al. (2014)	5	614.35	10.21	1	10.57
37	NSGPLVNP	Soy infant formula	Puchalska et al. (2014)	8	796.41	5.44	0	9.78
38	RDPIYS	Soy infant formula	Puchalska et al. (2014)	6	749.36	6.49	0	12.12
39	RPSYT	Soy infant formula	Puchalska et al. (2014)	5	622.31	9.59	1	9.85
40	ALPEEVIQHTFNLK	Soy infant formula	Puchalska et al. (2014)	14	1637.87	5.23	-1	17.01
41	DFYNPKA	Soy infant formula	Puchalska et al. (2014)	7	853.39	6.68	0	13.41
42	KNKPLVVQ	Soy infant formula	Puchalska et al. (2014)	8	924.57	10.58	2	13.09
43	NKNPFLFG	Soy infant formula	Puchalska et al. (2014)	8	935.48	10.16	1	9.02
44	NPFLFG	Soy infant formula	Puchalska et al. (2014)	6	693.34	5.41	0	5.37
45	RPSYTNGPQEIYIQQGKGIFG	Soy infant formula	Puchalska et al. (2014)	21	2352.18	9.60	1	18.40
46	VKNNNPFSFLVPPQE	Soy infant formula	Puchalska et al. (2014)	15	1728.88	6.71	0	12.94
47	VKNNNPFSFLVPPQESQRR	Soy infant formula	Puchalska et al. (2014)	19	2256.17	11.35	2	17.79
48	VSIIDTNSLENQLDQMPRR	Soy infant formula	Puchalska et al. (2014)	19	2228.11	4.27	-1	21.11
49	WWMYNNEDTPVVA	Soy infant formula	Puchalska et al. (2014)	13	1623.70	3.05	-2	11.28
50	NQLDQ	Soy infant formula	Puchalska et al. (2014)	5	616.28	3.02	-1	12.68
51	YNFREGDLLIAVPTG	Defatted soy flour protein	Coscueta et al. (2016)	14	1550.77	4.00	-1	15.77
52	IYNFREGDLLIAVPTG	Defatted soy flour protein	Coscueta et al. (2016)	15	1663.85	4.01	-1	14.65
53	YRAELSEQDIFVIPAG	Defatted soy flour protein	Coscueta et al. (2016)	16	1806.91	3.73	-2	17.76
54	FEITPEKNPQLRDLDIFLSI	Defatted soy flour protein	Coscueta et al. (2016)	20	2387.26	4.00	-2	19.13
55	INAENNQRNFLAGSQDNVISQIPSQV	Defatted soy flour protein	Coscueta et al. (2016)	26	2855.41	4.01	-1	20.74
56	FAIGINAENNQRNFLAGSQDNVISQIPSQV	Defatted soy flour protein	Coscueta et al. (2016)	30	3243.62	4.00	-1	19.56
57	TPRVF	Soy protein	Wang & de Mejia (2005)	5	618.34	10.90	1	7.93
58	RPLKPW	GM soy protein	Wang & de Mejia (2005)	6	795.47	11.53	2	9.45

*The characteristics of molecular weight, isoelectric point (pI), net charge, and hydrophobicity were obtained from www.tulane.edu/~biochem/WW/PepDraw/.

TABLE 3.3

Antioxidative Peptides from Soybeans and Derived Products: Sequences and Their Properties

No	Peptide Sequence	Sample	Reference	Peptide Length (Residues)	MW (Da)	pI	Net Charge	Hydrophobicity (kcal/mol)
1	TIIPLPV	Soybeans and soymilk	Capriotti et al. (2015)	7	751.48	5.36	0	4.48
2	LLPHHADADY	Soybeans and soymilk	Capriotti et al. (2015)	10	1,150.53	4.98	−2	17.77
3	LLPHH	Soybeans and soymilk	Capriotti et al. (2015)	5	615.34	7.84	0	10.20
4	LVNPHDHQN	Soybeans and soymilk	Capriotti et al. (2015)	9	1,072.50	6.05	−1	17.10
5	TTYY	Soybeans and soymilk	Capriotti et al. (2015)	4	546.23	5.28	0	6.98
6	LQSGDALRVPSGTTYY	Soybeans and soymilk	Capriotti et al. (2015)	16	1,726.84	6.47	0	14.10
7	HH	Soybeans	Chen et al. (1995)	2	292.12	7.84	0	12.56
8	QSGDALR	Soybeans and soy infant formula	Chen et al. (1995); Puchalska et al. (2014)	7	745.37	6.42	0	14.98
9	SGDALR	Soybeans	Chen et al. (1995)	6	617.31	6.42	0	14.21
10	RPSYT	Soy infant formula	Puchalska et al. (2014)	5	622.31	9.59	1	9.85
11	IRHFNEGDVLVIPPGVPY	Defatted soy flour protein	Coscueta et al. (2016)	18	2,021.07	5.19	−1	15.59
12	IRHFNEGDVLVIPPGVPYW	Defatted soy flour protein	Coscueta et al. (2016)	19	2,207.14	5.19	−1	13.50
13	YRAELSEQDIFVIPAG	Defatted soy flour protein	Coscueta et al. (2016)	16	1,806.91	3.73	−2	17.76
14	IYNFREGDLLAVPTG	Defatted soy flour protein	Coscueta et al. (2016)	15	1,663.85	4.01	−1	14.65
15	VSIIDTNSLENQLDQMPRR	Defatted soy flour protein	Coscueta et al. (2016)	19	2,228.12	4.27	−1	21.11

TABLE 3.4

Antimicrobial Peptides from Soybeans and Derived Products: Sequences and Their Properties

No	Peptide Sequence	Sample	Reference	Peptide Length (Residues)	MW (Da)	pI	Net Charge	Hydrophobicity (Kcal/mol)
1	FVLPVIRGNGGGIQVA	Soybeans and soymilk	Capriotti et al. (2015)	16	1,595.91	11.18	1	9.99
2	NVLKVIPAGSSSGAKKA	Soybeans and soymilk	Capriotti et al. (2015)	17	1,625.94	10.90	3	19.18
3	IIVVQGKGAIGF	Soybeans and soymilk	Capriotti et al. (2015)	12	1,200.72	9.93	1	9.43
4	ASRGIRVNGVAPGPVWTPIQPA	Soybeans and soymilk	Capriotti et al. (2015)	22	2,242.22	12.49	2	13.65
5	IIIAQGKGALGV	Soybeans and soymilk	Capriotti et al. (2015)	12	1,138.71	10.14	1	10.85
6	IVTVKGGLRVTAPA	Soybeans and soymilk	Capriotti et al. (2015)	14	1,380.84	11.56	2	12.70
7	KIGGIGTVPVGRVETGVLKPGMVV	Soybeans and soymilk	Capriotti et al. (2015)	24	2,362.37	10.63	2	19.70
8	SGGIKLPTDIISKISPLPVLKEI	Soybeans and soymilk	Capriotti et al. (2015)	23	2,417.44	9.81	1	18.11
9	LFVLSGRAIL	Soybeans and soymilk	Capriotti et al. (2015)	10	1,087.67	11.11	1	4.78
10	LAFPGSAKDIENLIKSQ	Soybeans and soymilk	Capriotti et al. (2015)	17	1,829.98	6.68	0	19.15
11	ITLAIPVNKPG	Soybeans and soymilk	Capriotti et al. (2015)	11	1,121.67	10.16	1	9.78
12	WAISKDISEGPPAIKL	Soybeans and soymilk	Capriotti et al. (2015)	16	1,723.94	6.92	0	17.42
13	SGGIKLPTDIISKISPLPV	Soybeans and soymilk	Capriotti et al. (2015)	19	1,934.14	9.93	1	14.05
14	MIIIAQGKGALGV	Soybeans and soymilk	Capriotti et al. (2015)	13	1,269.74	10.14	1	10.18
15	IIVVQGKGAIG	Soybeans and soymilk	Capriotti et al. (2015)	11	1,053.65	10.16	1	11.14
16	VDINEGALLLPHFNSKAIV	Soybeans and soymilk	Capriotti et al. (2015)	19	2,049.12	5.19	−1	16.13
17	VLSGRAILTLV	Soybeans and soymilk	Capriotti et al. (2015)	11	1,140.72	11.11	1	6.28
18	GKVKIGINGFGRIGRLV	Soybeans and soymilk	Capriotti et al. (2015)	17	1,783.09	12.50	4	16.48
19	IIYALNGRALVQV	Soybeans and soymilk	Capriotti et al. (2015)	13	1,428.84	9.91	1	7.11
20	IYALNGRALIQV	Soybeans and soymilk	Capriotti et al. (2015)	12	1,329.77	9.91	1	7.57
21	LAGSKDNVIRQIQKQVKEL	Soybeans and soymilk	Capriotti et al. (2015)	19	2,166.24	10.36	2	24.99
22	GIRVNGVAPGPVWTPIQPA	Soybeans and soymilk	Capriotti et al. (2015)	19	1,928.06	11.18	1	10.88
23	PGTAVFK	Soybeans	McClean et al. (2014)	7	718.4	9.8	1	10.57

TABLE 3.5

Bioactive Peptides Present in Tempeh (a Fungi-Fermented Soy Product): Sequences and Their Properties

No	Peptide Sequence	Bioactivity	Ref.*	Peptide Length (Residues)	MW (Da)	pI	Net Charge	Hydrophobicity (Kcal/mol)
1	AV	Antihypertensive	a	2	118.12	5.60	0	7.94
2	EL	Antihypertensive, antioxidant	a, b	2	260.14	3.09	−1	10.28
3	GF	Antihypertensive	a	2	222.10	5.53	0	7.34
4	PL	Antihypertensive	a, b	2	228.15	5.23	0	6.79
5	AF	Antihypertensive	a	2	236.12	5.53	0	6.69
6	DM	Antihypertensive	a	2	262.08	2.95	−1	10.87
7	DY	Antihypertensive	a	2	296.10	2.95	−1	10.83
8	IG	Antihypertensive	b	2	188.12	5.58	0	7.93
9	VP	Antihypertensive	b	2	214.13	5.69	0	7.58
10	LY	Antihypertensive, antioxidant	b	2	294.16	5.48	0	5.94
11	VW	Antihypertensive, antioxidant, antidiabetic	b	2	303.16	5.58	0	5.35
12	IP	Antihypertensive	b	2	228.15	5.62	0	6.92
13	KL	Antihypertensive	b	2	259.19	10.14	1	9.45
14	IY	Antihypertensive, antioxidant	b	2	294.16	5.48	0	6.07
15	FT	Antihypertensive	b	2	266.13	5.37	0	6.44
16	PAP	Antihypertensive	a	3	283.15	5.25	0	8.68
17	RIY	Antihypertensive	a	3	450.26	9.64	1	7.88
18	IAK	Antihypertensive	a	3	330.23	9.80	1	10.08
19	GEP	Antihypertensive	b	3	301.13	3.21	−1	12.82
20	ALEP	Antihypertensive	a	4	428.23	3.21	−1	10.92
21	VIKP	Antihypertensive	a	4	455.31	10.59	1	9.26
22	KP	Antioxidant	a, b	2	243.16	10.59	1	10.84
23	TY	Antioxidant	a	2	282.12	5.29	0	7.44
24	MY	Antioxidant	a	2	312.11	5.32	0	6.52
25	LK	Antioxidant	b	2	259.19	9.80	1	9.45
26	KD	Antioxidant	b	2	261.13	6.77	0	14.34
27	PS	Antidiabetic	a	2	202.10	5.18	0	8.50
28	SV	Antidiabetic	a	2	204.11	5.45	0	7.90
29	AE	Antidiabetic	a	2	218.09	3.21	−1	12.03
30	SI	Antidiabetic	a	2	218.13	5.46	0	7.24
31	EP	Antidiabetic	a	2	244.11	3.09	−1	11.67
32	HV	Antidiabetic	a	2	254.14	7.89	0	9.77
33	VH	Antidiabetic	a	2	254.14	7.69	0	9.77
34	PF	Antidiabetic	a	2	262.13	5.19	0	6.33
35	RN	Antidiabetic	a	2	288.15	10.60	1	10.56
36	NR	Antidiabetic	a	2	288.15	10.73	1	10.56
37	HF	Antidiabetic	a	2	302.14	7.68	0	8.52
38	KK	Antimicrobial	b	2	274.20	10.57	2	13.50
39	YG	Immunomodulatory	b	2	238.10	5.45	0	8.34
40	RKP	Immunomodulatory	b	3	399.26	11.71	2	12.65
41	GLF	Immunomodulatory	b	3	335.18	5.53	0	6.09
42	EAE	Immunomodulatory	b	3	347.13	2.92	−2	15.66
43	EF	Anticancer	a	2	294.12	3.09	−1	9.82
44	VVV	Anticancer	b	3	315.22	5.63	0	6.52
45	PG	Antithrombotic	b	2	172.08	5.23	0	9.19

*References: a = Tamam et al. (2019); b = Sitanggang et al. (2020).

3.3 CHARACTERISTICS OF BIOACTIVE PEPTIDES

The characteristics of bioactive peptides from soybeans and derived products can be revealed through the characteristics of peptide length, molecular weight, isoelectric point (pI), net charge, and hydrophobicity using the PepDraw application (www.tulane.edu/~biochem/WW/PepDraw/). The characteristics of antihypertensive peptides, antioxidative peptides, and antimicrobial peptides are shown in Tables 3.2, 3.3, and 3.4, respectively. Tamam et al. (2018) have studied the characteristics of peptides from different sources (not only soybeans) through a meta-analysis to recognize the bioactivities. Peptides with antimicrobial activity have a peptide length longer (14.8 amino acid residues on average) than peptides with antihypertensive and antioxidant activities (7.7 and 8.4 amino acid residues on average). Antimicrobial peptides could be affected by peptide length, molecular weight, pI, and net charge. Peptides with antioxidant activity are influenced by hydrophobicity with amino acid residues which could provide a proton donor or a metal ion binder. However, most soy bioactive peptides have a peptide length of two to five peptides. The molecular weights of them were below 1,000 Da.

Boman (2003) states that peptides that have antimicrobial properties generally have 15–45 amino acid residues and have a positive net charge. The antimicrobial activity of the peptides of this group is related to their low molecular weight, net charge, and formation of small channels in the lipid bilayer. These properties encourage the interaction between peptides and the membrane of microorganisms (Gobbetti et al., 2004).

Table 3.2 shows that antihypertensive peptides containing amino acids Ala, Phe, and His have been isolated from soybeans fermented by *Bacillus natto* (a natto product from Japan) and a chunggugjang product (from Korea) fermented by *Bacillus subtilis* (Korhonen & Pihlanto, 2003). Antihypertensive peptides (His-His-Leu) were also found in soybean paste (Shin et al., 2001) and soy sauce (Okamoto et al., 1995). Nakahara et al. (2010) reported that soy sauce fermented by *Aspergillus sojae* has produced antihypertensive peptides, namely Gly-Tyr, Ala-Phe, Val-Pro, Ala-Ile, and Val-Gly. Likewise, Inoue et al. (2009) reported that miso fermentation using the *Aspergillus oryzae* mold produced two antihypertensive peptides, namely the peptides Val-Pro-Pro and Ile-Pro-Pro. Wang and de Mejia (2005) have reviewed several antihypertensive peptides, such as Val-Ala-His-Ile-Asn-Val-Gly-Lys or Tyr-Val-Trp-Lys, from soybeans fermented with *Bifidobacterium natto* or *Bacillus subtilis*. Tsai et al. (2006) have reported on soy milk fermented with several strains of lactic acid bacteria such as *Lactobacillus casei, L. acidophilus, L. bulgaricus, Streptococcus thermophilus,* and *Bacillus longum*. The peptides produced have antihypertensive functional properties. Vallabha and Tiku (2014) revealed that the peptides Leu-Ile-Val-Thr-Lys and Leu-Ile-Val-Thr were produced by soy protein fermented by *Lactobacillus casei* spp.

Nakahara et al. (2010) have made a peptide-enriched seasoning called fermented soybean seasoning (FSS) by modifying the brewing process in making soy sauce. FSS has a total peptide concentration of 2.7 times higher than that of regular soy sauce, and the concentration of antihypertensive peptides is also higher than that of regular soy sauce.

The pI of antihypertensive peptides ranges from 3.02 to 11.53, and the net charge ranges from −2 to +2. The hydrophobicity of them ranges from 3.23 to 20.74 (Table 3.2). These wide properties make the typical characteristics of hypertensive peptides difficult to recognize.

Table 3.3 presented that the pI of antioxidative peptides ranges from 3.73 to 9.59, and net charge ranges from −2 to +1. The hydrophobicity of them ranges from 4.48 to 21.11. Meanwhile, the pI of antimicrobial peptides ranges from 5.19 to 12.50, and the net charge ranges from −1 to +4. The hydrophobicity of them ranges from 4.78 to 24.99 (Table 3.4).

Tempeh-derived bioactive peptides have pI ranged from 2.92 to 11.71, a net charge from −2 to +2, and hydrophobicity from 5.35 to 15.66 (Table 3.5). Tempeh has 21 antihypertensive peptides, ten antidiabetic peptides, ten antioxidant peptides (four of them are the same as antihypertensive peptides), four immunomodulatory peptides, two anticancer peptides, and one antithrombotic peptide (Sitanggang et al., 2020; Tamam et al., 2019).

3.4 PURIFICATION AND IDENTIFICATION OF BIOACTIVE PEPTIDES

Several purification methods have been reported to be used in the production of bioactive peptides. Chromatographic techniques are among the most widely used, such as high-performance liquid chromatography (HPLC) and ultra-HPLC (UHPLC) (Singh et al., 2014; Yang et al., 2015). Reversed-phase HPLC (RP-HPLC) can be used to separate peptides based on their hydrophobicity. Hydrophilic interaction liquid chromatography (HILIC) is used for the separation of hydrophilic substances. Gel electrophoresis and ultrafiltration are used as supporting techniques prior to the analysis of the structural and chemical composition of peptides (Roblet et al., 2012). Preparation of isolated peptide fractions by stepwise ultrafiltration, Sephadex G-15 or Sephadex G-25 gel filtration chromatography, followed by RP-HPLC separation, has been used in many studies (Kuba et al., 2003; Tamam et al., 2019; Tsou et al., 2013).

Liquid chromatography–mass spectrometry (LC-MS-MS) has been commonly used in the identification of peptide sequences (Singh et al., 2014; Sitanggang et al., 2020; Tamam et al., 2019; Tsou et al., 2013; Yang et al., 2015). Electrospray ionization (ESI) and matrix-assisted laser desorption ionization (MALDI) are types of mass spectrometry techniques with *time of flight* (TOF) or *triple quadrupole* (TQ) ion analyzers that are often used in the identification and characterization of bioactive peptides (Mamone et al., 2009; Tamam et al., 2019). The ESI is currently used in tandem with high-resolution mass spectrometry using quadrupole-orbitrap to provide peptide identification accurately with a more precise exact mass of ion (up to five decimals of ion mass). This system is supported with proteome discoverer software to process the peptide sequence accurately and fast (Sitanggang et al., 2020). However, the software could process only for sequences higher than five amino acid residues, but bioactive peptides with lower than six amino acid residues are abundantly found. Thus, the identification of peptides from the mass spectra data using UniProt KB or SwissProt SIB (www.expasy.org/resources/uniprotkb-swiss-prot) followed by FindPept (https://web.expasy.org/findpept/) and the BIOPEP database becomes a challenging effort (Minkiewicz et al., 2008; Tamam et al., 2019).

The current available sophisticated analyses have led bioactive peptides discovery to become a part of proteomic study. The steps to conduct a proteomic study are shown in Figure 3.1. Peptide separation using ultrafiltration followed by a gel filtration chromatography method and isolation by HPLC, prior to LC-MS or LC-MS-MS analysis, could be a choice in the bioactive peptide discovery (Amadou et al., 2013; Tamam et al., 2019). Bioinformatics using a peptide cutter on soy proteins (different soy protein sequences can be provided by UniProt) by complex enzymes derived by some microorganisms involved in soybean fermentation products (https://web.expasy.org/peptide_cutter/) and then connected to BIOPEP is also a challenge in bioactive peptide discovery by *in silico* method (Udinigwe, 2014). This procedure is also used to elucidate an allergenic property of some peptides, an issue of soy protein hydrolysate's safety (Schaafsma, 2009).

3.5 SUMMARY

Bioactive peptides can be produced from soybeans, soybean products (fermented and unfermented), and soy hydrolysates. The bioactivity of the peptides differs depend on the amino acid residue composition or sequence, peptide length, peptide net charge, and hydrophobicity. Peptides with antihypertensive activity have been highly investigated in many different soybean treatments and having various characteristics. Antioxidative peptides have been lesser studied, but many fermented soybeans have shown the antioxidant activities which might be imparted by the peptidic antioxidants. Bioactive peptide isolation and identification have let many researchers use chromatographic techniques linked to LC-MS-MS with and without the use of search engines for peptide bioactivities.

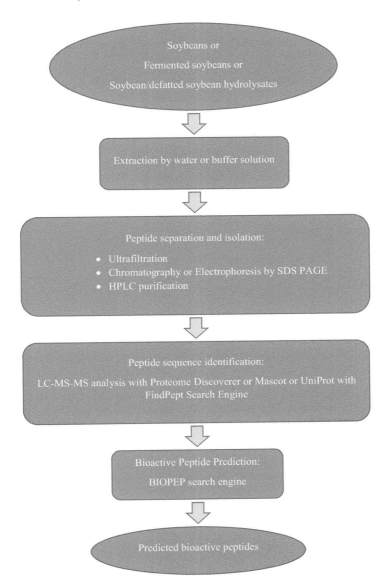

FIGURE 3.1 Bioactive peptides isolation and identification from soybeans and derived products, and their activity prediction through *in silico* by BIOPEP.

REFERENCES

Amadou, I., Le, G.W., Amza, T., Sun, J., Shi, Y.H. (2013). Purification and characterization of foxtail mil-let-derived peptides with antioxidant and antimicrobial activities. *Food Research International*, 51: 422–428.

Boman, H.G. (2003). Antimicrobial peptides: Basic facts and emerging concepts. *Journal of Internal Medicine*, 254(3): 197–215.

Capriotti, A.L. Caruso, G., Cavaliere, C., Saperi, R., Ventura, S., Chiozzi, R.Z., Lagana, A. (2015). Identification of potential bioactive peptides generated by simulated gastrointestinal digestion of soybean seeds and soy milk proteins. *Journal of Food Composition and Analysis*, 44: 205–213.

Carrasco-Castilla, J., Hernández-Álvarez, A.J., Jiménez-Martínez, C., Jacinto-Hernández, C., Alaiz, M., Girón-Calle, J. (2012). Antioxidant and metal chelating activities of *Phaseolus vulgaris* L var. Jamapa protein isolates, phaseolin and lectin hydrolysates. *Food Chemistry*, 131: 1157–1164.

Castro, R.J.S., Sato, H.H. (2015). Biologically active peptides: Processes for their generation, purification and identification and applications as natural additives in the food and pharmaceutical industries. *Food Research International*, 74: 185–198.

Chen, H.M., Muramoto, K., Yamauchi, F. (1995). Structural analysis of antioxidative peptides from soybean beta-conglycynin. *Journal of Agricultural and Food Chemistry*, 43: 574–578.

Chen, H.M., Muramoto, K., Yamauchi, F., Fujimoto, K., Nokihara, K. (1998). Antioxidative properties of histidine-containing peptides designed from peptide fragments found in the digests of a soybean protein. *Journal of Agricultural and Food Chemistry*, 46: 49–53.

Cheung, H.S., Wang, F.L., Ondetti, M.A., Sabo, E.F., Cushman, D.W. (1980). Binding of peptide substrates and inhibitors of angiotensin-converting enzyme. *Journal of Biological Chemistry*, 255: 401–407.

Coscueta, E.R., Amorim, M., Glenise, B.V., Bibiana, B.N., Guillermo, A.P., Manuela, E.P. (2016). Bioactive properties of peptides obtained from Argentinian defatted soy flour protein by Corolase PP hydrolysis. *Food Chemistry*, 198: 36–44.

Dashper, S.G., Liu, S.W., Reynolds, E.C. (2007). Antimicrobial peptides and their potential as oral therapeutic agents. *International Journal of Peptide Research and Therapeutics*, 13: 505–516.

Duan, X., Ocen, D., Wu, F., Li, M., Yang, N., Xu, J., Chen, H., Huang, L., Jin, Z., Xu, X. (2014). Purification and characterization of a natural antioxidant peptide from fertilized eggs. *Food Research International*, 56: 18–24.

Gallego, M., Mora, L., Toldra, F. (2018). Health relevance of antihypertensive peptides in foods. *Current Opinion in Food Science*, 19: 8–14.

Gibbs, B.F., Zougman, A., Masse, R., Mulligan, C. (2004). Production and characterization of bioactive peptides from soy hydrolysate and soy-fermented food. *International Food Research Journal*, 37: 123–131.

Gobbetti, M., Minervini, F., Rizzello, C.G. (2004). Angiotensin I-converting enzyme inhibitory and antimicrobial bioactive peptides. *International Journal of Dairy Technology*, 57: 173–188.

Gomes, L.S., Senna, R., Sandim, V., Silva-Neto, M.A.C., Pirales, J.E.A., Zingali, R.B., Soares, M., Fialho, E. (2014). Four conventional soybean (*Glycine max* (L) Merrill) seeds exhibit different protein profiles as revealed by proteomic analysis. *Journal of Agricultural and Food Chemistry*, 62: 1283–1293.

Haque, E., Chand, R. (2008). Antihypertensive and antimicrobial bioactive peptides from milk protein. *European Food Research Technology*, 227: 7–15.

Herman, E.M. (2014). Soybean seed proteome rebalancing. *Frontiers in Plant Science*, 5: 00437.

Huang, Y., Lai, Y., Chou, C. (2011). Fermentation temperature affects the antioxidant activity of the enzyme-ripened sufu, an oriental traditional fermented product of soybean. *Journal of Bioscience and Bioengineering*, 112: 49–53.

Inoue, K., Gotou, T., Kitajima, H., Mizuno, S., Nakazawa, T., Yamamoto, N. (2009). Release of antihypertensive peptides in miso paste during its fermentation, by the addition of casein. *Journal of Bioscience and Bioengineering*, 108: 111–115.

Iwai, K., Nakaya, N., Kawasaki, Y., Matsue, H. (2002). Antioxidative functions of natto, a kind of fermented soybean: Effect on LDL oxidation and lipid metabolism in cholesterol-fed rats. *Journal of Agricultural and Food Chemistry*, 50: 3597–3601.

Kodera, T., Nio, N. (2006). Identification of an angiotensin I-converting enzyme inhibitory peptides from protein hydrolysates by a soybean protease and the antihypertensive effects of hydrolysates in 4 spontaneously hypertensive model rats. *Journal of Food Science*, 71: 164–173.

Korhonen, H., Pihlanto, A. (2003). Food-derived bioactive peptides-opportunities for designing future foods. *Current Pharmaceutical Design*, 9: 1297–1308.

Kuba, M., Tanaka, K., Tawata, S., Takeda, Y., Yasuda, M. (2003). Angiotensin I-converting enzyme inhibitory peptides isolated from tofuyo fermented soybean food. *Bioscience, Biotechnology and Biochemistry*, 67: 1278–1283.

Mamone, G., Picariello, G., Caira, S., Addeo, F., Ferranti, P. (2009). Analysis of food proteins, and peptides by mass spectrometry-based techniques: Review. *Journal of Chromatography A*, 1216: 7130–7142.

Matsuzaki, K. (1999). Why and how are peptide-lipid interactions utilized for self defense? Magainins and tachyplesins as archetypes. *Biochimica et Biophysica Acta*, 1462: 1–10.

McClean, S., Beggs, L.B., Welch, R.W. (2014). Antimicrobial activity of antihypertensive food-derived peptides and selected alanine analogues. *Food Chemistry*, 146: 443–447.

Minkiewicz, P., Dziuba, J., Iwaniak, A., Dziuba, M., Darewicz, M. (2008). BIOPEP database and other pro-grams for processing bioactive peptide sequences. *Journal of AOAC International*, 91: 965–980.

Nakahara, T., Sano, A., Yamaguchi, H., Sugimoto, K., Chikata, H., Kinoshita, E., Uchida, R. (2010). Antihypertensive effect of peptide-enriched soy sauce-like seasoning and identification of its angio-tensin I-converting enzyme inhibitory substances. *Journal of Agricultural and Food Chemistry*, 58: 821–827.

Okamoto, A., Hanagata, H., Matsumoto, E., Kawamura, Y., Koizumi, Y., Yanadiga, F. (1995). Angiotensin I converting enzyme inhibitory activities of various fermented foods. *Bioscience, Biotechnology and Biochemistry*, 59: 1147–1149.

Ortiz-Martinez, M., Winkler, R., Garcia-Lara, S. (2014). Preventive and theurapeutical potential of peptides from cereals against cancer. *Journal of Proteomics*, 111: 165–183.

Puchalska, P., Garcia, M.C., Marina, M.L. (2014). Identification of native angiotensin-1 converting enzyme inhibitory peptides in commercial soybean based infant formulas using HPLC-Q-ToF-MS. *Food Chemistry*, 157: 62–69.

Roblet, C., Amiot, J., Lavigne, J., Marette, A., Lessard, M., Jean, J., Ramassamy, C., Moresoli, C., Bazinet, L. (2012). Screening of in vitro bioactivities of a soy protein hydrolysate separated by hollow fiber and spiral-wound ultrafiltration membranes. *International Food Research Journal*, 46: 237–249.

Rho, S.J., Lee, J.S., Chung, Y.I., Kim, Y.W., Lee, H.G. (2009). Purification and identification of an angiotensin I-converting enzyme inhibitory peptide from fermented soybean extract. *Process Biochemistry*, 44: 490–493.

Sanjukta, S. Rai, A.K. (2016). Production of bioactive peptides during soybean fermentation and their poten-tial health benefits. *Trends in Food Science and Technology*, 50: 1–10.

Sanjukta, S., Rai, A.K., Muhammed, A., Jeyaram, K., Talukdar, N.C. (2015). Enhancement of antioxidant properties of two soybean varieties of Sikkim Himalayan region by proteolytic *Bacillus subtilis* fermen-tation. *Journal of Functional Foods*, 14: 650–658.

Sarmadi, B.H., Ismail, A. (2010). Antioxidative peptides from food proteins: A review. *Peptides*, 31: 1949–1956.

Sato, K., Miyasaka, S., Tsuji, A., Tachi, H. (2018). Isolation and characterization of peptides with dipepti-dyl peptidase IV (DPPIV) inhibitory activity from natto using DPPIV from *Aspergillus oryzae*. *Food Chemistry*, 261: 51–56.

Schaafsma, G. (2009). Safety of protein hydrolysates, fractions thereof and bioactive peptides in human nutri-tion. *European Journal of Clinical Nutrition*, 63: 1161–1168.

Shin, Z.I., Yu, R., Park, S.A., Chung, D.K., Ahn, C.W., Nam, H.S., Kim, K.S., Lee, H.J. (2001). His-His-Leu, an angiotensin I converting enzyme inhibitory peptide derived from Korean soybean paste, exerts anti-hypertensive activity in vivo. *Journal of Agricultural and Food Chemistry*, 49: 3004–3009.

Singh, B.P., Vij, S., Hati, S. (2014). Functional significance of bioactive peptides derived from soybean. *Peptides*, 54: 171–179.

Sitanggang, A.B., Lesmana, M., Budijanto, S. (2020). Membrane-based preparative methods and bioactivities mapping of tempe-based peptides. *Food Chemistry*, 329: 127193.

Tamam, B., Syah, D., Lioe, H.N., Suhartono, M.T., Kusuma, W.A. (2018). Sequence-based identifiers to rec-ognize the functionality of bioactive peptides: An explorative study. *Journal of Food Technology and Industry (Jurnal Teknologi dan Industri Pangan)*, 29: 1–9.

Tamam, B., Syah, D., Suhartono, M.T., Kusuma, W.A., Tachibana, S., Lioe, H.N. (2019). Proteomic study of bioactive peptides from tempe. *Journal of Bioscience and Bioengineering*, 128: 241–248.

Tsai, J.S., Lin, Y.S., Pan, B.S., Chen, T.J. (2006). Antihypertensive peptides and -aminobutyric acid from prozyme facilitated lactic acid bacteria fermentation of soymilk. *Journal of Process Biochemistry*, 41: 1282–1288.

Tsou, M.J., Kao, F.J., Lu, H.C., Kao, H.C., Chiang, W.D. (2013). Purification and identification of lipolysis-stimulating peptides derived from enzymatic hydrolysis of soy protein. *Food Chemistry*, 138: 1454–1460.

Udinigwe, C.C. (2014). Bioinformatic approaches, prospects and challenges of food bioactive peptide research. *Trends in Food Science and Technology*, 36: 137–143.

Vallabha, V., Tiku, P.K. (2014). Antihypertensive peptides derived from soy protein by fermentation. *International Journal of Peptide Research and Therapeutics*, 20: 161–168.

Wang, D., Wang, L.J., Zhu, F.X., Zhu, J.Y., Chen, X.D., Zou, L., Saito, M., Li, L. (2008). In vitro and in vivo studies on the antioxidant activities of the aqueous extracts of Douchi (a traditional Chinese salt-fermented soybean food). *Food Chemistry*, 107: 1421–1428.

Wang, W., de Mejia, D.E.G. (2005). A new frontier in soy bioactive peptides that may prevent age-related chronic diseases. *Comprehensive Reviews in Food Science and Food Safety*, 4: 63–78.

Wijesekara, I., Qian, Z., Ryu, B., Ngo, D., Kim, S. (2011). Purification and identification of antihypertensive peptides from seaweed pipefish (*Syngnathus schlegeli*) muscle protein hydrolysate. *Food Research International*, 44: 703–707.

Yang, J.H., Mau, J.L., Ko, P.T., Huang, L.C. (2000). Antioxidant properties of fermented soybean broth. *Food Chemistry*, 71: 249–254.

Yang, Y., Boysen, R.I., Chowdhury, J., Alam, A., Hearn, M.T.W. (2015). Analysis of peptides and protein digests by reversed phase high performance liquid chromatography–electrospray ionisation mass spectrometry using neutral pH elution conditions. *Analytica Chimica Acta*, 872: 84–94.

4 Antioxidant Activity and Health Benefits of Anthocyanin of Black Soybeans

Ignasius Radix A.P. Jati

CONTENTS

4.1 INTRODUCTION

Soybean (*Glycine max* [L.] Merrill) is one of the world's most important agricultural commodities. The production of soybean globally reaches 362.76 million metric tons (USDA, 2020). The United States, Brazil, and Argentina are the highest producers of soybean. Approximately 68% of global soybean production is used to meet demand in the food sector.

Soybean belongs to the family *Fabaceae*, genus *Glycine*, and subgenus *Soja* (Moench). *Glycine soja* is the name of wild soybean founded in China and neighboring countries. The domesticated soybean is known as *Glycine max* (L.) Merrill. The seed coats have different colors, such as the most commonly grown yellow, green, and black seed coats.

According to Kumudini et al. (2008), the structures of the soybean plant leaf are characterized as the seed (cotyledon) leaves, the primary (unifoliolate) leaves, the trifoliolate leaves, and the prophylls. Meanwhile, the shape of the mature soybean seed is oval and consists of a seed coat surrounding a large embryo. Even though the planting properties are similar to those of the yellow soybean, in that it can be planted in various well-drained soils, favoring a slightly acidic soil (pH 6.0–6.5); needs a salinity threshold of approximately 5 ds/m; and needs the temperature between 10° C and 40° C during the growing season, thus being considered a short-day plant, the black soybean is reported to be more resistant to disease and environmental stress (Lee et al., 2020).

In recent years, the popularity of black soybean is increasing rapidly due to its health properties. Due to its similar characteristics with common yellow soybean, various products can be made from black soybean, for example, vegetable oil and its derivatives, such as margarine, salad dressing, and mayonnaise. In addition, black soybean can also be used as an alternative to meat or animal-based

DOI: 10.1201/9781003030294-4

protein, as a stabilizer in restructured products such as nuggets and sausages, and as a meat-mimicking food in the vegetarian diet.

4.2   UTILIZATION OF BLACK SOYBEAN

Black soybean is one of the soybean varieties and has a dark black seed coat color. Like the yellow soybean variety, black soybean originated from East Asia and began to be domesticated in the period of the Shang Dynasty (1700–1100 BC) in the northern part of China. Based on ancient inscriptions, soybean is one of the five sacred commodities with rice, wheat, millet, and adzuki beans. In ancient times, black soybean was not consumed as food; however, it was widely used as remedies by traditional healers to treat various diseases such as weakness, dizziness, headaches, and digestion problems. Therefore, besides its essential function for consumption, soybean is included as a sacred grain due to its additional value as a remedy to cure various diseases. In ancient times, black soybean could only be consumed by noble families as part of traditional ceremonies. A picture of black soybean is presented in Figure 4.1.

Similar to in China, the utilization of black soybean after spreading to South East Asia, especially Indonesia, is a part of worshipping gods, known as *sesajen*. *Sesajen* is a compulsory traditional gift that is believed to be given by humans to the one who possesses almighty power and rules all living creatures in the world. *Sesajen* consists of various grains, vegetables, fruits, and also animal-based foods. *Sesajen* is available in various ceremonies such as births, birthdays, weddings, funerals, and other socio-cultural ceremonies. The aims of preparing *sesajen* are to seek safety and protection from gods for all the members of society to live in harmony and prosperity. Through being domesticated, the popularity of black soybean has been increasing and is followed by a number of various black soybean–based products, both daily consumption food products and healthy diet food products. In general, black soybean food products are divided into two major categories, which are fermented and non-fermented products. Examples of fermented black soybean products are tempeh, natto, black soybean paste, and soy sauce. Meanwhile, some non-fermented black soybean products are tofu, soy milk, and soy protein isolate. Recent progress in research on the health benefits of black soybean has led to the development of various modern and innovative products such as black soybean tea, black soybean noodle, spaghetti, cookies, and also black soybean drink. Examples of black soybean-based products can be seen in Table 4.1. Meanwhile, pictures of

FIGURE 4.1   Black soybean (*Glycine max* [L.] Merrill).

TABLE 4.1

Black Soybean-Based Products

Groups	Example	Countries of Production
Traditional fermented product	• *In si, tau si* (Dried by-product of the mashed black soybean sauce fermented with *Aspergillus oryzae*)	China
	• Natto (Traditional Japanese soybean product fermented with *Bacillus subtilis*)	Japan
	• Soy sauce (Sauce fermented with *Aspergillus oryzae* and *Aspergillus soyae*, used as a condiment)	Asian countries
	• Tempeh (Traditional food from black or yellow soybean fermented with *Rhizopus oligosporus*)	Indonesia
	Cheonggukjang, doenjang (Steamed black soybeans fermented with *Bacillus* species)	Korea
Traditional non-fermented product	• Tofu (Protein gel-like product from soybean)	Asian countries
	• Soy milk (Soybean-based beverage made by soaking and grinding the soybean, boiling the mixture, and filtering the large-sized particles)	Worldwide
Newly developed commercial product	• Black soybean tea	Japan, Korea
	• Black soybean spaghetti	United States
	• Black soybean snack	Korea

Source: Modified from Harlen and Jati (2018).

black soybean products are presented in Figure 4.2. The rapid progress of the black soybean–based food products market is possibly due to the contribution of the black soybean's health properties. Consumers believe that consuming black soybean will provide a better health condition, which has been done for centuries by their ancestors. Moreover, the traditional belief has been supported by extensive research on the bioactive compounds of food plants, which can inhibit the onset of various degenerative diseases.

4.3 ANTHOCYANIN

Bioactive compounds are substances from food sources commonly consumed by animals and humans that are available in trace amounts and possess biologically active properties, which could affect physiological functions and cellular activities. Consuming bioactive compounds could give health benefits, both as food intake, which provides energy and other essential nutrients, and as remedial agents that contribute to the reduction of inflammation, decrease the rate of oxidative stress, and normalize metabolic disorder (Siriwardhana et al., 2013). The health effects of a high intake of bioactive compounds through the consumption of varieties of plant foods have long been known. For example, the most popular Mediterranean diet, which is based on traditional dietary and lifestyle habits in the Mediterranean region adapted to the new modern lifestyle diet, successfully exhibits potency in reducing the incidence of various degenerative diseases such as cancer, heart disease, stroke, Alzheimer's, diabetes, cataracts, and age-related functional degeneration (Hassimotto, Genovese, & Lajolo, 2009; Siriwardhana et al., 2013). The advancement of research and the awareness of a healthy diet has led to the discoveries and isolation of numerous bioactive compounds from plants such as polyphenolic compounds, including anthocyanin.

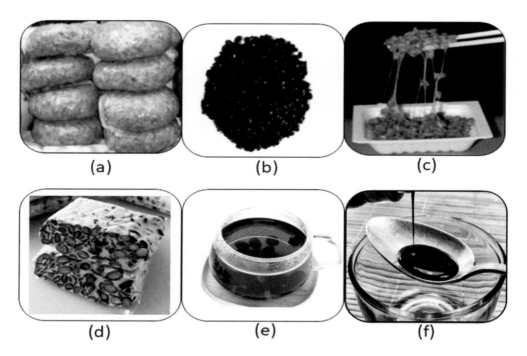

FIGURE 4.2 Examples of some commercial and non-commercial black soybean–based products: (a) *Cheonggukjang*, (b) *douchi*, (c) sweet soy sauce, (d) tempeh, (e) black soybean tea, (f) black soybean snack.

Anthocyanin is a water-soluble pigment containing the substances responsible for the formation of red, blue, and black colors in flowers and any other parts of the plant. Anthocyanin is a secondary plant metabolite included in the polyphenol group because it contains the phenolic ring in its chemical structure. The structures of anthocyanin and cyanidin 3 glucoside (C3G) as the most abundant anthocyanin found in plants are presented in Figure 4.3.

The production of anthocyanin by plants has several biological functions, such as attracting pollinators and frugivores. Anthocyanin plays a critical task in attracting pollinators and frugivores using their appealing color. Thus, pollination can be conducted. On the other hand, the anthocyanin colors also act as a repellent for herbivores and parasites. They provide a signal for herbivores and parasites that the plants contain toxic substances or signal a negative impression of unpalatable food. Anthocyanin could also contribute to plants as camouflage, a defensive mechanism to protect itself from insects and any other destructive organism.

Among several functions of anthocyanins in plants, the most investigated is the ability to act as an antioxidant with sunscreen properties due to the function of anthocyanins in protecting leaves in plants facing unfavorable conditions, such as various stressors. In a stress condition, the metabolism of the plant will be in an unbalanced state, thus resulting in an excessive oxidation rate. The anthocyanin plays an essential role as an antioxidant, which could help stabilize the reactive oxygen species due to its keen ability to act as an antioxidant in the plant system. Numerous investigators postulate that anthocyanin could also become a contributor to human health in the form of fruits, vegetables, and legumes, including black soybean rich in anthocyanins, which are consumed in the diet.

A study on anthocyanin formation in black soybean plants was first reported by Nagai (1921). Black soybeans contain a high content of anthocyanins in their seed coats. Various reports have been published in the determination of the anthocyanin content of black soybean seed coats. In agreement, the previously published research revealed that the anthocyanin content of black soybean seed coat is equal to other accessible sources of anthocyanin such as blueberry, blackberry, and

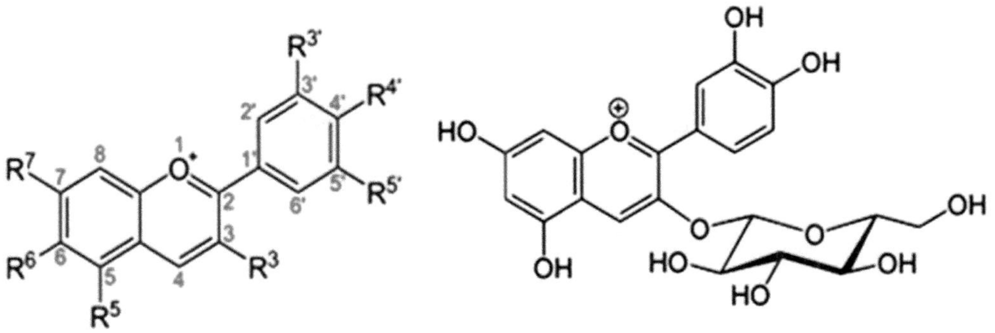

FIGURE 4.3   Chemical structure of anthocyanin (left) and cyanidin 3 glucoside (right).

TABLE 4.2

Anthocyanin Content of Black Soybean and Black Soybean Products

Black Soybean Varieties	Sources	Total Anthocyanin Content (mg/G)	References
Mallika	Indonesia	13.63	Astadi et al., 2009
Cikuray	Indonesia	14.68	Astadi et al., 2009
Cheongja 3	Korea	12.11	Jang et al., 2010
A3	Sichuan, China	3.95	Wu et al., 2017
QWT31	Yunnan, China	4.96	Wu et al., 2017
QWT5	Guizhou, China	3.01	Wu et al., 2017
JJ16	Chongqing, China	3.62	Wu et al., 2017
Black Tokyo	Serbia	1.92	Kalusevic et al., 2017
Cheongja 4 ho	Miryang, Korea	1.68	Ryu & Koh, 2018
852	Heilongjiang, China	6.96	Xie et al., 2018

Source: Jati (2020).

grapes. Some of the newest reports on the anthocyanin content of black soybean seed coats are presented in Table 4.2. It is shown that there were differences observed in anthocyanin content among black soybean varieties due to the variety of species, climatic conditions, and also geographical location. In addition to the anthocyanin content, a number of studies were performed to elucidate the individual anthocyanin of the black soybean seed coat. Such research mainly aims to investigate the prevalent individual anthocyanin found in the black soybean seed coat. Thus, in-depth exploration of the mechanism of anthocyanin's health properties, such as the capability of anthocyanin to inhibit the oxidation process and the role of anthocyanin in combatting degenerative diseases, could be investigated. The number of publications investigating the individual anthocyanin in black soybean seed coat is presented in Table 4.3. As shown in Table 4.3, the most common and abundantly found individual anthocyanin in black soybean seed coat is cyanidin. Meanwhile, other anthocyanins such as delphinidin, peonidin, malvidin, petunidin, and pelargonidin were also present in the black soybean seed coat. However, the concentration of individual anthocyanin depends on the black soybean plant varieties.

4.3.1   Cyanidin 3 Glucoside (C3G)

C3G is the most prominent anthocyanin found in black soybean. Besides its abundant presence, numerous studies have suggested that C3G is the main compound responsible for anthocyanin's

TABLE 4.3

Individual Anthocyanins of Black Soybean

Black Soybean Varieties	Source	Individual Anthocyanin	References
Cheongja 3	Korea	Cyanidin-3-O-glucoside, petunidin-3-O-glucoside, delphinidin-3-O-glucoside	Jang et al., 2010
A3	Sichuan, China	Cyanidin 3 glucoside, petunidin 3 glucoside, delphinidin 3 glucoside, peonidin 3 glucoside	Wu et al., 2017
Black Tokyo	Serbia	Cyanidin 3 glucoside, pelargonidin 3 glucoside, delphinidin 3 glucoside	Kalusevic et al., 2017
Cheongja 4 ho	Miryang, Korea	Cyanidin-3-O-glucoside, petunidin-3-O-glucoside, delphinidin-3-O-glucoside	Ryu & Koh, 2018
852	Heilongjiang, China	Cyanidin 3 glucoside	Xie et al., 2018

Source: Jati (2020).

beneficial health properties. Matsukawa et al. (2015) investigated the antidiabetes effect of C3G from black soybean on mice. It shows that exposure of adipocytes to C3G induces the differentiation of 3T3-L1 preadipocytes into smaller, insulin-sensitive adipocytes, which induced skeletal muscle metabolism. Another study on rats with breast cancer indicated that the isolate of C3G could inhibit cancer cells' development through the increase of apoptosis process activation (Cho et al., 2017). Meanwhile, dietary C3G significantly reduced body weight gain by enhancing energy expenditure, maintained glucose homeostasis, and increased insulin sensitivity in the obese mice by upregulating brown adipose tissue (BAT) mitochondrial function (You et al., 2017).

From previously published research, the beneficial health properties of C3G are postulated to be due to its radical scavenging capacity, epigenetic action, competitive protein-binding, and enzyme inhibition; thus it could act as an antioxidant and have several anti–degenerative disease capacities. The capability of C3G to act as an antioxidant is believed to be due to the two hydroxyl groups on the B ring that can donate their hydrogen atoms to stabilize free radicals (Khoo et al., 2017). Meanwhile, the activity of C3G to inhibit cancer formation and progression is due to its epigenetic action. C3G can perform epigenetic modification to regulate gene expression in various cancer cells (El-Ella & Bishayee, 2019). Moreover, the protein binding properties and enzyme inhibition capacity of C3G have been previously reported. C3G has a strong capability to bind with macromolecules, such as protein (Wiese et al., 2009). Therefore, in the metabolism system, C3G could act as an enzyme inhibitor by binding to the enzyme's active site (Balasuriya & Rupasinghe, 2011; Bräunlich et al., 2013; Sui et al., 2016). The capability to inhibit the work of enzymes is the reason behind the antidiabetic, hypolipidemic, antihypertension, and other metabolism-related disease inhibition capacities of C3G. For example, it uses glucosidase and amylase enzyme inhibition in converting carbohydrates to glucose and inhibiting hypertension-related enzymes such as an angiotensin-converting enzyme.

4.4   ANTIOXIDANT ACTIVITY AND HEALTH BENEFITS

The number of reports concerning antioxidant activity and its health benefits has been increasing rapidly in the last decades. This condition is related to the progressive rate of incidence of various diseases such as Alzheimer's, cancer, cardiovascular diseases, and diabetes. The changes of traditional healthy to modern unbalanced lifestyles in terms of workplace stress, quality of food intake, dietary habits, and environmental pollution are believed to play a crucial role in the occurrence of such diseases. An unhealthy lifestyle contributes to human metabolism by creating an unbalanced status and increasing susceptibility to the onset of various diseases. For example, oxygen metabolism, which is an ordinary process under normal circumstances to generate reactive oxygen species (ROS), could shift to excessive production of ROS as a response from the body to the abnormal oxidation process. The ROS production, which is usually used for cell signaling and homeostasis, had become uncontrollable. Therefore, it is also called free radicals.

Free radicals tend to attack other molecules in order to be stable. DNA, lipids, and proteins are the most vulnerable substances in the presence of free radicals. This process is suggested to be the start of various diseases' development. Free radicals can be stabilized by substances known as antioxidants through different pathways, such as donating their hydrogen to scavenge free radicals, known as a primary antioxidant, thus breaking the chain reaction, and also decomposing hydroperoxide radicals into non-reactive substances, known as the secondary antioxidant. The human metabolism system has its defense mechanism against free radicals through the numbers of enzymes with antioxidant capacity called indigenous antioxidants such as catalase, superoxide dismutase, and glutathione peroxidase.

The rate of ROS production in the human body due to environmental stress, however, could not be managed by indigenous antioxidants alone. Therefore, an exogenous antioxidant from various sources is needed. Intake of fruits, vegetables, and legumes, which for centuries have been known as healthy food, becomes the researcher's focus to explore the substances responsible for the health effects of such commodities. Among many plants, black soybean is rich in anthocyanin, which could act as a free radical scavenger, having anti-inflammatory, anticancer, and anti-atherosclerosis activity, the ability to prevent coronary heart disease, and antidiabetic and anti-obesity activity.

4.4.1 Free Radical Scavenging Activity

A free radical is defined as an unstable substance due to its unpaired electron configuration. The incidence of various diseases is believed to be caused by free radicals which reactively attack molecules in the human system such as DNA, protein, and lipids. Free radicals can be stabilized by antioxidants through the hydrogen atom donation or free radical scavenging process. Antioxidants can rapidly donate their hydrogen atoms to free radicals, stabilize, and thus terminate the chain reactions. Antioxidant compounds such as anthocyanin have high free radical scavenging activity. The action mechanism of anthocyanin as an antioxidant is available in Figure 4.4. It can be seen that anthocyanin can act as a hydrogen donor that could stabilize free radicals, and is thus called a radical scavenger. Also, anthocyanin could react with hydroperoxide to yield a non-radical product. Different methods have been developed to examine their activities due to the vital function of antioxidants as a free radical scavenger. The examples of the methods are DPPH (2,2-diphenyl-1-picrylhydrazyl), FRAP (ferric reducing antioxidant power), hydroxy radical scavenging activity, superoxide anion radical scavenging, and ABTS (2,2'-Azino-bis[3-ethylbenzthiazolin-6-sulfonic acid]).

Due to the importance of free radical scavenging assays, combined methods were commonly provided by researchers in their published reports in order to ensure that the substances examined were showing similar trends in free radical scavenging activities using different assay protocols. Numerous studies of black soybean free radical scavenging activities have been published. Such research spreads from the exploration of raw black soybean seeds to black soybean–based food products. Moreover, different processing methods, as well as geographical regions, were also widely investigated. A report by Astadi et al. (2009) examines the antioxidant activity of black soybean seed coat of the Mallika and Cikuray variety using the DPPH method. The result shows that the extract of both varieties could scavenge more than 90% of DPPH radicals. The black soybean Mallika variety is mainly utilized to produce sweet soy sauce products in Indonesia. A study from China by Zhang et al. (2011) on the radical scavenging capacity of 60 different varieties of black soybean revealed that antioxidant properties detected by DPPH, FRAP, and Oxygen Radical Absorbance Capacity (ORAC) methods all showed wide variations ranging from 4.8 to 65.3 µg/100

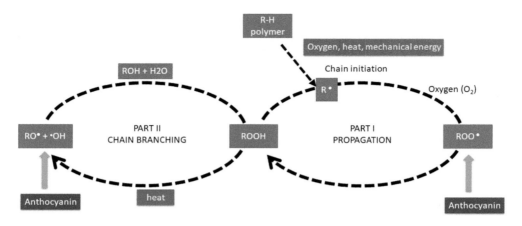

FIGURE 4.4   Antioxidative mechanism of anthocyanin.

mL (expressed as half-maximal effective concentration/EC_{50}), from 17.5 to 105.8 units/g, and from 42.5 to 1,834.6 µmol Trolox equivalent/g, respectively. Thus, this finding is scientifically supported by the traditional belief of the Chinese in using black soybean as an herb for the treatment of various diseases. The high content of the antioxidant compound could be the main reason that black soybean has beneficial properties for human health. Although black soybean is mostly popular in Asian countries, research on black soybean antioxidant activity is also reported from the black soybean grown in Central Europe. Two varieties of black soybean from Serbia were examined by Malencic et al. (2012) and compared with common yellow soybean and other colored soybeans. It is shown that both varieties of black soybean had higher total polyphenols, anthocyanin, and flavonoid contents than other varieties. In accordance, the free radical scavenging activity determination using the DPPH method exhibits a positive correlation between the total polyphenol content and the ability to scavenge DPPH radicals. Meanwhile, a recent study by Lee et al. (2020) on 172 samples of black soybean landrace in Korea shows that all of the samples exhibit free radical scavenging activity using DPPH, ABTS, and FRAP methods. The majority of published research with a wide variety of black soybean cultivars is from Korea. Compared to other countries, black soybean is a common food ingredient in Korea. Traditional food such as black soybean paste is a top-rated food product among Koreans. Meanwhile, in other countries, soybean-based food products mostly use yellow-colored soybean.

The exploration of the health benefits of soybean, especially the black-colored soybean, played an essential role in the increase of black soybean utilization as a top-rated product because soybean in Western countries was mainly explored for the oil content and also used for non-food applications. The continuous research on black soybean seeds' beneficial properties leads to the awareness of essential methods of processing used as one of the crucial steps before consumption. Black soybean processing procedures have been in the spotlight as a critical factor in maintaining the bioactive compound content and antioxidant capacity of black soybean products. A number of studies investigated the effect of different processing methods on the bioactive compounds and antioxidant activity of black soybeans, such as the production of soy milk, black soybean beverage, fermented products such as *cheonggukjang*, tempeh, natto, miso, yogurt, *petit suisse*, tofu, roasted black soybean, and germinated black soybean. Investigation of black soybean soy milk was performed by Xu and Chang (2009), which compared traditional processing with the modern ultrahigh temperature (UHT) method. The result suggested that the antioxidant activity examined by DPPH and FRAP methods was higher in the UHT and traditional processing methods than in the raw soy milk. This result could be due to the increase of the flavonoid content of black soybean soy milk resulting from the heating process.

On the other hand, the total phenolic content was decreased. Thus, it can be proposed that flavonoids, including anthocyanin, are responsible for black soybean soy milk antioxidant activity. Meanwhile, Ma and Huang (2014) investigated the bioactive compound and antioxidant activity of different soybean-based milk, including black soybean milk. The result revealed that black soybean milk had higher antioxidant activity and total phenolic content than yellow soybean milk. The black seed coat of black soybean contributes to the high content of phenolic as well as anthocyanin. Therefore, the antioxidant capacity was higher compared to the yellow soybean milk. A study by Tan et al. (2016) produced black soybean soy milk with different grinding methods. The result shows that even though black soybeans had gone through several steps of processing to become soy milk, the soy milk product maintained its antioxidant activity to scavenge free radicals. The heating process, which is widely accepted as a factor contributing to the significant decrease of various bioactive compounds and antioxidant activities of food products, only slightly reduced the ability to scavenge free radicals of cooked soy milk.

Another popular soybean-based food product is tofu, a yellow soybean–based product manufactured by the curdling process of soy proteins. Due to the increased focus on the health properties of black soybean, it is common to produce tofu using black soybean. An investigation by Shih et al. (2002) revealed that black soybean tofu exhibits higher antioxidant potential in inhibiting the formation of peroxide compared to the common yellow soybean. There was no significant difference observed in the antioxidant activity of the black soybean and black soybean tofu. This result is possibly due to the combined action by anthocyanin, phenolic, isoflavone, and other bioactive compounds, including peptides in tofu.

The development of black soybean products has been expanding from traditional products to new and popular products such as black soybean tea, spaghetti, and crackers. Research on the bioactive compounds and antioxidant activity of black soybean-based crackers shows that the bioactive compound content and antioxidant activity of black soybean crackers is higher than that of the yellow soybean crackers (Slavin et al., 2013). The high content of anthocyanin in black soybean played a crucial role in maintaining the antioxidant activity of crackers even though a decrease was observed on phenolic and anthocyanin. Moreover, the contribution to the antioxidant activity of crackers could be due to the isoflavone content, which was not significantly affected by the heating process. This research suggested that moderate temperature in crackers manufacture is needed to retain its phenolic and anthocyanin contents.

The effect of roasting in producing black soybean snacks and beverages was also investigated. Shen et al. (2019) performed research on the effect of roasting on the antioxidant activity of small black soybean. Small black soybean is known as a remedy or herb for the traditional treatment of diseases. Unlike the common fact that the heating process will reduce the bioactive compound and antioxidant activity, the roasting process increases the phenolic content and antioxidant activity measured by DPPH and ABTS. The release of phenolic compounds from its matrices by roasting is believed to be responsible for the result. Thus, the reaction with DPPH and ABTS resulted in higher antioxidant activity than the unroasted black soybean. On the other hand, a study by Zhou et al. (2017) revealed that the pre-treatment process of soaking before roasting on black soybean could decrease the antioxidant activity as measured by DPPH, ABTS, and FRAP methods. The soaking process is responsible for the leaching of anthocyanin and phenolic compounds, thus decreasing its antioxidant activity. Based on the fact that roasted black soybean is popular to be consumed as a snack and health food supplement, the preparation and cooking process should be done cautiously.

Another popular method to utilize black soybean, especially as a medicinal food, is to germinate the black soybean to become a black soybean sprout. Many studies show that germination increases the beneficial properties of legumes due to the increased rate of enzyme activities during germination, which leads to the leaching of various nutritional and functional compounds from their matrices. Research by Kumari et al. (2015) revealed that germination could increase the antioxidant activity of black soybean sprouts by approximately 403% due to the release of anthocyanin, phenolic, and isoflavone, as well as the vitamin C content of black soybean due to enzyme activity during germination. A similar trend was observed by Lee et al. (2018) using DPPH and ABTS methods. The seventh day of germination yielded the highest ability to act as an antioxidant. Nevertheless, it was suggested that the degraded polysaccharides with an additional hydroxyl group, especially uronic acid, were responsible for increasing antioxidant activity due to the capability to scavenge free radicals during germination.

Meanwhile, fermentation has long been known as one of the oldest methods of food processing. In the beginning, fermentation was conducted to preserve foods. Several published studies suggested a strong correlation between consuming fermented food products and the inhibition of various diseases (Pala et al., 2011; Kriss et al., 2018; Gille et al., 2018; Song & Giovannucci, 2018). The most recent development of soybean-based food products is fermented soy milk, and unlike the other traditional soy food products, fermented soybean milk is believed to be originated from Western countries (Shurtleff & Aoyagi, 2004). Lee et al. (2015) investigated the effect of fermentation of black soybean soy milk using *Lactobacillus acidophilus* ATCC 4356, *Lactobacillus plantarum* P8, and *Streptococcus thermophilus* S10 as starter cultures on the bioactive compound and antioxidant activity of soy milk. The result shows that the phenolic compound of fermented black soybean milk was higher compared to the non-fermented soy milk. Fermentation using *Streptococcus thermophilus* resulted in the highest phenolic compound compared to other starter cultures. In support of the increase of phenolic content by fermentation, the antioxidant activity of fermented soy milk was higher than that of non-fermented soy milk, as examined by the DPPH method. The higher content of phenolic compounds in fermented soy milk, which contributes to the higher antioxidant activity, could be because fermentation will decompose the substrate of the fermented product, which also breaks the matrix of foods, thus resulting in the release of phenolics content. Moreover, the antioxidant activity of fermented soy milk was also contributed by the isoflavone aglycone, which increased due to fermentation. This result is in line with previous work conducted by Cheng et al. (2013), which examined the antioxidant activity of black soybean milk extract using the DPPH method. It is reported that fermentation of black soybean milk by *Rhizopus oligosporous* exhibits higher capability in scavenging DPPH radicals due to the breakdown of the matrix in the soy milk. Therefore, the bioactive compound was released thus could donate its hydrogen atoms to stabilize the DPPH radicals.

Other black soybean food-based products reported their antioxidant activity in black soybean yogurt fermented by *L. delbrueckii* subsp. *bulgaricus* 1.1480 (Lb) (þ) and *S. thermophilus* ys14(St) (þ). The black soybean yogurt had higher antioxidant activity measured by DPPH and FRAP methods than cow milk (Ye et al., 2013). Moreover, Moraes-Filho et al. (2014) conducted research on the antioxidant activity of black soybean cheese manufactured using a mixture of *Lactobacillus acidophilus*, *Bifidobacterium animalis* subsp. *lactis*, *Lactobacillus delbrueckii* subsp. *bulgaricus*, and exopolysaccharide (EPS) producer *Streptococcus thermophilus*. It shows that black soybean milk and its cheese product had high phenolic and isoflavone content, which resulted in the high antioxidant activity examined by DPPH and ABTS methods. The fermentation process is believed to play an essential role in the hydrolysis of isoflavone and the release of the phenolic compound from the food matrix, thus affecting the antioxidant activity of the product.

Other black soybean fermented products that are scientifically reported for their antioxidant activity are black soybean fermented paste, which is commonly used in the daily dishes of Korea. Examples of black soybean fermented paste are *chunjang*, *doenjang*, *daemaekjang*, and *cheonggukjang*. A report by Kwak et al. (2007) indicated that black soybean paste *cheonggukjang* shows a stronger antioxidant activity in scavenge DPPH radicals and inhibits lipid peroxidation compared to unfermented steamed soybean. The antioxidant activity of black soybean paste was positively correlated with the increase of phenolic and isoflavone aglycone and malonylglycoside contents in the fermented paste. This finding was then confirmed by Hwang et al. (2013), which investigated two varieties of Korean black soybean as the main ingredients to produce *cheonggukjang*, using potential probiotic *Bacillus subtilis* CSY191 for their antioxidant activities. The result shows that fermentation could increase the free radical scavenging activities of *cheonggukjang* examined by

DPPH and ABTS methods. During fermentation, the isoflavone aglycone and malonylglycoside were increased, which is believed to contribute to the higher antioxidant activity of *cheonggukjang*. Meanwhile, a published study of *doenjang* was done by Kim et al. (2009). *Doenjang* was made from black soybean fermented using *Bacillus subtilis*. The result revealed that black soybean fermented paste *doenjang* exhibits higher antioxidant activity and phenolic compounds compared to the unfermented black soybean. The fermentation process is responsible for the increase in the beneficial properties of *doenjang*. The maximum level of phenolic and antioxidant activity is observed for 110 days of fermentation.

Due to the popularity of the food fermentation process, fermented soybean products are widely developed. In China, *douchi* is one of the traditional black soybean fermented products which are also commercially available. A comprehensive report conducted by Xu et al. (2015) investigates 28 commercially available soybean-based fermented products. Among all the samples, black soybean *douchi* products show the highest antioxidant activity examined by the DPPH method. This could be due to the conversion of isoflavone glycoside to their aglycone form. Moreover, it was presented that there was an increase of essential amino acids, which could be used as an indicator of the availability of bioactive peptides that can also act as antioxidants. Research from Japan, conducted by Jiang et al. (2019), investigated the antioxidant activity of black soybean supplemented in rice miso. The result shows that the products have high antioxidant activity and peptides content. The fermentation process could degrade the amino acids to their smaller peptides, which provides bioactive properties in the inhibition of the oxidation process.

The fact that black soybean and its products are provided free radical scavenging capacity is widely acknowledged. Besides, the promising abilities of black soybean seed and its products were also clearly observed. However, the *in vitro* free radical scavenging examinations using reagents are not sufficient to reach an agreement on the health benefit effects of the black soybean. An in-depth investigation is needed using various tests *in vitro* as well as *in vivo* using animal and human studies on health effects such as anti-inflammatory, anticancer, anti-atherosclerosis and coronary heart disease, antidiabetic, and anti-obesity activity.

4.4.2 Anti-Inflammatory and Anticancer Activity

Among several health property investigation methods, anti-inflammatory and anticancer activities from the natural compound are the most commonly examined. Inflammation has been widely investigated because it is associated with various types of diseases, for example, cancer, atherosclerosis, arthritis, and allergy. Early work on the anti-inflammatory effect of black soybean, especially its anthocyanin, was performed by Nizamutdinova et al. (2009), which suggested that anthocyanin plays an important role in the inhibition of pro-inflammatory cytokines and also stimulates wound healing in fibroblasts and keratinocytes. As postulated by Wang et al. (2013), inflammation is a natural biological process conducted by the human body in response to the abnormal condition of infection, irritation, or other injuries. The mechanism of anti-inflammatory activities of natural products is widely investigated since natural products or extracts have been commonly used to treat patients with inflammatory symptoms since ancient times. Inflammation is a process when the immune system responds to abnormal conditions by releasing pro-inflammatory cytokines such as interleukin (IL)-1b, IL-6, and tumor necrosis factor-alpha (TNF-a) sequentially. These pro-inflammatory cytokines' production should be inhibited to prevent or reduce the risk of inflammatory disease incidence. The inducible nitric oxide synthase (iNOS) and cyclooxygenase-2 (COX-2) are also inflammatory mediators involved in various inflammatory processes. The evidence can be seen in the presence of those inflammatory mediators in the inflammatory processes–related cells. Therefore, research has been conducted to suppress the activity or down-regulate the expression of inflammatory mediators using various plants containing the bioactive compound extract. Black soybean, rich in polyphenol, anthocyanin, isoflavone, and bioactive peptides, has also been investigated for potency as an anti-inflammatory agent. Research by Jeong et al. (2013) revealed that anthocyanins from black soybean were able to downregulate lipopolysaccharide-induced inflammatory responses in BV2 microglial cells. The anthocyanin mechanism in downregulating inflammatory response is by suppressing the NF-κB and Akt/MAPKs signaling pathways. Thus, anthocyanin from black soybean can be suggested to be used as therapeutic remedies for the condition of neurodegenerative disease. In this research, nitric oxide (NO) and prostaglandins E2, as well as TNF-α and interleukin (IL)-1β as LPS-induced pro-inflammatory mediators, were inhibited by anthocyanin. Downregulating the capability of anthocyanin was also shown in the expression of inducible NO synthase, cyclooxygenase-2, TNF-α, and IL-1β in LPS-stimulated BV2 cells. The ability of anthocyanin of black soybean as an anti-inflammatory agent is in agreement with other work conducted by Kim et al. (2013), which examined the activity of anthocyanin from black soybean to inhibit *Helicobacter pylori*–induced inflammation in human gastric epithelial AGS cells. *Helicobacter pylori* is well known for commonly infecting the gastric epithelial cells and leads to an inflammation process and various pathological incidences. Moreover, the infection of gastric epithelial cells by *Helicobacter pylori* will increase ROS, iNOS, COX-2, and IL-8 as inflammatory-associated gene expression. ROS plays a significant role in the oxidative damage of DNA, protein, and lipid. Anthocyanin of black soybean in this research can decrease the production of ROS. Besides, anthocyanin could inhibit the expression of iNOS and COX-2 as well as reduce the IL-8 production by 45.8%. Therefore, it can be suggested that anthocyanin could have a strong protective effect against gastric damage triggered by *Helicobacter pylori* infection. Both of these studies support a previous study conducted by Kim et al. (2008), which explored the capability of anthocyanin of black soybean to reduce the rate of inflammation *in vitro* using colorectal cancer cells and also *in vivo* with an animal model. The colorectal cancer cell was preferred because colorectal cancer is one of the most commonly observed cancer incidences in humans. Besides, previous research suggested that there was a negative correlation between the rate of consuming legumes and the incidence of colorectal cancer. The results show that iNOS and COX-2 expression were suppressed by anthocyanin, possibly by reducing the cellular oxidative stress. This ability is due to the hydroxyl group's presence at the 3 position of the B ring in the anthocyanin structure. Meanwhile, for the *in vivo* study, anti-inflammatory effects were also observed. Nevertheless, the anti-inflammatory properties could not solely be contributed by anthocyanin content because a similar result was also obtained for yellow soybean. A possible explanation was that the role of isoflavone and bioactive peptides in both black and yellow soybean could also act as an anti-inflammatory agent.

In support of this finding, Kim et al. (2017) investigated the downregulation of LPS-induced inflammatory markers of nitric oxide (25.01%), TNF-α (76.78%), IL-1β (58.99%), and IL-6 (84.48%) by the extract from germinated black soybean. The significant decreases of inflammatory markers were possibly due to the low molecular weight of peptides and free amino acids in the extract, which can suppress the inflammation process. However, the mechanism of low molecular peptides and free amino acids to inhibit the inflammatory process remains unclear. A more recent study conducted by Kim et al. (2018) on bioactive peptides which could act as anti-inflammatory agents revealed that black soybean could inhibit the gene expression of NO, TNF-α, IL- 1β, and IL-6 and showed that the germinated black soybean will release the smaller bioactive peptides which are readily available for antioxidative reaction. Thus, it could significantly reduce the expression of the pro-inflammatory cytokines.

Kim et al. (2017) revealed that black soybean anthocyanins significantly decreased LPS-stimulated production of ROS, inflammatory mediators such as nitric oxide (NO) and prostaglandin E2, and pro-inflammatory cytokines, including tumor necrosis factor α and interleukin-6. The capability of black soybean extract to decrease the production of ROS, PGE-2, and nitric oxide is due to the free radical scavenging activity of anthocyanin, especially cyanidin 3 glucoside, which is the most abundant anthocyanin found in the black soybean seed coat. Meanwhile, anthocyanin's mechanism in reducing the expression of NO and PGE-2 is possibly contributed by the ability of anthocyanin in inhibiting the expression of protein enzymes responsible for NO and PGE-2 production.

In addition to the *in vitro* study using cell culture, *in vivo* research was also widely conducted to examine the anti-inflammatory properties of the black soybean. Research by Kanamoto et al. (2011) shows that the administration of black soybean extract in high-fat diet-fed mice resulted in reduced gene expression of major inflammatory cytokines such as tumor necrosis factor-R and monocyte chemoattractant protein-1. Although the cellular mechanism is unclear, the results provide a promising potency of black soybean to be developed as a functional food, reducing the inflammatory process.

Meanwhile, the anthocyanin study in downregulating pro-inflammatory cytokine expression was conducted by Park et al. (2015). By consuming anthocyanin-rich extract from black soybean, the COX-2 expression in the normal-diet-fed mice was significantly reduced by the addition of anthocyanin extract. The prominent contributor of anthocyanin to inhibit the inflammatory process is cyanidin 3 glucoside. It is postulated that such a role was not merely attributed to the extract's antioxidant activity but also the ability of individual anthocyanin to interfere with a signaling pathway by a direct blockage. Besides, the lower serum concentration of PGE2 in mice was also observed by anthocyanin extract supplementation. PGE2 is known as a metabolite of COX-2. Thus, supplementation of anthocyanin from black soybean could have beneficial effects in reducing the inflammatory incidence, and therefore, black soybean could potentially be developed as a functional food.

Another *in vivo* study was also performed using rats induced with a high-fat diet to investigate the capability of consuming black soybean seed coat extract in the inhibition of obesity-related inflammatory processes (Kim et al., 2015). The result shows that administering black soybean seed coat extract could remarkably suppress the gene expression of TNF-α and IL-6, which are pro-inflammatory adipocytokines that play a role in the adipogenesis pathway. The gene suppression capability could be due to the cyanidin 3 glucoside in black soybean, which could activate the AMPK pathway by decreasing TNF-α expression and contribute to the significant decrease of body fat accumulation (Kwon et al., 2007). This result suggested that black soybean can be optimized as food for the diet in obesity prevention.

The inflammatory process is also closely related to the onset of various diseases, including cancer. As reported by the World Health Organization, cancer is the second leading cause of death globally (WHO, 2008). Therefore, significant research has been conducted in various fields, including medicine, pharmacology, and pharmacognosy, to elucidate the complex mechanism of cancer and investigate drugs and plant bioactive compounds that could potentially be used as drugs or remedies for cancer treatment. Black soybean, which has long been known as a traditional herb for various diseases, has also been investigated for its anticancer properties.

An early study was performed by Shon et al. (2007) on the anticancer activity of fermented black soybean extract on the HeLa, HepG-2, HT-29, and MCF-7 cancer cells. The result showed that fermented black soybean extract has strong potential as an anticancer agent contributed by the anthocyanin and phenolic content known to have high antioxidant activity. The anticancer result was positively correlated with the antioxidant activity measured by DPPH, ABTS, reducing power, and the inhibition of NO production. In agreement, research by Zou and Chang (2011) revealed that black soybean extract was capable of suppressing the proliferation of human AGS gastric cancer cells due to the polyphenol content in black soybeans such as phenolic acid, anthocyanin, isoflavone, and flavonols. Those bioactive compounds could induce the apoptosis process in cancer cells by altering the ratio of Bax to Bcl-2 and activation of caspase-3, followed by cleavage of PARP. Meanwhile, a study on the anticancer activity of black soybean paste *doenjang* was performed by Park et al. (2015). HT-29 human colon cancer cells were used to examine the anticancer activity of the extract. It was reported that black soybean extract exhibited an anticancer effect on HT-29 cells by MTT assay. It was suggested that this activity could be closely related to the ability of black soybean *doenjang* to reduce the inflammation process by downregulating the pro-inflammatory cytokines such as TNF-α, IL-6, and COX-2. Although the molecular mechanism of cancer cell growth inhibition remains unclear, it is strongly believed that bioactive compounds in black soybean *doenjang*, such as phenolic acid and anthocyanin, are playing a vital role in such accomplishment.

In addition to the famous polyphenol content, black soybean, similar to the conventional yellow soybean, is a rich source of protein. Six low molecular weights of the protein were found in black soybean and harmful amino acids (Chen et al., 2019), which is responsible for the antioxidant and anticancer activity of black soybean. The different protein content, the prevalence of acidic amino acids, and the limited content of hydrophobic amino acids are parameters responsible for inhibiting ovarian cancer cell growth. Meanwhile, a recent study by Chen et al. (2018) shows that bioactive peptides isolated from black soybean by-products could inhibit the growth of cancer cells using human liver (HepG2), lung (MCF-7), and cervical (Hela) cancer cell lines. From the extensive studies on the anti-inflammatory effect of black soybean, different polyphenols, including anthocyanin, flavonoids, and other phenolic compounds, were suggested to be responsible for such a significant effect. The mechanism of action of polyphenols to inhibit the inflammatory process is described by Zhang et al. (2019). There could be positive interaction among polyphenols available (Figure 4.5). Based on the various published research, black soybean in raw, fermented, and by-product forms

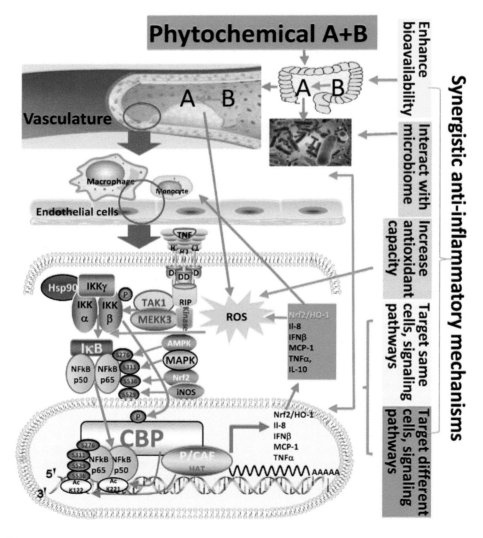

FIGURE 4.5 Mechanisms of the synergistic anti-inflammatory effect of combined phytochemicals. Source: Zhang et al. (2019).

could be used as functional food related to their anti-inflammatory and anticancer activities and polyphenols content, especially anthocyanin and the low molecular bioactive peptides available.

4.4.3   Anti-Atherosclerosis and Coronary Heart Disease

Coronary heart disease and related cardiovascular diseases have become the highest cause of mortality globally. The blockage of blood circulation causes this disease by the accumulation of plaque in the blood vessel, thus blocking delivery to the heart. The blockage of blood vessels is usually caused by a fat deposit. The process of depositing fat in the artery walls is known as atherosclerosis. Atherosclerosis is formed by three consecutive processes: Fatty streak formation, atheroma formation, and atherosclerotic plaque formation (Rafieian-Kopaei et al., 2014). One of the contributors to the formation of plaque is low-density lipoprotein (LDL) cholesterol. The oxidation of LDL increases the formation of foam cells, which then accumulate in the arteries. Therefore, the antioxidant substances play an important role in inhibiting the initial step of atherosclerosis by preventing the oxidation of LDL, as shown in Figure 4.6 (Moss et al., 2018). Anthocyanin is an antioxidant that could inhibit LDL oxidation and also reduce the incidence of inflammation, which, as a result, could decrease atherosclerotic formation in the blood vessel. Various studies of plant herbs and medicine capability on the inhibition of LDL oxidation have been published, including black soybeans. A study by Takahashi et al. (2005) investigated the antioxidant activity of black soybean and yellow soybean seed coat on the capability to inhibit LDL oxidation. The result shows that the extract of black soybean seed coat could prolong the lag time of LDL oxidation compared to the yellow soybean seed coat. This condition describes the ability of black soybean extract to delay the propagation phase after the initial phase. This result is probably due to the higher anthocyanin content in black soybean compared to yellow soybean. Moreover, hydrolyzing soybean with β-glucosidase has successfully increased the inhibition rate of LDL oxidation due to the fact that hydrolyzed soybeans are rich in aglycone which has higher antioxidant activity. Aglycone is also prominently found in fermented soybean. Therefore, the consumption of fermented soybean products such as tempeh, natto, miso, and soybean paste is recommended to decrease the risk of atherosclerotic formation. In agreement, Astadi et al. (2009) examined two local Indonesian varieties of black soybean, which were Mallika and Cikuray. Both black soybeans could decrease the LDL oxidation. The ability of black soybeans Mallika and Cikuray was higher than that of BHT, a synthetic antioxidant used as a positive control. The anthocyanin content of black soybean is believed to be responsible for the antioxidative action. The most dominant anthocyanin in black soybean, cyanidin 3 glucoside, is

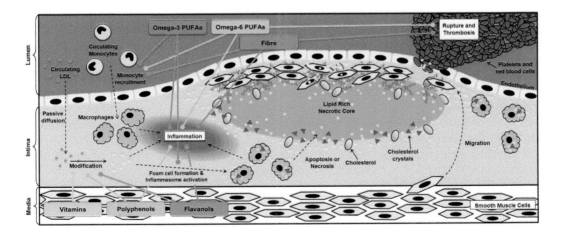

FIGURE 4.6   Role of anthocyanin in inhibiting atherosclerotic formation. Source: Moss et al. (2018).

reported to have potent antioxidant activity, preventing the oxidation of LDL. Published work by Chen et al. (2011) revealed that the fermentation process could increase the LDL oxidation inhibition by black soybean. Black soybean was fermented with *Aspergillus awamori*, and this process is usually used to make miso. The result shows that total phenolic content and amino nitrogen were significantly increased by fermentation due to the hydrolysis of the black soybean matrix and the release of bioactive compounds. This process is believed to be responsible for fermented black soybean's ability to inhibit LDL oxidation. It can be suggested that besides bioactive compounds such as anthocyanin, phenolic acid, isoflavone, and, amino nitrogen and other small peptides contribute to the beneficial health effects of black soybean and other legumes. Meanwhile, Kim et al. (2013) published their work on the antiplatelet aggregation of the black soybean. Platelet aggregation is one of the processes in atherosclerosis formation. Therefore, antiplatelet aggregation is an important property to decrease the risk of cardiovascular diseases. The inhibition of platelet activation is contributed by the anthocyanin, peptides, isoflavone, and adenosine. An *in vivo* study by Chao et al. (2013) was published on the effect of black soybean as prevention of the atherogenic process in the hypercholesterolemic rabbit. Black soybean could prolong the lag time of LDL oxidation, which can inhibit LDL oxidation. Moreover, the atheroma region in the aortic arch of the rabbit was significantly decreased by consuming a black soybean diet. The beneficial effects of the black soybean diet are several powerful bioactive compounds such as anthocyanin, phenolic, isoflavone aglycone, and small peptides available in black soybeans. Meanwhile, research on the prevention of atherosclerosis by black soybean was conducted by Lee et al. (2017). In this research, the monocyte-endothelial cell adhesion method, associated with atherosclerosis progression, was used. The result explained that black soybean could lower the monocyte-endothelial cell adhesion because of the isoflavone and proanthocyanidin content. Although the report is only *in vitro* study, several other works has been published on the correlation between consuming isoflavone and proanthocyanidin-rich foods with the decreased rate of coronary disease. Further study by Lee et al. (2018) revealed that black soybean could also suppress the TNF-α stimulated the expression of vascular cell adhesion molecule-1 and monocyte chemotactic protein-1 and phosphorylation of IκB kinase and IκBα involved in the initiation of atherosclerosis in HUVECs. This ability is due to the contribution of a daidzein metabolite, 7,8,4'-trihydroxyisoflavone (7,8,4'-THI), bioavailable in the blood of rats administered with soybean embryo extract and isoflavone.

4.4.4 Antidiabetic Activity

Diabetes mellitus (DM), primarily type 2, is a complex metabolic disease that is caused by insulin resistance to the increase of blood glucose level, known as hyperglycemia. This condition will lead to the development of various diseases such as blood vessel–related diseases (heart disease, stroke) and tissue dysfunction (liver, kidney, and pancreas). Nowadays, the number of cases of DM type 2 has increased rapidly and become infamously known as the mother of diseases. The main contributor to DM type 2 is an unbalanced diet and unhealthy lifestyles such as smoking, exposure to pollution, and excessive alcohol consumption. Various drugs have been developed for the prevention and treatment of DM type 2 in regulating blood glucose by direct blood glucose homeostasis intervention or by increasing the insulin response sensitivity. However, this is not an easy task due to the complexity of factors contributing to DM type 2 and the side effects of consuming the drugs that need to be taken into account.

Recently, a number of works were published on the potency of plant food for the prevention and treatment of DM type 2. The bioactive compound of a plant is believed to contribute positively to the signaling response of insulin, for example, anthocyanin. Published work by Chen et al. (2007) reported the ability of fermented black soybean *douchi* to inhibit the activity of the α-glucosidase enzyme. This enzyme is responsible for catalyzing the hydrolysis of carbohydrates to simple sugar, increasing blood glucose levels. Fermented black soybean has anti-α-glucosidase activity through its bioactive compounds, such as the aglycone form of isoflavone, and its anthocyanin compound by binding to the active side of α-glucosidase enzyme and preventing the enzyme from hydrolyzing the complex carbohydrate into glucose.

A similar conclusion was provided by Jang et al. (2010) who worked on both *in vitro* and *in vivo* investigations of black soybean peptides' capability to improve insulin resistance. The *in vitro* study focused on the effect of black soybean peptides on endoplasmic reticulum (ER) stress. ER stress contributed by obesity, which then affects insulin resistance, leads to DM type 2. The result shows that black soybean peptides could decrease the ER stress and, therefore, could ameliorate insulin resistance. Furthermore, the *in vivo* study using mice suggested that the intake of black soybean peptides could reduce blood glucose and improve the animal model's glucose tolerance. Bioactive peptides in black soybean are possibly improving the signaling pathway of insulin and inhibiting the glucosidase enzyme. Therefore the homeostasis of blood glucose could be maintained. Besides, a human clinical trial was performed by Kwak et al. (2010) on the ability of the peptide to improve glucose control in prediabetes, and newly diagnoses subjects with DM type 2. The result revealed that subjects with 12-week supplementation of black soybean peptides tended to have lower fasting glucose levels and a significant reduction in two hours post-load glucose compared to the placebo group. Although the mechanism of the decrease of blood glucose level by black soybean peptides is still unclear, it is suggested that peptides can be bound to various sites of α-glucosidase, which then inhibit their capacity to hydrolyze carbohydrates.

Meanwhile, work by Kurimoto et al. (2013) focused on the ability of black soybean seed coat extract, which is rich in polyphenol content, to improve the hyperglycemia condition and insulin sensitivity in diabetic mice. The result suggested that the intake of black soybean seed coat extract could ameliorate the hyperglycemia shown by the decrease in blood glucose levels. The insulin sensitivity was improved through the activation of AMP-activated protein kinase (AMPK) in the skeletal muscle and liver of the animal model. Besides, the upregulation of glucose transporter 4 in the skeletal muscle and the downregulation of gluconeogenesis in the liver were observed. The beneficial effects of black soybean seed coat extract were caused by cyanidin 3 glucoside and proanthocyanidin, which are abundantly available. In agreement, cyanidin 3 glucoside was reported to have antidiabetic activity via the initiation of differentiation of preadipocytes into a smaller size and improved insulin sensitivity (Matsukawa et al., 2015). The administration of black soybean seed coat extract reduces the body and the white adipose tissue weight and decreases the size of adipocytes in white adipose tissue. The mechanism was revealed using 3T3-Ll cells treated using black soybean seed coat extract and individual cyanidin 3 glucosides. The result shows that smaller adipocytes were observed as a result of 3T3-L1 differentiation. Furthermore, PPARγ and C/EBPα gene expressions and adiponectin secretion were increased. On the other hand, the tumor necrosis factor-α secretion was decreased. Meanwhile, the insulin signaling was activated and improved, and the glucose uptake was increased.

Besides peptides and anthocyanin, phenolic compounds could also contribute to the improvement of DM type 2 by inhibiting the work of α-amylase and α-glucosidase enzymes (Tan et al., 2017). All of the crude extracts, semi fractionated and fractionated, show better inhibition capacity than the commercial inhibitor. An interesting finding is that the fractionated extracts provide a different result for both enzymes. For example, myricetin could significantly inhibit α-amylase but shows no significant differences observed for α-glucosidase. Thus, it can be suggested that the synergistic effect among phenolic compounds is crucial in the inhibition of both enzymes.

4.4.5 Anti-Obesity

Obesity in recent years has become an international concern due to its progressive development rate. Obesity is not only prevalent in developed countries but also spread widely in developing countries. Diabetes, atherosclerosis, coronary heart diseases, and cancer are diseases that closely relate to obesity. It is believed that obesity is playing a significant role in the occurrence of such morbid diseases. Balancing the diet could contribute to the reduced risk of obesity along with physical activities. The development of biochemistry and genetic-related research leads to the elucidation of the mechanism behind the onset of obesity. It was reported that bioactive compounds such as anthocyanin and also peptides had anti-obesity effects. Previously published research has also investigated the ability of black soybean, a rich source of anthocyanin and peptides, to prevent the incidence of obesity *in vitro* and *in vivo*.

The anti-obesity and hypolipidemic effect of anthocyanin from black soybean seed coat was reported by Kwon et al. (2007) using high-fat diet-fed rats. The result shows that the intake of black soybean seed coat and black soybean anthocyanin extract lowered the body weight gain, suppressed liver weight gain, and decreased the weight of epididymal and perirenal fat pads. Consuming the black soybean extract could improve the rats' lipid profile, which includes lowering the triglyceride and cholesterol level and increasing the high-density lipoprotein content. The anti-obesity and hypolipidemic effect of black soybean could be contributed by the ability of bioactive compounds to interfere with the gene expression responsible for lipid metabolism. Meanwhile, the fecal excretion rate was also increased. Furthermore, anthocyanin could also take part in starch digestion by inhibiting α-glucosidase enzyme activity, thus reducing glucose metabolism. Anthocyanin can probably affect the triglyceride synthesis in the metabolism and downregulate the mRNA expression of the lipolytic enzyme for lipid hydrolysis.

Besides its anthocyanin content, black soybean is rich in bioactive peptides, suggesting adipogenesis inhibitory activity (Kim et al., 2007). The presence of bioactive peptides identified as tripeptide (isoleucine, glutamine, asparagine) could suppress the differentiation of the 3T3-L1 preadipocyte cells. Thus, it can be postulated that the peptides in black soybean affect the gene expression in adipose tissue, which in result regulates adipogenesis effectively.

Meanwhile, a double-blind, randomized, controlled study in overweight and obese human subjects was performed by Kwak et al. (2012) to investigate the weight reduction effect of black soybean peptides supplementation. After completing the study period, there was a significant reduction in weight, body mass index, and body fat mass in the test group. Moreover, a lower fasting blood glucose level was observed. Furthermore, the supplementation of black soybean peptides lowered the leptin level in the subjects. Leptin is a critical adipose-derived hormone that plays a crucial role in energy intake and energy expenditure and regulates appetite and metabolism, which is usually found at a high level in obese people. Black soybean peptides are suggested to affect the leptin pathway, which then downregulates the energy and lipid metabolism along with the decrease of appetite. Thus, it decreased the body weight, body fat mass, and body mass index of the subjects.

A study by Kim et al. (2012) shows that anthocyanins could reduce adipose tissue mass by acting directly on adipocytes using 3T3-L1 preadipocyte cell line exposed to anthocyanin from black soybean. Moreover, anthocyanin could inhibit the proliferation of pre-confluent preadipocytes and mature post-confluent adipocytes and reduce the number of viable cells. Furthermore, the accumulation of lipids was decreased, and black soybean anthocyanin was able to downregulate the peroxisome proliferator-activated receptor γ, a main transcription factor for the adipogenic gene.

Meanwhile, research by Jung et al. (2013) revealed that black soybean intake could reduce hepatic cholesterol accumulation in high-fat-diet-induced non-alcoholic fatty liver disease rats. Non-alcoholic fatty liver disease is a condition of excess fat in the liver. The published research shows that the intake of black soybean powder could reduce the liver's cholesterol and triglyceride levels. The mechanism of the ability of black soybean to reduce the cholesterol and triglyceride levels, thus potentially reducing the risk of diabetes, liver disease, as well as metabolic disorders, is presented in Figure 4.7 (Jung et al., 2013). The expression of SREBP2 as an indicator of cholesterol metabolism was suppressed by black soybean supplementation, and it can be pointed out that black soybean could decrease the HMG CoA reductase expression. Moreover, black soybean supplementation could increase the work of superoxide dismutase, catalase, and glutathione peroxidase antioxidant enzymes, and thus could balance the oxidation process in the body, resulting in the lower production of ROS and in the long term reducing the risk of atherosclerosis.

Research on the effect of fermentation of black soybean on the anti-obesity capacity was published by Lee et al. (2015). Using *Monascus pilosus* as a culture starter of fermented black soybean,

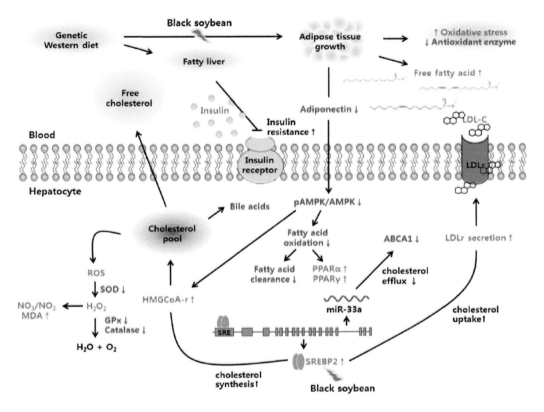

FIGURE 4.7 The mechanism of black soybean against non-alcoholic fatty liver disease. Source: Jung et al. (2013).

high-fat-diet-induced obese mice were supplemented by both the extract and the powder of fermented black soybean. The anti-obesity capacity was also performed in adipocytes. The result shows that fermented black soybean could decrease the body weight gain of mice and also suppress the mRNA expression of adipogenesis-related genes such as peroxisome proliferator-activated receptor γ (PPAR γ), fatty acid-binding protein 4 (FABP4), and fatty acid synthase (FAS). Meanwhile, the lipid accumulation of 3T3-L1 adipocytes was also decreased by the presence of fermented black soybean. Fermentation could increase the level of isoflavone glycoside, which is known to play an essential role in lipid metabolism and is also responsible for part of hydrolysis of protein to smaller peptides that also could give beneficial effects for the obesity condition. Furthermore, anthocyanin and other phenolic compounds available in black soybean also contribute to the ability of black soybean to improve obesity in the animal model.

In line with previously published studies, anthocyanin of black soybean seed coat was proven to contribute to the improvement of lipid profile, level of abdominal fat, and also low-density lipoprotein content of Korean overweight/obese adults in a randomized controlled trial (Lee et al., 2016). The result suggested that anthocyanin from the black soybean seed coat is responsible for reducing body weight and lipid accumulation. Its essential task is to activate the AMPK pathway in the white adipose tissue, skeletal muscle, and liver, and thus promote the catabolic and inhibit the anabolic pathways of lipids.

Meanwhile, work by Jing et al. (2018) explored the effect of black soybean intake on the lipid and also gut microbiome profile of high-fat-diet-induced mice. The result shows that the improvement of the lipid profile was observed, and a significant decrease of triglyceride, total cholesterol, and low-density lipoprotein content was found in mice in the black soybean supplemented diet group. Moreover, the short-chain fatty acid level, especially propionate and butyrate in the feces, was improved by the black soybean diet. Propionate and butyrate are known to be antiobesogenic and able to reduce visceral and liver fat. In agreement with several previously published works, the anti-obesity capacity of black soybean is contributed by their peptides and anthocyanin content. Additionally, this research suggested that the fiber content of black soybean is also giving beneficial effects by altering the gut microbiome profile, which leads to the improvement of lipid metabolism and inhibits lipid accumulation.

4.5 FUTURE POTENCY

Research on the health benefit of black soybean has been widely established. Black soybean is reported to have antioxidant, anti-inflammatory, anticancer, anti-atherosclerosis and coronary heart disease, antidiabetic, and anti-obesity capacity due to certain compounds such as anthocyanin, isoflavone, peptides, fiber, and other polyphenol compounds. Due to various health benefits, the extract of black soybean can be produced and used as remedies for various diseases. Meanwhile, the processing of black soybean to make various products is proven to only slightly reduce some bioactive compounds, and on the other hand, it increases other bioactive compounds and thus could retain the antioxidant properties. Therefore, black soybean is a potential commodity to be developed as a functional food. The already established and well-known black soybean health properties can be essential in producing black soybean food products that can be widely accepted. Thus, it could contribute to the health improvement of society by providing black soybean–based healthy food products.

4.6 SUMMARY

Black soybean is a rich source of anthocyanin, isoflavone, phenolic compounds, bioactive peptides, and fiber. Such health-promoting compounds are responsible for antioxidant activity, as well as anti-inflammatory, anticancer, anti-atherosclerosis and coronary heart diseases, antidiabetes, and anti-obesity capacity. Black soybean, both as a whole seed and the seed coat only, was proven to have health benefit properties as measured by *in vitro* chemical and cell culture assays, *in vivo* animal models, and also clinical trials with humans. The method of processing as an essential factor in consuming black soybean such as soaking, grinding, boiling, and roasting did not significantly affect the antioxidant activities. Moreover, the fermentation of black soybean could increase the bioavailability of the bioactive compounds as well as their beneficial properties. Based on its characteristics, black soybean has the potential to be developed as a functional food.

REFERENCES

Astadi, I. R., Astuti, M., Santoso, U., & Nugraheni, P. S. (2009). In vitro antioxidant activity of anthocyanins of black soybean seed coat in human low density lipoprotein (LDL). *Food Chemistry*, *112*(3), 659–663.

Balasuriya, B. N., & Rupasinghe, H. V. (2011). Plant flavonoids as angiotensin converting enzyme inhibitors in regulation of hypertension. *Functional Foods in Health and Disease*, *1*(5), 172–188.

Bräunlich, M., Slimestad, R., Wangensteen, H., Brede, C., Malterud, K. E., & Barsett, H. (2013). Extracts, anthocyanins and procyanidins from Aronia melanocarpa as radical scavengers and enzyme inhibitors. *Nutrients*, *5*(3), 663–678.

Chao, P. Y., Chen, Y. L., Lin, Y. C., Hsu, J. I., Lin, K. H., Lu, Y. F., ... & Yang, C. M. (2013). Effects of black soybean on atherogenic prevention in hypercholesterolemic rabbits and on adhesion molecular expression in cultured HAECs. *Food and Nutrition Sciences*, *4*, 8A.

Chen, J., Cheng, Y. Q., Yamaki, K., & Li, L. T. (2007). Anti-α-glucosidase activity of Chinese traditionally fermented soybean (douchi). *Food Chemistry*, *103*(4), 1091–1096.

Chen, Y. F., Lee, S. L., & Chou, C. C. (2011). Fermentation with *Aspergillus awamori* enhanced contents of amino nitrogen and total phenolics as well as the low-density lipoprotein oxidation inhibitory activity of black soybeans. *Journal of Agricultural and Food Chemistry*, *59*(8), 3974–3979.

Chen, Z., Li, W., Santhanam, R. K., Wang, C., Gao, X., Chen, Y., … & Chen, H. (2019). Bioactive peptide with antioxidant and anticancer activities from black soybean [Glycine max (L.) Merr.] byproduct: Isolation, identification and molecular docking study. *European Food Research and Technology*, *245*(3), 677–689.

Chen, Z., Wang, J., Liu, W., & Chen, H. (2017). Physicochemical characterization, antioxidant and anticancer activities of proteins from four legume species. *Journal of Food Science and Technology*, *54*(4), 964–972.

Cheng, K. C., Wu, J. Y., Lin, J. T., & Liu, W. H. (2013). Enhancements of isoflavone aglycones, total phenolic content, and antioxidant activity of black soybean by solid-state fermentation with *Rhizopus* spp. *European Food Research and Technology*, *236*(6), 1107–1113.

Cho, E., Chung, E. Y., Jang, H. Y., Hong, O. Y., Chae, H. S., Jeong, Y. J., … & Park, K. H. (2017). Anti-cancer effect of cyanidin-3-glucoside from mulberry via caspase-3 cleavage and DNA fragmentation in vitro and in vivo. *Anti-Cancer Agents in Medicinal Chemistry (Formerly Current Medicinal Chemistry-Anti-Cancer Agents)*, *17*(11), 1519–1525.

El-Ella, D. M. A., & Bishayee, A. (2019). The epigenetic targets of berry anthocyanins in cancer prevention. In *Epigenetics of Cancer Prevention*, Bishayee. A. and Bhatia, D. (eds), (pp. 129–148). Academic Press.

Gille, D., Schmid, A., Walther, B., & Vergères, G. (2018). Fermented food and non-communicable chronic diseases: A review. *Nutrients*, *10*(4), 448.

Harlen, W. C., & Jati, I. R. A. (2018). Antioxidant activity of anthocyanins in common legume grains. In *Polyphenols: Mechanisms of Action in Human Health and Disease*, Watson, R., Preedy, V., and Zibadi, S. (eds), (pp. 81–92). Academic Press.

Hassimotto, N. M. A., Genovese, M. I., & Lajolo, F. M. (2009). Antioxidant capacity of Brazilian fruit, vegetables and commercially-frozen fruit pulps. *Journal of Food Composition and Analysis*, *22*(5), 394–396.

Hwang, C. E., Seo, W. T., & Cho, K. M. (2013). Enhanced antioxidant effect of black soybean by Cheonggukjang with potential probiotic Bacillus subtilis CSY191. *Korean Journal of Microbiology*, *49*(4), 391–397.

Jang, E. H., Ko, J. H., Ahn, C. W., Lee, H. H., Shin, J. K., Chang, S. J., … & Kang, J. H. (2010). In vivo and in vitro application of black soybean peptides in the amelioration of endoplasmic reticulum stress and improvement of insulin resistance. *Life Sciences*, *86*(7–8), 267–274.

Jang, H., Ha, U. S., Kim, S. J., Yoon, B. I., Han, D. S., Yuk, S. M., & Kim, S. W. (2010). Anthocyanin extracted from black soybean reduces prostate weight and promotes apoptosis in the prostatic hyperplasia-induced rat model. *Journal of Agricultural and Food Chemistry*, *58*(24), 12686–12691.

Jati, I. R. A. (2020). Black soybean seed: Black soybean seed antioxidant capacity. In *Nuts and Seeds in Health and Disease Prevention*, Preedy, V. and Watson, R. (eds), (pp. 147–159). Academic Press.

Jeong, J. W., Lee, W. S., Shin, S. C., Kim, G. Y., Choi, B. T., & Choi, Y. H. (2013). Anthocyanins downregulate lipopolysaccharide-induced inflammatory responses in BV2 microglial cells by suppressing the NF-κB and Akt/MAPKs signaling pathways. *International Journal of Molecular Sciences*, *14*(1), 1502–1515.

Jiang, C., Ci, Z., & Kojima, M. (2019). α-Glucosidase inhibitory activity in rice miso supplementary with black soybean. *American Journal of Food Science and Technology*, *7*(1), 27–30.

Jing, C., Wen, Z., Zou, P., Yuan, Y., Jing, W., Li, Y., & Zhang, C. (2018). Consumption of black legumes glycine soja and glycine max lowers serum lipids and alters the gut microbiome profile in mice fed a high-fat diet. *Journal of Agricultural and Food Chemistry*, *66*(28), 7367–7375.

Jung, J. H., & Kim, H. S. (2013). The inhibitory effect of black soybean on hepatic cholesterol accumulation in high cholesterol and high fat diet-induced non-alcoholic fatty liver disease. *Food and Chemical Toxicology*, *60*, 404–412.

Kalušević, A., Lević, S., Čalija, B., Pantić, M., Belović, M., Pavlović, V., … & Nedović, V. (2017). Microencapsulation of anthocyanin-rich black soybean coat extract by spray drying using maltodextrin, gum Arabic and skimmed milk powder. *Journal of Microencapsulation*, *34*(5), 475–487.

Kanamoto, Y., Yamashita, Y., Nanba, F., Yoshida, T., Tsuda, T., Fukuda, I., … & Ashida, H. (2011). A black soybean seed coat extract prevents obesity and glucose intolerance by up-regulating uncoupling proteins and down-regulating inflammatory cytokines in high-fat diet-fed mice. *Journal of Agricultural and Food Chemistry*, *59*(16), 8985–8993.

Khoo, H. E., Azlan, A., Tang, S. T., & Lim, S. M. (2017). Anthocyanidins and anthocyanins: Colored pigments as food, pharmaceutical ingredients, and the potential health benefits. *Food & Nutrition Research*, *61*(1), 1361779.

Kim, H. G., Kim, G. W., Oh, H., Yoo, S. Y., Kim, Y. O., & Oh, M. S. (2011). Influence of roasting on the antioxidant activity of small black soybean (Glycine max L. Merrill). *LWT-Food Science and Technology*, *44*(4), 992–998.

Kim, H. J., Bae, I. Y., Ahn, C. W., Lee, S., & Lee, H. G. (2007). Purification and identification of adipogenesis inhibitory peptide from black soybean protein hydrolysate. *Peptides*, *28*(11), 2098–2103.

Kim, H. K., Kim, J. N., Han, S. N., Nam, J. H., Na, H. N., & Ha, T. J. (2012). Black soybean anthocyanins inhibit adipocyte differentiation in 3T3-L1 cells. *Nutrition Research, 32*(10), 770–777.

Kim, J. M., Kim, K. M., Park, E. H., Seo, J. H., Song, J. Y., Shin, S. C., ... & Youn, H. S. (2013). Anthocyanins from black soybean inhibit *Helicobacter pylori*-induced inflammation in human gastric epithelial AGS cells. *Microbiology and Immunology, 57*(5), 366–373.

Kim, J. M., Kim, J. S., Yoo, H., Choung, M. G., & Sung, M. K. (2008). Effects of black soybean [Glycine max (L.) Merr.] seed coats and its anthocyanidins on colonic inflammation and cell proliferation in vitro and in vivo. *Journal of Agricultural and Food Chemistry, 56*(18), 8427–8433.

Kim, J. N., Han, S. N., Ha, T. J., & Kim, H. K. (2017). Black soybean anthocyanins attenuate inflammatory responses by suppressing reactive oxygen species production and mitogen activated protein kinases signaling in lipopolysaccharide-stimulated macrophages. *Nutrition Research and Practice, 11*(5), 357–364.

Kim, K., Lim, K. M., Shin, H. J., Seo, D. B., Noh, J. Y., Kang, S., ... & Bae, O. N. (2013). Inhibitory effects of black soybean on platelet activation mediated through its active component of adenosine. *Thrombosis Research, 131*(3), 254–261.

Kim, M. Y., Jang, G. Y., Lee, S. H., Kim, K. M., Lee, J., & Jeong, H. S. (2018). Preparation of black soybean (Glycine max L) extract with enhanced levels of phenolic compound and estrogenic activity using high hydrostatic pressure and pre-germination. *High Pressure Research, 38*(2), 177–192.

Kim, M. Y., Jang, G. Y., Oh, N. S., Baek, S. Y., Lee, S. H., Kim, K. M., ... & Jeong, H. S. (2017). Characteristics and in vitro anti-inflammatory activities of protein extracts from pre-germinated black soybean [Glycine max (L.)] treated with high hydrostatic pressure. *Innovative Food Science & Emerging Technologies, 43*, 84–91.

Kim, S. Y., Son, H. S., & Oh, S. H. (2009). Characteristics of Korean soybean paste (Doenjang) prepared by the fermentation of black soybeans. *Journal of Food Science and Nutrition, 14*(2), 134–141.

Kim, S. Y., Wi, H. R., Choi, S., Ha, T. J., Lee, B. W., & Lee, M. (2015). Inhibitory effect of anthocyanin-rich black soybean testa (Glycine max (L.) Merr.) on the inflammation-induced adipogenesis in a DIO mouse model. *Journal of Functional Foods, 14*, 623–633.

Kriss, J. L., Ramakrishnan, U., Beauregard, J. L., Phadke, V. K., Stein, A. D., Rivera, J. A., & Omer, S. B. (2018). Yogurt consumption during pregnancy and preterm delivery in M exican women: A prospective analysis of interaction with maternal overweight status. *Maternal & Child Nutrition, 14*(2), e12522.

Kumari, S., Krishnan, V., & Sachdev, A. (2015). Impact of soaking and germination durations on antioxidants and anti-nutrients of black and yellow soybean (Glycine max. L) varieties. *Journal of Plant Biochemistry and Biotechnology, 24*(3), 355–358.

Kumudini, S., Prior, E., Omielan, J., & Tollenaar, M. (2008). Impact of *Phakopsora pachyrhizi* infection on soybean leaf photosynthesis and radiation absorption. *Crop Science, 48*(6), 2343–2350.

Kurimoto, Y., Shibayama, Y., Inoue, S., Soga, M., Takikawa, M., Ito, C., ... & Tsuda, T. (2013). Black soybean seed coat extract ameliorates hyperglycemia and insulin sensitivity via the activation of AMP-activated protein kinase in diabetic mice. *Journal of Agricultural and Food Chemistry, 61*(23), 5558–5564.

Kwak, J. H., Ahn, C. W., Park, S. H., Jung, S. U., Min, B. J., Kim, O. Y., & Lee, J. H. (2012). Weight reduction effects of a black soy peptide supplement in overweight and obese subjects: Double blind, randomized, controlled study. *Food & Function, 3*(10), 1019–1024.

Kwak, J. H., Lee, J. H., Ahn, C. W., Park, S. H., Shim, S. T., Song, Y. D., ... & Chae, J. S. (2010). Black soy peptide supplementation improves glucose control in subjects with prediabetes and newly diagnosed type 2 diabetes mellitus. *Journal of Medicinal Food, 13*(6), 1307–1312.

Kwak, C. S., Lee, M. S., & Park, S. C. (2007). Higher antioxidant properties of Chungkookjang, a fermented soybean paste, may be due to increased aglycone and malonylglycoside isoflavone during fermentation. *Nutrition Research, 27*(11), 719–727.

Kwon, S. H., Ahn, I. S., Kim, S. O., Kong, C. S., Chung, H. Y., Do, M. S., & Park, K. Y. (2007). Anti-obesity and hypolipidemic effects of black soybean anthocyanins. *Journal of Medicinal Food, 10*(3), 552–556.

Lee, A. L., Yu, Y. P., Hsieh, J. F., Kuo, M. I., Ma, Y. S., & Lu, C. P. (2018). Effect of germination on composition profiling and antioxidant activity of the polysaccharide-protein conjugate in black soybean [Glycine max (L.) Merr.]. *International Journal of Biological Macromolecules, 113*, 601–606.

Lee, C. C., Dudonné, S., Dubé, P., Desjardins, Y., Kim, J. H., Kim, J. S., ... & Lee, C. Y. (2017). Comprehensive phenolic composition analysis and evaluation of Yak-Kong soybean (*Glycine max*) for the prevention of atherosclerosis. *Food Chemistry, 234*, 486–493.

Lee, C. C., Dudonné, S., Kim, J. H., Kim, J. S., Dubé, P., Kim, J. E., ... & Lee, C. Y. (2018). A major daidzin metabolite 7, 8, 4′-trihydroxyisoflavone found in the plasma of soybean extract-fed rats attenuates monocyte-endothelial cell adhesion. *Food Chemistry, 240*, 607–614.

Lee, K. J., Baek, D. Y., Lee, G. A., Cho, G. T., So, Y. S., Lee, J. R., ... & Hyun, D. Y. (2020). Phytochemicals and antioxidant activity of Korean black soybean (Glycine max L.) landraces. *Antioxidants, 9*(3), 213.

Lee, K. J., Lee, J. R., Ma, K. H., Cho, Y. H., Lee, G. A., & Chung, J. W. (2016). Anthocyanin and isoflavone contents in Korean black soybean landraces and their antioxidant activities. *Plant Breeding and Biotechnology, 4*(4), 441–452.

Lee, M., Hong, G. E., Zhang, H., Yang, C. Y., Han, K. H., Mandal, P. K., & Lee, C. H. (2015). Production of the isoflavone aglycone and antioxidant activities in black soymilk using fermentation with Streptococcus thermophilus S10. *Food Science and Biotechnology, 24*(2), 537–544.

Lee, M., Sorn, S. R., Park, Y., & Park, H. K. (2016). Anthocyanin rich-black soybean testa improved visceral fat and plasma lipid profiles in overweight/obese Korean adults: A randomized controlled trial. *Journal of Medicinal Food, 19*(11), 995–1003.

Lee, Y. S., Choi, B. K., Lee, H. J., Lee, D. R., Cheng, J., Lee, W. K., ... & Suh, J. W. (2015). Monascus pilosus-fermented black soybean inhibits lipid accumulation in adipocytes and in high-fat diet-induced obese mice. *Asian Pacific Journal of Tropical Medicine, 8*(4), 276–282.

Ma, Y., & Huang, H. (2014). Characterisation and comparison of phenols, flavonoids and isoflavones of soymilk and their correlations with antioxidant activity. *International Journal of Food Science & Technology, 49*(10), 2290–2298.

Malenčić, D., Cvejić, J., & Miladinović, J. (2012). Polyphenol content and antioxidant properties of colored soybean seeds from Central Europe. *Journal of Medicinal Food, 15*(1), 89–95.

Matsukawa, T., Inaguma, T., Han, J., Villareal, M. O., & Isoda, H. (2015). Cyanidin-3-glucoside derived from black soybeans ameliorate type 2 diabetes through the induction of differentiation of preadipocytes into smaller and insulin-sensitive adipocytes. *The Journal of Nutritional Biochemistry, 26*(8), 860–867.

Moraes Filho, M. L. D., Hirozawa, S. S., Prudencio, S. H., Ida, E. I., & Garcia, S. (2014). Petit suisse from black soybean: Bioactive compounds and antioxidant properties during development process. *International Journal of Food Sciences and Nutrition, 65*(4), 470–475.

Moss, J. W., Williams, J. O., & Ramji, D. P. (2018). Nutraceuticals as therapeutic agents for atherosclerosis. *Biochimica et Biophysica Acta (BBA)-Molecular Basis of Disease, 1864*(5), 1562–1572.

Nagai, I. (1921). A genetico-physiological study on the formation of anthocyanin and brown pigments in plants. *Tokyo University College of Agriculture Journal., 8*, 1–92.

Nizamutdinova, I. T., Kim, Y. M., Chung, J. I., Shin, S. C., Jeong, Y. K., Seo, H. G., ... & Kim, H. J. (2009). Anthocyanins from black soybean seed coats stimulate wound healing in fibroblasts and keratinocytes and prevent inflammation in endothelial cells. *Food and Chemical Toxicology, 47*(11), 2806–2812.

Pala, V., Sieri, S., Berrino, F., Vineis, P., Sacerdote, C., Palli, D., ... & Giurdanella, M. C. (2011). Yogurt consumption and risk of colorectal cancer in the Italian European prospective investigation into cancer and nutrition cohort. *International Journal of Cancer, 129*(11), 2712–2719.

Park, E. S., Lee, J. Y., & Park, K. Y. (2015). Anticancer effects of black soybean doenjang in HT-29 human colon cancer cells. *Journal of the Korean Society of Food Science and Nutrition, 44*(9), 1270–1278.

Rafieian-Kopaei, M., Setorki, M., Doudi, M., Baradaran, A., & Nasri, H. (2014). Atherosclerosis: Process, indicators, risk factors and new hopes. *International Journal of Preventive Medicine, 5*(8), 927.

Ryu, D., & Koh, E. (2018). Application of response surface methodology to acidified water extraction of black soybeans for improving anthocyanin content, total phenols content and antioxidant activity. *Food Chemistry, 261*, 260–266.

Shen, Y., Song, X., Li, L., Sun, J., Jaiswal, Y., Huang, J., ... & Guan, Y. (2019). Protective effects of p-coumaric acid against oxidant and hyperlipidemia-an in vitro and in vivo evaluation. *Biomedicine & Pharmacotherapy, 111*, 579–587.

Shih, M. C., Yang, K. T., & Kuo, S. J. (2002). Quality and antioxidative activity of black soybean tofu as affected by bean cultivar. *Journal of Food Science, 67*(2), 480–484.

Shon, M. Y., Lee, S. W., & Nam, S. H. (2007). Antioxidant and anticancer activities of glycine semen germinatum fermented with germinated black soybean and some bacteria. *Korean Journal of Food Preservation, 14*(5), 538–544.

Shurtleff, W., & Aoyagi, A. (2012). History ofsoy yoghurt, soy acidophilus milk, and other cultured soymilks. *History of Soybeans and Soyfoods, 001-590.*

Siriwardhana, N., Kalupahana, N. S., Cekanova, M., LeMieux, M., Greer, B., & Moustaid-Moussa, N. (2013). Modulation of adipose tissue inflammation by bioactive food compounds. *The Journal of Nutritional Biochemistry, 24*(4), 613–623.

Slavin, M., Lu, Y., Kaplan, N., & Yu, L. L. (2013). Effects of baking on cyanidin-3-glucoside content and antioxidant properties of black and yellow soybean crackers. *Food Chemistry, 141*(2), 1166–1174.

Song, M., & Giovannucci, E. (2018). Substitution analysis in nutritional epidemiology: Proceed with caution. *European Journal of Epidemiology*, *33*(2), 137–140.

Sui, X., Zhang, Y., & Zhou, W. (2016). In vitro and in silico studies of the inhibition activity of anthocyanins against porcine pancreatic α-amylase. *Journal of Functional Foods*, *21*, 50–57.

Takahashi, R., Ohmori, R., Kiyose, C., Momiyama, Y., Ohsuzu, F., & Kondo, K. (2005). Antioxidant activities of black and yellow soybeans against low density lipoprotein oxidation. *Journal of Agricultural and Food Chemistry*, *53*(11), 4578–4582.

Tan, Y., Chang, S. K., & Zhang, Y. (2016). Innovative soaking and grinding methods and cooking affect the retention of isoflavones, antioxidant and antiproliferative properties in soymilk prepared from black soybean. *Journal of Food Science*, *81*(4), H1016–H1023.

Tan, Y., Chang, S. K., & Zhang, Y. (2017). Comparison of α-amylase, α-glucosidase and lipase inhibitory activity of the phenolic substances in two black legumes of different genera. *Food Chemistry*, *214*, 259–268.

USDA. (2020). World agricultural supply and demand estimates. *USDA Reports*. Available at: https://usda .library.cornell.edu/concern/publications/3t945q76s?locale=en

van den Berg, R., Haenen, G. R., van den Berg, H., & Bast, A. A. L. T. (1999). Applicability of an improved Trolox equivalent antioxidant capacity (TEAC) assay for evaluation of antioxidant capacity measurements of mixtures. *Food Chemistry*, *66*(4), 511–517.

Wang, Q., Kuang, H., Su, Y., Sun, Y., Feng, J., Guo, R., & Chan, K. (2013). Naturally derived anti-inflammatory compounds from Chinese medicinal plants. *Journal of Ethnopharmacology*, *146*(1), 9–39.

Wiese, S., Gärtner, S., Rawel, H. M., Winterhalter, P., & Kulling, S. E. (2009). Protein interactions with cyanidin-3-glucoside and its influence on α-amylase activity. *Journal of the Science of Food and Agriculture*, *89*(1), 33–40.

World Health Organization, & Research for International Tobacco Control. (2008). *WHO Report on the Global Tobacco Epidemic, 2008: The MPOWER Package*. World Health Organization.

Wu, H. J., Deng, J. C., Yang, C. Q., Zhang, J., Zhang, Q., Wang, X. C., … & Liu, J. (2017). Metabolite profiling of isoflavones and anthocyanins in black soybean [Glycine max (L.) Merr.] seeds by HPLC-MS and geographical differentiation analysis in Southwest China. *Analytical Methods*, *9*(5), 792–802.

Xie, Y., Zhu, X., Li, Y., & Wang, C. (2018). Analysis of the ph-dependent fe (iii) ion chelating activity of anthocyanin extracted from black soybean [glycine max (l.) merr.] coats. *Journal of Agricultural and Food Chemistry*, *66*(5), 1131–1139.

Xu, B., & Chang, S. K. (2009). Isoflavones, flavan-3-ols, phenolic acids, total phenolic profiles, and antioxidant capacities of soy milk as affected by ultrahigh-temperature and traditional processing methods. *Journal of Agricultural and Food Chemistry*, *57*(11), 4706–4717.

Xu, L., Du, B., & Xu, B. (2015). A systematic, comparative study on the beneficial health components and antioxidant activities of commercially fermented soy products marketed in China. *Food Chemistry*, *174*, 202–213.

Ye, M., Ren, L., Wu, Y., Wang, Y., & Liu, Y. (2013). Quality characteristics and antioxidant activity of hickory-black soybean yogurt. *LWT-Food Science and Technology*, *51*(1), 314–318.

You, Y., Yuan, X., Liu, X., Liang, C., Meng, M., Huang, Y., … & Zhang, Q. (2017). Cyanidin-3-glucoside increases whole body energy metabolism by upregulating brown adipose tissue mitochondrial function. *Molecular Nutrition & Food Research*, *61*(11), 1700261.

Zhang, L., Virgous, C., & Si, H. (2019). Synergistic anti-inflammatory effects and mechanisms of combined phytochemicals. *The Journal of Nutritional Biochemistry*, *69*, 19–30.

Zhang, R. F., Zhang, F. X., Zhang, M. W., Wei, Z. C., Yang, C. Y., Zhang, Y., … & Chi, J. W. (2011). Phenolic composition and antioxidant activity in seed coats of 60 Chinese black soybean (Glycine max L. Merr.) varieties. *Journal of Agricultural and Food Chemistry*, *59*(11), 5935–5944.

Zhou, R., Cai, W., & Xu, B. (2017). Phytochemical profiles of black and yellow soybeans as affected by roasting. *International Journal of Food Properties*, *20*(12), 3179–3190.

Zou, Y., & Chang, S. K. (2011). Effect of black soybean extract on the suppression of the proliferation of human AGS gastric cancer cells via the induction of apoptosis. *Journal of Agricultural and Food Chemistry*, *59*(9), 4597–4605.

5 Soybean Oil
Chemical Properties and Benefits for Health

Jane Mara Block, Renan Danielski,
Gerson Lopes Teixeira, and Itaciara Larroza Nunes

CONTENTS

5.1 INTRODUCTION

It has been reported that the expected production of soybean (*Glycine max* [L.] Merrill) and soybean oil worldwide in 2019/2020 will be 337.137 and 56.520 (in 1,000 metric tons), respectively. Currently, the largest world producer of soybeans is Brazil, with a production of 133 million tons and a planted area of 36.945 million hectares with a productivity of 3.272 kg/ha. The production in the USA, which has been the leading producer of soybeans in the last decade, dropped by almost 20% in relation to the previous harvest, falling from 120.52 in 2018/2019 to 96.84 million tons in the

DOI: 10.1201/9781003030294-5

2019/2020 harvest. In this period, the planted area in the USA was 3a0.332 million hectares with a productivity of 3.187 kg/ha. The production of soybeans in the USA increased to 113.5 million metric tons in 2020/2021 (CONAB, 2020; Farmnews, 2020; STATISTA, 2021a). These numbers show the commercial and economic importance of the soy complex in the world.

Soybeans are used in the animal food industry, for human consumption, and for non-food products as well. Soybean meal is mainly used as poultry and livestock feed since the value of its protein is much greater than that of other oilseeds. A small amount of soybean meal produced in the West (about 3% in the USA) is used in food products such as protein alternatives, soymilk, soy sauce, tofu, soy-fortified pasta, breakfast cereals and bars, beverages, and whipped toppings (Hammond, Johnson, Su, Wang, & White, 2005).

Soybean oil, which currently is the second-largest source of edible oil, is used for cooking, frying, and baking foods and as an ingredient in foods such as salad dressings, margarine, and shortenings. The global production of soybean oil in 2020/2021 was 60.26 million tons (STATISTA, 2021b), which represents 28.81% of the 209.14 million tons of total vegetable oils produced in the world. On the other hand, the global consumption of soybean oil reached 59.48 million tons in 2020/2021 (STATISTA, 2021c). Despite Brazil and the USA leading the soybean production, China is the largest soybean oil producer in the world, with a total volume of 17.74 million tons in 2020/2021 marketing year (STATISTA, 2021d). The production of soybean oil in Brazil reached 9.55 million tons in 2020 and is expected to maintain this trend in 2021. The domestic consumption was 8.53 million tons; about 1.11 million tons was exported (ABIOVE, 2021).

The primary component in soybean oil is triacylglycerol, which accounts for 94.4–97% of the lipid phase. The high level of linoleic acid (C18:2) and relatively high levels of linolenic acid (C18:3) make soybean oil susceptible to oxidation reactions. Genetic engineering methods have been used to modify the composition of soybean oil and increase its stability. In addition to triacylglycerols, soybean oil is rich in phospholipids, which can reach from 1.5% to 3.7% in crude oil (Hammond, Johnson, Su, Wang, & White, 2005; Perkins, 1995). It is used for producing commercial lecithin, a valuable co-product obtained from the refining of soybean oil.

The unsaponifiable matter, which represents 1.3–1.6% of the soybean oil, is composed of 2% squalene, 16% phytosterols, and 8.5% tocopherols (Hammond, Johnson, Su, Wang, & White, 2005). These minor compounds are considered bioactive and can be recovered from the deodorizer distillate during the last step of the soybean oil refining. The tocopherols, which are concentrated in the deodorizer distillate from 11% to 13%, can be used as vitamin E and as an antioxidant in the food industry. On the other hand, the sterols are used in the production of hormones, and the squalene in cosmetic preparations (Mendes et al., 2002).

The composition of fatty acids and triacylglycerols and the non-glyceride components of soybean oil and modified soybean oil will be presented in this chapter. Their importance in human nutrition, bioactivity, bioaccessibility, and their relationship to human health will also be discussed.

5.2   COMPOSITION OF SOYBEAN OIL

Soybean oil (Figure 5.1) has a major role in the development of many oil-based products worldwide (Biermann et al., 2011) mainly because it is produced in larger amounts than any other vegetable oil (Gunstone, 2005). The composition of vegetable oils is dependent on many factors such as plant variety, crop conditions, agriculture practices, as well as the quality of the seeds. On the other hand, the chemical and physical properties of the oil have a direct impact on the formulation and production of foods (Hammond, 2003).

Table 5.1 shows that crude refined soybean oil is composed of many chemical components, mostly neutral lipids, tri-, di-, and mono-acylglycerols, free fatty acids (FFAs), hydrocarbons, sterols, tocopherols, polar lipids such as phospholipids, and traces of metals (Gerde et al., 2020; Wang, 2011). The triacylglycerols represent the main component (> 94%) in soybean oil, followed by phospholipids (3.7%). The unsaponifiable matter represents about 1.45% of its composition,

FIGURE 5.1   Soybean oil. Source: United Soybean Board/Flickr.

TABLE 5.1
Composition of Crude and Refined Soybean Oil

Component	Crude	Refined
Triacylglycerol (%)	94.4–97	> 99
Phospholipids (%)	1.5–3.7	0.003–0.045
Unsaponifiable matter (%)	1.3–1.6	0.3
Sterols (mg/100 g)	236–330	130
Campesterol	59–68	47
Stigmasterol	54	47
β-Sitosterol	123–183	123
Δ^5-Avesnasterol	5	1
Δ^7-Stigmasterol	5	1
Δ^7-Avenasterol	2	< 0.5
Tocopherols (mg/100 g)	113–145	77
α	9–12	6–9
β	18	–
γ	74–102	45–50
δ	24–30	19–22
Hydrocarbons (%)	0.38	–
Free fatty acids (%)	0.3–0.7	–

Source: Hammond et al. (2005); Perkins (1995).

and it is composed of 26.20% hydrocarbons, 16.28% sterols, and 8.48% tocopherols. Other minor and non-identified compounds are among the remaining 49% of the unsaponifiable matter in soybean oil.

5.2.1 FATTY ACIDS OF SOYBEAN AND THEIR HEALTHFUL PROPERTIES

The fatty acid (FA) composition of seed oils has a direct relationship with the plant source. Differences in the content and distribution of each fatty acid are attributed to many factors, such as climatic or seasonal conditions (Fedeli & Jacini, 1971). Also, the FA composition of vegetable oils affects their main chemical and physical properties, especially their stability against oxidation and consistency, which define their applications in foods (Hammond, 2003). The chemical composition can also be used for authenticity purposes and for detecting fraud in oils such as soybean, sunflower, rapeseed, maize, and especially olive oil (Aparicio et al., 2018).

Table 5.2 shows that soybean oil is rich in linoleic (omega-6), oleic, palmitic, linolenic (omega-3), and stearic acids. Other FAs, such as myristic, palmitoleic, arachidic, gondoic, behenic, and lignoceric acids, are also reported in percentages lower than 0.5%. Most commercial soybean varieties usually present a typical composition with the same major FAs (Gerde et al., 2020; Gunstone, 2005). Important progress related to good yield and better quality of the oil is associated with plant breeding as well as other agronomic properties. As a result, the composition of soybean oil can be changed according to the objectives of the producer (Gerde et al., 2020; Gerde & White, 2008; Gerde, Hardy, Fehr, & White, 2007; Warner & Gupta, 2003). Besides using classical breeding, mutagenic agents and genetic engineering techniques can be used for the development of varieties with a modified fatty acid composition (Homrich et al., 2012).

Despite the wide variation in the FA profile of soybean oil, the soybean seeds do not present great variations in the yield in the oil content. However, it has been reported that varieties with elevated or low content of palmitic acid, compared to regular soybean, commonly present low oil content (Hartmann et al., 1996; Horejsi et al., 1994; Ndzana et al., 1994). Also, low yield and poor

---

TABLE 5.2

Fatty Acid Composition of Different Soybean Oil Samples

Fatty Acid	[A]	[B]	[C]	[D]	[E]	[F]*	[G]*	Range (%)**
Myristic (C14:0)	–	–	0.1	0.1	–	–	–	tr–0.03
Palmitic (C16:0)	10.0	10.0	10.3	11.6	11	15.9	15.6	3.2–26.4
Palmitoleic (C16:1)	–	0.2	0.2	0.1	–	–	–	tr–0.7
Stearic (C18:0)	4.1	3.5	3.8	3.4	–	4.1	4.5	2.6–32.6
Oleic (C18:1)	25.2	21.0	22.8	24.1	23	24.0	22.8	8.6–79.0
Linoleic (C18:2)	53.4	55.3	51.0	53.9	53	37.2	46.5	35.2–64.8
Linolenic (C18:3)	7.3	9.2	6.8	6.4	7	18.8	10.6	1.7–19
Arachidic (C20:0)	–	0.5	–	0.3	–	–	–	tr–0.7
Gondoic (C20:1)	–	–	0.2	–	–	–	–	tr–0.7
Behenic (C22:0)	–	0.3	–	–	–	–	–	tr–1.0
Lignoceric (C24:0)	–	–	–	–	–	–	–	0.1–0.5
ω–6/ω–3	7:1	6:1	7:1	8:1	7:1	2:1	4:1	–

Source: A = Neff and Byrdwell (1995); B = Hammond (2003); C = Dupont (2003); D = Kalo and Kemppinen (2003); E = Gunstone (2005); F and G = Yeom et al. (2020).

Notes

*High α-linolenic transgenic soybean oil.

**Gerde et al. (2020); tr = trace.

germination are reported for lines with a high content of stearic acid. The soil management, as well as the soil treatment, were described as important factors that can significantly affect the content of oleic acid in soybean oil (Marro et al., 2020).

Oleic acid is considered one of the most important FAs in vegetable oils, as it plays a major role in many chemical properties (Biermann et al., 2011). Although presenting 23% oleic acid on average, genetically engineered soybean oils for food applications with 79% of this FA were reported (Kinney, 1996). Traditional plant breeding can be used for the development of high-oleic soybean seeds, but the content of oleic acid can show a high variation. It can happen because of the growing environment, and it has a negative influence on its market value (Fehr & Hammond, 1997). The maturity of seeds and seed oil deposition also can have an impact on the FA composition of soybean oil. The palmitic and linolenic acids tend to decrease with maturity. On the other hand, linoleic acid increases (Fehr et al., 1971; Graef et al., 1985; Ishikawa et al., 2001). The FA composition of all vegetable oils has a strong relationship with its physicochemical properties, shelf-life, and health (Hammond, 2003; Yin et al., 2011).

The content of FFAs in crude soybean oil is lower than 0.1% (Evans et al., 1974; Wang, 2011) and it is strongly related to the harvest, quality of the initial raw material, and extraction process (Wang & Johnson, 2001). The FFAs are reduced considerably by refining processes, during the neutralization or deodorization in the classical or physical refining, respectively (O'Brien, 2008; Wang, 2011). The FFAs in the oil can negatively affect characteristics such as heat and moisture transfer properties, causing several problems in frying products (Hammond, 2003). Thus, the percentage of FFAs and unsaponifiable matter in fully refined oil must be lower than 0.05% and 1.5%, respectively (Gerde et al., 2020).

The refining of soybean oil is mandatory for improving its quality and sensory properties. The lecithin obtained after the degumming step is used as a surfactant. Some minor compounds from soybean oil, such as tocopherols and sterols, are recovered during refining processes because they have industrial applications (Greyt & Kellens, 2005; Gunstone, 2005).

5.2.2 Essential Fatty Acids

Polyunsaturated fatty acids (PUFAs) can be classified as ω–3 and ω–6 FAs. Linoleic acid (LA – ω–6) and alpha-linolenic acid (ALA – ω–3) are known as essential fatty acids. These FAs cannot be synthesized by humans but are indispensable for cell membrane synthesis and maintenance of optimal health (Das, 2006; Schmitz & Ecker, 2008; Simopoulos, 1999; Singh, 2005). Besides, ω–3 and ω–6 FAs are precursors of signaling molecules named "eicosanoids" that present opposing effects and modulate membrane microdomain composition, receptor signaling, and gene expression (Patterson et al., 2012; Schmitz & Ecker, 2008). When consumed, these PUFAs undergo a series of metabolic processes within mammalian cells to be further converted into very-long-chain FAs (VLCFAs). VLCFAs are constituents of cellular lipids such as sphingolipids and glycerophospholipids, as well as precursors of lipid mediators (Sassa & Kihara, 2014).

Figure 5.2 shows the main steps of the metabolism of ω–3 and ω–6 FAs. In this process, desaturase- and elongase-enzymes convert LA (18:2) to γ-linolenic acid (18:3), and dihomo-γ-linolenic acid (20:3), as well as arachidonic acid (AA) (20:4), all from the ω–6 family. The latter is a key intermediate to docosapentaenoic acid (22:5) or eicosanoids. Likewise, ALA undergoes the same enzyme pathway on the metabolization of stearidonic acid (SDA) (18:4) and eicosatetraenoic acid (20:4) to further produce eicosapentaenoic acid (EPA) (20:5). Then, EPA is further metabolized to docosahexaenoic acid (DHA) (22:6) or eicosanoids. Both AA and EPA have a crucial role in the production of VLCFAs, which are also vital for numerous human metabolic processes (Calder, 2006; Haag, 2003; Patterson et al., 2012; Schmitz & Ecker, 2008).

LA and ALA are involved in several biological activities, such as the development and maintenance of the brain (Haag, 2003; Singh, 2005; Uauy et al., 2001; Uauy & Dangour, 2008), optimal functioning of the retina (Connor et al., 2009; Uauy et al., 2001), prevention of chronic and

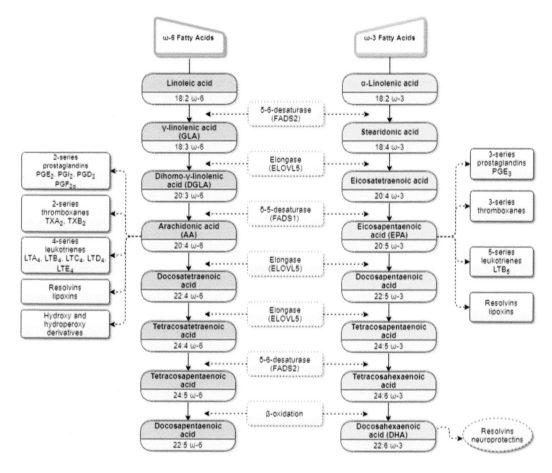

FIGURE 5.2 Metabolism of ω−6 and ω−3 polyunsaturated fatty acids. Source: adapted from Patterson et al. (2012).

heart diseases (Pizzini et al., 2018; Simopoulos, 1999), and regulation of the inflammatory process (Calder, 2006). However, the competition among ω−6 and ω−3 groups for the same series of enzymes on human metabolism causes a significant decrease in the conversion of the other (Schmitz & Ecker, 2008). In addition, the FAs derived from ω−6 and ω−3 present different activities in human metabolism. Generally, the eicosanoids produced from ω−3 PUFAs, such as EPA and DHA, present anti-inflammatory properties, while those derived from ω−6 PUFAs, such as AA, show pro-inflammatory activity (Calder, 2006; Das, 2006; DiNicolantonio & O'Keefe, 2018a; Patterson et al., 2012; Simonetto et al., 2019). Thus, great attention has been given to the ω−3 and ω−6 ratio in the diet, as it strongly affects the production of autacoids such as leukotrienes, prostaglandins, and thromboxanes (Calder, 2006; Das, 2006; Harris, 2018). It has been reported that the ingestion of a low ω−6/ω−3 ratio has several positive effects on human metabolism (DiNicolantonio & O'Keefe, 2018). Elevated production of anti-inflammatory autacoids can be achieved with an increased intake of ω−3 FAs in contrast to low consumption of ω−6. As a result, the synthesis of inflammatory eicosanoids from ω−6 AA is reduced (Schmitz & Ecker, 2008). Several metabolic inter-relationships are involved in the biotransformation of FAs in the human body. As shown in Figure 5.2, LA and ALA undergo a complex series of desaturation and elongation reactions to be converted into their higher unsaturated derivatives (Candela et al., 2011; Russo, 2009; Simopoulos, 2010). The enzymes cyclooxygenase and lipooxygenase are responsible for converting AA and EPA into eicosanoids as prostaglandins, leukotrienes, and thromboxanes that play a crucial role in cellular signaling and inflammation (Le et al., 2009). The ω−3 PUFAs such as ALA are highly important in human metabolic processes because they can decrease the development of those pro-inflammatory eicosanoids by competing with ω−6 PUFAs and displacing AA in phospholipids membrane (Prasad et al., 2020; Simonetto et al., 2019).

Some reports suggest that the ancient human diet included the consumption of an equilibrated ω−6/ω−3 ratio near to 1:1 (Simopoulos, 2002, 2016, 2020). However, changes in dietary patterns over the last 150 years and the increased consumption of ω−6 in Western diets caused an imbalance in the ω−6/ω−3 ratio. The ideal 1:1 proportion then raised to 20–30:1 in 2010 (Candela et al., 2011) and dropped to approximately 16:1 nowadays (Simopoulos, 2020). Since most populations have different eating habits, mostly with low consumption of ω−3 FAs, an ideal 1:1 ω−6/ω−3 ratio is difficult to accomplish. However, it has been reported that an intake ratio of ω−6/ω−3 between 4 and 5:1 is satisfactory to achieve positive effects and maintain a healthy diet (Martin et al., 2006). In soybean oil ratios of ω−6/ω−3 from 2 to 8:1 were reported (Table 5.2). The diets which contain a low ω−6/ω−3 ratio have a significant and positive impact on biological processes involved in the maintenance of normal brain function (Haag, 2003). Also, the consumption of a balanced ratio of ω−6/ω−3 between 2.5 and 5:1 has beneficial effects for patients with asthma and rheumatoid arthritis and for the prevention of cardiovascular diseases (Simopoulos, 2002). On the contrary, excessive consumption of ω−6 PUFAs and high ω−6/ω−3 ratio represents a risk for overweight and obesity (Simopoulos, 2016) and can cause several illnesses such as cancer, inflammatory processes, cardiovascular and autoimmune diseases (Simopoulos, 2002). Sublette et al. (2006) also support that low levels of DHA and high proportions of ω−6/ω−3 are associated with major depression and, possibly, suicidal behavior.

Although considered a useful metric for decades, there is controversial evidence reinforcing that the ω−6/ω−3 ratio is of little value regarding cardiovascular diseases, and may be imprecise, non-specific, and scientifically outdated (Griffin, 2008; Harris, 2007; Pizzini et al., 2018). Instead, the "Omega-3 Index", which considers the content of EPA + DHA acids in the red blood cells, is suggested as a more accurate tool in clinical evaluations (Harris, 2007, 2018; Harris et al., 2008; Whelan, 2009). According to Harris (2018), chronic diseases in Western cultures may not be a problem related to the high consumption of LA and ALA, but the low intake of EPA and DHA, a reason why the Omega-3 Index could be more useful as an emerging cardiovascular risk marker. Furthermore, a report found that the absolute content of ALA and LA in the diet influences the conversion of ALA in humans but not their ratio (Goyens et al., 2006).

Countries with high consumption of ω−6 PUFA such as LA, usually have major challenges, because the excessive consumption of oils rich in those FAs favors many health issues, such as oxidative stress, the risk of heart diseases, as well as chronic low-grade inflammation and atherosclerosis (Dinicolantonio & O'Keefe, 2018b). The lower cost and higher availability and consumption of ω−6 PUFAs oils, compared to fish oils rich in ω−3 FAs, are correlated with the higher occurrence of cardiovascular diseases (CVD) in developing countries (Singh, 2005). Particularly in India, some health problems such as the high rates of CVD are associated with the elevated intake of SFA, trans-fatty acids (TFA), and other dietary habits (Misra et al., 2011; Prabhakaran et al., 2016, 2018). The Asian country is the second-largest consumer and the major importer of vegetable oils (OECD/FAO, 2019), as well as the main importer of soybean oil worldwide (India, 2019). Nevertheless, India has a high consumption of ω−6 PUFAs, presenting a very high ω−6/ω−3 ratio of about 38:1 (Prasad et al., 2020).

Currently, CVD is the main cause of mortality among citizens in India (Prabhakaran et al., 2016, 2018). The high consumption of TFA and SFA, especially from coconut oil, which accounts for 80% of the fat intake among Southern Indians, in contrast with the low intake of PUFAs and monounsaturated FAs, contributes to higher high CVD rates in India (Misra et al., 2011; Prabhakaran et al., 2018). As the increasing intake of plant-based ω−3 FA may be associated with the reducing risk of certain diseases, the adoption of healthier oils in the diets of such populations may be an alternative option to handling disorders related to an imbalanced consumption of FAs (Prasad et al., 2020). Hence, soybean oil with increased content of ω−3 ALA in contrast to the low content of ω−6 LA is more desirable (Yeom et al., 2020). Foods with high levels of ALA usually have a reduced shelf life, and the addition of antioxidants is necessary, as this FA is highly susceptive to oxidation (Dubois et al., 2007).

5.2.3 TRIACYLGLYCEROL COMPOSITION

The major neutral lipids in soybean oil are triacylglycerols (TAG). Most TAG molecules in soybean oil present at least two unsaturated FAs due to their high content of monounsaturated and polyunsaturated FAs (Wang, 2011). The content and distribution of each TAG species in soybean oil change with the FA composition (Neff et al., 1999; Neff & Byrdwell, 1995). Table 5.3 shows the major acyl groups and the triacylglycerol composition of soybean oil. The TAG profile of soybean oil is mostly composed of LLL, LLO, and LLP (30%, 27%, and 21%, respectively). Gunstone (2005) reported a TAG composition for soybean oil mostly distributed by LLO, LLL, and LLS. The content of an acyl group in a specific position is associated with the content of that acyl group in the whole TAG structure (Harp & Hammond, 1998).

The distribution of the acyl group in the TAG structure in soybean oil also has an impact on its oxidative stability. The higher levels of unsaturated FAs in the sn-2 position of the soybean oil TAG may enhance its oxidative stability when compared to TAGs with unsaturated FAs on the sn-1 and sn-3 positions (Neff & List, 1999; Wang, 2011). In most vegetable oils, the occurrence of unsaturated FA is higher in the sn-2 position rather than the sn-1 or sn-3 positions, which usually shows SFAs (Kubow, 1996). The fatty acids palmitic and stearic are preferentially in the sn-1 position in soybean, sunflower, peanut, and high-erucic rapeseed oils (Harp & Hammond, 1998; Kalo & Kemppinen, 2003). The position of the fatty acid in the TAG has a key role in the process of digestion and absorption and determined whether FAs are absorbed as 2-monoacylglycerol (2-MAG) or FFA (Ramírez et al., 2001). Pancreatic and lipoprotein lipases preferentially target the FAs in the sn-1 and sn-3 positions of TAG, resulting in FFAs and 2-MAG (Karupaiah & Sundram, 2007; Kubow, 1996). It was suggested that pancreatic lipase presents a higher affinity for ester bonds in the sn-1 position, compared to the sn-3 position (Rogalska et al., 1990). The rate of hydrolysis at the sn-2 position is very slow, and the FA in the sn-2 position remains as 2-MAG during digestion and absorption. Long-chain SFAs have digestion and absorption less readily than shorter-chain or more highly unsaturated FAs. It is also reported that ω−3 PUFAs occupying the sn-2 position are preserved during absorption (Kubow, 1996).

The content of LA in conventional soybean oil, which exceeds 50%, is found in high proportions in the TAGs, decreasing its oxidative stability. Thus, breeders, farmers, and research institutes are still seeking to develop enhanced soybean traits aiming at solving problems related to physical and chemical parameters such as stability and nutritional profiles (Monsanto, 2020; Pham et al., 2012; Qualisoy, 2020; Warner & Gupta, 2006).

5.2.4 SOYBEAN OIL OF MODIFIED FATTY ACID COMPOSITION AND ITS HEALTH BENEFITS

The commodity of soybean oil is mostly intended for food purposes (Biermann et al., 2011). However, its use in foods has raised several questions over the past few years. Recently, after the recognition by the World Health Organization (WHO) of the connection between the consumption of trans fat and the incidence of heart attacks and other heart diseases (Morin & Lees, 2018; WHO, 2018), many countries decided to completely ban or reduce considerably the use of partially hydrogenated oils (PHO) in their food products. Brazil, Canada, and the USA approved regulations in 2018 seeking to ban PHO, which is the major source of industrial trans fat in foods. In addition, a total of 23 countries from different parts of the world are committed to creating new policies and regulations on trans fat in foods (WHO, 2018). The main producers of soybean for decades,

TABLE 5.3

Triacylglycerol (TAG) Composition of Soybean Oil (%)

TAG Species	[A][1]	[B][2]	[C][3]	[D][4]							[E]	Range
LnLnLn	0.2	0.1	0.4	–	–	–	–	–	–	–	–	0.1–0.4
LnLnL	1.1	1.2	2.0	–	–	–	–	–	–	–	–	1.1–2.0
LLLn	6.0	4.8	1.5	–	–	–	–	–	–	–	–	1.5–6.0
LnLL	–	–	–	2.0	1.2	1.4	3.2	2.6	2.08	0.83	7	2.0–3.2
LnLnO	1.4	1.3	4.8	0.0	0.1	0.1	0.2	0.1	0.1	0.07	–	0–4.8
LnLnP	0.5	0.4	2.3	0.0	0.0	0.4	0.6	0.1	0.22	0.27	–	0–2.3
LLL	17.3	9.0	6.8	30.0	11.5	3.7	9.6	6.5	12.26	10.35	15	3.7–30
LnLO	5.1	4.9	2.4	1.7	1.5	0.8	2.0	1.9	1.58	0.48	5	0.8–5.1
PLLn	3.1	3.8	13.6	–	–	–	–	–	–	–	–	3.1–13.6
LLO	17.2	9.2	4.5	26.9	14.4	3.6	8.7	7.1	12.14	9.13	16	3.6–26.9
LnOO	1.3	1.1	0.0	0.4	0.5	0.1	0.3	0.3	0.32	0.15	–	0.1–1.3
LLP	12.1	10.6	19.4	6.4	20.7	17.5	21.4	11.6	15.52	6.40	–	6.4–21.4
LnOP	1.4	1.6	2.6	0.1	1.0	1.4	2.1	0.4	1.0	0.8	–	0.1–2.6
LnPP	–	–	–	0.1	0.0	1.5	2.0	0.1	0.74	0.94	–	0–2.0
PPLn	0.3	0.3	2.5	–	–	–	–	–	–	–	–	0.3–2.5
LOO	–	–	–	13.9	7.3	1.3	3.1	2.5	5.62	5.15	8	1.3–13.9
OOL	8.4	3.0	1.2	–	–	–	–	–	–	–	–	1.2–8.4
LLS	3.0	9.7	0.0	3.6	2.6	7.7	2.0	13.0	5.78	4.61	13	2.0–13.0
LOP	–	–	–	3.7	16.3	7.4	12.2	6.4	9.2	5.02	–	7.4–16.3
POL	8.3	7.3	12.4	–	–	–	–	–	–	–	–	7.3–12.4
PLP	–	–	–	0.8	8.6	13.8	14.8	2.0	8.0	6.48	–	0.8–14.8
PPL	1.5	2.2	16.0	–	–	–	–	–	–	–	–	1.5–16.0
OOO	2.9	1.4	0.6	4.6	2.1	1.0	0.8	1.1	1.92	1.58	–	0.6–4.6
OOS	0.2	1.2	0.1	–	–	–	–	–	–	–	5	0.1–5.0
LOS	–	–	–	2.6	2.3	3.9	1.3	11.8	4.38	4.25	8	1.3–11.8
SOL	2.6	8.3	0.5	–	–	–	–	–	–	–	–	0.5–8.3
POO	–	–	–	0.9	3.2	0.6	1.0	0.5	1.24	1.11	–	0.5–3.2
SLP	1.0	–	–	0.8	2.2	0.0	0.0	0.0	0.60	0.96	–	0–2.2
LnSS	–	5.6	–	0.0	0.0	16.0	3.0	8.8	5.56	6.85	–	0–16.0
POP	–	–	–	0.2	1.7	1.2	1.5	0.3	0.98	0.69	–	0.2–1.7
SOO	–	–	–	0.7	0.5	0.3	0.2	2.1	0.76	0.77	–	0.2–2.1
SOP	0.4	1.2	0.4	0.1	0.4	1.3	0.2	1.4	0.68	0.62	–	0.2–1.4
PPP	0.1	0.4	0.1	0.0	0.0	0.6	0.0	3.4	0.8	1.48	–	0–3.4
SOS	0.2	0.7	0.4	0.0	0.0	0.6	0.0	0.1	0.14	0.26	–	0–0.7
SSS	0.1	0.1	0.1	0.0	0.0	0.2	0.0	0.1	0.06	0.09	–	0–0.2

Sources: A, B, and C = Neff et al. (1999); D = Neff and Byrdwell (1995); E = Gunstone (2005).

Notes

P = palmitic; S = stearic; O = oleic; L = linoleic; Ln = linolenic.

[1] Regular soybean.

[2] High stearic soybean.

[3] High palmitic soybean.

[4] Transgenic soybean.

countries such as the USA, Brazil, Argentina, and China, have an important role in changes in the legislation (Gunstone, 2005).

The concerns on the consumption of trans fat from PHO and its association with harmful health effects also encouraged many studies on the modification and development of soybean varieties with an altered FA composition (Clemente & Cahoon, 2009; Homrich et al., 2012; Lee et al., 2019; Neff & List, 1999; Primomo et al., 2002; Yang et al., 2018; Yeom et al., 2020). These new soybean varieties can improve chemical (FA composition and oxidative stability), physical (melting point and consistency), and nutritional properties of the soybean oil (Cahoon, 2003; Kalo & Kemppinen, 2003; Kinney, 1996; Uauy & Dangour, 2008). Research on soybean seeds with a specific composition and a balanced ω−6/ω−3 ratio in order not to change the composition of the oil by chemical, physical, or enzymatic methods has been done and is still being developed (Demorest et al., 2016; Homrich et al., 2012; Lee et al., 2019; Marro et al., 2020; Pham et al., 2012; Qualisoy, 2020; Yang et al., 2018; Yeom et al., 2020). Classical techniques, as well as bioengineering, have been successfully used over the last decades in the development of soybean seeds with an enhanced composition (Clemente & Cahoon, 2009; Kinney, 1996; O'Brien, 2008). Most of the changes made in the soybean oil were focused on reducing the concentration of PUFA (Gerde & White, 2008; Kanai et al., 2019; Primomo et al., 2002; Yang et al., 2018; Yeom et al., 2020) and increasing the oleic acid content (Demorest et al., 2016; Kim et al., 2015; Kinney, 1996; Pham et al., 2010; Pham et al., 2012). Soybean cultivars with low content of palmitic acid (Cardinal et al., 2014; Cherrak et al., 2003) or palmitic and linolenic acids were developed, focusing on enhancing its nutritional quality and oxidative stability (Primomo et al., 2002). Studies have reported the development of soybeans with high α-linolenic acid content and a low ω−6/ω−3 ratio (Yeom et al., 2020), as well as oleic-acid-enriched seeds (Demorest et al., 2016; Pham et al., 2010; Pham et al., 2012), through genetic modification.

The FA composition of a specific oil or fat is critical for its performance. The high saturated and high stearic acid soybean oils can be used for margarine and shortening applications as they require particular attributes related to plasticity and melting (Gerde & White, 2008). On the other hand, frying processes require high resistance to oxidation. Therefore, it is recommended to use oils with reduced concentrations of PUFA, especially linolenic acid, for this application (Gerde et al., 2007; Warner & Gupta, 2003). Oils with a higher content of oleic acid are considered more resistant to oxidation and thermal oxidation during the frying process (Cahoon, 2003; Nill, 2015).

Table 5.4 shows that soybean oils with modified FA include a high concentration of stearic, or oleic, acids, a medium and high concentration of oleic acid, as well as a low concentration of linolenic acid. Soybean traits with a high amount of stearidonic acid are reported as well.

The high-oleic soybean oil should present more than 70% of oleic acid (Huth et al., 2015). Moreover, some genetically engineered seeds may achieve up to 90% of oleic acid (Cahoon, 2003). The first experiments using HOSO showed high stability under frying conditions but caused changes in the flavor of fried products (Warner et al., 1997). However, the HOSO currently commercialized has a neutral flavor, and it is being used for the formulation of fats for frying, bakery, and ice cream, without changing the sensorial characteristics for these products (Qualisoy, 2020).

Soybean seeds rich in oleic acid and reduced PUFA started to be commercialized in the USA a few years ago. Plenish™, Vistive® Gold, SOYLEIC™, and Calyno™ were developed by DuPont-Pioneer (now Corteva Agriscience), Monsanto-Bayer, University of Missouri, and Calyxt, respectively (Calyxt, 2020; DuPont Pioneer, 2020; Missouri Soybean, 2020; Monsanto, 2020). Vistive® Gold, Plenish™, SOYLEIC™, and Calyno™ present up to 72%, 76%, 82%, and 84% oleic acid, respectively. These oils show a reduced content of SFA and a fatty acid composition similar to olive oil. The use of HOSO has many advantages over hydrogenated oils since it can be used in deep-frying and baking, reducing the intake of TFA (Huth et al., 2015). Warner and Gupta (2006) reported that fried potato chips showed higher shelf life when prepared with HOSO as compared to cottonseed and low linolenic acid soybean oils.

Recently the USA Food and Drug Administration (FDA) has approved the health claim for oleic acid in edible oils after supporting evidence that consuming oleic-acid-rich oils from different

TABLE 5.4

Fatty Acid Profile of Oils from Soybeans with Modified Fatty Acid Compositions

Oil	Composition (%)						Reference
	16:0	18:0	18:1	18:2	18:3	18:4	
Soybean oil	9.5–13.5	3.0–6.1	17.7–28.5	46.2–57.1	5.5–11.0	–	Firestone (2013)
High palmitic	25.8	3.5	13.6	43.1	13.8	–	Neff and Byrdwell (1995)
High stearic	9.3	27.2	16.7	38.6	6.5	–	Warner (2008)
	8.9	17.2	16.7	47.2	10.0	–	Neff and Byrdwell (1995)
High saturates	23.3	20.0	10.5	39.7	6.5	–	Warner (2008)
Low saturates	4.6	3.8	22.4	62.0	7.1	–	Wang and Briggs (2002)
Mid oleic	8.9	5.9	52.1	31.2	1.0	–	Warner and Fehr (2008)
High oleic	6	3	86	2	2	–	Liu (1999)
	7.6	2.9	85.9	1.9	1.8	–	Pham et al. (2012)
	7.9	3.7	82.2	2.3	3.9	–	Pham et al. (2010)
	7.3	3.4	85.1	1.3	2.0	–	Warner and Gupta (2006)
	6.7	3.8	79.2	7.2	3.1	–	Wang and Briggs (2002)
	9	3	79	3	6	–	Wilson (1999)
	7.4	3.0	77.8	4.9	6.8	–	Kim et al. (2015)
Plenish™	6.5	3.9	79.4	8.1	2.2	–	Seemamahannop et al. (2019)
	11	4	76	7	2	–	Zhang et al. (2018)
	6.1	4.1	71.9	9.1	2.7	–	1Tarté et al. (2020)
	6.3	3.9	75.7	6.7	1.6	–	Napolitano et al. (2018)
Vistive Gold®	2.5	3.5	72	16	3	–	Monsanto (2020)
	2.9	3.0	71.5	19.8	2.7	–	Seemamahannop et al. (2019)
SOYLEIC®	7.8	6.0	82	4.9	2.7	–	Missouri Soybean (2020)
Low linolenic	10.8	4.5	26.1	55.4	3.0	–	Warner (2008)
	11.3	4.7	22.4	59.7	2.0	–	Warner and Gupta (2003)
	10.7	4.5	25.0	56.6	3.0	–	Wang and Briggs (2002)
	11.4	4.4	25.0	55.3	3.7	–	Warner and Mounts (1993)
Ultra-low linolenic	11.5	4.6	24.8	58.2	0.8	–	Warner and Gupta (2003)
	9.9	5.4	27.8	54.9	1.1	–	Warner and Fehr (2008)
	9.9	6.0	26.0	56.6	1.5	–	Gerde et al. (2007)
High stearidonic	11	4	20	24	10	20	Mermelstein (2010)
	11.2	4.6	18.1	31.5	10.6	16.6	Harris et al. (2008)
	12.7	4.3	14.5	23.7	10.9	25.0	Kleiner et al. (2012)

sources may reduce the risk of coronary heart disease – CHD (FDA, 2018). Likewise, the FDA also supported the qualified health claim for the unsaturated FA from soybean oil on reducing the risk of CHD. The daily consumption of about 20 g of soybean oil has positive effects on human health, decreasing the risk of CHD (FDA, 2017). Therefore, the recent research on the development of soybeans with modified FA composition was mostly targeted at oleic-acid-rich seeds. The commercial HOSO currently available is produced from soybean seeds grown exclusively in the USA. It is among the main oils with modified FA composition marketed in the United States (United Soybean Board, 2020). The production of high oleic soybean in the USA is increasing, and it is expected to reach an estimated nine billion pounds by 2027 (Qualisoy, 2020). However, there is an increasing concern about the growth and consumption of plants produced by direct genetic modification using genes from different species, often referred to as "genetically modified organisms – GMO". The United Soybean Board, a farmers' organization in the USA, is promoting the production of high-oleic soybean varieties engineered with approaches different from those of conventional GMO techniques. Thus, some producers refer to their soybean as "gene-edited non-GMO" traits (United Soybean Board, 2020). Calixt claims that their high-oleic soybeans are "non-GMO", as they do not add any foreign DNA into the product, but instead, they edit the soybean gene (Calyxt, 2020; Wilke, 2019). This action was also a response to the concern of the consumers regarding foods with trans fat. The planted acreage of high-oleic soybean varieties achieved 650,000 acres (263,045 hectares) in 2017. Farmers from 13 states in the USA were able to grow high-oleic soybeans in 2018 (United Soybean Board, 2020). Projections are expected to increase the planted acreage to 16 billion pounds and 647,497,027 hectares (Qualisoy, 2020).

SDA, an ω–3 FA with 18 carbons and four double bonds (Decker et al., 2012), is an intermediate in the biosynthesis of the very-long-chain polyunsaturated ω–3 fatty acids eicosapentaenoic acid (EPA) and docosahexaenoic acid (DHA) (Clemente & Cahoon, 2009). Although rarely occurring naturally among vegetable oils (Gunstone, 2006), soybean varieties enriched with SDA have been successfully developed in order to improve their nutritional profile (Decker et al., 2012; Harris et al., 2008; Mermelstein, 2010). The reasons to produce SDA-enriched soybean oils are mainly due to the low consumption of marine-based (ω–3) PUFA in many countries such as the USA, India, and Europe (Banz et al., 2012; Prasad et al., 2020). Also, because the conversion of ALA into SDA via $\Delta6$ desaturase to further be transformed to EPA in the human metabolism has limitations, since ALA may undergo oxidation during the process. In contrast, SDA does not need $\Delta6$ desaturase to be converted to EPA and is thus considered a "pro-EPA" FA (Banz et al., 2012; James et al., 2003; Whelan, 2009). Therefore, potential health benefits for SDA have been reported (James et al., 2003; Prasad et al., 2020; Ruiz-López et al., 2009; Whelan, 2009), and positive claims have been attributed to this SDA-enriched soybean oil. The consumption of SDA is considered a strategy for increasing tissue EPA and DHA levels in humans, reducing the dependency on fish oil for the intake of omega-3 fatty acids (Hammond et al., 2008; Harris et al., 2008; James et al., 2003; Lemke et al., 2010; Nill, 2015).

Clinical trials showed that vegetable oils containing SDA would be more effective in increasing tissue EPA concentrations than ALA-rich oils, suggesting SDA as a dietary source of ω–3 FA. The study also recommended the use of SDA oils in food products as an alternative source of EPA in the human diet (James et al., 2003). The health benefits of consuming SDA-enriched soybean oil were also clinically proven, and the trials indicated that supplementation with this oil increases the percentage of red blood cells EPA and the omega-3 index (Lemke et al., 2010).

Whittinghill and Welsby (2010) reported that several foods such as bagels, breakfast bars, pastries, cookies, icings, and chocolate coatings formulated with SDA-enriched soybean oil ($\approx$ 23% SDA) showed 7 points on average in a 9-point hedonic scale. Good consumer acceptance also has been reported by Decker et al. (2012). Qualisoy® is an independent, third-party collaboration that promotes the development of and builds the market for the latest soybean traits in the USA. The organization is currently promoting the "SDA omega-3 soybean oil", which was engineered to show between 18% and 20% SDA (Qualisoy, 2020). The safety of soybean oil rich in SDA has been confirmed by toxicological studies, and it was considered safe for consumption by humans without present adverse effects (Hammond et al., 2008). In addition, the FDA has endorsed SDA with the status of Generally Recognized as Safe (GRAS) for use in foods and beverages. Despite the several advantages of SDA-enriched soybean oil, the available traits are still being improved in the USA, and the oil is not produced at a commercial level (Qualisoy, 2020).

Soybean oil with 1.9–2.5% and with 1% of LA (18:3) are referred to as low-linolenic oil (Fehr & Hammond, 1997) and ultra-low linolenic oil (Warner & Gupta, 2003), respectively. These oils were developed to replace hydrogenated soybean oil for reducing the consumption of TFA, as well as improving the stability of processed foods (DiRienzo et al., 2006). Although presenting better performance under frying conditions (Dupont, 2003; Warner & Gupta, 2003), the commercial introduction of low-linolenic and ultra-low-linolenic soybean oils has been hindered by the availability of corn oil, which has a composition very similar to those oils (Tan & Shahidi, 2012). The main company that produced ultra-low-linolenic soybean oil (Asoyia Inc.) was formed by 25 farmers from Iowa (USA) and it started its operations in 2004. However, the company did not succeed and closed its doors after working for only six years (AOCS, 2010). Therefore, the current research and production of modified soybean are mostly targeted at seeds enriched with oleic and stearidonic acids, as they produce oils with more positive claims for human consumption.

5.3 NONACYLGLYCEROL COMPONENTS OF SOYBEAN OIL: STRUCTURE, OCCURRENCE, AND HEALTH-PROMOTING EFFECTS

5.3.1 PHOSPHOLIPIDS

5.3.1.1 Structure and Technological Relevance

Phospholipids are a group of complex lipids composed of two fatty acids esterified to a glycerol molecule (positions sn-1 and sn-2), containing a phosphate group that can be attached to an alcohol component (e.g. choline, serine, inositol, and ethanolamine). Phospholipids that do not possess an alcohol group are called phosphatides, and they can be found in the cell membranes in small quantities. Due to their structural characteristics, phospholipids are amphipathic molecules. In other words, they carry hydrophobic (fatty acids) and hydrophilic (phosphate and alcohol) moieties, which makes it possible for them to interact with both polar and non-polar substances (Gundermann, Kuenker, Kuntz, & Droździk, 2011).

Due to their amphipathic nature, phospholipids are major components of cell membranes, forming lipid bilayers, where their hydrophobic tails (fatty acids) face each other and form a membrane of hydrophilic heads interacting with the water. This configuration allows polar and non-polar molecules to diffuse through the membranes (Gundermann et al., 2011).

Soybean is the richest source of phospholipids among oilseeds. The commercial soybean lecithin is composed of around 48% phospholipids, and it is obtained as a co-product of the oil degumming process, the first step of soybean oil refining. Crude lecithin is typically composed of 18% phosphatidylcholine (PC), 14% phosphatidylethanolamine (PE), 9% phosphatidylinositol (PI), 5% (phosphatidic acid), 2% minor phospholipids (PL), 11% glycolipids, 5% complex sugars, and 37% neutral oil. Afterward, lecithin is defatted in order to achieve higher purity, increasing the phospholipid levels (23% PC, 20% PE, 14% PI, 8% PA, and 8% minor PL) (Gulkari et al., 2017).

Phosphatidylcholine presents a choline group linked to the glycerophosphoric acid. The hydrophobic portion of PC is generally composed of a saturated (e.g. palmitic, hexadecanoic acid) and an unsaturated (e.g. oleic acid) fatty acid. After the ingestion, the phosphatic acid and soluble choline are released from PC into the cytosol by the enzyme phospholipase D. Phosphatidylethanolamine is structurally similar to PC. However, the polar group in the molecule is ethanolamine, instead of choline. In living organisms, PE accounts for 25% of all phospholipids, being found in nervous tissues in humans. Stearic acid (sn-1 position) and arachidonic acid (sn-2 position) are the most common fatty acids in the PE and PI molecules. The polarity of PI is due to the presence of an inositol group, and this phospholipid is involved in lipid and cell signaling, as well as membrane trafficking (Scholfield, 1981). The structures of soybean phospholipids are shown in Figure 5.3.

Lecithin is considered a valuable co-product, largely used in the industry as an emulsifier, decreasing surface tension in emulsion-based products such as fat spreads. It also improves the functional properties of powdered food products (e.g. cocoa, milk) and baked goods. Lecithin is also used in the pharmaceutical industry to stabilize encapsulated compounds and in animal feed to improve palletization (List, 2015). Over the decades, lecithin has become as important as soybean oil. According to a 2019 projection (Market Watch, 2019), the worldwide market for soy lecithin is expected to grow 3% in the next five years, going from US$ 150 million in 2019 to US$ 179 million in 2024.

Besides the classical functional properties of lecithin, this compound has been recently used as a component of a novel chemically synthesized category of molecules with increased antioxidant

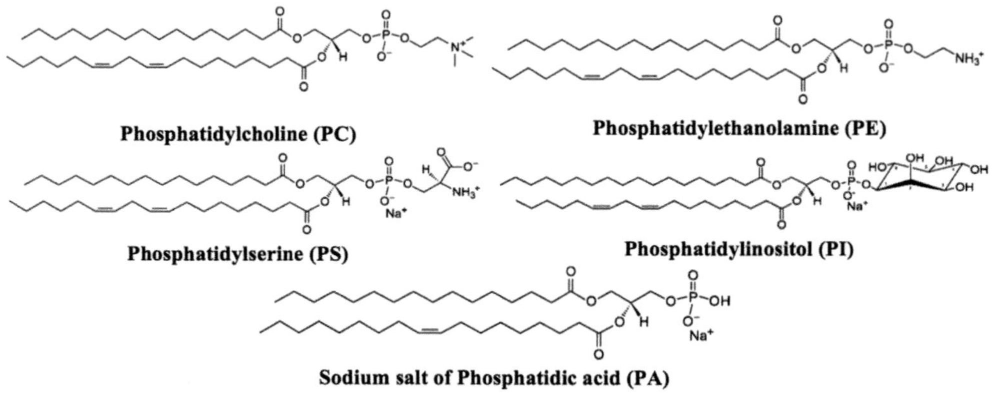

FIGURE 5.3 Phospholipids present in soybean oil. Source: adapted from Wang (2008).

capacity known as phenolipids. It has been reported that hydrophobic antioxidants are more efficient than hydrophilic ones in systems of high surface/volume ratio, such as emulsions due to a higher affinity for the oil/water interface. This mechanism is known as the antioxidant polar paradox and suggests that lipid oxidation in more polar food systems, such as oil-in-water emulsions or liposomes, would be more efficiently prevented with the use of lipophilic antioxidants. On the other hand, hydrophilic antioxidants would be more effective in less polar food systems such as bulk oils (Zhong & Shahidi, 2012). However, according to Zhong and Shahidi (2012), the polar paradox theory presents some limitations, such as the effect of the hydrophobic chain length and the antioxidant concentration in the system. According to the authors, at low concentrations, fat-soluble antioxidants would act more efficiently in lipophilic environments, while the opposite effect would dominate at high concentrations. Therefore, the existence of a threshold value for antioxidant concentration is hypothesized, where the solubility effect overcomes the interfacial phenomenon.

Several studies have incorporated fatty acids into hydrophilic phenolics in order to increase their hydrophobicity. A quercetin-lecithin complex (1:99 and 3:97, w/w tested) could retard the oxidation of sunflower oil through radical scavenging more effectively than when using quercetin or lecithin alone, showing the existence of a synergistic interaction between the two compounds (Ramadan, 2012).

Lecithin can also be used to create delivery systems for target substances. Rashidinejad et al. (2014) reported the incorporation of catechin and epigallocatechin gallate into soy lecithin liposomes. The application of 0.25% w/v of such structures to low-fat hard cheese showed retention considered satisfactory, with no catechin and epigallocatechin gallate identified in the whey, as determined by HPLC, indicating that the compounds were all retained in the cheese. The authors concluded that besides being a protein and calcium source, the cheese also became a source of antioxidant compounds, increasing the product's functionality and nutritional quality.

5.3.1.2 Bioavailability and Health Effects

The absorption of phospholipids starts in the intestinal lumen, where they have their ester bonds hydrolyzed in the positions *sn*-1 and *sn*-2 by the enzymes phospholipase A1 or A2. The hydrolysis results in a lysophospholipid and a free fatty acid. These molecules are further regenerated and incorporated into the surface of chylomicrons and to a lesser extent into very-low-density lipoproteins (VLDL). Subsequently, the chylomicrons and VLDL are released into the lymph and are ready to reach the circulatory system. In the bloodstream, the chylomicrons are decomposed by the lipoprotein lipase, resulting in free fatty acids that originate in the triacylglycerols and ultimately reach the tissues. On the other hand, the phospholipids liberated upon chylomicrons hydrolysis are hydrolyzed by endothelial lipase. The non-hydrolyzed phospholipid portion is incorporated into high-density lipoproteins (HDL). Therefore, phospholipids play a substantial role in the composition of lipoproteins, which influences the lipid composition and functionality of the tissues receiving them (Robert, Couëdelo, Vaysse, & Michalski, 2020).

Moreover, lecithin, which has the ability to interact with both polar and non-polar substances, has been widely used as a carrier system for compounds that present low bioavailability. The fat-soluble monophenols such as the tocopherols are better absorbed when solubilized in lipophilic systems. Kimura et al. (1989) dispersed d-α-tocopherol acetate in soy lecithin containing medium-chain triacylglycerols and orally administered the preparation to male Wistar rats. The transport of d-α-tocopherol acetate was remarkably improved, being absorbed through the lymphatic route and showing a slow and prolonged absorption pattern.

Curcumin, a polyphenol with antioxidant and anti-inflammatory activities, is poorly absorbed in the gastrointestinal (GI) tract due to its lack of stability under GI's pH conditions. Takahashi et al. (2009) reported an increase in curcumin bioavailability and a higher plasma antioxidant activity for encapsulated curcumin (2.5%) in liposomes composed of soy lecithin (5%). The capsules were orally administered to Sprague-Dawley rats at a dose of 100 mg curcumin/kg body weight. The authors concluded that the encapsulation in liposomes was efficient for increasing the bioactivity of the polyphenol.

It has been reported that phospholipids can impact the content of several lipid classes in the plasma. The administration of soybean lecithin (23% PC) to hypercholesterolemic rabbits was demonstrated to significantly reduce total plasma cholesterol (Polichetti et al., 2000). Wilson et al. (1998) reported that hypercholesterolemic monkeys and hamsters consuming a diet supplemented with PC had their plasma cholesterol concentration lowered. On the other hand, the concentration of HDL levels was not modified. The reduction in the serum LDL/HDL ratio has been reported as an important marker of metabolic syndrome. Cohn et al. (2010) reported that HDL levels can increase when lecithin is incorporated into the diet due to the affinity of phospholipids for this specific lipoprotein. Moreover, it has been reported that the interaction of phospholipids with the membranes of enterocytes decreases the binding capacity of cholesterol by competition. The degree of saturation and the chain length of fatty acids bound to phospholipids influence the binding of cholesterol. The highly unsaturated and long-chain fatty acids have a more effective action for inhibiting the absorption of cholesterol since the saturated fatty acids have less affinity for phospholipase A2. Since saturated PC are not good substrates for phospholipase A2, they are not readily hydrolyzed, leading to their accumulation, which would hinder the access of pancreatic lipase to the core TG of the emulsion. This slows down the formation and diffusion of micelles, as well as the uptake of cholesterol by the enterocytes, reducing their absorption (Jiang, Noh, & Koo, 2001).

The effects of lecithin on lipid metabolism have been associated with the potential prevention and treatment of obesity. Lee et al. (2014) reported that supplementation in the diet of obese mice decreased the impact of complications related to obesity. In addition, a decrease in plasma triglycerides, cholesterol, and LDL/HDL ratio also was observed. It has been reported that the fatty acids omega-3 EPA and DHA decrease dyslipidemia related to obesity. This action is associated with the reduction of hepatic fatty acid synthesis and enhancement of β-oxidation and serum adiponectin levels (Shirouchi et al., 2007). Narce and Poisson (2002) reported that PC with omega-3 fatty acids in its structure acts on the suppression of gene expression of nuclear receptors participating in the regulation of lipid and energy homeostasis.

5.3.2 Tocopherols

5.3.2.1 Structure and Occurrence

The term tocopherol(s) should be used as a generic descriptor for all mono-, di-, and trimethyltocols. They are lipophilic monophenols found in mitochondrial membranes, animal and human adipose tissue, lipoproteins, oilseeds, and vegetable oils. These compounds present antioxidant activity *in vivo* and *in vitro*. Since they can mitigate oxidative stress, tocols have been associated with

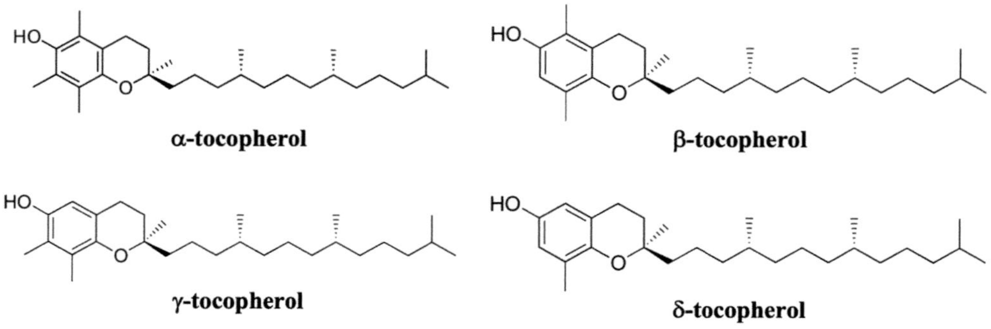

α-tocopherol

β-tocopherol

γ-tocopherol

δ-tocopherol

FIGURE 5.4   Structures of tocopherol homologs. Source: adapted from Shahidi and de Camargo (2016).

health-promoting effects, such as the reduced risk of cardiovascular diseases, some types of cancer, diabetes, and obesity (Maciel et al., 2019).

The tocopherols consist of a group of eight molecules with a phenoxyl ring linked to a heterocyclic ring named chromanol, which is bound to a 16-carbon side chain. The pattern of double bonds across this side chain determines the type of tocol. Tocopherols present a completely saturated side chain, while tocotrienols possess three double bonds located at C-3', C-7', and C-11' (Figure 5.4). Moreover, four homologs of tocopherols and tocotrienols are possible, depending on the number and position of methyl groups located on the chromanol head – α (three methyl groups), β and γ (two methyl groups), and δ (one methyl group). Additionally, tocopherols carry three chiral centers (C-2', C-4', and C-8'), making possible the existence of eight stereoisomers for each one of them. Different isomers can present distinct stability and/or bioactivity. RRR-α-tocopherol is the only stereoisomer found in nature. Nevertheless, chemically synthesized α-tocopherol is a racemic mixture of its eight possible stereoisomers. In the literature, it is not uncommon to find tocopherols being referred to as vitamin E, a generic term that includes all forms of tocopherols and tocotrienols (Combs, 2008). However, not all tocopherols display significant vitamin E activity and therefore this term cannot be used as a synonym with the term vitamin E. This type of bioactivity will vary according to their affinity for the α-tocopherol transfer protein (α-TTP) and will be addressed in more detail in Section 5.3.2.3.

Tocopherols and tocotrienols can be found in plant food containing high amounts of unsaturated fatty acids (e.g. nuts, wheat germ oil, cottonseed oil, sunflower oil). Tocopherols represent a large proportion of soybean oil's minor components, being one of the richest sources of γ-tocopherol among commercial oils. However, soybean oil does not present tocotrienols (Shahidi & De Camargo, 2016). The composition of tocopherols in soybean oil and other edible oils is presented in Table 5.5.

Although α-tocopherol is the most common homolog encountered in most of the commercial oils, soybean oil is richer in δ- and γ-tocopherol. The amount of γ-tocopherol in soybean oil is seven times higher than α-tocopherol. Soybean, palm, and canola oil, which are the oils most consumed in the world, are rich in γ-tocopherol. Therefore, several studies reported the potential of the health benefits of γ-tocopherol. Additionally, during the deodorization step of soybean oil refining, tocopherols can be recovered and used as antioxidants in oils and other lipid-rich foods (Hwang & Winkler-Moser, 2017).

5.3.2.2   Antioxidant Mechanism

Oxidative stress can be defined as an imbalance between free radicals and other reactive oxygen species (ROS) and antioxidants in the cells. The excessive propagation of these chemical species can damage cellular macromolecules (e.g. lipids, protein, DNA) and has been associated with an increased risk of developing a myriad of chronic diseases (de Camargo et al., 2018). Besides health issues, oxidation also represents a problem for lipid-containing foods, especially the ones with a

TABLE 5.5

Tocopherol Content (mg/100 g of Oil) of Edible Oils

Oil	α-Tocopherol	β-Tocopherol	γ-Tocopherol	δ-Tocopherol
Soybean	9.53–12.0	1.0–1.31	61.0–69.9	23.9–26.0
Wheat germ	151–192	31.2–65.0	tr–52.3	nd–0.55
Sunflower	32.7–59.0	tr–2.4	1.40–4.5	0.27–0.5
Cottonseed	30.5–57.3	0.04–0.3	10.5–31.7	tr
Safflower	36.7–47.7	nd–2.1	tr–2.56	tr–0.65
Palm	6.05–42.0	nd–0.42	tr–0.02	nd–0.02
Sesame	0.24–36.0	0.28–0.8	16.0–57.0	0.17–13.0
Corn	18.0–25.7	0.95–1.1	44.0–75.2	2.20–3.25
Peanut	8.86–30.4	nd–0.38	3.5–19.2	0.85–3.1
Rapeseed	18.9–24.0	nd–tr*	37–51	0.98–1.9
Barley	14.2–20.1	0.6–1.9	3.5–15.1	0.9–4.6
Olive	11.9–17.0	nd–0.27	0.89–1.34	nd–tr
Rice bran	0.73–15.9	0.19–2.5	0.26–8.0	0.03–2.7
Coconut	0.2–1.82	tr–0.25	tr–0.12	nd–0.39
Linseed	0.54–1.2	nd–tr	52.0–57.3	0.75–0.95

Source: adapted from Shahidi and de Camargo (2016).

*nd = not detected; tr = traces.

high concentration of polyunsaturated fatty acids (e.g. vegetable oils). The attack of free radicals can lead to the development of secondary oxidation products, including ketones, aldehydes, alcohol, and carboxylic acids, causing the formation of off-flavors and reducing the product's shelf-life (Alamed et al., 2009).

The autoxidation of polyunsaturated fatty acids is initiated by the formation of a free radical through exposure to light, high temperatures, ionizing radiation, lipoxygenases, or transition metals. The highly reactive free radical ($R^•$) is produced when the polyunsaturated fatty acid (RH) loses a hydrogen atom, in what is known as the initiation step. Then, the free radical reacts with oxygen, forming peroxide radical ($ROO^•$), which is able to subtract another hydrogen atom from the polyunsaturated fatty acid, producing hydroperoxide radical (ROOH), a primary oxidation product, and a new free radical ($R^•$). A chain reaction or propagation step happens until peroxide radicals start to accumulate and react with each other. In the last stage of the reaction, the termination step, stable radicals are formed, and the reaction stops. The hydroperoxides which were formed further decompose into secondary oxidation products of low molecular weight, which are volatiles and responsible for rancidity in lipid-rich foods. The hydroperoxides, although not responsible for causing off-flavors, represent a hazard to the human body, once they can damage macromolecules leading to a possible higher incidence of chronic diseases (Shahidi & Ambigaipalan, 2015).

Tocopherols that have antioxidant activity can donate a hydrogen atom from their hydroxyl group to free radicals, stabilizing them and interrupting the propagation of oxidation chain reactions. This mechanism is represented in Figure 5.5. After losing its hydrogen atom, the tocopherol is transformed into a tocopheryl radical ($T^•$), which is stabilized by resonance, being significantly less reactive than the free radicals. The tocopheryl radical can be regenerated by a hydrogen-donor compound, such as ascorbic acid, through a redox reaction. The efficiency of tocopherols as antioxidants will vary depending on the methylation pattern (number and position) in the molecule, since it will influence their capacity to donate hydrogen atoms. This fact explains why each tocopherol homolog shows different levels of antioxidant activity (Martin-Rubio et al., 2018).

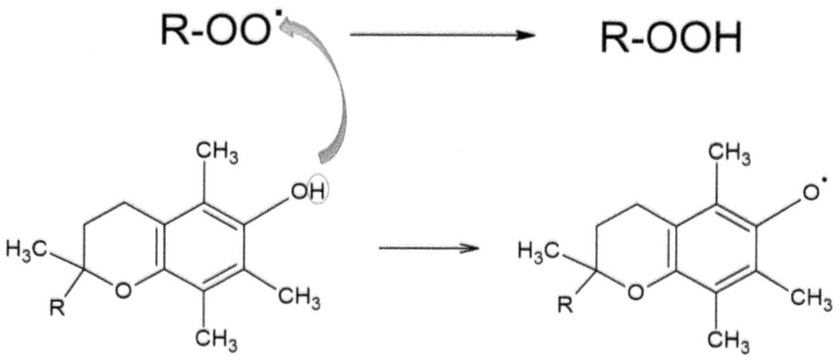

FIGURE 5.5 Antioxidant mechanism of α-tocopherol. Source: adapted from Shahidi and de Camargo (2016).

A synergistic effect with phospholipids can increase the ability of tocopherols in scavenging free radicals of the tocopherols. Due to phospholipids' amphiphilic nature, they can form microemulsions in organic solvent systems where tocopherols are solubilized. In this configuration, the polar hydroxyl groups of tocopherols are located close to the membrane surface, where they can easily scavenge aqueous peroxyl radicals (Xu et al., 2019).

The temperature also can influence the antioxidant activity of tocopherols. Wagner et al. (2001) observed that under low-to-mild temperatures and optimum levels of tocopherols, α-tocopherol had the highest antioxidant potential, followed by β-, γ-, and δ-tocopherol. However, at frying conditions, δ-tocopherols showed the highest protection of coconut fat against oxidation, followed by γ-, and α-tocopherol. A concentration of 5,000 mg/kg showed the highest antioxidant efficiency at 160° C, followed by 2,000, 1,000, and 100 mg/kg, indicating a dose-dependent effect at high temperatures.

Tocopherols can present a prooxidant activity at concentrations of 700 mg/kg or higher. At high concentrations, the tocopheryl radical can absorb hydrogen atoms either from polyunsaturated fatty acids or hydrogen peroxide. Additionally, α-tocopherol may react with hydrogen peroxide, decreasing its antioxidant capacity (Martin-Rubio et al., 2018). Health Canada recommends a daily intake of 15 mg/day of α-tocopherol for healthy adults, which is equivalent to 22 international units (IU) of natural-source vitamin E or 33 IU from synthetic sources. The same daily amount is recommended by the Food and Drug Administration (FDA).

5.3.2.3 Bioavailability

The antioxidant activity and other biological functions in the body of the tocopherols are dependent on their absorption, transport through the bloodstream, and distribution to the tissues. After their ingestion, the tocopherols, similarly to other fat-soluble vitamins, are transported to the small intestine by lipoproteins, reaching the intestinal microvillus through passive diffusion. Once in the enterocytes, the tocopherols, along with other lipophilic compounds (e.g. triacylglycerol, phospholipids, cholesterol), are packed into chylomicrons. By the action of lipoprotein lipase, the release compounds are distributed to adipose, hepatic, and muscular tissues where they are taken up by lipoprotein receptors. However, this mechanism has been questioned recently due to the discovery of broadly specific vitamin E intestinal transporters, revealing that the interaction between tocopherols and dietary lipids may have a crucial role in tocopherol's absorption process. The existence of other unidentified protein transporters involved in tocopherol absorption has been questioned but this mechanism still remains unclear (Reboul, 2017).

Another possible route has been described where a small proportion of tocopherols remains in the surface of the chylomicrons, being transferred to high-density lipoproteins (HDL) and further reaching the liver along with tocopherols located in the remaining chylomicrons. From there, α-tocopherol is preferentially transported to the plasma, while the largest proportion of the other homologs are not absorbed. This can be explained by the affinity of α-tocopherol for α-tocopherol transfer protein (α-TTP), which incorporates α-tocopherol into very-low-density lipoproteins (VLDL) before being released into the blood circulation. Therefore, the ability to reach the bloodstream gives α-tocopherol a competitive advantage regarding the vitamin E activity, once this form is securely present in the tissues in order to display the protection against oxidative damage. Comparatively, β-tocopherol presents vitamin E activity of 50% in relation to α-tocopherol's bioactivity, followed by γ- (10%) and δ-tocopherol (3%), which is due to their relative affinity for α-TTP (Maciel et al., 2019).

5.3.2.4 Biological Activity and Health-Promoting Effects

Several types of bioactivity associated with health-promoting benefits have been reported for tocopherols. The α-tocopherol carries the highest vitamin E activity and the other tocopherol homologs have shown strong antioxidant activities acting by different pathways in order to promote health benefits. The α-, γ-, and δ-tocopherol added to stripped soybean oil showed synergistic effects with arginine and lysine at frying temperature (180 C), enhancing the antioxidant capacity of such amino acids (Hwang & Winkler-Moser, 2017). The γ-tocopherol was verified to be a more potent antioxidant than α- and δ-tocopherol under these conditions (Thompson & Cooney, 2020), which is consistent with what other authors have reported.

It has been reported that the antioxidant activity of tocopherols *in vivo* may be linked to the prevention of cardiovascular diseases. The accumulation of oxidized low-density lipoprotein cholesterol (LDL-c) promotes the formation of atheromatous plaques associated with an increased risk of atherosclerosis, leading to the development of coronary heart diseases. As phenolic antioxidants, tocopherols act as chain breakers through the inhibition of lipid peroxidation. Besides, their lipophilicity represents an advantage; once increased, hydrophobicity in phenolics has been related to enhancing ROS scavenging activity. The capacity of tocopherols of acting as free radical scavengers is related to their demonstrated ability to prevent *in vitro* LDL-c cholesterol oxidation (Wallert et al., 2019).

A higher inhibition of LDL-c oxidation by α-tocopherol compared to the other homologs has been confirmed by human trials (Shahidi & De Camargo, 2016). Moreover, a meta-analysis study concluded that α-tocopherol plays an important role in reducing myocardial infarction in randomized clinical trials, decreasing the risk by 20% (Loffredo et al., 2015). In another study, the administration of 400 IU/day of α-tocopherol lowered the levels of oxidized LDL-c in healthy male subjects (Jialal et al., 1995).

Tocopherols have also been suggested as adjuvants in cancer treatment and prevention by inducing apoptosis and cell cycle arrest of carcinogenic cells (Abid-Essefi et al., 2003). The δ- and γ-tocopherol exhibit the highest apoptosis capacity among tocopherol homologs, by acting through caspase-9 and -3 activation. The γ-tocopherol has also been demonstrated to possess an inhibitory capacity of colon cell lines (Jiang, 2019). In addition, α-tocopherol was able to reduce the formation of colon carcinogenesis biomarkers induced by the consumption of cured meat (Pierre et al., 2013). The ingestion of tocopherol was associated with reduced mortality caused by prostate cancer in a long-term (18 years) and post-intervention study (Virtamo et al., 2014).

Damages to the DNA molecule can lead to mutagenesis and cancer development. Therefore, many *in vitro* studies (Donnelly et al., 1999; Placzek et al., 2005) have used free radical–induced DNA strand scission to assess the capacity of tocopherols in preventing such damages by preventing ROS-induced oxidation. A mixture of α-tocopherol and other homologs have demonstrated inhibition of high-fat-diet–induced DNA strand breakage in an animal study using C57BL/6J male mice (Remely et al., 2017).

5.3.3 PHYTOSTEROLS

5.3.3.1 Structure and Occurrence

Phytosterols are plant sterols, with more than 250 different compounds reported in the literature. Phytosterols and cholesterol have a similar chemical structure to a C-5 double bond and a 3 β-hydroxyl group. On the other hand, there are structural modifications of the C-24 side chain. Cholesterol has 27 C without the methyl, ethyl, or unsaturation in its side carbon chain. Phytosterols have 28 or 29 C, with a linkage of a methyl, methylene, ethyl, or ethylidene group in carbon 24 (Moreau et al., 2018). In addition, the unsaturation pattern of the side chain and rings can also be different (Moreau et al., 2018). Based on the presence of a methyl group in C4 the phytosterol can be divided into three groups: -desmethyl sterols (without any group); -monomethyl sterols (one group) and 4.4-dimethyl sterols (two groups). Phytosterols are not synthesized by humans and they derive from the diet. The -desmethyl sterols include sitosterol (29 carbons), stigmasterol (29 carbons), and campesterol (28 carbons), accounting for between 95% and 98% of the phytosterols in the diet. The campesterol and β-sitosterol, which are the 24-methyl and the 24-ethyl analogs of cholesterol, respectively, are the main phytosterols present at low concentrations in plasma. On the other hand, phytostanols are not abundant in nature (Calpe-Berdiel et al., 2010).

Usually, free phytosterols possess a double bond between carbons 5 and 6, or in most plants, between carbons 7 and 8 in the B ring. Stanols are phytosterols with a fully saturated ring, being considered a subgroup of phytosterols. Phytosterols exist in plant material either in the free or conjugated (esterified to fatty acids or via glycosidic linkage to glucose) state. Phytosterols exist in plant material either in the free or conjugated (esterified to fatty acids or via glycosidic linkage to glucose) state. Glycosylated sterols play a role as structural components of biological membranes. This includes acting as transport molecules for lipid precursors in cell walls. They are also believed to be part of cellulose biosynthesis (Schrick et al., 2012).

Crude rapeseed oil is reported to have the highest content of phytosterols among oilseeds (4,500–14,000 mg/kg). Table 5.6 shows that soybean oil is rich in phytosterols and the main compounds found in this raw material are β-sitosterol, campesterol, and stigmasterol (Figure 5.6).

5.3.3.2   Biological Role of Phytosterols

The presence of additional carbon atoms in the chain increases the hydrophobicity of the phytosterols when compared with cholesterol. The apparently small difference in structure has a substantial impact on the absorption of both compounds. Cholesterol exhibits a high absorption rate (around 50%), and on the other hand, the absorption of phytosterols is very low (around 5%) (Cohen, 2008; Ros, 2000). In order to be absorbed, sterols need to be incorporated into mixed micelles composed of bile acid salts, monoacylglycerols, free fatty acids, lysophospholipids, phospholipids, and free cholesterol, with the objective to reach the enterocytes.

Cholesterol, which presents an amphipathic character, is a cellular membrane component that acts as a regulation cofactor for receptors. In addition, cholesterol acts as a precursor for steroid hormones and is an essential sterol for humans. Nevertheless, the accumulation of normal and oxidized cholesterol in the serum and other tissues can lead to deleterious effects, such as cytotoxicity,

TABLE 5.6

Phytosterol Composition of Soybean Oil

Phytosterol	Crude Oil (mg/kg)	Refined (mg/kg)
β-Sitosterol	1,250–2,360	1,230
Campesterol	620–1,310	470
Stigmasterol	470–770	470
Δ^5 – Avenasterol	50	10
Δ^7 – Stigmasterol	50	10
Δ^7 – Avenasterol	20	5

Source: adapted from Wang (2008).

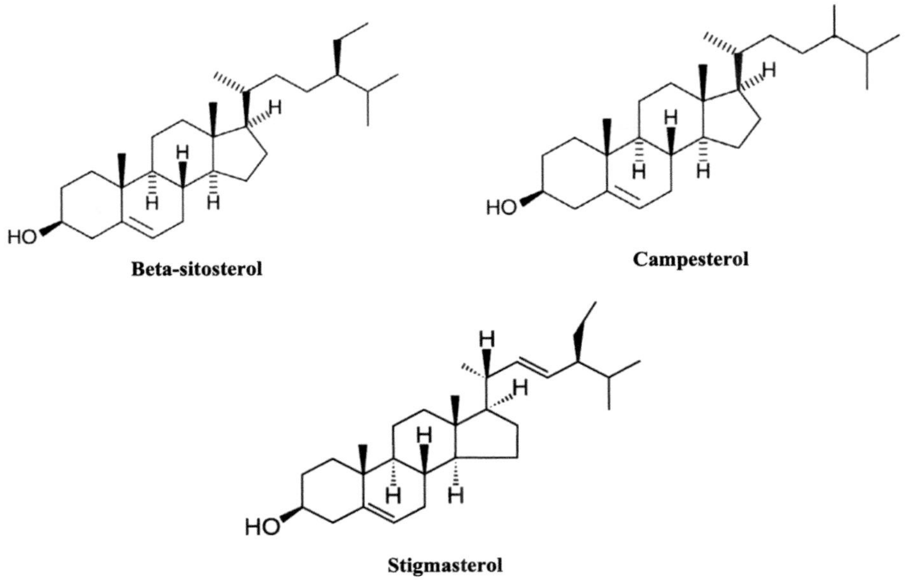

FIGURE 5.6 Major phytosterols in soybean oil. Source: adapted from Wang (2008).

atherogenesis, and coronary heart diseases. It has been reported since the 1950s that phytosterols inhibit the accumulation of cholesterol, positively impacting human health (Brauner et al., 2012; Lin et al., 2010; Rozner & Garti, 2006).

The bioactivity of the phytosterols has been related to a mechanism of competition. Phytosterols, which are more hydrophobic than cholesterol, may present a higher affinity for transporting micelles, and therefore, take cholesterol's place with a favorable free energy change. This would prevent excessive cholesterol from reaching the enterocytes and further reaching the bloodstream (Rozner & Garti, 2006). On the other hand, phytosterols, after being linked to the micelles, will be excreted since they cannot be absorbed by the human body (Rozner & Garti, 2006).

Government health agencies from several countries recognize phytosterols intake as a natural means to reduce blood cholesterol. Health Canada, the Food and Drug Administration (FDA), and the European Food Safety Authority (EFSA) have approved claims of the ability to reduce the risk of cardiovascular diseases through the reduction in cholesterol levels for phytosterol-enriched foods (Moreau et al., 2018). It has been reported that relatively small quantities of naturally occurring phytosterols (around 500 mg/d) could be sufficient for reducing LDL cholesterol, with a 20% reduction (Lin et al., 2010; Ostlund et al., 2002; Sanclemente et al., 2012). However, a consensus about the optimum dosage of this compound has not been reached yet (Doornbos et al., 2006; Plat et al., 2000; Ras et al., 2014), showing the need for more reliable human trials.

5.3.4 OTHER COMPONENTS

5.3.4.1 Sphingolipids

Table 5.7 shows the content of other minor compounds found in soybean oil. Sphingolipids are important components of the cell membrane, although they are present to a lesser extent than phospholipids. They consist of a sphingoid long-chain (18 carbon atoms) dihydroxyl base along with an α-hydroxyl fatty acyl chain connected to the base through an amide bond. The major sphingolipids found in soybean are ceramides and cerebrosides. The cerebroside content of soybean varies from 122 to 389 nmol/g on a dry wet basis, depending on the genotype (Wang, 2008).

TABLE 5.7

Lutein, β-Carotene, and Sphingolipids in Soybean Oil

Minor Compound	Content (mg/kg)
Lutein	8.95–21.19
β-Carotene	2.91–4.91
Sphingolipids	193

Source: adapted from Wang (2008).

The sphingolipids exert biological activity, acting as cell growth and differentiation mediators, as well as programmed cell death. An animal study (Dillehay et al., 1994) found that the inclusion of sphingomyelin (0.05%) into mice diet led to a decrease of 50% in the amount of an early biomarker of tumor development compared with the control diet. In a similar study (Schmelz et al., 1996), the administration of 0.1% of sphingomyelin in the diet of mice promoted a 70% reduction of the same tumor biomarker.

5.3.4.2 Carotenoids

Carotenoids are isoprenoid compounds classified as natural colorants encountered in fruits, vegetables, and some marine sources responsible for proving shades of yellow, orange, and red. Carotenoids can be further divided into two subgroups, based on structural characteristics. Carotenes (e.g., lycopene, β-carotene) are solely composed of hydrocarbons while xanthophylls (e.g. lutein, astaxanthin) contain oxygen (Rodriguez-Amaya, 2019).

Carotenoids display bioactivity due to their antioxidant and pro-vitamin A activities. They are especially efficient at preventing photooxidation by quenching singlet oxygen, a highly reactive oxygen form responsible for attacking molecules susceptible to oxidative damage. Additionally, carotenoids yielding β-ionone rings, such as β- and α-carotenes, exhibit pro-vitamin A activity due to the ability to be converted in retinol, a form of vitamin A crucial for normal vision, immune system, and reproduction (Rodriguez-Amaya, 2019).

Soybean oil has a relatively low concentration of carotenoids (Table 5.7), with the major ones being lutein and β-carotene. These compounds are usually removed by the oil refining during the bleaching step, being considered as a by-product of the process (Wang, 2008).

5.4 COMPOSITION AND OXIDATIVE STABILITY OF SOYBEAN OIL

The oxidative stability of oils and fats represents their resistance to oxidation which, in turn, depends on their history, composition, and the presence of anti- and pro-oxidant compounds. Oils rich in unsaturated fatty acids are more susceptible to oxidation. On the other hand, raw materials which present a high concentration of saturated fatty acids are more stable. Tocopherols and phenolic compounds can increase the oxidative stability of these raw materials. However, the presence of catalysts of the oxidation, such as metals and chlorophyll, make oils and fats very susceptible to oxidation. The oxidative state of fats and oils represents their current state and it is the result of their composition, conditions of handling, processing, and storage, or in other words, the history of the product.

5.4.1 Factors That Influence the Oxidative Stability

Soybean oil is composed mainly of unsaturated fatty acids such as linoleic (46.2–59%), oleic (17–30%), and linolenic (4.5–11%) acids (Codex Alimentarius, 2019; Firestone, 2013). The presence of about 80% unsaturation and a relatively high content of linolenic acid makes the soybean oil highly susceptible to oxidation reactions and one of the least stable edible oils for deterioration (Bockish, 1998). Therefore, soybean oil shows a short shelf-life, even when stored at low temperatures, and it is not recommended for frying and cooking food at high temperatures (Naeli et al., 2017).

The lipid oxidation of soybean oil can be delayed by the natural antioxidants present in the oil, as well as the natural and synthetic antioxidants added to it. Refined soybean oil presents 61–69.9 and 23.9–26 mg/100 g of γ-tocopherol and δ-tocopherol, respectively, and low levels of α- and β-tocopherol (Shahidi & de Camargo, 2016). Evans et al. (2002) reported that natural soybean oil tocopherol mixture (ratio of 1:13:5 for α-, γ-, and δ-tocopherol, respectively) between 340 and 660 ppm was the optimal concentration for inhibiting soybean oil oxidation. On the other hand, when the tocopherol mixture was above their optimal concentrations, the authors reported a pro-oxidation effect which increased at 40° to 60° C. Jung and Min (1990) reported that the oxidative stability of soybean oil was increased when the concentrations of α -, γ -, and δ-tocopherols in soybean oil was 100, 250, and 500 ppm, respectively. In addition, a significant pro-oxidant effect was observed at higher concentrations. Martin-Rubio et al. (2018) reported that at concentrations above 700 mg/kg, tocopherols can present a pro-oxidant activity. The presence of tocopherols in oilseeds, when they are consumed in their natural form, without processing, is considered important for providing the minimum vitamin E in the diet of humans and animals.

The main antioxidants in soybean oil are the tocopherols, classified as primary antioxidants because they can inactivate the free radicals formed during the initiation or propagation of the reaction, through the donation of hydrogen atoms to these molecules, interrupting the chain reaction (Figure 5.5).

In refined soybean oil, the addition of citric acid is common, which is used as a synergist with primary natural or synthetic antioxidants. Citric acid is classified as a secondary antioxidant due to the chelation of metals, and in soybean oil, it can form chelates with metal ions at concentrations of 0.005–0.2% (Dziezak, 1986).

Pro-oxidants are substances such as pigments and metals that reduce the level of activation energy necessary for the occurrence of the lipid oxidation reaction. Among the pigments present in soybean, chlorophyll can absorb light energy and transfer this energy to the molecules involved in photooxidation reactions (McClements & Decker, 2008). The chlorophyll in crude soybean oil (0.30 ppm) is practically eliminated from the oil during refining. After the bleaching step using clay and/ or silica the remaining chlorophyll is around 0.08 ppm (Hammond, Johnson, Su, Wang, & White, 2005).

Transition metal ions, especially Fe^{2+} and Fe^{3+}, are major pro-oxidants, accelerating the decomposition of primary lipid oxidation products, giving rise to other radicals, or acting in the formation of singlet oxygen (McClements & Decker, 2008). Other transitional metals such as copper, magnesium, and manganese, which present pro-oxidant activity, should be removed from soybean oil during refining (Hammond, Johnson, Su, Wang, White, & Gerde, 2005; Szyczewski et al., 2016).

Radiation (i.e. ultraviolet radiation) is a physical pro-oxidant factor that facilitates the transmission of energy to molecules to promote the rupture of the carbon-hydrogen bonds, formation of singlet oxygen, or decomposition of hydroperoxides and peroxides (McClements & Decker, 2008).

Autoxidation of edible oils and fats can also be catalyzed by other factors such as exposure to air, high temperature, and moisture. This process is a free radical chain reaction, leading to an increase in reactive radicals, which initiate further reactions (Choe & Min, 2006; Hammond, Johnson, Su, Wang, White, & Gerde, 2005; Taghvaei & Jafari, 2015). These parameters will be discussed in Section 5.4.2.

5.4.2 How to Prevent Oil Oxidation and Determine the Oxidative Stability

Oxidation may lead to the formation of toxic compounds and off-flavors in soybean oil, decreasing its quality and nutritional value (Naz et al., 2004). Loss has been reported of essential fatty acids and bioactive lipids, vitamins, and amino acids in oxidized oils. The consumption of these oils can contribute to the development of cardiovascular diseases, cancer, and neurodegenerative diseases (Dobarganes & Márquez-Ruiz, 2003).

The oxidation of fats and oils can be inhibited by eliminating oxygen access (reduction of oxygen pressure, vacuum packaging, packing under inert gas or modified atmosphere); inactivating enzymes involved in oxidation catalysis; adding chelating agents; using lower temperature; and packaging with barriers suitable for light, oxygen (McClements & Decker, 2008), and humidity (Rodsamran & Sothornvit, 2019).

Another method of protection against oxidation is to use antioxidants (Kiran et al., 2015). Synthetic antioxidants (e.g. tertiary-butylhydroquinone – TBHQ) are commonly used to increase the stability of soybean oil. However, studies with animals show the possibility that they may have toxic effects (Whysner et al., 1994). Therefore, there is a growing interest in replacing these antioxidants with extracts obtained from different vegetable raw materials and their by-products (Rodsamran & Sothornvit, 2019).

The effects of natural compounds such as carotenoids and tocopherols (Kaur et al., 2015); nano-encapsulated olive leaf extract (Mohammadi et al., 2016); aromatic plants (Saoudi et al., 2016); and lime peel extract (Rodsamran & Sothornvit, 2019) have been reported in the literature with promising results.

Lycopene and β-carotene showed pro-oxidant activity when the concentration was increased from 100 to 200 ppm. On the other hand, γ-tocopherol acted as an antioxidant at 100 ppm, and an increase in the concentration to 200 ppm did not result in changes in PV (Kaur et al., 2015).

Oxidation stability is one of the most important quality parameters of fats and oils for predicting their shelf-life. Several methods are used to determine the oxidative stability of oils. The most reliable test is the storage test, but it takes a long time to be done. The Rancimat test and Schaal oven Test determine the oxidation stability of oils in a noticeably short time. The Rancimat is carried out at high temperature (50–150 C) and intensive aeration, which change the nature of the oxidation process. On the other hand, the Schaal oven test uses low temperatures (30–63 C) without intense aeration. Therefore, the test mimics the changes that occur during regular storage conditions, but the results are obtained in at least 15 days (Maszewska et al., 2018).

The oxidative state of the fats and oils can be determined by the determination of the products of primary oxidation such as peroxides and hydroperoxides by peroxide value. In addition, the products of secondary oxidation, such as epoxides, volatile and non-volatile compounds, are usually determined by ρ-anisidine value (ρ-AV); 2-thiobarbituric acid (TBA) and volatile compounds using gas chromatography by mass spectrophotometry (CG-MS) (Silva et al., 1999). Chemiluminescence, Raman spectroscopy (Carmona et al., 2014), and differential scanning calorimetry (DSC) methods are also used. DSC is the most used method to determine chemical properties and thermo-oxidative decomposition of oils (Tengku et al., 2016; Thurgood et al., 2007). These methods should be used in combination since a single parameter of oxidation is not sufficient to explain the oxidative changes in fats and oils in the various stages of the reaction and at different conditions (Goyary et al., 2015; Naz et al., 2004).

5.5   FINAL CONSIDERATIONS

Health agencies worldwide have been promoting campaigns for a low fat intake, mainly saturated fatty acids and cholesterol, to prevent cardiovascular diseases. The relationship between the intake of lipids and the appearance of cardiovascular diseases has been alerting consumers of fast food, meat products, and other products with high-fat content to increase their attention to the type of fat present in the food. A high intake of saturated fats has been associated with several diseases such as cardiovascular disease and cancer. On the other hand, an increase in the intake of monounsaturated and polyunsaturated fatty acids has been recommended. Soybean oil is composed mainly of unsaturated fat and is one of the few sources of omega-3 fatty acids in vegetables, which may help to reduce the risk of both cardiovascular disease and cancer. Other minor bioactive compounds such as tocopherols and phytosterols are also present in soybean oil and are important for good health. The oxidative instability of the soybean can be modified through genetic engineering and it is an area of science that can still be exploited. The projections for soybean production indicate that the production of this raw material will continue to grow in the coming years and will be present in the diet of the population around the world.

REFERENCES

Abid-Essefi, S., Baudrimont, I., Hassen, W., Ouanes, Z., Mobio, T. A., Anane, R., Creppy, E. E., et al. (2003). DNA fragmentation, apoptosis and cell cycle arrest induced by zearalenone in cultured DOK, Vero and Caco-2 cells: Prevention by vitamin E. *Toxicology, 192*:2–3, 237–248.

ABIOVE. (2021). Brazilian Association of Vegetable Oil Industries. *Monthly statistics*. Retrieved October 07, 2021, from https://abiove.org.br/en/statistics/

Alamed, J., Chaiyasit, W., McClements, D. J., & Decker, E. A. (2009). Relationships between free radical scavenging and antioxidant activity in foods. *Journal of Agricultural and Food Chemistry, 57*:7, 2969–2976.

AOCS. (2010). The ultra-low-linolenic soybean market. *INFORM magazine*. Retrieved June 4, 2020, from https://www.aocs.org/stay-informed/inform-magazine/featured-articles/the-ultra-low-linolenic-soybean-market-april-2010

Aparicio, R., García González, D. L., & Aparicio-Ruiz, R. (2018). Vegetable oils. In *FoodIntegrity Handbook* (pp. 349–370). Eurofins Analytics France. Retrieved from https://foodintegrity.eu/foodintegrity/index.cfm?sectionid=83

Banz, W. J., Davis, J. E., Clough, R. W., & Cheatwood, J. L. (2012). Stearidonic acid: Is there a role in the prevention and management of type 2 diabetes mellitus? *The Journal of Nutrition, 142*:3, 635S–640S. Retrieved from https://academic.oup.com/jn/article/142/3/635S/4631007

Biermann, U., Bornscheuer, U., Meier, M. A. R. R., Metzger, J. O., & Schäfer, H. J. (2011). Oils and fats as renewable raw materials in chemistry. *Angewandte Chemie International Edition, 50*:17, 3854–3871. Retrieved from http://doi.wiley.com/10.1002/anie.201002767

Bockish, M. (1998). Vegetable fats and oils. In *Fats and Oils Handbook* (pp. 174–344). Champaign, IL: AOCS Press. Retrieved from https://linkinghub.elsevier.com/retrieve/pii/B9780981893600500093

Brauner, R., Johannes, C., Ploessl, F., Bracher, F., & Lorenz, R. L. (2012). Phytosterols reduce cholesterol absorption by inhibition of 27-hydroxycholesterol generation, liver X receptor α activation, and expression of the basolateral sterol exporter atp-binding cassette A1 in caco-2 enterocytes. *Journal of Nutrition, 142*:6, 981–989. Retrieved from https://academic.oup.com/jn/article/142/6/981/4689081

Cahoon, E. B. (2003). Genetic enhancement of soybean oil for industrial uses. *AgbioForum, 6*:1–2, 11–13. Retrieved from http://www.unitedsoybean.org/

Calder, P. C. (2006). Polyunsaturated fatty acids and inflammation. *Prostaglandins Leukotrienes and Essential Fatty Acids, 75*:3, 197–202.

Calpe-Berdiel, L., Méndez-González, J., Llaverias, G., Escolà-Gil, J. C., & Blanco-Vaca, F. (2010). Plant sterols, cholesterol metabolism and related disorders. In L. Avigliano, L., Rossi (Ed.), *Biochemical Aspects of Human Nutrition* (pp. 223–242). Transworld Research Network.

Calyxt. (2020). First commercial sale of Calyxt High Oleic Soybean Oil on the U.S. Market. *Calyno™ High Oleic Soybean Oil*. Retrieved June 12, 2020, from https://calyxt.com/first-commercial-sale-of-calyxt-high-oleic-soybean-oil-on-the-u-s-market/

de Camargo, A. C., Schwember, A. R., Parada, R., Garcia, S., Maróstica Júnior, M. R., Franchin, M., Regitano-D'arce, M. A. B., et al. (2018). Opinion on the hurdles and potential health benefits in value-added use of plant food processing by-products as sources of phenolic compounds. *International Journal of Molecular Sciences, 19*:11.

Candela, C. G., López, L. M. B., & Kohen, V. L. (2011). Importance of a balanced omega 6/omega 3 ratio for the maintenance of health. Nutritional recommendations. *Nutricion Hospitalaria, 26*:2, 323–329.

Cardinal, A. J., Whetten, R., Wang, S., Auclair, J., Hyten, D., Cregan, P., Bachlava, E., et al. (2014). Mapping the low palmitate fap1 mutation and validation of its effects in soybean oil and agronomic traits in three soybean populations. *Theoretical and Applied Genetics, 127*:1, 97–111. Retrieved from http://link.springer.com/10.1007/s00122-013-2204-8

Carmona, M. Á., Lafont, F., Jiménez-Sanchidrián, C., & Ruiz, J. R. (2014). Raman spectroscopy study of edible oils and determination of the oxidative stability at frying temperatures. *European Journal of Lipid Science and Technology, 116*:11, 1451–1456. Retrieved from http://doi.wiley.com/10.1002/ejlt.201400127

Cherrak, C. M., Pantalone, V. R., Meyer, E. J., Ellis, D. L., Melton, S. L., West, D. R., & Mount, J. R. (2003). Low-palmitic, low-linolenic soybean development. *Journal of the American Oil Chemists' Society*, *80*:6, 539–543. Retrieved from http://doi.wiley.com/10.1007/s11746-003-0734-9

Choe, E., & Min, D. B. (2006). Mechanisms and factors for edible oil oxidation. *Comprehensive Reviews in Food Science and Food Safety*, *5*:4, 169–186. Retrieved from http://doi.wiley.com/10.1111/j.1541-4337 .2006.00009.x

Clemente, T. E., & Cahoon, E. B. (2009). Soybean oil: Genetic approaches for modification of functionality and total content. *Plant Physiology*, *151*:3, 1030–1040. Retrieved from http://www.plantphysiol.org/ lookup/doi/10.1104/pp.109.146282

CODEX ALIMENTARIUS. (2019). *International Food Standards. Standard for named vegetable oils. CXS 210-1999.*

Cohen, D. E. (2008). Balancing cholesterol synthesis and absorption in the gastrointestinal tract. *Journal of Clinical Lipidology*, *2*:2, S1–S3. Retrieved from https://linkinghub.elsevier.com/retrieve/pii/ S1933287408000081

Cohn, J. S., Kamili, A., Wat, E., Chung, R. W. S., & Tandy, S. (2010). Dietary phospholipids and intestinal cholesterol absorption. *Nutrients*, *2*:2, 116–127. Retrieved from http://www.mdpi.com/2072-6643/2/2/116

Combs, G. F. (2008). *The Vitamins: Fundamental Aspects in Nutrition and Health* (3rd ed.). Boston, MA: Elsevier.

CONAB. (2020). 1º Levantamento - Safra 2020/1. *Boletim da safra de grãos - Companhia Nacional de Abastecimento* - Brazilian National Supply Company. Retrieved July 10, 2020, from https://www.conab .gov.br/info-agro/safras/graos/boletim-da-safra-de-graos

Connor, W. E., Neuringer, M., & Reisbick, S. (2009). Essential fatty acids: The importance of n-3 fatty acids in the retina and brain. *Nutrition Reviews*, *50*:4, 21–29. Retrieved from https://academic.oup.com/nutri- tionreviews/article-lookup/doi/10.1111/j.1753-4887.1992.tb01286.x

Das, U. (2006). Essential fatty acids - A review. *Current Pharmaceutical Biotechnology*, *7*:6, 467–482.

Decker, E. A., Akoh, C. C., & Wilkes, R. S. (2012). Incorporation of (n-3) fatty acids in foods: Challenges and opportunities. *The Journal of Nutrition*, *142*:3, 610S–613S. Retrieved from https://academic.oup.com/ jn/article/142/3/610S/4630980

Demorest, Z. L., Coffman, A., Baltes, N. J., Stoddard, T. J., Clasen, B. M., Luo, S., Retterath, A., et al. (2016). Direct stacking of sequence-specific nuclease-induced mutations to produce high oleic and low linolenic soybean oil. *BMC Plant Biology*, *16*:1, 225. Retrieved from http://bmcplantbiol.biomedcentral.com/ articles/10.1186/s12870-016-0906-1

Dillehay, D. L., Webb, S. K., Schmelz, E. M., & Merrill, A. H. (1994). Dietary sphingomyelin inhibits 1,2-dimethylhydrazine-induced colon cancer in CF1 mice. *Journal of Nutrition*, *124*:5, 615–620. Retrieved from https://academic.oup.com/jn/article/124/5/615/4724256

Dinicolantonio, J. J., & O'Keefe, J. H. (2018a). Omega-6 vegetable oils as a driver of coronary heart disease: The oxidized linoleic acid hypothesis. *Open Heart*, *5*:2, e000898. Retrieved from http://openheart.bmj .com/lookup/doi/10.1136/openhrt-2018-000898

DiNicolantonio, J. J., & O'Keefe, J. H. (2018b). Importance of maintaining a low omega–6/omega–3 ratio for reducing inflammation. *Open Heart*, *5*:2, e000946. Retrieved from http://openheart.bmj.com/lookup/ doi/10.1136/openhrt-2018-000946

DiRienzo, M. A., Astwood, J. D., Petersen, B. J., & Smith, K. M. (2006). Effect of substitution of low linolenic acid soybean oil for hydrogenated soybean oil on fatty acid intake. *Lipids*, *41*:2, 149–157. Retrieved from http://doi.wiley.com/10.1007/s11745-006-5083-9

Dziezak, J.D. (1986). Preservatives Antioxidants. The Ultimate Answer to Oxidation. *Food Technology*, *40*:9, 94–102.

Dobarganes, C., & Márquez-Ruiz, G. (2003). Oxidized fats in foods. *Current Opinion in Clinical Nutrition and Metabolic Care*, *6*:2, 157–163. Retrieved from http://journals.lww.com/00075197-200303000-00004

Donnelly, E. T., McClure, N., & Lewis, S. E. M. (1999). The effect of ascorbate and α-tocopherol supplementation in vitro on DNA integrity and hydrogen peroxide-induced DNA damage in human spermatozoa. *Mutagenesis*, *14*:5, 505–511. Retrieved from https://academic.oup.com/mutage/article-lookup/doi/10 .1093/mutage/14.5.505

Doornbos, A. M. E., Meynen, E. M., Duchateau, G. S. M. J. E., van der Knaap, H. C. M., & Trautwein, E. A. (2006). Intake occasion affects the serum cholesterol lowering of a plant sterol-enriched single-dose yoghurt drink in mildly hypercholesterolaemic subjects. *European Journal of Clinical Nutrition*, *60*:3, 325–333.

Dubois, V., Breton, S., Linder, M., Fanni, J., & Parmentier, M. (2007). Fatty acid profiles of 80 vegetable oils with regard to their nutritional potential. *European Journal of Lipid Science and Technology, 109*:7, 710–732. Retrieved from http://doi.wiley.com/10.1002/ejlt.200700040

Dupont, J. (2003). Vegetable oils| dietary importance. In B. Caballero (Ed.), *Encyclopedia of Food Sciences and Nutrition* (pp. 5921–5925). Elsevier. Retrieved from https://linkinghub.elsevier.com/retrieve/pii/B012227055X012293

DuPont Pioneer. (2020). Plenish® High Oleic Soybeans. *Pioneer Products*. Retrieved May 10, 2020, from https://www.pioneer.com/home/site/about/products/product-traits-technology/plenish/

Evans, C. D., List, G. R., Beal, R. E., & Black, L. T. (1974). Iron and phosphorus contents of soybean oil from normal and damaged beans. *Journal of the American Oil Chemists Society, 51*:10, 444–448. Retrieved from http://doi.wiley.com/10.1007/BF02635151

Evans, J. C., Kodali, D. R., & Addis, P. B. (2002). Optimal tocopherol concentrations to inhibit soybean oil oxidation. *JAOCS, Journal of the American Oil Chemists' Society, 79*:1, 47–51. Retrieved from http://doi.wiley.com/10.1007/s11746-002-0433-6

Farmnews. (2020). Main soy producing countries in 2020. *Main soy producing countries in 2020 – April data (2020)*. Retrieved July 10, 2020, from https://www.farmnews.com.br/mercado/principais-paises-produtoras-de-soja/

FDA. (2017). Soybean oil and reduced risk of coronary heart disease. *Qualified Health Claim Petition - Soybean Oil and Reduced Risk of Coronary Heart Disease (Docket No. FDA-2016-Q-0995)*. Retrieved June 10, 2020, from https://www.fda.gov/media/106649/download

FDA. (2018). FDA completes review of qualified health claim petition for oleic acid and the risk of coronary heart disease. *CFSAN Constituent Updates*. Retrieved June 10, 2020, from https://www.fda.gov/food/cfsan-constituent-updates/fda-completes-review-qualified-health-claim-petition-oleic-acid-and-risk-coronary-heart-disease

Fedeli, E., & Jacini, G. (1971). Lipid composition of vegetable oils (pp. 335–382). Retrieved from https://linkinghub.elsevier.com/retrieve/pii/B9780120249091500144

Fehr, W. R., Thorne, J. C., & Hammond, E. G. (1971). Relationship of fatty acid formation and chlorophyll content in soybean seed. *Crop Science, 11*:2, 211–213. Retrieved from https://onlinelibrary.wiley.com/doi/abs/10.2135/cropsci1971.0011183X001100020013x

Fehr, Walter R., & Hammond, E. G. (1997). Soybeans having low linolenic acid content and method of production. *Biotechnology Advances, 15*:1, 275–276. Retrieved from https://linkinghub.elsevier.com/retrieve/pii/S0734975097885606

Firestone, D. (2013). *Physical and Chemical Characteristics of Oils, Fats, and Waxes* (D. Firestone, Ed.). Urbana, IL: AOCS Press.

Gerde, J. A., Hammond, E. G., Johnson, L. A., Su, C., Wang, T., & White, P. J. (2020). Soybean oil. In F. Shahidi (Ed.), *Bailey's Industrial Oil and Fat Products Volume 2* (7th ed., pp. 1–68). John Wiley & Sons, Ltd. Retrieved from https://doi.org/10.1002/047167849X.bio041.pub2

Gerde, J. A., & White, P. J. (2008). Lipids. *Soybeans* (pp. 193–227). Elsevier. Retrieved from https://linkinghub.elsevier.com/retrieve/pii/B978189399764650010X

Gerde, J., Hardy, C., Fehr, W., & White, P. J. (2007). Frying performance of no-trans, low-linolenic acid soybean oils. *JAOCS, Journal of the American Oil Chemists' Society, 84*:6, 557–563. Retrieved from http://doi.wiley.com/10.1007/s11746-007-1066-0

Goyary, J., Kumar, A., & Nayak, P. K. (2015). Changes of quality parameters of soybean oil in deep fat frying: A review. *Journal of Basic Application Research International, 2*, 951–954.

Goyens, P. L., Spilker, M. E., Zock, P. L., Katan, M. B., & Mensink, R. P. (2006). Conversion of α-linolenic acid in humans is influenced by the absolute amounts of α-linolenic acid and linoleic acid in the diet and not by their ratio. *The American Journal of Clinical Nutrition, 84*:1, 44–53. Retrieved from https://academic.oup.com/ajcn/article/84/1/44/4633063

Graef, G. L., Miller, L. A., Fehr, W. R., & Hammond, E. G. (1985). Fatty acid development in a soybean mutant with high stearic acid. *Journal of the American Oil Chemists' Society, 62*:4, 773–775. Retrieved from http://doi.wiley.com/10.1007/BF03028752

Greyt, W. De, & Kellens, M. (2005). Deodorization. In Fereidoon Shahidi (Ed.), *Bailey's Industrial Oil and Fat Products Volume 5* (6th ed., pp. 341–383). John Wiley & Sons, Inc.

Griffin, B. A. (2008). How relevant is the ratio of dietary n-6 to n-3 polyunsaturated fatty acids to cardiovascular disease risk? Evidence from the OPTILIP study. *Current Opinion in Lipidology, 19*:1, 57–62. Retrieved from http://journals.lww.com/00041433-200802000-00011

Gulkari, V. D., Chaple, D. R., & Dubey, A. L. (2017). Behavioral effect of phosphatidylcholine isolated from soy lecithin in streptozotocin induced experimental alzheimers model. *Journal of Pharmacognosy and Phytochemistry*, 6:3, 702–709.

Gundermann, K.-J., Kuenker, A., Kuntz, E., & Droździk, M. (2011). Activity of essential phospholipids (EPL) from soybean in liver diseases. *Pharmacological Reports*, *63*(3), 643–659. https://doi.org/10.1016/S1734 -1140(11)70576-X

Gunstone, F. D. (2005). Vegetable oils. In Fereidoon Shahidi (Ed.), *Bailey's Industrial Oil and Fat Products Volume 1* (6th ed., pp. 213–267). John Wiley & Sons, Inc.

Gunstone, F. D. (2006). Vegetable sources of lipids. *Modifying Lipids for Use in Food* (pp. 11–27). Oxford: Blackwell Publishing.

Haag, M. (2003). Essential fatty acids and the brain. *Canadian Journal of Psychiatry*, *48*:3, 195–203. Retrieved from http://journals.sagepub.com/doi/10.1177/070674370304800308

Hammond, B. G., Lemen, J. K., Ahmed, G., Miller, K. D., Kirkpatrick, J., & Fleeman, T. (2008). Safety assessment of SDA soybean oil: Results of a 28-day gavage study and a 90-day/one generation reproduction feeding study in rats. *Regulatory Toxicology and Pharmacology*, *52*:3, 311–323. Retrieved from https:// linkinghub.elsevier.com/retrieve/pii/S0273230008001918

Hammond, E. G., Johnson, L. A., Su, C., Wang, T. T., White, P. J., & Gerde, J. A. (2005). Soybean oil. In F. Shahidi (Ed.), *Bailey's Industrial Oil and Fat Products Volume 2* (6th ed., pp. 577–653). John Wiley & Sons, Inc. Retrieved from https://doi.org/10.1002/047167849X.bio041.pub2

Hammond, E. G., Johnson, L. A., Su, C., Wang, T., & White, P. J. (2005). Soybean oil. *Bailey's Industrial Oil and Fat Products* (pp. 577–653). Wiley. Retrieved from https://onlinelibrary.wiley.com/doi/abs/10.1002 /047167849X.bio041

Hammond, E. W. (2003). Vegetable oils | types and properties. In B. Caballero (Ed.), *Encyclopedia of Food Sciences and Nutrition* (pp. 5899–5904). Elsevier. Retrieved from https://linkinghub.elsevier.com/ retrieve/pii/B012227055X012256

Harp, T. K., & Hammond, E. G. (1998). Stereospecific analysis of soybean triacylglycerols. *Lipids*, *33*:2, 209–216. Retrieved from http://doi.wiley.com/10.1007/s11745-998-0197-7

Harris, W. S. (2007). The omega-6/omega-3 ratio and cardiovascular disease risk: Uses and abuses. *Current Cardiovascular Risk Reports*, *1*:1, 39–45.

Harris, W. S. (2018). The Omega-6:Omega-3 ratio: A critical appraisal and possible successor. *Prostaglandins, Leukotrienes and Essential Fatty Acids*, *132*, 34–40. Retrieved from https://linkinghub.elsevier.com/ retrieve/pii/S095232781830067X

Harris, W. S., Lemke, S. L., Hansen, S. N., Goldstein, D. A., DiRienzo, M. A., Su, H., Nemeth, M. A., et al. (2008). Stearidonic acid-enriched soybean oil increased the Omega-3 index, an emerging cardiovascular risk marker. *Lipids*, *43*:9, 805–811. Retrieved from http://doi.wiley.com/10.1007/s11745-008-3215-0

Hartmann, R. B., Fehr, W. R., Welke, G. A., Hammond, E. G., Duvick, D. N., & Cianzio, S. R. (1996). Association of elevated palmitate content with agronomic and seed traits of soybean. *Crop Science*, *36*:6, 1466–1470. Retrieved from https://onlinelibrary.wiley.com/doi/abs/10.2135/cropsci1996.0011183 X003600060007x

Homrich, M. S., Wiebke-Strohm, B., Weber, R. L. M., & Bodanese-Zanettini, M. H. (2012). Soybean genetic transformation: A valuable tool for the functional study of genes and the production of agronomically improved plants. *Genetics and Molecular Biology*, *35*:4 suppl 1, 998–1010. Retrieved from http://www .scielo.br/scielo.php?script=sci_arttext&pid=S1415-47572012000600015&lng=en&tlng=en

Horejsi, T. F., Fehr, W. R., Welke, G. A., Duvick, D. N., Hammond, E. G., & Cianzio, S. R. (1994). Genetic control of reduced palmitate content in soybean. *Crop Science*, *34*:2, 331–334. Retrieved from https:// onlinelibrary.wiley.com/doi/abs/10.2135/cropsci1994.0011183X003400020003x

Huth, P. J., Fulgoni, V. L., & Larson, B. T. (2015). A systematic review of high-oleic vegetable oil substitutions for other fats and oils on cardiovascular disease risk factors: Implications for novel high-oleic soybean oils. *Advances in Nutrition*, *6*:6, 674–693. Retrieved from https://academic.oup.com/advances/article/6 /6/674/4555146

Hwang, H. S., & Winkler-Moser, J. K. (2017). Antioxidant activity of amino acids in soybean oil at frying temperature: Structural effects and synergism with tocopherols. *Food Chemistry*, *221*:, 1168–1177. Retrieved from https://linkinghub.elsevier.com/retrieve/pii/S0308814616318659

India. (2019). Commodity profile of edible oil for September −2019. Department of Agriculture Cooperation & Farmers Welfare. Retrieved June 10, 2020, from http://agricoop.gov.in/sites/default/files/Edible-oil -Profile-21-11-2019.pdf

Ishikawa, G., Hasegawa, H., Takagi, Y., & Tanisaka, T. (2001). The accumulation pattern in developing seeds and its relation to fatty acid variation in soybean. *Plant Breeding, 120*:5, 417–423. Retrieved from http://doi.wiley.com/10.1046/j.1439-0523.2001.00631.x

James, M. J., Ursin, V. M., & Cleland, L. G. (2003). Metabolism of stearidonic acid in human subjects: Comparison with the metabolism of other n-3 fatty acids. *American Journal of Clinical Nutrition, 77*:5, 1140–1145. Retrieved from https://academic.oup.com/ajcn/article/77/5/1140/4689812

Jialal, I., Fuller, C. J., & Huet, B. A. (1995). The effect of α-tocopherol supplementation on LDL oxidation: A dose-response study. *Arteriosclerosis, Thrombosis, and Vascular Biology, 15*:2, 190–198.

Jiang, Q. (2019). Natural Forms of Vitamin E and Metabolites—Regulation of Cancer Cell Death and Underlying Mechanisms. *IUBMB Life 71*:4, 495–506.

Jiang, Y., Noh, S. K., & Koo, S. I. (2001). Egg Phosphatidylcholine Decreases the Lymphatic Absorption of Cholesterol in Rats. *The Journal of Nutrition, 131*:9, 2358–2363. https://doi.org/10.1093/jn/131.9.2358

Jung, M. Y., & Min, D. B. (1990). Effects of α-, γ-, and δ-tocopherols on oxidative stability of soybean oil. *Journal of Food Science, 55*:5, 1464–1465. Retrieved from http://doi.wiley.com/10.1111/j.1365-2621.1990.tb03960.x

Kalo, P., & Kemppinen, A. (2003). Triglycerydes | structures and properties. *Encyclopedia of Food Sciences and Nutrition* (pp. 5857–5868). Elsevier. Retrieved from https://linkinghub.elsevier.com/retrieve/pii/B012227055X012098

Kanai, M., Yamada, T., Hayashi, M., Mano, S., & Nishimura, M. (2019). Soybean (Glycine max L.) triacylglycerol lipase GmSDP1 regulates the quality and quantity of seed oil. *Scientific Reports, 9*:1, 8924. Retrieved from http://www.nature.com/articles/s41598-019-45331-8

Karupaiah, T., & Sundram, K. (2007). Effects of stereospecific positioning of fatty acids in triacylglycerol structures in native and randomized fats: A review of their nutritional implications. *Nutrition & Metabolism, 4*:1, 16. Retrieved from http://nutritionandmetabolism.biomedcentral.com/articles/10.1186/1743-7075-4-16

Kaur, D., Sogi, D. S., & Wani, A. A. (2015). Oxidative stability of soybean triacylglycerol using carotenoids and y-tocopherol. *International Journal of Food Properties, 18*:12, 2605–2613. Retrieved from http://www.tandfonline.com/doi/full/10.1080/10942912.2013.803118

Kim, H. J., Ha, B. K., Ha, K. S., Chae, J. H., Park, J. H., Kim, M. S., Asekova, S., et al. (2015). Comparison of a high oleic acid soybean line to cultivated cultivars for seed yield, protein and oil concentrations. *Euphytica, 201*:2, 285–292.

Kimura, T., Fukui, E., Kageyu, A., Kurohara, H., Kurosaki, Y., Nakayama, T., … Ohsawa, S. (1989). Enhancement of Oral Bioavailability ofd-a-Tocopherol d-a-Tocopherol Acetate by Lecithin-Dispersed Aqueous Preparation Containing Medium-Chain Triglycerides in Rats. *Chemical and Pharmaceutical Bulletin, 37*(2), 439–441. https://doi.org/10.1248/cpb.37.439

Kinney, A. J. (1996). Development of genetically engineered soybean oils for food applications. *Journal of Food Lipids, 3*:4, 273–292. Retrieved from http://doi.wiley.com/10.1111/j.1745-4522.1996.tb00074.x

Kleiner, L., Vázquez, L., & Akoh, C. C. (2012). Lipase-catalyzed concentration of stearidonic acid in modified soybean oil by partial hydrolysis. *JAOCS, Journal of the American Oil Chemists' Society, 89*:11, 1999–2010.

Kubow, S. (1996). The influence of positional distribution of fatty acids in native, interesterified and structure-specific lipids on lipoprotein metabolism and atherogenesis. *Journal of Nutritional Biochemistry, 7*:10, 530–541. Retrieved from https://linkinghub.elsevier.com/retrieve/pii/S0955286396001064

Le, H. D., Meisel, J. A., de Meijer, V. E., Gura, K. M., & Puder, M. (2009). The essentiality of arachidonic acid and docosahexaenoic acid. *Prostaglandins, Leukotrienes and Essential Fatty Acids, 81*:2–3, 165–170. Retrieved from https://linkinghub.elsevier.com/retrieve/pii/S0952327809000842

Lee, H. S., Nam, Y., Chung, Y. H., Kim, H. R., Park, E. S., Chung, S. J., … Jeong, J. H. (2014). Beneficial effects of phosphatidylcholine on high-fat diet-induced obesity, hyperlipidemia and fatty liver in mice. *Life Sciences, 118*:1, 7–14. https://doi.org/10.1016/j.lfs.2014.09.027

Lee, K. J., Lee, J. R., Shin, M. J., Cho, G. T., Ma, K. H., Chung, J. W., & Lee, G. A. (2019). Selection and molecular characterization of soybeans with high oleic acid from plant germplasm of genebank. *Journal of Crop Science and Biotechnology, 22*:4, 323–333.

Lemke, S. L., Vicini, J. L., Su, H., Goldstein, D. A., Nemeth, M. A., Krul, E. S., & Harris, W. S. (2010). Dietary intake of stearidonic acid–enriched soybean oil increases the omega-3 index: randomized, double-blind clinical study of efficacy and safety. *The American Journal of Clinical Nutrition, 92*:4, 766–775. Retrieved from https://academic.oup.com/ajcn/article/92/4/766/4597510

Lin, X., Racette, S. B., Lefevre, M., Spearie, C. A., Most, M., Ma, L., & Ostlund, R. E. (2010). The effects of phytosterols present in natural food matrices on cholesterol metabolism and LDL-cholesterol: A controlled feeding trial. *European Journal of Clinical Nutrition, 64*:12, 1481–1487.

List, G. R. (2015). Soybean lecithin: Food, industrial uses, and other applications. *Polar Lipids* (pp. 1–33). Elsevier. Retrieved from https://linkinghub.elsevier.com/retrieve/pii/B9781630670443500054

Liu, K. (1999). Soy oil modification: Products and applications. *INFORM, 10*:, 868–878.

Loffredo, L., Perri, L., Di Castelnuovo, A., Iacoviello, L., De Gaetano, G., & Violi, F. (2015). Supplementation with vitamin E alone is associated with reduced myocardial infarction: A meta-analysis. *Nutrition, Metabolism and Cardiovascular Diseases, 25*:4, 354–363. Retrieved from https://linkinghub.elsevier.com/retrieve/pii/S0939475315000241

Maciel, L. G., Lima, R. da S., & Block, J. M. (2019). Tocoferóis. In C. V. de M. B. Pimentel, M. F. Elias, & S. T. Philippi (Eds.), *Alimentos Funcionais e Compostos Bioativos* (1st ed., pp. 333–378). Barueri: Manole.

Market Watch. (2019). Market watch. *Soy Lecithin*. Retrieved July 10, 2020, from https://www.marketwatch.com/

Marro, N., Cofré, N., Grilli, G., Alvarez, C., Labuckas, D., Maestri, D., & Urcelay, C. (2020). Soybean yield, protein content and oil quality in response to interaction of arbuscular mycorrhizal fungi and native microbial populations from mono- and rotation-cropped soils. *Applied Soil Ecology, 152*:, 103575. Retrieved from https://linkinghub.elsevier.com/retrieve/pii/S0929139319313253

Martin-Rubio, A. S., Sopelana, P., Ibargoitia, M. L., & Guillén, M. D. (2018). Prooxidant effect of α-tocopherol on soybean oil. Global monitoring of its oxidation process under accelerated storage conditions by 1H nuclear magnetic resonance. *Food Chemistry, 245*:, 312–323. Retrieved from https://linkinghub.elsevier.com/retrieve/pii/S0308814617317375

Martin, C. A., Almeida, V. V. de, Ruiz, M. R., Visentainer, J. E. L. J. V., Matshushita, M., Souza, N. E. de, & Visentainer, J. E. L. J. V. (2006). Ácidos graxos poliinsaturados ômega-3 e ômega-6: Importância e ocorrência em alimentos. *Revista de Nutrição, 19*:6, 761–770. Retrieved from http://www.scielo.br/scielo.php?script=sci_arttext&pid=S1415-52732006000600011&lng=pt&nrm=iso&tlng=pt

Maszewska, M., Florowska, A., Dłuzewska, E., Wroniak, M., Marciniak-Lukasiak, K., & Zbikowska, A. (2018). Oxidative stability of selected edible oils. *Molecules, 23*:7, 1746. Retrieved from http://www.mdpi.com/1420-3049/23/7/1746

McClements, D. J.; Decker, E. A. (2008). Lipids. In Damodaran, S.; Parkin, K. L.; Fennema, O. R. (Ed.), *Fennema's Food Chemistry* (4th ed.). CRC Press.

Mendes, M. F., Pessoa, F. L. P., & Uller, A. M. C. (2002). An economic evaluation based on an experimental study of the vitamin E concentration present in deodorizer distillate of soybean oil using supercritical CO_2. *Journal of Supercritical Fluids, 23*:3, 257–265. Retrieved from https://linkinghub.elsevier.com/retrieve/pii/S0896844601001401

Mermelstein, N. H. (2010, August). Improving soybean oil. *Food Technology Magazine*, 72–76. Chicago, IL. Retrieved from https://www.ift.org/news-and-publications/food-technology-magazine/issues/2010/august/columns/food-safety-and-quality

Misra, A., Singhal, N., Sivakumar, B., Bhagat, N., Jaiswal, A., & Khurana, L. (2011). Nutrition transition in India: Secular trends in dietary intake and their relationship to diet-related non-communicable diseases. *Journal of Diabetes, 3*:4, 278–292. Retrieved from http://doi.wiley.com/10.1111/j.1753-0407.2011.00139.x

Missouri Soybean. (2020). SOYLEIC™. *Missouri Soybean Technology*. Retrieved June 12, 2020, from https://mosoy.org/wp-content/uploads/2019/05/MoSoy_HighOleic_TechSheet_102717.pdf

Mohammadi, A., Jafari, S. M., Esfanjani, A. F., & Akhavan, S. (2016). Application of nano-encapsulated olive leaf extract in controlling the oxidative stability of soybean oil. *Food Chemistry, 190*, 513–519. Retrieved from https://linkinghub.elsevier.com/retrieve/pii/S0308814615008596

Monsanto. (2020). Vistive® Gold Composition. *About Vistive® Gold*. Retrieved April 10, 2020, from https://www.vistivegold.com/About

Moreau, R. A., Nyström, L., Whitaker, B. D., Winkler-Moser, J. K., Baer, D. J., Gebauer, S. K., & Hicks, K. B. (2018). Phytosterols and their derivatives: Structural diversity, distribution, metabolism, analysis, and health-promoting uses. *Progress in Lipid Research, 70*, 35–61. Retrieved from https://linkinghub.elsevier.com/retrieve/pii/S0163782717300620

Morin, J.-F., & Lees, M. (Eds.). (2018). *FoodIntegrity Handbook*. Eurofins Analytics France. Retrieved from https://foodintegrity.eu/foodintegrity/index.cfm?sectionid=83

Naeli, M. H., Farmani, J., & Zargaraan, A. (2017). Rheological and physicochemical modification of trans-free blends of palm stearin and soybean oil by chemical interesterification. *Journal of Food Process Engineering, 40*:2, e12409. Retrieved from http://doi.wiley.com/10.1111/jfpe.12409

Napolitano, G. E., Ye, Y., & Cruz-Hernandez, C. (2018). Chemical characterization of a high-oleic soybean oil. *JAOCS, Journal of the American Oil Chemists' Society*, *95*:5, 583–589.

Narce, M., & Poisson, J. P. (2002). Lipid metabolism: Regulation of lipid metabolism gene expression by peroxisome proliferator-activated receptor α and sterol regulatory element binding proteins. *Current Opinion in Lipidology*, *13*:4, 445–447. https://doi.org/10.1097/00041433-200208000-00013

Naz, S., Sheikh, H., Siddiqi, R., & Asad Sayeed, S. (2004). Oxidative stability of olive, corn and soybean oil under different conditions. *Food Chemistry*, *88*:2, 253–259. Retrieved from https://linkinghub.elsevier.com/retrieve/pii/S0308814604001104

Ndzana, X., Fehr, W. R., Welke, G. A., Hammond, E. G., Duvick, D. N., & Cianzio, S. R. (1994). Influence of reduced palmitate content on agronomic and seed traits of soybean. *Crop Science*, *34*:3, 646–649. Retrieved from https://onlinelibrary.wiley.com/doi/abs/10.2135/cropsci1994.0011183X003400030008x

Neff, W. E., & Byrdwell, W. C. (1995). Soybean oil triacylglycerol analysis by reversed-phase high-performance liquid chromatography coupled with atmospheric pressure chemical ionization mass spectrometry. *Journal of the American Oil Chemists' Society*, *72*:10, 1185–1191. Retrieved from http://doi.wiley.com/10.1007/BF02540986

Neff, W. E., List, G. R., & Byrdwell, W. C. (1999). Quantitative composition of high palmitic and stearic acid soybean oil triacylglycerols by reversed phase High Performance Liquid Chromatography: Utilization of evaporative light scattering and flame ionization detectors. *Journal of Liquid Chromatography & Related Technologies*, *22*:11, 1649–1662. Retrieved from https://www.tandfonline.com/doi/full/10.1081/JLC-100101758

Neff, William E., & List, G. R. (1999). Oxidative stability of natural and randomized high-palmitic-and high-stearic-acid oils from genetically modified soybean varieties. *Journal of the American Oil Chemists' Society*, *76*:7, 825–831. Retrieved from http://doi.wiley.com/10.1007/s11746-999-0072-9

Nill, K. (2015). *Soy Beans: Properties and Analysis. Encyclopedia of Food and Health* (1st ed.). Elsevier Ltd. Retrieved from http://dx.doi.org/10.1016/B978-0-12-384947-2.00642-5

O'Brien, R. D. (2008). Soybean oil purification. *Soybeans: Chemistry, Production, Processing, and Utilization* (pp. 377–408). Elsevier. Retrieved from https://linkinghub.elsevier.com/retrieve/pii/B9781893997646500159

OECD/FAO. (2019). OECD-FAO Agricultural Outlook (Edition 2019). *OECD Agriculture Statistics (database)*. Retrieved July 7, 2020, from https://www.oecd-ilibrary.org/content/data/eed409b4-en

Ostlund, R. E., Racette, S. B., Okeke, A., & Stenson, W. F. (2002). Phytosterols that are naturally present in commercial corn oil significantly reduce cholesterol absorption in humans. *American Journal of Clinical Nutrition*, *75*:6, 1000–1004. Retrieved from https://academic.oup.com/ajcn/article/75/6/1000/4689428

Patterson, E., Wall, R., Fitzgerald, G. F., Ross, R. P., & Stanton, C. (2012). Health implications of high dietary omega-6 polyunsaturated fatty acids. *Journal of Nutrition and Metabolism*, *2012*, 1–16. Retrieved from http://www.hindawi.com/journals/jnme/2012/539426/

Perkins, E. G. (1995). Composition of soybeans and soybean products. *Practical Handbook of Soybean Processing and Utilization* (pp. 9–28). AOCS Press. Retrieved from https://linkinghub.elsevier.com/retrieve/pii/B9780935315639500061

Pham, A.-T., Lee, J.-D., Shannon, J. G., & Bilyeu, K. D. (2010). Mutant alleles of FAD2-1A and FAD2-1B combine to produce soybeans with the high oleic acid seed oil trait. *BMC Plant Biology*, *10*:1, 195. Retrieved from http://bmcplantbiol.biomedcentral.com/articles/10.1186/1471-2229-10-195

Pham, A. T., Shannon, J. G., & Bilyeu, K. D. (2012). Combinations of mutant FAD2 and FAD3 genes to produce high oleic acid and low linolenic acid soybean oil. *Theoretical and Applied Genetics*, *125*:3, 503–515. Retrieved from http://link.springer.com/10.1007/s00122-012-1849-z

Pierre, F. H. F., Martin, O. C. B., Santarelli, R. L., Taché, S., Naud, N., Guéraud, F., Audebert, M., et al. (2013). Calcium and α-tocopherol suppress cured-meat promotion of chemically induced colon carcinogenesis in rats and reduce associated biomarkers in human volunteers. *American Journal of Clinical Nutrition*, *98*:5, 1255–1262.

Pizzini, A., Lunger, L., Sonnweber, T., Weiss, G., & Tancevski, I. (2018). The role of omega-3 fatty acids in the setting of coronary artery disease and COPD: A review. *Nutrients*, *10*:12, 1864. Retrieved from http://www.mdpi.com/2072-6643/10/12/1864

Placzek, M., Gaube, S., Kerkmann, U., Gilbertz, K. P., Herzinger, T., Haen, E., & Przybilla, B. (2005). Ultraviolet B-induced DNA damage in human epidermis is modified by the antioxidants ascorbic acid and D-α-tocopherol. *Journal of Investigative Dermatology*, *124*:2, 304–307. Retrieved from https://linkinghub.elsevier.com/retrieve/pii/S0022202X15321588

Plat, J., van Onselen, E. N. M., van Heugten, M. M. A., & Mensink, R. P. (2000). Effects on serum lipids, lipoproteins and fat soluble antioxidant concentrations of consumption frequency of margarines and shortenings enriched with plant stanol esters. *European Journal of Clinical Nutrition, 54*:9, 671–677. Retrieved from http://www.nature.com/articles/1601071

Polichetti, E., Janisson, A., de la Porte, P. L., Portugal, H., Léonardi, J., Luna, A., … Chanussot, F. (2000). Dietary polyenylphosphatidylcholine decreases cholesterolemia in hypercholesterolemic rabbits. *Life Sciences, 67*(21), 2563–2576. https://doi.org/10.1016/S0024-3205(00)00840-7

Prabhakaran, D., Jeemon, P., & Roy, A. (2016). Cardiovascular diseases in India: Current epidemiology and future directions. *Circulation, 133*:16, 1605–1620. Retrieved from https://www.ahajournals.org/doi/10.1161/CIRCULATIONAHA.114.008729

Prabhakaran, D., Singh, K., Roth, G. A., Banerjee, A., Pagidipati, N. J., & Huffman, M. D. (2018). Cardiovascular diseases in India compared with the United States. *Journal of the American College of Cardiology, 72*:1, 79–95. Retrieved from https://linkinghub.elsevier.com/retrieve/pii/S0735109718346643

Prasad, P., Anjali, P., & Sreedhar, R. V. (2020). Plant-based stearidonic acid as sustainable source of omega-3 fatty acid with functional outcomes on human health. *Critical Reviews in Food Science and Nutrition*, 1–13. Taylor & Francis. Retrieved from https://doi.org/10.1080/10408398.2020.1765137

Primomo, V. S., Falk, D. E., Ablett, G. R., Tanner, J. W., & Rajcan, I. (2002). Inheritance and interaction of low palmitic and low linolenic soybean. *Crop Science, 42*:1, 31–36. Retrieved from https://onlinelibrary.wiley.com/doi/abs/10.2135/cropsci2002.3100

QUALISOY. (2020). Soybean innovations. *Increased Omega-3 Soybean Oil.* Retrieved May 18, 2020, from https://www.qualisoy.com/soybean-innovations

Ramadan, M. F. (2012). Antioxidant characteristics of phenolipids (quercetin-enriched lecithin) in lipid matrices. *Industrial Crops and Products, 36*(1), 363–369. https://doi.org/10.1016/j.indcrop.2011.10.008

Ramírez, M., Amate, L., & Gil, A. (2001). Absorption and distribution of dietary fatty acids from different sources. *Early Human Development, 65*:SUPPL. 2.

Ras, R. T., Geleijnse, J. M., & Trautwein, E. A. (2014). LDL-cholesterol-lowering effect of plant sterols and stanols across different dose ranges: A meta-analysis of randomised controlled studies. *British Journal of Nutrition, 112*:2, 214–219.

Rashidinejad, A., Birch, E. J., Sun-Waterhouse, D., & Everett, D. W. (2014). Delivery of green tea catechin and epigallocatechin gallate in liposomes incorporated into low-fat hard cheese. *Food Chemistry, 156*, 176–183. https://doi.org/10.1016/j.foodchem.2014.01.115

Reboul, E. (2017). Vitamin e bioavailability: Mechanisms of intestinal absorption in the spotlight. *Antioxidants, 6*:4, 95. Retrieved from http://www.mdpi.com/2076-3921/6/4/95

Remely, M., Ferk, F., Sterneder, S., Setayesh, T., Kepcija, T., Roth, S., Noorizadeh, R., et al. (2017). Vitamin e modifies high-fat diet-induced increase of DNA strand breaks, and changes in expression and DNA methylation of DNMT1 and MLH1 in C57BL/6J male mice. *Nutrients, 9*:6.

Robert, C., Couëdelo, L., Vaysse, C., & Michalski, M.-C. (2020). Vegetable lecithins: A review of their compositional diversity, impact on lipid metabolism and potential in cardiometabolic disease prevention. *Biochimie, 169*, 121–132. https://doi.org/10.1016/j.biochi.2019.11.017

Rodriguez-Amaya, D. B. (2019). Update on natural food pigments - A mini-review on carotenoids, anthocyanins, and betalains. *Food Research International, 124*, 200–205.

Rodsamran, P., & Sothornvit, R. (2019). Lime peel pectin integrated with coconut water and lime peel extract as a new bioactive film sachet to retard soybean oil oxidation. *Food Hydrocolloids, 97*, 105173. Retrieved from https://linkinghub.elsevier.com/retrieve/pii/S0268005X1930757X

Rogalska, E., Ransac, S., & Verger, R. (1990). Stereoselectivity of lipases. II. Stereoselective hydrolysis of triglycerides by gastric and pancreatic lipases. *Journal of Biological Chemistry, 265*:33, 20271–20276.

Ros, E. (2000). Intestinal absorption of triglyceride and cholesterol. Dietary and pharmacological inhibition to reduce cardiovascular risk. *Atherosclerosis, 151*, 357–379.

Rozner, S., & Garti, N. (2006). The activity and absorption relationship of cholesterol and phytosterols. *Colloids and Surfaces A: Physicochemical and Engineering Aspects, 282–283*, 435–456. Retrieved from https://linkinghub.elsevier.com/retrieve/pii/S0927775705009787

Ruiz-López, N., Haslam, R. P., Venegas-Calerón, M., Larson, T. R., Graham, I. A., Napier, J. A., & Sayanova, O. (2009). The synthesis and accumulation of stearidonic acid in transgenic plants: A novel source of 'heart-healthy' omega-3 fatty acids. *Plant Biotechnology Journal, 7*:7, 704–716. Retrieved from http://doi.wiley.com/10.1111/j.1467-7652.2009.00436.x

Russo, G. L. (2009). Dietary n−6 and n−3 polyunsaturated fatty acids: From biochemistry to clinical implications in cardiovascular prevention. *Biochemical Pharmacology, 77*:6, 937–946. Retrieved from https://linkinghub.elsevier.com/retrieve/pii/S0006295208007776

Sanclemente, T., Marques-Lopes, I., Fajó-Pascual, M., Cofán, M., Jarauta, E., Ros, E., Puzo, J., et al. (2012). Naturally-occurring phytosterols in the usual diet influence cholesterol metabolism in healthy subjects. *Nutrition, Metabolism and Cardiovascular Diseases*, 22:10, 849–855. Retrieved from https://linkinghub.elsevier.com/retrieve/pii/S0939475311000378

Saoudi, S., Chammem, N., Sifaoui, I., Bouassida-Beji, M., Jiménez, I. A., Bazzocchi, I. L., Silva, S. D., et al. (2016). Influence of Tunisian aromatic plants on the prevention of oxidation in soybean oil under heating and frying conditions. *Food Chemistry*, 212:, 503–511. Retrieved from https://linkinghub.elsevier.com/retrieve/pii/S0308814616308743

Sassa, T., & Kihara, A. (2014). Metabolism of very long-chain fatty acids: Genes and pathophysiology. *Biomolecules and Therapeutics*, 22:2, 83–92. Retrieved from http://www.biomolther.org/journal/DOIx.php?id=10.4062/biomolther.2014.017

Scholfield, C. R. (1981). Composition of soybean lecithin. *Journal of the American Oil Chemists' Society*, 58(10), 889–892. https://doi.org/10.1007/BF02659652

Schmelz, E. M., Dillehay, D. L., Webb, S. K., Reiter, A., Adams, J., & Merrill, A. H. (1996). Sphingomyelin consumption suppresses aberrant colonic crypt foci and increases the proportion of adenomas versus adenocarcinomas in CF1 mice treated with 1,2-dimethylhydrazine: Implications for dietary sphingolipids and colon carcinogenesis. *Cancer Research*, 56:21, 4936–4941.

Schmitz, G., & Ecker, J. (2008). The opposing effects of n−3 and n−6 fatty acids. *Progress in Lipid Research*, 47:2, 147–155. Retrieved from https://linkinghub.elsevier.com/retrieve/pii/S0163782707000574

Schrick, K., DeBolt, S., & Bulone, V. (2012). Deciphering the molecular functions of sterols in cellulose biosynthesis. *Frontiers in Plant Science*, 3:MAY.

Seemamahannop, R., Bilyeu, K., He, Y., Kapila, S., Tumiatti, V., & Pompili, M. (2019). Assessment of oxidative stability and physical properties of high oleic natural esters. *2019 IEEE 20th International Conference on Dielectric Liquids (ICDL)* (pp. 1–6). IEEE. Retrieved from https://ieeexplore.ieee.org/document/8796627/

Shahidi, Fereidoon, & Ambigaipalan, P. (2015). Phenolics and polyphenolics in foods, beverages and spices: Antioxidant activity and health effects – A review. *Journal of Functional Foods*, 18:Part B, 820–897. Retrieved from https://linkinghub.elsevier.com/retrieve/pii/S1756464615003023

Shahidi, Fereidoon, & De Camargo, A. C. (2016). Tocopherols and tocotrienols in common and emerging dietary sources: Occurrence, applications, and health benefits. *International Journal of Molecular Sciences*, 17:10.

Shirouchi, B., Nagao, K., Inoue, N., Ohkubo, T., Hibino, H., & Yanagita, T. (2007). Effect of dietary omega 3 phosphatidylcholine on obesity-related disorders in obese Otsuka Long-Evans Tokushima fatty rats. *Journal of Agricultural and Food Chemistry*, 55:17, 7170–7176. https://doi.org/10.1021/jf071225x

Silva, F. A. M., Borges, M. F. M., & Ferreira, M. A. (1999). Métodos para avaliação do grau de oxidação lipídica e da capacidade antioxidante. *Química Nova*, 22:1, 94–103.

Simonetto, M., Infante, M., Sacco, R. L., Rundek, T., & Della-Morte, D. (2019). A novel anti-inflammatory role of omega-3 PUFAs in prevention and treatment of atherosclerosis and vascular cognitive impairment and dementia. *Nutrients*, 11:10, 2279. Retrieved from https://www.mdpi.com/2072-6643/11/10/2279

Simopoulos, A.P. (2002). The importance of the ratio of omega-6/omega-3 essential fatty acids. *Biomedicine & Pharmacotherapy*, 56:8, 365–379. Retrieved from https://linkinghub.elsevier.com/retrieve/pii/S0753332202002536

Simopoulos, Artemis P. (1999). Essential fatty acids in health and chronic disease. *The American Journal of Clinical Nutrition*, 70:3 Supplement, 560s–569s. Retrieved from https://academic.oup.com/ajcn/article/70/3/560s/4715011

Simopoulos, Artemis P. (2010). The omega-6/omega-3 fatty acid ratio: Health implications. *Oléagineux, Corps gras, Lipides*, 17:5, 267–275. Retrieved from http://www.ocl-journal.org/10.1051/ocl.2010.0325

Simopoulos, Artemis P. (2016). An increase in the omega-6/omega-3 fatty acid ratio increases the risk for obesity. *Nutrients*, 8:3, 128. Retrieved from http://www.mdpi.com/2072-6643/8/3/128

Simopoulos, Artemis P. (2020). Omega-6 and omega-3 fatty acids: Endocannabinoids, genetics and obesity. *OCL*, 27, 7. Retrieved from https://www.ocl-journal.org/10.1051/ocl/2019046

Singh, M. (2005). Essential fatty acids, DHA and human brain. *The Indian Journal of Pediatrics*, 72:3, 239–242. Retrieved from http://link.springer.com/10.1007/BF02859265

STATISTA. (2021a). Leading soybean producing countries worldwide from 2012/13 to 2020/21. Retrieved October 07, 2021, from https://www.statista.com/statistics/263926/soybean-production-in-selected-countries-since-1980/

STATISTA. (2020b). Production volume of soybean oil worldwide from 2012/13 to 2020/2021. Retrieved October 07, 2021, from https://www.statista.com/statistics/620477/soybean-oil-production-volume -worldwide

STATISTA. (2021c). Consumption of vegetable oils worldwide from 2013/14 to 2020/2021, by oil type. Retrieved October 07, 2021, from https://www.statista.com/statistics/263937/vegetable-oils-global -consumption/

STATISTA. (2021d). Production of soybean oil worldwide from 2013/14 to 2020/21, by country. Retrieved October 07, 2021, from https://www.statista.com/statistics/612557/soybean-oil-production-worldwide -by-country/

Sublette, M. E., Hibbeln, J. R., Galfalvy, H., Oquendo, M. A., & Mann, J. J. (2006). Omega-3 polyunsaturated essential fatty acid status as a predictor of future suicide risk. *American Journal of Psychiatry*, *163*:6, 1100–1102. Retrieved from http://psychiatryonline.org/doi/abs/10.1176/ajp.2006.163.6.1100

Szyczewski, P., Frankowski, M., Zioła-Frankowska, A., Siepak, J., Szyczewski, T., & Piotrowski, P. (2016). A comparative study of the content of heavy metals in oils: Linseed oil, rapeseed oil and soybean oil in technological production processes. *Archives of Environmental Protection*, *42*:3, 37–40.

Taghvaei, M., & Jafari, S. M. (2015). Application and stability of natural antioxidants in edible oils in order to substitute synthetic additives. *Journal of Food Science and Technology*, *52*:3, 1272–1282. Retrieved from http://link.springer.com/10.1007/s13197-013-1080-1

Takahashi, M., Uechi, S., Takara, K., Asikin, Y., & Wada, K. (2009). Evaluation of an oral carrier system in rats: Bioavailability and antioxidant properties of liposome-encapsulated curcumin. *Journal of Agricultural and Food Chemistry*, *57*:19, 9141–9146. Retrieved from https://pubs.acs.org/doi/10.1021/ jf9013923

Tan, Z., & Shahidi, F. (2012). Phytosterols, phytostanols, and their conjugates in foods: structural diversity, quantitative analysis, and health-promoting uses. *Journal of the American Oil Chemists' Society*, *89*:4, 457–500.

Tarté, R., Paulus, J. S., Acevedo, N. C., Prusa, K. J., & Lee, S.-L. (2020). High-oleic and conventional soybean oil oleogels structured with rice bran wax as alternatives to pork fat in mechanically separated chicken-based bologna sausage. *LWT*, 109659. Retrieved from https://linkinghub.elsevier.com/retrieve/ pii/S0023643820306484

Tengku, T. M., Birch, E. J., Tengku-Rozaina, T. M., & Birch, E. J. (2016). Thermal oxidative stability analysis of hoki and tuna oils by Differential Scanning Calorimetry and Thermogravimetry. *European Journal of Lipid Science and Technology*, *118*:7, 1053–1061.

Thompson, M. D., & Cooney, R. V. (2020). The potential physiological role of γ-tocopherol in human health: A qualitative review. *Nutrition and Cancer*, *72*:5, 808–825.

Thurgood, J., Ward, R., & Martini, S. (2007). Oxidation kinetics of soybean oil/anhydrous milk fat blends: A differential scanning calorimetry study. *Food Research International*, *40*:8, 1030–1037. Retrieved from https://linkinghub.elsevier.com/retrieve/pii/S0963996907000877

Uauy, R., & Dangour, A. D. (2008). Nutrition in brain development and aging: Role of essential fatty acids. *Nutrition Reviews*, *64*:, S24–S33. Retrieved from https://academic.oup.com/nutritionreviews/article -lookup/doi/10.1111/j.1753-4887.2006.tb00242.x

Uauy, R., Hoffman, D. R., Peirano, P., Birch, D. G., & Birch, E. E. (2001). Essential fatty acids in visual and brain development. *Lipids*, *36*:9, 885–895. Retrieved from http://doi.wiley.com/10.1007/s11745-001 -0798-1

United Soybean Board. (2020). United soybean board. *High oleic soy*. Retrieved May 18, 2020, from https:// unitedsoybean.org/topics/high-oleic-soy

Virtamo, J., Taylor, P. R., Kontto, J., Männistö, S., Utriainen, M., Weinstein, S. J., Huttunen, J., et al. (2014). Effects of α-tocopherol and β-carotene supplementation on cancer incidence and mortality: 18-Year postintervention follow-up of the Alpha-Tocopherol, Beta-Carotene Cancer Prevention Study. *International Journal of Cancer*, *135*:1, 178–185. Retrieved from http://www.embase.com/search/results?subaction =viewrecord&from=export&id=L372904382%0Ahttp://dx.doi.org/10.1002/ijc.28641

Wagner, K.-H., Wotruba, F., & Elmadfa, I. (2001). Antioxidative potential of tocotrienols and tocopherols in coconut fat at different oxidation temperatures. *European Journal of Lipid Science and Technology*, *103*:11, 746–751.

Wallert, M., Ziegler, M., Wang, X., Maluenda, A., Xu, X., Yap, M. L., Witt, R., et al. (2019). α-Tocopherol preserves cardiac function by reducing oxidative stress and inflammation in ischemia/reperfusion injury. *Redox Biology*, 26.

Wang, T. (2008). Minor constituents and phytochemicals of soybeans. *Soybeans* (pp. 297–329). Elsevier. Retrieved from https://linkinghub.elsevier.com/retrieve/pii/B9781893997646500135

Wang, T. (2011). Soybean oil. In Frank D. Gunstone (Ed.), *Vegetable Oils in Food Technology* (2nd ed., pp. 59–105). Oxford, UK: Wiley-Blackwell. Retrieved from http://doi.wiley.com/10.1002/9781444339925.ch3

Wang, T., & Briggs, J. L. (2002). Rheological and thermal properties of soybean oils with modified FA compositions. *JAOCS, Journal of the American Oil Chemists' Society*, *79*:8, 831–836.

Wang, T., & Johnson, L. A. (2001). Refining normal and genetically enhanced soybean oils obtained by various extraction methods. *Journal of the American Oil Chemists' Society*, *78*:8, 809–815. Retrieved from http://doi.wiley.com/10.1007/s11746-001-0347-3

Warner, K. A. (2008). Food uses for soybean oil and alternatives to trans fatty acids in foods. *Soybeans* (pp. 483–498). Elsevier. Retrieved from https://linkinghub.elsevier.com/retrieve/pii/B9781893997646500184

Warner, K., & Fehr, W. (2008). Mid-oleic/ultra low linolenic acid soybean oil: A healthful new alternative to hydrogenated oil for frying. *Journal of the American Oil Chemists' Society*, *85*:10, 945–951. Retrieved from http://doi.wiley.com/10.1007/s11746-008-1275-1

Warner, K., & Gupta, M. (2003). Frying quality and stability of low-and ultra-low-linolenic acid soybean oils. *JAOCS, Journal of the American Oil Chemists' Society*, *80*:3, 275–280. Retrieved from http://doi.wiley.com/10.1007/s11746-003-0689-x

Warner, K., Orr, P., & Glynn, M. (1997). Effect of fatty acid composition of oils on flavor and stability of fried foods. *JAOCS, Journal of the American Oil Chemists' Society*, *74*:4, 347–356. https://doi.org/10.1007/s11746-997-0090-4

Warner, K., & Mounts, T. L. (1993). Frying stability of soybean and canola oils with modified fatty acid compositions. *Journal of the American Oil Chemists' Society*, *70*:10, 983–988.

Warner, Kathleen, & Gupta, M. (2006). Potato chip quality and frying oil stability of high oleic acid soybean oil. *Journal of Food Science*, *70*:6, s395–s400. Retrieved from http://doi.wiley.com/10.1111/j.1365-2621.2005.tb11462.x

Whelan, J. (2009). Dietary stearidonic acid is a long chain (n-3) polyunsaturated fatty acid with potential health benefits. *The Journal of Nutrition*, *139*:1, 5–10. Retrieved from https://academic.oup.com/jn/article/139/1/5/4750894

Whittinghill, J., & Welsby, D. (2010). Use of SDA soybean oil in bakery applications. *Lipid Technology*, *22*:9, 203–205.

WHO. (2018). Policies to eliminate industrially-produced trans fat consumption. *Replace trans fat*. Retrieved May 25, 2020, from https://www.who.int/docs/default-source/documents/replace-transfats/replace-act-information-sheet.pdf?ua=1

Whysner, J., Wang, C. X., Zang, E., Iatropoulos, M., & Williams, G. (1994). Dose response of promotion by butylated hydroxyanisole in chemically initiated tumours of the rat forestomach. *Food and Chemical Toxicology*, *32*:3, 215–222. Retrieved from https://linkinghub.elsevier.com/retrieve/pii/0278691594901937

Wilke, C. (2019). Gene-edited soybean oil makes restaurant debut. The Scientist. Retrieved June 10, 2020, from https://www.the-scientist.com/news-opinion/gene-edited-soybean-oil-makes-restaurant-debut-65590

Wilson, R. F. (1999). Alternatives to genetically modified soybean – The Better Bean Initiative. *Lipid Technology*, *10*:, 107–110.

Wilson, T. A., Meservey, C. M., & Nicolosi, R. J. (1998). Soy lecithin reduces plasma lipoprotein cholesterol and early atherogenesis in hypercholesterolemic monkeys and hamsters: beyond linoleate. *Atherosclerosis*, *140*:1, 147–153. Retrieved from https://linkinghub.elsevier.com/retrieve/pii/S0021915098001324

Xu, N., Shanbhag, A. G., Li, B., Angkuratipakorn, T., & Decker, E. A. (2019). Impact of phospholipid-tocopherol combinations and enzyme-modified lecithin on the oxidative stability of bulk oil. *Journal of Agricultural and Food Chemistry*, *67*:28, 7954–7960.

Yang, J., Xing, G., Niu, L., He, H., Guo, D., Du, Q., Qian, X., et al. (2018). Improved oil quality in transgenic soybean seeds by RNAi-mediated knockdown of GmFAD2-1B. *Transgenic Research*, *27*:2, 155–166. Retrieved from http://link.springer.com/10.1007/s11248-018-0063-4

Yeom, W. W., Kim, H. J., Lee, K.-R., Cho, H. S., Kim, J.-Y., Jung, H. W., Oh, S.-W., et al. (2020). Increased production of α-linolenic acid in soybean seeds by overexpression of lesquerella FAD3-1. *Frontiers in Plant Science*, *10*:. Retrieved from https://www.frontiersin.org/article/10.3389/fpls.2019.01812/full

Yin, H., Xu, L., & Porter, N. A. (2011). Free radical lipid peroxidation: Mechanisms and analysis. *Chemical Reviews, 111*:10, 5944–5972.

Zhang, X., Burchell, J., & Mosier, N. S. (2018). Enzymatic epoxidation of high oleic soybean oil. *ACS Sustainable Chemistry and Engineering, 6*:7, 8578–8583.

Zhong, Y., & Shahidi, F. (2012). Antioxidant behavior in bulk oil: Limitations of polar paradox theory. *Journal of Agricultural and Food Chemistry, 60*:1, 4–6. Retrieved from https://pubs.acs.org/doi/10.1021/jf204165g

6 Polyamines in Soybean Food and Their Potential Benefits for the Elderly

Nelly C. Muñoz-Esparza, Oriol Comas-Basté,
Natalia Toro-Funes, M. Luz Latorre-Moratalla,
M. Teresa Veciana-Nogués, and M. Carmen Vidal-Carou

CONTENTS

6.1 INTRODUCTION

Polyamines are a group of aliphatic molecules that are ubiquitously distributed in all living organisms. These compounds were firstly described in 1678 by Antoni van Leeuwenhoek and later named spermidine and spermine due to their presence in particularly high amounts in human semen (Miller-Fleming et al., 2015). In humans, polyamines are involved in the regulation of several cellular processes, including cell proliferation, signal transduction, and membrane stabilization (Hunter & Burrit, 2012). Polyamines have important implications in human health, playing an important role in the prevention of diseases associated with age and thus favoring lifespan (Muñoz-Esparza et al., 2019).

These compounds are synthesized endogenously, starting from the decarboxylation of ornithine, or exogenously supplied through diet. It is known that their synthesis decreases with age, which is why polyamines from food acquire greater importance in the elderly population (Nishimura et al., 2006). In this sense, soybean food can be considered a good source of polyamines (Nishibori et al., 2007; Toro-Funes et al., 2015). This chapter reviews the content of polyamines found in soybeans and in soy-derived foods, as well as the implications of polyamines during aging and in the prevention of chronic diseases.

DOI: 10.1201/9781003030294-6

6.2   CHEMICAL AND BIOCHEMICAL PROPERTIES OF POLYAMINES

Spermidine (N-[3-aminopropyl]-1,4-butanediamine), spermine (N,N-bis[3-aminopropyl]-1,4-butane diamine), and putrescine (1,4-butanediamine) are low-molecular-weight aliphatic polycations that are widely distributed in nature, including microorganisms, plants, and animals (Lenis et al., 2017). Spermidine and spermine possess three and four amino groups, respectively, while putrescine may be considered a diamine (Figure 6.1). Polyamines are relatively stable compounds, capable of resisting acidic and alkaline conditions, and can establish hydrogen bonds with hydroxyl solvents such as water and alcohol. Likewise, they can strongly bind to biomolecules such as DNA, RNA, proteins, and phospholipids, stabilizing their negative charges and, in many cases, modulating their function (Gómez-Gallego et al., 2017; Handa et al., 2018; Hirano et al., 2021).

Polyamines are involved in several important cellular processes, especially in cell proliferation and differentiation. They also participate in the synthesis of proteins, the modulation of the immune response, and the regulation of ion channels, particularly by blocking potassium channels (Gómez-Gallego et al., 2017; Tofalo et al., 2019). Another important aspect for which polyamines have been studied is their antioxidant capacity, spermine being the polyamine with the greatest antioxidant potential due to the higher amount of positive charges in its chemical structure. The main antioxidant mechanism would be its ability to chelate metals, which prevents the formation of hydroperoxides and secondary oxidation compounds (Lovaas, 1991; Toro-Funes et al., 2013). In addition, polyamines also act as DNA protectors from oxidative damage by eliminating free radicals, especially in lipophilic media (Farriol et al., 2003; Toro-Funes et al., 2013).

Intracellular polyamine levels are mainly regulated by *de novo* synthesis (Figure 6.2). Putrescine is synthesized from the amino acid ornithine by the action of the enzyme ornithine decarboxylase. Spermidine is subsequently obtained from putrescine by the action of the enzyme spermidine-synthase, which catalyzes the addition of a propylamine group coming from the decarboxylation of S-adenosyl-methionine. Finally, the enzyme spermine-synthase is in charge of transforming

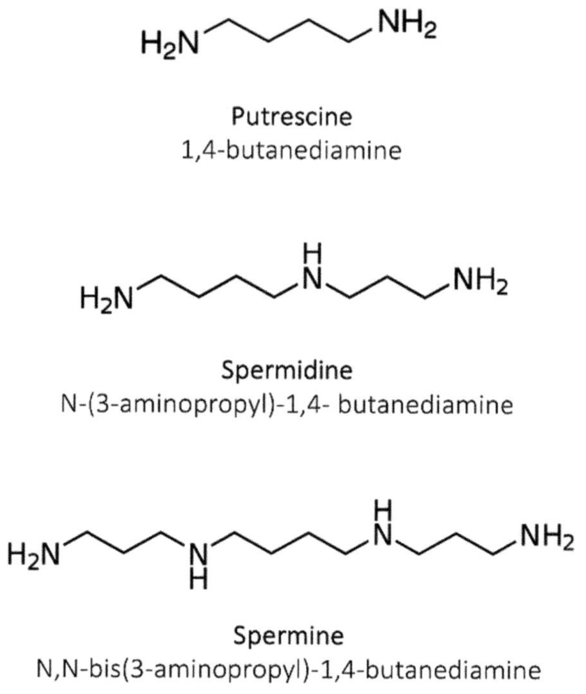

Putrescine
1,4-butanediamine

Spermidine
N-(3-aminopropyl)-1,4- butanediamine

Spermine
N,N-bis(3-aminopropyl)-1,4-butanediamine

FIGURE 6.1   Chemical structure of polyamines.

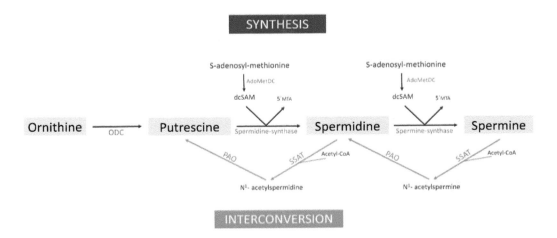

FIGURE 6.2 Synthesis and interconversion of polyamines. ODC = ornithine-decarboxylase; AdoMetDC = S-adenosyl-L-methionine-decarboxylase; dcSAM = decarboxylated S-adenosyl-methionine; 5′MTA = 5′-methylthioadenosine; SSAT = spermidine/spermine N1-acetyl-transferase; Acetyl-CoA = Acetyl coenzyme-A; PAO = polyamine-oxidase. Source: adapted from Munoz-Esparza et al. (2019).

spermidine into spermine by the addition of a second propylamine group (Miller-Fleming et al., 2015; Muñoz-Esparza et al., 2019).

In addition to *de novo* formation, interconversion among polyamines is also possible. This cyclical process controls the turnover of polyamines and regulates their intracellular homeostasis. The interconversion begins with the acetylation of either spermine or spermidine, which is catalyzed by an N-acetyl-transferase enzyme with the participation of acetyl coenzyme-A. Then, the enzyme polyamine oxidase removes a propylamine group. In this way, putrescine and spermidine may be obtained from the acetylated metabolite of spermidine and spermine, respectively (Larqué et al., 2007; Ruíz-Cano et al., 2012; Kalač, 2014).

In humans, polyamines can also have an exogenous origin mainly proceeding from food and/or the intestinal microbiota (Ramos-Molina et al., 2019; Tofalo et al., 2019). However, the contribution of the colonic microbiota to the global pool of polyamines is still uncertain and more information is needed on the polyamine-forming capacity of the intestinal microbiota and the possible biosynthetic pathways (Ramos-Molina et al., 2019). These bioactive compounds from the diet are absorbed in the small intestine, mainly in the duodenum and in the first segment of the jejunum, through transcellular mechanisms (i.e. passive diffusion and transporters) and by paracellular pathways (Pérez Cano et al., 2010; Ruíz-Cano et al., 2012; Hirano et al., 2021). Polyamines reaching systemic circulation are distributed to different tissues, such as adipose tissue, brain, liver, heart, pancreas, and thymus, in which they can be directly used by the cell or undergo interconversion (Larqué et al., 2007; Ramos-Molina et al., 2019).

6.3 POLYAMINES AND HEALTH IN THE ELDERLY

Polyamines are involved in a wide range of vital cellular processes, playing essential roles in cell growth and differentiation, metabolism, and physiological functions that may contribute to the inhibition of age-associated pathological changes (Minois et al., 2011; Soda et al., 2009a).

No age-related changes in polyamine levels in the blood have been reported so far, either in humans or animals. In blood cells, three separate mouse experiments did not find aging-associated changes in blood polyamine concentrations (Soda et al., 2009a; Soda et al., 2009b; Soda et al., 2013). In human blood, an age-related decrease in polyamines has not been observed either (Elworthy & Hitchcock, 1989; Soda et al., 2005). Similarly, Pucciarelli et al. (2012) reported that healthy human nonagenarians and centenarians retain whole-blood polyamines concentration in comparison with middle-aged individuals.

However, it has been evidenced that the levels of polyamines present in some tissues tend to decrease with age. In this sense, Nishimura et al. (2006) found that polyamine concentrations in various organs were significantly lower in 10- and 26-week-old mice than in 3-week-old mice. Similarly, Liu et al. (2008) observed a decrease in polyamine concentrations in different age-related memory-associated structures of the brain in rats. In human brains, Vivó et al. (2001) reported a negative correlation between spermidine and spermine content and age in several areas of the basal ganglia. The decline of polyamine concentrations with age in tissues may result from a decrease in the biosynthetic activities of polyamine-producing enzymes, especially ornithine decarboxylase (Yoshinaga et al., 1993).

Overall, to overcome the potential decrease in the synthesis of polyamines with aging, the exogenous supply of these compounds is crucial for the elderly population. In fact, it has been observed that the enrichment of the diet with polyamines during this stage can increase the concentration of these compounds in the blood and, hence, help reduce the risk of age-associated pathologies (Soda et al., 2009a; Soda et al., 2013).

6.3.1 Mechanism of Action of Polyamines in Aging

Several age-associated pathologies, including cancer, neurodegeneration, and cardiovascular diseases, are directly connected to the intracellular accumulation of toxic debris, and its removal by autophagy constitutes a well-documented pathway for protection against aging and disease (Choi et al., 2013; Frake et al., 2015). Many of the antiaging properties of polyamines, in particular spermidine, have been causally linked to their capacity to stimulate cytoprotective autophagy (Yin et al., 2016; Madeo et al., 2018).

The induction of autophagy by polyamines is independent of certain well-characterized pathways, such as sirtuins and mTOR (Morselli et al., 2011). Morselli et al. (2011) reported that spermidine could induce short-term autophagy regulating cytoplasmic (de)acetylations without transcription of new proteins. More recently, it has been reported that spermidine induces autophagy through the inhibition of several acetyltransferases, including that of E1A-associated protein p300, which is one of the main negative regulators of autophagy (Mariño et al., 2014; Pietrocola et al., 2015). In fact, the potency of spermidine has been quantified to be equivalent to that of rapamycin, an FDA-approved immunosuppressant with protective and autophagy-stimulatory properties (Harrison et al., 2009; Du Toit et al., 2018).

Moreover, a new mechanism of polyamines' regulation of autophagy in memory immune cells has been described (Puleston et al., 2019; Zhang et al., 2019). Autophagy levels are specifically reduced in mature lymphocytes, leading to compromised memory cell responses in old individuals (Phadwal et al., 2012). Zhang et al. (2019) demonstrated that spermidine induces autophagy *in vivo* and rejuvenates memory immune cell responses. These results reveal an unexpected autophagy regulatory mechanism, which can be harnessed to reverse immune senescence in humans (Zhang et al., 2019; Green, 2020).

Beyond the anti-inflammatory effects of autophagy itself, the antioxidant and anti-inflammatory activities of polyamines have been described as complementary mechanisms of action of polyamines in age-associated diseases (Minois, 2014a; Hussain et al., 2017). Polyamines' effect against oxidative stress has been studied both *in vitro* and *in vivo* with inflammation models. In all experiments, polyamines significantly inhibited the production of pro-inflammatory mediators, such as nitric oxide, prostaglandin E2, and cytokines. Moreover, polyamines also reduced the accumulation of intracellular reactive oxygen species (Choi & Park, 2012; Morón et al., 2013; Yang et al., 2016; Jeong et al., 2018).

Furthermore, other mechanisms have been ascribed to the action of polyamines, including the regulation of ion channels, DNA and RNA stability, regulation of DNA methylases, and protein acetylation. Soda (2020) recently reviewed the relation between polyamine metabolism and DNA methylation, as changes in DNA methylation status play an important role in lifespan and aging-associated pathologies. The authors suggested that the synthesis of polyamines acts by inhibiting aging-associated aberrant DNA methylation. In this sense, it has been shown, both *in vitro* and in experimental animals, that this beneficial effect of polyamines mainly acts by maintaining the activity of DNA-methyltransferase in a sustained manner (Soda, 2020).

6.3.2   HEALTH EFFECTS OF POLYAMINES RELATED TO AGING-ASSOCIATED PATHOLOGIES

Polyamines exert various beneficial effects on aging and age-related diseases. A recent prospective population-based study over a period of 15 years reported for the first time that a diet rich in spermidine was associated with a decrease in the risk of all-cause mortality in humans (Kiechl et al., 2018). In this study, spermidine showed the strongest inverse relation with mortality among 146 nutrients investigated. This effect was dose-dependent, and the authors stated that spermidine effectively induced autophagy and could reduce the acetylation of histones, which are critical processes for cell homeostasis in aging. However, more studies are still needed to clarify the role of polyamines in lifespan and to further unravel the age-associated effects of polyamines in humans. Here, a summary of the polyamine main effects on aging-associated diseases is presented.

6.3.2.1   Neuroprotection by Polyamines

Polyamines, in particular spermidine, seem to exert neuroprotective effects according to various *in vivo* studies performed in animal models. In flies, spermidine feeding protects from age-induced memory impairment (Gupta et al., 2013) and loss of locomotor activity (Minois et al., 2014b). These effects have been explained by a rejuvenation, dependent on autophagy, which maintains synaptic flexibility and plasticity (Gupta et al., 2016). In a mouse model for multiple sclerosis, spermidine attenuates disease progression and improves visual functions through reduced demyelination of the optic nerve and spinal cord and a decreased loss of retinal ganglion cells (Guo et al., 2011; Yang et al., 2016). Similarly, spermidine promotes optic nerve regeneration and blunts retinal degeneration in a mouse model of normal-tension glaucoma (Noro et al., 2015a; Noro et al., 2015b). On the other, it has been reported that spermine may improve the recognition memory deficit, spatial learning, and memory capabilities in rodent models (Velloso et al., 2009; Kibe et al., 2014). Overall, these studies suggest a wide range of neuroprotective effects of exogenously applied polyamines with relevance to several neurodegenerative motor disorders and dementias.

6.3.2.2   Polyamines in Cardiovascular and Metabolic Syndromes

The effect of polyamines in the protection from cardiac aging and cardiovascular pathologies has been widely described both in animal and human studies. Wang et al. (2020) reported the ability of spermidine to attenuate cardiac aging through activation of mitochondrial biogenesis. Eisenberg et al. (2016) showed that spermidine extends the lifespan of old mice and exerts cardioprotective effects, reducing cardiac hypertrophy and blood pressure, preserving diastolic function, and delaying the transition to heart failure. Moreover, spermidine reversed age-induced arterial stiffness with a reduction in oxidative damage of endothelial cells and alleviated the formation of atherosclerotic plaques in mice (LaRocca et al., 2013; Michiels et al., 2016). Finally, six-week supplementation of spermine and spermidine in mice reduced age-associated changes in myocardial morphology and inhibited cellular apoptosis of the heart (Zhang et al., 2017).

In humans, an increase in dietary polyamines inversely correlates with blood pressure and the incidence of cardiovascular disease and death (Soda et al., 2012; Eisenberg et al., 2016). However, it is necessary to perform interventional studies in humans to elucidate the exact mechanism of polyamines in the prevention and incidence of cardiovascular disease.

Regarding type 2 diabetes, polyamines could function as antiglycan agents due to their chemical structure, delaying the accumulation of advanced glycation end-products, which are associated with diabetes complications (Moinard et al., 2005; Bjelakovic et al., 2010). The interaction of the free amino groups of polyamines with the highly reactive carbonyl compounds may explain this protective effect (Gugliucci & Menini, 2003). *In vitro* studies have demonstrated that spermine can protect DNA and histones from glycation in the cell nucleus (Gugliucci & Menini, 2003). Whether polyamines might be useful in the treatment of obesity and type 2 diabetes is an important topic for future research.

In mice fed with hypercaloric diets, polyamines attenuated weight gain and the comorbidities of obesity induced by hypercaloric regimens, correlating with autophagy induction in white adipose tissue (Fernández et al., 2017). Moreover, polyamines prevented adiposity, improved glucose tolerance, led to increased energy expenditure, and conferred resistance to obesity-associated complications (Bonhoure et al., 2015; Sadasivan et al., 2014; Kraus et al., 2014).

6.3.2.3   Polyamines in Muscle-Related Diseases

One of the hallmarks of aging is the decline in stem cell function, resulting in impaired tissue regeneration and immunosenescence (López-Otín et al., 2013). In old mice, spermidine reversed the age-associated defect of autophagy and mitophagy in muscle stem cells, preventing their senescence and improving muscle regeneration (García-Prat et al., 2016). Spermidine application (or spermidine combined with exercise) successfully inhibited skeletal muscle atrophy in rats, concomitant with induction of autophagy and mitochondrial improvements (Fan et al., 2017). The spermidine-mediated ultrastructural and functional improvement of mitochondria from skeletal muscle stem cells further support the potential utility of spermidine in the treatment of muscle-related disorders (García-Prat et al., 2016; Fan et al., 2017).

6.3.2.4   Effects of Polyamines on Tumorigenesis

Polyamines are essential for cell proliferation and growth and, hence, a dysregulation of their metabolism is implicated in many tumor types. Although increased polyamine concentrations caused by enhanced biosynthesis have been found in skin, breast, colon, lung, and prostate cancers, the exact role of polyamines in cancer is still unclear (Nowotarski et al., 2013).

Despite that polyamines have shown *in vitro* pro-carcinogenic properties on cultured human cancer cells, *in vivo* animal studies have not demonstrated the same effect (Gerner & Meyskens, 2004). In fact, polyamines seem to reduce tumorigenesis, as demonstrated in skin tumors (Matsumoto et al., 2011), hepatocellular carcinomas (Yue et al., 2017), or colorectal tumors (Vargas et al., 2012; Miao et al., 2016). Moreover, polyamines reduce the growth of transplantable tumors in mice treated with chemotherapies by the stimulation of immunosurveillance. Concretely, spermidine improves the antitumor efficacy of chemotherapy *in vivo* by enhancing the anticancer immune response (Pietrocola et al., 2016). The role of polyamines in the stimulation of the immune system explains why they reduce tumorigenesis *in vivo*, although it enhances the proliferation of cancer cells *in vitro*.

6.4   POLYAMINES IN SOYBEAN AND SOY PRODUCTS

Polyamines are found in foods of both animal and plant origin, although their content and distribution depend on the type of product. The occurrence of spermidine, spermine, and putrescine in foods mainly has a physiological origin (as they proceed from animal and plant tissues) (Kalač et al., 2005; Sánchez-Pérez et al., 2018; Muñoz-Esparza et al., 2021). In addition, high levels of putrescine may be achieved by the action of fermentative and/or spoilage microorganisms (Bover-Cid et al., 2001; Kozová et al., 2009). Moreover, some authors have also associated the formation of spermidine and spermine with a bacterial activity, especially in fermented products (Atiya-Ali et al., 2011; Kalač, 2014; Kobayashi et al., 2016).

In foods of plant origin, spermidine is the main polyamine, usually followed by spermine, with the exception of fruits, in which putrescine is the highest polyamine, especially in citrus fruits (Sánchez-Pérez et al., 2018; Muñoz-Esparza et al., 2021). On the other hand, foods of animal origin tend to have a higher content of spermine and, in some cases, significant levels of putrescine can be achieved (Kalač et al., 2005; Nishimura et al., 2006; Nishibori et al., 2007; Kozová et al., 2009; Bover-Cid et al., 2014).

6.4.1 Soybean and Non-Fermented Soy Products

Soybean and soy products are a good source of polyamines, although their content shows great variability, even within the same type of food. Table 6.1 summarizes the polyamine content in soybean and non-fermented soy products reported in the literature. Due to its plant origin, the predominant polyamine in soybeans was spermidine, with mean values ranging from 106 to 218 mg/kg, followed by spermine (36–82 mg/kg) and putrescine (17–41 mg/kg). For non-fermented soy products, a similar polyamine distribution profile has been reported, although in notably lower amounts (with approximately a ten-fold reduction). Regarding soy drink, the fact that it is produced by the hydration and grinding of the soybeans may help explain the lower polyamine occurrence observed. Moreover, a subsequent heating process is applied to improve the taste and flavor of soy drink and achieve its stabilization. The application of both ultra-high temperature (UHT) and ultra-high-pressure homogenization (UHPH) treatments in the manufacturing of soy drinks did not modify its polyamine content (Toro-Funes et al., 2014). In the case of tofu, obtained by the coagulation of soy drink, a very similar distribution profile of polyamines has been reported.

In soybean sprouts, putrescine was the predominant polyamine, with a mean level of 44.7 mg/kg according to data reported by Toro-Funes et al. (2015), which may be attributed to the germination process in which putrescine acts as a growth factor (Kusano et al., 2008). In soybeans, the germination process leads to an accumulation of all polyamines, finding maximum values at 48 h of germination, followed by slightly lower values after 96 h (Glória et al., 2005). In this sense, Kralj-Cigic et al. (2020) reported that germination of lentils only increased the content of putrescine while spermine and spermidine remained practically unchanged. The conditions (e.g. luminosity, temperature, humidity) and the period of germination of pulses may modify the polyamine levels achieved in the sprouts (Gloria et al., 2005; Mao-Jun et al., 2005; Ponce de Leon et al., 2013; Menéndez et al., 2019).

Overall, the occurrence of polyamines in soybean and, subsequently, its derivatives, could be affected by the cultivar, the geographical location, certain environmental factors, and the specific cultivation and/or harvesting conditions (Glória et al., 2005; Kobayashi et al., 2017; Sagara et al., 2017). In plants, apart from the involvement of polyamines in the germination process, they also play a role in the response to stress to various environmental factors, such as drought or the presence of extreme temperatures during the harvest period (Toro-Funes et al., 2015; Shao et al., 2015; Dawood & Abeed, 2020). Actually, it has been observed that polyamine levels in plants, particularly those of putrescine, increase in response to stress during cultivation (Shao et al., 2015; Menéndez et al., 2019). Likewise, some studies have shown that the application of polyamines can also compensate for the negative effects of cold or drought, thus favoring the germination, growth, or survival of plants (Kusano et al., 2007; Luna-Esquivel et al., 2014; Chen et al., 2019; Menéndez et al., 2019). In addition, fertilization can also influence the levels of polyamines. On this topic, Losák et al. (2018) reported that sulfur and nitrogen fertilization increased the levels of spermidine present in soybeans.

6.4.2 Fermented Soy Products

Nowadays, a range of fermented soybean derivatives are available worldwide (i.e. natto, sufu, tempeh, tamari, soy sauce, and miso). These products are all obtained through the fermentation of the soybean, but they differ in the specific treatment applied to the raw material (e.g. dehulled, soaking, boiling, roasting, and steaming) and in the addition of several ingredients (e.g. sesame oil, wheat, and rice). Overall, the occurrence of polyamines in fermented soy products was highly variable among them, both from a quantitative and qualitative point of view (Table 6.2). This heterogeneity

TABLE 6.1

Polyamine Content in Soybean and Non-Fermented Soy Products (mg/Kg)

Soybean and Soy Products	n	Putrescine			Spermidine			Spermine			References
		Mean	SD	Range	Mean	SD	Range	Mean	SD	Range	
Soybean	3	–	–	1.6–6.5	–	–	33.2–62.1	–	–	29.7–34.3	Bardócz et al. (1993)
	1	17.0	–	–	128.0	–	–	–	–	–	Ziegler et al. (1994)
	2	41.0	–	–	207.0	–	–	69.0	–	–	Okamoto et al. (1997)
	13	–	–	3.7–16.8	–	–	99.2–389.0	–	–	27.8–114.0	Gloria et al. (2005)
	4	30.9	15.5	16.3–57.0	180.0	82.7	90.8–305.0	–	–	7.2–19.1	Kalač et al. (2005)
	2	–	–	35.2–57.2	158.0	–	–	58.6	–	–	Nishimura et al. (2006)
	5	17.1	–	6.4–24.2	106.0	–	88.2–125.0	36.6	–	30.3–41.6	Nishibori et al. (2007)
	6	18.1	0.7	–	218.0	9.2	–	82.6	2.4	–	Hou et al. (2019)*
	1	3.18	0.13	–	86.04	3.17	–	5.97	0.24	–	Tan et al. (2019)
Soy drink	2	2.1	–	–	16.3	–	–	2.8	–	–	Nishimura et al. (2006)
	8	1.3	0.4	0.7–2.1	10.1	0.5	9.1–10.9	2.2	0.6	1.6–3.2	Toro-Funes et al. (2015)
Tofu	4	nd	–	–	0.2	–	nd–0.2	nd	–	–	Nishibori et al. (2007)
	19	2.6	1.4	nd–5	23.6	5.6	15–35.3	7.6	4.1	4.1–19.2	Byun et al. (2013)
	2	1.8	–	–	15.5	–	–	6.8	–	–	Nishimura et al. (2006)
	8	0.8	0.6	nd–1.5	20.8	6.3	3.4–30.2	4.5	3.0	nd–8.1	Toro-Funes et al. (2015)
	32	44.6	2.4	–	9.1	0.3	–	–	–	–	Yue et al. (2019)
Soy sprouts	3	44.7	3.2	41.1–47.4	11.2	0.6	10.5–1.7	0.5	0.4	0.3–0.9	Toro-Funes et al. (2015)

Notes

nd = not detected; – = not reported by authors.

*mg per Kg of dry weight.

TABLE 6.2

Polyamine Content in Fermented Soy Products (mg/Kg)

Fermented Soy Products	n	Putrescine			Spermidine			Spermine			References
		Mean	SD	Range	Mean	SD	Range	Mean	SD	Range	
Natto	2	11.4	–	–	87.2	–	–	17.8	–	–	Nishimura et al. (2006)
	4	15.4	–	7.1–23.5	33.6	–	21.8–45.5	3.8	–	2.0–5.3	Nishibori et al. (2007)
	3	7.4	1.9	5.8–9.5	66.4	9.2	56.9–75.2	10.1	1.0	9.2–11.2	Toro-Funes et al. (2015)
Sufu	10	–	–	21.2–47.3	–	–	nd–27.1	–	–	nd–22.9	Tang et al. (2011)
	38	–	–	0.5–316.9	–	–	nd–4.0	–	–	nd–6.9	Guan et al. (2013)
	3	14.2	4.7	9.3–18.8	0.5	0.8	nd–1.4	1.8	0.8	0.9–2.2	Toro-Funes et al. (2015)
	3	–	–	13.8–36.4	–	–	0.6–6.5	–	–	nd–9.4	Yang et al. (2020)
Tempeh	2	45.3	–	–	85.6	–	–	13.6	–	–	Nishimura et al. (2006)
	3	23.2	7.0	17.5–31.1	108.9	13.7	97.3–124.0	12.7	8.3	6.1–21.8	Toro-Funes et al. (2015)
Tamari	3	15.3	2.0	13.1–17.1	34.8	4.6	29.45–38.02	4.4	2.10	2.8–6.8	Toro-Funes et al. (2015)
Douchi	30	–	–	nd–276.0	–	–	nd–74.92	–	–	nd	Yang et al. (2014)
	6	–	–	4.47–33.36	–	–	7.59–18.0	–	–	nd	Tan et al. (2019)
Soy sauce	10	26.8	–	–	12.9	–	–	nd	–	–	Nishimura et al. (2006)
	6	61.4	–	29.8–136.4	11.9	–	6.3–16.7	2.0	–	0.2–3.9	Nishibori et al. (2007)
	3	7.2	0.4	6.9–7.6	21.9	1.7	20.0–22.9	1.9	0.3	1.5–2.1	Toro-Funes et al. (2015)
	45	22.9	60.3	0.3–229.0	3.9	3.3	nd–10.4	–	–	–	Deetae et al. (2017)
	3	31.75	0.52	–	73.86	2.83	–	nd	–	–	Li et al. (2019)
Miso	6	26.1	–	19.8–34.3	1.7	–	0.4–5.9	1.0	–	0–6.3	Nishibori et al. (2007)
	3	11.7	7.8	2.7–17.8	8.4	1.3	7.5–9.9	2.9	0.6	2.5–3.5	Toro-Funes et al. (2015)

Notes

nd = not detected; – = not reported by authors.

is mainly due to the different occurrence of polyamines in the raw soybeans used in their manufacturing, as well as to the contribution from the different added ingredients (Toro-Funes et al., 2015; Kobayashi et al., 2017; Park et al., 2019; Yang et al., 2020; Liu et al., 2020).

Moreover, the fermentation process could, partly, influence the polyamine content, especially explaining the high putrescine values observed for some products, such as soy sauce and sufu (Tang et al., 2011; Yang et al., 2020). In this topic, it has been described that putrescine production may depend on the strains of the fermentative microorganisms, as well as on the fermentation conditions, such as temperature and period (Park et al., 2019; Tan et al., 2019; Lan et al., 2020; Yang et al., 2020). For example, Tan et al. (2019) evaluated the addition of two different starter cultures (*Aspergillus* and *Mucor*) during the douchi fermentation process. Their results showed an increase in putrescine content with respect to its initial content (raw); particularly the *Aspergillus* strains increased putrescine levels up to ten times, while in the *Mucor* strains this increase was lower (up to three times more).

During the production of these types of soy products, a traditional fermentation with naturally occurring microorganisms has been historically used (Tan et al., 2019; Park et al., 2019). In this case, the fermentation process takes longer and is generally carried out in open environments where random spontaneous microorganisms could be responsible for a higher formation of amines than in soy products fermented with starter cultures (Li et al., 2019; Tan et al., 2019). Thus, some authors related the occurrence of higher levels of putrescine in fermented soy products with the presence of spoilage microorganisms with decarboxylase activity (i.e. *Enterococcus*, *Bacillus*, or *Staphylococcus*) (Li et al., 2019; Yue et al., 2019; Hou et al., 2019; Yang et al., 2020).

Despite the fact that most of the studies dealing with soy fermented products do not link the formation of spermine and spermidine to bacterial activity, Kobayashi et al. (2016) reported the capacity of different strains of *Bacillus subtilis* to form spermidine during the production of natto. These authors suggested that the selection of starter cultures with spermidine production potential could contribute to improved polyamine levels in natto, reinforcing the health benefits of this traditional Japanese fermented food (Kobayashi et al., 2016).

6.5   IMPACT OF POLYAMINE INTAKE FROM SOYBEAN FOOD ON THE ELDERLY POPULATION

It has been argued that, during aging, polyamines evolve to the status of anti-aging vitamins, and thus its dietary intake should be promoted to secure the maintenance of homeostasis (Madeo et al., 2019). Although there are no specific recommendations on the daily intake of polyamines, some authors have shown that long-term intake of foods rich in polyamines could have a positive impact on the prevention of age-associated diseases. Specifically, Soda et al. (2009b) observed in a Japanese population that the continuous intake of soy and soy derivatives, such as natto, increased the blood levels of polyamines, which could potentially have a positive impact on cardiovascular health. Likewise, Binh et al. (2011) and Soda et al. (2012) have pointed out that a higher intake of polyamines in the frame of a Mediterranean dietary pattern could promote greater longevity and provide a protective effect against cardiovascular diseases. More recently, Kiechl et al. (2018) also observed that a diet rich in spermidine, mainly from plant origin foods, was associated with a lower risk of mortality in a follow-up cohort for 20 years.

The daily intake of polyamines has been estimated in various geographical areas, showing highly variable results (Muñoz-Esparza et al., 2019). Different studies performed in the adult European population showed ranges of polyamine intake from 140 to 390 µmol/day, depending on the country (Ralph et al., 1999; Atiya Ali et al., 2011; Buyukuslu et al., 2014). For the USA, the daily intake of polyamines in adults aged 40–80 years old was estimated at 250 µmol. In Japan, a slightly lower average intake of polyamines was estimated for children and adults (200 µmol/day) (Nishibori et al., 2007).

In recent years, the consumption of soy and its derivatives has increased in Western countries, mainly through the use of soy drink and a variety of traditional fermented products (Guan et al., 2013; Losák et al., 2018). This consumption behavior may be due not only to the potential health benefits, but also to their use as a substitute for animal protein or, in the case of soy drink, as an alternative to cow's milk for people who are lactose intolerant, allergic to milk protein, or avoid milk for other reasons (Toro-Funes et al., 2014). However, the estimation of the polyamine intake only derived from the consumption of soy-derived products in Western countries is still challenging due to the scarcity of their consumption data. Here, a first attempt has been made to assess the contribution of polyamines per serving of some of the most popular soy products by Western population (Table 6.3). The soy-derivative product that showed the highest contribution of polyamines per serving was tempeh (129.0 μmol), mainly due to its outstanding polyamine occurrence. Soy drink, tofu, soybean sprouts, natto, and sufu provided a lower polyamine contribution (10.4–29.3 μmol), spermidine being the predominant compound. As tamari, soy sauce, and miso are usually consumed as condiments, their expected polyamine contribution was very low.

Nowadays, cow milk is frequently substituted by several plant-based beverages, soy drink being one of the most largely consumed. Then, taking into account the cow milk polyamine content as reported by Bover-Cid et al. (2014), a 60-fold intake of polyamines from soy drink is obtained in comparison with cow milk. It is also noteworthy that the typical replacement of certain animal protein foods (i.e. meat and fish) by high-protein soy derivatives (i.e. tempeh or tofu) would also provide a greater polyamine intake. For example, the consumption of one serving of tofu would provide double the polyamine intake than one of fish, while tempeh would provide a 6- and 12-times higher intake in comparison with meat and fish, respectively. On the contrary, the consumption of sufu, a matured soy-product that is often consumed as cheese, would lead to a lower intake of polyamines.

Some authors have suggested that the daily intake of polyamines should be approximately 540 μmol/day (Atiya-Ali et al., 2011). In this sense, in a hypothetical scenario with daily consumption of two glasses of soy drink and a serving of sufu, tofu, and tempeh, the intake of dietary polyamines would achieve 207.7 μmol, covering almost half of the above-mentioned recommendation. The inclusion of other foods, such as vegetables, fruits, cereals, legumes, and foods of animal origin in the daily diet could help to reach the suggested intake of polyamines.

TABLE 6.3
Estimation of the Polyamine Contribution (μmol) per Serving of Different Soy Products*

	Serving Size (g)	Total Polyamines (μmol)	Putrescine (μmol)	Spermidine (μmol)	Spermine (μmol)
Soybean sprouts	50	29.3	25.4	3.8	0.1
Soy drink	250	23.7	3.6	17.5	2.7
Tofu	120	20.9	1.0	17.2	2.7
Natto	30	17.7	2.5	13.7	1.5
Sufu	60	10.4	9.7	0.2	0.5
Tempeh	120	129.0	31.5	90	7.4
Tamari	15	6.5	2.6	3.6	0.3
Soy sauce	15	3.6	1.2	2.3	0.1
Miso	15	3.0	1.9	0.9	0.2

Notes

*Content data from Toro-Funes et al. (2015) and serving sizes according to CESNID Food Composition Tables (CESNID, 2008) were used to estimate the polyamine contribution. The following equivalences were used: soy drink as a substitute for cow's milk, tempeh and tofu as substitutes for meat and fish, sufu as aged cheese, and miso, tamari and soy sauce as condiments.

In conclusion, polyamines are associated with the prevention of a range of age-related processes and pathologies, thus promoting life expectancy. Thus, the dietary intake of polyamines becomes of relevance for the elderly population. Considering that soybeans and soy products are a good source of dietary polyamines, their consumption, instead of animal protein foods, could help to significantly increase the intake of polyamines.

6.6 SUMMARY

Spermidine, spermine, and putrescine, known as polyamines, are involved in different aspects of human health. They are associated with the maturation and maintenance of the gastrointestinal tract, as well as with the prevention of age-associated pathologies, such as cardiovascular diseases and metabolic syndrome, and thus favoring the lifespan. Polyamines are also related to neuroprotection and to the stimulation of the antineoplastic immune response. This protective effect against aging and aging-associated pathologies is mainly related to the fact that spermidine can induce autophagy and reduce the acetylation of histones, which are critical processes for cell homeostasis in aging. In addition to endogenous synthesis, polyamines can also be exogenously supplied through food. It is known that during aging the endogenous synthesis of polyamines decreases, which is why dietary polyamines acquire greater importance for the elderly population. Polyamines can be found in all types of food, although with different levels and distribution depending on the type of product. Soybean is a plant-based food that stands out for having a high content of polyamines, and due to its origin, spermidine is the majority polyamine. Soy-derived products also show a high occurrence of polyamines, although their contents are highly variable, both from a quantitative and qualitative point of view. This heterogeneity is mainly due to the different occurrence of polyamines in the raw soybeans used in their manufacturing, as well as the contribution from the different added ingredients. Therefore, the regular consumption of both soybeans and soy products could help to increase the intake of polyamines and achieve the requirements of these compounds, especially in the elderly.

REFERENCES

Atiya-Ali, M., Poortvliet, E., Strömberg, R., Yngve, A. (2011). Polyamines in foods: Development of a food database. *Food and Nutrition Research*, 55: 5572. doi:10.3402/fnr.v55i0.5572

Bardócz, S., Grant, G., Brown, D.S., Ralph, A., Pusztai, A. (1993). Polyamines in food Implications for growth and health. *Journal Nutrition of Biochemistry*, 4: 66–71.

Binh, P.N.T., Soda K., Kawakami, M. (2011). Mediterranean diet and polyamine intake: possible contribution of increased polyamine intake to inhibition of age-associated disease. *Nutrition and Dietary Supplements*, 3. doi:10.2147/NDS.S15349

Bjelakovic, G., Beninati, S., Bjelakovic, N., Sokolovic, D., Jevtovic, T., Stojanovic, I., Rossi, S., Tabolacci, C., Kocić, G., Pavlovic, D., Saranac, L., Zivic, S. (2010). Does polyamine oxidase activity influence the oxidative metabolism of children who suffer of diabetes mellitus?. *Molecular and Cellular Biochemistry*, 341: 79–85.

Bonhoure, N., Byrnes, A., Moir, R.D., Hodroj, W., Preitner, F., Praz, V., Marcelin, G., Chua, S.C. Jr., Martinez-Lopez, N., Singh, R., Moullan, N., Auwerx, J., Willemin, G., Shah, H., Hartil, K., Vaitheesvaran, B., Kurland, I., Hernandez, N., Willis, I.M. (2015). Loss of the RNA polymerase III repressor MAF1 confers obesity resistance. *Genes & Development*, 29(9): 934–947.

Bover-Cid, S., Hugas, M., Izquierdo-Pulido, M., Vidal-Carou, M.C. (2001). Amino acid-descarboxylase activity of bacteria isolated from fermented pork sausages. *International Journal of Food Microbiology*, 66: 185–189.

Bover-Cid, S., Latorre-Moratalla, M.L., Veciana-Nogués, M.T., Vidal-Carou, M.C. (2014). Processing contaminants: Biogenic amines. In Y. Motarjemi, G.G. Moy, E.C.D. Todd (Eds.), *Encyclopedia of Food Safety* (Vol. 2, pp. 381–391). Burlington, MA: Elsevier Inc.

Buyukuslu N., Hizli H., Esin K., Garipagaoglu M. (2014). A cross-sectional study: Nutritional polyamines in frequently consumed foods of the turkish population. *Foods*, 3: 541–557. doi:10.3390/foods3040541

Byun, B.Y., Bai, X., Mah J.H. (2013). Occurrence of biogenic amines in Doubanjiang and Tofu. *Food Science of Biotechnology*, 22: 55–62. doi:10.1007/s10068-013-0008-x

CESNID. 2008.*Tablas de composición de alimentos por medidas caseras de consume habitual en España*. Ed. McGraw Hill Americana.

Chen, D., Shao, Q., Yin L., Younis, A., Zheng, B. (2019). Polyamine functions in plants: Metabolism, regulation on development and roles in abiotic stress responses. *Frontiers in Plant Science*, 9: 1945. doi:10.3389/fpls.2018.01945

Choi, Y.H., Park, H.Y. (2012) Anti-inflammatory effects of spermidine in lipopolysaccharide-stimulated BV2 microglial cells. *Journal of Biomedical Science*, 19: 31.

Choi, A.M.K., Ryter, S.W., Levine, B. (2013). Autophagy in human health and disease. *New England Journal of Medicine*, 368: 651–662.

Dawood, M.F.A., Abeed, A.H.A. (2020). Spermine-priming restrained water relations and biochemical deteriorations prompted by water deficit on two soybean cultivars. *Heliyon*, 6: e04038. https://doi.org/10.1016/j.heliyon.2020.e04038

Deetae, P., Jamnong, P., Assavanig, A., Lertsiri, S. (2017). Occurrence of biogenic amines in Thai soy sauces and soy bean pastes and their health concern. *International Food Research Journal*, 24: 1575–1578.

Du Toit, A., De Wet, S., Hofmeyr, J.S., Müller-Nedebock, K.K., Loos, B. (2018). The precision control of autophagic flux and vesicle dynamics—A micropattern approach. *Cells*, 7: 94.

Eisenberg, T., Abdellatif, M., Schroeder, S.E. (2016). Cardioprotection and lifespan extension by the natural polyamine spermidine. *Nature Medicine*, 22(12): 1428–1438.

Elworthy, P., Hitchcock, E. (1989). Polyamine levels in red blood cells from patient groups of different sex and age. *Biochimica et Biophysica Acta*, 993: 212–216.

Fan, J., Yang, X., Li, J., Shu, Z., Dai, J., Liu, X., Li, B., Jia, S., Kou, X., Yang, Y., Chen, N. (2017). Spermidine coupled with exercise rescues skeletal muscle atrophy from D-gal-induced aging rats through enhanced autophagy and reduced apoptosis via AMPK-FOXO3a signal pathway. *Oncotarget*, 8: 17475–17490.

Farriol, M., Segovia-Silvestre, T., Venereo, Y., Orta, X. (2003). Antioxidant effect of polyamines on erythrocyte cell membrane lipoperoxidation after free-radical damage. *Phytotherapy Research*, 17: 44–47. doi:10.1002/ptr.1073

Fernández, Á.F., Bárcena, C., Martínez-García, G.G., Tamargo-Gómez, I., Suárez, M.F., Pietrocola, F., et al. (2017). Autophagy couteracts weight gain, lipotoxicity and pancreatic b-cell death upon hypercaloric pro-diabetic regimens. *Cell Death & Disease*, 8: e2970.

Frake, R.A., Ricketts, T., Menzies, F.M., Rubinsztein, D.C. (2015). Autophagy and neurodegeneration. *Journal of Clinical Investigation*, 125: 65–74.

García-Prat, L., Martínez-Vicente, M., Perdiguero, E., Ortet, L., Rodríguez-Ubreva, J., Rebollo, E., et al. (2016). Autophagy maintains stemness by preventing senescence. *Nature*, 529: 37–42.

Gerner, E.W., Meyskens F.L. Jr. (2004). Polyamines and cancer: Old molecules, new understanding. *Nature Reviews Cancer*, 4: 781–792.

Gómez-Gallego C., Kumar, H., García-Mantrana, I., du Toit, E., Suomela, J.P., Linderborg, K.M., Zhang Y., Isolauri E., Yang B., Salminen S., Collado M.C. (2017) Breast milk polyamines and microbiota interactions: Impact of mode of delivery and geographical location. *Annals of Nutrition and Metabolism*, 70: 184–190. doi:10.1159/000457134

Glória, M.B.A., Tavares-Neto, J., Labanca, R.A., (2005). Influence of cultivar and germination on bioactive amines in soybeans (Glycine mas L. Merril). *Journal of Agricultural and Food Chemistry*, 53: 7480–7485.

Green D.R. (2020). Polyamines and aging: A CLEAR connection? *Molecular Cell*, 76: 5–7.

Guan, R.F., Liu Z.F., Zhang, J.J., Wei, Y.X., Wahab, S., Liu, D.H., Ye, X.Q. (2013). Investigation of biogenic amines in sufu (furu): A Chinese traditional fermented soybean food product. *Food Control*, 31: 345–352. http://dx.doi.org/10.1016/j.foodcont.2012.10.033

Gugliucci, A., Menini, T. (2003). The polyamines spermine and spermidine protect proteins from structural and functional damage by AGE precursors: A new role for old molecules? *Life Science*, 72: 2603–2616.

Guo, X., Harada, C., Namekata, K., Kimura, A., Mitamura, Y., Yoshida, H., Matsumoto, Y., Harada, T. (2011). Spermidine alleviates severity of murine experimental autoimmune encephalomyelitis. *Investigate Opthalmology & Visual Science*, 52: 2696–2703.

Gupta, V.K., Scheunemann, L., Eisenberg, T., Mertel, S., Bhukel, A., Koemans, T.S., et al. (2013). Restoring polyamines protects from age-induced memory impairment in an autophagy-dependent manner. *Nature Neuroscience*, 16: 1453–1460.

Gupta, V.K., Pech, U., Bhukel, A., Fulterer, A., Ender, A., Mauermann, S.F., et al. (2016). Spermidine suppresses age-associated memory impairment by preventing adverse increase of presynaptic active zone size and release. *PLoS Biology*, 14: e1002563.

Handa, A.k., Fatima, T., Mattoo, A.K. (2018). Polyamines: Bio-molecules with diverse functions in plant and human health and disease. *Frontiers in Chemistry*, 2–10. doi:10.3389/fchem.2018.00010

Harrison, D.E., Strong, R., Sharp, Z.D., Nelson, J.F., Astle, C.M., Flurkey, K., et al. (2009). Rapamycin fed late in life extends lifespan in genetically heterogeneous mice. *Nature*, 460: 392–395.

Hirano, R., Shirasawa, H., Kurihara, S. (2021). Health-promoting effects of dietary polyamines. *Medical Sciences*, 9, 8. https://doi.org/10.3390/medsci9010008

Hou, Y., He, W., Hu, S., Wu, G. (2019). Composition of polyamines and amino acids in plant-source foods for human consumption. *Amino Acids*, 51: 1153–1165. https://doi.org/10.1007/s00726-019-02751-0

Hunter, D.C., Burrit, D.J. (2012). Polyamines of plant origin - An important dietary consideration for human health. In V. Rao (Ed.), *Phytochemicals as Nutraceuticals - Global Approaches to Their Role in Nutrition and Health*. InTech. ISBN: 978-953-51-0203-8. Retrieved from: http://www.intechopen.com/books/phytochemicalsas-nutraceuticals-global-approaches-to-their-role-innutrition-and-health/polyamines-of-plant-origin-an-important-dietary-consideration-for-human-health

Hussain, T., Tan, B., Ren, W., Rahu, N., Dad, R., Kalhoro, D.H., Yin, Y. (2017). Polyamines: Therapeutic perspectives in oxidative stress and inflammatory diseases. *Amino Acids*, 49(9): 1457–1468.

Jeong, J.W., Cha, H.J., Han, M.H., Hwang, S.J., Lee, D.S., Yoo, J.S., et al. (2018). Spermidine protects against oxidative stress in inflammation models using macrophages and zebrafish. *Biomolecules & Therapeutics (Seoul)*, 26(2): 146–156.

Kalač, P., Krízek, M., Pelikánová, T., Langová, M., Veskrna, O. (2005) Contents of polyamines in selected foods. *Food Chemistry*, 90: 561–564. doi:10.1016/j.foodchem.2004.05.019

Kalač, P. (2014). Health effects and occurrence of dietary polyamines: A review for the period 2005–mid 2013. *Food Chemistry*, 161: 27–39. doi:10.1016/j.foodchem.2014.03.102

Kibe, R., Kurihara, S., Sakai, Y., Suzuki, H., Ooga, T., Sawaki, E., et al. (2014). Upregulation of colonic luminal polyamines produced by intestinal microbiota delays senescence in mice. *Scientific Reports*, 4: 4548.

Kiechl, S., Pechlaner, R., Willeit, P., Notdurfter, M., Paulweber, B., Willeit, K., et al. (2018). Higher spermidine intake is linked to lower mortality: A prospective population-based study. *American Journal of Clinical Nutrition*, 108: 371–80. doi:10.1093/ajcn/nqy102

Kobayashi, K., Shimojo, S., Watanabe, S. (2016). Contribution of a fermentation process using Bacillus subtilis (natto) to high polyamine contents of natto, a traditional Japanese fermented soy food. *Food Science and Technology Research*, 22: 153–157. doi:10.3136/fstr.22.153

Kobayashi, K., Horii Y., Watanabe, S., Kubo, Y., Koguchi, K., Hoshi, Y., Matsumoto, K., Soda, K. (2017). Comparison of soybean cultivars for enhancement of the polyamine contents in the fermented soybean natto using Bacillus subtilis (natto). *Bioscience, Biotechnology and Biochemistry*, 81: 587–594. doi:10.1080/09168451.2016.1270738

Kozová M., Kalač P., Pelikánová T. (2009). Contents of biological active polyamines in chicken meat, liver, heart and skin after slaughter and their changes during meat storage and cooking. *Food Chem*, 116: 419–425. doi:10.1016/j.foodchem.2009.02.057

Kralj-Cigic, I., Rupnik, S., Rijavec, T., Ulrih, N.P., Cigic, B. (2020). Accumulation of agmatine, spermidine and spermine in sprouts and microgreens of alfalfa, fenugreek, lentil and daikon radish. *Foods*, 9: 547. doi:10.3390/foods9050547

Kraus, D., Yang, Q., Kong, D., Banks, A.S., Zhang, L., Rodgers, J.T., et al. (2014). Nicotinamide N-methyltransferase knockdown protects against diet-induced obesity. *Nature*, 508: 258–262.

Kusano, T., Yamaguchi, K., Berberich, T., Takahashi, Y. (2007). Advances in polyamine research in 2007. *Journal in Plants Research*, 120: 345–350. doi:10.1007/s10265-007-0074-3

Kusano, T., Berberich, T., Tateda, C., Takahashi, Y. (2008). Polyamines: Essential factors for growth and survival. *Planta*, 228: 367–381. DOI 10.1007/s00425-008-0772-7

Lan, G., Li, C., He, L., Zeng, X., Zhu, Q. (2020). Effects of different strains and fermentation method on nattokinase activity, biogenic amines and sensory characteristics of natto. *Journal of Food Science and Technology*, https://doi.org/10.1007/s13197-020-04478-3

LaRocca, T.J., Gioscia-Ryan, R.A. Hearon C.M. Jr., Seals, D.R. (2013). The autophagy enhancer spermidine reverses arterial aging. *Mechanisms of Ageing and Development*, 134: 314–320.

Larqué, E., Sabater-Molina, M., Zamora, S. (2007). Biological significance of dietary polyamines. *Nutrition*, 23: 87–95. doi:10.1016/j.nut.2006.09.006

Lenis, Y., Elmetwally, M., Maldonado-Estrada, J., Bazer, F. (2017). Physiological importance of polyamines. *Zygote*, 25: 244–255. doi:10.1017/S0967199417000120

Li, J., Huang, H., Feng, W., Guan, R., Zhou, L., Cheng, H., Ye, X. (2019). Dynamic changes in biogenic amine content in the traditional brewing process of soy sauce. *Journal of Foof Protection*, 82: 1539–1545. https://doi.org/10.4315/0362-028X.JFP-19-035

Liu, B., Cao, Z., Qin, L., Li, J., Lian, R., Wang, C. (2020). Investigatión of the synthesis of biogenic amines and quality during high-salt liquid-state soy sauce fermentation. *LWT Food Science and Technology*, 133: 109835. https://doi.org/10.1016/j.lwt.2020.109835

Liu, P., Gupta, N., Jing, Y., Zhang, H. (2008). Age-related changes in polyamines in memory-associated brain structures in rats. *Neuroscience*, 155: 789–796.

Lošák, T., Ševčík, M., Plchová, R., von Bennewitz, E., Hlušek, J., Elbl, J., et al. (2018). Nitrogen and sulphur fertilisation affecting soybean seed spermidine content. *Journal of Elementology*, 23: 581–588. doi:10.5601/jelem.2017.22.3.1516

López-Otín, C., Blasco, M.A., Partridge, L., Serrano, M., Kroemer, G. (2013). The hallmarks of aging. *Cell*, 153: 1194–1217.

Lovaas, E. (1991). Antioxidative effects of polyamines. *Journal of the American Oil Chemists´ Soceaty*, 68: 353–457.

Luna-Esquivel, E.N., Ojeda-Barrios, D.L., Guerrero-Prieto, V.M., Ruíz-Achondo, T., Martínez-Téllez, J.J. (2014). Poliaminas como indicadores de estrés en plantas. *Revista Chapingo Serie Horticultura*, 20: 283–295. http://dx.doi.org/10.5154/r.rchsh.2013.05.019

Madeo, F., Eisenberg, T., Pietrocola F., Kroemer G. (2018). Spermidine in health and disease. *Science*, 359(6374): 2788.

Madeo, F., Bauer, M.A., Carmona-Gutierrez, D., Kroemer, G. (2019). Spermidine: A physiological autophagy inducer acting as an anti-aging vitamin in humans? *Autophagy*, 15(1): 165–168.

Mao-Jun, X., Ju-Fang, D., Mu-Yuan, Z. (2005). Effects of germination conditions on ascorbic acid level and yield of soybean sprouts. *Journal of the Science of Food and Agriculture*, 85: 943–947. doi:10.1002/jsfa.2050.

Mariño, G., Pietrocola, F., Eisenberg, T., Kong, Y., Malik, S.A., Andryushkova, A. et al. (2014). Regulation of autophagy by cytosolic acetyl-coenzyme A. *Molecular Cell*, 53: 710–725.

Matsumoto, M., Kurihara, S., Kibe, R., Ashida, H., Benno, Y. (2011). Longevity in mice is promoted by probiotic-induced suppression of colonic senescence dependent on upregulation of gut bacterial polyamine production. *PLoS One*, 6: e23652.

Menéndez, A.B., Calzadilla, P.I., Sansberro, P.A., Espasandin, F.D., Gazquez, A., Bordenave, C.D., et al. (2019) Polyamines and legumes: Joint stories of stress, nitrogen fixation and environment. *Frontiers in Plant Science*, 10: 1415. doi:10.3389/fpls.2019.01415

Miao, H., Ou, J., Peng, Y., Zhang, X., Chen, Y., Hao, L., et al. (2016). Macrophage ABHD5 promotes colorectal cancer growth by suppressing spermidine production by SRM. *Nature Communications*, 7: 11716.

Michiels, C.F., Kurdi, A., Timmermans, J.P., De Meyer, G.R.Y., Martinet, W. (2016). Spermidine reduces lipid accumulation and necrotic core formation in atherosclerotic plaques via induction of autophagy. *Atherosclerosis*, 251: 319–327.

Miller-Fleming, L., Olin-Sandoval, V., Campbell, K., Ralser M. (2015). Remaining mysteries of molecular biology: The role of polyamines in the cell. *Journal of Molecular Biology*, 427: 3389–3406. http://dx.doi.org/10.1016/j.jmb.2015.06.020

Minois, N., Carmona-Gutierrez, D., Madeo, F. (2011). Polyamines in aging and disease. *Aging*, 3: 716–732. doi:10.18632/aging.100361

Minois, N. (2014a). Molecular basis of the 'anti-aging' effect of spermidine and other natural polyamines - A mini-review. *Gerontology*, 60(4): 319–326.

Minois, N., Rockenfeller, P., Smith, T.K., Carmona-Gutierrez, D. (2014b). Spermidine feeding decreases age-related locomotor activity loss and induces changes in lipid composition. *PLoS One*, 9: e102435.

Moinard, C., Cynober, L., de Bandt, J.P. (2005). Polyamines: Metabolism and implications in human diseases. *Clinical Nutrition*, 24: 184–97. doi:10.1016/j.clnu.2004.11.001

Morón, B., Spalinger, M., Kasper, S., Atrott, K., Frey-Wagner, I., Fried, M., McCole, D. F., Rogler, G. and Scharl, M. (2013). Activation of protein tyrosine phosphatase non-receptor type 2 by spermidine exerts anti-inflammatory effects in human THP-1 monocytes and in a mouse model of acute colitis. *PLoS One*, 8: e73703.

Morselli, E., Mariño, G., Bennetzen, M.V., Eisenberg, T., Megalou, E., Schroeder, S., et al. (2011). Spermidine and resveratrol induce autophagy by distinct pathways converging on the acetylproteome. *Journal of Cell Biology*, 192: 615–629.

Muñoz-Esparza, N.C., Latorre-Moratalla, M.L, Comas-Basté, O., Toro-Funes, N., Veciana-Nogués, M.T, Vidal-Carou, M.C. (2019) Polyamines in food. *Frontiers in Nutrition*, 6: 108. doi:10.3389/fnut.2019.00108

Muñoz-Esparza, N.C., Costa-Catala, J., Comas-Basté, O., Toro-Funes, N., Latorre-Moratalla, M.L., Veciana-Nogués, M.T., Vidal-Carou, M.C. (2021) Occurrence of Polyamines in Foods and the Influence of Cooking Processes. *Foods*, 10, 1752. https://doi.org/10.3390/foods10081752

Nishibori, N., Fujihara, S., Akatuki, T. (2007). Amounts of polyamines in foods in Japan and intake by Japanese. *Food Chemistry*, 100: 491–497. doi:10.1016/j.foodchem.2005.09.070

Nishimura, K., Shiina, R., Kashiwagi, K., Igarashi, K. (2006). Decrease in polyamines with aging and their ingestion from food and drink. *Journal of Biochemistry*, 139: 81–90. doi:10.1093/jb/mvj003

Noro, T., Namekata, K., Kimura, A., Guo, X., Azuchi, Y., Harada, C., Nakano, T., Tsuneoka, H., Harada T. (2015a). Spermidine promotes retinal ganglion cell survival and optic nerve regeneration in adult mice following optic nerve injury. *Cell Death & Disease*, 6(4): e1720.

Noro, T., Namekata, K., Azuchi, Y., Kimura, A., Guo, X., Harada, C., Nakano, T., Tsuneoka, H., Harada T. (2015b). Spermidine ameliorates neurodegeneration in a mouse model of normal tension glaucoma. *Investigative Ophthalmology Visual Science*, 56: 5012–5019.

Nowotarski, S.L., Woster, P.M., Casero R.A. Jr. (2013). Polyamines and cancer: Implications for chemotherapy and chemoprevention. *Expert Reviews in Molecular Medicine*, 15: e3.

Okamoto, A., Sugi, E., Koizumi, Y., Yanadiga, F., Udaka, S. (1997). Polyamine content of ordinary foodstuffs and various fermented foods. *Bioscience, Biotechnology and Biochemistry*, 61: 1582–1584. https://doi.org/10.1271/bbb.61.1582

Park, Y.K., Lee, J.H., Mah, J.H. (2019). Occurrence and reduction of biogenic amines in traditional Asian fermented soybeans foods: A review. *Food Chemistry*, 278: 1–9. https://doi.org/10.1016/j.foodchem.2018.11.045

Pérez-Cano, F.J., González-Castro, A., Castellote, C., Franch, A., Castell, M. (2010). Influence of breast milk polyamines on suckling rat immune system maturation. *Development and Comparative Immunology*, 34: 210–218. doi:10.1016/j.dci.2009.10.001

Phadwal, K., Alegre-Abarrategui, J., Watson, A.S., Pike, L., Anbalagan, S., Hammond, E.M., Wade-Martins, R., McMichael, A., Klenerman, P., Simon, A.K. (2012). A novel method for autophagy detection in primary cells: Impaired levels of macroautophagy in immunosenescent T cells. *Autophagy*, 8: 677–689.

Pietrocola, F., Lachkar, S., Enot, D.P., Niso-Santano, M., Bravo-San Pedro, J.M., Sica, V., Izzo, V., Maiuri, M.C., Madeo, F., Mariño, G., Kroemer, G. (2015). Spermidine induces autophagy by inhibiting the acetyltransferase EP300. *Cell Death & Differentiation*, 22: 509–516.

Pietrocola, F., Pol, J., Vacchelli, E., Rao, S., Enot, D.P., Baracco, E.E., et al. (2016). Caloric restriction mimetics enhance anticancer immunosurveillance. *Cancer Cell*, 30: 147–160.

Ponce de Leon, C., Torija, M.E., Matallana, M.C. (2013). Use in the alimentation of some germinated seeds: Mung bean and wheat sprouts. *Boletín de la Real Sociedad Española de Historia Natural Sección Biología*, 107: 47–55.

Pucciarelli, S., Moreschini, B., Micozzi, D., De Fronzo, G.S., Carpi, F.M., Polzonetti, V., Vincenzetti, S., Mignini, F., Napolioni, V. (2012). Spermidine and spermine are enriched in whole blood of nona/centenarians. *Rejuvenation Research*, 15: 590–595.

Puleston, D., Buck M., Pearce, E. (2019). Polyamines and eIF5A hypusination modulate mitochondrial respiration and macrophage activation. *Cell Metabolism*, 30(2): 352–363.

Ralph A., Englyst K., Bardócz S. (1999). Polyamine content of the human diet. In Bardócz S., White A. (Eds.), *Polyamines in Health and Nutrition* (pp. 123–137). London: Kluwer Acad Publ.

Ramos-Molina, B., Queipo-Ortuño, M.I., Lambertos, A., Tinahones, F.J., Peñafiel, R. (2019). Dietary and gut microbiota polyamines in obesity- and age-related diseases. *Frontiers in Nutrition*, 2: 24. doi:10.3389/fnut.2019.00024

Ruíz-Cano, D., Pérez-Llamas, F., Zamora, S. (2012). Implicaciones de las poliaminas en la salud infantil. *Archivos Argentinos de Pediatría*, 110: 244–250. http://dx.doi.org/10.5546/aap.2012.244

Sadasivan, S.K., Vasamsetti, B., Singh, J., Marikunte, V.V., Oommen, A.M., Jagannath, M.R., Pralhada Rao, R. (2014). Exogenous administration of spermine improves glucose utilization and decreases bodyweight in mice. *European Journal of Pharmacology*, 729: 94–99.

Sagara, T., Fiechter, G., Pachner, M., Mayer, H.K., Vollmann, J. (2017). Soybean Spermidine concentration: Genetic and environmental variation of a potential anti-aging constituent. *Journal of Food Composition and Analysis*, 56: 11–17. http://dx.doi.org/10.1016/j.jfca.2016.11.008

Sánchez-Pérez, S., Comas-Basté, O., Rabell-González, J., Veciana-Nogués, M.T., Latorre-Moratalla, M.L., Vidal-Carou, M.C. (2018). Biogenic amines in plant-origin foods: Are they frequently underestimated in low-histamine diets? *Foods*, 7: 205. doi:10.3390/foods7120205

Shao, CG, Wang, H, Yu-Fen, B.I. (2015). Relationship between endogenous polyamines and tolerance in Medicago sativa L.under heat stress. *Acta Agrestia Sinica*, 23: 1214–1219.

Soda, K., Kano, Y., Nakamura, T., Kasono, K., Kawakami, M., Konishi, F. (2005). Spermine, a natural polyamine, suppresses LFA-1 expression on human lymphocyte. *Journal of Immunology*, 175: 237–245.

Soda, K., Dobashi, Y., Kano, Y., Tsujinaka, S., Konishi, F. (2009a). Polyamine-rich food decreases age-associated pathology and mortality in aged mice. *Experimental Gerontology*, 44: 727–732.

Soda, K., Kano, Y., Sakuragi, M., Takao, K., Lefor, A., Konishi, F. (2009b). Long-term oral polyamine intake increases blood polyamines concentrations. *Journal of Nutritional Science and Vitaminology*, 55: 361–366. doi:10.3177/jnsv.55.361

Soda, K., Kano, Y., Chiba, F., (2012). Food polyamine and cardiovascular disease, an epidemiological study. *Global Journal of Health Science*, 4: 170–178. doi:10.5539/gjhs.v4n6p170

Soda, K., Kano, Y., Chiba, F., Koizumi, K., Miyaki, Y. (2013). Increased polyamine intake inhibits age-associated alteration in global DNA methylation and 1,2-dimethylhydrazine-induced tumorigenesis. *PLoS One*, 8: e64357. doi:10.1371/journal.pone.0064357

Soda, K. (2020). Spermine and gene methylation: A mechanism of lifespan extension induced by polyamine-rich diet. *Amino Acids*, 52: 213–224.

Tan, Y., Zhang, R., Chen, G., Wang, S., Li, C., Xu, Y., Kan, J. (2019). Effect of different starter cultures on the control of biogenic amines and quality change of douche by rapid fermentation. *LWT Food Science and Technology*, 109: 395–405. https://doi.org/10.1016/j.lwt.2019.04.041

Tang, T., Qian, K., Shi, T., Wang, F., Li, J., Cao, Y., Hu, Q. (2011). Monitoring the contents of biogenic amines in sufu by HPLC with SPE and pre-column derivatization. *Food Control*, 22: 1203–1208. doi:10.1016/j.foodcont.2011.01.018

Tofalo, R., Cocchi, S., Suzzi, G. (2019). Polyamines and gut microbiota. *Frontiers in Nutrition*, 6: 16. doi:10.3389/fnut.2019.00016

Toro-Funes, N., Bosch-Fusté, J., Veciana-Nogués, MT., Izquierdo-Pulido, M., Vidal-Carou, M.C. (2013). In vitro antioxidant activity of dietary polyamines. *Food Research International*, 51: 141–147. http://dx.doi.org/10.1016/j.foodres.2012.11.036

Toro-Funes, N., Bosch-Fusté, J., Veciana-Nogués, MT., Izquierdo-Pulido, M., Vidal-Carou, M.C. (2014). Effect of ultra high pressure homogenization treatment on the bioactive compounds of soya milk. *Food Chemistry*, 152: 597–602. http://dx.doi.org/10.1016/j.foodchem.2013.12.015

Toro-Funes, N., Bosch-Fusté, J., Latorre-Moratalla, M.L., Veciana-Nogués, M.T., Vidal-Carou, M.C. (2015). Biologically active amines in fermented and non-fermented commercial soybean products from the Spanish market. *Food Chemistry*, 173: 1119–1124. http://dx.doi.org/10.1016/j.foodchem.2014.10.118

Vargas, A.J., Wertheim, B.C., Gerner, E.W., Thomson, C.A., Rock, C.L., Thompson, P.A. (2012). Dietary polyamine intake and risk of colorectal adenomatous polyps. *American Journal of Clinical Nutrition*, 96: 133–141.

Velloso, N.A., Dalmolin, G.D., Gomes, G.M., Rubin, M.A., Canas, P.M., Cunha, R.A., Mello, C.F. (2009). Spermine improves recognition memory deficit in a rodent model of Huntington's disease. *Neurobiology of Learning and Memory*, 92: 574–580.

Vivó, M., de Vera, N., Cortés, R., Mengod, G., Camón, L., Martínez, E. (2001). Polyamines in the basal ganglia of human brain. Influence of aging and degenerative movement disorders. *Neuroscience Letters*, 304: 107–111.

Wang, J., Li, S., Wang, J., Wu, F., Chen, Y., Zhang, H., Guo, Y., Lin, Y., Li, L., Yu, X., Liu, T., Zhao, T. (2020). Spermidine alleviates cardiac aging by improving mitochondrial biogenesis and function. *Aging*, 12(1): 650–671.

Yang, B., Tan, Y., Kan, J. (2020). Regulation of quality and biogenic amine production during suf fermentation by pure Mucor strains. *LWT- Food Science and Technology*, 117: 18637. https://doi.org/10.1016/j.lwt.2019.108637

Yang, J., Ding, X., Qin, Y., Zeng, Y. (2014). Safety assessment of the biogenic amines in fermented soya beans and fermented bean curd. *Journal of Agricultural and Food Chemistry*, 62: 7947–7954.

Yang, Q., Zheng, C., Cao, J., Cao, G., Shou, P., Lin, L., Velletri, T., Jiang, M., Chen, Q., Han, Y., Li, F., Wang, Y., Cao, W., Shi, Y. (2016). Spermidine alleviates experimental autoimmune encephalomyelitis through inducing inhibitory macrophages. *Cell Death & Differentiation*, 23: 1850–1861.

Yin, Z., Pascual, C., Klionsky, D.J. (2016). Autophagy: Machinery and regulation. *Microbial Cell*, 3(12): 588–596.

Yoshinaga, K., Ishizuka, J., Evers, B.M., Townsend, C.M. Jr, Thompson, J.C. (1993). Age-related changes in polyamine biosynthesis after fasting and refeeding. *Experimental Gerontology*, 28: 565–572.

Yue, C.S., Ng, Q.N., Lim, A.K., Lam, M.H., Chee, K.N. (2019). Biogenic amine content in various types of tofu: Occurrence, validation and quantification. *International Food Research Journal*, 26: 999–1009.

Yue, F., Li, W., Zou, J., Jiang, X., Xu, G., Huang, H., Liu, L. (2017). Spermidine prolongs lifespan and prevents liver fibrosis and hepatocellular carcinoma by activating MAP1S-mediated autophagy. *Cancer Research*, 77: 2938–2951.

Zhang H., Wang J., Li L., Chai N., Chen Y., Wu F., et al. (2017). Spermine and spermidine reversed age-related cardiac deterioration in rats. *Oncotarget*, 8: 64793–64808.

Zhang H., Alsaleh G., Feltham J., Sun Y., Napolitano G., Riffelmacher T., Charles P., Frau L., Hublitz P., Yu Z., Mohammed, S., Ballabio, A., Balabanov, S., Mellor, J., Simon, A.K. (2019). Polyamines control eIF5A hypusination, TFEB translation, and autophagy to reverse B cell senescence. *Molecular Cell*, 76: 110–125.

Ziegler, W., Hahn, M., Wallnöfer, P.R. (1994). Changes in biogenic amine contents during processing of several plant foods. *Deutsche Lebensmittel Rundschau*, 90: 108–112.

Zoumas-Morse, C., Rock, C.L., Quintana, E.L., Neuhouser M.L., Gerner E.W., Meyskens Jr F.L. (2007). Development of a polyamine database for assessing dietary intake. *Journal of the American Dietetic Association*, 107: 1024–1027. doi:10.1016/j.jada.2007.03.012.

7 Soybean Glyceollins and Human Health

Ikenna C. Ohanenye, Innocent U. Okagu,
Oluwaseyi A. Ogunrinola, Precious E. Agboinghale,
and Chibuike C. Udenigwe

CONTENTS

7.1 INTRODUCTION

There is an increasing interest in the use of natural compounds from a variety of sources for health and biomedical applications. The human body is constantly exposed to several harmful agents and, for survival, the body is equipped with several defense mechanisms against these agents, including the immune system, metabolic regulations, and antioxidant defense mechanisms. Unfortunately, situations abound when the burden of external toxicants overwhelms the body's natural defense systems, resulting in dysregulation of metabolism and deficiency in the ability to restore normalcy. Knowledge gained from human interaction with the environment, such as ingestion of plant-based diets rich in health-promoting compounds, is crucial for managing the aberrant processes. Many plant-based diets contain compounds that activate the body's natural defense mechanisms or possess the direct ability to control abnormal physiological processes. Among the plant-derived compounds, phytoalexins are generating a lot of interest as bioactive compounds; these are inducible compounds produced by different plant species, including legumes (Ejike et al., 2013; Nwachukwu et al., 2013). Specifically, soybean (*Glycine max*) produces a structurally related group of phytoalexins in response to exposure to microbial components (Boué et al., 2008, 2012) and abiotic factors, including radiation, physical damage, low temperature, and inorganic chemicals (Zimmermann et al., 2010; Boué et al., 2012). The major soybean phytoalexins, glyceollins, have been recognized to offer several bioactivities and potential health benefits, such as modulation of lipid and glucose metabolism, anti-inflammatory, anti-estrogenic, and anticancer activities, inhibition of melanogenesis, antioxidant and anti-lipid peroxidation effects (Kim et al., 2010; Kim et al., 2011; Seo et al.,

DOI: 10.1201/9781003030294-7

2017; Son et al., 2019; Jeong et al., 2019). This chapter discusses the biosynthesis, chemical structures, and inducers of glyceollin production in soybean, as well as the transport, bioavailability, and metabolism of glyceollins. The chapter also discusses evidence of the health-promoting potentials of the soybean phytoalexins and their prospects in the formulation of nutraceuticals and functional foods.

7.2 CHEMICAL STRUCTURE AND BIOSYNTHESIS OF SOYBEAN GLYCEOLLINS

Glyceollin isomers have the molecular formula $C_{20}H_{18}O_5$ and a molecular weight of 338.4 g/mol. The soybean phytoalexins are members of the family of pterocarpans, which are prenylated derivatives of isoflavonoids. Glyceollins have simple pterocarpanoid skeletons that are linked to a cyclic ether ring originating from C5 prenyl substitutions (Akashi et al., 2009). With two chiral centers at C6α and C11α, glyceollins each have four possible enantiomers (RR, RS, SR, SS), but only the SS enantiomer occurs in nature (Zimmerman et al., 2010; Nwachukwu et al., 2013). Glyceollins are categorized as major phytoalexins of soybeans produced when exposed to certain fungi and abiotic elicitors, such as wounding, UV light, aluminum chloride, freezing, or methyl jasmonate, and possess tremendous potential as bioactive components for pharmaceutical development (Selvam et al., 2017). Although many types of glyceollins are biosynthesized following the exposure of soybean to elicitors (Simons et al., 2011), glyceollin I, II, and III are the three primary forms (Figure 7.1). Other forms of glyceollins are produced directly from the glyceollidins and glyceollin III in challenged soybean, including glyceollin IV, glyceollin V, glyceollin VI, and glyceofuran (Simons et al., 2011; Nwachukwu et al., 2013). The proportion of the glyceollins varies in soybean tissues and depends on the type, amount, and duration of exposure of the external factors. A comprehensive discussion of the biosynthesis of soybean glyceollins has been published (Nwachukwu et al., 2013). As summarized in Figure 7.1, glycinol, the parent non-prenylated 6α-hydroxypterocarpan, is the precursor of glyceollins through a reaction catalyzed by prenyltransferases to form the glyceollidins. Thereafter, NADPH-dependent glyceollin synthase catalyzes the cyclization reaction to transform the prenylated precursors into glyceollins.

7.3 ELICITORS OF GLYCEOLLIN PRODUCTION

Elicitation is a technique used to stimulate the plant defense system and elicitors are factors that induce the stress response (Patel et al., 2020). Elicitors can be biotic or abiotic. Biotic elicitors are biological materials from plants or microbial sources (Patel & Krishnamurthy, 2013). Abiotic elicitors are chemical or physical factors involved in triggering phytochemical biosynthesis in plants (Owolabi et al., 2018). Treatment with different elicitors induces the production of different intensities of the compounds in soybean tissues (Boue et al., 2008). For instance, soybean seeds treated with silver nitrite produced lower amounts of glyceollins than those treated with a wall glycan elicitor derived from soybean pathogen, *Phytophthora sojae*. As shown in Figure 7.2, a combination of the biotic and abiotic elicitors resulted in an additive effect on glyceollin accumulation. This was proposed to be because of the distinct mechanisms through which the elicitors acted, with the wall glycan elicitor activating the accumulation of phytoalexin biosynthetic gene whereas silver nitrite inhibited glyceollin degradation and promoted the generation of glyceollin precursor, 6''-O-malonyldaidzin (Farrell et al., 2017). Moreover, glyceollins can be produced during food processing. Glyceollins I–VI and glyceofuran were reported to be produced when soybean was subjected to fermentation and UV irradiation (Chen et al., 2012; Nwachukwu et al., 2013).

The elicitation of glyceollin can be done by an extensive variety of fungal pathogens. In addition to the whole cells, accumulation of glyceollins can be induced using various components of the fungal cell wall (Table 7.1). A comparison of the elicitation features of some biotic elicitors indicates that the highest amount of glyceollin I was induced by a fungal cell wall extract (Farrell et al., 2017). This could be because of the increased accessibility of the elicitor within the extract compared to

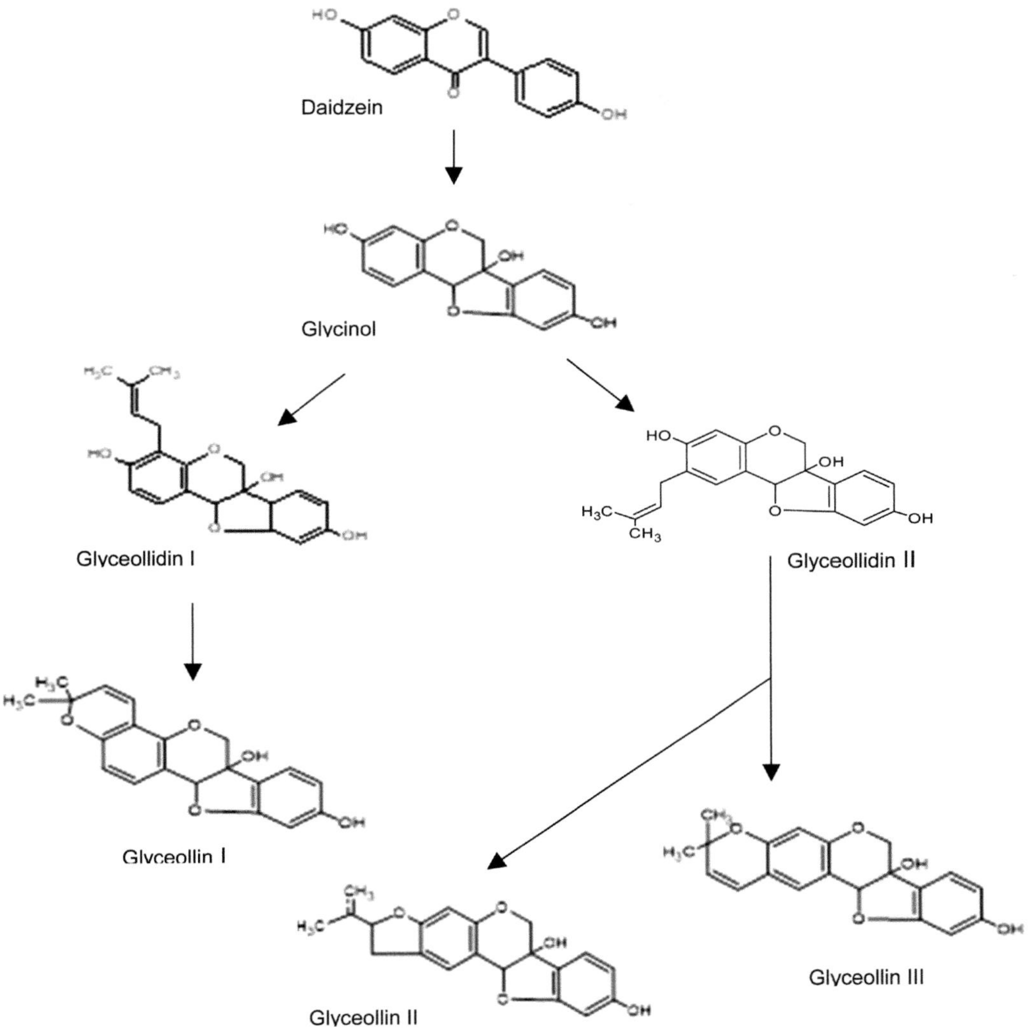

FIGURE 7.1 Pathway for the biosynthesis of glyceollins I–III from daidzein.

whole organisms. For the abiotic elicitors, exposure of a soybean variety (Aga 3) to UV radiation for 15 min gave rise to relatively higher glyceollin accumulation compared to aluminum chloride (Park et al., 2017); however, acidity stress (pH 3) elicited a larger quantity of glyceollins in soybean (Jahan et al., 2019). Considering that glyceollins are produced through the same biosynthesis pathway, the differences in their accumulation can be attributed to differences in elicitor receptors, stress signaling, exposure amount and duration, and relative stability of the phytoalexins within the soybean seed matrices and surrounding environments.

7.4 ABSORPTION AND BIOAVAILABILITY OF GLYCEOLLINS

The beneficial health effects of any dietary supplement or drug depend on its intestinal transport, systemic absorption when administered orally, and bioavailability. Intestinal transport and metabolism of glyceollin are rapid in human small intestinal Caco-2 cells and cellular uptake is dominated by the passive diffusion mechanism (Bamji & Corbitt, 2017). The permeability of glyceollins

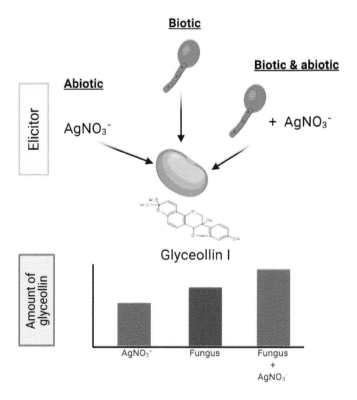

FIGURE 7.2 Effect of biotic and abiotic elicitors on the amount of soy glyceollin. The biotic factor is a wall glycan elicitor derived from the soybean pathogen, *Phytophthora sojae*. Source: image created with Biorender based on data from Farrell et al. (2017).

TABLE 7.1

Elicitors and Elicitor Types for Glyceollin Accumulation in Soybean Seeds

Elicitors	Elicitor Type	References
AgNO$_3$	Abiotic elicitor	Park et al., 2017
CuSO$_4$	Abiotic elicitor	Park et al., 2017
CuCl$_2$	Abiotic elicitor	Farrell et al., 2017
Fungal extracellular metabolite	Biotic elicitor	Park et al., 2017
Fungal cell wall	Biotic elicitor	Park et al., 2017
Rhizopus oligosporous	Biotic elicitor	Park et al., 2017
Aspergillus sojae	Biotic elicitor	Park et al., 2017
Triton X-100	Abiotic elicitor	Park et al., 2017
UV light	Abiotic elicitor	Park et al., 2017
Jasmonate	Biotic elicitor	Park et al., 2017
Acidity stress pH 3.0	Aboitic elicitor	Jahan et al., 2019

examined using Caco-2 cell culture is high and comparable to other compounds with 100% absorption; in addition, glyceollins can be directly metabolized to sulfate and glucuronide conjugates in the small intestine (Chimezie et al., 2014; Chimezie et al., 2016; Oh et al., 2019). The transport of glyceollins across the intestinal and blood-brain barrier enhanced the delivery of other nutraceutical compounds, such as genistein (Bamji & Corbitt, 2017). Glyceollin permeability was reported to be

~2 × 10^{-4} cm/s (Chimezie et al., 2014). Glyceollins do not appear to affect the P-glycoprotein (Pgp) function activity or induce Pgp expression in Caco-2 cells; Pgp plays a protective role by limiting the uptake of orally consumed xenobiotics in the small intestine (Bamji & Corbitt, 2017). In connection with the high permeability of glyceollins, Chimezie et al. (2014) suggested that intestinal and hepatic metabolism may have contributed to the low oral bioavailability. Moreover, glyceollins appeared to inhibit the metabolism of the genistein, thus improving its intestinal transport *in vitro* (Bamji & Corbitt, 2017).

Although the health-promoting benefits and mechanisms of glyceollin have been well established, there are limited studies on their metabolism. Phytochemicals ingested are typically metabolized by phases I and II enzymes in the liver before being excreted in the urine and bile (Oh et al., 2019). Phase II metabolic reactions are conjugative, usually occurring after phase I redox reaction, but can also happen directly. An understanding of the ingested glyceollins was possible due to the detection of their conjugated derivatives by ion scanning using liquid chromatography coupled with electrospray ionization tandem mass spectrometry (LC-ESI-MS/MS) (Pham et al., 2019). Using this technique, glyceollin metabolites derived from phase I (hydroxylated) and phase II metabolism (sulfated, glutathione, or glucuronide conjugates) were found in the blood, urine, and feces of rats after oral administration of the phytoalexins (Pham et al., 2019). Glyceollins I, II, and III were also detected intact in the feces (Pham et al., 2019).

Despite their poor systemic bioavailability, glyceollins have been demonstrated to manifest biologically beneficial effects *in vivo* (Nwachukwu et al., 2013; Pham et al., 2019; Oh et al., 2019). The low bioavailability of phytochemicals may be due to their sulfation and glucuronidation by intestinal enzymes. Although the conjugates formed facilitate elimination and decrease the bioavailability of the parent compound, they may also be beneficial physiologically (Pham et al., 2019; Oh et al., 2019). It is also likely that glyceollins or their metabolites act indirectly by binding receptors in the gastrointestinal tract to transmit signals to target organs and tissues throughout the body (Oh et al., 2019). This type of effect could also be mediated by the gut microbiota and their metabolites and should be explored in future studies. Furthermore, glyceollins could impact the metabolism of other compounds by regulating the expression of genes encoding cytochrome P450 phase I and other detoxification enzymes (Pham et al., 2019). Glyceollins decrease the activity of the efflux transporters ABCC2 (MRP2) and ABCG2 (BCRP) without major modification of their expression. This interaction can have major effects on the effectiveness of drugs by inhibiting apical excretion and increasing basolateral efflux, thus potentially increasing systemic transmission (Pham et al., 2019).

7.5 HEALTH-PROMOTING PROPERTIES OF GLYCEOLLIN

Glyceollins, as with other phytoalexins, are not usually stored in large quantities but produced under stress and as part of the plant defense mechanism against pathogenic attacks. For instance, it was reported that the overexpression of coenzyme A ligase (a soybean 4-coumaric acid) caused increases in daidzein, genistein, and glyceollin production that led to enhanced resistance to *Phytophthora sojae*, a destructive root and stem disease in soybeans (Chen et al., 2019). Further implication of glyceollin in fungal resistance has previously been reported (Lygin et al., 2010). Although primarily produced for the protection of the plants, glyceollin elicits several health benefits by producing anti-estrogenic and anticancer, antioxidant, anti-inflammatory, antidiabetic, hypolipidemic, and bone health-enhancing effects.

7.5.1 Anti-Estrogenic and Anticancer Activities

Estrogens are hormones with multifunctional roles that include cell differentiation, development, and growth with regulatory functions in the reproductive and central nervous systems (Klinge, 2000). In relation to cancer, estrogen receptor (ER)-positive cancer is the most common with ERα and ERβ as the only two subtypes (Lecomte et al., 2017); thus, ERα and ERβ are central to hormonal therapy. The cancer-related activities of ER involve a cascade of events that include the binding to estrogen-responsive elements, stimulation of activator protein 1, and modulation of MAPK and PI3K/AKT signaling pathways (Levin & Pietras, 2008; Lecomte, 2017). ER binds E2, such that E2 acts as an ER modulator overseeing the survival and growth of breast epithelial cells but can cause the initiation and growth of breast cancer (Pham, Lecomte, Efstathiou, Ferriere, & Pakdel, 2019). Typically, ER activities are mainly through the chromatin remodeling-linked transcription factors or through direct interaction with estrogen-responsive elements (ERE) by altering the transcription of the target genes (Klinge, 2000). However, E2 target genes lack ERE in their promoters (Pham et al., 2019); therefore, for ER to circumvent this challenge it modulates its transcription by interacting with specificity proteins 1 (Sp1) and activator proteins 1 (AP1) (Safe & Kim, 2004).

Natural products of plant origin such as glyceollin, known as phytoestrogens, have been revealed to elicit some chemo-preventive and anticancer properties (Hasanpourghadi, Pandurangan, & Mustafa, 2018; Salvo et al., 2006). For instance, glyceollin was reported to cause the upregulation of tumor suppressor genes and oncogenes by altering miRNA expression that selectively inhibited the growth of tumor cells (Ahmed et al., 2020). Glyceollin also functions by removing reactive oxygen species (ROS) via induction of the nuclear factor (erythroid-derived 2)-like 2 (Nrf2) signaling pathways (Jeong et al., 2019). As an antimetastatic agent, glyceollin inhibits the epithelial-mesenchymal transition through the blocking of the transforming growth factor-β receptors (TGFBR) types I and II signaling pathways (Hermanto, Rifa'I, & Widodo, 2019).

Globally, breast cancer is the most prevalent cancer in women, and accounted for 2.3 million (24.5%) of all the reported cancer cases in women in 2020 (WHO, 2021). Cytotoxicity chemotherapy is a common cancer treatment that is accompanied by several adverse effects. Hormonal therapy was identified as an alternative treatment with less destructive outcomes. Anti-estrogenic activities are part of the hormonal therapy that has emerged as the preferred treatment option for patients (Van Weelden, Massuger, Pijnenborg, & Romano, 2019). It is equally effective, safer, and more cost-effective when compared to cytotoxic chemotherapy (Carlson, Thiel, & Leslie, 2014). An *in vivo* study conducted in ovariectomized athymic mice highlighted the anti-estrogenic activities of glyceollin, where it was reported to suppress E2-initiated tumor cell growth of BG-1 and MCF-7 by 73.1% and 53.4%, respectively (Salvo et al., 2006). The capability of glyceollin to elicit hormonal responses was thought to be due to its structural similarities with E2. Glyceollin binds ER, causing the inhibition of cell proliferation (Lecomte et al., 2017; Pham et al., 2019), thus presenting glyceollin as an anti-estrogenic agent with therapeutic promises in cancer treatments. Moreover, glyceollin actions on ER function and estrogen-dependent tumor growth coincided with lower E2-induced progesterone receptor expression in tumors while showing no estrogen-agonist effects on uterine morphology (Salvo et al., 2006). Although all glyceollin types bind ERα and ERβ, they show a stronger affinity for ERα. Moreover, each of the glyceollin types I, II, and III vary in their affinities for ER with type I having the strongest affinity to ER when compared to glyceollin II and glyceollin III (Zimmermann et al., 2010). Other phytoestrogens such as coumestrol, daidzein, and genistein are known and they act as ER agonists under low estrogen conditions but become antagonists of ER at high estrogen concentrations. In contrast, glyceollin antagonizes ER activities only under low estrogen concentrations (Burow et al., 2001; Pham et al., 2019; Zimmermann et al., 2010).

Indeed, the majority of glyceollin activities is through its induction of Nrf2; however, the safety of this mechanism has been questioned. A recent study reported that glyceollin increased the growth of existing colon tumors through a cascade of events that involved the scavenging of p53-regulated ROS produced by the Warburg effect, which enhanced drug metabolism that led to resistance to chemotherapy (Jeong et al., 2019). Nonetheless, the study reported that the increases in tumor growth were only found in cases where high doses of glyceollin were administered in the presence of the wild-type p53 HCT116 xenografts, and no such growths were found in HCT116 xenografts devoid of p53. Moreover, in all cases where low dosages of glyceollin were used, there were no increases in tumor growth (Jeong et al., 2019). Previously, glyceollins were found to induce apoptosis in hepatoma cell culture at high concentrations through ROS generation, as evidenced by the decrease in Bcl-2 and increase in pro-apoptotic p21 and p27, among others (Kim et al., 2014). Glyceollins were also proposed to regulate tumor growth via inhibition of the synthesis and stability of hypoxia-inducing factor-1α, which is involved in tumor growth and invasion (Lee et al., 2015). Taken together, the reevaluation of the types of cancer and corresponding glyceollin doses, as well as the physiological effect *in vivo*, is crucial in assessing the beneficial effects.

7.5.2 Anti-Inflammatory Activities

Inflammation is a defense mechanism that involves the cells of the immune system through complex reactions and leads to protection from damages caused by external factors. On exposure to elicitors, the cells of the immune system produce inflammatory mediators that initiate a cascade of reactions aimed at destroying the response initiator. However, the hyper-activation of the immune cells can cause cellular damage. Inflammation has been implicated in many health conditions such as inflammatory bowel disease, intestinal ulcerative colitis, osteo- and rheumatoid arthritis, vasculitis, pelvic inflammatory disease, asthma, atherosclerosis, and lupus (Glass, Saijo, Winner, Marchetto, & Gage, 2010; Kazemi, Shirzad, & Rafieian-Kopaei, 2018; Strohacker & McFarlin, 2020; Tabas, 2010). Cardiovascular disease, which is a leading cause of death in developed countries, is also characterized by inflammation (Huang et al., 2013). Consequently, compounds that modulate inflammatory events (anti-inflammatory drugs) are widely used in various disease treatments (Ogbu, Okagu, & Nwodo, 2020; Wongrakpanich, Wongrakpanich, Melhado, & Rangaswami, 2018).

Using an *in vitro* model, soybean glyceollin induced by *Phytophthora sojae* and *Aspergillus sojae* suppressed lipopolysaccharide (LPS)-induced inflammation in macrophages, including production of nitric oxide and pro-inflammatory cytokines (Kim, Sung, & Kim, 2011; Yoon, Kim, Cui, Kim, & Lee, 2012). A reduction in nitric oxide concentration downregulates the expression of mRNA of inducible nitric oxide synthase (*iNOS*) gene leading to reduced activity (Fan et al., 2009; Liu et al., 2015). In macrophages and related immune cells, LPS increases the activity of cyclooxygenase (COX)-2, the enzyme central to prostaglandins (PGs) synthesis. The PGs are anti-inflammation mediators; thus, the inhibition of COX-2 has been targeted in the design of non-steroidal anti-inflammatory drugs. Soybean glyceollins were demonstrated to potently downregulate COX-2 gene expression by suppressing the nuclear factor (NF)-κβ-mediated activation of inflammation pathways (Yoon et al., 2012). Glyceollins also inhibited NF-κβ by phosphorylation, thus preventing its downstream activation of inflammation. In addition to mediating inflammation via NO and COX-2, the NF-κβ signaling pathway regulates pro-inflammatory cytokines, such as IL-1β, IL-6, and IL-8, involved in inflammation (Baeuerle & Baltimore, 1996).

Further evidence of the anti-inflammatory role of glyceollin was demonstrated in mice exposed to 12-O-tetradecanoylphorbol-13-acetate, which elicits skin inflammatory events such as leukocyte infiltration of cutaneous tissues, enhanced production of arachidonic acid from membrane phospholipids by phospholipase A2, synthesis of prostaglandin E2, and increased levels of pro-inflammatory cytokines (Kim et al., 2011). However, treatment with soybean glyceollins halted all the observed inflammatory responses (Kim et al., 2011). Furthermore, soybean glyceollins were reported to prevent colitis in a murine model by suppressing lipid peroxidation (malondialdehyde), oxidative stress (8-hydroxy-2-guanosine), and inflammation (NO and IL-6 via NF-κβ pathway), while activating the Nrf2 signaling pathway (Seo et al., 2017). Ulcerative colitis is a chronic inflammatory condition of the intestinal mucosa and, although the colitis was experimentally induced, the findings highlighted a potential role for the soybean phytoalexins in colitis treatment. Steamed soybean wastewater rich in glyceollin was reported to strongly suppress inflammation by inhibiting the production of nitric oxide and pro-inflammatory cytokines and expression of iNOS and COX-2 genes in activated macrophages, which prevented the mucosal inflammation in a murine model of colitis (Jeong et al., 2019). These anti-inflammatory properties in cell culture and animal models of inflammation strengthen the potential of glyceollins as natural anti-inflammatory agents.

Inflammation is a characteristic of cardiovascular disease (CVD). Therefore, compounds with anti-inflammatory and anticholesterol activities are thought to play positive roles in CVD. The anti-inflammatory activity of glyceollins positions them as strong candidates with potential in CVD treatments. For instance, a study in a golden Syrian hamster model showed that diets supplemented with glyceollin caused significant reductions in hepatic cholesterol esters, plasma VLDL, and total lipid content by altering the genes in the liver involved in cholesterol metabolism (Huang et al., 2013). In another study, glyceollin was revealed to prevent the phosphorylation and activation of the transcription nuclear factor-kappa B (NF-κB), which in turn inhibited the pro-inflammatory cytokines from proliferating of inflammation (Pham et al., 2019). Taken together, these strengthened the argument for glyceollin as a significant mitigator of CVD. That said, future studies would be needed to ascertain the potency of glyceollin in CVD conditions.

7.5.3   Antioxidant Activities

The human cells are constantly exposed to exogenous sources of free radicals and reactive oxygen species, such as cigarette, industrial, and domestic smokes, alcohol, drugs, radiation, and environmental chemicals such as heavy metals and other agricultural and industrial pollutants (Żukowski, Maciejczyk, & Waszkiel, 2018). These reactive species can increase oxygen demand and inefficient consumption of oxygen leads to mitochondrial oxidative stress, which can cause certain enzymes such as NADPH oxidase, cytochrome P450 enzyme system, diamine oxidases, myeloperoxidase, and thiol oxidase to elicit an uncontrolled response (Dizdaroglu & Jaruga, 2012; Phaniendra, Jestadi, & Periyasamy, 2015). Having unpaired electrons, free radicals are highly unstable; thus, they are known to cause damages to cells by picking off electrons from their nearest cellular component such as proteins, nucleic acid, and lipids (Santo, Zhu, & Li, 2016). The damaging effects of free radicals are mitigated by endogenous antioxidants, such as glutathione, superoxide dismutase, and catalase known to scavenge free radicals and quench their chain reactions to prevent cellular damages. Dietary antioxidants such as vitamins A, C, and E, phytochemicals, and some minerals (selenium, copper, and zinc that serve as cofactors to the endogenous antioxidants) directly scavenge free radicals and boost the activities of endogenous antioxidant machinery (Ighodaro & Akinloye, 2018).

Soybean glyceollins are produced in response to stresses such as oxidative stress, which suggests that glyceollins may be among compounds produced to mitigate the oxidative effects. Consequently, a glyceollin-rich mixture of soybean isoflavones was reported to scavenge free radicals (DPPH, ABTS⁺, hydroxyl radical), quench singlet oxygen species *in vitro*, and halt lipid peroxidation in mouse hepatoma treated with ROS (Boué et al., 2008; Kim et al., 2010). Furthermore, glyceollins were reported to exhibit their antioxidative effects by inhibiting oxidative effects induced by glutathione depletion in mouse hepatoma cells via nuclear translocation and activation of Nrf2 signaling and expression of endogenous antioxidant enzymes, including γ-glutamylcysteine synthetase, heme oxygenase 1, and glutathione reductase (Kim et al., 2011; Jung et al., 2013). Collectively, these activities show that glyceollins possess both direct and indirect antioxidant properties. However, the low bioavailability and biotransformation of glyceollins *in vivo* suggest that future studies need to focus on their metabolites as the more physiologically relevant active compounds when elucidating molecular mechanisms and health benefits.

7.5.4   Effects on Glucose Metabolism

Blood glucose level, whether high (hyperglycemia) or low (hypoglycemia), is implicated in several health conditions, such as diabetes, insulin resistance, stroke, and aging (Silva, Ferreira, & de Pinho, 2017; Lee & Halter, 2017). Therefore, glucose metabolism is central to health and wellness. The blood glucose level is tightly regulated such that excess glucose is stored as reserve energy and alteration of this balance may result in type 2 diabetes and other health conditions (Marcovecchio,

Lucantoni, & Chiarelli, 2011). For instance, sustained hyperglycemia generates oxidative stress and other deleterious reactions that impair normal cardiovascular and metabolic functions, potentially resulting in CVD, nephropathy, retinopathy, neuropathy, diabetic foot ulcer, cancer, and aging (Giri et al., 2018; Roberts & Porter, 2013; Spinelli et al., 2020). Currently, therapeutic strategies to maintain normal glucose levels include the use of insulin therapy, hypoglycemic drugs, dipeptidyl peptidase-4 inhibitors, and insulin secretagogues, as well as lifestyle modifications, such as consuming low-calorie diets and engaging in physical activity (Singh, Paramanick, Sharma, & Pundir, 2020; Tiwari, 2015).

Soybean phytoalexins have been shown to improve glucose control by promoting postprandial glucose tolerance largely by increasing insulin release and glucose uptake in cultured 3T3-L1 adipocytes (Park, Kim, Kim, Kim, & Kim, 2012). In addition, glyceollins were found to modulate glucose metabolism under healthy and disease conditions. The mechanism behind the improved glucose uptake was reported to be independent of the peroxisome proliferator-activated receptor (PPAR)-γ pathway, suggesting that glyceollins might have acted by increasing mRNA expression of glucose transporters known to shuttle glucose into the cells. Glyceollins were also shown to protect pancreatic β-cells exposed to endoplasmic reticulum stress and attenuate β-cell apoptosis in Min6 cells, probably via their antioxidative mechanism (Kim et al., 2010) as well as ER-mediated β-cell regeneration (Soo, Jin, & Park, 2005). Furthermore, in a murine model of type 2 diabetes, glyceollin-rich fermented soybean was reported to reduce blood glucose level by increasing insulin release via an ER-dependent mechanism (Burow et al., 2001; Park et al., 2012). In addition to glucose homeostasis, glyceollin-rich diet also suppressed hepatic *de novo* glucose synthesis from non-carbohydrate sources (gluconeogenesis) by promoting the phosphorylation of phosphoenolpyruvate carboxykinase (PEPCK) (Wood et al., 2012). This effect, however, resulted from the activation (by phosphorylation of specific serine residues) of protein kinase B (Akt), an indication of enhanced insulin signaling in the liver. PEPCK is the rate-limiting enzyme of gluconeogenesis and phosphorylation renders it inactive (Cherrington, Edgerton & Ramnanan, 2009). Notwithstanding, glyceollins activated the AMP-activated protein kinase (AMPK), a serine/threonine kinase that mediates insulin signaling, which led to glucose homeostasis by inhibiting gluconeogenesis and glycogenolysis (Wood et al., 2012).

Furthermore, glyceollins inhibited the activity of glucose-metabolizing enzymes of the gastrointestinal tract (Son et al., 2019). α-Glucosidase hydrolyzes the α-(1-4) glycosidic bonds of dietary starch to release glucose in the intestinal brush border for absorption. Thus, it contributes significantly to an increase in blood glucose level after consumption of carbohydrate-rich diets (i.e. glycemic response). Thus, α-glucosidase inhibition has been explored as a major strategy for controlling hyperglycemia (Okuyama, Saburi, Mori, & Kimura, 2016; Di Stefano et al., 2018). Glyceollins have been reported to inhibit α-glucosidase activity as well as hepatic glucose synthesis and glycogen breakdown under normal and hyperglycemic conditions (Park et al., 2012). The dual mechanisms may be responsible for glyceollin regulation of glucose metabolism in diet-induced hyperglycemia. Glyceollins also induced glucose homeostasis by protecting pancreatic β-cells and improving their insulin-producing capacity, as well as via insulin signaling to mediate glucose uptake into the cells (Park et al., 2010). Glyceollins led to increased secretion of incretin glucagon-like peptide-1 in enteroendocrinal cells (Park et al., 2010), and this effect was likely mediated through the inhibition of dipeptidyl peptidase-4, a major regulator of insulin secretion and glucose metabolism and widely targeted enzyme for managing type 2 diabetes. Therefore, further studies would be needed to ascertain whether dipeptidyl peptidase-4 is involved in the roles of glyceollin in glucose metabolism.

7.5.5   Effects on Lipid Metabolism

A glyceollin-rich diet was shown to influence the mRNA expression profiles of some genes involved in the regulation of lipid metabolism. For instance, a study of the serum of female postmenopausal monkeys fed with a glyceollin-rich diet revealed the increased expression of peroxisome proliferator-activated receptor (PPAR)-γ, adiponectin, leptin, lipin-1, and lipoprotein lipase genes. These genes are involved in the reduction of lipid synthesis and storage, and the effects of glyceollins resulted in a reduction in the serum lipid profile (total cholesterol, low-density lipoproteins, and triacylglycerol levels) of the animals (Wood et al., 2012). Glyceollins were also reported to reduce the synthesis and accumulation of fat in both normal and diabetic conditions (Park et al., 2012). Similarly, a study with hyperlipidemic hamsters reported that a glyceollin-supplemented diet significantly suppressed plasma and hepatic lipid profiles and altered the regulation of lipogenic and lipolytic genes (Huang et al., 2013). Notably, the expression of genes of the major enzyme involved in cholesterol biosynthesis, 3-hydroxy-3-methylglutaryl coenzyme-A (HMG-CoA) reductase, were all suppressed. Furthermore, glyceollins downregulated the gene expression and enzymatic activity of acetyl coenzyme A carboxylase (ACC), which catalyzes the rate-limiting step in lipogenesis and lipid storage. This activity of glyceollins may have been involved in lowering the level of triacylglycerol in a murine model (Park et al., 2012).

Proposed mechanisms behind the hypocholesterolemic effects of glyceollin supplementation involve a decrease in cholesterol absorption and transport into circulation and elevation in its clearance. This was supported by upregulation of mRNA expression of ATP-binding cassette sub-family G members 5 and 8 (ABCG5 and ABCG8), which are known to suppress intestinal absorption and increase the biliary clearance of sterols (Hegele & Robinson, 2005). Furthermore, glyceollin supplementation increased the expression of *CYP7A1*, the gene for the major enzyme in bile acid biosynthesis, microsomal cytochrome P-450 cholesterol 7α-hydrolase (Huang et al., 2013). Taken together, these findings highlight the need for further studies on the role of glyceollins in the management of hyperlipidemia and related cardiovascular diseases.

7.5.6 Effects on Melanogenesis

Glyceollin plays a role in melanin synthesis, which may be of interest to some consumers and the cosmeceutical industry. Melanin is the dark skin pigment that is responsible for shielding the cells against UV radiation. Melanogenesis is the synthesis of melanin, and the stem cell factor (SCF)/ SCF receptor signaling pathway plays a major role in this process during embryo development (dos Santos Videira, Moura & Magina, 2013). The interaction of SCF with its receptor induces autophosphorylation of the tyrosine kinase receptor, which activates the signaling pathway for tyrosinase activation (Shin & Lee, 2013). Downstream the cascade of events is the activation of mitogen-activated protein kinase (MAPK) and extracellular responsive kinase (ERK), culminating in melanin production. Tyrosinase activity is needed for the maturation and protection of melanocytes, which are melanin-producing cells of the skin epidermis (Galibert, Carreira, & Goding, 2001; Kim et al., 2007). Moreover, active MAPK is known to increase the activity of adenylyl cyclase, which converts adenosine triphosphate to cyclic adenosine monophosphate (cAMP). The binding of cAMP to its response element-binding protein (CREB) upregulates the microphthalmia-associated transcription factor (MITF) expression involved in melanocyte development and maturation (Hirata et al., 2007).

A study on the potential application of glyceollins in the cosmeceutical industry reported that the exposure of zebrafish embryos to glyceollins markedly suppressed melanin synthesis, tyrosinase activity and expression of pigment cell-specific gene (Sox10) compared to control (Shin & Lee, 2013). Western blot analysis further demonstrated that B16F10 melanoma cells treated with glyceollins have reduced SCF-induced melanogenesis, which was achieved by inhibition of tyrosinase gene expression and its activity via inhibition of MAPK activity (Shin & Lee, 2013). Similarly, phytoalexins were reported to decrease melanin synthesis via hormone-mediated signaling in B16 melanocytes (Lee et al., 2010). Taken together, glyceollins inhibit melanogenesis via SCF/SCF receptor signaling that activates adenylyl cyclase, p38 MAPK, and ERK/MAPK pathways. Although a promising additive for the cosmeceutical industry in some parts of the world, the general concept of inhibiting melanogenesis for esthetic skin lightening poses a significant health risk and may cause skin cancer, especially with the loss of skin protection against UV irradiation.

7.5.7 EFFECTS ON BONE HEALTH

Plant-derived compounds, including glyceollins, have shown some promising effects in promoting bone health. Osteoporosis is the most prevalent bone disease in humans and is characterized by an imbalance between bone resorption and bone regeneration, leading to fragility and susceptibility to fracture (Wood, Register, & Cline, 2007). The disease presents a great burden to the most vulnerable population such as older adult women (Su et al., 2019). The prevention and management of osteoporosis towards improving the quality of life of the patients rely on drugs such as alendronic acid, risedronate, zoledronic acid, and ibandronic acid. However, these drugs have low oral bioavailability and elicit undesirable side effects such as headache and muscle pain, gastrointestinal disturbance, and nephrotoxicities (Rodríguez-Merchán, 2021).

Estrogen level declines significantly in post-menopausal women leading to primary osteoporosis (González-Chávez, Quiñonez-Flores, & Pacheco-Tena, 2016). Thus, estrogen receptor ligands may play beneficial roles in the prevention and management of osteoporosis. E2 is the most active form of estrogen; however, its use in osteoporosis treatment is discouraged due to its role in the initiation of breast and endometrial carcinomas (Pham et al., 2019). As alternative treatments, phytoestrogens were reported to block bone resorption, induce osteogenesis, and improve the maturation of bone marrow–generated stem cells and osteoblasts in both animal models and humans (Marini et al., 2007). Notably, the presence of glyceollins as phytoestrogens improved calcium deposition and osteogenesis in adipose-derived and bone marrow–derived stromal cells (Bateman et al., 2017). On further investigation, it was reported that ER was involved in glyceollin-induced osteogenesis, as an ER antagonist (fulvestrant) blocked the activities of the phytoalexins in the treated cells. Glyceollin II was shown to have a higher upregulatory effect on genes related to osteogenesis, including runt-related transcription factor 2 (RUNX2), FBJ murine osteosarcoma viral oncogene homolog (c-FOS), ABCG1, and apolipoprotein A-I (APOA1) compared to glyceollin I, which contrasts previous reports on the higher affinity of the latter for ER. Nonetheless, glyceollin I showed a higher upregulation of ABCG8 expression compared to glyceollin II. Despite these prospects, confirmation of these bioactivities in animal models and humans is warranted.

7.6 CONCLUSION AND FUTURE DIRECTION

This chapter presented the potential of glyceollins as naturally occurring compounds in soybean with a range of biological activities. There is tremendous opportunity in exploring these phytoalexins further for nutraceutical and functional food applications. First, the inducible nature of glyceollins using different biotic and abiotic elicitors in culture conditions would facilitate their sustainable production in large amounts for industrial scale-up, compared to naturally constitutive phytochemicals that are often obtained in low yields. Second, the multifunctional effects of glyceollins make them excellent candidates for targeting several diseases or health issues with multiple pathophysiologies. Nonetheless, there are some limitations impeding the practical uses of glyceollins as nutraceuticals or functional food components, such as the limited information on the oral pharmacokinetics and biological effects of the phytoalexins in humans. Furthermore, there is a dearth of information on the toxicological profile and safety of dietary consumption of different doses of soybean glyceollins. Future studies are needed to address these knowledge gaps in order to facilitate the application of soybean glyceollins in human health promotion.

REFERENCES

Ahmed, F., Ijaz, B., Ahmad, Z., Farooq, N., Sarwar, M. B., & Husnain, T. (2020). Modification of miRNA expression through plant extracts and compounds against breast cancer: Mechanism and translational significance. *Phytomedicine*, 68(August 2019), 153168. https://doi.org/10.1016/j.phymed.2020.153168.

Akashi, T., Sasaki, K., Aoki, T., Ayabe, S.-I., & Yazaki, K. (2009). Molecular cloning and characterization of a cdna for pterocarpan 4-dimethylallyltransferase catalyzing the key prenylation step in the biosynthesis of glyceollin, a soybean phytoalexin. *Plant Physiology*, 149, 683–693.

Baeuerle, P. A., & Baltimore, D. (1996). Nf-κB: Ten years after. *Cell*, 87(1), 13–20. https://doi.org/10.1016/S0092-8674(00)81318-5.

Bamji, S. F., & Corbitt, C. (2017). Glyceollins: Soybean phytoalexins that exhibit a wide range of health promoting effects. *Journal of Functional Foods*, 34, 98–105. https://doi.org/10.1016/j.jff.2017.04.020.

Bateman, M. E., Strong, A. L., Hunter, R. S., Bratton, M. R., Komati, R., Sridhar, J., … Bunnell, B. A. (2017). Osteoinductive effects of glyceollins on adult mesenchymal stromal/stem cells from adipose tissue and bone marrow. *Phytomedicine*, 27, 39–51. https://doi.org/10.1016/j.phymed.2017.02.003.

Boué, S. M., Isakova, I. A., Burow, M. E., Cao, H., Bhatnagar, D., Sarver, J. G., Shinde, K. V., Erhardt, P. W., & Heiman, M. L. (2012). Glyceollins, soy isoflavone phytoalexins, improve oral glucose disposal by stimulating glucose uptake. *Journal of Agricultural and Food Chemistry*, 60(25), 6376–6382.

Boué, S. M., Shih, F. F., Shih, B. Y., Daigle, K. W., Carter-Wientjes, C. H., & Cleveland, T. E. (2008). Effect of biotic elicitors on enrichment of antioxidant properties and induced isoflavones in soybean. *Journal of Food Science*, 73(4). https://doi.org/10.1111/j.1750-3841.2008.00707.x.

Burow, M. E., Boue, S. M., Collins-Burow, B. M., Melnik, L. I., Duong, B. N., Carter-Wientjes, C. H., … McLachlan, J. A. (2001). Phytochemical glyceollins, isolated from soy, mediate antihormonal effects through estrogen receptor a and b.pdf. *The Journal of Clinical Endocrinology & Metabolism*, 86(4), 1750–1758. https://doi.org/10.1210/jcem.86.4.7430.

Carlson, M. J., Thiel, K. W., & Leslie, K. K. (2014). Past, present, and future of hormonal therapy in recurrent endometrial cancer. *International Journal of Women's Health*, 6, 429–435. https://doi.org/10.2147/IJWH.S40942.

Chen, K. I., Erh, M. H., Su, N. W., Liu, W. H., Chou, C. C., & Cheng, K. C. (2012). Soyfoods and soybean products: From traditional use to modern applications. *Journal of Applied Microbiology and Technology*, 96(1), 9–22.

Chen, X., Fang, X., Zhang, Y., Wang, X., Zhang, C., Yan, X., … Zhang, S. (2019). Overexpression of a soybean 4-coumaric acid: Coenzyme A ligase (GmPI4L) enhances resistance to *Phytophthora sojae* in soybean. *Functional Plant Biology*, 46(4), 304–313. https://doi.org/10.1071/FP18111.

Cherrington, A. D., Edgerton, D. S., & Ramnanan, C. J. (2009). The role of insulin in the regulation of PEPCK and gluconeogenesis in vivo. *US Endocrinology*, 5(1), 34–39. https://doi.org/10.17925/USE.2009.05.1.34.

Chimezie, C., Ewing, A. C., Quadri, S. S., Cole, R. B., Boue, S. M., Omari, C. F., Bratton, M., Glotser, E., Skripnikova, E., Townley, I., & Stratford Jr, R. E. (2014). Glyceollin transport, metabolism, and effects on P-glycoprotein function in caco-2 cells. *Journal of Medicinal Food*, 17(4), 462–471. https://doi.org/10.1089/jmf.2013.0115.

Chimezie, C., Ewing, A., Schexnayder, C., Bratton, M., Glotser, E., Skripnikova, E., Sá, P., Boué, S., & Stratford Jr, R. E. (2016). Glyceollin effects on MRP2 and BCRP in Caco-2 cells, and implications for metabolic and transport interactions. *Journal of Pharmaceutical Sciences*, 105(2), 972–981. https://doi.org/10.1002/jps.24605.

Di Stefano, E., Oliviero, T., & Udenigwe, C. C. (2018). Functional significance and structure–activity relationship of food-derived α-glucosidase inhibitors. *Current Opinion in Food Science*, 20, 7–12. https://doi.org/10.1016/j.cofs.2018.02.008.

Dizdaroglu, M., & Jaruga, P. (2012). Mechanisms of free radical-induced damage to DNA. *Free Radical Research*, 46(4), 382–419. https://doi.org/10.3109/10715762.2011.653969.

dos Santos Videira, I. F., Moura, D. F. L., & Magina, S. (2013). Mechanism regulating melanogenesis. *Anais Brasileiros de Dermatologia*, 88(1), 76–83. https://doi.org/10.1590/S0365-05962013000100009.

Ejike, C. E., Gong, M., & Udenigwe, C. C. (2013). Phytoalexins from the Poaceae: Biosynthesis, function and prospects in food preservation. *Food Research International*, 52(1), 167–177. https://doi.org/10.1016/j.foodres.2013.03.012.

Fan, G. W., Gao, X. M., Wang, H., Zhu, Y., Zhang, J., Hu, L. M., … Zhang, B. L. (2009). The anti-inflammatory activities of Tanshinone IIA, an active component of TCM, are mediated by estrogen receptor activation and inhibition of iNOS. *Journal of Steroid Biochemistry and Molecular Biology*, 113(3–5), 275–280. https://doi.org/10.1016/j.jsbmb.2009.01.011.

Farrell, K., Jahan, M. A., & Kovinich, N. (2017). Distinct mechanisms of biotic and chemical elicitors enabled additive elicitation of the anticancer phytoalexin glyceollin 1. *Journal of Synthetic Chemistry and Natural Product Chemistry*, 22, 1261.

Galibert, M. D., Carreira, S., & Goding, C. R. (2001). The Usf-1 transcription factor is a novel target for the stress-responsive p38 kinase and mediates UV-induced tyrosinase expression. *EMBO Journal*, 20(17), 5022–5031. https://doi.org/10.1093/emboj/20.17.5022.

Giri, B., Dey, S., Das, T., Sarkar, M., Banerjee, J., & Dash, S. K. (2018). Chronic hyperglycemia mediated physiological alteration and metabolic distortion leads to organ dysfunction, infection, cancer progression and other pathophysiological consequences: An update on glucose toxicity. *Biomedicine and Pharmacotherapy*, 107(April), 306–328. https://doi.org/10.1016/j.biopha.2018.07.157.

Glass, C. K., Saijo, K., Winner, B., Marchetto, M. C., & Gage, F. H. (2010). Mechanisms underlying inflammation in neurodegeneration. *Cell*, 140(6), 918–934. https://doi.org/10.1016/j.cell.2010.02.016.

González-Chávez, S. A., Quiñonez-Flores, C. M., & Pacheco-Tena, C. (2016). Molecular mechanisms of bone formation in spondyloarthritis. *Joint Bone Spine*, 83(4), 394–400. https://doi.org/10.1016/j.jbspin.2015.07.008.

Hasanpourghadi, M., Pandurangan, A. K., & Mustafa, M. R. (2018). Modulation of oncogenic transcription factors by bioactive natural products in breast cancer. *Pharmacological Research*, 128, 376–388. https://doi.org/10.1016/j.phrs.2017.09.009.

Hegele, R. A., & Robinson, J. F. (2005) ABC transporters and sterol absorption. *Current Drug Targets Cardiovascular & Hematological Disorders*, 5(1), 31–37. https://doi.org/10.2174/1568006053005029.

Hermanto, F. E., Rifa'I, M., & Widodo. (2019). Potential role of glyceollin as anti-metastatic agent through transforming growth factor-β receptors inhibition signaling pathways: A computational study. *AIP Conference Proceedings*, 2155. https://doi.org/10.1063/1.5125539.

Hirata, N., Naruto, S., Ohguchi, K., Akao, Y., Nozawa, Y., Iinuma, M., & Matsuda, H. (2007). Mechanism of the melanogenesis stimulation activity of (-)-cubebin in murine B16 melanoma cells. *Bioorganic and Medicinal Chemistry*, 15(14), 4897–4902. https://doi.org/10.1016/j.bmc.2007.04.046.

Huang, H., Xie, Z., Boue, S. M., Bhatnagar, D., Yokoyama, W., Yu, L., & Wang, T. T. Y. (2013). Cholesterol-lowering activity of soy-derived glyceollins in the golden syrian hamster model. *Journal of Agricultural and Food Chemistry*, 61(24), 5772–5782. https://doi.org/10.1021/jf400557p.

Ighodaro, O. M., & Akinloye, O. A. (2018). First line defence antioxidants-superoxide dismutase (SOD), catalase (CAT) and glutathione peroxidase (GPX): Their fundamental role in the entire antioxidant defence grid. *Alexandria Journal of Medicine*, 54(4), 287–293. https://doi.org/10.1016/j.ajme.2017.09.001.

Jahan, M. A., Harris, B., Lowery, M., Coburn, K., Infante, A. M., Percifield, R. J., Ammer, A. G., & Kovinich N. (2019). The NAC family transcription factor GmNAC42–1 regulates biosynthesis of the anticancer and neuroprotective glyceollins in soybean. *BMC Genomics*, 20, 149.

Jahan, M. A., & Kovinich, Nik. (2019). Acidity stress for the systemic elicitation of glyceollin phytoalexins in soybean plants. *Plant Signaling and Behavior*, 14(7).

Jeong, G., Oh, J., & Kim, J.-S. (2019). Glyceollins modulate tumor development and growth in a mouse xenograft model of human colon cancer in a p53-dependent manner. *Journal of Medicinal Food*, 22(5), 521–528. https://doi.org/10.1089/jmf.2018.4290.

Jung, C. L., Kim, H. J., Park, J. H. Y., Kong, A. N. T., Lee, C. H., & Kim, J. S. (2013). Synergistic activation of the Nrf2-signaling pathway by glyceollins under oxidative stress induced by glutathione depletion. *Journal of Agricultural and Food Chemistry*, 61(17), 4072–4078. https://doi.org/10.1021/jf303948c.

Kazemi, S., Shirzad, H., & Rafieian-Kopaei, M. (2018). Recent findings in molecular basis of inflammation and anti-inflammatory plants. *Current Pharmaceutical Design*, 24(14), 1551–1562. https://doi.org/10.2174/1381612824666180403122003.

Kim, D. S., Park, S. H., Kwon, S. B., Na, J. I., Huh, C. H., & Park, K. C. (2007). Additive effects of heat and p38 MAPK inhibitor treatment on melanin synthesis. *Archives of Pharmacal Research*, 30(5), 581–586. https://doi.org/10.1007/BF02977652.

Kim, H. J., di Luccio, E., Kong, A. N. T., & Kim, J. S. (2011). Nrf2-mediated induction of phase 2 detoxifying enzymes by glyceollins derived from soybean exposed to *Aspergillus sojae*. *Biotechnology Journal*, 6(5), 525–536. https://doi.org/10.1002/biot.201100010.

Kim, H. J., Jung, C. L., Jeong, Y. S., & Kim, J. S. (2014). Soybean-derived glyceollins induce apoptosis through ROS generation. *Food & Function*, 5(4), 688–695. https://doi.org/10.1039/C3FO60379B.

Kim, H. J., Suh, H. J., Kim, J. H., Park, S., Joo, Y. C., & Kim, J. S. (2010). Antioxidant activity of glyceollins derived from soybean elicited with *Aspergillus sojae*. *Journal of Agricultural and Food Chemistry*, 58(22), 11633–11638. https://doi.org/10.1021/jf102829z.

Kim, H. J., Sung, M. K., & Kim, J. S. (2011). Anti-inflammatory effects of glyceollins derived from soybean by elicitation with *Aspergillus sojae*. *Inflammation Research*, 60(10), 909–917. https://doi.org/10.1007/s00011-011-0351-4.

Klinge, C. M. (2000). Estrogen receptor interaction with co-activators and co-repressors. *Steroids*, 65(5), 227–251. https://doi.org/10.1016/S0039-128X(99)00107-5.

Lecomte, S., Chalmel, F., Ferriere, F., Percevault, F., Plu, N., Saligaut, C., ... Pakdel, F. (2017). Glyceollins trigger anti-proliferative effects through estradiol-dependent and independent pathways in breast cancer cells. *Cell Communication and Signaling*, 15(1), 1–18. https://doi.org/10.1186/s12964-017-0182-1.

Lee, P. C., & Halter, J. B. (2017). The pathophysiology of hyperglycemia in older adults: Clinical considerations. *Diabetes Care*, 40, 444–452. http://dx.doi.org/10.2337/dc16-1732.

Lee, S. H., Jee, J. G., Bae, J. S., Liu, K. H., & Lee, Y. M. (2015). A group of novel HIF-1α inhibitors, glyceollins, blocks HIF-1α synthesis and decreases its stability via inhibition of the PI3K/AKT/mTOR pathway and Hsp90 binding. *Journal of Cellular Physiology*, 230(4), 853–862. https://doi.org/10.1002/jcp.24813.

Lee, Y. S., Kim, D. W., Kim, S., Choi, H. I., Lee, Y., Kim, C. D., ... Lee, Y. H. (2010). Downregulation of NFAT2 promotes melanogenesis in B16 melanoma cells. *Anatomy & Cell Biology*, 43(4), 303. https://doi.org/10.5115/acb.2010.43.4.303.

Levin, E. R., & Pietras, R. J. (2008). Estrogen receptors outside the nucleus in breast cancer. *Breast Cancer Research and Treatment*, 108:3, 351–361. https://doi.org/10.1007/s10549-007-9618-4

Liu, X., Guo, C. Y., Ma, X. J., Wu, C. F., Zhang, Y., Sun, M. Y., ... Yin, H. J. (2015). Anti-inflammatory effects of tanshinone IIA on atherosclerostic vessels of ovariectomized ApoE mice are mediated by estrogen receptor activation and through the ERK signaling pathway. *Cellular Physiology and Biochemistry*, 35(5), 1744–1755. https://doi.org/10.1159/000373986.

Lygin, A. V., Hill, C. B., Zernova, O. V., Crull, L., Widholm, J. M., Hartman, G. L., & Lozovaya, V. V. (2010). Response of soybean pathogens to glyceollin. *Genetics and Resistance*, 100(9), 897–903. https://doi.org/10.1094/PHYTO-100-9-0897.

Marcovecchio, M. L., Lucantoni, M., & Chiarelli, F. (2011). Role of chronic and acute hyperglycemia in the development of diabetes complications. *Diabetes Technology & Therapeutics*, 13(3), 389–394. https://doi.org/10.1089/dia.2010.0146.

Marini, J. C., Forlino, A., Cabral, W. A., Barnes, A. M., San Antonio, J. D., Milgrom, S., ... Byers, P. H. (2007). Consortium for osteogenesis imperfecta mutations in the helical domain of type I collagen: Regions rich in lethal mutations align with collagen binding sites for integrins and proteoglycans. *Human Mutation*, 28(3), 209–221. https://doi.org/10.1002/humu.20429.

Nwachukwu, I., Luciano, F. B., & Udenigwe, C. (2013). The inducible soybean glyceollin phytoalexins with multifunctional health promoting properties. *Journal of Food Research International*, 54, 1208–1216.

Ogbu, C. P., Okagu, I. U., & Nwodo, O. F. C. (2020). Anti-inflammatory activities of crude ethanol extract of *Combretum zenkeri* Engl. & Diels leaves. *Comparative Clinical Pathology*, 29(2), 397–409. https://doi.org/10.1007/s00580-019-03072-0.

Oh, J., Jang, C. H., & Kim, J. (2019). Soy-derived phytoalexins: Mechanism of in vivo biological effectiveness in spite of their low availability. *Food Science Biotechnology*, 28(1), 1–6. https://doi.org/10.1007/s10068-018-0498-7.

Okuyama, M., Saburi, W., Mori, H., & Kimura, A. (2016). α-Glucosidases and α-1,4-glucan lyases: Structures, functions, and physiological actions. *Cellular and Molecular Life Sciences*, 73(14), 2727–2751. https://doi.org/10.1007/s00018-016-2247-5.

Owolabi, L. O., Yupanqui, C. T., & Siripongvutikorn, S. (2018). Enhancing secondary metabolites (emphasis on phenolics and antioxidants) in plants through elicitation and metabolomics. *Pakistan Journal of Nutrition*, 17, 411–420.

Park, S., Ahn, I. S., Kim, J. H., Lee, M. R., Kim, J. S., & Kim, H. J. (2010). Glyceollins, one of the phytoalexins derived from soybeans under fungal stress, enhance insulin sensitivity and exert insulinotropic actions. *Journal of Agricultural and Food Chemistry*, 58(3), 1551–1557. https://doi.org/10.1021/jf903432b.

Park, S., Kim, D. S., Kim, J. H., Kim, J. S., & Kim, H. J. (2012). Glyceollin-containing fermented soybeans improve glucose homeostasis in diabetic mice. *Nutrition*, 28(2), 204–211. https://doi.org/10.1016/j.nut.2011.05.016.

Park, I. S., Kim, J. H., Jeong, S. Y., Kim, W. K., & Kim, S. J. (2017). Differential abilities of Korean soybean varieties to biosynthesize glyceollins by biotic and abiotic elicitors. *Food Science Biotechnology*, 26(1), 255–261.

Patel, H., & Krishnamurthy, R. (2013). Elicitors in plants tissue culture. *Journal of Pharmacognosy and Phytochemistry*, 2, 60–65.

Patel, Z. M., Mahapatra, R., & Mohan Jampala, S. S. (2020). Role of fungal elicitors in plant defense mechanism. In Eds. Sharma, V., Salwan, R., & Tawfeeq Al-Ani, L. K., *Molecular Aspects of Plant Beneficial Microbes in Agriculture* (pp. 143–158). https://doi.org/10.1016/B978-0-12-818469-1.00012-2

Pham, T. H., Lecomte, S., Efstathiou, T., Ferriere, F., & Pakdel, F. (2019). An update on the effects of glyceollins on human health: Possible anticancer effects and underlying mechanisms. *Nutrients*. https://doi.org/10.3390/nu11010079.

Phaniendra, A., Jestadi, D. B., & Periyasamy, L. (2015). Free radicals: Properties, sources, targets, and their implication in various diseases. *Indian Journal of Clinical Biochemistry*, 30(1), 11–26. https://doi.org /10.1007/s12291-014-0446-0.

Roberts, A. C., & Porter, K. E. (2013). Cellular and molecular mechanisms of endothelial dysfunction in diabetes. *Diabetes and Vascular Disease Research*. https://doi.org/10.1177/1479164113500680.

Rodríguez-Merchán, E. C. (2021). A review of recent developments in the molecular mechanisms of bone healing. *International Journal of Molecular Sciences*, 22(2), 1–14. https://doi.org/10.3390/ijms22020767.

Safe, S., & Kim, K. (2004). Nuclear receptor-mediated transactivation through interaction with Sp proteins 1 stephen safe and I. Sp1 protein. *Progress in Nucleic Acid Research and Molecular*, 77, 1–36. https://doi .org/10.1016/S0079-6603(04)77001-4.

Salvo, V., Boue, S. M., Fonseca, J. P., Elliott, S., Corbitt, C., Collins-Burow, B. M., … Burow, M. E. (2006). Antiestrogenic glyceollins suppress human breast and ovarian carcinoma tumorigenesis. *Clinical Cancer Research*, 12(23), 7159–7164. https://doi.org/10.1158/1078-0432.CCR-06-1426.

Santo, A., Zhu, H., & Li, Y. R. (2016). Free radicals: From health to disease. *Reactive Oxygen Species*, 2(4), 245–263. https://doi.org/10.20455/ros.2016.847.

Selvam, C., Jordan, B. C., Prakash, S., Mutisya, D., & Thilagavathi, R. (2017). Pterocarpan scaffold: A natural lead molecule with diverse pharmacological properties. *European Journal of Medicinal Chemistry*, 128, 219–236.

Seo, H., Oh, J., Hahn, D., Kwon, C.-S., Lee, J. S., & Kim, J.-S. (2017). Protective effect of glyceollins in a mouse model of dextran sulfate sodium-induced colitis. *Journal of Medicinal Food*, 20(11), 1055–1062. https://doi.org/10.1089/jmf.2017.3960.

Shin, S. H., & Lee, Y. M. (2013). Glyceollins, a novel class of soybean phytoalexins, inhibit SCF-induced melanogenesis through attenuation of SCF/c-kit downstream signaling pathways. *Experimental and Molecular Medicine*, 45(2), 1–9. https://doi.org/10.1038/emm.2013.20.

Silva, E. F. F., Ferreira, C. M. D., & de Pinho, L. (2017). Risk factors and complications in type 2 diabetes outpatients. *Revista da Associacao Medica Brasileira*, 63(7), 621–627. http://dx.doi.org/10.1590/1806 -9282.63.07.621.

Simons, R., Vincken, J. P., Roidos, N., Bovee, T. F. H., Van Iersel, M., & Verbruggen, M. A. (2011). Increasing soy isoflavonoid content and diversity by simultaneous malting and challenging by a fungus to modulate estrogenicity. *Journal of Agricultural and Food Chemistry*, 59, 6748–6758.

Singh, V. K., Paramanick, D., Sharma, V., & Pundir, A. (2020). A comprehensive review on new approaches for management of diabetes. *Archives of Medicine*, 12(5:23), 1–9. https://doi.org/10.36648/1989-5216.12.

Son, H. U., Yoon, E. K., Yoo, C. Y., Park, C. H., Bae, M. A., Kim, T. H., Lee, C. H., Lee, K. W., Seo, H., Kim, K. J., & Lee, S. H. (2019). Effects of synergistic inhibition on α-glucosidase by phytoalexins in soybeans. *Biomolecules*, 9, 828. https://doi.org/10.3390/biom9120828.

Soo, B. C., Jin, S. J., & Park, S. (2005). Estrogen and exercise may enhance β-cell function and mass via insulin receptor substrate 2 induction in ovariectomized diabetic rats. *Endocrinology*, 146(11), 4786–4794. https://doi.org/10.1210/en.2004-1653.

Spinelli, R., Parrillo, L., Longo, M., Florese, P., Desiderio, A., Zatterale, F., … Beguinot, F. (2020). Molecular basis of ageing in chronic metabolic diseases. *Journal of Endocrinological Investigation*, 43(10), 1373–1389. https://doi.org/10.1007/s40618-020-01255-z.

Strohacker, K., & McFarlin, B. K. (2020). Influence of obesity, physical inactivity, and weight cycling on chronic inflammation. *Frontiers in Bioscience (Elite Ed)*, 1(2), 98–104. https://doi.org/10.2741/e70.

Su, N., Yang, J., Xie, Y., Du, X., Chen, H., Zhou, H., & Chen, L. (2019). Bone function, dysfunction and its role in diseases including critical illness. *International Journal of Biological Sciences*, 15(4), 776–787. https://doi.org/10.7150/ijbs.27063.

Tabas, I. (2010). Macrophage death and defective inflammation resolution in atherosclerosis. *Nature Reviews Immunology*, 10(1), 36–46. https://doi.org/10.1038/nri2675.

Tiwari, P. (2015). Recent trends in therapeutic approaches for diabetes management: A comprehensive update. *Journal of Diabetes Research*, 2015, 340838. https://doi.org/10.1155/2015/340838

Van Weelden, W. J., Massuger, L. F. A. G., Pijnenborg, J. M. A., & Romano, A. (2019). Anti-estrogen treatment in endometrial cancer: A systematic review. *Frontiers in Oncology*, 9(MAY). https://doi.org/10.3389/ fonc.2019.00359.

WHO. (2021). Estimated number of new cases in 2020, worldwide, female, all ages. Retrieved from http://gco .iarc.fr/

Wongrakpanich, S., Wongrakpanich, A., Melhado, K., & Rangaswami, J. (2018). A comprehensive review of non-steroidal anti-inflammatory drug use in the elderly. *Aging and Disease*, 9(1), 143–150. https://doi .org/10.14336/AD.2017.0306.

Wood, C. E., Boue, S. M., Collins-Burow, B. M., Rhodes, L. V., Register, T. C., Cline, J. M., Dewi, F. N., & Burow, M. E. (2012). Glyceollin-elicited soy protein consumption induces distinct transcriptional effects compared to standard soy protein. *Journal of Agricultural and Food Chemistry*, 60(1), 81–86. https://doi.org/10.1021/jf2034863.

Wood, C. E., Register, T. C., & Cline, J. M. (2007). Soy isoflavonoid effects on endogenous estrogen metabolism in postmenopausal female monkeys. *Carcinogenesis*, 28(4), 801–808. https://doi.org/10.1093/carcin/bgl163.

Yoon, E. K., Kim, H. K., Cui, S., Kim, Y. H., & Lee, S. H. (2012). Soybean glyceollins mitigate inducible nitric oxide synthase and cyclooxygenase-2 expression levels via suppression of the NF-κB signaling pathway in RAW 264.7 cells. *International Journal of Molecular Medicine*, 29(4), 711–717. https://doi.org/10.3892/ijmm.2012.887.

Zimmermann, M. C., Tilghman, S. L., Boué, S. M., Salvo, V. A., Elliott, S., Williams, K. Y., Williams, K. Y., Skripnikova, E. V., Ashe, H., Payton-Stewart, F., Vanhoy-Rhodes, L., & Burow, M. E. (2010). Glyceollin I, a novel antiestrogenic phytoalexin isolated from activated soy. *Journal of Pharmacology and Experimental Therapeutics*, 332(1), 35–45. https://doi.org/10.1124/jpet.109.160382.

Żukowski, P., Maciejczyk, M., & Waszkiel, D. (2018). Sources of free radicals and oxidative stress in the oral cavity. *Archives of Oral Biology*, 92(January), 8–17. https://doi.org/10.1016/j.archoralbio.2018.04.018.

8 Soybean Allergens

Caterina Villa, Joana Costa, and Isabel Mafra

CONTENTS

8.1 INTRODUCTION

Soybean (*Glycine max*) belongs to the Leguminosae (Fabaceae) family and is popularly known as the miracle crop, being the major source of protein and oil in the world. Native to China, presently soybean is used worldwide for producing meal for livestock feeding and vegetable oil, with a growing rate used for food production due to its favorable nutritional composition, health benefits, and functional characteristics (Watanabe et al., 2018). Water and fat absorption, emulsification, foaming, gelation, and binding are useful properties for the application of soybean proteins as ingredients in a wide variety of food formulations, meat products, bakery, dairy, and pastry products, as well as edible spreads (L'Hocine & Boye, 2007). In 2018, soybean ranked first in the production of oilseed crops (46.2%), followed by rapeseed (9.9%) and palm (9.5%) (FAOSTAT, 2020). In the same year, the worldwide production of soybean was above 348 million tons, the USA being the main producer with almost 124 million tons (35.6% of total world production), followed by Brazil with 118 million tons (34% of total world production). Europe produced 12 million tons of soybean in 2018, Russia being its main producer (FAOSTAT, 2020). Currently, a great part of soybean production is assured by biotech crops, from which 95.9 million hectares corresponded to genetically modified soybean, representing 76.8% of total soybean area production and 50% of total biotech crops (FAOSTAT, 2020; ISAAA, 2018).

Soybean is also considered as one of the eight groups of allergenic foods responsible for about 90% of all food allergies, with an estimated prevalence ranging between 0.3–0.4% (L'Hocine & Boye, 2007), depending on local consumption habits and exposure (Dean et al., 2013). Among children with food allergies until one year of age, 16.6% are allergic to soybean (Jiang et al., 2020). The health risk associated with the ingestion of products containing soybean proteins is high, with the occurrence of mild to potentially severe and lethal systemic allergic reactions in sensitized

DOI: 10.1201/9781003030294-8

individuals. In this context, the European Union has implemented regulations that define the mandatory discrimination of 14 groups of foods, including soybean, from the rest of the list of ingredients present on the labels of pre-packaged food, regardless of its amount (Directive 2007/68/EC; Regulation (EU) No 1169/2011).

This chapter intends to provide an overview of soybean as an allergenic food, describing the molecular characterization of its allergens and cross-reactions associated with the ingestion of related legume species. Additionally, the application of different types of food processing technologies is referred to, focusing on the reduction of soybean allergenicity, as well as its correlation with digestibility, and genetic modification. Finally, recent analytical methodologies aiming at detecting soybean in processed foods are summarized and discussed.

8.2   SOYBEAN ALLERGY

The prevalence of soybean allergy in the general population is not exact, although it has been estimated around 0.3% and 0.4% (L'Hocine & Boye, 2007), with an incidence of 3.2–3.6% in European countries (Italy, Germany, and France) and 4.7% in the USA (Burney et al., 2010). In children, the prevalence of soybean allergy is around 0.4–0.6%, being 4.7–8.3% in children with a food allergy and 14% with cow's milk allergy (Gupta et al., 2018). However, it seems that children are able to develop a state of tolerance with the progression of age, with 50% outgrowing the allergy by the age of one year old, 67% by two years old, and 45% by six years old (Jiang et al., 2020; Sampson & McCaskill, 1985; Sampson & Scanlon, 1989; Savage et al., 2010).

Similar to other food allergies, individuals with soybean allergy present a wide range of IgE- and non-IgE-mediated clinical symptoms, the first occurring within one to two hours after ingestion and the latter with a more delayed onset beyond two hours after soybean ingestion (Jiang et al., 2020). These manifestations can mainly affect the cutaneous, cardiovascular, and respiratory systems, as well as the gastrointestinal tract, ranging from severe enterocolitis to atopic eczema and immediate IgE-mediated systemic reactions, such as anaphylaxis (L'Hocine & Boye, 2007; Verma et al., 2013). Fatal reactions related to the ingestion of soybean are rare, compared to peanuts, tree nuts, or fish, though some cases have been described (Bock et al., 2001; Foucard & Malmheden-Yman, 1999; Vidal et al., 1997). In Figure 8.1, the route of soybean allergic response is represented. The threshold levels for developing an adverse reaction in soybean allergic individuals have been reported to vary between 0.0013 mg and 500 mg (Becker et al., 2004). In the study developed by Ballmer-Weber et al. (2007) regarding soybean allergy in Europe, the minimum amount of protein causing (subjective) symptoms in allergic individuals was calculated to be 5.3 mg of soybean. Clearly, the amount of total protein eliciting objective symptoms can vary to a large degree, depending on the allergenic proteins contributing to the total protein content in the consumed food, the challenge procedure, and the characteristics of the allergic individual (Selb et al., 2017). Diagnostic tests for a suspected soybean allergy involve skin-prick testing, atopy patch test, soybean specific IgE and IgG dosing, and basophil activation test. Ultimately, the oral food challenges in open, closed, or double-blind placebo-controlled food challenge (DBPCFC) (the gold standard method) formats are performed to confirm the diagnosis of soybean allergy (Verma et al., 2013).

As for other food allergens, patients have to strictly avoid soybean-containing foods and, in the case of unintended ingestion, emergency treatments such as epinephrine auto-injectors are required. Therefore, the production of hypoallergenic soybean products has been investigated by several researchers, including thermal, enzymatic, and chemical modification of allergenic proteins or conventional breeding and genetic modification. According to the American Academy of Pediatrics and the European Society for Paediatric Gastroenterology, Hepatology, and Nutrition (ESPGHAN), a hypoallergenic food has been shown, with 95% confidence, not to cause allergic reactions in 90% of patients with a confirmed diagnosis of food allergy (American Academy of Pediatrics, 2000).

The sensitization or the development of allergic reactions to soybean may also be triggered by different routes rather than ingestion. The inhalation of soybean flour or dust can be responsible for

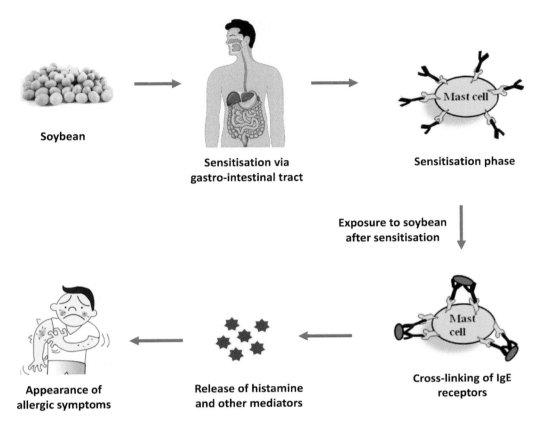

Soybean

Sensitisation via gastro-intestinal tract

Sensitisation phase

Exposure to soybean after sensitisation

Cross-linking of IgE receptors

Appearance of allergic symptoms

Release of histamine and other mediators

FIGURE 8.1 Route of soybean allergic response. Source: Adapted from Verma et al. (2013).

occupational asthma in food industry workers, although the allergens responsible for such reactions are different from the ones triggering food allergy (Ballmer-Weber & Vieths, 2008).

Soybean is also considered the most important genetically modified (GM) crop. Therefore, it cannot be excluded that genetic transformation or transgenic modification can adversely affect human health and increase the allergenic potential of this crop. The safety of GM plants is strictly regulated by international guidelines (Codex Alimentarius, 2009) and EU regulations (Regulation [EU] No 503/2013), which stated that endogenous allergens, after genetic modification of plants considered to be allergenic, such as soybean, liable to any potential changes in allergenicity have to be assessed before placement on the European market. In this context, there are at least three categories of risks for the allergic individuals that must be carefully evaluated, namely: 1) The potential for transferring a specific allergen or a cross-reactive one to the new GM crop; 2) the expression of novel proteins with *de novo* sensitization capacity (development of new food allergy); and 3) the potential for increasing the allergenicity of a GM crop (e.g. soybean) versus its non-GM counterpart (Ladics, 2019). Studies using sera from soybean-allergic patients to evaluate the IgE-binding capacity of GM soybean versus conventional soybean have been reported. Matsuo et al. (2020) verified that the GM soybean events containing the CP4-EPSPS gene (glyphosate-tolerant trait) have similar allergen abundance levels and allergen reactivity to the non-GM counter-part. Additionally, the expressed proteins from the CP4-EPSPS gene did not bind to the IgE of patients' sera (Hoff et al., 2007; Matsuo et al., 2020; Sten et al., 2004). Likewise, Lu et al. (2018) and Tsai et al. (2017a) concluded that there were no significant differences regarding the IgE-binding capacity and histamine release of the allergenic proteins from the GM lines compared with the non-GM soybean. The amounts of allergens expressed in the GM soybean lines were similar to the conventional crop, therefore not affecting their allergenic potential.

8.3   MOLECULAR CHARACTERIZATION AND CLINICAL RELEVANCE OF SOYBEAN ALLERGENS

Soybean contains approximately 20% oil and 37% proteins, from which about 90% are salt-soluble storage globulins, with the remaining 10% being water-soluble albumins, including γ-conglycinin (minor globulin), and relatively large amounts of proteins such as lipoxygenase, β-amylase, lectin, and Kunitz trypsin inhibitors. Four protein fractions have been revealed, based on the sedimentation coefficient, namely 2S, 7S, 11S, and 15S, representing approximately 8–22%, 35%, 31–52%, and 5% of protein content, respectively (Luthria et al., 2018; Nishinari et al., 2014; Verma et al., 2013). At least 16 IgE-reactive proteins have been suggested as soybean allergens (ALLERGOME, 2020), most of them displaying storage, protective, and metabolic/regulatory functions (L'Hocine & Boye, 2007; Matricardi et al., 2016). However, only eight are recognized as allergens by the World Health Organization/International Union of Immunological Societies (WHO/IUIS) Allergen Nomenclature Subcommittee (WHO/IUIS, 2020). Table 8.1 summarizes the known allergenic proteins of soybean.

8.3.1   GLY M 1 AND GLY M 2 (SOYBEAN HULL PROTEINS)

Gly m 1 and Gly m 2 allergens are components of the soybean hull and are considered aeroallergens because of their association with respiratory symptoms through their inhalation, as reported in Barcelona, Spain (Ballmer-Weber & Vieths, 2008; Codina et al., 1999). However, they do not seem to be correlated with occupational asthma caused by soybean flours in bakers, where high molecular weight (MW) allergens are involved (L'Hocine & Boye, 2007).

Gly m 1 possesses two isoforms (Gly m 1.0101 and Gly m 1.0102) with MW of 7.5 and 7.0 kDa, respectively, identified as soybean hydrophobic proteins that belong to the non-specific lipid transfer protein (nsLTP) family, also named pathogenesis-related (PR)-14 proteins from the prolamin superfamily (Gonzalez et al., 1992; Gonzalez et al., 1995). PR-14 are monomeric (between 7 and 9 kDa) proteins with highly conserved motifs having disulfide bonds. These proteins present a very compact and stable tertiary structure with a central cavity possessing a lipid-binding site, whose main function is the phospholipid transportation from liposomes to the mitochondria. They show high resistance to gastrointestinal proteolysis, harsh pH changes, and high temperatures, frequently associated with systemic and even anaphylactic reactions in sensitized individuals (Breiteneder & Radauer, 2004; Scala et al., 2018).

Gly m 2 proteins belong to the PR-12/defensin family with a MW of 8 kDa and a possible defense role in soybean plants due to their N-terminal sequence, showing homologies of 71% and 64% with a storage protein from the cotyledon of cow pea and a disease response protein from green pea, respectively (Codina et al., 1997; L'Hocine & Boye, 2007).

8.3.2   GLY M 3 AND GLY M 4 (BIRCH-POLLEN RELATED ALLERGENS)

Gly m 3 proteins belong to the profilin family, characterized as cytosolic proteins found in all eukaryotic cells, which participate in the polymerization of actin filaments by binding to monomeric actin (Breiteneder & Radauer, 2004). Gly m 3 were produced as recombinant proteins, showing that their IgE-binding reactivity is mediated by conformational epitopes. The recombinant form of Gly m 3 demonstrated an amino-acid sequence homology of 73% with the well-characterized rBet v 2 birch pollen-related allergen, suggesting the presence of IgE-binding epitopes (Rihs et al., 1999). These proteins are very heat-labile, susceptible to enzymatic degradation and their clinical relevance is most commonly associated with oral allergy syndrome (OAS) (L'Hocine & Boye, 2007).

TABLE 8.1

Identified Soybean Allergens and Their Isoforms/Variants According to Their Biochemical Classification, Biological Function, and Clinical Relevance

Allergen	Isoallergen	Isoforms or Variants	Biochemical Classification	MW (kDa)	Biological Function	Clinical Relevance	Nucleotide* (NCBI)	Protein* (NCBI)	Protein* (UniProt)	PDB
Gly m 1	Gly m 1.01	Gly m 1.0101	nsLTP (Prolamin superfamily/PR-14)	7.5	Hydrophobic hull seed protein	Major allergen. Associated to systemic and anaphylactic reactions. Asthmatic symptoms.	–	Q9S8F3	Q9S8F3	–
		Gly m 1.0102		7.0			–	Q9S8F2	Q9S8F2	–
Gly m 2	Gly m 2.01	Gly m 2.0101	Defensin (Defensin/myotoxin-like superfamily/PR-12)	8.0	Defense	Major allergen. Associated to systemic and anaphylactic reactions. Asthmatic symptoms.	A57106	–	–	–
Gly m 3	Gly m 3.01	Gly m 3.0101	Profilin	14	Actin-binding protein	Minor allergen. Oral allergy syndrome	AJ223982	CAA11756	O65809	–
		Gly m 3.0102					AJ223981	CAA11755	O65810	–
Gly m 4	Gly m 4.01	Gly m 4.0101	PR-10 (Bet v 1-homologous) (Pathogenesis-related protein family)	16.6	Defense	Major allergen. Oral allergy syndrome, mild to severe systemic and anaphylactic reactions.	X60043	CAA42646	P26987	2K7H
Gly m 5	Gly m 5.01	Gly m 5.0101	Beta-conglycinin (Vicilin, 7S globulin) (Cupin superfamily)	67	Storage protein	Major allergen. Severe allergic reactions.	AB008678	BAA23360	O22120	–
	Gly m 5.02	Gly m 5.0201		71			AB008680	BAA74452	Q9FZP9	–
	Gly m 5.03	Gly m 5.0301		50			S44893	AAB23463	P25974	1IPJ
		Gly m 5.0302						–	P25974	1IPK
Gly m 6	Gly m 6.01	Gly m 6.0101	Glycinin (Legumin, 11S globulin) (Cupin superfamily)	53.6	Storage protein	Minor allergen. Severe allergic reactions.	M36686 AB113349	AAA33966 BAC78522	P04776	1FXZ
	Gly m 6.02	Gly m 6.0201		52.4			D00216	BAA00154	P04405	–
	Gly m 6.03	Gly m 6.0301	(Cupin superfamily)	52.2			X15123	CAA33217	P11828	–
	Gly m 6.04	Gly m 6.0401		61.2			AB004062	BAA74953	Q9SB11	–
	Gly m 6.05	Gly m 6.0501		55.4			AB049440	BAB15802	Q7GC77	2D5F 2D5H
Gly m 7	Gly m 7.01	Gly m 7.0101	Seed biotinylated protein (LEA)	76.2	Plant germination	Minor allergen	GQ168590	ACS49840	C6K8D1	–

(continued)

TABLE 8.1 (CONTINUED)

Identified Soybean Allergens and Their Isoforms/Variants According to Their Biochemical Classification, Biological Function, and Clinical Relevance

Allergen	Isoallergen	Isoforms or Variants	Biochemical Classification	MW (kDa)	Biological Function	Clinical Relevance	Nucleotide* (NCBI)	Protein* (NCBI)	Protein* (UniProt)	PDB
Gly m 8	Gly m 8.01	Gly m 8.0101	2S albumin (Prolamin superfamily)	28	Storage proteins	Major allergen. Severe allergic reactions.	AF005030	AAB71140	P19594	–
Gly m Bd28K	–	–	7S Vicilin (Cupin superfamily)	26	Storage proteins	Major allergen	AB046874.2	–	Q9AVK8	–
Gly m Bd30K0	–	–	Thiol protease (Papain superfamily)	34	Storage proteins. Papain-like protease	Major allergen. Atopic dermatitis	–	ACO55749.1	O64458 P22895	–
Gly m KTI	–	–	Kunitz Trypsin Inhibitor	21	Protease inhibitor. Defense	Minor allergen. Anaphylaxis, respiratory symptoms.	–	AAB23464.1	B1ACD5 P01070 Q9LLX2	–
Gly m Bd39Kd	–	–	Soybean lecithin	39	Proteins from soybean oil fraction	Minor allergen. Anaphylaxis, chronic diarrhea, respiratory symptoms.	–	–	–	–
Gly m agglutinin	–	–	Lectin (Agglutinin)	14.5	Carbohydrate-binding protein	Not reported	–	ACZ74649.1	E3W7C3	–
Gly m Bd50Kd	–	–	17AA nonterminal residue	50		Major allergen. Bronchial asthma, allergic rhinitis	–	–	P82947	–
Gly m CPI	–	–	Cystatin squash aspartic acid proteinase inhibitor	25	Cysteine protease inhibitors. Defense	Not reported	–	–	–	–
Soybean lipoxygenase	–	–	Lipoxygenase	94	Plant growth and development. Cell signaling	Major allergen	–	–	–	–

* Accession members from Uniprot database (www.uniprot.org/), NCBI database (www.ncbi.nlm.nih.gov/), and PDB (www.rcsb.org/).

Gly m 4 (also designated as SAM22) belongs to a group of proteins responsible for the defense system of a plant, namely the PR-10 family. It possesses a tertiary structure with seven-stranded anti-parallel β-sheets and three α-helices, whose expression is greatly influenced by biotic and abiotic factors. It is known that Gly m 4 shares important sequence and structure similarities with the major birch pollen allergen Bet v 1, which is the protein responsible for the sensitization to birch pollen in more than 95% of birch-pollen sensitized/allergic patients (Cabanillas et al., 2018; L'Hocine & Boye, 2007; Scala et al., 2018). It was also demonstrated that 96% of children allergic to soybean were IgE-reactive to Gly m 4; thus it is considered a major soybean allergen (Mittag et al., 2004). Sensitization to this protein has been associated with mild to severe allergic symptoms, including oral allergy syndrome and anaphylactic reactions (Kleine-Tebbe et al., 2002; Matricardi et al., 2016).

8.3.3   GLY M 5 (β-CONGLYCININ) AND GLY M 6 (GLYCININ)

Gly m 5 and Gly m 6 are seed storage proteins from the cupin superfamily, classified as 7S (vicilins) and 11S (legumins) globulins and accounting for about 30% and 40% of the total seed proteins, respectively. Both share conserved sequence motifs containing similar amino-acid residues and a tertiary structure composed of anti-parallel β-sheets associated with α-helices and forming a β-barrel cavity where is located a binding site for hydrophobic ligands (L'Hocine & Boye, 2007). Gly m 5 is a 180 kDa trimeric glycoprotein, whose mature form is located in the vacuole as trimers composed of three isoforms, namely Gly m 5.0101 (α), Gly m 5.0201 (α'), and Gly m 5.0301/Gly m 5.0302 (β), all considered allergenic (Krishnan et al., 2009; Luthria et al., 2018). The α-subunit can also be called of Gly m Bd 60K (Yang et al., 2011). Gly m 6, a 360 kDa hexameric glycoprotein, is synthesized in seeds during embryogenesis and encoded by five non-allelic genes that produce the five allergenic isoforms of glycinin: G1 (Gly m 6.0101), G2 (Gly m 6.0201), G3 (Gly m 6.0301), G4 (Gly m 6.0401), and G5 (Gly m 6.0501). Each subunit consists of at least one acidic and one basic polypeptide chain linked by a disulfide bond (Holzhauser et al., 2009; Luthria et al., 2018).

According to Holzhauser et al. (2009), both proteins are considered minor allergens since 43% and 36% of soybean-allergic patients have specific IgE to Gly m 5 and Gly m 6, respectively. However, 86% of these patients had anaphylaxis, while only 36% showed mild subjective symptoms, suggesting that these allergens are potential diagnostic markers for severe allergic reactions to soybean (Holzhauser et al., 2009). Gly m 5 has also been associated with food-dependent exercise-induced anaphylaxis in a patient without previous allergy to soybean, whose challenge with tofu prior to exercise lead to the development of severe urticaria and face swelling (Adachi et al., 2009).

8.3.4   GLY M 7 AND GLY M 8

Gly m 7 is a biotinylated protein only found within the seed embryo, accounting for less than 0.01% of the total seed protein content. It plays an important role as a source of biotin during plant germination and in desiccation and stress tolerance in many plant seeds, allowing them to survive the dry storage phase. Its primary structure has also been linked to the late embryogenesis abundant (LEA) family of proteins, which have been recently described as allergens (Batista et al., 2007; Gagnon et al., 2010). It was reported to be IgE-reactive to approximately 73% of the sera of soybean-allergic patients, and that glycosylation could possibly be involved in its allergenicity. In terms of potency, Gly m 7 seems comparable to Ara h 1 (peanut), but more potent than Gly m 5. Gly m 7 is most likely to induce severe allergic responses, despite the small amount of this protein in soybean seeds and its rather recent identification, since its clinical significance is still difficult to evaluate (Riascos et al., 2016).

Gly m 8 are 2S albumins (28 kDa) that belong to the prolamin superfamily. Typically, these proteins are heterodimeric with two polypeptide chains held together by four disulfide bonds. Additionally, they show a conserved tri-dimensional structure, high stability to gastrointestinal digestion, and high resistance to thermal processing (Breiteneder & Radauer, 2004; Ebisawa et al., 2013). Gly m 8 is considered a major soybean allergen and a potential marker for severe reactions in soybean-allergic patients, both in children and adults (Ebisawa et al., 2013; Kattan & Sampson, 2015; Klemans et al., 2013).

8.3.5 OTHER SOYBEAN ALLERGENS

In addition to the above eight mentioned soybean allergens registered in the IUIS/WHO list of allergens, eight other proteins have demonstrated their capacity for triggering adverse reactions in sensitized individuals, which lead to their inclusion in the ALLERGOME database (ALLERGOME, 2020; WHO/IUIS, 2020).

Gly m 28 K is a glycoprotein from the 7S globulin fraction (vicilin) that belongs to the cupin superfamily, possessing high resistance to thermal denaturation and proteolysis. It was demonstrated to be highly homologous to the MP27/MP32 proteins of pumpkin seeds and carrot globulin-like protein (Tsuji et al., 2001). Gly m Bd 30 K, also known as P34, is a 34 kDa oil body-associated seed protein from the vacuoles of soybean. It has high sequence similarity with the thiol proteinases of the papain family, including the allergenic thiol proteinase Der p 1 from house dust mites (L'Hocine & Boye, 2007; Ogawa et al., 1993) and possesses a notable resistance to proteolysis in the gastrointestinal tract (Watanabe et al., 2018). Both Gly m 28 K and Gly m 30 K have been identified and characterized as major soybean allergens, the latter recognized by 65% of soybean-sensitized patients with atopic dermatitis (Hiemori et al., 2004; Ogawa et al., 1991; Ogawa et al., 1993; Tsuji et al., 2001).

Gly m TI (Kunitz trypsin inhibitor) belongs to one of the many families of proteinase inhibitors of the plant defense system, such as serine proteinases, thiol proteinases, or aspartic proteinases (Breiteneder & Radauer, 2004). This protein possesses an anti-parallel β-sheet protein containing two disulfide bridges, being highly resistant to chemical and thermal denaturation (Roychaudhuri et al., 2004). It is also considered an occupational aeroallergen since respiratory symptoms have been reported in bakers and workers of a soybean mill after the inhalation of soybean flour (L'Hocine & Boye, 2007).

Gly m 39kD (soybean lecithin) was also considered a source of allergenic proteins, with the identification of an IgE-binding protein termed P39. Soybean lecithin is derived from the oil fraction and is essentially composed of phospholipids, namely phosphatides, phytoglycolipids, triglycerides, phytosterols, tocopherols, and free fatty acids (Gu et al., 2001; Xiang et al., 2008). Its allergenicity can be associated with the development of respiratory symptoms, chronic diarrhea, and even anaphylaxis (Palm et al., 1999).

Gly m agglutinin (lectin), also known as Gly m Bd 50Kd, is a heat-labile carbohydrate-binding protein found in soybean seeds, being considered potentially allergenic (L'Hocine & Boye, 2007). It is a 50 kDa protein from the soybean hull, which was found to be homologous with the chlorophyll A-B binding protein precursors from tomato, spinach, and petunia. Its N-terminal amino acid sequence (first 17 amino acids) showed 51.7% of IgE-binding capacity in soybean-allergic patients, thus being considered as a major allergen (Codina et al., 2002). Gly m CPI (cystatin) is a cysteine proteinase inhibitor that was also found to be allergenic (Batista et al., 2007), as well as the soybean lipoxidase that was IgE-reactive with 50% of the sera from patients with soybean allergy, being classified as an aeroallergen (Baur et al., 1996).

8.4 IgE CROSS-REACTIVITY OF SOYBEAN ALLERGENS WITH OTHER SPECIES

Cross-reactivity among legumes is frequent because of their structurally homologous proteins and shared common epitopes. Thus, IgE-mediated soybean allergy can result from primary sensitization as well as from cross-reactivity with other members of the legume family. Extensive studies about this topic have been performed on different legume species. However, the clinical relevance of cross-reactivity among legumes is still unclear since the existing studies were merely based on soybean-sensitized subjects with unclear clinical reactivity (Bernhisel-Broadbent & Sampson, 1989; Eigenmann et al., 1996). A study developed by Bernhisel-Broadbent & Sampson (1989) revealed that, from 69 patients with legume-hypersensitivity, 43% were sensitized to soybean, 87% to peanut, 26% to pea, and 22% to green bean. Of these patients, 6.5% of the peanut-allergic children had a concomitant soybean allergy. In another study, the rate of patients simultaneously allergic to peanut and soybean was 15% (Skolnick et al., 2001). Later, in a study involving 140 peanut-allergic patients, 7% were found to be allergic to soybean as assessed by a combination of clinical history, serum-specific IgE testing, SPT, and OFC (Green et al., 2007). The highest rate was verified by Savage et al. (2010) in a study carried out with 122 soybean-allergic patients from the USA, from which 88% had a concomitant peanut allergy. However, despite the high degree of cross-sensitization evidenced by serum IgE-testing and SPT, the real clinical reactivity between soybean and peanut is in fact rare. When patients with positive SPT and serological testing towards soybean are submitted to DBPCFC, the actual clinical reactivity ranges from 1% to 6.5% (Bock & Atkins, 1989; L'Hocine & Boye, 2007).

At the molecular level, high homology and sequence identity have been reported between allergenic proteins of soybean and peanut. IgE-binding epitopes of Gly m 6.0101 and Gly m 6.0201 (subunit G1 and G2 of soybean glycinin) were revealed to be homologous to Ara h 3 (peanut), suggesting that their position might be related to a potential allergenic domain within the legume family (Beardslee et al., 2000; Xiang et al., 2002). Chruszcz et al. (2011) demonstrated a sequence identity of 51% between the Gly m 5 (soybean β-conglycinin) and Ara h 1 (peanut vicilin). However, the common sequences are not involved in IgE-binding, thus indicating a low frequency of cross-reactivity between these proteins (Verma et al., 2013). As suggested by Hurlburt et al. (2013), Ara h 3 shows 68% sequence identity and 84% sequence similarity with Gly m 4, the birch-pollen related protein from soybean.

Similarly, the well-characterized birch profilin Bet v 2 shares important IgE-binding epitopes with the Gly m 3 (soybean profilin) and an amino acid homology of 73% (Rihs et al., 1999). The sequence identity of Gly m 4 and Bet v 1 was also demonstrated by Berkner et al. (2009). The authors reported sequence similarities of 66–69% between Gly m 4 and Bet v 1, Mal d 1 and Pru av 1, showing significant cross-reactivity between these species. The high molecular homology between Bet v 1 and Gly m 4 seems to be well supported by the clinical data since 82 out of 138 birch pollen–sensitized patients (59%) reacted to soybean in a DBPCFC (Treudler et al., 2017). Contrarily, the homologies between Gly m 4 and Api g 1, and Gly m 4 and Dau c 1 (58% and 56%, respectively) did not lead to any clinical cross-reactivity. Moreover, Lup l 4 (PR-10 protein) from yellow lupine demonstrates to be highly homologous to Gly m 4 with sequence similarities of 74–84%. The allergenic protein α-conglutin from lupine revealed an extremely high sequence similarity with Gly m 6.0101. The homology was lower for the allergenic lupine protein β-conglutin and Gly m Bd 28 K (soybean vicilin), evidencing 26.1% and 58.8% of sequence identity and similarity, respectively (Guillamón et al., 2010). Despite the high molecular homology between soybean allergens and other allergenic proteins, their clinical relevance is still to be determined for most of the legume species.

Several authors have reported cross-reactivity between soybean and cow's milk proteins after the ingestion of soybean formulas commonly used as milk substitutes by cow's milk–allergic individuals (Villa et al., 2018). Bovine caseins share epitopes with Gly m Bd 30 K and Gly m Bd 28 K, the soybean thiol protease and vicilin, respectively (Candreva et al., 2016; Candreva et al., 2015; Smaldini et al., 2012). Likewise, the α-subunit of soybean β-conglycinin (Gly m 5.0101) showed important cross-reactive epitopes with bovine α-casein (Candreva et al., 2017; Curciarello et al., 2014), while Gly m 6.0401 shares sequential common epitopes with α-, β-, and κ-caseins involved in cross-reactions between soybean and cow's milk (Rozenfeld et al., 2002; Smaldini et al., 2012). Smaldini et al. (2012) used a murine food allergy model to test the potential cross-reactivity of cow's milk with soybean proteins. When the cow's milk–sensitized mice were challenged with soybean proteins by oral intake, they evidenced clear hypersensitivity reactions, suggesting that such immunochemical cross-reactivity might be clinically relevant.

8.5 STRATEGIES TO MITIGATE SOYBEAN ALLERGENICITY

Soybean proteins can be present in uncounted food products, not only due to their important technological characteristics, but also because of potential cross-contamination during product manufacture and as a result mislabeling, or, on the contrary, be included in precautionary labeling. Allergic individuals are thus forced to exclude all the products suspected of containing soybean from the diet in order to avoid the risk of triggering adverse reactions. To improve the quality of life of these individuals, several methods have been investigated to reduce the allergenicity of soybean proteins and produce hypoallergenic products. Several food processing technologies, including thermal or high-pressure treatments, fermentation, and enzymatic hydrolysis have demonstrated interesting effects on soybean allergenicity. The use of chemical modification, such as polysaccharide conjugation or genetic modification by gene silencing, has also proved to be effective. Tables 8.2 and 8.3 summarize studies regarding the application of processes to reduce the allergenicity of soybean products based on processing technologies and biotechnology, respectively.

8.5.1 EFFECT OF FOOD PROCESSING

Soybean can undertake several processing treatments to obtain different derived technological products such as soybean flour, soybean protein concentrates (SPC), soybean protein isolates (SPI), soybean hydrolysates, and fermented products. All these products may undergo structural modifications that can affect the structural properties of soybean proteins, which can result in changes in their allergenicity. The destruction of conformational epitopes or the occurrence of chemical reactions with proteins, fat, or sugars can reduce the availability of the protein to the immune system, while the formation of neoepitopes or the decrease in protein digestibility can lead to an increase of IgE-binding (Costa et al., 2020).

Heat treatment is one of the most commonly used processing technologies in food production, including boiling, autoclaving, or oven cooking. Heat treatments can lead to protein denaturation, significantly altering their structure or even destroying existing epitopes in native molecules. In fact, there are some studies reporting a reduction in the IgE-binding capacity of soybean proteins when submitted to treatments involving high temperatures (Burks et al., 1991; Ohishi et al., 1994; van Boxtel et al., 2008; van de Lagemaat et al., 2007). However, some authors have reported no effect or even an increase in protein immunoreactivity (Burks et al., 1992; Codina et al., 2002; Shibasaki et al., 1980), suggesting that thermal treatment may also create new allergenic epitopes, which are uncovered after the denaturation of the protein. Different heat treatments during different periods of time can also display dissimilar results as suggested by Wilson et al. (2008), who reported an enhanced immunoreactivity of Gly m Bd 30 K (P34) after 5 min at 100° C, but a significant reduction after 60 min. More recently, Villa et al. (2020) demonstrated a clear reduction of the immunoreactivity of the Gly m TI (soybean trypsin inhibitor) by the application of autoclaving and mild oven-cooking to sausages and cooked hams, respectively. Different conditions of heating, namely temperature and period of time, may lead to distinct protein immunoreactivity changes (Villa et al., 2020), as well as the intrinsic properties of the proteins and the physicochemical conditions of its environment, such as the food matrix (L'Hocine & Boye, 2007; Villa et al., 2020).

Fermentation and enzymatic hydrolysis can also play important roles in reducing soybean allergenicity. The production of small peptides by proteases from different bacteria strains or fungi, or by the use of other proteolytic enzymes, was extensively studied as an efficient way to obtain hypoallergenic soybean. The type of bacteria used in fermentation can be fundamental in determining which proteins are hydrolyzed and whether the formed peptides retain allergenic potential. Generally, the use of proteases from *Bacillus*, *Aspergyllus*, *Enterococcus*, and *Lactobacillus* species

TABLE 8.2
Reported Studies on the Reduction of Soybean Allergenicity by Food Processing Technologies and Correlation with Digestibility

Allergen	Treatment	Allergenicity/Immunoreactivity	Digestibility	Reference
Soybean proteins	Heat treatment (80° C, 30 min)	Increased IgE-binding in 2S fraction; Reduced IgE-binding in 11S and 7S fractions	–	Shibasaki et al. (1980)
Soybean proteins	Heat treatment (80° C, 100° C, 120° C, 60 min)	Reduced IgE-binding	Reduced IgE-binding (pepsin, trypsin, chymotrypsin, intestinal mucosal peptidases)	Burks et al. (1991)
Soybean proteins	Heat treatment (100° C)	Unaffected IgE-binding	–	Burks et al. (1992)
Soybean proteins	Heat treatment (< 66° C)	Reduced immunoreactivity (0.1%)	–	Ohishi et al. (1994)
Gly m Bd 30 K	Proteolysis (proteases from *Rhizopus niveus, B. subtilis, Aspergyllus oryzae, A. melleus*) Heat treatment (autoclaving)	Reduced allergenicity	–	Yamanishi et al. (1995); Yamanishi et al (1996)
Gly m Bd 30 K	Polysaccharide conjugation or TGase cross-linking	Unaffected allergenicity (TGase) Reduced allergenicity (Galactomannan conjugation)	Unaffected allergenicity (chymotrypsin digestion)	Babiker et al. (1998)
Soybean hull proteins	Heat treatment (37° C, 55° C, 80° C)	Increased allergenicity	–	Codina et al. (1998)
Glycinin; β-conglycinin; Gly m Bd 30 K	Enzymatic hydrolysis (*Bacillus* protease), Heat treatment (70° C)	Reduced immunoreactivity (β-conglycinin and Gly m Bd 30 K) Glycinin (unaffected immunoreactivity)	β-conglycinin (More pronounced digestion at 70° C) Glycinin (more pronounced digestion at low pH)	Tsumura (2009); Tsumura et al. (1999)
Soybean proteins	Twin-screw extrusion	Reduced immunoreactivity	–	Saitoh et al. (2000)
Gly m Bd 30 K	Polysaccharide conjugation (yeast expression system)	Reduced IgE (70%) and IgG (125%) production in mice	–	Arita et al. (2001)
Soybean proteins	Fermentation (*L. lactis* subsp. *lactis, A. oryzae,* and *B. subtilis*) Heat treatment (steaming)	Reduced immunoreactivity	–	Lee et al. (2004)
Soybean proteins	Polysaccharide conjugation (chitosan)	Reduced allergenicity (Gly m Bd 30 K)		Usui et al. (2004)
Soybean proteins	Enzymatic proteolysis (alcalase, corolase, neutrase) HP (100, 200, 300 MPa)	Reduced immunoreactivity	Increased hydrolysis at 200 and 300 MPa	Peñas et al. (2006)

(continued)

TABLE 8.2 (CONTINUED)

Reported Studies on the Reduction of Soybean Allergenicity by Food Processing Technologies and Correlation with Digestibility

Allergen	Treatment	Allergenicity/Immunoreactivity	Digestibility	Reference
Glycinin	Enzymatic hydrolysis (pepsin and chymotrypsin)	Reduced allergenicity	–	Lee et al. (2007)
Glycinin	Ionic strength and pH	High IgG-binding (pH 2.2 and 7.2) Low IgG-binding (pH 3-6)	–	L'Hocine et al. (2007)
Glycinin and β-conglycinin	Heat treatment (95° C, 1 h), Glycation (fructooligosaccharides)	Reduced immunoreactivity	–	van de Lagemaat et al. (2007)
Soybean proteins	Fermentation (B. subtilis, L. plantarum, B. lactis, S. cereviseae)	Reduced IgE-binding	–	Frias et al. (2008); Song et al. (2008)
Glycinin	Heat treatment (100° C, 10 min) Pepsin digestion	Decreased IgE-binding	Slightly decreased with heat treatment Reduced IgE-binding after digestion	van Boxtel et al. (2008)
Gly m Bd 30 K	Heat treatment (100° C)	Increased immunoreactivity (5 min) Reduced immunoreactivity (60 min)	–	Wilson et al. (2008)
Glycinin and β-conglycinin	Pulsed UV light (4 min)	Reduced IgE-binding (44%)	–	Yang et al. (2010)
Glycinin and β-conglycinin	Enzymatic hydrolysis (pepsin and trypsin)	Immunoreactivity positively related with stability of proteins to digestion	Glycinin (more pronounced digestion with pepsin and trypsin) β-conglycinin (more pronounced digestion with trypsin)	Zhao et al. 2010
Soybean proteins	Controlled pressure drop (6 bar, 3 min)	Reduced immunoreactivity	–	Cuadrado et al. (2011); Takács et al. (2014)
Soybean proteins	HHP (300 MPa, 15 min, 40° C)	Reduced immunoreactivity	–	Peñas et al. (2011)
Glycinin and β-conglycinin	Enzymatic hydrolisys (proteases from B. subtilis)	Reduced immunoreactivity	–	Han et al. 2012)
Gly m KTI	Heat treatment (100° C), HP (30 min)	Reduced immunoreactivity (higher with high pressure)	Higher digestibility with high pressure	Shen et al. (2013)
Glycinin and β-conglycinin	Enzymatic hydrolysis (alcalase and trypsin)	Reduced immunoreactivity (higher with alcalase)	–	Wang et al. (2014)

(continued)

TABLE 8.2 (CONTINUED)

Reported Studies on the Reduction of Soybean Allergenicity by Food Processing Technologies and Correlation with Digestibility

Allergen	Treatment	Allergenicity/Immunoreactivity	Digestibility	Reference
Glycinin and β-conglycinin	Glycation (fructooligosaccharides)	Reduced immunoreactivity (78% and 82%)	Different rates of digestion along the digestive tract	Wang et al. (2015a)
Soybean proteins	Fermentation (Virginiamycin-added fungal fermentation)	Reduced immunoreactivity	–	Chen et al. (2015)
Soybean proteins	Enzymatic hydrolysis (alcalase, pepsin, and papain)	Complete degradation of major soybean allergens	–	Meinlschmidt et al. (2016a)
β-conglycinin	UV light, (CAPP), and gamma-irradiation	Reduced immunoreactivity, more pronounced with CAPP	–	Meinlschmidt et al. (2016b)
β-conglycinin	Fermentation (L. helveticus)	Reduced IgE-binding (100%)	–	Meinlschmidt et al. (2016c)
Glycinin and β-conglycinin	Fermentation (Enterococcus faecalis)	Reduced immunoreactivity	Complete hydrolysis	Biscola et al. (2017)
Glycinin	Glycation (xylose, 55° C)	Reduced immunoreactivity (18%)	–	Bu et al. (2017)
Soybean proteins	Fermentation (moromi), heat treatment (85° C), and filtration	Reduced allergenicity	–	Magishi et al. (2017)
β-conglycinin	HP (400 MPa, 50° C, 15 min), Enzymatic hydrolysis	Reduced immunoreactivity	–	Meinlschmidt et al. (2017)
Glycinin	Microbial transglutaminase (TGase) cross-linking Heat treatment (100° C)	Increased immunoreactivity	Decreased	Yang et al. (2017)
β-conglycinin	Enzymatic hydrolysis (Neutrase and Flavourzyme)	Reduced immunoreactivity (64.46%)	–	Huang et al. (2018)
Soybean proteins	HPP (300 MPa, 15 min)	Reduced immunoreactivity (45.5%)	–	Li et al. (2018); Li et al. (2012)
Soybean proteins	Fermentation (L. casei, yeast, and B. subtilis)	Reduced allergenicity	–	Yang et al. (2018b)
β-conglycinin	Enzymatic deglycosylation (PNGase F) Ultrasound (40 kHz, 300 W)	Reduced IgE-binding (enzymatic deglycosylation) Increased IgE-binding (with ultra-sound pre-treatment)	–	Yang et al. (2018a)
Soybean proteins in soymilk	Fermentation (L. brevis and L. sp.)	Reduced allergenicity	–	Xia et al. (2019)
Gly m TI	Heat treatment (oven cooking and autoclaving)	Reduced immunoreactivity	–	Villa et al. (2020)
Glycinin and β-conglycinin	Fermentation (24 h) Enzymatic hydrolysis (alcalase, 15 min)	Reduced immunoreactivity	–	Yang et al. (2020)

TABLE 8.3

Reported Studies on the Genetic Modification of Soybean for Reducing Its Allergenic Potential

Allergen	Germplasm	Genetic Modification	Effects on Soybean Proteins	Reference
Gly m 5	Mo-shi-dou Gong 503	Germplasm screening	Reduced allergen level	Kitamura & Kaizuma (1981)
Gly m 5.0101	Kari-kei 434	Germplasm screening	Reduced allergen level	Takahashi et al. (1994)
Gly m Bd 30K, Gly m Bd 28K, Gly m 5.0101	Tohuku 124	Germplasm screening	Removal rates of Gly m Bd 30K: 99.8%, α-subunit of Gly m 5: 100%, Gly m Bd 28K: 100%	Samoto et al. (1997); Samoto et al. (1996)
Gly m 5	Wild Soybean Line QT2	Germplasm screening	Reduced allergen level	Hajika et al. (1998)
Gly m Bd 30K	WT soybean	Single-site amino acid substitution	IgE-binding reduction	Helm et al. (2000)
Gly m Bd 28K	Fukuyutaka, Enrei, Tachinagaha, Tamahomare, Murayutaka	Cultivar screening	Reduced allergen level	Kanegae et al. (2001)
Gly m Bd 30 K	Transgenic line	Gene silencing	Suppression of Gly m Bd 30 K-related peptides	Herman et al. (2003)
Gly m 5 and Gly m Bd 28K	Yumeminori	Gamma-ray irradiation	Reduced allergen level	Takahashi et al. (2004)
Gly m Bd 30 K	PI 567476, PI 603570A	Germplasm screening	Reduced allergen level	Joseph et al. (2006)
Gly m Bd 28K and Gly m Bd 30K	Various genotypes from Chinese core collection	Germplasm screening	Reduced allergen level	Zhang et al. (2006)
Gly m 5, Gly m 6, Gly m KTI	Mutant lines of VLSoy-2	Gamma-ray irradiation	Reduced allergen level	Manjaya et al. (2007)
Gly m 5.0101, Gly m 5.0102	Nagomimaru	Backcrossing Yumeminori into Tachinagaha	Reduced allergen level	Hajika et al. (2009)
Gly m 5 and Gly m KTI	Mutant lines of Korean landraces and cultivars	Gamma-ray irradiation	Reduced allergen level	Lee et al. (2011)
Gly m Bd 30 K	PI 603570A and PI567476	Insertion mutation in promoter region	Down-regulation of P34 translation	Koo et al. (2013)
Gly m Bd 30 K	PI 567476 × Hwanggeum	Selection of segregants	Reduced allergen level	Jeong et al. (2013)
Gly m Bd 30 K	Transgenic line	Gene silencing	Reduced allergen level	Liu et al. (2013b)
Gly m 5.0101	R iB × D ongnong47	Selection of segregants	Reduced allergen level	Song et al. (2014)
Gly m KTI, Gly m agglutinin and Gly m Bd 30 K	*Triple Null*	Selection of segregants	Reduced allergen level	Schmidt et al. (2015)
Gly m 4 and Gly m Bd 30 K	Non-GM and genetically modified (glyphosate tolerant) soybean	GM-soybean	Unaffected allergenicity	Tsai et al. (2017b)
Gly m Bd 30 K	PI 567476 × OAC Erin lines	Selection of segregants	Reduced immunoreactivity	Watanabe et al. (2017)

contribute to reducing soybean protein allergenicity, such as the case of Gly m Bd 30 K, Gly m 5, and Gly m 6 (Biscola et al., 2017; Frias et al., 2008; Han et al., 2012; Lee et al., 2004; Meinlschmidt et al., 2016c; Song et al., 2008; Tsumura, 2009; Tsumura et al., 1999; Xia et al., 2019; Yamanishi et al., 1995; Yamanishi et al., 1996; Yang et al., 2018b). The extent of the hydrolysis can affect the residual allergenicity of each protein, which can be affected by the combination of other treatments or the addition of specific proteases or substances. Some studies reported that fermentation coupled to enzymatic hydrolysis with alcalase (Yang et al., 2020), heat treatment (Lee et al., 2004; Magishi et al., 2017), or the addition of virginiamycin (Chen et al., 2015) may improve the proteolysis of soybean allergens. Enzymatic hydrolysis using specific proteases, such as pepsin, trypsin, papain, or alcalase (Huang et al., 2018; Lee et al., 2007; Meinlschmidt et al., 2016a; Wang et al., 2014; Zhao et al., 2010) and in combination with high-pressure (Meinlschmidt et al., 2017; Peñas et al., 2006) and heat treatments (Tsumura et al., 1999; van Boxtel et al., 2008) has also demonstrated to be effective in reducing soybean immunoreactivity. For example, Tsumura et al. (1999) were able to selectively hydrolyze Gly m Bd 30 K and Gly m 5, but not Gly m 6 with a *Bacillus* protease (Proleather FG-F), based on different temperatures of denaturation of each protein at neutral pH. Therefore, the IgE-binding of both Gly m Bd 30 K and Gly m 5 was considerably reduced in combination with a more rapid digestion at 70° C. Proteolysis seems to be more effective in the destruction of conformational epitopes by disrupting the complex protein 3D-structure, but linear epitopes may remain intact even after the process (Lee et al., 2007). In this context, it is important to know the structural character-istics of the allergen to combine the most efficient strategy to decrease its allergenicity.

Chemical modification of proteins by means of glycation (Maillard reaction/conjugation with carbohydrates) is described as a powerful alternative to reduce soybean allergenicity. This process often leads to the formation of protein aggregates that may be responsible for masking the existing allergenic epitopes, which become unavailable to the immune system (L'Hocine & Boye, 2007). The first successful study was performed by Babiker et al. (1998), who showed that conjugating soy-bean protein with galactomannan through spontaneous Maillard reactions was effective in reducing Gly m Bd 30 K allergenicity. Usui et al. (2004) demonstrated that chitosan was more effective than galactomannan in decreasing Gly m Bd 30 K allergenicity. Posteriorly, glycation with fructooligo-saccharides (van de Lagemaat et al., 2007; Wang et al., 2015a) and xylose (Bu et al., 2017) was also revealed to be effective in reducing the IgE-binding capacity of soybean proteins. In some cases, heat treatment seemed to improve this effect as studied by Bu et al. (2017) and van de Lagemaat et al. (2007), who observed an immunoreactivity reduction of soybean glycated products processed at high temperatures. In opposition, transglutaminase (TG) cross-linking seems to have no effect or even increase soybean immunoreactivity, as demonstrated by Babiker et al. (1998) and Yang et al. (2017).

More recently, some novel food processing technologies have been applied in this field, namely pulsed UV light (Yang et al., 2010), controlled pressure drop (DIC) (Cuadrado et al., 2011; Takács et al., 2014), high hydrostatic pressure (Li et al., 2018; Li et al., 2012; Peñas et al., 2011; Shen et al., 2013), ultrasound (Yang et al., 2018a), and UV light combined with cold-atmospheric pressure plasma (CAPP) and gamma-irradiation (Meinlschmidt et al., 2016b). For example, Cuadrado et al. (2011) showed an almost complete abolishment of the IgE-binding capacity of soybean proteins after treatment with DIC at 6 bar for 3 min. An interesting approach applying a pre-treatment by ultrasound (40 kHz, 300 W) followed by enzymatic deglycosylation with PNGase F of soybean 7S globulins showed that the individual process of deglycosylation could reduce the IgE-binding capacity of proteins. However, when applying the ultrasound pre-treatment, the IgE-binding capac-ity of soybean 7S globulins increased significantly (Cuadrado et al., 2011).

It seems that promising results have been obtained by the application of food technologies to produce hypoallergenic soybean products. However, the majority of these studies do not perform *in vivo* tests (skin prick tests or oral challenges) to ensure that these products do not elicit positive responses in sensitized individuals. It is important to evaluate these aspects in order to guarantee the safety of consumers and avoid unintended adverse reactions.

8.5.2 DIGESTIBILITY

During the digestion process, most of the proteins are hydrolyzed into single amino-acid or small peptides, facilitating their absorption by the intestinal mucosa. However, some allergenic proteins can resist proteolytic digestion of the gastrointestinal tract, triggering the sensitization of the mucosal immune system after their absorption since their IgE-binding epitopes still remain intact (Moreno, 2007). In this context, the resistance to the digestion process is strictly connected with the occurrence of adverse reactions in allergic patients (Bøgh & Madsen, 2016).

Some studies have been performed about the digestibility of soybean proteins associated with their allergenicity (Amigo-Benavent et al., 2011; De Angelis et al., 2017; Lee et al., 2007; Li et al., 2013; Schafer et al., 2016; Shim et al., 2010; Wang et al., 2017a). De Angelis et al. (2017) performed *in vitro* simulated gastrointestinal digestion experiments directly on ground soybean seeds, showing that Gly m 5 and Gly m TI are highly resistant to the gastric and duodenal phases, while Gly m 6 is less resistant. The resistant peptides from Gly m 5 and Gly m 6 demonstrated to retain their allergenic epitopes, being potentially able to trigger an immunoreaction upon crossing the intestinal mucosa. Amigo-Benavent et al. (2011) investigated the resistance to gastrointestinal digestion of Gly m 5 and its deglycosylated form. The authors showed that Gly m 5 partially survives the digestive process, while its deglycosylated form was completely removed, supporting the idea that structural glycans affect the rate of proteolysis and the amount of intact protein available to reach the intestinal tract, which will develop the allergenic response. Digestibility of GM soybean is one of the parameters that should be carefully considered in the risk assessment of newly expressed proteins (König et al., 2004). Several authors have performed this evaluation in the last years (Harrison et al., 1996; Kim et al., 2006; Okunuki et al., 2002; Schafer et al., 2016; Shim et al., 2010), suggesting that GM proteins are rapidly digested by the gastrointestinal tract, thus not posing additional allergenic risk comparing to their non-GM counterparts.

The evaluation of the digestibility of soybean allergens after the application of a food processing technology with the aim of reducing allergenicity is also extremely important. Table 8.2 summarizes the reports about the effects of food processing on soybean allergenicity combined with digestibility assays. As an example, Wang et al. (2015a) studied the effect of glycation on Gly m 5 and Gly m 6 followed by simulated gastrointestinal digestion. The results showed the loss of the immunoreactivity of glycated proteins along the digestive tract, with the fastest rate in the duodenum and middle jejunum. Moreover, the reduction rate of immunoreactive Gly m 6 was a little higher than Gly m 5, which indicates that glycated Gly m 6 was more easily degraded than glycated Gly m 5 (Wang et al., 2015a).

The clear objective of a digestion model is to simulate food degradation as closely as possible to the human physiological process. However, most of the studies about soybean digestibility are performed under very different experimental conditions, making their results difficult or even impossible to compare. Some aspects are important to consider: the nature and the processing of the raw material; the composition of the digestive fluids, pH, or buffers; the ratio enzyme-substrate; the number of steps and the incubation time during the gastrointestinal digestion; as well as the mechanical aspects associated to the process (Picariello et al., 2013). Recently, a harmonized *in vitro* digestion method for food analysis was proposed by Brodkorb et al. (2019) and Mulet-Cabero et al. (2020) in the scope of INFOGEST action, aiming at producing more comparable data among researchers and laboratories.

8.5.3 GENETIC MODIFICATION

Recently, breeding and mutagenesis have been considered successful technologies to reduce soybean allergenicity based on the identification of various soybean mutant genes, which control the production of enzymes, allergenic proteins, and storage proteins (L'Hocine & Boye, 2007). These strategies aim at lowering or removing the content of specific allergy-related proteins by

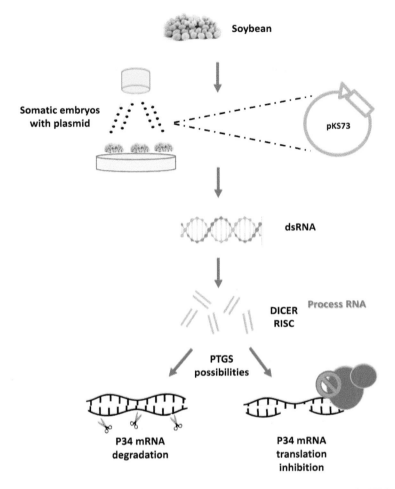

FIGURE 8.2 Schematic representation of a co-suppression technology used to lessen the P34 protein expression in soybean (gene knockout). Briefly, somatic embryos were transformed with plasmids containing the complete P34 open-reading-frame, in sense, under the control of a Gly m 5 promoter. The formed dsRNA activates the PTGS, which involves the processing of dsRNA by proteins such as RISC and DICER, leading to P34 degradation or stopping the P34 mRNA translation. Legend: dsRNA = double-strand RNA; PTGS = post-transcriptional gene silencing; RISC = RNA-induced silencing complex. Source: Adapted from Shamoon et al. (2016).

screening germplasm lines or silencing the native genes that encode allergenic proteins by genetic transformation (Shamoon et al., 2016). Table 8.3 summarizes the studies about the use of genetic modification as a successful approach to reduce soybean allergenicity. An interesting co-suppression technology developed by Herman et al. (2003) and represented in Figure 8.2 was applied to lessen the expression of Gly m Bd 30K in soybean cotyledons, being completely removed from transgenic somatic embryos. Immunoblotting using sera from six soybean allergic patients showed no significant differences between transgenic and wild-type soybeans, though P34 was absent in transgenic lines. Similarly, Helm et al. (2000) have reported a single-site amino-acid substitution of the IgE-binding epitopes of P34 with alanine, resulting in the reduction of the IgE-binding activity. Low-P34 lines selected from a cross-population (PI 567476 × OAC Erin lines) showed reduced concentrations (around 50–70%) of this protein as identified with a polyclonal antibody and compared to P34-containing controls (Watanabe et al., 2017). Despite the reported interesting findings, it remains elusive whether the low expression of Gly m Bd 30 K would be tolerated by soybean-sensitized patients. Takahashi et al. (2004) developed a Gly m 5 and Gly m 6 deficient soybean variety, QF2, by crossing a Gly m 5 deficient natural mutant with a Gly m 6 absent inbred line, QY2 and EnB1. A marker-assisted background selection (MABS) was used in backcross breeding to incorporate cgy-2, a null phenotype version of the gene encoding the α-subunit of β-conglycinin (Gly m 5.0101), from the donor line RiB into the genetic background of the Chinese cultivar Dongnong47 (DN47), a popular high-oil superfine seed soybean cultivar from China. All of the advanced backcrossing breeding lines lacked the referred soybean allergen (Song et al., 2014). Although it seems possible to reduce or remove some of the known soybean allergens with genetic modification, it remains unclear their real hypoallergenization. More studies applying sera from allergic individuals for IgE-binding evaluation, as well as clinical tests, should be performed in order to assess the allergenic potential of such transgenic lines. Moreover, the structural composition and physiological development of the seeds could be affected, having negative consequences on the functional and nutritional characteristics of soybean (L'Hocine & Boye, 2007).

8.6   TRACEABILITY OF SOYBEAN ALLERGENS IN PROCESSED FOODS

Currently, several regulations have been implemented regarding the labeling of food products containing allergenic ingredients with the aim of protecting the sensitized/allergic individuals. In the European Union, the Directive 2007/68/EC and Regulation (EU) No. 1169/2011 clearly state the mandatory labeling of 14 groups of potentially allergenic ingredients in processed foods, including soybean, regardless of their amount. In order to ensure compliance with food label regulations and minimize the risks for consumers from exposure to undeclared allergens in the food supply, the development of strategies for detecting allergenic foods is extremely important. Food allergens might be present at trace amounts due to possible cross-contamination during product manufacture and, in some cases, can be hidden by the food matrix. Therefore, analytical methodologies should have high sensitivity/specificity. Several methods have been developed and employed in soybean traceability, and have been based on the direct detection of the allergenic protein or the indirect identification of a DNA sequence from the target species (Costa et al., 2017b). In Table 8.4, the most recent and innovative methods (2010–2020) for the specific detection of soybean in processed foods are summarized.

Protein-based methods mostly rely on immunochemical techniques using the recognition of the antigen by a specific antibody, such as lateral flow devices (LFD), ELISA (enzyme-linked immunosorbent assay), immunosensors, and other relevant immunoassays. Alternatively, mass spectrometry (MS) platforms have also been applied to detect allergenic proteins. ELISA is one of the most common methods in soybean detection because it provides quantitative results with high sensitivity, precision, simple handling, and good potential for standardization (Costa et al., 2017b; L'Hocine & Boye, 2007). Several formats have been developed aiming at detecting soybean proteins in processed foodstuffs, including indirect, sandwich, and competitive ELISA able to reach sensitivities ranging from 10 pg/mL to 20 ng/mL (Geng et al., 2015; Hei et al., 2012; Liu et al., 2013a; Liu et al., 2012; Ma et al., 2010; Morishita et al., 2014; Segura-Gil et al., 2018; Ueberham et al., 2019; Xi & Shi, 2016; Zhang et al., 2006). However, in some cases, ELISA can be prone to false-positive and negative results due to food matrix interferences and food processing, respectively (Costa et al., 2017b). In this context, MS-based methodologies are useful alternatives because they are independent of biological interaction between the antibody and the target allergen, avoiding cross-reactivity problems often related to immunoassays and allowing the unequivocal identification of the target. However, the selection of proper markers in both raw and processed matrices is an important step that requires additional attention since it is known that processing can affect differently the structure of proteins (Verhoeckx et al., 2015). In recent years, numerous methods using MS coupled to liquid chromatography (LC) have been successfully employed for soybean detection in processed

TABLE 8.4

Reported Analytical Methods Applied to Detect Soybean Allergens in Food Products

Method	Matrix	Target Allergen	Limit of Detection	References
Immunochemical Assays				
Competitive ELISA	Soybean and soybean products	Gly m 6	0.3 ng/mL	Ma et al. (2010)
Sandwich ELISA	Soybean products	Gly m 5	1.63 ng/mL	Hei et al. (2012)
Competitive ELISA	Soybean products	Gly m 5	0.65 ng/mL	Liu et al. (2012)
Indirect ELISA	Soybean products	Gly m Bd 28K	NR	Liu et al. (2013a)
Sandwich ELISA	Fermented soybean products	Gly m Bd 30K	10 μg/g	Morishita et al. (2014)
Sandwich ELISA	Soybean varieties	Gly m 4	2.1 ng/mL	Geng et al. (2015)
Lateral Flow Colloidal Gold Immunoassay Strip	Powdered milk	Gly m 6	0.69 mg/kg	Wang et al. (2015b)
Indirect competitive ELISA	Soybean products	Gly m Bd 28K	0.235 μg/L	Xi & Shi (2016)
Direct competitive ELISA	Soybean samples	Gly m 5	20 ng/mL	Zhang et al. (2016)
LFD-AuNPs	Bovine milk	Soy proteins	1.75%	Gautam et al. (2017)
Lateral flow colloidal gold immunoassay	Powdered milk	Gly m 5	1.66 mg/kg	Wang et al. (2017c)
Indirect competitive and sandwich ELISA	UHT milk, bread, sausages, cookies	Soybean protein isolate Gly m 6	SPI in sausages: 0.005% SPI in bread: 0.05% Soy drink in UHT milk: 0.05% Soy seeds in cookies: 0.005%	Segura-Gil et al. (2018)
Sandwich ELISA	Muffins, sausages, cookies	Gly m 8	LOD: 10 pg/mL LOQ: 65 pg/mL	Ueberham et al. (2019)
SERS-based lateral flow immunoassay	Purified protein	Gly m 5	1 μg/mL	Xi & Yu (2020)
MS Platforms				
2DLC-UV/MS	Commercial soybean lines	Gly m 4	12.5 μg/mL	Julka et al. (2012)
UHPLC–MS/MS	Chocolate, ice cream, processed cookies, and sauce	Gly m 5, Gly m 6, Gly m 8	5 mg/kg	Planque et al. (2016)
HPLC-MS/MS	Sausages	Gly m 6	4 mg/kg	Hoffmann et al. (2017)
LC-Q-TOF-MS/MS	Poultry meat products	Gly m 5	NR	Montowska & Fornal (2018)
LC-MS/MS	Cookies	Soybean ingredients	10 mg/kg	Chen et al. (2019)
LC-MS/MS	Breads and corn flours	Soybean	NR	Croote et al. (2019)
LC-MS/MS	Soybean seeds	Gly m 5.0101	0.48 ng/mL	Jia et al. (2019)
UHPLC–MS	Sausages	Gly m 6.0101 Gly m 5.0401 Gly m 5	0.45%	Montowska et al. (2019)
LC-MS/MS	Bread, cookie, fried fish, and frozen pasta	Gly m KTI (4 peptides)	NR	Ogura et al. (2019)
DNA-Based Methods				

(continued)

TABLE 8.4 (CONTINUED)

Reported Analytical Methods Applied to Detect Soybean Allergens in Food Products

Method	Matrix	Target Allergen	Limit of Detection	References
End-point PCR	Meat, fish, and bakery processed foods	Soybean lectin DNA	End-point PCR: 0.0625%	Espiñeira et al. (2010)
Real-time PCR with TaqMan probe			Real-time PCR: 0.05%	
End-point PCR	Meat products	Soybean lectin DNA	15 g/kg	Soares et al. (2010)
DNA microarray-DVD	Jam and frozen ready meal	Soybean DNA	1 µg/g	Tortajada-Genaro et al. (2012)
Real-time PCR with EvaGreen dye	Sausages	Soybean lectin DNA	0.001%	Safdar & Abasıyanık (2013)
Tetraplex PCR	Sausages	Soybean lectin DNA	0.01%	Safdar et al. (2014)
RPA-ELISA	Cookies, chocolate, soup	Soybean lectin DNA	RPA: 2.01 µg/g	Santiago-Felipe et al. (2014)
PCR-ELISA			PCR: 1.47 µg/g	
Real-time PCR with TaqMan probe	Meat products	Soybean lectin DNA	LOD: 0.0125% LOQ: 0.0125%	Soares et al. (2014)
Real-time PCR with TaqMan probe	Meat products	Soybean lectin DNA	0.1 g/kg	Šmíd et al. (2015)
Decaplex PCR	Bakery products, beverages, noodles, chocolate, among others	Gly m Bd 28K gene of soybean	0.005%	Cheng et al. (2016)
Real-time PCR with TaqMan probe	Cooked ham and mortadella	Soybean lectin DNA	LOD: 10 mg/kg LOQ: 10 mg/kg	Costa et al. (2017a)
Real-time PCR with TaqMan probe	Sausages	Soybean lectin DNA	5-10 mg/kg	Waiblinger et al. (2017)
Real-time PCR with TaqMan probe	Rice cookies, Hollandaise sauce powder, and sausage	Mitochondrial soybean DNA	1 ppm	Ladenburger et al. (2019)
ddPCR	Meat products, flour, milk, and fatty creams	Soybean DNA	LOD: 0.16 mg/kg LOQ: 0.60 mg/kg	Mayer et al. (2019)
Other Methods				
Optical thin-film biosensor chips	Chocolate chips, dark chocolate, fried mud carps with fermented soybean	Soybean lectin DNA	NR	Wang et al. (2011)
Turbidimetric method	Powdered milk	SPI	LOD: 211 µg/mg LOQ: 639 µg/mg	Scholl et al. (2014)
Capillary electrophoresis	Canned tuna samples	Soybean lectin DNA	NR	Rodríguez-Ramírez et al. (2015)
Label-free biochip	Cream cake, cereal bar, chocolate biscuits, and pineapple cake	Gly m Bd 30K	1 mg/L	Wang et al. (2017b)
LAMP-microfluidic chip	Biscuits and candy	Gly m Bd 28K	0.4 ng/µL	Yuan et al. (2018)
HCR and AuNPs	Biscuits and sweets	Soybean DNA	0.5 nM	Yuan et al. (2019a)
GO-HCR-FRET	Soybean	Soybean DNA	1 nM	Yuan et al. (2019b)

AuNPs = gold nanoparticles; dd = digital droplet; DVD = digital versatile disc; ELISA = enzyme-linked immunosorbent assay; GO-HCR-FRET = graphene-oxide hybridization chain reaction fluorescence resonance energy transfer; HPLC = high-performance liquid chromatography; LAMP = loop-mediated isothermal amplification; LFD = lateral flow device; MS = mass spectrometry; PCR = polymerase chain reaction; Q-TOF = quadrupole time-of-flight; RPA = recombinase polymerase amplification; SERS = surface enhanced Raman spectroscopy; UHPLC = ultra-HPLC.

foodstuffs, reaching sensitivities down to 0.48–12.5 µg/mL (Chen et al., 2015; Croote et al., 2019; Hoffmann et al., 2017; Jia et al., 2019; Julka et al., 2012; Montowska & Fornal, 2018; Montowska et al., 2019; Ogura et al., 2019; Planque et al., 2016). Comparing the lowest sensitivity obtained by a sandwich ELISA (Ueberham et al., 2019) with an LC–MS/MS method (Jia et al., 2019), the latter proved to be more sensitive for soybean detection.

LFD and biosensors are also considered useful approaches for food allergen detection, considering their low cost, portability, simple operation, high-speed execution, and automation feasibility. Several studies have reported the use of LFD (Allgöwer et al., 2020; Gautam et al., 2017; Wang et al., 2015b; Wang et al., 2017c; Xi & Yu, 2020) and biosensors (Wang et al., 2011; Wang et al., 2017b; Yuan et al., 2019b; Yuan et al., 2018) to detect soybean allergens in processed foodstuffs. The first relies on the direct recognition of the antigen by a specific antibody used to confirm the presence of the target, providing qualitative or semi-quantitative results that can be interpreted visually (Costa et al., 2017b; Villa et al., 2018). Biosensors provide quantitative results with high sensitivity by the measurement of a signal produced by the biological interaction between a receptor (antibody or probes) and a target molecule (protein or DNA) (Prado et al., 2016).

DNA-based techniques usually rely on the amplification of gene sequences encoding allergenic proteins or other specific regions from the genome of the target species. They are mainly based on the polymerase chain reaction (PCR), generally real-time PCR, which has proven to be very useful for food allergen detection and quantification, including soybean. PCR-based methods are independent of possible biological effects associated with antibody production, they are highly specific and sensitive, taking advantage of the relatively high thermal stability of DNA molecules, compared with proteins (Costa et al., 2017b; Prado et al., 2016). Despite the referred advantages and wide applicability to food allergen analysis, these approaches are not well accepted by some researchers because of the indirect identification of the allergenic food (Holzhauser, 2018). Recent studies about soybean detection reported sensitivities ranging from 0.16 mg/kg to 15 g/kg (Table 8.4) (Chen et al., 2019; Cheng et al., 2016; Costa et al., 2017a; Espiñeira et al., 2010; Ladenburger et al., 2019; Mayer et al., 2019; Safdar & Abasıyanık, 2013; Safdar et al., 2014; Santiago-Felipe et al., 2014; Šmíd et al., 2015; Soares et al., 2014; Soares et al., 2010; Waiblinger et al., 2017).

8.7   CONCLUSIONS

Soybean allergy has attracted worldwide attention due to its constant prevalence increase and to the severity/frequency of the reported allergic reactions. Soybean allergic individuals need to adopt an elimination diet to avoid adverse reactions, which can attain systemic and life-threatening episodes, such as anaphylaxis. Soybean proteins have high homology with other legumes' proteins, mainly from peanuts, but also with milk proteins, which can induce cross-reactivity phenomena and increase the risk of developing allergic reactions in sensitized individuals. Conversely, the clinical relevance of cross-reactivity between legumes is still unclear, thus demanding more clinical tests, such as double-blind placebo-controlled food challenges. Several soybean allergens have been identified and characterized. However, not all of them have been clinically confirmed and their allergenic epitopes have not yet been fully elucidated. The advanced knowledge of the physicochemical properties of soybean allergens and how they may impact their allergenicity is imperative for the development of technological strategies able to mitigate soybean allergenicity. Currently, many efforts have been made in this field based on the application of food processing technologies or genetic modifications as efficient strategies to produce soybean hypoallergenic formulas. Numerous analytical methods have also been developed for soybean traceability in processed foods, aiming at detecting very low amounts of soybean allergens, as effective tools to verify labeling compliance and protect sensitized individuals from suffering adverse reactions. So far, innovative progresses, both in protein and DNA analysis, have been advanced, which can be useful in supporting the industrial management of allergenic foods, always with the final goal of increasing the quality of life of allergic consumers.

ACKNOWLEDGMENTS

This work was supported by Fundação para a Ciência e Tecnologia under the Partnership Agreement UID/QUI/50006/2020 and by the projects AlleRiskAssess – PTDC/BAA-AGR/31720/2017 and NORTE-01-0145-FEDER-000052.

Caterina Villa is grateful for the FCT grant (PD/BD/114576/2016) financed by POPH-QREN (subsidized by FSE and MCTES).

REFERENCES

Adachi, A., Horikawa, T., Shimizu, H., Sarayama, Y., Ogawa, T., Sjolander, S., Tanaka, A., & Moriyama, T. (2009). Soybean β-conglycinin as the main allergen in a patient with food-dependent exercise-induced anaphylaxis by tofu: Food processing alters pepsin resistance. *Clinical & Experimental Allergy, 39*(1), 167–173.

ALLERGOME. (2020). Allergome - A database of allergenic molecules, Allergy Data Laboratories, Latina, Italy. Accessed on July 17, 2020. Retrieved from http://www.allergome.org/

Allgöwer, S. M., Hartmann, C. A., & Holzhauser, T. (2020). The development of highly specific and sensitive primers for the detection of potentially allergenic soybean (*Glycine max*) using Loop-Mediated Isothermal Amplification combined with Lateral Flow Dipstick (LAMP-LFD). *Foods, 9*(4), 423.

American Academy of Pediatrics (2000). American Academy of Pediatrics: Committee on Nutrition. Hypoallergenic infant formulae. *Pediatrics, 106*, 346–349.

Amigo-Benavent, M., Clemente, A., Ferranti, P., Caira, S., & del Castillo, M. D. (2011). Digestibility and immunoreactivity of soybean β-conglycinin and its deglycosylated form. *Food Chemistry, 129*(4), 1598–1605.

Arita, K., Babiker, E. E., Azakami, H., & Kato, A. (2001). Effect of chemical and genetic attachment of polysaccharides to proteins on the production of IgG and IgE. *Journal of Agricultural and Food Chemistry, 49*(4), 2030–2036.

Babiker, E. f. E., Hiroyuki, A., Matsudomi, N., Iwata, H., Ogawa, T., Bando, N., & Kato, A. (1998). Effect of polysaccharide conjugation or transglutaminase treatment on the allergenicity and functional properties of soy protein. *Journal of Agricultural and Food Chemistry, 46*(3), 866–871.

Ballmer-Weber, B. K., Holzhauser, T., Scibilia, J., Mittag, D., Zisa, G., Ortolani, C., Oesterballe, M., Poulsen, L. K., Vieths, S., & Bindslev-Jensen, C. (2007). Clinical characteristics of soybean allergy in Europe: A double-blind, placebo-controlled food challenge study. *The Journal of Allergy and Clinical Immunology, 119*(6), 1489–1496.

Ballmer-Weber, B. K., & Vieths, S. (2008). Soy allergy in perspective. *Current Opinion in Allergy and Clinical Immunology, 8*(3), 270–275.

Batista, R., Martins, I., Jeno, P., Ricardo, C. P., & Oliveira, M. M. (2007). A proteomic study to identify soya allergens--the human response to transgenic versus non-transgenic soya samples. *International Archives of Allergy and Immunology, 144*(1), 29–38.

Baur, X., Pau, M., Czuppon, A., & Fruhmann, G. (1996). Characterization of soybean allergens causing sensitization of occupationally exposed bakers. *Allergy, 51*(5), 326–330.

Beardslee, T. A., Zeece, M. G., Sarath, G., & Markwell, J. P. (2000). Soybean glycinin G1 acidic chain shares IgE epitopes with peanut allergen Ara h 3. *International Archives of Allergy and Immunology, 123*(4), 299–307.

Becker, W., Brasseur, D., Bresson, J., Flynn, A., Jackson, A., Lagiou, P., Mingrone, G., Moseley, B., Palou, A., & Przyrembel, H. (2004). Opinion of the scientific panel on dietetic products, nutrition and allergies on a request from the commission relating to the evaluation of allergenic foods for labelling purposes (Request nr EFSA-Q-2003-016). *EFSA Journal, 32*, 1–197.

Berkner, H., Neudecker, P., Mittag, D., Ballmer-Weber, B. K., Schweimer, K., Vieths, S., & Rösch, P. (2009). Cross-reactivity of pollen and food allergens: Soybean Gly m 4 is a member of the Bet v 1 superfamily and closely resembles yellow lupine proteins. *Bioscience Reports, 29*(3), 183–192.

Bernhisel-Broadbent, J., & Sampson, H. A. (1989). Cross-allergenicity in the legume botanical family in children with food hypersensitivity. *Journal of Allergy and Clinical Immunology, 83*(2), 435–440.

Biscola, V., de Olmos, A. R., Choiset, Y., Rabesona, H., Garro, M. S., Mozzi, F., Chobert, J.-M., Drouet, M., Haertlé, T., & Franco, B. (2017). Soymilk fermentation by *Enterococcus faecalis* VB43 leads to reduction in the immunoreactivity of allergenic proteins β-conglycinin (7S) and glycinin (11S). *Beneficial Microbes, 8*(4), 635–643.

Bock, S. A., & Atkins, F. M. (1989). The natural history of peanut allergy. *Journal of Allergy and Clinical Immunology, 83*(5), 900–904.

Bock, S. A., Muñoz-Furlong, A., & Sampson, H. A. (2001). Fatalities due to anaphylactic reactions to foods. *Journal of Allergy and Clinical Immunology, 107*(1), 191–193.

Bøgh, K. L., & Madsen, C. B. (2016). Food allergens: Is there a correlation between stability to digestion and allergenicity? *Critical Reviews in Food Science and Nutrition, 56*(9), 1545–1567.

Breiteneder, H., & Radauer, C. (2004). A classification of plant food allergens. *Journal of Allergy and Clinical Immunology, 113*(5), 821–830.

Brodkorb, A., Egger, L., Alminger, M., Alvito, P., Assunção, R., Ballance, S., Bohn, T., Bourlieu-Lacanal, C., Boutrou, R., Carrière, F., Clemente, A., Corredig, M., Dupont, D., Dufour, C., Edwards, C., Golding, M., Karakaya, S., Kirkhus, B., Le Feunteun, S., Lesmes, U., Macierzanka, A., Mackie, A. R., Martins, C., Marze, S., McClements, D. J., Ménard, O., Minekus, M., Portmann, R., Santos, C. N., Souchon, I., Singh, R. P., Vegarud, G. E., Wickham, M. S. J., Weitschies, W., & Recio, I. (2019). INFOGEST static *in vitro* simulation of gastrointestinal food digestion. *Nature Protocols, 14*(4), 991–1014.

Bu, G., Zhu, T., & Chen, F. (2017). The structural properties and antigenicity of soybean glycinin by glycation with xylose. *Journal of the Science of Food and Agriculture, 97*(7), 2256–2262.

Burks, A. W., Williams, L. W., Helm, R. M., Thresher, W., Brooks, J. R., & Sampson, H. A. (1991). Identification of soy protein allergens in patients with atopic dermatitis and positive soy challenges; determination of change in allergenicity after heating or enzyme digestion. *Advances in Experimental Medicine and Biology, 289*, 295–307.

Burks, A. W., Williams, L. W., Thresher, W., Connaughton, C., Cockrell, G., & Helm, R. M. (1992). Allergenicity of peanut and soybean extracts altered by chemical or thermal denaturation in patients with atopic dermatitis and positive food challenges. *Journal of Allergy and Clinical Immunology, 90*(6), 889–897.

Burney, P., Summers, C., Chinn, S., Hooper, R., Van Ree, R., & Lidholm, J. (2010). Prevalence and distribution of sensitization to foods in the European Community Respiratory Health Survey: A EuroPrevall analysis. *Allergy, 65*(9), 1182–1188.

Cabanillas, B., Jappe, U., & Novak, N. (2018). Allergy to peanut, soybean, and other legumes: Recent advances in allergen characterization, stability to processing and ige cross-reactivity. *Molecular Nutrition & Food Research, 62*(1), 1700446.

Candreva, Á., M., Smaldini, P. L., Curciarello, R., Fossati, C. A., Docena, G. H., & Petruccelli, S. (2016). The major soybean allergen Gly m Bd 28K induces hypersensitivity reactions in mice sensitized to cow's milk proteins. *Journal of Agricultural and Food Chemistry, 64*(7), 1590–1599.

Candreva, Á. M., Ferrer-Navarro, M., Bronsoms, S., Quiroga, A., Curciarello, R., Cauerhff, A., Petruccelli, S., Docena, G. H., & Trejo, S. A. (2017). Identification of cross-reactive B-cell epitopes between Bos d 9.0101 (*Bos Taurus*) and Gly m 5.0101 (*Glycine max*) by epitope mapping MALDI-TOF MS. *Proteomics, 17*(15–16), 1700069.

Candreva, A. M., Smaldini, P. L., Curciarello, R., Cauerhff, A., Fossati, C. A., Docena, G. H., & Petruccelli, S. (2015). Cross-reactivity between the soybean protein P34 and bovine caseins. *Allergy, Asthma & Immunology Research, 7*(1), 60–68.

Chen, L., Vadlani, P. V., Madl, R. L., Wang, W., Shi, Y., & Gibbons, W. R. (2015). The investigation of virginiamycin-added fungal fermentation on the size and immunoreactivity of heat-sensitive soy protein. *International Journal of Polymer Science, 2015*(5), 1–7.

Chen, S., Yang, C. T., & Downs, M. L. (2019). Detection of six commercially processed soy ingredients in an incurred food matrix using parallel reaction monitoring. *Journal of Proteome Research, 18*(3), 995–1005.

Cheng, F., Wu, J., Zhang, J., Pan, A., Quan, S., Zhang, D., Kim, H., Li, X., Zhou, S., & Yang, L. (2016). Development and inter-laboratory transfer of a decaplex polymerase chain reaction assay combined with capillary electrophoresis for the simultaneous detection of ten food allergens. *Food Chemistry, 199*, 799–808.

Chruszcz, M., Maleki, S. J., Majorek, K. A., Demas, M., Bublin, M., Solberg, R., Hurlburt, B. K., Ruan, S., Mattison, C. P., Breiteneder, H., & Minor, W. (2011). Structural and immunologic characterization of Ara h 1, a major peanut allergen. *The Journal of Biological Chemistry, 286*(45), 39318–39327.

Codex Alimentarius. (2009). Foods derived from modern biotechnology. Codex Alimentarius Commission, Joint FAO/WHO Food Standards Programme, Rome.

Codina, R., Ardusso, L., Lockey, R. F., Crisci, C. D., Jaén, C., & Bertoya, N. H. (2002). Identification of the soybean hull allergens involved in sensitization to soybean dust in a rural population from Argentina and N-terminal sequence of a major 50 KD allergen. *Clinical & Experimental Allergy, 32*(7), 1059–1063.

Codina, R., Lockey, R. F., Fernández-Caldas, E., & Rama, R. (1997). Purification and characterization of a soybean hull allergen responsible for the Barcelona asthma outbreaks. II. Purification and sequencing of the Gly m 2 allergen. *Clinical & Experimental Allergy*, 27(4), 424–430.

Codina, R., Lockey, R. F., Fernández-Caldas, E., & Rama, R. (1999). Identification of the soybean hull allergens responsible for the Barcelona asthma outbreaks. *International Archives of Allergy and Immunology*, 119(1), 69–71.

Codina, R., Oehling Jr, A. G., & Lockey, R. F. (1998). Neoallergens in heated soybean hull. *International Archives of Allergy and Immunology*, 117(2), 120–125.

Costa, J., Amaral, J. S., Grazina, L., Oliveira, M. B. P. P., & Mafra, I. (2017a). Matrix-normalised real-time PCR approach to quantify soybean as a potential food allergen as affected by thermal processing. *Food Chemistry*, 221, 1843–1850.

Costa, J., Bavaro, S. L., Benedé, S., Diaz-Perales, A., Bueno-Diaz, C., Gelencser, E., Klueber, J., Larré, C., Lozano-Ojalvo, D., Lupi, R., Mafra, I., Mazzucchelli, G., Molina, E., Monaci, L., Martín-Pedraza, L., Piras, C., Rodrigues, P. M., Roncada, P., Schrama, D., Cirkovic-Velickovic, T., Verhoeckx, K., Villa, C., Kuehn, A., Hoffmann-Sommergruber, K., & Holzhauser, T. (2020). Are Physicochemical Properties Shaping the Allergenic Potency of Plant Allergens? *Clinical Reviews in Allergy & Immunology*. doi: 10.1007/s12016-020-08810-9

Costa, J., Fernandes, T. J., Villa, C., Oliveira, M., & Mafra, I. (2017b). Advances in food allergen analysis. *Food safety: Innovative Analytical Tools for Safety Assessment*, 305–360.

Croote, D., Braslavsky, I., & Quake, S. R. (2019). Addressing complex matrix interference improves multiplex food allergen detection by targeted LC–MS/MS. *Analytical Chemistry*, 91(15), 9760–9769.

Cuadrado, C., Cabanillas, B., Pedrosa, M. M., Muzquiz, M., Haddad, J., Allaf, K., Rodriguez, J., Crespo, J. F., & Burbano, C. (2011). Effect of instant controlled pressure drop on IgE antibody reactivity to peanut, lentil, chickpea and soybean proteins. *International Archives of Allergy and Immunology*, 156(4), 397–404.

Curciarello, R., Smaldini, P. L., Candreva, A. M., Gonzalez, V., Parisi, G., Cauerhff, A., Barrios, I., Blanch, L. B., Fossati, C. A., Petruccelli, S., & Docena, G. H. (2014). Targeting a cross-reactive Gly m 5 soy peptide as responsible for hypersensitivity reactions in a milk allergy mouse model. *PLoS One*, 9(1), e82341.

De Angelis, E., Pilolli, R., Bavaro, S. L., & Monaci, L. (2017). Insight into the gastro-duodenal digestion resistance of soybean proteins and potential implications for residual immunogenicity. *Food & Function*, 8(4), 1599–1610.

Dean, T., MacKenzie, H., Kilburn, S., Moonesinghe, H., Lee, K., Maslin, K., & Venter, C. (2013). Literature searches and reviews related to the prevalence of food allergy in Europe. *EFSA supporting publication*, 2013: EN-506.

Directive 2007/68/EC of 27 November 2007 amending Annex IIIa to Directive 2000/13/EC of the European Parliament and of the Council as regards certain food ingredients. *Official Journal of the European Union*, L310/11.

Ebisawa, M., Brostedt, P., Sjölander, S., Sato, S., Borres, M. P., & Ito, K. (2013). Gly m 2S albumin is a major allergen with a high diagnostic value in soybean-allergic children. *Journal of Allergy and Clinical Immunology*, 132(4), 976-978.e971-975.

Eigenmann, P. A., Burks, A. W., Bannon, G. A., & Sampson, H. A. (1996). Identification of unique peanut and soy allergens in sera adsorbed with cross-reacting antibodies. *Journal of Allergy and Clinical Immunology*, 98(5), 969–978.

Espiñeira, M., Herrero, B., Vieites, J. M., & Santaclara, F. J. (2010). Validation of end-point and real-time PCR methods for the rapid detection of soy allergen in processed products. *Food Additives & Contaminants: Part A*, 27(4), 426–432.

FAOSTAT. (2020). Food and Agriculture Organization of the United Nations - Statistics Division. Retrieved 2020 July 20, http://www.fao.org/faostat/en/#home

Foucard, T., & Malmheden-Yman, I. (1999). A study on severe food reactions in Sweden–is soy protein an underestimated cause of food anaphylaxis? *Allergy*, 54(3), 261–265.

Frias, J., Song, Y., Martínez-Villaluenga, C., De Mejia, E. G., & Vidal-Valverde, C. (2008). Fermented soyabean products as hypoallergenic food. *Proceedings of the Nutrition Society*, 67(OCE1).

Gagnon, C., Poysa, V., Cober, E. R., & Gleddie, S. (2010). Soybean allergens affecting north american patients identified by 2D gels and mass spectrometry. *Food Analytical Methods*, 3(4), 363–374.

Gautam, P. B., Sharma, R., Lata, K., Rajput, Y. S., & Mann, B. (2017). Construction of a lateral flow strip for detection of soymilk in milk. *Journal of Food Science and Technology*, 54(13), 4213–4219.

Geng, T., Liu, K., Frazier, R., Shi, L., Bell, E., Glenn, K., & Ward, J. M. (2015). Development of a sandwich ELISA for quantification of Gly m 4, a soybean allergen. *Journal of Agricultural and Food Chemistry*, 63(20), 4947–4953.

Gonzalez, R., Polo, F., Zapatero, L., Caravacaj, F., & Carreira, J. (1992). Purification and characterization of major inhalant allergens from soybean hulls. *Clinical & Experimental Allergy, 22*(8), 748–755.

Gonzalez, R., Varela, J., Carreira, J., & Polo, F. (1995). Soybean hydrophobic protein and soybean hull allergy. *The Lancet, 346*(8966), 48–49.

Green, T. D., LaBelle, V. S., Steele, P. H., Kim, E. H., Lee, L. A., Mankad, V. S., Williams, L. W., Anstrom, K. J., & Burks, A. W. (2007). Clinical characteristics of peanut-allergic children: Recent changes. *Pediatrics, 120*(6), 1304–1310.

Gu, X., Beardslee, T., Zeece, M., Sarath, G., & Markwell, J. (2001). Identification of IgE-binding proteins in soy lecithin. *International Archives of Allergy and Immunology, 126*(3), 218–225.

Guillamón, E., Rodríguez, J., Burbano, C., Muzquiz, M., Pedrosa, M. M., Cabanillas, B., Crespo, J. F., Sancho, A. I., Mills, E. N. C., & Cuadrado, C. (2010). Characterization of lupin major allergens (*Lupinus albus* L.). *Molecular Nutrition & Food Research, 54*(11), 1668–1676.

Gupta, R. S., Warren, C. M., Smith, B. M., Blumenstock, J. A., Jiang, J., Davis, M. M., & Nadeau, K. C. (2018). The public health impact of parent-reported childhood food allergies in the United States. *Pediatrics, 142*(6).

Hajika, M., Takahashi, K., Yamada, T., Komaki, K., Takada, Y., Shimada, H., Sakai, T., Shimada, S., Adachi, T., & Tabuchi, K. (2009). Development of a new soybean [*Glycine max*] cultivar for soymilk 'Nagomimaru'. *Bulletin of the National Institute of Crop Science (Japan), 48*(4), 383–386.

Hajika, M., Takahashi, M., Sakai, S., & Matsunaga, R. (1998). Dominant inheritance of a trait lacking beta-conglycinin detected in a wild soybean line. *Japanese Journal of Breeding, 48*(4), 383–386.

Han, X., Nagano, H., Phromraksa, P., Tsuji, M., Shimoyamada, M., Kasuya, S., Suzuki, T., & Khamboonruang, C. (2012). Hydrolysis of soybean 7S and 11S globulins using *Bacillus subtilis*. *Food Science and Technology Research, 18*(5), 651–657.

Harrison, L. A., Bailey, M. R., Naylor, M. W., Ream, J. E., Hammond, B. G., Nida, D. L., Burnette, B. L., Nickson, T. E., Mitsky, T. A., Taylor, M. L., Fuchs, R. L., & Padgette, S. R. (1996). The expressed protein in glyphosate-tolerant soybean, 5-enolpyruvylshikimate-3-phosphate synthase from *Agrobacterium* sp. strain CP4, is rapidly digested *in vitro* and is not toxic to acutely gavaged mice. *Journal of Nutrition, 126*(3), 728–740.

Hei, W., Li, Z., Ma, X., & He, P. (2012). Determination of beta-conglycinin in soybean and soybean products using a sandwich enzyme-linked immunosorbent assay. *Analytica Chimica Acta, 734*, 62–68.

Helm, R. M., Cockrell, G., Connaughton, C., West, C. M., Herman, E., Sampson, H. A., Bannon, G. A., & Burks, A. W. (2000). Mutational analysis of the IgE-binding epitopes of P34/Gly m Bd 30K. *Journal of Allergy and Clinical Immunology, 105*(2), 378–384.

Herman, E. M., Helm, R. M., Jung, R., & Kinney, A. J. (2003). Genetic modification removes an immunodominant allergen from soybean. *Plant Physiology, 132*(1), 36–43.

Hiemori, M., Ito, H., Kimoto, M., Yamashita, H., Nishizawa, K., Maruyama, N., Utsumi, S., & Tsuji, H. (2004). Identification of the 23-kDa peptide derived from the precursor of Gly m Bd 28K, a major soybean allergen, as a new allergen. *Biochimica et Biophysica Acta (BBA) - General Subjects, 1675*(1), 174–183.

Hoff, M., Son, D.-Y., Gubesch, M., Ahn, K., Lee, S.-I., Vieths, S., Goodman, R. E., Ballmer-Weber, B. K., & Bannon, G. A. (2007). Serum testing of genetically modified soybeans with special emphasis on potential allergenicity of the heterologous protein CP4 EPSPS. *Molecular Nutrition & Food Research, 51*(8), 946–955.

Hoffmann, B., Münch, S., Schwägele, F., Neusüß, C., & Jira, W. (2017). A sensitive HPLC-MS/MS screening method for the simultaneous detection of lupine, pea, and soy proteins in meat products. *Food Control, 71*, 200–209.

Holzhauser, T. (2018). Protein or no protein? Opportunities for DNA-based detection of allergenic foods. *Journal of Agricultural and Food Chemistry, 66*(38), 9889–9894.

Holzhauser, T., Wackermann, O., Ballmer-Weber, B. K., Bindslev-Jensen, C., Scibilia, J., Perono-Garoffo, L., Utsumi, S., Poulsen, L. K., & Vieths, S. (2009). Soybean (*Glycine max*) allergy in Europe: Gly m 5 (beta-conglycinin) and Gly m 6 (glycinin) are potential diagnostic markers for severe allergic reactions to soy. *The Journal of Allergy and Clinical Immunology, 123*(2), 452–458.

Huang, T., Bu, G., & Chen, F. (2018). The influence of composite enzymatic hydrolysis on the antigenicity of β-conglycinin in soy protein hydrolysates. *Journal of Food Biochemistry, 42*(5), e12544.

Hurlburt, B. K., Offermann, L. R., McBride, J. K., Majorek, K. A., Maleki, S. J., & Chruszcz, M. (2013). Structure and function of the peanut panallergen Ara h 8. *Journal of Biological Chemistry, 288*(52), 36890–36901.

ISAAA. (2018). Global status of commercialized Biotech/GM Crops in 2018: Biotech crops continue to help meet the challenges of increased population and climate change. ISAAA Brief No. 54. ISAAA: Ithaca, NY.

Jeong, K.-H., Choi, M. S., Lee, S.-K., Seo, M.-j., Hwang, T.-Y., Yun, H.-T., Kim, H.-s., Kim, J., Kwon, Y.-U., & Kim, Y.-H. (2013). Development of low Gly m Bd 30K (P34) allergen breeding lines using molecular marker in soybean [Glycine max (L.) Merr.]. *Plant Breeding and Biotechnology, 1*, 298–306.

Jia, H., Zhou, T., Zhu, H., Shen, L., & He, P. (2019). Quantification of Gly m 5.0101 in soybean and soy products by liquid chromatography-tandem mass spectrometry. *Molecules, 24*(1), 68.

Jiang, J., Warren, C. M., & Gupta, R. S. (2020). Epidemiology and racial/ethnic differences in food allergy pediatric food allergy. In R.S. Gupta (Ed.), *Pediatric Food Allergy* (pp. 3–16). Switzerland: Springer.

Joseph, L. M., Hymowitz, T., Schmidt, M. A., & Herman, E. M. (2006). Evaluation of glycine germplasm for nulls of the immunodominant allergen P34/Gly m Bd 30k. *Crop Science, 46*(4), 1755–1763.

Julka, S., Kuppannan, K., Karnoup, A., Dielman, D., Schafer, B., & Young, S. A. (2012). Quantification of Gly m 4 protein, a major soybean allergen, by two-dimensional liquid chromatography with ultraviolet and mass spectrometry detection. *Analytical Chemistry, 84*(22), 10019–10030.

Kanegae, R., Obata, A., Matsunaga, R., Takahashi, M., & Komatsu, K. (2001). Analysis of Gly m Bd 28K, major soybean allergen, in several leading varieties in Japan. *Journal of the Japanese Society for Food Science and Technology, 48*(5): 344–348.

Kattan, J. D., & Sampson, H. A. (2015). Clinical reactivity to soy is best identified by component testing to Gly m 8. *Journal of Allergy and Clinical Immunology: In Practice, 3*(6), 970-972.e971.

Kim, J.-H., Lieu, H.-Y., Kim, T.-W., Kim, D.-O., Shon, D.-H., Ahn, K.-M., Lee, S.-I., & Kim, H.-Y. (2006). Assessment of the potential allergenicity of genetically modified soybeans and soy-based products. *Food Science and Biotechnology, 15*(6), 954–958.

Kitamura, K., & Kaizuma, N. (1981). Mutant strains with low level of subunits of 7S globulin in soybean (*Glycine max* Merr.) seed. *Japanese Journal of Breeding, 31*(4), 353–359.

Kleine-Tebbe, J., Wangorsch, A., Vogel, L., Crowell, D. N., Haustein, U.-F., & Vieths, S. (2002). Severe oral allergy syndrome and anaphylactic reactions caused by a Bet v 1 related PR-10 protein in soybean, SAM22. *Journal of Allergy and Clinical Immunology, 110*(5), 797–804.

Klemans, R. J., Knol, E. F., Michelsen-Huisman, A., Pasmans, S. G., de Kruijf-Broekman, W., Bruijnzeel-Koomen, C. A., van Hoffen, E., & Knulst, A. C. (2013). Components in soy allergy diagnostics: Gly m 2S albumin has the best diagnostic value in adults. *Allergy, 68*(11), 1396–1402.

König, A., Cockburn, A., Crevel, R. W. R., Debruyne, E., Grafstroem, R., Hammerling, U., Kimber, I., Knudsen, I., Kuiper, H. A., Peijnenburg, A. A. C. M., Penninks, A. H., Poulsen, M., Schauzu, M., & Wal, J. M. (2004). Assessment of the safety of foods derived from genetically modified (GM) crops. *Food and Chemical Toxicology, 42*(7), 1047–1088.

Koo, S. C., Seo, J. S., Park, M. J., Cho, H. M., Park, M. S., Choi, C. W., Jung, W.-H., Lee, K. H., Jin, B. J., Kim, S. H., Shim, S. I., Chung, J.-S., Chung, J. I., & Kim, M. C. (2013). Identification of molecular mechanism controlling P34 gene expression in soybean. *Plant Biotechnology Reports, 7*(3), 331–338.

Krishnan, H. B., Kim, W. S., Jang, S., & Kerley, M. S. (2009). All three subunits of soybean beta-conglycinin are potential food allergens. *Journal of Agricultural and Food Chemistry, 57*(3), 938–943.

L'Hocine, L., & Boye, J. I. (2007). Allergenicity of soybean: new developments in identification of allergenic proteins, cross-reactivities and hypoallergenization technologies. *Critical Reviews in Food Science and Nutrition, 47*(2), 127–143.

L'Hocine, L., Boye, J. I., & Jouve, S. (2007). Ionic strength and pH-induced changes in the immunoreactivity of purified soybean glycinin and its relation to protein molecular structure. *Journal of Agricultural and Food Chemistry, 55*(14), 5819–5826.

Ladenburger, E.-M., Dehmer, M., Grünberg, R., Waiblinger, H.-U., Stoll, D., & Bergemann, J. (2019). Highly sensitive matrix-independent quantification of major food allergens peanut and soy by competitive real-time PCR targeting mitochondrial DNA. *Journal of AOAC International, 101*(1), 170–184.

Ladics, G. S. (2019). Assessment of the potential allergenicity of genetically-engineered food crops. *Journal of Immunotoxicology, 16*(1), 43–53.

Lee, H.-W., Keum, E.-H., Lee, S.-J., Sung, D.-E., Chung, D.-H., Lee, S.-I., & Oh, S. (2007). Allergenicity of proteolytic hydrolysates of the soybean 11S globulin. *Journal of Food Science, 72*(3), C168–C172.

Lee, J., Lee, S., Cho, S., Oh, C., & Ryu, C. (2004). A new technique to produce hypoallergenic soybean proteins using three different fermenting microorganism. *Journal of Allergy and Clinical Immunology, 113*(2), S239.

Lee, K. J., Kim, J.-B., Kim, S. H., Ha, B.-K., Lee, B.-M., Kang, S.-Y., & Kim, D. S. (2011). Alteration of seed storage protein composition in soybean [*Glycine max* (L.) Merrill] mutant lines induced by γ-irradiation mutagenesis. *Journal of Agricultural and Food Chemistry, 59*(23), 12405–12410.

Li, H., Jia, Y., Peng, W., Zhu, K., Zhou, H., & Guo, X. (2018). High hydrostatic pressure reducing allergenicity of soy protein isolate for infant formula evaluated by ELISA and proteomics via Chinese soy-allergic children's sera. *Food Chemistry*, *269*, 311–317.

Li, H., Zhu, K., Zhou, H., & Peng, W. (2012). Effects of high hydrostatic pressure treatment on allergenicity and structural properties of soybean protein isolate for infant formula. *Food Chemistry*, *132*(2), 808–814.

Li, H., Zhu, K., Zhou, H., Peng, W., & Guo, X. (2013). Comparative study about some physical properties, *in vitro* digestibility and immunoreactivity of soybean protein isolate for infant formula. *Plant Foods for Human Nutrition*, *68*(2), 124–130.

Liu, B., Teng, D., Wang, X., & Wang, J. (2013a). Detection of the soybean allergenic protein Gly m Bd 28K by an indirect enzyme-linked immunosorbent assay. *Journal of Agricultural and Food Chemistry*, *61*(4), 822–828.

Liu, B., Teng, D., Yang, Y., Wang, X., & Wang, J. (2012). Development of a competitive ELISA for the detection of soybean α subunit of β-conglycinin. *Process Biochemistry*, *47*(2), 280–287.

Liu, S., Chen, G., Yang, L., Gai, J., & Zhu, Y. (2013b). Production of transgenic soybean to eliminate the major allergen Gly m Bd 30K by RNA interference-mediated gene silencing. *Journal of Pure Applied Microbiology*, *7*, 589–599.

Lu, M., Jin, Y., Ballmer-Weber, B., & Goodman, R. E. (2018). A comparative study of human IgE binding to proteins of a genetically modified (GM) soybean and six non-GM soybeans grown in multiple locations. *Food and Chemical Toxicology*, *112*, 216–223.

Luthria, D. L., Maria John, K. M., Marupaka, R., & Natarajan, S. (2018). Recent update on methodologies for extraction and analysis of soybean seed proteins. *Journal of the Science of Food and Agriculture*, *98*(15), 5572–5580.

Ma, X., Sun, P., He, P., Han, P., Wang, J., Qiao, S., & Li, D. (2010). Development of monoclonal antibodies and a competitive ELISA detection method for glycinin, an allergen in soybean. *Food Chemistry*, *121*(2), 546–551.

Magishi, N., Yuikawa, N., Kobayashi, M., & Taniuchi, S. (2017). Degradation and removal of soybean allergen in Japanese soy sauce. *Molecular Medicine Reports*, *16*(2), 2264–2268.

Manjaya, J. G., Suseelan, K. N., Gopalakrishna, T., Pawar, S. E., & Bapat, V. A. (2007). Radiation induced variability of seed storage proteins in soybean [*Glycine max* (L.) Merrill]. *Food Chemistry*, *100*(4), 1324–1327.

Matricardi, P. M., Kleine-Tebbe, J., Hoffmann, H. J., Valenta, R., Hilger, C., Hofmaier, S., Aalberse, R. C., Agache, I., Asero, R., Ballmer-Weber, B., Barber, D., Beyer, K., Biedermann, T., Bilò, M. B., Blank, S., Bohle, B., Bosshard, P. P., Breiteneder, H., Brough, H. A., Caraballo, L., Caubet, J. C., Crameri, R., Davies, J. M., Douladiris, N., Ebisawa, M., EIgenmann, P. A., Fernandez-Rivas, M., Ferreira, F., Gadermaier, G., Glatz, M., Hamilton, R. G., Hawranek, T., Hellings, P., Hoffmann-Sommergruber, K., Jakob, T., Jappe, U., Jutel, M., Kamath, S. D., Knol, E. F., Korosec, P., Kuehn, A., Lack, G., Lopata, A. L., Mäkelä, M., Morisset, M., Niederberger, V., Nowak-Węgrzyn, A. H., Papadopoulos, N. G., Pastorello, E. A., Pauli, G., Platts-Mills, T., Posa, D., Poulsen, L. K., Raulf, M., Sastre, J., Scala, E., Schmid, J. M., Schmid-Grendelmeier, P., Hage, M., Ree, R., Vieths, S., Weber, R., Wickman, M., Muraro, A., & Ollert, M. (2016). EAACI molecular allergology user's guide. *Pediatric Allergy and Immunology*, *27*(S23), 1–250.

Matsuo, A., Matsushita, K., Fukuzumi, A., Tokumasu, N., Yano, E., Zaima, N., & Moriyama, T. (2020). Comparison of various soybean allergen levels in genetically and non-genetically modified soybeans. *Foods*, *9*(4), 522.

Mayer, W., Schuller, M., Viehauser, M. C., & Hochegger, R. (2019). Quantification of the allergen soy (*Glycine max*) in food using digital droplet PCR (ddPCR). *European Food Research and Technology*, *245*(2), 499–509.

Meinlschmidt, P., Brode, V., Sevenich, R., Ueberham, E., Schweiggert-Weisz, U., Lehmann, J., Rauh, C., Knorr, D., & Eisner, P. (2017). High pressure processing assisted enzymatic hydrolysis – An innovative approach for the reduction of soy immunoreactivity. *Innovative Food Science & Emerging Technologies*, *40*, 58–67.

Meinlschmidt, P., Sussmann, D., Schweiggert-Weisz, U., & Eisner, P. (2016a). Enzymatic treatment of soy protein isolates: Effects on the potential allergenicity, technofunctionality, and sensory properties. *Food Science & Nutrition*, *4*(1), 11–23.

Meinlschmidt, P., Ueberham, E., Lehmann, J., Reineke, K., Schlüter, O., Schweiggert-Weisz, U., & Eisner, P. (2016b). The effects of pulsed ultraviolet light, cold atmospheric pressure plasma, and gamma-irradiation on the immunoreactivity of soy protein isolate. *Innovative Food Science & Emerging Technologies*, *38*, 374–383.

Meinlschmidt, P., Ueberham, E., Lehmann, J., Schweiggert-Weisz, U., & Eisner, P. (2016c). Immunoreactivity, sensory and physicochemical properties of fermented soy protein isolate. *Food Chemistry*, *205*, 229–238.

Mittag, D., Vieths, S., Vogel, L., Becker, W. M., Rihs, H. P., Helbling, A., Wüthrich, B., & Ballmer-Weber, B. K. (2004). Soybean allergy in patients allergic to birch pollen: Clinical investigation and molecular characterization of allergens. *The Journal of Allergy and Clinical Immunology*, *113*(1), 148–154.

Montowska, M., & Fornal, E. (2018). Detection of peptide markers of soy, milk and egg white allergenic proteins in poultry products by LC-Q-TOF-MS/MS. *LWT*, *87*, 310–317.

Montowska, M., Fornal, E., Piątek, M., & Krzywdzińska-Bartkowiak, M. (2019). Mass spectrometry detection of protein allergenic additives in emulsion-type pork sausages. *Food Control*, *104*, 122–131.

Moreno, F. J. (2007). Gastrointestinal digestion of food allergens: effect on their allergenicity. *Biomedicine & Pharmacotherapy*, *61*(1), 50–60.

Morishita, N., Matsumoto, T., Morimatsu, F., & Toyoda, M. (2014). Detection of soybean proteins in fermented soybean products by using heating extraction. *Journal of Food Science*, *79*(5), T1049–T1054.

Mulet-Cabero, A.-I., Egger, L., Portmann, R., Ménard, O., Marze, S., Minekus, M., Le Feunteun, S., Sarkar, A., Grundy, M. M. L., Carrière, F., Golding, M., Dupont, D., Recio, I., Brodkorb, A., & Mackie, A. (2020). A standardised semi-dynamic *in vitro* digestion method suitable for food – An international consensus. *Food & Function*, *11*(2), 1702–1720.

Nishinari, K., Fang, Y., Guo, S., & Phillips, G. O. (2014). Soy proteins: A review on composition, aggregation and emulsification. *Food Hydrocolloids*, *39*, 301–318.

Ogawa, T., Bando, N., Tsuji, H., Okajima, H., Nishikawa, K., & Sasaoka, K. (1991). Investigation of the IgE-binding proteins in soybeans by immunoblotting with the sera of the soybean-sensitive patients with atopic dermatitis. *Journal of Nutritional Science and Vitaminology*, *37*(6), 555–565.

Ogawa, T., Tsuji, H., Bando, N., Kitamura, K., Zhu, Y.-L., Hirano, H., & Nishikawa, K. (1993). Identification of the soybean allergenic protein, Gly m Bd 30K, with the soybean seed 34-kDa oil-body-associated protein. *Bioscience, Biotechnology, and Biochemistry*, *57*(6), 1030–1033.

Ogura, T., Clifford, R., & Oppermann, U. (2019). Simultaneous detection of 13 allergens in thermally processed food using targeted LC–MS/MS approach. *Journal of AOAC International*, *102*(5), 1316–1329.

Ohishi, A., Watanabe, K., Urushibata, M., Utsuno, K., Ikuta, K., Sugimoto, K., & Harada, H. (1994). Detection of soybean antigenicity and reduction by twin-screw extrusion. *Journal of the American Oil Chemists' Society*, *71*(12), 1391–1396.

Okunuki, H., Teshima, R., Shigeta, T., Sakushima, J., Akiyama, H., Goda, Y., Toyoda, M., & Sawada, J. (2002). Increased digestibility of two products in genetically modified food (CP4-EPSPS and Cry1Ab) after preheating. *Shokuhin Eiseigaku Zasshi*, *43*(2), 68–73.

Palm, M., Moneret-Vautrin, D. A., Kanny, G., Denery-Papini, S., & Frémont, S. (1999). Food allergy to egg and soy lecithins. *Allergy*, *54*(10), 1116–1117.

Peñas, E., Gomez, R., Frias, J., Baeza, M. L., & Vidal-Valverde, C. (2011). High hydrostatic pressure effects on immunoreactivity and nutritional quality of soybean products. *Food Chemistry*, *125*(2), 423–429.

Peñas, E., Restani, P., Ballabio, C., Préstamo, G., Fiocchi, A., & Gómez, R. (2006). Assessment of the residual immunoreactivity of soybean whey hydrolysates obtained by combined enzymatic proteolysis and high pressure. *European Food Research and Technology*, *222*(3–4), 286–290.

Picariello, G., Mamone, G., Nitride, C., Addeo, F., & Ferranti, P. (2013). Protein digestomics: Integrated platforms to study food-protein digestion and derived functional and active peptides. *TrAC Trends in Analytical Chemistry*, *52*, 120–134.

Planque, M., Arnould, T., Dieu, M., Delahaut, P., Renard, P., & Gillard, N. (2016). Advances in ultra-high performance liquid chromatography coupled to tandem mass spectrometry for sensitive detection of several food allergens in complex and processed foodstuffs. *Journal of Chromatography A*, *1464*, 115–123.

Prado, M., Ortea, I., Vial, S., Rivas, J., Calo-Mata, P., & Barros-Velázquez, J. (2016). Advanced DNA- and protein-based methods for the detection and investigation of food allergens. *Critical Reviews in Food Science and Nutrition*, *56*(15), 2511–2542.

Regulation (EU) 503/2013 of 3 April 2013 on applications for authorisation of genetically modified food and feed in accordance with Regulation (EC) No 1829/2003 of the European Parliament and of the Council and amending Commission Regulations (EC) No 641/2004 and (EC) No 1981/2006. *Official Journal of the European Union*, L157:1.

Regulation (EU) 1169/2011 of 25 October 2011 on the provision of food information to consumers, amending Regulations (EC) No 1924/2006 and (EC) No 1925/2006 of the European Parliament and of the Council, and repealing Commission Directive 87/250/EEC, Council Directive 90/496/EEC, Commission Directive 1999/10/EC, Directive 2000/13/EC of the European Parliament and of the Council, Commission Directives 2002/67/EC and 2008/5/EC and Commission Regulation (EC) No 608/2004. *Official Journal of the European Union*, L304: 18–63.

Riascos, J. J., Weissinger, S. M., Weissinger, A. K., Kulis, M., Burks, A. W., & Pons, L. (2016). The seed bio-tinylated protein of soybean (*Glycine max*): A boiling-resistant new allergen (Gly m 7) with the capac-ity to induce IgE-mediated allergic responses. *Journal of Agricultural and Food Chemistry 64*(19), 3890–3900.

Rihs, H. P., Chen, Z., Ruëff, F., Petersen, A., Rozynek, P., Heimann, H., & Baur, X. (1999). IgE binding of the recombinant allergen soybean profilin (rGly m 3) is mediated by conformational epitopes. *The Journal of Allergy and Clinical Immunology, 104*(6), 1293–1301.

Rodríguez-Ramírez, R., Vallejo-Cordoba, B., Mazorra-Manzano, M. A., & González-Córdova, A. F. (2015). Soy detection in canned tuna by PCR and capillary electrophoresis. *Analytical Methods, 7*(2), 530–537.

Roychaudhuri, R., Sarath, G., Zeece, M., & Markwell, J. (2004). Stability of the allergenic soybean Kunitz trypsin inhibitor. *Biochimica et Biophysica Acta, 1699*(1–2), 207–212.

Rozenfeld, P., Docena, G. H., Añón, M. C., & Fossati, C. A. (2002). Detection and identification of a soy pro-tein component that cross-reacts with caseins from cow's milk. *Clinical and Experimental Immunology, 130*(1), 49–58.

Safdar, M., & Abasıyanık, M. F. (2013). Development of fast multiplex real-time PCR assays based on EvaGreen fluorescence dye for identification of beef and soybean origins in processed sausages. *Food Research International, 54*(2), 1652–1656.

Safdar, M., Junejo, Y., Arman, K., & Abasıyanık, M. F. (2014). A highly sensitive and specific tetraplex PCR assay for soybean, poultry, horse and pork species identification in sausages: Development and valida-tion. *Meat Science, 98*(2), 296–300.

Saitoh, S., Urushibata, M., Ikuta, K., Fujimaki, A., & Harada, H. (2000). Antigenicity in soybean hypocotyls and its reduction by twin-screw extrusion. *Journal of the American Oil Chemists' Society, 77*(4), 419.

Samoto, M., Fukuda, Y., Takahashi, K., Tabuchi, K., Hiemori, M., Tsuji, H., Ogawa, T., & Kawamura, Y. (1997). Substantially complete removal of three major allergenic soybean proteins (Gly m Bd 30K, Gly m Bd 28K, and the α-subunit of conglycinin) from soy protein by using a mutant soybean, Tohoku 124. *Bioscience, Biotechnology, and Biochemistry, 61*(12), 2148–2150.

Samoto, M., Takahashi, K., Fukuda, Y., Nakamura, S., & Kawamura, Y. (1996). Substantially complete removal of the 34kDa allergenic soybean protein, Gly m Bd 30 K, from soy milk of a mutant lacking the alpha- and alpha'-subunits of conglycinin. *Bioscience, Biotechnology, and Biochemistry, 60*(11), 1911–1913.

Sampson, H. A., & McCaskill, C. C. (1985). Food hypersensitivity and atopic dermatitis: evaluation of 113 patients. *The Journal of Pediatrics, 107*(5), 669–675.

Sampson, H. A., & Scanlon, S. M. (1989). Natural history of food hypersensitivity in children with atopic dermatitis. *The Journal of Pediatrics, 115*(1), 23–27.

Santiago-Felipe, S., Tortajada-Genaro, L. A., Puchades, R., & Maquieira, A. (2014). Recombinase polymerase and enzyme-linked immunosorbent assay as a DNA amplification-detection strategy for food analysis. *Analytica Chimica Acta, 811*, 81–87.

Savage, J. H., Kaeding, A. J., Matsui, E. C., & Wood, R. A. (2010). The natural history of soy allergy. *Journal of Allergy and Clinical Immunology, 125*(3), 683–686.

Scala, E., Villalta, D., Meneguzzi, G., Giani, M., & Asero, R. (2018). Storage molecules from tree nuts, seeds and legumes: Relationships and amino acid identity among homologue molecules. *European Annals of Allergy and Clinical Immunology, 50*(4), 148–155.

Schafer, B. W., Embrey, S. K., & Herman, R. A. (2016). Rapid simulated gastric fluid digestion of in-seed/grain proteins expressed in genetically engineered crops. *Regulatory Toxicology and Pharmacology, 81*, 106–112.

Schmidt, M. A., Hymowitz, T., & Herman, E. M. (2015). Breeding and characterization of soybean Triple Null; a stack of recessive alleles of Kunitz Trypsin Inhibitor, Soybean Agglutinin, and P34 allergen nulls. *Plant Breeding, 134*(3), 310–315.

Scholl, P. F., Farris, S. M., & Mossoba, M. M. (2014). Rapid turbidimetric detection of milk powder adultera-tion with plant proteins. *Journal of Agricultural and Food Chemistry, 62*(7), 1498–1505.

Segura-Gil, I., Nicolau-Lapeña, I., Galán-Malo, P., Mata, L., Calvo, M., Sánchez, L., & Pérez, M. D. (2018). Development of two ELISA formats to determine glycinin. Application to detect soy in model and com-mercial processed food. *Food Control, 93*, 32–39.

Selb, R., Wal, J. M., Moreno, F. J., Lovik, M., Mills, C., Hoffmann-Sommergruber, K., & Fernandez, A. (2017). Assessment of endogenous allergenicity of genetically modified plants exemplified by soybean – Where do we stand? *Food and Chemical Toxicology, 101*, 139–148.

Shamoon, M., Sajid, M. W., Safdar, W., Haider, J., Omar, M., Ammar, A., Sharif, H. R., Khalid, S., & Randhawa, M. A. (2016). An update on hypoallergenicity of peanut and soybean: Where are we now? *RSC Advances, 6*(82), 79185–79195.

Shen, J.-D., Cai, Q.-F., Liu, G.-M., Zhang, L., Zhang, L.-J., & Cao, M.-J. (2013). A comparison study of the impact of boiling and high pressure steaming on the stability of soybean trypsin inhibitor. *International Journal of Food Science & Technology*, *48*(9), 1877–1883.

Shibasaki, M., Suzuki, S., Tajima, S., Nemoto, H., & Kuroume, T. (1980). Allergenicity of major component proteins of soybean. *International Archives of Allergy and Immunology*, *61*(4), 441–448.

Shim, S.-M., Choi, M.-H., Park, S.-H., Gu, Y.-U., Oh, J.-M., Kim, S., Kim, H.-Y., Kim, G.-H., & Lee, Y. (2010). Assessing the digestibility of genetically modified soybean: Physiologically based *in vitro* digestion and fermentation model. *Food Research International*, *43*(1), 40–45.

Skolnick, H. S., Conover-Walker, M. K., Koerner, C. B., Sampson, H. A., Burks, W., & Wood, R. A. (2001). The natural history of peanut allergy. *Journal of Allergy and Clinical Immunology*, *107*(2), 367–374.

Smaldini, P., Curciarello, R., Candreva, A., Rey, M. A., Fossati, C. A., Petruccelli, S., & Docena, G. H. (2012). *In vivo* evidence of cross-reactivity between cow's milk and soybean proteins in a mouse model of food allergy. *International Archives of Allergy and Immunology*, *158*(4), 335–346.

Šmíd, J., Godálová, Z., Piknová, Ľ., Siekel, P., & Kuchta, T. (2015). Semi-quantitative estimation of soya protein-based additives in meat products using real-time polymerase chain reaction. *Journal of Food & Nutrition Research*, *54*(2), 165–170.

Soares, S., Amaral, J. S., Oliveira, M. B. P. P., & Mafra, I. (2014). Quantitative detection of soybean in meat products by a TaqMan real-time PCR assay. *Meat Science*, *98*(1), 41–46.

Soares, S., Mafra, I., Amaral, J. S., & Oliveira, M. B. P. (2010). A PCR assay to detect trace amounts of soybean in meat sausages. *International Journal of Food Science & Technology*, *45*(12), 2581–2588.

Song, B., Shen, L., Wei, X., Guo, B., Tuo, Y., Tian, F., Han, Z., Wang, X., Li, W., & Liu, S. (2014). Marker-assisted backcrossing of a null allele of the α-subunit of Soybean (*Glycine max*) β-conglycinin with a Chinese soybean cultivar (a). The development of improved lines. *Plant Breeding*, *133*(5), 638–648.

Song, Y. S., Frias, J., Martinez-Villaluenga, C., Vidal-Valdeverde, C., & de Mejia, E. G. (2008). Immunoreactivity reduction of soybean meal by fermentation, effect on amino acid composition and antigenicity of commercial soy products. *Food Chemistry*, *108*(2), 571–581.

Sten, E., Skov, P. S., Andersen, S. B., Torp, A. M., Olesen, A., Bindslev-Jensen, U., Lars K., B.-J., & Bindslev-Jensen, C. (2004). A comparative study of the allergenic potency of wild-type and glyphosate-tolerant gene-modified soybean cultivars. *APMIS - Journal of Pathology, Microbiology and Immunology*, *112*(1), 21–28.

Takács, K., Guillamon, E., Pedrosa, M. M., Cuadrado, C., Burbano, C., Muzquiz, M., Haddad, J., Allaf, K., Maczó, A., & Polgár, M. (2014). Study of the effect of instant controlled pressure drop (DIC) treatment on IgE-reactive legume-protein patterns by electrophoresis and immunoblot. *Food and Agricultural Immunology*, *25*(2), 173–185.

Takahashi, K., Banba, H., Kikuchi, A., Ito, M., & Nakamura, S. (1994). An induced mutant line lacking the alpha-subunit of beta-conglycinin in soybean [*Glycine max* (L.) Merrill]. *Japanese Journal of Breeding*, *44*(1), 65–66.

Takahashi, K., Shimada, S., Shimada, H., Takada, Y., Sakai, T., Kono, Y., Adachi, T., Tabuchi, K., Kikuchi, A., & Yumoto, S. (2004). A new soybean cultivar" Yumeminori" with low allergenicity and high content of 11S globulin. *Bulletin of the National Agricultural Research Center for Tohoku Region*, *102*, 23–39.

Tortajada-Genaro, L. A., Santiago-Felipe, S., Morais, S., Gabaldón, J. A., Puchades, R., & Maquieira, Á. (2012). Multiplex DNA detection of food allergens on a digital versatile disk. *Journal of Agricultural and Food Chemistry*, *60*(1), 36–43.

Treudler, R., Franke, A., Schmiedeknecht, A., Ballmer-Weber, B., Worm, M., Werfel, T., Jappe, U., Biedermann, T., Schmitt, J., Brehler, R., Kleinheinz, A., Kleine-Tebbe, J., Brüning, H., Ruëff, F., Ring, J., Saloga, J., Schäkel, K., Holzhauser, T., Vieths, S., & Simon, J. C. (2017). BASALIT trial: double-blind placebo-controlled allergen immunotherapy with rBet v 1-FV in birch-related soya allergy. *Allergy*, *72*(8), 1243–1253.

Tsai, J.-J., Chang, C.-Y., & Liao, E.-C. (2017a). Comparison of allergenicity at Gly m 4 and Gly m Bd 30K of soybean after genetic modification. *Journal of Agricultural and Food Chemistry*, *65*(6), 1255–1262.

Tsai, J. J., Chang, C. Y., & Liao, E. C. (2017b). Comparison of allergenicity at Gly m 4 and Gly m Bd 30 K of soybean after genetic modification. *Journal of Agricultural and Food Chemistry*, *65*(6), 1255–1262.

Tsuji, H., Hiemori, M., Kimoto, M., Yamashita, H., Kobatake, R., Adachi, M., Fukuda, T., Bando, N., Okita, M., & Utsumi, S. (2001). Cloning of cDNA encoding a soybean allergen, Gly m Bd 28K. *Biochimica et Biophysica Acta, 1518*(1-2), 178–182.

Tsumura, K. (2009). Improvement of the physicochemical properties of soybean proteins by enzymatic hydrolysis. *Food Science and Technology Research*, *15*(4), 381–388.

Tsumura, K., Kugimiya, W., Bando, N., Hiemori, M., & Ogawa, T. (1999). Preparation of hypoallergenic soybean protein with processing functionality by selective enzymatic hydrolysis. *Food Science and Technology Research*, *5*(2), 171–175.

Ueberham, E., Spiegel, H., Havenith, H., Rautenberger, P., Lidzba, N., Schillberg, S., & Lehmann, J. r. (2019). Simplified tracking of a soy allergen in processed food using a monoclonal antibody-based sandwich ELISA targeting the soybean 2S albumin Gly m 8. *Journal of Agricultural and Food Chemistry*, *67*(31), 8660–8667.

Usui, M., Tamura, H., Nakamura, K., Ogawa, T., Muroshita, M., Azakami, H., Kanuma, S., & Kato, A. (2004). Enhanced bactericidal action and masking of allergen structure of soy protein by attachment of chitosan through Maillard-type protein-polysaccharide conjugation. *Food/Nahrung*, *48*(1), 69–72.

van Boxtel, E. L., van den Broek, L. A. M., Koppelman, S. J., & Gruppen, H. (2008). Legumin allergens from peanuts and soybeans: Effects of denaturation and aggregation on allergenicity. *Molecular Nutrition & Food Research*, *52*(6), 674–682.

van de Lagemaat, J., Manuel Silván, J., Javier Moreno, F., Olano, A., & Dolores del Castillo, M. (2007). *In vitro* glycation and antigenicity of soy proteins. *Food Research International*, *40*(1), 153–160.

Verhoeckx, K. C. M., Vissers, Y. M., Baumert, J. L., Faludi, R., Feys, M., Flanagan, S., Herouet-Guicheney, C., Holzhauser, T., Shimojo, R., van der Bolt, N., Wichers, H., & Kimber, I. (2015). Food processing and allergenicity. *Food and Chemical Toxicology*, *80*, 223–240.

Verma, A. K., Kumar, S., Das, M., & Dwivedi, P. D. (2013). A comprehensive review of legume allergy. *Clinical Reviews in Allergy & Immunology*, *45*(1), 30–46.

Vidal, C., Pérez-Carral, C., & Chomón, B. (1997). Unsuspected sources of soybean exposure. *Annals of Allergy, Asthma & Immunology*, *79*(4), 350–352.

Villa, C., Costa, J., Oliveira, M. B. P. P., & Mafra, I. (2018). Bovine milk allergens: A comprehensive review. *Comprehensive Reviews in Food Science and Food Safety*, *17*(1), 137–164.

Villa, C., Moura, M. B., Costa, J., & Mafra, I. (2020). Immunoreactivity of lupine and soybean allergens in foods as affected by thermal processing. *Foods*, *9*(3), 254.

Waiblinger, H.-U., Boernsen, B., Geppert, C., Demmel, A., Peterseil, V., & Koeppel, R. (2017). Ring trial validation of single and multiplex real-time PCR methods for the detection and quantification of the allergenic food ingredients mustard, celery, soy, wheat and rye. *Journal of Consumer Protection and Food Safety*, *12*(1), 55–72.

Wang, R., Edrington, T. C., Storrs, S. B., Crowley, K. S., Ward, J. M., Lee, T. C., Liu, Z. L., Li, B., & Glenn, K. C. (2017a). Analyzing pepsin degradation assay conditions used for allergenicity assessments to ensure that pepsin susceptible and pepsin resistant dietary proteins are distinguishable. *PLoS One*, *12*(2), e0171926.

Wang, T., Tan, Z., & Sun, Z. (2015a). Low immunoreactive glycated soybean antigen proteins production: system-wide analysis of their immunogenicity *in vitro* and *in vivo*. *Food and Agricultural Immunology*, *26*(5), 703–716.

Wang, W., Han, J., Wu, Y., Yuan, F., Chen, Y., & Ge, Y. (2011). Simultaneous detection of eight food allergens using optical thin-film biosensor chips. *Journal of Agricultural and Food Chemistry*, *59*(13), 6889–6894.

Wang, W., Zhu, X., Teng, S., Fan, Q., & Qian, H. (2017b). Label-free biochips for rapid detection of soybean allergen GlymBd 30K (P34) in foods. *Tropical Journal of Pharmaceutical Research*, *16*(4), 755–760.

Wang, Y., Deng, R., Zhang, G., Li, Q., Yang, J., Sun, Y., Li, Z., & Hu, X. (2015b). Rapid and sensitive detection of the food allergen glycinin in powdered milk using a lateral flow colloidal gold immunoassay strip test. *Journal of Agricultural and Food Chemistry*, *63*(8), 2172–2178.

Wang, Y., Li, Z., Pei, Y., Li, Q., Sun, Y., Yang, J., Yang, Y., Zhi, Y., Deng, R., Hou, Y., & Hu, X. (2017c). Establishment of a lateral flow colloidal gold immunoassay strip for the rapid detection of soybean allergen β-conglycinin. *Food Analytical Methods*, *10*(7), 2429–2435.

Wang, Z., Li, L., Yuan, D., Zhao, X., Cui, S., Hu, J., & Wang, J. (2014). Reduction of the allergenic protein in soybean meal by enzymatic hydrolysis. *Food and Agricultural Immunology*, *25*(3), 301–310.

Watanabe, D., Adányi, N., Takács, K., Maczó, A., Nagy, A., Gelencsér, É., Pachner, M., Lauter, K., Baumgartner, S., & Vollmann, J. (2017). Development of soybeans with low P34 allergen protein concentration for reduced allergenicity of soy foods. *Journal of the Science of Food and Agriculture*, *97*(3), 1010–1017.

Watanabe, D., Lošák, T., & Vollmann, J. (2018). From proteomics to ionomics: Soybean genetic improvement for better food safety. *Genetika*, *50*(1), 333–350.

WHO/IUIS. (2020). World Health Organization/International Union of Immunological Societies (WHO/IUIS) Allergen Nomenclature Sub-committee. Accessed on March 17, 2020. Retrieved from http://www.allergen.org/.

Wilson, S., Martinez-Villaluenga, C., & De Mejia, E. G. (2008). Purification, thermal stability, and antigenicity of the immunodominant soybean allergen P34 in soy cultivars, ingredients, and products. *Journal of Food Science*, *73*(6), T106–T114.

Xi, J., & Shi, Q. (2016). Development of an indirect competitive ELISA kit for the detection of soybean allergenic protein Gly m Bd 28K. *Food Analytical Methods*, *9*(11), 2998–3005.

Xi, J., & Yu, Q. (2020). The development of lateral flow immunoassay strip tests based on surface enhanced Raman spectroscopy coupled with gold nanoparticles for the rapid detection of soybean allergen β-conglycinin. *Spectrochimica Acta Part A: Molecular and Biomolecular Spectroscopy*, *241*, 118640.

Xia, J., Zu, Q., Yang, A., Wu, Z., Li, X., Tong, P., Yuan, J., Wu, Y., Fan, Q., & Chen, H. (2019). Allergenicity reduction and rheology property of *Lactobacillus*-fermented soymilk. *Journal of the Science of Food and Agriculture*, *99*(15), 6841–6849.

Xiang, P., Baird, L. M., Jung, R., Zeece, M. G., Markwell, J., & Sarath, G. (2008). P39, a novel soybean protein allergen, belongs to a plant-specific protein family and is present in protein storage vacuoles. *Journal of Agricultural and Food Chemistry*, *56*(6), 2266–2272.

Xiang, P., Beardslee, T. A., Zeece, M. G., Markwell, J., & Sarath, G. (2002). Identification and analysis of a conserved immunoglobulin E-binding epitope in soybean G1a and G2a and peanut Ara h 3 glycinins. *Archives of Biochemistry and Biophysics*, *408*(1), 51–57.

Yamanishi, R., Huang, T., Tsuji, H., Bando, N., & Ogawa, T. (1995). Reduction of the soybean allergenicity by the fermentation with *Bacillus natto*. *Food Science and Technology International*, *1*(1), 14–17.

Yamanishi, R., Tsuji, H., Bando, N., Yamada, Y., Nadaoka, Y., Huang, T., Nishikawa, K., Emoto, S., & Ogawa, T. (1996). Reduction of the allergenicity of soybean by treatment with proteases. *Journal of Nutritional Science and Vitaminology*, *42*(6), 581–587.

Yang, A., Deng, H., Zu, Q., Lu, J., Wu, Z., Li, X., Tong, P., & Chen, H. (2018a). Structure characterization and IgE-binding of soybean 7S globulin after enzymatic deglycosylation. *International Journal of Food Properties*, *21*(1), 171–182.

Yang, A., Xia, J., Gong, Y., Deng, H., Wu, Z., Li, X., Tong, P., & Chen, H. (2017). Changes in the structure, digestibility and immunoreactivities of glycinin induced by the cross-linking of microbial transglutaminase following heat denaturation. *International Journal of Food Science & Technology*, *52*(10), 2265–2273.

Yang, A., Zuo, L., Cheng, Y., Wu, Z., Li, X., Tong, P., & Chen, H. (2018b). Degradation of major allergens and allergenicity reduction of soybean meal through solid-state fermentation with microorganisms. *Food & Function*, *9*(3), 1899–1909.

Yang, H., Qu, Y., Li, J., Liu, X., Wu, R., & Wu, J. (2020). Improvement of the protein quality and degradation of allergens in soybean meal by combination fermentation and enzymatic hydrolysis. *LWT*, *128*, 109442.

Yang, W. W., Chung, S.-Y., Ajayi, O., Krishnamurthy, K., Konan, K., & Goodrich-Schneider, R. (2010). Use of pulsed ultraviolet light to reduce the allergenic potency of soybean extracts. *International Journal of Food Engineering*, *6*(3).

Yang, W. W., De Mejia, E. G., Zheng, H., & Lee, Y. (2011). Soybean Allergens: Presence, detection and methods for mitigation. In H. El-Shemy (Ed.), *Soybean and Health* (pp. 433–464). Rijeka, Croatia: InTek.

Yuan, D., Fang, X., Liu, Y., Kong, J., & Chen, Q. (2019a). A hybridization chain reaction coupled with gold nanoparticles for allergen gene detection in peanut, soybean and sesame DNAs. *Analyst*, *144*(12), 3886–3891.

Yuan, D., Kong, J., Fang, X., & Chen, Q. (2019b). A graphene oxide-based paper chip integrated with the hybridization chain reaction for peanut and soybean allergen gene detection. *Talanta*, *196*, 64–70.

Yuan, D., Kong, J., Li, X., Fang, X., & Chen, Q. (2018). Colorimetric LAMP microfluidic chip for detecting three allergens: Peanut, sesame and soybean. *Scientific Reports*, *8*(1), 1–8.

Zhang, S., Cao, K., Liu, D., Gaowa, N., Bao, N., & Zhao, Y. (2016). Development of dcELISA method for rapid detection of β-conglycinin in soybean. *International Journal of Food Engineering*, *12*(5), 461.

Zhang, Y., Guan, R., Liu, Z., Chang, R., Yao, Y., & Qiu, L. (2006). Identification of Gly m Bd 28K and Gly m Bd 30K lacking soybean by using random sampling of core collection in soybean. *Zuo wu xue bao*, *32*(3), 324–329.

Zhao, Y., Qin, G. X., Sun, Z. W., Zhang, B., & Wang, T. (2010). Stability and immunoreactivity of glycinin and β-conglycinin to hydrolysis *in vitro*. *Food and Agricultural Immunology*, *21*(3), 253–263.

9 Isolation, Bioactivity, Identification, and Commercial Application of Soybean Bioactive Peptides

Akhunzada Bilawal, Zhanmei Jiang, Yang Li, and Baokun Qi

CONTENTS

9.1 INTRODUCTION TO SOYBEAN BIOACTIVE PEPTIDES

Recently, several studies have reported that the derivation of peptides from precursor proteins, normally extracted from edible proteins, are also present in independent form and with maternal proteins (Wang & De Mejia, 2005). During the process of intestinal digestion or directly from processed foods, the peptides are extracted from the maternal protein. The peptides do the same function as hormones do in the body during the processing and regulating of food (Korhonen & Pihlanto, 2003). For animals and humans, the edible protein has a key and important role in the conversion of energy from these items (Aoyama, Fukui, Takamatsu, Hashimoto, & Yamamoto, 2000b). Various important bioactive peptides are known for their specific functions (Korhonen & Pihlanto, 2003). As of now, more than 1,500 various types of bioactive peptides have been identified according to the BIOPEP database (Dziuba et al., 1999). According to their antihypertensive activities, dipeptidyl peptidase IV inhibitors and particularly angiotensin-converting enzymes are more importantly known for their specific function. Opioid agonistic, antagonistic, antioxidative, anticancer, and immunomodulatory actions are some specific biological functions of peptides (Wang et al., 2005). Currently, peptides and antihypertensive peptides are really hot and important topics for research scientists. Peptides have the ability to maintain blood pressure with the help of constraining the angiotensin-converting enzymes (ACE) (Natesh et al., 2003). Figure 9.1 shows a schematic diagram of bioactive peptides.

In the last few years, the attention toward biologically active peptides formed through enzymatic hydrolysis and fermentation has been increased (Moldes & Vecino, 2017). Over 3,200 bioactive peptides are reported in the BIOPEP database in several studies. Bioactive peptides are a series of peptides within a protein that exhibit valuable effects on body utilities beyond the recognized dietary importance (Walther & Sieber, 2011). Proteins have structural and functional properties ascribed by biologically active peptides (Blomstrand & Newsholme, 1992). During enzymatic hydrolysis, the inactive proteins become active (peptides). Most bioactive peptides have standard features, including structures. Nearly all peptides contain two to nine amino acids and are mainly hydrophobic (Kitts & Weiler, 2003). According to Korhonen and Pihlanto (2003), the range can

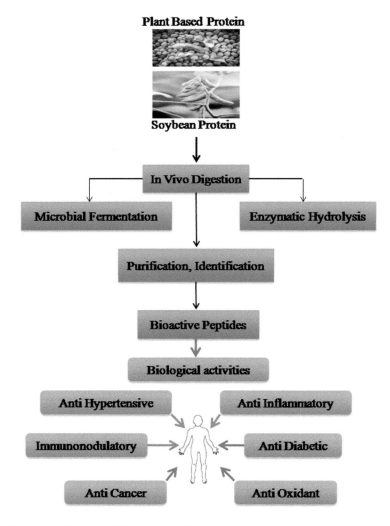

FIGURE 9.1   Schematic diagram of bioactive peptides.

sometimes be prolonged to 20 or more amino acids. Lunasin is a food-derived bioactive peptide and has been proven to have strong anticancer bioactivity; it has 43 amino acids with MW 5,400 kDa (Jeong et al., 2003). Most bioactive peptide activities are controlled by a series of amino acids that act together with other body proteins to regulate normal processes (Moller, Scholz-Ahrens, Roos, & Schrezenmeir, 2008). Based on their structure and composition of amino acids, these peptides can normalize essential body utilities through many bioactivities such as immunomodulatory, antihypertensive, antithrombotic, antimicrobial, and antioxidant (Singh, Vij, & Hati, 2014). Many bioactive peptides reveal specific bioactivity even though other peptides possess multifunctional properties (Andre-Frei et al., 1999).

Sánchez and Vázquez (2017) found that peptides could be extracted from different sources, such as dairy products, seafood, animals, and plants. Milk and different dairy products are the best sources of bioactive peptides (Floris et al., 2003). Some of the most utilized animal proteins include eggs, milk, and meat (Lassoued et al., 2015), whereas peptides have also been isolated and characterized from soy, canola, wheat, hemp, flax seeds, oat, and pulses (Singh et al., 2014). Marine animals like fish, squid, salmon, oysters, seahorses, and snow crab are sources of bioactive peptides. Of these, milk proteins are the most isolated bioactive compounds (Aoyama, Fukui, et al., 2000a). Plant-derived food is a significant source of bioactive peptides and protein hydrolysates. Soy is an essential marginal source of bioactive compounds. Soybean (*Glycine Max*) comprises no less than 35% protein, 20% lipids, 9% nutritional fiber, and 8.5% moisture content (Chatterjee, Gleddie, & Xiao, 2018). However, the composition of bioactive peptides of soybeans depends on the geography and environmental conditions of planting.

Soybean is one of the essential ancient foodcrops of the Far East. Soy has been domesticated in Asia for almost 5,000 years and was considered a daily bread in East Asia (Kamran & Reddy, 2018; Chatterjee et al., 2018; Agyei, 2015). Soybean originated in Southeast Asia but was cultivated for the first time by Chinese farmers around 1100BC, and later grown in Japan and other countries in the first century AD. Soybean was formally presented in Europe and the United States in the eighteenth century and nineteenth century, respectively. It has remained the main economic crop in the US since the 1940s. In 1765, a British colony from Georgia cultivated soybean seeds in China, and by 1851, it was disseminated to other crofters in Illinois and Corn Belt states. Figure 9.2 presents the soybean plant and seeds. Significant progress in the use of soybean in the US was made in 1999 after the Food and Drug Administration (FDA) sanctioned soybeans following the health assertion that soy proteins reduce the possibility of coronary heart syndrome (Chang, Lee, & Hungria, 2014). The popularity of soy foods increased, particularly in North America. Today, the US is ranked as the top manufacturer of soybeans, accounting for more than 30% of worldwide productions (Vollmann, 2016). Soy is a significant source of high-grade proteins that contain essential amino acids. From the time when civilization began, a lot of people have utilized soybean as a rich source of nutritious protein and oil. Soybean has proven to be an excellent source of protein for vegetarians and a significant source of micronutrients and fiber as it can be processed into a variety of food products through various technological and biotechnological incursions. Even so, the demographic depletion of soybean differs geographically. Asians consume nearly 20 to 80 g each day in the form of conventionally fermented soy foods while Americans utilize 1to 3 g per day. The increasing interest in soy foods is owed to the emerging evidence of the potential curative functions for various disorders such as cardiovascular syndrome, cancer, osteoporosis, and menopausal symptoms (Lemes et al., 2016).

The key components of soybean that have biological activities include proteins, protease inhibitors, saponins, and isoflavones. The two most important soy proteins are β-conglycinin (βCG, 7S) and glycinin (11S) which entail 80–90% of the overall composition of protein in soybean. The components of βCG are α', α, and β while those for glycinin are acidic (A) and basic (B) including A1aB2, A2B1a, and A3B4 as sub-elements (Maet al., 2016). Minor protein components in soybean are 2S, 9S, and 15S, lectin, Kunitz, and protease-blocking enzymes such as Bowman-Birk (BBI). The different compositions elucidate the variances in utility properties

(a) Soybean Plant   (b) Soybean Seeds

FIGURE 9.2   Soybean plant and seeds.

in terms of value, harvest, and quality during production. Soybean consists of 40% protein that conforms toa complex mixture of different types of proteins. Conglycinin and glycinin are the main sources of soy proteins, which are 50–70% of the total seed proteins. Glycinin is a hexamer with a molecular weight (MW) of 320–375 kDa and five major subunits (G1, G2, G3, G4, and G5) composed of an acidic and basic chain about 40 and 20 kDa respectively that are used by disulfide bonds. The first three distinct components (G1, G2, G3) are grouped together since they have the same homologies sequence, unlike the last two (G4 and G5). On the other hand, conglycinin has an MW of 50–200 kilodaltons (kDa) and with three subunits; a, a', and 3(6). Each subunit consists of a high degree of homology and regions of extensions since it acts as a trimer (Raman et al., 2018).

Soy isoflavone is a major soybean bioactive compound, denoted as phytoestrogens since they physically look like 17β-estradiol (Cohen, 2000). Isoflavone has estrogenic and anti-estrogenic characteristics, with more affinity for ERβ since they can equally fasten estrogen receptors (ER) α and β. The hormonal and antioxidant properties of soy isoflavones contribute to their health benefits and physiological functions that vary depending on the level of each isoflavone in soybean (Raman et al., 2018). Subsequently, soy saponins are a minor bioactive peptide component with compound and varied "oleanane triterpenoid glycosides". They are amphiphilic molecules with polar sugar chains attached to a non-polar pentacyclic ring structure. Saponins have anti-inflammatory, anticarcinogenic, antimicrobial, hepatic-, and cardio-protective effects. Soybean also has many enzymes, including lipoxygenase, chalcone synthase, catalase, and urease. However, only a few of them exceed 1% of total protein seeds. As already mentioned above, soy proteins are usually digested into peptides upon ingestion into the body by gastrointestinal enzymes.

Lunasin is one of the bioactive soybean peptides that can be identified through preservation. It is a novel soybean bioactive peptide that consists of 43 amino acids of 5.5 k Da molecular weight and nine aspartic acid filtrates on its carboxyl end. Lunasin has a cell linkage motif that has arginine-glycine-aspartic acid filtrates, and a foreseen twist within its structural homology that has a sealed section of chromatin-binding proteins (Wang, Dia, et al., 2008). Lunasin is absorbed intact to the target tissues but may not be wholly digested in the gastrointestinal system. Lunasin has a monomer peptide (de Mejia & Dia, 2009).

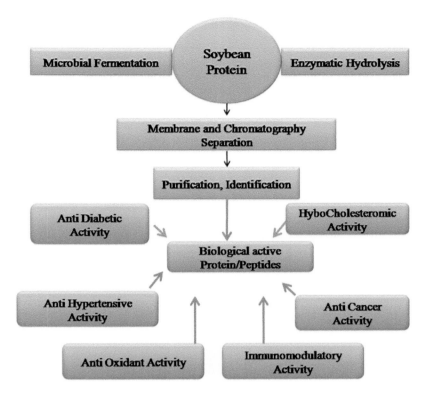

FIGURE 9.3 Flow chart of industrial production and separation process.

9.1.1 PREPARATION, ISOLATION, AND IDENTIFICATION OF SOYBEAN BIOACTIVE PEPTIDES

Bioactive soy proteins are small protein fragments that are prepared through fermentation, food processing, enzymatic hydrolysis, and gastrointestinal digestion of larger soybean proteins. Figure 9.3 shows a flow chart of the industrial production and separation process. Table 9.1 shows the preparation of soybean peptides.

9.1.1.1 Gastrointestinal Digestion

Soybean hydrolysates produce many peptides that are prepared through gastrointestinal digestion. It is the process where bioactive peptides are formed during absorption and consumption of soybean by acid and digestive enzymes such as pepsin, trypsin, pancreatin, and chymotrypsin present in the stomach, small intestine, and pancreas. The released peptides are absorbed in the bloodstream through the walls of the small intestines to enable them to exert systematic effects in the human body. Capriotti et al. (2015) prepared protein samples and subjected them to gastrointestinal digestion to produce bioactive peptides. They found that soybeans passed through a degradation process during gastrointestinal digestion which generated a significant amount of bioactive peptides, with biological and antimicrobial activity. Further, peptides were prepared through gastroduodenal digestion of six days germinated beans that yielded robust reactions (González-Montoya et al., 2018).

9.1.1.2 Enzymatic Hydrolysis

The *in vitro* enzymatic hydrolysis process is a peptides producer that occurs commercially in more substantial volumes that are effective and have better quality control to acquire specific peptides. The process generates a peptide by digesting soy proteins in an ideal pH and temperature condition using specific and non-specific proteases including pepsin, trypsin, chymotrypsin, papain, and

TABLE 9.1

Preparation of Soybean Peptides

Peptides	Preparation	Origin	Action	Reference
Ultra-filtration separated peptides with different molecular weights	Trypsin, Pepsin and Porcine	Soy concentrate	Regulation of LDL by cell receptors	Arnold et al., 2001
PGTAVFK	Bacillus Subtillis Protease	Soybean	Antihypertensive	Kodera & Nia, 2002
Thr-Pro-Arg-Val-Phe;0 Pro-Asn- Asn- Lys-Pro-Phe-Gin	Protease D3	Soy protein	Hypotensive	Kitts & Wettler, 2003
X-Met-Leu-Pro-ser-Tyr-Ser-Pro-Tyr	Thermolase	Defatted soy protein	Anticancer	Kim et al., 2000
Soluble peptides	Pronase	Soy flour	Growth promotion	Frank et al., 2000
HHL	Fermentation	Korean soybean paste	Peptide (hypertensive)	Shin et al., 2001
Catlionic resin	Alcatase	Defatted soy meal	Hypotensive	
RPLKPW;LLPHH	Proteinase S	Soybean protein (genetically modified)	Antihypertensive and antioxidative	Korhonen & Pilhanto, 2002
LPYPR	Peptide	Soybean glycinin	Hypocholesterolemic	

peptidase (Gu & Wu, 2013). Soy proteins are hydrolyzed with enzymes like trypsin, pepsin, chymotrypsin, bromelain, ficain, or papain. Enzymatic hydrolysis has numerous advantages, including getting rid of toxic filtrate chemicals and organic solvents in the final product. The cost of production rises when enzymes are used on an industrial scale. The impact can be lowered by using cheap enzymes such as the by-products of animals. During an *in vitro* enzymatic hydrolysis, a mixture of proteins can be generated relative to a composite of the starting material which impacts the purification process making it arduous and puzzling. The peptide may sometimes require a multifaceted purification progression. Even so, naturally occurring peptides are safer than those produced through enzymatic hydrolysis, but the commercialization of bioactive soybean peptides is economical because it does not involve expensive purification.

Soybean is a promising source of bioactive peptides. Peptides are obtained mainly by hydrolysis during fermentation or formed by the microorganisms associated with fermentation. Specific bioactive peptides are produced because of the hydrolysis of soybean proteins (glycinin and β-conglycinin). Enzymatic hydrolysis involves subjecting proteins to enzymes at a specific temperature and pH to yield peptides at different size ranges. The method is easier and quicker than microbial fermentation. The enzymatic hydrolysis is followed by an isolation technique performed through ultra-filtration or by cationic exchange resins which result in individual microbial strains that contribute to the formation of specific bioactive peptides with respective biological activities and health benefits depending on the difference in amino acids composition. The biological activities of peptides vary depending on enzymes used, processing conditions, and initial protein sources. Soy protein's exposure to different enzymes for hydrolysis provides peptides that have anticancer properties (Kim et al., 2000), antioxidant peptides (Pena-Ramos & Xiong, 2002), or hypotensive properties (Wu & Ding, 2001).

9.1.1.3   Food Processing

Food processing is also one of the rich sources of bioactive peptides due to the alteration of the structures of molecules. The most typical food processing procedures are high temperature, change in pH, protein isolation, extreme high-pressure processing (HPP), and storage conditions.

9.1.1.4 Fermentation

There is a rising interest in fermented soybean products because they generate health-promoting bioactive compounds usually formed during microbial fermentation. Asians, mainly Koreans, Chinese, and Japanese, have been utilizing fermented soybean food, including soy sauce, natto, paste, and tempeh. Fermentation is one of the most economical processes for producing bioactive peptides. A protein is hydrolyzed through a microbial process or bacterial enzymatic process. A microbe is secluded from the fermented soybean products to create many bioactive constituents including polyphenols, peptides, gamma-aminobutyric acid, antimicrobial composites, fibrinolytic enzymes, and exopolysaccharides (Rai et al., 2017). Lactic acid originating from the upper gastro-intestinal pathway is the most utilized bacteria during fermentation to produce peptides. Fermented soybean products comprise a wide range of bioactive compounds that are either created during the modification of soybean compounds into their bioactive form or by a starter culture. The types of bioactive peptides generated during the fermentation of a soybean depend on the microbiological culture which performs fermentation at the intraspecific level to precisely hydrolyze soybean proteins. Occasionally, soybean proteins are not fully hydrolyzed with covalent and enzymatic modification and R-group interactions. Enzymes such as pronase, trypsin, and plasma proteases are required to generate minor peptides with enhanced bioactivity. The bioactive peptide derived from soybean during fermentation demonstrates several health benefits, dependent on the dimensions and structures of amino acids (Sanjukta & Rai, 2016). Table 9.1 lists the preparation of soybean peptides.

9.1.2 ISOLATION OF SOYBEAN

Proteins play a primitive role in any living organism and control all mechanisms of action in the body. Proteins can be short or long chains of peptides, and so the chance to isolate the active peptides requires the development of automated and continuous systems (Dia, Wang, Oh, de Lumen, & de Mejia, 2009). Numerous technologies and energy-efficient systems are used for bioactive peptides' isolation and purification. Different methods and techniques have been used such as ion exchange, retrogressing high-yield liquid chromatography, fastening liquid proteins, affinity chromatography, membrane, and gel filtration. Today, isolation of bioactive peptides is commonly carried out by using chromatography techniques that have replaced the use of batch methods and solvent extraction (Kitts & Weiler, 2003). The highest bioactive compounds are isolated from soybeans and soy food products including isoflavones, peptides, flavonoids, soy lipids, soy phytoalexins, soya saponins, soy toxins, and vitamins (Ahn et al., 2000). Capriotti et al. (2015) applied liquid chromatography combined with mass spectrometry to isolate and enable peptide sequencing. The separation of bioactive soybean peptides involves the use of A-25 ion-exchange chromatography when the DH value is reached. The process begins during enzymatic reactions where the DH values are separated. Veličković et al. (2012) observed that the enzymatic reactions end at $24.64 \pm$ DH values. The SPI excerpts are separated by using A-25 ion-exchange chromatography into three segments. One of the fractions exhibits a potent antioxidant activity that is then separated using a Sephadex G-10 gel filtration column into two sections. One of these new sub-units has a potent antioxidant activity.

González-Montoya et al. (2018) isolated peptides through ultra-filtration. Concisely, 100 g of soybeans were drenched in 600 mL purified water and left to grow for six days in a germination chamber and in darkness at a temperature of 30° C. Later, the germinated seeds were dried for a day at a temperature of 40° C. The protein isolate extracted from the germinated soybean was prepared by isoelectric precipitation of pH 4.5 and alkaline extraction at pH 9.0. The generated fractions were analyzed through ultra-filtration to evaluate their ability to suppress dipeptidyl peptidase IV (DPP-IV), α-amylase, and α-glycosidase. The DPP-IV activity was inhibited by the six-day germinated seeds in a dose-dependent manner.

Successively, different chromatographic techniques are merged to ensure a successful separation of peptides due to their complexity, particularly those derived from fermented foods and during digestion (Acquah, Chan, Pan, Agyei, & Udenigwe, 2019). Angiotensin-converting enzyme

(ACE) inhibitory peptides can be obtained from soy proteins through physiological gastrointestinal absorption and enzyme action. Gao et al. (2003) used gel filtration chromatography (GFC) to purify and isolate ACE inhibitory peptides derived from douchi, a Chinese fermented soybean product. The separation revealed that peptides composed of phenylalanine, isoleucine, and glycine showed a 1:2:5 ratio of highest ACE inhibitory activity. Multiple studies have been done to analyze the existence of ACE inhibitory peptides in soy hydrolysate and fermented foods. The studies have found that many oligopeptides with complex biological activities have been produced from endoproteases' lower meticulousness.

Kuba et al. (2003) also used GFC and high-performance liquid chromatography (HPLC) to isolate ACE inhibitor peptides from tofu extract. The reverse phased HPLC and ion-pair chromatographies merged with matrix-assisted laser desorption/ionization to separate the ACE inhibitory peptides found in glycinin enzymatic hydrolysates. They found Val-Leu-Ile-Val-Pro, a strong ACE inhibitory peptide in its protease P hydrolysate. Ogawa et al. (2007) combined electrospray ionization with GFC and RP-HPLC to identify the protein hydrolysates derived from protease D3. Kuba et al. (2003) isolated four ACE inhibitors from soybean seed storage protein using an ultra-filtration process to purify the active elements of glycinin and 3-conglycinin. Lee et al. (2013) established mediators for intercellular calcium from the peptides that were obtained from the isolated soybean protein. Usually, the molecular weight of the proteins was measured by using gel filtration chromatography, which revealed that the phosphor peptides of soybeans could be strong calcium carriers and can be used to inhibit the low absorption of dietary calcium in animals. In consequence, Wang et al. (2008) used size-exclusion chromatography (SEC) to approve the amino acid structure quantification derived from the black soybean. GFC was used to analyze the distribution of the molecular weight, which showed that 80% of black soybean had less than 0.05 MW, unlike those of male rats.

9.2   IDENTIFICATION OF SOYBEAN BIOACTIVE PEPTIDES

Saz and Marina (2008) briefly explained the use of HPLC and capillary electrophoresis to identify, isolate, purify, and characterize soybean proteins. Identification of soybean peptides can be performed through western blot, chromatography, and enzyme-linked immunosorbent assay. The defatted soybean protein excerpt is made through extraction in ion-exchange chromatography at the speed of Tris-HCl, and Tris-HCl plus 2 M NaCl buffer. The extraction process occurs at a PH of 8.4 while the fractions are collected at 10 mL/min through step gradient elution (Saz et al., 2008). Capriotti et al. (2015) used specific databases to clarify the bioactive active presence in soybean seed samples. They identified the peptides in the soybean seeds as 1,173. The investigation process was aided by the biological and probable antimicrobial activity of soy protein.

In consequence, Ma et al. (2016) used liquid chromatography with tandem mass spectrometry to purify and identify the polypeptide chain soybean protein. The process began with eluting fraction 1 from a low to high (10–90% acetonitrile) organic mobile phase to purify the monomer peptide (FDPAL). The molecular-ion mass ($M+H^+$) of the peptide was determined using the Triple TOF mass spectrum system. The process of separating antioxidant peptides from SPI uses Sephadex A-25 through elution using a 10 mM phosphate buffer (pH 9.0) and a liner ingredient of 0–5 M NaCl at a flow rate of 1.0 Ml/min for 8 h.

9.3   BIOACTIVITY VERIFICATION OF SOYBEAN PEPTIDES

The bioactivity of peptides can be tested through biochemical analysis of their cell structures and clinical trials or *in vivo* studies. The focus on the identification and characterization of soybeans has shifted. Most peptides were reported to have functioned as ACE inhibitors. The soybean has been considered an important source of bioactive peptides, possessing different biological properties such as antioxidant, anticancer, anti-inflammatory antidiabetic, neuromodulator, and antihypertensive properties. Català-Clariana (2013) confirmed that bioactive soybean peptides have the same sequence with other biological activities including anticancer, antidiabetic, hypotensive, antioxidant, anti-inflammatory, and hypocholesterolemic activities.

9.3.1 Hypocholesterolemic

Most soy peptides have been identified as hypolipidemic which has been proven to reduce the cholesterol levels and triglycerides in the blood. The bioactivity of soy peptides can suppress fat synthesis and storage. LPYPR, a hypocholesterolemic peptide, is the first subunit of soybean from the glycinin to be revealed (Yoshikawa et al., 2000). When a dosage of 50 mg per kg of peptides is administered in the body, minus isoflavones for two days, the total amount of serum and "low-density lipoprotein" (LDL) cholesterol will be lowered in rats by 25%. More studies have proven that LPYPR is hypocholesterolemic and functions as a viable inhibitor of 3-hydroxy-3-methylglutaryl CoA reductase (HMGR), which is a significant inhibiting rate for enzymes during cholesterol biosynthesis. It can also increase the absorption of LDL in the cultured liver cells by stimulating the LDL receptor (LDLR) which acts as a steroid alcohol monitory element that binds the two protein pathways. IAVPGEVA and IAVPTGVA are also cholesterol-lowering peptides that are glycinin consequent. YVVNPDNDEN and YVVNPDNNEN are hypocholesterolemic peptides derived from soy β-conglycinin (βCG) that regulate cholesterol using an identical mechanism. Numerous hypotriglyceridemic di-peptides such as KA, VK, and SY all from glycinin, βCG, and lipoxygenase that are identified as soybean constituents. In the experiment performed in rats, the βCG has proven to prevent the synthesis of fatty acid in the liver, and therefore, contributed to lowering the level of serum triglycerides. More studies have identified peptides such as KNPQLR, EITPEKNPQLR, and RKQEEDEDEEQQRE to subdue the activities of fatty acid synthase (FAS). Equally, the subunits of α'βCG such as KNPQLR, EITPEKNPQLR, and RKQEEDEDEEQQRE act as FAS inhibitors, and have helped in reducing the concentration of lipid in mouse 3T3-L1 adipocytes by encoding more lively peptides in the cells than glycinin subunits.

9.3.2 Antidiabetic

Most soy peptides with hypolipidemic functions have anti-diabetic properties from diverse trialsimulations. Hypocholesterolemia soy peptides, including LPYP, IAVPGEVA, and IAVPTGVA, have proven to advance the uptake of glucose by increasing its absorption in the hepatic cells through the biological membrane. IAVPTGVA particularly has been identified in *in vitro* studies as an effective moderator of dipeptidyl peptidase IV (DPP-IV) which is also a serine exopeptidase. DPP-IV hydrolyzes glucagon-like and glucose-dependent insulinotropic peptides, which regulates the level of blood sugar (Chatterjee et al., 2018). Soybean protein and isoflavones function as diabetic rodent models because they lower the serum and glucose levels as well as increase insulin emission and reduce plasma glucose. Occasionally, fermented soybean produces peptides such as natto and *cheonggukjang* that prevent the inception of type II diabetes in humans and mice (Azuma, Machida, Saeki, Kanamoto, & Iwami, 2000). Therefore, when a person with diabetes, particularly women aged 18–40, consumes a soy protein for more than a month, they will have better insulin resistance capability. Also, the levels of serum, insulin, and plasma glucose in their blood will be lowered.

9.3.3 Antihypertensive

High blood pressure and hypertension are the main participating sources of heart disease. Antihypertensive peptides are bioactive peptides to be extensively studied (De Mejia & Ben, 2006). The bioactivity of soybean proteins is exhibited by the suppression of angiotensin-converting enzyme (ACE). Soy properties have shown increased antihypertensive activities such as inhibiting angiotensin-converting enzyme (ACE) and the renin-angiotensin-aldosterone system (RAAS) to control blood pressure (Yoshikawa et al., 2000). ACE is a proteolytic enzyme that is found to enhance the elimination of dipeptides in the renin-angiotensin system (RAS). The hormonal system is responsible for blood pressure regulation, electrolyte balance, and systematic vascular resistance (Gaoet al., 2003). Angiotensinogen is released by the liver and is then converted into the deca-peptide angiotensin I by the plasma resin from the renal blood. The dipeptidyl carboxypeptidase activity of ACE then converts angiotensin I into the octapeptide angiotensin II, a vasoconstrictive peptide that narrows the blood vessels, causing an increase in blood pressure. ACE suppression therefore produces an antihypertensive effect (Nateshet al., 2003).

Hydrolysates of soy proteins have demonstrated diverse ACE-suppressing bioactive peptides. Peptide fractions detached from soybean proteins' peptic digests by ion-exchange chromatography suppressed ACE activity. Yoshikawa et al. (2002) employed genetic engineering and presented an advanced corporeal function into soy protein. RPLKPW, a strong antihypertensive peptide was given to three analogous organs and when the fourth string was introduced, the antihypertensive activity was enhanced. Kodera and Nio (2002) created pleasant-tasting peptides with ACE-suppressing activity derived from soybeans. Soybean protein digestion with protease produced five peptides, Pro-Asn-Asn-LysPro-Phe-Gln, Tyr-Val-Val-Phe-Lys, Ile-Pro-Pro-Gly-Val, Pro-Tyr-Trp-Thr,Thr-Pro-Arg-Val-Phe, and Asn-Trp-Gly-Pro-Leu-Val (Kodera et al., 2002).

The fermented soybean manufacturing products also generate ACE-suppressing bioactive peptides such as Tyr-Val-Trp-Lys during the fermentation process (Kimura et al., 2000). Fermented soybean paste, which is popular in Korea, when treated with chymotrypsin consists of the hypotensive tripeptide HHL (Shin et al., 2001). Many Asian fermented soybean goods like natto and tempeh contain inhibitory peptides for ACE, which helps with the regulation of blood (Borer, 2007). Similarly, Korean fermented soybean pastes have been linked with chymotrypsin and are rich in hypotensive tripeptide (Wang & De Mejia, 2005). The soybean fermented with *Bacillus subtilis* or *Bacillus natto* has been identified to have two ACE antioxidant peptides, VAHINVGZK and YVWK (Sanjukta & Rai, 2016). The fermented soybean seasoning contains SY and GY peptides that lower hypertension in salt-sensitive dahl, suppress the renin-angiotensin system, and lower the amount of serum aldosterone, hence creating more ACE inhibitory activity, unlike soy sauce (García et al., 2013). Additional antihypertensive soy peptides are NWGPLV, PGTAVFK, PNNKPFQ, LLF, LSW, LNF, LEF, TPRVF, YVVFK, and IVF. It has been demonstrated that bioactive peptides with blood antihypertensive properties have some structural similarities. They are basically resistant to deterioration by digestive enzymes due to Pro or hydroxyl-Pro at the C-terminal. Interestingly, the dipeptides with the C-terminal Tyr recorded increased antihypertension effect over dipeptides with C-terminal Phe (Kimura et al., 2000).

9.3.4 ANTICANCER

Soy isoflavones have proven to control cancer in the past years. Soy peptides have anticancer properties. For example, soy protein is a main source of hydrophobic peptide (X-MLPSYSPY) to control the cell cycle movement of murine lymphoma cells during a mitotic G2 phase (Chatterjee et al., 2018). Subsequent studies have proven that lunasin and Bowman–Birk inhibitor (BBI) soy peptides are richer in anticancer properties. BBI is a small protein granule that acts as an anti-nutrient that can inhibit the action of trypsin and chymotrypsin. BBI acts as a nontoxic chemoprophylaxis agent for cancer that inhibits and suppresses the initiation of cancer formation and transformation using. It usually inhibits 800–2,000 chymotrypsin units. BBI applies its anticancer properties using an apoptosis mechanism which damages mitochondria and results in reactive oxygen species and proteasomal and angiogenesis inhibition. Lunasin has been proven in SENCAR skincare for mice to inhibit skin tumor conditions. Lunasin, a soybean-derived peptide, generates extensive anti-inflammatory effects such as repression of NF-κB movement, less appearance of cytokine, and lower prevalence of cyclooxygenase-2 (COX-2) as well as the antioxidant and anticarcinogenic compositions (Chatterjee et al., 2018). Lunasin promotes group subdual of carcinogens and viral oncogenes induced in mammalian cells. It contains arginyl-glycyl-aspartic acid (RGD) which starts the anti-inflammatory effects that cause integrin signals and downstream pro-inflammatory cascades (Dia & de Mejia, 2010). When a soybean is utilized, a peptide is produced through different methods degraded by enzymes and bacteria. Relatively, they may impact anti-inflammatory effects on macrophage cells. A good example is *cheonggukjang*, which is a fermented soybean item for consumption with peptides. A domestic study in Korea has proven *cheonggukjang* to have anti-inflammatory effects in breast cancer cells since it negatively impacts cytokine/chemokines formation and activates the transformational growth factor (TGF)-beta signaling. Therefore, the study concluded that isoflavones and flavones present in soy can aid in cancer prevention; nonetheless, more investigations need to be carried out to confirm the interactions (Caro et al., 1989; Birt et al., 2001). A study indicated that a high-soybean diet changed the profile of the colonic gene and increased somatostatin, a growth inhibitive agent for colon cancer cells, thus inhibiting tumors (Xiao et al., 2005). Frequent consumption of soy proteins lowered the occurrence of tumors in rats' colons (Hakkak et al., 2001). Part of these anticancer activities may be because of bioactive peptides acquired from soy proteins. The theoretical and experimental works that use a derivation of peptides from soy may possess anticancer properties. Soy isoflavones are phytochemicals that have been associated with health in humans because of antioxidant properties inhibiting diabetes, heart disease, prostate and breast cancers, and osteoporosis.

9.3.5 Antioxidant and Anti-Inflammatory

Soybean bioactive peptides not only show anticancer activities but exert anti-inflammatory and antioxidant effects. Carcinogenesis relies on irregular tissue growth due to pro-oxidant, pro-inflammatory, and immunosuppressive systems. Both proteins and peptides having anticancer properties frequently present anti-inflammatory and antioxidant properties as well (Xue et al., 2016). Soybean peptides have proven to have anti-inflammatory effects and diminish oxidative stress with or without isoflavones. Isoflavones are phytochemicals and bioactive compounds affiliated with soy proteins (Amigo-Benavent et al., 2014). They are also referred to as phytoestrogens because they anatomically resemble estradiol and can bond with estrogen receptors (Park et al., 2010). The cell culture has shown that soybean peptides have estrogenic and anti-estrogenic effects. They occur naturally in soybeans as genistin, glycetin, glycosides, and daidzin (Pyo et al., 2006). The soy protein exhibits its anti-inflammatory properties by suppressing nuclear factor-kappa and obstructing the secretion of pro-inflammatory cytokines which have an adverse effect on other cells.

Amino acids such as Typ and Tyr are said to have antioxidants properties (Saito et al., 2003). Several amino acids are considered antioxidants and when the soy protein pattern is altered during hydrolysis, it exposes a more vigorous amino acid R group (Vernaza et al., 2012). Thus, soybean peptides may have increased antioxidative effects than intact proteins (Chen et al., 1998). When glycinin and con-glycinin underwent enzyme digestion, radical scavenging effects were increased three to five times. Heat did not influence the protein activity, demonstrating that peptide formation was more crucial than protein structure preserving (Matoba, 2002).

Similar results have been found in male rats; the intake of soy peptide resulted in the reduction of neurotoxicant paraquat (PQ)-induced oxidative stress, inhibiting the increase in lung weight caused by paraquat (Takenaka et al., 2003). It was not specified whether isoflavones were present. When 28 anatomically related peptides were isolated from soybean protein digests, Pro-His-His was established as a dynamic core. Thus, it was deemed that His- comprising peptides may perform as active-oxygen quenches, metal-ion bonding agents, and radical scavengers that support the antioxidative effect of peptides (Saito et al., 2003).

Chen et al. (1996) reported that soy proteins have anti-inflammatory properties and function as oxidative stress regulators; this may be attributed to the fact that soy proteins impede nuclear factor-kappa B (NF-κB). They block pro-inflammatory cytokines release and oxidative stress that may cause tissue damage during inflammation. The significance of the property has been proven in rat, hyperlipidemic mouse, and human models (Chen et al., 1998). Lunasin (SKWQHQQDSCRKQKQ) functions effectively as an anticancer agent because of its strong anti-inflammatory and antioxidative properties (Singh et al., 2014).

9.3.6   IMMUNOMODULATORY

Immunomodulatory peptides are closely linked with anticancer, antioxidant, and anti-inflammatory peptides. They boost the cell immune system by regulating cytokines and cell activities. Immunomodulatory peptides are found in hydrolyzed soy proteins which are enzymatically digested. Soy protein hydrolysates rich in alkali prevent the multiplication of more murine splenic lymphocytes. A good example is HCGAPA and GAPA peptides from the glycinin component which inhibits phagocytosis effects of peritoneal macrophages. Some therapeutic properties of soy bioactive peptides have been listed in Table 9.2.

9.4   COMMERCIAL APPLICATION OF SOYBEAN PEPTIDES

9.4.1   NUTRACEUTICAL

The application of bioactive peptides is an emerging field in food science. In all bio-agents, the endogenous peptides have significantly functioned as hormones and neurotransmitters. They regulate the metabolism such as nutrients, water, and minerals through their hormone-receptor interactions and signaling cascades. In performing physiological roles, endogenous peptides control the secretion of glands, adjust blood pressure, and impact body growth. Endogenous peptides impact the central nervous system (CNS) which then affects the sleeping patterns, memory, learning, sexual behavior, pain, appetite, and stress of humans (Wang & De Mejia, 2005). In living organisms, most peptide hormones are involved in the hypothalamus hormones cascade. A good example is corticoliberin (CRH, 41-peptide amide) and thyroliberin (TRH, pGlu-His-ProNH2) that has the function of hypothalamic hormones. They can stimulate the emission of two pituitary hormones, corticotropin (acth, 39aa), and thyrotropin (glycoprotein, chain 96aa, chain 112aa). Sewald and Jakubke (2002) found that peptides such as gastroentero pancreatic peptides that include insulin and glucagons have proven significant effects in regulating metabolism.

Lee et al. (2013) observed that bioactive soybean peptides are associated with numerous health benefits including the prevention of certain diseases. Numerous *in vitro* studies on animal trials and epidemiological observations have linked the consumption of soy phytochemicals, particularly isoflavones and peptides, with health benefits including reduced mortality rates connected to prostate, breast, and endometrial cancer (Abdollahi et al., 2018). Bioactive peptides are active during enzymatic reactions, gastrointestinal (GI) digestion, food processing, and fermentation but are inactive when they are involved in the practical process of producing amino acids. Approximately 2–20 amino acids are absorbed into the blood by the intestines to generate efficient or indigenous biological effects in target tissues. Regarding that, several studies have proven that when 11S peptides are administered in the human intestinal wall, they produce higher concentrations of amino acids in the venous blood compared to when 11S globulin is absorbed as a beverage in the blood system (Chatterjee et al., 2018). Those results proved that a hydrolyzed soy protein is faster and can be quickly absorbed into the blood system during circulation.

Soybean is an excellent source of nutraceutical; it has physiological functions including antihypertensive, anticholesterol, immunomodulatory, antioxidant, and anticancer properties. The high intake of soybean among the Asian population has helped in reducing the palpable effects of these conditions. The health effects increase when soy proteins are being processed into peptides in the gestational intestines. Proteins such as lunasin and soymorphins have more than one of these properties that aid them to inhibit the distribution of many chronic diseases (Demejia & De Lumen, 2006). Among these peptides, lunasin is the most promising and active anticancer agent associated with the most accessible component of healthy living (Lule et al., 2015). Other potential benefits

TABLE 9.2

Properties of Soybean Bioactive Peptides

Source of Soybean Protein	Verification Procedure	Properties of Soybean Peptide	Soy Bioactive Peptide
BCG	HepG2 human liver cells	Hypocholesterolemic	YVVNPDNDEN YVVNDPNNEN
	Inhibitory activity of ACE assay	ACE inhibiting	LAIPVNKP LPHF
βCG (α-subunit)	Fatty acid synthase (FAS) inhibitor studies; 3T3-L1mouse adipocyte	Fatty acid synthase (FAS) inhibition	KNPQLR. RKQEEDEDEEQQRE. EITPEKNPQLR.
βCG (β-subunit)	Ileum assay in guinea pig; opioid activity indiabetic mice (KKAʸ)	Immunostimulating; antidiabetic intestinal	Soymorphin-5: YPFVV
	Maze test in male ddY mice	transit	Soymorphin-6: YPFVVN
	Male ddY and BALB/c mice	triglyceride-lowering	Soymorphin-7: YPFVVN
Glycinin (A4 and A5)	Mice HMGR activity assay (50 mg/kg for 2 days)	Hypocholesterolemic	LPYPR
Lunasin	Suppression of skin papilloma in SENCAR mice; improved the tumor activity of natural killer cells both *in vitro* and *in vivo*; anti-apoptotic in MCF-7 breast cancer cells, acts as antioxidant in Caco-2 cells	Hypocholesterolemic; anti-inflammatory; anticancer; antioxidative	GVNLTPCEKHIMEKIQ SKWQHQQDSCR KQKQGRGDDD DDDDDD
BBI (Bowman-Birk inhibitory)	Inhibited ROS in prostate cancer cells including 267B1/Ki-ras, BRF-55T,PC-3 cells) and LNCaP; chemopreventive in chromosome abnormalities	Anticancer proteinase inhibitor chemoprevention	
Glycinin, βCG-α, βCG-α', βCG-β, trypsin inhibitor, and lipoxygenase	Long-Evans TokushimaOtsuka fatty male rats; HepG2cells; Wistar male rats	Lowering triglycerides	KA
Glycinin, trypsin inhibitor, and lipoxygenase			VK
Glycinin, lipoxygenase, βCG-α', βCG-α, & βCG-β			SY
Defatted protein from soybeans	Arrest of G2/M step P388DI mouse monocytemacrophage to block cell cycle progression	Anticancer disease	X-MLPSYSPY
Soybean protein	Postmenopausal women; ApoE knockout mice	Anti-inflammatory	WGAPSL; VAWWMY; FVVNATSN
	Rats; HepG2 cells	Hypocholesterolemia	
Chymotrypsin fermented Korean soybean paste	Hypertensive rats randomly	Hypotensive	HHL
Soybean protein genetically engineered		Antioxidative; antihypertensive	LLPHH; RPLKPW
Protein black soybean	Adipocytes of the 3T3-L1 mouse	Inhibition by adipogenesis	IQN

include the prevention of chronic diseases such as obesity, type II diabetes, and immune disorders as well as cardiovascular diseases. Wang and De Mejia (2005) report that soybean-derived peptides play a crucial role in their physiological properties and help in the prevention of chronic diseases. The physiological roles of soy proteins are attributed to their bioactive peptides after they are processed (Singh et al., 2014). These results have attributed to the sanction of health claims of soy proteins to reduce the risk of coronary heart disease in America and approval of soy proteins to lower cholesterol levels in Canada (Chatterjee et al., 2018). Bioactive soybean peptides have proven efficient in their pharmaceutical application.

9.4.2 FUNCTIONAL FOOD INGREDIENTS

Recently, soybean has been commonly used to generate a series of functional foods from its peptide extract, protein complexes, and dietary supplements (El Sohaimy, 2012; Lee et al., 2013; Li-Chan, 2015). Soybean peptides are generated through classical ways before their bioactivity is confirmed. Bioactive peptides isolated from food include the concoction of peptides and are nothing like artificial drug molecules which are single units (Barrios et al., 2014). Bioactive peptides are hydrophobic, and thus are less soluble in higher concentrations. However, the purified ones enhance their cost level and reduce yields by excluding any valuable synergistic effects present in the whole hydrolysate. The two major potential soy protein ingredients used in food commercials to stimulate cholesterol are lunasin (peptide extract) and LunaSoy™ (protein complex). Lunasin is one of the most promising dietary compounds (Hsieh et al., 2017). Numerous lunasin products have been made commercially available. Lunasin is most used by Relive International Company to produce LunaRich, which is a dietary complement for heart and cellular well-being. Agro-Mercantil Ltd, a commercial company, is working on the commercialization of soybean for human consumption, particularly those that are to be exported (Singh et al., 2014). The cultivation process depends on the grain size, which yet again impacts the levels of proteins to be extracted.

9.5 SUMMARY

The present evaluation has revealed that peptides resulting from soybean possess outstanding multifunctional health advantages like immune-stimulatory, anti-inflammatory, anticancer, hypolipidemic, antihypertensive, antidiabetic, neuromodulatory, and antioxidant properties established in various models. Soybean bioactive peptides are fragments of small protein obtained by hydrolysis of enzymes, fermentation, processing of food, and gastrointestinal digestion of bigger soybean proteins. The soybean peptide can be prepared and produced by various methods and is influenced by the bacteria or enzymes and the type of soy proteins that are used in the isolation processes. Soybean bioactive peptides are currently isolated through several energy-efficient schemes and technological methods such as gel filtration, affinity chromatography, ion exchange, fastening liquid proteins, membrane, and retrogressing high yield liquid chromatography. Consumption of soybean has been allied with numerous health welfares in lessening chronic diseases such as cardiovascular disease, obesity, insulinresistance/type II diabetes, immune disorders, and certain types of cancers. The studies proved the physiological health benefits accredited to soy proteins, either intact or from bioactive peptides resulting from soybean processing. Software-based methods, also known as *in silico* analysis, are used for identifying and predicting cryptic peptides. Commercial application of bioactive peptides includes their use in cell culture media as bio-pharmaceuticals with hypocholesterolemic properties.

REFERENCES

Abdollahi, M. R., Zaefarian, F., Gu, Y., Xiao, W., Jia, J., & Ravindran, V. (2018). Influence of soybean bioactive peptides on performance, foot pad lesions and carcass characteristics in broilers. *Journal of Applied Animal Nutrition, 6e3.*

Acquah, C., Chan, Y. W., Pan, S., Agyei, D., & Udenigwe, C. C. (2019). Structure-informed separation of bioactive peptides. *Journal of Food Biochemistry, 43*(1), e12765.

Agyei, D. (2015). Bioactive proteins and peptides from soybeans. *Recent Patents on Food, Nutrition & Agriculture, 7*(2), 100–107.

Ahn, S. W., Kim, K. M., Yu, K. W., Noh, D. O., & Suh, H. J. (2000). Isolation of angiotensin I converting enzyme inhibitory peptide from soybean hydrolysate. *Food Science Biotechnology, 9*(3), 378–381.

Amigo-Benavent, M., Clemente, A., Caira, S., Stiuso, P., Ferranti, P., & del Castillo, M. D. (2014). Use of phytochemomics to evaluate the bioavailability and bioactivity of antioxidant peptides of soybean β-conglycinin. *Electrophoresis, 35*(11), 1582–1589.

Andre-Frei, V., Perrier, E., Augustin, C., Damour, O., Bordat, P., Schumann, K., Förster, T., & Waldmann-Laue, M. (1999). A comparison of biological activities of a new soya biopeptide studied in an in vitro skin equivalent model and human volunteers. *International Journal of Cosmetic Science, 21*(5), 299–311.

Aoyama, T., Fukui, K., Nakamori, T., Hashimoto, Y., Yamamoto, T., Takamatsu, K., & Sugano, M. (2000a). Effect of soy and milk whey protein isolates and their hydrolysates on weight reduction in genetically obese mice. *Bioscience, Biotechnology, and Biochemistry, 64*(12), 2594–2600.

Aoyama, T., Fukui, K., Takamatsu, K., Hashimoto, Y., & Yamamoto, T. (2000b). Soy protein isolate and its hydrolysate reduce body fat of dietary obese rats and genetically obese mice (Yellow KK). *Nutrition, 16*(5), 249–254.

Arnoldi, A., D'Agostina, A., Boschin, G., Lovati, M. R., Manzoni, C., & Sirtori, C. R. (2001). Soy protein components active in the regulation of cholesterol homeostasis. Biologically active phytochemicals in food. *Royal Society of Chemistry, 269*, 103–106.

Azuma, N., Machida, K., Saeki, T., Kanamoto, R., & Iwami, K. (2000). Preventive effect of soybean resistant proteins against experimental tumorigenesis in rat colon. *Journal of Nutritional Science and Vitaminology, 46*(1), 23–29.

Birt, D. F., Hendrich, S., & Wang, W. (2001). Dietary agents in cancer prevention: Flavonoids and isoflavonoids. *Pharmacology and Therapeutics, 90*(2/3), 157–177.

Blomstrand, B., & Newsholme, E. A. (1992). Effect of branched-chain amino acid supplementation on the exercise-induced change in aromatic amino acid concentration in human muscle. *Acta Physiologica Scandinavica, 146*(3), 293–298.

Borer, J. S. (2007). Angiotensin-converting enzyme inhibition: A landmark advance in treatment for cardiovascular diseases. *European Heart Journal Supplements, 9*, E2–E9.

Capriotti, A. L., Caruso, G., Cavaliere, C., Samperi, R., Ventura, S., Chiozzi, R. Z., & Laganà, A. (2015). Identification of potential bioactive peptides generated by simulated gastrointestinal digestion of soybean seeds and soy milk proteins. *Journal of Food Composition and Analysis, 44*, 205–213.

Caro, J. F., Dohm, L. G., Pories, W. J., & Sinha, M. K. (1989). Cellular alterations in liver, skeletal muscle, and adipose tissue responsible for insulin resistance in obesity and type II diabetes. *Diabetes/Metabolism Reviews, 5*(8), 665–689. doi: 10.1002/dmr.5610050804.

Català-Clariana, S., Benavente, F., Giménez, E., Barbosa, J., Sanz-Nebot, V. (2013) Identification of bioactive peptides in hypoallergenic infant milk formulas by CE-TOF-MS assisted by semiempirical model of electromigration behavior. *Electrophoresis, 34*(13), 1886–1894.

Chang, W.-S., Lee, H.-I., & Hungria, M. (2014). Soybean production in the Americas. In *Principles of Plant-Microbe Interactions* (pp. 393–400). doi: 10.1007/978-3-319-08575-3_41.

Chatterjee, C., Gleddie, S., & Xiao, C.-W. (2018). Soybean bioactive peptides and their functional properties. *Nutrients, 10*(9), 1211. doi: 10.3390/nu10091211.

Chen, H.-M., Muramoto, K., Yamauchi, F., Fujimoto, K., & Nokihara, K. (1998). Antioxidative properties of histidine-containing peptides designed from peptide fragments found in the digests of a soybean protein. *Journal of Agricultural and Food Chemistry, 46*(1), 49–53. doi: 10.1021/jf970649w.

Chen, H.-M., Muramoto, K., Yamauchi, F., & Nokihara, K. (1996). Antioxidant activity of designed peptides based on the antioxidative peptide isolated from digests of a soybean protein. *Journal of Agricultural and Food Chemistry, 44*(9), 2619–2623. doi: 10.1021/jf950833m.

Cohen, L. A. (2000). Effect of intact and isoflavone-depleted soy protein on NMU-induced rat mammary tumorigenesis. *Carcinogenesis, 21*(5), 929–935. doi: 10.1093/carcin/21.5.929.

De Mejia, E., & Delumen, B. (2006). Soybean bioactive peptides: A new horizon in preventing chronic diseases. *Sexuality, Reproduction and Menopause, 4*(2), 91–95. doi: 10.1016/j.sram.2006.08.012.

De Mejia, E. G., & Dia, V. P. (2009). Lunasin and lunasin-like peptides inhibit inflammation through suppression of NF-κB pathway in the macrophage. *Peptides, 30*(12), 2388–2398. doi: 10.1016/j.peptides.2009.08.005.

Dia, V. P., & Mejia, E. G. de. (2010). Lunasin promotes apoptosis in human colon cancer cells by mitochondrial pathway activation and induction of nuclear clusterin expression. *Cancer Letters*, *295*(1), 44–53. doi: 10.1016/j.canlet.2010.02.010.

Dia, V. P., Wang, W., Oh, V. L., Lumen, B. O. d., & de Mejia, E. G. (2009). Isolation, purification and characterisation of lunasin from defatted soybean flour and in vitro evaluation of its anti-inflammatory activity. *Food Chemistry*, *114*(1), 108–115. doi: 10.1016/j.foodchem.2008.09.023.

Dziuba, J., Minkiewicz, P., Nałecz, D., & Iwaniak, A. (1999). Database of biologically active peptide sequences. *Nahrung/Food*, *43*(3), 190–195. doi: 10.1002/(sici)1521-3803(19990601)43:3<190::aid-food190>3.0.co;2-a.

ElSohaimy, S. A. (2012). Functional foods and nutraceuticals-modern approach to food science. *World Applied Sciences Journal*, *20*(5), 691–708. doi: 10.5829/idosi.wasj.2012.20.05.66119.

Floris, R., Recio, I., Berkhout, B., & Visser, S. (2003). Antibacterial and antiviral effects of milk proteins and derivatives thereof. *Current Pharmaceutical Design*, *9*(16), 1257–1275. doi: 10.2174/1381612033454810.

Franek, F., Hohenwarter, O., & Katinger, H. (2000). Plant protein hydrolysates: Preparation of defined peptide fractions promoting growth and production in animal cells cultures. *Biotechnology Progress*, *16*(5), 688–692. doi: 10.1021/bp0001011.

Gao, Y., Chen, C., Zhang, P., Chai, Z., He, W., & Huang, Y. (2003). Detection of metalloproteins in human liver cytosol by synchrotron radiation X-ray fluorescence after sodium dodecyl sulphate polyacrylamide gel electrophoresis. *Analytica Chimica Acta*, *485*(1), 131–137. doi: 10.1016/s0003-2670(03)00347-7.

García, M. C., Puchalska, P., Esteve, C., & Marina, M. L. (2013). Vegetable foods: A cheap source of proteins and peptides with antihypertensive, antioxidant, and other less occurrence bioactivities. *Talanta*, *106*, 328–349. doi: 10.1016/j.talanta.2012.12.041.

González-Montoya, M., Hernández-Ledesma, B., Mora-Escobedo, R., & Martínez-Villaluenga, C. (2018). Bioactive peptides from germinated soybean with anti-diabetic potential by inhibition of dipeptidyl peptidase-IV, α-amylase, and α-glucosidase enzymes. *International Journal of Molecular Sciences*, *19*(10), 2883. doi: 10.3390/ijms19102883.

Gu, Y., & Wu, J. (2013). LC–MS/MS coupled with QSAR modeling in characterising of angiotensin I-converting enzyme inhibitory peptides from soybean proteins. *Food Chemistry*, *141*(3), 2682–2690. doi: 10.1016/j.foodchem.2013.04.064.

Hakkak, R., Korourian, S., Ronis, M. J., Johnston, J. M., & Badger, T. M. (2001). Soy protein isolate consumption protects against azoxymethane-induced colon tumors in male rats. *Cancer Letters*, *166*(1), 27–32. doi: 10.1016/s0304-3835(01)00441-4.

Hsieh, C.-C., Martínez-Villaluenga, C., de Lumen, B. O., & Hernández-Ledesma, B. (2017). Updating the research on the chemopreventive and therapeutic role of the peptide lunasin. *Journal of the Science of Food and Agriculture*, *98*(6), 2070–2079. doi: 10.1002/jsfa.8719.

Jeong, H. J., Park, J. H., Lam, Y., & de Lumen, B. O. (2003). Characterization of lunasin isolated from soybean. *Journal of Agricultural and Food Chemistry*, *51*(27), 7901–7906. doi: 10.1021/jf034460y.

Kamran, F., & Reddy, N. (2018). Bioactive peptides from legumes: Functional and nutraceutical potential. *Recent Advances in Food Science*, *1*(3), 134–149.

Kim, S. E., Kim, H. H., Kim, J. Y., Kang, Y. I., Woo, H. J., & Lee, H. J. (2000). Anticancer activity of hydrophobic peptides from soy proteins. *BioFactors*, *12*(1–4), 151–155.

Kimura, A., Takada, A., Okada, T., & Yamada, H. (2000). Microbial manufacture of angiotensin I-converting enzyme inhibiting peptides. Japan: Toyo Hatsuko K.K.

Kitts, D., & Weiler, K. (2003). Bioactive proteins and peptides from food sources. Applications of bioprocesses used in isolation and recovery. *Current Pharmaceutical Design*, *9*(16), 1309–1323. doi: 10.2174/1381612033454883.

Kodera, T., & Nio, N. (2002). *Angiotensin Converting Enzyme Inhibitors*. PCT International Application (WO 2002055546 A1 18 43 p. KokaiTokkyoKoho).

Korhonen, H., & Pihlanto, A. (2003). Food-derived bioactive peptides - Opportunities for designing future foods. *Current Pharmaceutical Design*, *9*(16), 1297–1308. doi: 10.2174/1381612033454892.

Kuba, M., Tanaka, K., Tawata, S., Takeda, Y., & Yasuda, M. (2003). Angiotensin I-converting enzyme inhibitory peptides isolated from Tofuyo fermented soybean food. *Bioscience, Biotechnology, and Biochemistry*, *67*(6), 1278–1283. doi: 10.1271/bbb.67.1278.

Lassoued, I., Mora, L., Barkia, A., Aristoy, M.-C., Nasri, M., & Toldrá, F. (2015). Bioactive peptides identified in thornback ray skin's gelatin hydrolysates by proteases from *Bacillus subtilis* and *Bacillus amyloliquefaciens*. *Journal of Proteomics*, *128*, 8–17. doi: 10.1016/j.jprot.2015.06.016.

Lee, J. K., Li-Chan, E. C. Y., Jeon, J.-K., & Byun, H.-G. (2013). Development of functional materials from seafood by-products by membrane separation technology. *Seafood Processing By-Products*, 4: 35–62. doi: 10.1007/978-1-4614-9590-1_4.

Lemes, A., Sala, L., Ores, J., Braga, A., Egea, M., & Fernandes, K. (2016). A review of the latest advances in encrypted bioactive peptides from protein-rich waste. *International Journal of Molecular Sciences*, *17*(6), 950. doi: 10.3390/ijms17060950.

Li-Chan, E. C. (2015). Bioactive peptides and protein hydrolysates: Research trends and challenges for application as nutraceuticals and functional food ingredients. *Current Opinion in Food Science*, *1*, 28–37. doi: 10.1016/j.cofs.2014.09.005.

López-Barrios, L., Gutiérrez-Uribe, J. A., & Serna-Saldívar, S. O. (2014). Bioactive peptides and hydrolysates from pulses and their potential use as functional ingredients. *Journal of Food Science*, *79*(3), R273–R283. doi: 10.1111/1750-3841.12365.

Lule, V. K., Garg, S., Pophaly, S. D., Hitesh, & Tomar, S. K. (2015). Potential health benefits of lunasin: A multifaceted soy-derived bioactive peptide. *Journal of Food Science*, *80*(3), R485–R494. doi: 10.1111/1750-3841.12786.

Ma, H., Liu, R., Zhao, Z., Zhang, Z., Cao, Y., Ma, Y., ... Xu, L. (2016). A novel peptide from soybean protein isolate significantly enhances resistance of the organism under oxidative stress. *PLOS ONE*, *11*(7), e0159938. doi: 10.1371/journal.pone.0159938.

Matoba, T. (2002). How does the radical-scavenging activity of soy protein food change during heating? *DaizuTanpakushitsuKenkyu*, *5*, 47–50.

Moldes, A. B., Vecino, X., & Cruz, J. M. (2017). Nutraceuticals and food additives. *Current Developments in Biotechnology and Bioengineering*, 6: 143–164. doi: 10.1016/b978-0-444-63666-9.00006-6.

Möller, N. P., Scholz-Ahrens, K. E., Roos, N., & Schrezenmeir, J. (2008). Bioactive peptides and proteins from foods: Indication for health effects. *European Journal of Nutrition*, *47*(4), 171–182. doi: 10.1007/s00394-008-0710-2.

Natesh, R., Schwager, S. L. U., Sturrock, E. D., & Acharya, K. R. (2003). Crystal structure of the human angiotensin-converting enzyme–lisinopril complex. *Nature*, *421*(6922), 551–554. doi: 10.1038/nature01370.

Ogawa, J., Yamanaka, H., Mano, J., Doi, Y., Horinouchi, N., Kodera, T., ... Shimizu, S. (2007). Synthesis of 4-hydroxyisoleucine by the aldolase–transaminase coupling reaction and basic characterization of the aldolase from arthrobactersimplexAKU 626. *Bioscience, Biotechnology, and Biochemistry*, *71*(7), 1607–1615. doi: 10.1271/bbb.60655.

Park, S. Y., Lee, J.-S., Baek, H.-H., & Lee, H. G. (2010). Purification and characterization of antioxidant peptides from soy protein hydrolysate. *Journal of Food Biochemistry*, *34*, 120–132. doi: 10.1111/j.1745-4514.2009.00313.x.

Penta-Ramos, E. A., & Xiong, Y. L. (2002). Antioxidant activity of soy protein hydrolysates in a liposomal system. *Journal of Food Science*, *67*(8), 2952–2956. doi: 10.1111/j.1365-2621.2002.tb08844.x.

Pyo, Y.-H., Lee, T.-C., & Lee, Y.-C. (2006). Effect of lactic acid fermentation on enrichment of antioxidant properties and bioactive isoflavones in soybean. *Journal of Food Science*, *70*(3), S215–S220. doi: 10.1111/j.1365-2621.2005.tb07160.x.

Rai S. N., Hareram, B., Singh, S. S., Zahra, W., Patil, R. R., Jadhav, J. P., Gedda, M. R., Singh, S. P. (2017) Mucuna pruriens Protects against MPTP Intoxicated Neuroinflammation in Parkinson's Disease through NF-κB/pAKT Signaling Pathways[J]. *Frontiers in Aging Neuroscience*, 9, 421.

Raman, P., Ekant, T., Aurovind, A., Su, C-H., Gopinath, S. C. B., Chen, Y., Velusamy, P. (2018) Separation and identification of bioactive peptides from stem of Tinospora cordifolia (Willd.) Miers[J]. *Plos One*, *13*(3), e0193717.

Saito, K., Jin, D.-H., Ogawa, T., Muramoto, K., Hatakeyama, E., Yasuhara, T., & Nokihara, K. (2003). Antioxidative properties of tripeptide libraries prepared by the combinatorial chemistry. *Journal of Agricultural and Food Chemistry*, *51*(12), 3668–3674. doi: 10.1021/jf021191n.

Sánchez, A., & Vázquez, A. (2017). Bioactive peptides: A review. *Food Quality and Safety*, *1*(1), 29–46. doi: 10.1093/fqsafe/fyx006.

Sanjukta, S., & Rai, A. K. (2016). Production of bioactive peptides during soybean fermentation and their potential health benefits. *Trends in Food Science & Technology*, *50*, 1–10. doi: 10.1016/j.tifs.2016.01.010.

Saz, J. M., & Marina, M. L. (2008). Application of micro- and nano-HPLC to the determination and characterization of bioactive and biomarker peptides. *Journal of Separation Science*, *31*(3), 446–458. doi: 10.1002/jssc.200700589.

Sewald, N., & Jakubke, H. D. (2002). *Peptides: Chemistry and Biology*. Weinheim, Germany: Wiley-Vch. Verlag GmbH, p. 562.

Shin, Z.-I., Yu, R., Park, S.-A., Chung, D. K., Ahn, C.-W., Nam, H.-S., ... Lee, H. J. (2001). His-His-Leu, an angiotensin I converting enzyme inhibitory peptide derived from Korean soybean paste, exerts antihypertensive activity in vivo. *Journal of Agricultural and Food Chemistry*, *49*(6), 3004–3009. doi: 10.1021/jf001135r.

Singh, B. P., Vij, S., & Hati, S. (2014). Functional significance of bioactive peptides derived from soybean. *Peptides*, *54*, 171–179. doi: 10.1016/j.peptides.2014.01.022.

Takenaka, A., Annaka, H., Kimura, Y., Aoki, H., & Igarashi, K. (2003). Reduction of paraquat-induced oxidative stress in rats by dietary soy peptide. *Bioscience, Biotechnology, and Biochemistry*, *67*(2), 278–283. doi: 10.1271/bbb.67.278.

Veličković, D. T., Ristić, M. S., Milosavljević, N. P., Karabegović, I. T., Stojičević, S. S., Lazić, M. L. (2012) Chemical composition of the essential oils of Salvia austriaca Jacq. and Salvia amplexicaulis Lam. from Serbia[J]. *Agro Food Industry Hi Tech*, *23*(3), 8–10.

Vernaza, M. G., Dia, V. P., Gonzalez de Mejia, E., & Chang, Y. K. (2012). Antioxidant and antiinflammatory properties of germinated and hydrolysed Brazilian soybean flours. *Food Chemistry*, *134*(4), 2217–2225. doi: 10.1016/j.foodchem.2012.04.037.

Vollmann, J. (2016). Soybean versus other food grain legumes: A critical appraisal of the United Nations International year of pulses 2016. *Die Bodenkultur: Journal of Land Management, Food and Environment*, *67*(1), 17–24. doi: 10.1515/boku-2016-0002.

Walther, B., & Sieber, R. (2011). Bioactive proteins and peptides in foods. *International Journal for Vitamin and Nutrition Research*, *81*, 181–191.

Wang, W., & de Mejia, E. G. (2005). A new frontier in soy bioactive peptides that may prevent age-related chronic diseases. *Comprehensive Reviews in Food Science and Food Safety*, *4*(4), 63–78. doi: 10.1111/j.1541-4337.2005.tb00075.x.

Wang, W., Dia, V. P., Vasconez, M., de Mejia, E. G., & Nelson, R. L. (2008). Analysis of soybean protein-derived peptides and the effect of cultivar, environmental conditions, and processing on lunasin concentration in soybean and soy products. *Journal of AOAC International*, *91*(4), 936–946. doi: 10.1093/jaoac/91.4.936.

Wu, J., & Ding, X. (2001). Hypotensive and physiological effect of angiotensin converting enzyme inhibitory peptides derived from soy protein on spontaneously hypertensive rats. *Journal of Agricultural and Food Chemistry*, *49*(1), 501–506. doi: 10.1021/jf000695n.

Xiao, R., Badger, T. M., & Simmen, F. A. (2005). Dietary exposure to soy or whey proteins alters colonic global geneexpression profiles during rat colon tumorigenesis. *Molecular Cancer, 4(1)*, 1. doi: 10.1186/1476-4598-4-1.

Xue, Z., Wang, C., Zhai, L., Yu, W., Chang, H., Kou, X., & Zhou, F. (2016). Bioactive compounds and antioxidant activity of mung bean (*Vigna radiata* L.), soybean (*Glycine max* L.) and black bean (*Phaseolus vulgaris* L.) during the germination process. *Czech Journal of Food Sciences*, *34*(1), 68–78. doi: 10.17221/434/2015-cjfs.

Yoshikawa, M., Fujita, H., Matoba, N., Takenaka, Y., Yamamoto, T., Yamauchi, R., … Takahata, K. (2000). Bioactive peptides derived from food proteins preventing lifestyle-related diseases. *BioFactors*, *12*(1–4), 143–146. doi: 10.1002/biof.5520120122.

Yoshikawa, M., Yamada, Y., Matoba, N., Utsumi, S., Maruyama, N., & Onishi, K. (2002). Study on introducing new physiological function into soy protein by genetic engineering. *Daizu Tanpakushitsu Kenkyu*, *5*, 26–30.

10 Revisiting Side Streams of Soy Product Processing

Jian-Yong Chua, Weng Chan Vong, and Shao-Quan Liu

CONTENTS

10.1 INTRODUCTION

Soymilk, tofu (soybean curd), and soy protein isolate are common consumer products, but less well-known are the side streams arising from their production – okara (soybean pulp) and soy whey (Figure 10.1). To produce soymilk, soaked soybeans are blended with water, then filtered. The filtrate is the soymilk, while the finely-ground residue that remains is known as okara or soy pulp, a solid side stream. In tofu-making, the soymilk is boiled, then further coagulated by adding various coagulants (e.g. calcium salts or acids), and the curds are pressed to the desired firmness and shape. During the process, soy whey, a light-yellow liquid, is pressed out as another side stream. Soy whey is also generated during soy protein isolate production (Figure 10.1). Given the great popularity of soy foods, such side streams are also generated in huge volumes by the food industry, and their disposal leads to economic losses and environmental concerns.

More than four million tons of okara are generated each year globally, with the top producers being China (accounting for about 70% of the global okara output), Japan (about 20%), and Korea (less than 10%). Despite the large volumes of okara generated by the food industry, most of it is discarded as okara contains about 75–80% water and is very perishable.

In the past 20 years, there has been growing interest in ways of reusing the okara. Okara may be dried for use as an ingredient in baked products. However, the high moisture content of okara leads to high costs of drying. This investment may not be justified, especially if the dried okara is

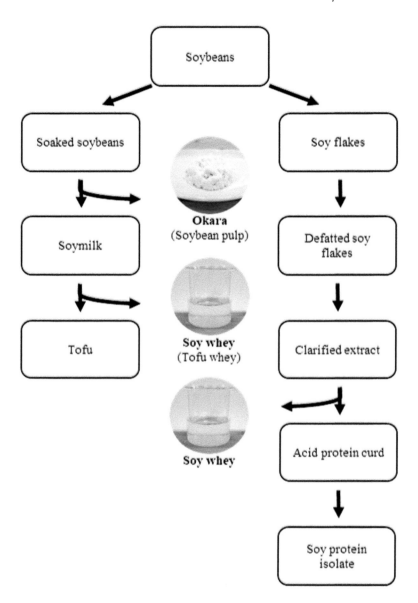

FIGURE 10.1   Overview of generation of okara and soy whey side streams from soymilk, tofu, and soy protein isolate production. Source: modified from Chua and Liu (2019).

used in products with low profit margins. Okara is also susceptible to putrefaction due to its high protein content; it must be refrigerated, frozen, or dried within hours after production if further use is desired. Moreover, the high amount of insoluble dietary fiber in okara gives a gritty mouthfeel, while its particulate nature makes it hard to bind together. These features limit the amount of okara that can be incorporated into food products, as adding too much of it will cause the final product to have a crumbly, fragile texture with poor palatability.

Due to these issues, okara is usually used as compost or animal feed, or simply disposed of. None of these approaches realize or recover the true value of okara. While the extraction of soy protein and/or dietary fiber (largely insoluble) from okara is possible, the spent cake constitutes yet another waste stream.

On the other hand, soy whey is also disposed of as sewage during tofu or soy protein isolate production. However, as soy whey still contains some carbohydrates, nitrogenous compounds, and minerals, it can support microbial growth and causes organic pollution of water if discharged directly. Some negative effects include bad odors and pollution to surface and groundwater. Therefore, soy whey needs to be treated before disposal, further increasing costs for tofu producers.

Some current uses of soy whey include using it as a microbiological nutrient media for the production of nisin (a natural antimicrobial compound), citric acid and biohydrogen, or its extraction for functional food additives. However, for most of these technologies, significant amounts of residual and/or new wastes are still generated eventually; technologies that can utilize soy whey entirely are limited.

Bioprocessing using enzymes and/or microorganisms is one approach to valorizing these soy side streams. While okara and soy whey spoil easily, this also means that that microbial growth is well-supported by both side streams. The key then lies in using the right microorganisms and/or enzymes to achieve targeted, desirable changes so as to improve their composition and facilitate their utilization.

Due to the differences in form and composition, okara and soy whey are discussed separately in this chapter. Selected studies are highlighted to exemplify the potential of these side streams for high-value applications in food and functional foods/supplements.

10.2   OKARA VALORIZATION

10.2.1   An Overview of Okara Composition and Bioactive Components

The composition of okara differs depending on the cultivar of soybeans, the method of soymilk processing, and the amount of water-soluble components extracted from the ground soybeans. Table 10.1 shows the general composition of okara.

Fiber, mainly insoluble fiber in the form of cellulose and hemicellulose, makes up the bulk of the dry matter content at 40–60% (Redondo-Cuenca, Villanueva-Suárez, & Mateos-Aparicio, 2008; van der Riet, Wight, Cilliers, & Datel, 1989). In comparison, free carbohydrates (such as arabinose, glucose, galactose, fructose, sucrose, raffinose, and stachyose) are present in low amounts at 4–5% (dry basis). The comparatively low content of fermentable carbohydrates limits efficient microbial growth in okara (Redondo-Cuenca et al., 2008). Okara also contains 1.4% stachyose and raffinose, which may cause bloat and flatulence in some people.

Protein makes up 15.2–33.4% of okara (dry basis), with the two main proteins being basic 7S globulin and 11S globulin (Singh, Meena, Kumar, Dubey, & Hassan, 2015). Okara protein isolates contain all essential amino acids, but have low solubility in water (Chan & Ma, 1999; Ma, Liu, Kwok, & Kwok, 1996). Okara protein is resistant to complete digestion by the gastrointestinal pepsin and pancreatin. Out of these peptides, the low molecular weight fraction (less than 1 kDa) is most potent in inhibiting angiotensin-converting enzymes and displays the greatest antioxidant activity, possibly due to its high proportion of hydrophobic amino acids (Jiménez-Escrig, Alaiz, Vioque, & Rupérez, 2010). Trypsin inhibitors make up about 5.2–14.4% of the okara protein, but they can be almost completely inactivated by heat (Stanojevic, Barac, Pesic, Jankovic, & Vucelic-Radovic, 2013).

There is also a considerable amount of lipids in okara, at 8.3–10.9% (dry basis). Most of the fatty acids are mono- or polyunsaturated, and they consist of linoleic acid (54.1% of total fatty acids), oleic acid (20.4%), palmitic acid (12.3%), linolenic acid (8.8%), and stearic acid (4.7%) (Mateos-Aparicio, Redondo-Cuenca, Villanueva-Suárez, Zapata-Revilla, & Tenorio-Sanz, 2010). During grinding of soybeans, soy lipoxygenase and hydroperoxide lyase react with the unsaturated fatty acids, primarily linoleic acid, leading to the formation of odor-active compounds such as hexyl and heptyl aldehydes and alcohols. These odorants with low detection thresholds cause the off-flavors in raw soymilk.

TABLE 10.1

General Composition of Okara

Macro Components	Amount (g/100 g Dry Matter)
Carbohydrate	3.8–5.3
Protein	15.2–33.4
Fat	8.3–10.9
Dietary fiber	42.4–58.1
Insoluble dietary fiber	40.2–50.8
Soluble dietary fiber	4.2–14.6
Ash	3.0–4.5

Micro Components	Amount (mg/100 g Dry Matter)
Thiamine (B_1)	0.48–0.59
Riboflavin (B_2)	0.03–0.04
Niacin (B_3)	0.82–1.04
Potassium	936–1350
Sodium	16–96
Calcium	260–428
Magnesium	130–165
Iron	0.6–11
Copper	0.1–1.2
Magnesium	0.2–3.1
Zinc	0.3–3.5

Phytochemicals	Amount (g/100 g Dry Matter)
Aglycones	5.41
Glucosides	10.3
Malonyl glucosides	19.7
Acetyl glucosides	0.32
Phytic acid	0.5–1.2
Saponin	0.10

Sources: Van der Riet et al. (1989), Anderson and Wolf (1995), Jackson et al. (2002), Redondo-Cuenca et al. (2008), Li, Qiao et al. (2012).

Okara also contains a fair amount of isoflavones: About 12–30% of the isoflavones in soybeans are retained in okara during soymilk production (Jackson et al., 2002; Wang & Murphy, 1996). Glycosides and aglycones are the main isoflavones (at 28.9% and 15.4% respectively), with a smaller proportion being acetyl genistin (0.89%) (Jackson et al., 2002). Among these isoflavones, the aglycone forms have the greatest bioavailability in humans (Izumi et al., 2000).

Antinutritional factors, such as saponin and phytate are also present in okara, limiting its use in animal feeds. Okara also contains a trypsin inhibitor, although this can be eliminated by heat treatment (Anderson & Wolf, 1995). Fermentative microorganisms may metabolize these factors and other macromolecules, such as proteins and allergens, resulting in a final fermented product with improved nutrition and digestibility (Anderson & Wolf, 1995; Wang, Qin, Sun, & Zhao, 2014).

Untreated okara can be used as a flour replacement in baked goods (such as bread or cereal bars) to reduce calories and boost fiber content (Radočaj & Dimić, 2013). Okara can also act as a meat replacement in burger patties. Turhan, Temiz, and Sagir (2007) demonstrated that up to 22.5% wet okara can be added to beef patties without adversely affecting sensory properties. The cholesterol content of okara-added beef patties was reduced by 9–42% in cooked beef patties. Wang et al. (2015) also used okara as an extender in pork meat gel products, although sensory results showed that okara with smaller particle size in meatballs was preferred for their better mouthfeel. Untreated okara can therefore boost nutrients and cut costs in such products, although the amount added will be limited by any adverse sensorial impact on the final product. As such, pre-treatments to improve the mouthfeel and functionalities of okara become important in the creation of a versatile ingredient.

10.2.2 Enzyme Treatment

Since the bulk of okara is insoluble dietary fiber (IDF), carbohydrases are often used for digesting okara. Early research focuses on selecting the right carbohydrases and conditions to optimize the conversion of okara insoluble fiber into soluble dietary fiber (SDF). The okara cell wall is complex and cannot be completely digested simply by carbohydrases. Kasai et al. (2004) showed that the primary cell wall of okara could be digested by cellulases, while the secondary cell wall – a complex composition of galacturonic acid, neutral sugars, and protein – could only be digested by pectinases. Still, despite serial digestion by cellulase and pectinase, about 15–17% of okara remained undigested. These residues were oil body complexes in the soybean cells and fiber-like organs between the okara cells.

Carbohydrase digestion of okara offers several benefits. As IDF is converted into SDF, the okara becomes less gritty and more soluble, enhancing its use as an ingredient in foods and (especially) beverages. Functional properties such as oil and water retention capacity and swelling capacity are also improved due to greater surface area (smaller particle size) for adsorption (Villanueva-Suárez, Pérez-Cózar, & Redondo-Cuenca, 2013). Carbohydrase-treated okara can thus be added into dough and meat matrices without them breaking apart easily.

Moreover, okara treated by carbohydrase Ultraflo® improved *in vitro* fermentability by the probiotic *Bifidobacterium bifidus* (Villanueva-Suárez et al., 2013) and also significantly reduced the serum and liver triglyceride levels in rats fed a high-cholesterol diet (Villanueva-Suarez, Perez-Cozar, Mateos-Aparicio, & Redondo-Cuenca, 2016). Similarly, the carbohydrases Celluclast®1.5L (containing cellulase) and Viscozyme®L (a multi-enzyme cocktail of cellulases and hemicellulases) converted the okara IDF into SDF (Vong, Lim, & Liu, 2017). These carbohydrases (side activities of these enzyme preparations) also converted glycosidic isoflavones (genistin, daidzin, and glycitin) in okara into their aglycone form (genistein, daidzein, and glycitein respectively) almost completely within four hours.

Furthermore, carbohydrase treatment can aid in the extraction of protein from okara. De Figueiredo et al. (2018) first treated okara with a carbohydrase under the optimized conditions of 53° C, pH 6.2, and 4% of Viscozyme®L concentration. Under these conditions, a protein content of 56% (dry weight basis) and a recovery of 28% were obtained, representing an increase of 17% and 86%, respectively, compared to the sample with no enzymatic pre-treatment.

On the other hand, physical processing can also enhance the enzymatic digestion of okara IDF. Simple heat treatment at 121° C broke down the okara primary cell wall such that enzymes could access the secondary cell wall more easily (Kasai et al., 2004). Simultaneous treatment of okara with β-glucanase Ultraflo® and high hydrostatic pressure treatment (pressure of 600 MPa) at 40° C for 30 min increased the SDF content by 58.2% (30% higher than when only Ultraflo® was used) (Pérez-López, Cela, Costabile, Mateos-Aparicio, & Rupérez, 2016).

Compared to carbohydrase studies, protease treatment of okara is less well studied. Proteases are usually applied to the okara protein isolates, not to the native okara, for improvement of functional properties and production of bioactive peptides. Okara protein isolates have poor solubility, at about 5% at pH 6. Trypsin hydrolyzed okara protein isolates by 5% to 14% (degree of hydrolysis = 4.9–13.7) and increased the solubility to about 7–30% at pH 6; more extensively hydrolyzed okara protein isolates showed greater solubility (Chan & Ma, 1999). On the other hand, Yokomizo et al. (2002) digested okara with seven individual proteases, then isolated and analyzed the antioxidative peptides. These peptides were composed of two and three amino acid residues, with sequences Ala-Tyr, Gly-Tyr-Tyr, Ala-Asp-Phe, and Ser-Asp-Phe. Notably, these peptides all have an aromatic amino acid at the C-terminal end.

There are even fewer studies on the lipase treatment of okara, despite the fair amount of lipids in okara. Vahvaselkä and Laakso (2010) hydrolyzed okara using lipolytically active oat flour at a water activity of 0.70. The resulting free linoleic acid was then isomerized predominantly to *cis*-9, *trans*-11-conjugated linoleic acid by resting cells of *Propionibacterium freudenreichii* ssp. *shermanii* in 5% aqueous okara slurries. After 21 days, the resultant yield was 1.1 mg/mL (78 mg/g of total lipids) and 22 mg/g dry matter. In another study, Vong and Liu (2018) treated okara directly with a commercial lipase to release fatty acids for their bioconversion into esters by yeast *Lindnera saturnus*. This study is elaborated in Section 10.2.4.

10.2.3 FERMENTATION

Selected enzymes can efficiently break down (often) single, targeted polymeric fractions in okara, but microorganisms produce a mix of extracellular enzymes which can simultaneously transform two or more components in okara. The composition of okara can thus be improved in several ways and may be a more wholesome product. Three classes of microorganisms – fungi, bacteria, and yeasts – have been used in okara fermentation.

10.2.3.1 Fungi

Okara particles offer a physical surface for fungal adherence and growth. Extracellular fungal enzymes can break down okara particles for growth while the fermented okara has enhanced digestibility. Table 10.2 summarizes several studies on the fungal fermentation of okara.

While numerous studies have shown that fungal-fermented okara has higher antioxidant activity using chemical assays, it is more informative if the actual bioactive compound(s) can be identified and monitored during fermentation. For instance, 8-hydroxydaidzein, an inhibitor of aldose reductase and tyrosinase, has been isolated from okara fermented with *Aspergillus* sp. HK-388, with a yield of 30 mg/kg okara (wet basis) (Fujita, Funako, & Hayashi, 2004). Monacolin K, which helps lower cholesterol, has also been extracted from okara fermented by *Monascus purpureus* with a yield of 192 mg/kg okara (dry basis) (Wongkhalaung, Leelawatcharamas, & Japakaset, 2009). Zhu and colleagues (2015) used okara for cultivating various types of edible fungi (mushrooms) (Table 10.2) for the extraction of fungal polysaccharides. Another functional ingredient is γ-aminobutyric acid (GABA). Ochi et al. (2016) reported that in okara fermented by *Aspergillus oryzae*, the GABA content increased from 2.05 to 7.66 mg/100 mL. Higher levels of GABA could have accounted for the higher levels of ACE inhibitory activity observed in the fermented okara.

More recently, Chan et al. (2019) screened four different fungi for the one that showed the greatest extent of α-glucosidase inhibition using chemical assays. Okara fermented by *Eurotium cristatum* was the most potent; this fungus is usually used for making Fuzhuan tea, a type of fermented tea in China. Following that, a metabolomics approach coupled with instrumental analysis revealed the exclusive presence of anthraquinones (emodin and physcion) in the *E. cristatum*–fermented okara. When mice on a normal or high-fat diet were fed the *E. cristatum*–fermented okara 15 min before being fed corn starch, their postprandial blood glucose levels were significantly reduced relative to those fed unfermented okara. Emodin and physicon were suggested to have increased insulin sensitivity in mice. Subsequently, a novel crispy snack suitable for diabetics was developed using the *E. cristatum*–fermented okara as a form of preload intervention therapy.

The consumption of fungal-fermented okara as food is not entirely new. In Hubei Province, China, one indigenous food is *meitauza*, fungal-fermented okara. Traditionally, okara is first shaped into blocks, cooled, and covered with rice straws at below 20° C for about 8–14 days for natural fermentation (Li & Ma, 2014). The major microorganisms involved include a fungus, *Actinomucor elegans*, and a bacterium, *Zymomonas mobilis* (Xu, Liu, & Zhou, 2012), and the cooked *meitauza* has

TABLE 10.2

Summary of Fungal Fermentation of Okara

Products	Fungi	Reference
Bioactive compounds		
8-Hydroxydaidzein	*Aspergillus* sp. HK-388	Fujita et al. (2004)
Monacolin K	*Monascus purpureus*	Wongkhalaung et al. (2009)
Oyster mushroom polysaccharide	*Pleurotus ostreatus*	Shi, Yang et al. (2011)
Enokidake mushroom polysaccharide	*Flammulina velutipes*	Shi, Yang et al. (2012)
Shiitake mushroom polysaccharide	*Lentinus edodes*	Shi, Yang et al. (2012)
Lingzhi polysaccharide	*Ganoderma lucidum*	Shi, Yang et al. (2013); Shi, Yang et al. (2014)
Common morel mushroom polysaccharide	*Morchella esculenta*	Li, Sang et al. (2013); Li, Chen et al. (2016)
Wood decay fungus polysaccharide	*Wolfiporia extensa*	Li, Wang et al. (2014)
Fungal polysaccharide, adenosine, and ergosterol	*Preussia aemulans*	Li, Meng et al. (2015)
Maitake mushroom polysaccharide	*Grifola frondosa*	Zhu et al. (2015)
γ-Aminobutyric acid	*Aspergillus oryzae*	Ochi et al. (2016)
Emodin and physcion	*Eurotium cristatum*	Chan et al. (2019)
Foodstuffs		
Meitauza[a]	*Actinomucor elegans, Zymomonas mobilis*	Xu et al. (2012)
Meitauza[a]	*Actinomucor elegans* DYC-1	Guan, Wang et al. (2017)
Tempeh	*Rhizopus oligosporus*	Yogo et al. (2011)
Okara-koji flour	*Aspergillus oryzae*	Matsuo (1999)
Okara-miso	*Neurospora intermedia, Aspergillus oryzae*	Matsuo & Takeuchi (2003); Matsuo (2004); Matsuo (2006)
Okara-meju	*Aspergillus oryzae, Monascus pilosus*	Lee et al. (2013)

[a] *Meitauza:* A traditional fermented Chinese food.

a chewy texture with a mild flavor. Okara can also be used to make tempeh, a traditional Indonesian food, which is normally made from whole soybeans fermented by *Rhizopus oligosporus*. Yogo et al. (2011) observed that consuming okara-tempeh improved good microflora, such as *Bifidobacterium*, in dogs, possibly due to soy oligosaccharides in okara.

Similarly, okara may replace soybeans in making *koji* (mold-fermented soybean mass, typically using *Aspergillus oryzae*) and provide additional benefits. When used as a flour substitute, okara-koji extended the shelf-life of high-fat baked goods by reducing lipid oxidation and starch retrogradation (Matsuo, 1999). Alternatively, when okara replaced part of the soybeans in making the Japanese condiment miso (salted soybean paste), *in vivo* studies showed that rats fed okara-miso had significantly lower serum cholesterol levels, and lower serum and liver thiobarbituric acid reactive substances values compared to rats fed normal miso (Matsuo, 1997, 2004).

Lee et al. (2013) also observed similar hypocholesterolemic and hypolipidemic effects in mice fed okara-soybean meju, a traditional Korean soybean cake fermented mainly by *A. oryzae* and *M. pilosus*. Mice on a high-fat diet had significantly lower serum LDL-cholesterol, higher levels of glutathione, and lower levels of hepatic triglycerides and lipid peroxide in liver tissue when fed fermented okara, relative to the group fed unfermented okara. It has been suggested that the hypocholesterolemic and hypolipidemic effects in crude okara are contributed by the dietary fiber and protein, but not isoflavones (Villanueva, Yokoyama, Hong, Barttley, & Rupérez, 2011).

10.2.3.2   Bacteria

Most studies on bacterial fermentation of okara involved *Bacillus* species (Table 10.3). Bacilli produce extracellular alkaline proteases and are typically used in soybean fermentation (Bhunia, Basak, & Dey, 2012). On the other hand, okara has also been investigated for prebiotic potential using probiotic bacteria.

10.2.3.2.1   Production of Bioactive Compounds

Natto is a traditional Japanese food with a distinctive pungent smell and slimy texture; it is usually fermented by *B. subtilis* var. *natto* and *B. subtilis*. Replacing soybeans with okara, *B. natto*–fermented okara extracts also displayed scavenging effects on free radicals and reduced inflammation in *in vivo* experiments (Yokota, Hattori, Ohami, Ohishi, & Watanabe, 1996; Yokota, Hattori, Ohishi, Ohami, & Watanabe, 1996).

On the other hand, *B. subtilis* produces extracellular alkaline proteases, which can hydrolyze okara proteins to give bioactive compounds, such as γ-polyglutamic acid (Oh, Jang, Seo, Ryu, & Lee, 2007), bioactive peptides (Zhu, Fan, Cheng, & Li, 2008), and nattokinase, a fibrinolytic enzyme that is also found in natto (Oh, Kim, & Lee, 2006; Zu et al., 2010). In particular, one strain of *B. subtilis* isolated by Zhu et al. (2010) (*B. subtilis* B2) could produce 1-deoxynojirimycin during okara fermentation. 1-deoxynojirimycin is a potential antidiabetic agent usually extracted from mulberry leaves.

10.2.3.2.2   Use of Dried Okara as Prebiotics

Due to its high fiber content, okara may also serve as a source of prebiotics; prebiotics selectively stimulate the growth and/or activity of certain gut microbes that confer health benefits to the host when consumed. To this end, the effect of okara on probiotic viability has been studied.

Espinosa-Martos and Rupérez (2009) noted that okara particles provided a surface for cell adherence by probiotics *Bifidobacterium bifidum* and *Lactobacillus acidophilus*, thereby

TABLE 10.3
Summary of Bacterial Fermentation of Okara

Products	Bacteria	Reference
Bioactive compounds		
Fermented okara with higher antioxidant activity	*Bacillus subtilis* var. *natto*	Yokota, Hattori et al. (1996)
γ-Polyglutamic acid	*Bacillus subtilis* GT-D, KU-A	Oh et al. (2007)
1-Deoxynojirimycin	*Bacillus. subtilis* B2	Zhu et al. (2010)
Nattokinase	*Bacillus subtilis* var. *natto*	Zu et al. (2010)
Conjugated linoleic acid	*Propionibacterium freudenreichii* ssp. *shermanii*	Vahvaselkä & Laakso (2010)
Dried okara as prebiotics		
Soy-okara yogurt	*Lactobacillus delbrueckii* subsp. *delbrueckii*	Kitawaki et al. (2007); Kitawaki et al. (2009)
Prebiotics	*Lactobacillus acidophilus*, *Bifidobacterium bifidum*	Espinosa-Martos & Rupérez (2009); Bedani, Rossi et al. (2013)
Prebiotics	*Lactobacillus bulgaricus*, *Streptococcus thermophilus*	Tu et al. (2014)
Soy-okara yogurt (with inulin)	*Lactobacillus acidophilus*, *Bifidobacterium animalis* subsp. *lactis*, *Streptococcus thermophilus*	Bedani, Campos et al. (2014), Bedani et al. (2015)
Probiotic creamy sauce (savory)	*Lactobacillus acidophilus* LA3	Lima de Moraes Filho, Busanello et al. (2018)

enhancing substrate uptake and cell growth. Similar effects were also observed when yogurt cultures *Streptococcus thermophilus* and *L. delbrueckii* subsp. *bulgaricus* were used (Tu et al., 2014).

Consequently, the health benefits of such okara probiotic products have been investigated. One group of researchers prepared soy-okara yogurt by mixing soymilk and dried okara powder (ratio of 2:1) and fermenting it with *Lactobacillus delbrueckii* subsp. *bulgaricus* (Kitawaki et al., 2009, 2007). Regardless of a high-cholesterol diet or cholesterol-free diet, the consumption of soymilk-okara yogurt significantly and consistently reduced the plasma total cholesterol in rats compared to the control group and another group fed soymilk yogurt only.

These effects may be extended to humans too. The consumption of soy-okara yogurt (with inulin) by 36 men with normal cholesterol levels for eight weeks reduced their LDL-cholesterol level and LDL-cholesterol to HDL-cholesterol ratio by 10.3% and 11.6% respectively (Bedani et al., 2015). The hypocholesterolemic effect could have been exerted by okara fiber, which facilitates the excretion of bile acids through their adsorption to fecal matter. Still, the confounding effect from the live probiotics in the yogurts cannot be overlooked.

Another innovative probiotic okara product is a creamy sauce developed by Busanello and colleagues. The fermentation of okara and soymilk by *Lactobacillus* was first optimized using a central composite rotational design (Moraes Filho, Busanello, & Garcia, 2016). A creamy sauce was then developed by fermenting soymilk containing 3% okara flour with *Lactobacillus acidophilus* LA3 for 48 h (Moraes Filho, Busanello, Prudencio, & Garcia, 2018). Sensory tests showed that the sauce was considered "satisfactory". After 30 days of refrigerated storage, the sauce still yielded a viable probiotic count of > 6 log CFU/g after simulated gastrointestinal treatment, further highlighting its probiotic potential.

Lactic acid fermentation may also be one way to extend the shelf life of okara. Hiwatashi et al. (2015) fermented okara with *Lactobacillus casei* and *Weissella paramesenteroides* under optimized conditions (an inoculum of 6 log CFU/g, 30° C, semi-anaerobic conditions) for rapid lactic acid bacterial growth and high organic acid production. After seven days, no spore-forming bacteria, coliforms, or fungi were detected in the okara.

10.2.3.3 Yeasts

Studies on yeast fermentation of okara began around the 2010s, relatively later than those using fungi and bacteria (Table 10.4). Early studies focused on changes in nutrition (proximate composition) and aroma (volatile profile).

Rashad et al. (2011) conducted solid-state fermentation of okara using several yeasts (*Candida albicans*, *C. guilliermondii*, *Kluyveromyces marxianus*, *Pichia pinus*, and *Saccharomyces cerevisiae*). Individual yeasts were added to sterilized okara and fermented for three days. Generally, yeast fermentation increased the amounts of protein and ash and reduced the crude fiber, carbohydrate, and lipid contents. Hu et al. (2019) also showed that *Kl. marxianus* increased the okara soluble fiber content by 158% and reduced its phytic acid content (61.7%) and trypsin inhibitor activity (92.7%). In addition, the authors noticed a reduction in the overall beany odor.

This positive change in off-odor modulation was studied more extensively by Vong and Liu (2017), who inoculated individual yeasts typically associated with fermented dairy products ("dairy yeast") or wines ("wine yeast"). The four dairy yeasts (*Geotrichum candidum*, *Yarrowia lipolytica*, *Debaryomyces hansenii*, and *Kl. lactis*) had high proteolytic and lipolytic capabilities. They broke down the lipid fraction in okara to produce mainly methyl ketones to give a musty, cheese-like flavor. On the other hand, the six wine yeasts (*S. cerevisiae*, *Lanchancea thermotolerans*, *Metschnikowia pulcherrima*, *P. kluyveri*, *Torulaspora delbrueckii*, and *Lindnera saturnus*) converted short-chain aliphatic aldehydes (which have an undesirable grassy note) in okara into more pleasant-smelling esters (which have a fruity note). In particular, okara fermented by *L. saturnus* contained the greatest amount and widest variety of esters and smelled perceptibly fruity.

The mechanism and rate-limiting enzymes behind this bioconversion were further investigated. The generation of C7 esters followed the authors' hypothesized pathway: Linoleic and linolenic

TABLE 10.4

Summary of Yeast Fermentation of Okara

Products	Yeasts	Reference
Erythritol	*Yarrowia lipolytica* M53, with pre-treatment by fungus *Mucor flavus*	Liu et al. (2017) and Liu, Yu et al. (2018)
Hexyl and heptyl esters	*Lindnera saturnus*	Vong & Liu (2018)
Foodstuffs		
Fermented okara with improved nutrition and higher antioxidant activity	*Candida albicans, C. guilliermondii, Kluyveromyces marxianus* NRRL Y-7571 and NRRL Y-8281, *Pichia pinus, Saccharomyces cerevisiae*	Rashad et al. (2011)
	Yarrowia lipolytica	Vong, Au Yang et al. (2016)
Fermented okara with improved aroma	*Geotrichum candidum, Yarrowia lipolytica, Debaryomyces hansenii, K. lactis, Saccharomyces cerevisiae, Lanchancea thermotolerans, Metschnikowia pulcherrima, Pichia kluyveri, Torulaspora delbrueckii, Lindnera saturnus*	Vong & Liu (2016)
Fermented okara with improved nutritional and processing qualities	*Kluyveromyces marxianus*	Hu et al. (2019)
Fermented okara with improved nutritional qualities	*Saccharomyces cerevisiae*	Queiroz Santos et al. (2018)
Co-cultured probiotic beverage	*Lindnera saturnus*	Vong & Liu (2019)

acids were first oxidized by endogenous soy lipoxygenase in okara, then broken down into C7 aldehyde (heptanal) by hydroperoxide lyase (Vong & Liu, 2018). Heptanal was then converted into heptyl esters by the yeast *L. saturnus*, likely by reaction with highly reactive yeast metabolite intermediates, fatty acyl-CoA, and acetyl-CoA. The production of C6 esters, however, resulted from yeast *L. saturnus* bioconversion from okara lipids or *de novo* synthesis, although the mechanism remains unclear. *L. saturnus* fermentation of okara could therefore be a simple one-pot set-up for the production of C6 and C7 esters; okara provides the substrate and enzymes while yeast provides the cofactors and metabolite intermediates.

10.2.4 USING MULTIPLE BIOCATALYSTS FOR OKARA UTILIZATION

As could be seen, enzymes and microorganisms can break down or convert specific fractions of okara to improve its nutrition, aroma, or functional properties. Yet, given the complex composition of okara, its complete utilization via bioprocessing will require two or more biocatalysts.

For instance, carbohydrase or fungal pre-treatments can be applied to release simple sugars for enhanced fermentation subsequently. Liu et al. (2017) first digested the fiber in okara using fungi *Mucor flavus* or *Trichoderma reesei*. Next, the *Mucor*-fermented okara was selected to be sterilized (thereby inactivating the fungus) and then fermented by yeast *Yarrowia lipolytica* for erythritol production. In a pilot scale of 5-L fermentation, an erythritol titer of 14.7 g/L, with a yield of 0.49 g/g okara, was obtained. The authors further experimented with the set-up by switching to solid-state fermentation and adding buckwheat husk, also an agro-waste, as inert support in okara. Maximum erythritol production (143.3 mg/g dried solids) was obtained from okara–buckwheat husk mixture (5:2, w/w) supplemented with 0.01 g/g dried solids NaCl, with an initial moisture content of 60% and pH of 4.0 for 192 h. This is equivalent to a yield of 0.102 g/g okara, about 80% lower than that obtained in the liquid fermentation set-up.

Vong et al. (2018) also first fermented okara with a fungus *R. oligosporus*, then with yeast *Y. lipolytica*. In contrast to Liu's work (Liu et al., 2017), the fungus-fermented okara was not sterilized.

Instead, the yeast *Y. lipolytica* was added directly for further solid-state fermentation, allowing for interaction between the fungus and yeast enzymes. After 72 h, relative to the monoculture-fermented okara (i.e. either with only *R. oligosporus* or *Y. lipolytica*), the coculture-fermented okara yielded more acids and volatile phenols such as guaiacol, syringol, and 4-ethylguaiacol. These compounds all contribute to a smoke-like, spicy aroma.

In some cases, a mix of enzymes and mixed microbial cultures may be used. Vong and Liu (2019) first treated okara with carbohydrase Viscozyme®L, breaking down the bulk fraction of insoluble fiber. Then, probiotic *Lb. paracasei* and yeast *L. saturnus* were added for further fermentation. The final okara probiotic beverage contained higher amounts of soluble fiber, free amino acids, and aglycone isoflavones, with a natural fruity aroma. Moreover, the inclusion of both the probiotic and yeast in the okara medium extended probiotic viability during ambient storage; the viable probiotic count remained > 7 log CFU/mL when stored at 25° C for eight weeks. This positive interaction was only observed in specific probiotic–yeast combinations (Liu & Vong, 2019).

As evident, understanding the nature of okara allows for the selection of suitable enzymes, microorganisms, and set-ups with specific end-goals or target compounds in mind. While the research done thus far provides some guidance, there still remains much scope for creative permutations and combinations of biocatalysts for complete okara valorization.

10.3 SOY WHEY VALORIZATION

10.3.1 OVERVIEW OF THE COMPOSITION OF SOY WHEY

Table 10.5 shows the general composition of soy whey from both tofu and soy protein isolate production. As compared to the nutritional composition of okara (Table 10.1), soy whey from both tofu and soy protein isolate (SPI) contained significantly lower amounts of nutrients despite being able to support the growth of different types of microorganisms as reflected in Table 10.6. The nutrients present in soy whey represent the nutrients that are lost during the main product manufacturing process (i.e. nutrients that are lost from tofu and SPI production). Therefore, it is wasteful to discard soy whey while there are still valuable nutrients in soy whey to valorize.

As seen from Table 10.5, soy whey consists mainly of residual carbohydrates, proteins, fats, and minerals that are left behind after tofu and SPI production. The composition of soy whey depends heavily on the soybeans as well as the processing methods (e.g. soaking of beans, cooking process, types of coagulants used, etc.) and therefore, a range of values are reported for each of the components. Soy whey from tofu production contains various carbohydrates ranging from oligosaccharides

---

TABLE 10.5

Composition of Soy Whey

Components[&]	Soy Whey (Tofu)	Soy Whey (SPI)
Carbohydrates (g/L)	8.50	9.50
Stachyose	5.48–6.40	No data available
Raffinose	0.30–1.80	No data available
Sucrose	2.87–11.30	No data available
Fructose	0.36–1.10	No data available
Glucose	0.14–1.20	No data available
Proteins (g/L)	1.33–8.20	0.3–3.00
Fats (g/L)	3.9–10.0	Not present in SPI whey
Mineral (ash; g/L)	3.9–4.6	1.93

[&]A comparison between the composition of milk or dairy whey and soy whey has been reviewed by Chua and Liu (2019).

TABLE 10.6
Summary of Enzymatic and Fermentation Processes of Soy Whey Valorization

Products	Biocatalysts	References
Enzymatic processes		
Peptides	Trypsin, chymotrypsin, pepsin	Peñas, Préstamo, & Gomez (2004)
	Alcalase, Neutrase, Corolase 7089 and Corolase PNL	Peñas, Préstamo, Polo, & Gomez (2006)
	Protease from *Aspergillus awamori* nakazawa	Singh & Banerjee (2013)
Prebiotic (lactosucrose)	Levansucrase SacB from *Bacillus subtilis* CECT 39	Corzo-Martinez, Luscher, De Las Rivas, Muñoz, & Moreno (2015)
Prebiotic (Fructooligosaccharides)	Pectinex Ultra SP-L	Corzo-Martínez, García-Campos, Montilla, & Moreno (2016)
Fermentation		
Fungal mycelial protein	*Tricholoma nudum* and *Boletus indecisus*	Falanghe, Smith, & Rackis (1964)
Citric acid	*Aspergillus niger*	Khare, Jha, & Gandhi (1994)
Hydrogen*	*Rhodobacter sphaeroides*	Zhu, Suzuki, Tsygankov, Asada, & Miyake (1999)
Lactic acid bacteria growth media	*Lactobacillus paracasei* ssp. *paracasei* LG3	Thi, Champagne, Lee, & Goulet (2003)
Probiotic growth media	*Lactobacillus plantarum* LB17	Ben Ounis, Champagne, Makhlouf, & Bazinet (2008)
	Lactobacillus acidophilus FTCC 0291, *Lactobacillus casei* FTCC 0442, *Lactobacillus fermentum* FTD 13, *Bifidobacterium bifidum* BB12	Fung, Woo, & Liong (2008) Fung & Liong (2010)
Nisin	*Lactococcus lactis* subsp. *lactis* ATCC 7962	Mitra et al. (2010)
Endo-polysaccharides	*Hericium erinaceus*	Zhang et al. (2012)
Tuberculosis detection assay*	*Mycobacterium tuberculosis* strain H37Rv	Masangkay (2012)
Biodiesel source*	*Chlorella vulgaris*	Dianursanti, Rizkytata, Gumelar, & Abdullah (2014)
	Chlorella pyrenoidosa FACHB-9	Wang, Wang, Miao, & Tian (2018)
	Nostoc muscorum	Rusydi & Yakupitiyage (2019)
Functional beverages fermented with lactic acid bacteria /probiotics	*Lactobacillus plantarum* B1-6	Xiao et al. (2015)
	Lactobacillus bulgaricus and *Streptococcus thermophilus*	Benedetti et al. (2016)
	Lactobacillus amylolyticus L6	Fei et al. (2017)
	Lactobacillus paracasei 22709, *Leuconostoc mesenteroides* 22264, *Lactobacillus rhamnosus* GG, and *Lactobacillus plantarum*	Zhu, Wang, & Zhang (2019)
Soy whey alcoholic beverages	*Saccharomyces cerevisiae* Merit, *S. cerevisiae* EC1118, *S. cerevisiae* R2, and *S. cerevisiae* 71B	Chua, Lu, & Liu (2017)
	Torulaspora delbrueckii Biodiva; *Lachancea thermotolerans* Concerto; *Metschnikowia pulcherrima* Flavia; *Pichia kluyveri* Frootzen, and *Williopsis saturnus* NCYC2251	Chua, Lu, & Liu (2018)
Kombucha	Dominant yeasts: *Pichia*, *Dekkera* Dominant bacteria: *Acetobacter*, *Lactobacillus*.	Tu, Tang, Azi, Hu, & Dong (2019)
Water kefir beverage	Water kefir grains (yeasts, lactic acid bacteria, and acetic acid bacteria)	Tu, Azi, Huang, Xu, Xing, & Dong (2019); Azi, Tu, Rasheed, & Dong (2019)
Tofu acidic coagulant	*Lactobacillus plantarum* JMC-1	Li et al. (2017)
	Lactobacillus casei YQ336	Xu et al. (2019)
Biosurfactant (rhamnolipids)	*Pseudomonas fluorescens* FNCC 0063	Suryanti, Handayani, Marliyana, & Suratmi (2017)
Biological control aquaculture	*Flavobacterium sp.* and *Bacillus sp.*	Satyantini, Pratiwi, Sahidu, & Nindarwi (2019)
Biocementation*	*Bacillus cereus* NS4	Fang, He, Achal, & Plaza (2019)
Bacteriocin-like substance	*Bacillus* spp. JY1	Zhang et al. (2019)
Docosahexaenoic acid	*Schizochytrium* sp. S31 (ATCC 20888)	Wang, Wang, Tian, & Cui (2020)

*These applications are non-food-related.

(stachyose and raffinose) to simple sugars (sucrose, fructose, and glucose). Proteins and fats are also detected in soy whey from tofu production. Minerals such as calcium and magnesium ions are commonly found in soy whey due to the coagulants used to make tofu (calcium salt or magnesium salt). As for soy whey from SPI production, not much information nor data is available as seen in Table 10.5; soy whey also contains carbohydrates, proteins, and minerals. Fat is not detected in soy whey from SPI production due to the defatting process carried out at the earlier steps of the production (Figure 10.1).

In the tofu manufacturing industry, approximately up to 9 kg of soy whey can be generated from every 1 kg of soybeans used to make tofu (Fei et al., 2017). As for SPI, approximately 20 kg of soy whey is generated from every 1 kg of SPI produced (Wang, Wu, Zhao, Liu, & Gao, 2013). Therefore, a substantial amount of soy whey is generated daily around the world by both tofu and SPI production and the amount of soy whey discarded is too massive to be ignored. Direct disposal of soy whey not only results in the loss of the valuable nutrients present in the liquid side stream, but it also can lead to foul odor generation and pollution of surface and groundwater (Belén, Sánchez, Hernández, Auleda, & Raventós, 2012). The typical method for wastewater treatment either uses aerobic or anaerobic digestion to treat the wastewater to discharge standards but this process is complicated and requires specialized equipment and high operating cost (Wang & Serventi, 2019). To overcome the wastewater treatment cost and to reduce the impact of soy whey disposal, numerous studies have been carried out to evaluate methods to utilize soy whey for different applications.

The components in soy whey can be extracted to be used for various applications. The residual whey soy protein from SPI production can be recovered via a continuous technology of successive foaming and defoaming to precipitate the proteins out of the soy whey wastewater (Li, Xintong, Youshuang, Yuran, & Jing, 2018). Oligosaccharides can be recovered via first concentrating the oligosaccharides content using reverse osmosis or nanofiltration, followed by dehydration and purification (Matsubara, Iwasaki, Nakajima, Nabetani, & Nakao, 1996). Magnesium ions found in the soy (tofu) whey can be recovered up to 65% via a sequential process of electrodialysis followed by bipolar membrane electro-acidification (Bazinet, Ippersiel, & Lamarche, 1999). Recovery of the components from soy whey provides manufacturers with an opportunity to recover the nutrients from the soy whey while at the same time reducing the COD and BOD of the soy whey, but the recovery process does not add value to the soy whey. Bio-valorization of the soy whey via enzymatic treatment or microbial fermentation serves as a better solution for manufacturers to value-add to their waste stream. Table 10.6 summarizes the research associated with the bio-valorization of soy whey using enzymatic treatment and fermentation process.

10.3.2 Enzymatic Treatment

As seen from Table 10.6, enzymatic treatments were utilized to produce peptides and prebiotics from soy whey. Peñas et al. (2004) reported that high pressure of 100 MPa enhanced the hydrolysis of the soybean whey protein in soy (tofu) whey by trypsin, chymotrypsin, and pepsin into peptides (lower than 14 kDa). The peptides generated via this hydrolysis method can be used as a base in certain diets. The same researchers then went on to use several proteases, Alcalase, Neutrase, Corolase 7089, and Corolase PNL to evaluate the hydrolysis of the soy whey protein at high pressure (Peñas, Préstamo, Polo, & Gomez, 2006). Using high pressure combined with enzymatic treatment aided in the reduction of the allergen Gly m 1 in the soy whey and therefore allowed the creation of peptide sources with low antigenicity. Singh and Banerjee (2013) used proteases from *Aspergillus awamori* nakazawa to hydrolyze soy (tofu) whey into bioactive peptides with a radical scavenging ability of 40–50% in normal conditions. The bioactive peptides can then be added as additives for food preparation and formulation.

Besides peptides, prebiotics were also created from soy whey via enzymatic processes. Corzo-Martinez et al. (2015) reported that a high value-added prebiotic called lactosucrose could be developed using the enzyme levansucrase SacB from *Bacillus subtilis* CECT 39. The process involved the transfructosylation of lactose contained in cheese whey by using sucrose, raffinose, and stachyose in soy (tofu) whey. This study carried out by Corzo-Martinez et al. (2015) used both the liquid side stream from cheese and tofu production to create lactosucrose. Subsequently, Corzo-Martinez et al. (2016) used soy whey from tofu production to produce prebiotic fructooligosaccharides involving Pectinex Ultra SP-L to transfructosylate sucrose, raffinose, and stachyose. Through these two experiments, soy whey (as well as cheese whey) can be used to create high-value products through enzymatic processes.

10.3.3   FERMENTATION

The amount of research work done on the fermentation of soy whey is more extensive than on the enzymatic treatment processes on soy whey as reflected in Table 10.6. The fermentation of soy whey can be further characterized into two segments: Food fermentation and fermentation for non-food application.

10.3.3.1   Food Fermentation

As seen from Table 10.6, different microorganisms have been utilized in the fermentation of soy whey, ranging from fungi to yeasts to bacteria.

10.3.3.1.1   Fungal and Yeast Fermentation

The first soy whey research dated back to 1964 where Falanghe et al. (1964) successfully cultivated *Tricholoma nudum* and *Boletus indecisus* for the production of mycelial protein in soy whey obtained from SPI production. This research presented a solution to utilizing soy whey, a low-value side stream, to create a product of high protein content. Zhang et al. (2012) went on to cultivate *Hericium erinaceus* (common name: Lion's mane mushroom) in soy (tofu) whey to obtain endo-polysaccharides from the mycelia of the fungi. The endo-polysaccharides in one of the fractions were found to have strong activity on antioxidant *in vitro* and potent hepatoprotective effect *in vivo*. Besides cultivating edible mushrooms, *Aspergillus niger* immobilized in agarose beads was used in the production of citric acid from soy (tofu) whey supplemented with 10% (sucrose) (Khare, Jha, & Gandhi, 1994). A maximum yield of 27 g/L of citric acid could be obtained after ten days of fermentation. From these examples, soy whey can be seen as a low-cost starting material to produce new substances through fermentation.

Aside from mushroom fermentation, yeast fermentation has also been explored in soy whey valorization. Soy (tofu) whey, supplemented with sucrose, has been fermented with different strains of *Saccharomyces cerevisiae* (Chua, Lu, & Liu, 2017) and different genera of non-*Saccharomyces* yeasts (Chua, Lu, & Liu, 2018). From these two studies, different yeasts were able to reduce the endogenous grassy and beany aromatic compounds to low or trace levels with the formation of new aroma compounds, mainly secondary alcohols and esters. Isoflavone glucosides were also hydrolyzed to their respective aglycones, resulting in an increase in the antioxidant capacity in all the samples. Hence, fermenting soy whey into soy alcoholic beverages not only improves the flavor profile of the soy whey, but also improves the functionality of the product.

10.3.3.1.2   Bacterial Fermentation

As seen from Table 10.6, the research work done using lactic acid bacteria was more extensive as compared to research on other microorganisms. Soy (tofu) whey supplemented with yeast extracts, salts (phosphate, citrates, magnesium, and manganese), glucose, and Tween was able to support the growth of *Lactobacillus paracasei* ssp. *paracasei* LG3 to a maximum population of 2.9×10^9 CFU/mL (Thi, Champagne, Lee, & Goulet, 2003). Soy whey supplemented with the necessary nutrients can be used as a low-cost material for lactic starters. Ben Ounis et al. (2008) went on to demineralize and deproteinate soy (tofu) whey prior to using the demineralized skimmed soy whey for culturing *Lactobacillus plantarum* LB17. This method allows the recovery of coagulant reagents and proteins to be reused in tofu production to increase yield while at the same time generating a low-cost growth medium for lactobacilli. In the same year, Fung et al. (2008) reported on the growth optimization of *Lactobacillus acidophilus* FTCC 0291, a probiotic strain, in soy (tofu) whey supplemented with seven nitrogen sources. Subsequently, Fung and Liong (2010) evaluated the proteolytic and angiotensin-I converting enzyme (ACE)-inhibitory activities of *L. acidophilus* FTCC 0291 in the optimized soy whey formulated in the previous study. The work showed that peptides liberated during the fermentation of the soy whey had *in vitro* antihypertensive properties. Therefore, soy whey can be used as a carrier for probiotics with enhanced functional properties.

Besides using soy whey as a cheap alternative growth medium for lactic acid bacteria and probiotics, soy whey can also be fermented by the same groups of microorganisms to produce probiotic beverages or functional lactic acid beverages. Xiao et al. (2015) fermented soy whey with *Lactobacillus plantarum* B1-6 and resulted in an increase in the total phenolic and isoflavone aglycone contents. The outcomes enhanced the antioxidant capacity of the soy whey and created a functional soy beverage that may contribute to the health and nutritional status improvement of consumers. Benedetti et al. (2016) first concentrated the soy (tofu) whey to a volume reduction factor of 4.5 using a nanofiltration process. The concentrated soy whey was added to milk at 10% and 20% ratios and fermented with *Lactobacillus bulgaricus* and *Streptococcus thermophilus*. Higher percentages of concentrated soy whey usage in the fermented lactic beverage production contributed to the higher content of the total isoflavone content in the beverage, allowing the creation of a functional fermented beverage. Fei et al. (2017) utilized *Lactobacillus amylolyticus* L6 to ferment soy (tofu) whey to produce a tofu coagulant (due to a large amount of lactic acid produced) and also a functional beverage with enhanced antioxidant activities due to hydrolysis of isoflavone glycosides to aglycones. A mixture of *Lactobacillus rhamnosus* GG and *Lactobacillus paracasei* was found to have great potential for efficient enrichment of bioactive aglycones in soy whey (Zhu, Wang, & Zhang, 2019). The fermentation process using these two lactic bacteria also resulted in the metabolism of the undesirable aroma compounds to trace or undetectable levels, particularly the aldehydes and alcohols. Hence, from the examples above, soy whey has been shown to be a promising ingredient for the production of new functional beverages using lactic acid bacteria.

The lactic acid formed by lactic acid bacteria in soy whey can also be used as coagulants for tofu production. As seen in the previous paragraph, Fei et al. (2017) utilized *L. amylolyticus* L6 to produce a tofu coagulant (due to high lactic acid production from homofermentative pathway) in conjunction with a functional beverage with enhanced antioxidant activity. The acidified soy whey can be used to coagulate the soy milk to produce traditional Chinese tofu. Li et al. (2017) isolated a lactic bacterium from naturally fermented soy whey and identified it as *Lactobacillus plantarum* JMC-1. Lactic acid and acetic acid were the major organic acids present in the soy whey fermented by *L. plantarum* JMC-1 and the tofu prepared by this lactic bacterium was of better quality in terms of sensory attributes, texture, yield, and retention capacity when compared to tofu made from the use of calcium sulphate and magnesium chloride. Xu et al. (2019) did high-throughput sequencing of naturally fermented soy whey and identified *Lactobacillus casei* YQ336 as a lactic bacterium with high coagulating ability and lactic acid production. The acid slurry produced by *Lb. casei* YQ336 contained organic acids, metal ions, and protease that promoted the coagulation of the protein in the soymilk, resulting in the formation of tofu. Hence, acidic soy whey itself can be reutilized as a coagulant for tofu production with the right selection of lactic bacteria.

Soy whey from SPI production was used to culture *Lactococcus lactis* subsp. *lactis* for nisin production (Mitra et al., 2010). The production of nisin using soy whey was slightly lower when compared to culturing the lactococcus in MRS broth (619 mg/L versus 672 mg/L), but it demonstrated that a low-value side stream can be used to produce a high-value fermentation end product.

Other than lactic bacteria, other bacterial species have also been used in soy whey fermentation for food-related applications. *Pseudomonas fluorescens* FNCC 0063 was fermented in soy (tofu) whey containing 8 g/L nutrient broth and 5 g/L sodium chloride for two days to produce rhamnolipids, a type of biosurfactants (Suryanti, Handayani, Marliyana, & Suratmi, 2017). This biosurfactant has a critical micelle concentration (CMC) value of 638 mg/L and surface tension of 54 mN/m and this biosurfactant is comparable to commercially available surfactants such as Triton X-100 and Tween 80.

Bacillus spp. JY-1, isolated from Chinese traditional fermented soybean product (douchi), was fermented in soy whey to produce a bacteriocin-like substance that has a broad inhibitory spectrum against some food-borne pathogens such as *Staphylococcus aureus* and *Listeria monocytogenes* (Zhang et al., 2019). The bacteriocin-like substance was stable between pH 2–10 with heat resistance (65–105° C) and may be used as a potential alternative to control food pathogens in food production.

Bacillus sp. and *Flavobacterium* sp. were cultivated together in culture media containing 0%, 10%, 20%, and 30% of the soy whey to be used as a biological control agent in aquaculture (Satyantini, Pratiwi, Sahidu, & Nindarwi, 2019). *Bacillus* sp. has been reported to produce antibiotics that can inhibit *Vibrio parahaemolyticus* and *Flavobacterium* sp. has been reported to produce antibacterial compounds that suppress the growth of other bacteria. From the study, culture media containing 10% soy whey provided the best growth for the two bacteria out of the four concentrations of soy whey used.

The production of docosahexaenoic acid (DHA) by *Schizochytrium* sp. S31 in soy whey supplemented with 70 g/L glucose and 15 g/L sea salt has also been reported (Wang, Wang, Tian, & Cui, 2020). The purpose of fermenting *Schizochytrium* in soy whey is to reduce the cost of the medium which will lead to the reduction of the cost of DHA. The use of soy whey showed better culture performance as compared to traditional media commonly used for the growth of *Schizochytrium* and it also costs less than a third of that of traditional media.

10.3.3.1.3   Mixed Culture Fermentation

Besides the use of single cultures in the fermentation of soy whey, mixed culture fermentation has also been exploited in recent years. Soy whey has been fermented into kombucha using commercial kombucha culture (Tu, Tang, Azi, Hu, & Dong, 2019). The kombucha culture consisted of yeasts (dominantly *Pichia* and *Dekkera*) and bacteria (mainly *Acetobacter* and *Lactobacillus* genus). The fermentation of soy whey into kombucha enhanced the antioxidant capacity of the soy whey and the resulting beverage has antibacterial activity against *Staphylococcus aureus*, *Bacillus subtilis*, and *Escherichia coli*. Other than soy whey kombucha, Tu et al. (2019) also fermented soy whey into water kefir. The water kefir culture consisted mainly of yeasts (mainly *S. cerevisiae*) and bacteria (mainly *Acetobacter* and *Lactobacillus*). Fermentation with the water kefir culture enhanced the antioxidant capacity of the soy whey and improved the sensory quality of the soy whey through new aromatic volatile production. Azi et al. (2019) went on to use three different water kefir microbiota for the soy whey fermentation. Through this study, the researchers found that *Lactobacillus* dominated water kefir microbiota and produced soy whey beverages with high phenolic acids, isoflavone aglycones, and antioxidant activity. Hence, the use of mixed cultures in soy whey fermentation has been proven useful in converting soy whey into novel functional beverages.

10.3.3.2   Fermentation for Non-Food Applications

Besides fermenting soy whey for food applications, soy whey has also been used in non-food applications. Soy whey has been shown to be able to produce biological hydrogen through the use of an anoxygenic phototrophic bacterium *Rhodobacter sphaeroides* immobilized in agar gels (Zhu, Suzuki, Tsygankov, Asada, & Miyake, 1999). The hydrogen generated through this process can be used for energy production. Other than producing hydrogen as a fossil fuel alternative, soy whey has also been used for the cultivation of biodiesel resources such as microalgae *Chlorella vulgaris*, *Chlorella pyrenoidosa*, and *Nostoc muscorum*. Dianursanti et al. (2014) reported optimum cultivation of *C. vulgaris* in 20–30% of soy whey diluted with water and approximately 23% of lipids could be obtained from this cultivation method. *C. pyrenoidosa* has also been shown to grow well in soy whey with just a simple pH adjustment to slightly alkaline conditions (pH 9) (Wang, Wang, Miao,

& Tian, 2018). Rusydi and Yakupitiyage (2019) reported that soy whey diluted to 40% of its concentration and supplemented with a small amount of nitrate and phosphate supported good growth of *Nostoc muscorum*. Even though the above-mentioned studies involved cultivating microalgae as a source of biodiesel, microalgae can also be used to produce edible nutrients, pharmaceutical and cosmetic products, natural dyes, and many more products if the liquid medium used for the microalgae cultivation is sufficiently clean (i.e. lack of toxic compounds), for example, soy whey (Wang et al., 2018).

Soy whey has also been demonstrated to be a suitable low-cost alternative to the standard media (Middlebrook 7H9 medium) for the propagation of *Mycobacterium tuberculosis* strain H37Rv (Masangkay, 2012). The use of soy whey for testing *M. tuberculosis* allows the creation of cheap alternative test kits to be used in poorly resourced countries.

Last but not least, soy whey has also been reported in biocementation, an eco-friendly process in sustainable construction (Fang, He, Achal, & Plaza, 2019). Soy whey was able to support the growth of a ureolytic bacterium *Bacillus cereus*, and the use of soy whey for the biocementation process improved the compressive strength of sandstones and mortars when compared to control sandstones and control mortars (mixture containing no microbial culture). Biocemetation depends on the urease activity of the bacteria to break down urea to produce CO_3^{2-} ions, which then react with calcium ions to form calcium carbonate precipitation. The improvement of the compressive strength in the microbial mortars could be partially attributed to the microbial biomass grown inside the mortar matrix.

10.4   FUTURE PROSPECTS

Numerous studies have been carried out to determine methods to bio-valorize okara and soy whey into value-added products. However, the translation of the research work from the lab to the market in this area has been rather limited. Two Singapore-based start-ups (Soynergy and SinFooTech) are now commercializing the technologies underlying a co-cultured probiotic okara beverage (described in Section 10.2.4; Figure 10.2) and fermented alcoholic soy whey beverage (described in Section 10.3.3.1.1; Figure 10.2). Ultimately, bridging this gap between benchwork and actual application in the marketplace is needed for reducing the disposal of okara and soy whey. Prior to the translation of technology, a cost–benefit analysis must be carried out to weigh the value of the final product against the cost of the technology, so as to ensure the commercial viability of the technology.

Despite efforts in exploiting different technologies for various applications of these soy side-streams, several such technologies still generate certain types of waste at the end of the processes. For example, fermenting okara for the extraction of 8-hydroxydaidzein and monacolin K (Table 10.2) will still result in the extracted okara being discarded as waste. A similar problem exists in the extraction of citric acid and nisin from fermented soy whey (Table 10.6). In recent years, research attention has shifted to focus on using the entirety of the okara and soy whey for the creation of new

FIGURE 10.2   Examples of soy side-streams and their valorized products. Okara is converted into a co-cultured probiotic beverage (left picture). Soy whey (in the conical flask) is converted into an alcoholic beverage (in the wine glass) (right picture).

products ("zero waste"). For example, okara has been added to yogurt, sauce, or even beverages to improve the functionality of the food product (Table 10.3 and Table 10.4). The entirety of soy whey has also been fermented into alcoholic beverages, kombucha, and kefir (Table 10.6). The complete use of okara and soy whey reduces the waste generation at the end of the processes, therefore creating a zero-waste solution for the food processing industry.

10.5   CONCLUSION

Side streams (okara and soy whey) of soymilk, tofu, and soy protein isolates are more than just waste products generated from food production. Using specific enzymes and microorganisms, new products or compounds can be produced from these side streams. Discarding the side streams as waste not only causes environmental problems in the long run, but also results in the loss of valuable resources to generate high-value products. More recently, the global research direction shifted to developing solutions to use okara and soy whey entirely, instead of simply using them as substrates for extraction. Such a scientific feat prompts the logical next step: Translating such technologies from lab to commercial scale in order to harness the true benefits of the technologies for the food industry.

REFERENCES

Anderson, R. L., & Wolf, W. J. (1995). Compositional changes in trypsin inhibitors, phytic acid, saponins and isoflavones related to soybean processing. *Journal of Nutrition, 125*(3) Supplement, 518S–588S.

Azi, F., Tu, C., Rasheed, H. A., & Dong, M. (2019). Comparative study of the phenolics, antioxidant and metagenomic composition of novel soy whey-based beverages produced using three different water kefir microbiota. *International Journal of Food Science and Technology, 55*(4), 1689–1697.

Bazinet, L., Ippersiel, D., & Lamarche, F. (1999). Recovery of magnesium and protein from soy tofu whey by electrodialytic configurations. *Journal of Chemical Technology and Biotechnology, 74*(7), 663–668.

Bedani, R., Campos, M. M., Castro, I. A., Rossi, E. A., & Saad, S. M. (2014). Incorporation of soybean by-product okara and inulin in a probiotic soy yoghurt: texture profile and sensory acceptance. *Journal of the Science of Food and Agriculture, 94*(1), 119–125.

Bedani, R., Rossi, E. A., Cavallini, D. C. U., Pinto, R. A., Vendramini, R. C., Augusto, E. M., Abdalla, D. S. P., & Saad, S. M. I. (2015). Influence of daily consumption of synbiotic soy-based product supplemented with okara soybean by-product on risk factors for cardiovascular diseases. *Food Research International, 73*, 142–148.

Bedani, R., Rossi, E. A., & Saad, S. M. I. (2013). Impact of inulin and okara on Lactobacillus acidophilus La–5 and Bifidobacterium animalis Bb–12 viability in a fermented soy product and probiotic survival under in vitro simulated gastrointestinal conditions. *Food Microbiology, 34*(2), 382–389.

Belén, F., Sánchez, J., Hernández, E., Auleda, J. M., & Raventós, M. (2012). One option for the management of wastewater from tofu production: Freeze concentration in a falling-film system. *Journal of Food Engineering, 110*(3), 364–373.

Ben Ounis, W., Champagne, C. P., Makhlouf, J., & Bazinet, L. (2008). Utilization of tofu whey pre-treated by electromembrane process as a growth medium for *Lactobacillus plantarum* LB17. *Desalination, 229*(1–3), 192–203.

Benedetti, S., Prudencio, E. S., Muller, C. M. O., Verruck, S., Mandarino, J. M. G., Leite, R. S., & Petrus, J. C. C. (2016). Utilization of tofu whey concentrate by nanofiltration process aimed at obtaining a functional fermented lactic beverage. *Journal of Food Engineering, 171*, 222–229.

Bhunia, B., Basak, B., & Dey, A. (2012). A review on production of serine alkaline protease by *Bacillus* spp. *Journal of Biochemical Technology, 3*(4), 448–457.

Chan, L. Y., Takahashi, M., Lim, P. J., Aoyama, S., Makino, S., Ferdinandus, F., Ng, S. Y. C., Arai, S., Fujita, H., Tan, H. C., Shibata, S., & Lee, C.-L. K. (2019). *Eurotium cristatum* fermented okara as a potential food ingredient to combat diabetes. *Scientific Reports, 9*(1), 17536.

Chan, W., & Ma, C. (1999). Modification of proteins from soymilk residue (okara) by trypsin. *Journal of Food Science, 64*(5), 781–786.

Chan, W. M., & Ma, C. Y. (1999). Acid modification of proteins from soymilk residue (okara). *Food Research International, 32*(2), 119–127.

Chua, J.-Y., & Liu, S.-Q. (2019). Soy whey: More than just wastewater from tofu and soy protein isolate industry. *Trends in Food Science & Technology*, *91*(June), 24–32.

Chua, J.-Y., Lu, Y., & Liu, S.-Q. (2017). Biotransformation of soy whey into soy alcoholic beverage by four commercial strains of *Saccharomyces cerevisiae*. *International Journal of Food Microbiology*, *262*(July), 14–22.

Chua, J.-Y., Lu, Y., & Liu, S.-Q. (2018). Evaluation of five commercial non-*Saccharomyces* yeasts in fermentation of soy (tofu) whey into an alcoholic beverage. *Food Microbiology*, *76*(April), 533–542.

Corzo–Martínez, M., García–Campos, G., Montilla, A., & Moreno, F. J. (2016). Tofu whey permeate is an efficient source to enzymatically produce prebiotic fructooligosaccharides and novel fructosylated α–galactosides. *Journal of Agricultural and Food Chemistry*, *64*(21), 4346–4352.

Corzo–Martinez, M., Luscher, A., De Las Rivas, B., Muñoz, R., & Moreno, F. J. (2015). Valorization of cheese and tofu whey through enzymatic synthesis of lactosucrose. *PLoS ONE*, *10*(9), 1–19.

de Figueiredo, V. R. G., Yamashita, F., Vanzela, A. L. L., Ida, E. I., & Kurozawa, L. E. (2018). Action of multi-enzyme complex on protein extraction to obtain a protein concentrate from okara. *Journal of Food Science and Technology*, *55*(4), 1508–1517.

de Moraes Filho, M. F., Busanello, M., Prudencio, S. H., & Garcia, S. (2018). Soymilk with okara flour fermented by *Lactobacillus acidophilus*: Simplex-centroid mixture design applied in the elaboration of probiotic creamy sauce and storage stability. *LWT-Food Science and Technology*, *93*, 339–345.

Dianursanti, Rizkytata, B. T., Gumelar, M. T., & Abdullah, T. H. (2014). *Industrial tofu wastewater as a cultivation medium of microalgae Chlorella vulgaris*. *Energy Procedia*, *47*, 56–61

Espinosa-Martos, I., & Rupérez, P. (2009). Indigestible fraction of okara from soybean: Composition, physicochemical properties and *in vitro* fermentability by pure cultures of *Lactobacillus acidophilus* and *Bifidobacterium bifidum*. *European Food Research and Technology*, *228*(5), 685–693.

Falanghe, H., Smith, A. K., & Rackis, J. J. (1964). Production of fungal mycelial protein in submerged culture of soybean whey. *Applied Microbiology*, *12*(4), 330–334.

Fang, C., He, J., Achal, V., & Plaza, G. (2019). Tofu wastewater as efficient nutritional source in biocementation for improved mechanical strength of cement mortars. *Geomicrobiology Journal*, *36*(6), 515–521.

Fei, Y., Liu, L., Liu, D., Chen, L., Tan, B., Fu, L., & Li, L. (2017). Investigation on the safety of *Lactobacillus amylolyticus* L6 and its fermentation properties of tofu whey. *LWT - Food Science and Technology*, *84*, 314–322.

Fujita, T., Funako, T., & Hayashi, H. (2004). 8-Hydroxydaidzein, an aldose reductase inhibitor from okara fermented with *Aspergillus* sp. HK-388. *Bioscience, Biotechnology, and Biochemistry*, *68*(7), 1588–1590.

Fung, W.-Y., & Liong, M.-T. (2010). Evaluation of proteolytic and ACE-inhibitory activity of *Lactobacillus acidophilus* in soy whey growth medium via response surface methodology. *LWT - Food Science and Technology*, *43*(3), 563–567.

Fung, W. Y., Woo, Y. P., & Liong, M. T. (2008). Optimization of growth of *Lactobacillus acidophilus* FTCC 0291 and evaluation of growth characteristics in soy whey medium: A response surface methodology approach. *Journal of Agricultural and Food Chemistry*, *56*(17), 7910–7918.

Guan, Y., Wang, J., Wu, J., Wang, L., Rui, X., Xing, G., & Dong, M. (2017). Enhancing the functional properties of soymilk residues (okara) by solid–state fermentation with Actinomucor elegans. *CyTA–Journal of Food*, *15*(1), 155–163.

Hiwatashi, M., Kano, S., & Kato, T. (2015). Development of a preservation method for okara using lactic acid fermentation. *Nippon Shokuhin Kagaku Kogaku Kaishi*, *62*(12), 572–578.

Hu, Y., Piao, C., Chen, Y., Zhou, Y., Wang, D., Yu, H., & Xu, B. (2019). Soybean residue (okara) fermentation with the yeast *Kluyveromyces marxianus*. *Food Bioscience*, *31*, 100439.

Izumi, T., Piskula, M. K., Osawa, S., Obata, A., Tobe, K., Saito, M., Kataoka, S., Kubota, Y., & Kikuchi, M. (2000). Soy isoflavone aglycones are absorbed faster and in higher amounts than their glucosides in humans. *Journal of Nutrition*, *130*(7), 1695–1699.

Jackson, C. J. C., Dini, J. P., Lavandier, C., Rupasinghe, H. P. V, Faulkner, H., Poysa, V., Buzzell, D., & DeGrandis, S. (2002). Effects of processing on the content and composition of isoflavones during manufacturing of soy beverage and tofu. *Process Biochemistry*, *37*(10), 1117–1123.

Jiménez-Escrig, A., Alaiz, M., Vioque, J., & Rupérez, P. (2010). Health-promoting activities of ultra-filtered okara protein hydrolysates released by in vitro gastrointestinal digestion: Identification of active peptide from soybean lipoxygenase. *European Food Research and Technology*, *230*(4), 655–663.

Kasai, N., Murata, A., Inui, H., Sakamoto, T., & Kahn, R. I. (2004). Enzymatic high digestion of soybean milk residue (okara). *Journal of Agricultural and Food Chemistry*, *52*(18), 5709–5716.

Khare, S. K., Jha, K., & Gandhi, A. P. (1994). Use of agarose-entrapped *Aspergillus niger* cells for the production of citric acid from soy whey. *Applied Microbiology and Biotechnology*, *41*(5), 571–573.

Kitawaki, R., Nishimura, Y., Takagi, N., Iwasaki, M., Tsuzuki, K., & Fukuda, M. (2009). Effects of *Lactobacillus* fermented soymilk and soy yogurt on hepatic lipid accumulation in rats fed a cholesterol-free diet. *Bioscience, Biotechnology and Biochemistry, 73*(7), 1484–1488.

Kitawaki, R., Takagi, N., Iwasaki, M., Asao, H., Okada, S., & Fukuda, M. (2007). Plasma cholesterol-lowering effects of soymilk and okara treated by lactic acid fermentation in rats. *Journal of the Japanese Society for Food Science and Technology (Japan), 54*(8), 379–382.

Lee, S.-I., Lee, Y.-K., Kim, S.-D., Lee, J.-E., Choi, J., Bak, J.-P., Lim, J.-H., Suh, J.-W., & Lee, I.-A. (2013). Effect of fermented soybean curd residue (FSCR; SCR-meju) by *Aspergillus oryzae* on the anti-obesity and lipids improvement. *Journal of Nutrition and Health, 46*(6), 493–502.

Li, B., Qiao, M., & Lu, F. (2012). Composition, nutrition, and utilization of okara (soybean residue). *Food Reviews International, 28*(3), 231–252. https://doi.org/10.1080/87559129.2011.595023

Li, C., Rui, X., Zhang, Y., Cai, F., Chen, X., & Jiang, M. (2017). Production of tofu by lactic acid bacteria isolated from naturally fermented soy whey and evaluation of its quality. *LWT - Food Science and Technology, 82*, 227–234.

Li, L. T., & Ma, Y. L. (2014). Diversity of plant-based food products involving alkaline fermentation. In P. K. Sarkar & M. J. Robert Nout (Eds.), *Handbook of Indigenous Foods Involving Alkaline Fermentation* (pp. 78–87). CRC Press.

Li, R., Xintong, J., Youshuang, Z., Yuran, Z., & Jing, G. (2018). Precipitation of proteins from soybean whey wastewater by successive foaming and defoaming. *Chemical Engineering and Processing - Process Intensification, 128*(January), 124–131.

Li, S., Chen, Y., Li, K., Lei, Z., & Zhang, Z. (2016). Characterization of physicochemical properties of fermented soybean curd residue by Morchella esculenta. *International Biodeterioration & Biodegradation, 109*, 113–118.

Li, S., Wang, L., Song, C., Hu, X., Sun, H., Yang, Y., ... & Zhang, Z. (2014). Utilization of soybean curd residue for polysaccharides by Wolfiporia extensa (Peck) Ginns and the antioxidant activities in vitro. *Journal of the Taiwan Institute of Chemical Engineers, 45*(1), 6–11.

Li, Y., Meng, S., Wang, L., & Zhang, Z. (2015). Optimum fermentation condition of soybean curd residue and rice bran by Preussia aemulans using solid–state fermentation method. *International Journal of Biology, 7*(3), 66.

Liu, S.-Q., & Vong, W. C. (2019). An okara-based beverage. Patent No. WO 2019177536. World Intellectual Property Organisation.

Liu, X., Yu, X., Xia, J., Lv, J., Xu, J., Dai, B., Xu, X., & Xu, J. (2017). Erythritol production by *Yarrowia lipolytica* from okara pretreated with the in-house enzyme pools of fungi. *Bioresource Technology, 244*(Part 1), 1089–1095.

Liu, X., Yu, X., Zhang, T., Wang, Z., Xu, J., Xia, J., ... & Xu, J. (2018). Novel two–stage solid–state fermentation for erythritol production on okara–buckwheat husk medium. *Bioresource technology, 266*, 439–446.

Ma, C. Y., Liu, W. S., Kwok, K. C., & Kwok, F. (1996). Isolation and characterization of proteins from soymilk residue (okara). *Food Research International, 29*(8), 799–805.

Masangkay, F. R. (2012). The performance of tofu-whey as a liquid medium in the propagation of *Mycobacterium tuberculosis* strain H37Rv. *International Journal of Mycobacteriology, 1*(1), 45–50.

Mateos-Aparicio, I., Redondo-Cuenca, A., Villanueva-Suárez, M.-J., Zapata-Revilla, M.-A., & Tenorio-Sanz, M.-D. (2010). Pea pod, broad bean pod and okara, potential sources of functional compounds. *LWT - Food Science and Technology, 43*(9), 1467–1470.

Matsubara, Y., Iwasaki, K., Nakajima, M., Nabetani, H., & Nakao, S. (1996). Recovery of oligosaccharides from steamed soybean waste water in tofu processing by reverse osmosis and nanofiltration membranes. *Bioscience, Biotechnology and Biochemistry, 60*(3), 421–428.

Matsuo, M. (1997). *In vivo* antioxidant activity of okara koji, a fermented okara, by *Aspergillus oryzae*. *Bioscience, Biotechnology, and Biochemistry, 61*(12), 1968–1972.

Matsuo, M. (1999). Application of okara koji, okara fermented by *Aspergillus oryzae*, for cookies and cup-cakes. *Journal of Home Economics of Japan, 50*(10), 1029–1034.

Matsuo, M. (2004). Low-salt O-miso produced fromkoji fermentation of oncom improves redox state and cho-lesterolemia in rats more than low-salt soybean-miso. *Journal of Nutritional Science and Vitaminology, 50*(5), 362.

Matsuo, M., & Takeuchi, T. (2003). Preparation of low salt miso–like fermented seasonings using soy–oncom and okara–oncom (fermented soybeans and okara with Neurospora intermedia) and their antioxidant activity and antimutagenicity. *Food science and technology research, 9*(3), 237–241.

Mitra, D., Pometto, A. L., Khanal, S. K., Karki, B., Brehm-Stecher, B. F., & Van Leeuwen, J. (2010). Value-added production of nisin from soy whey. *Applied Biochemistry and Biotechnology*, *162*(7), 1819–1833.

Moraes Filho, M L, Busanello, M., & Garcia, S. (2016). Optimization of the fermentation parameters for the growth of *Lactobacillus* in soymilk with okara flour. *LWT-Food Science and Technology*, *74*, 456–464.

Ochi, H., Mizutani, M., Matsuura, Y., Furuichi, K., & Hayashi, S. (2016). Characteristics of fermented products made from okara koji. *Nippon Shokuhin Kagaku Kogaku Kaishi*, *63*(6), 274–279.

Oh, S., Kim, C., & Lee, S. (2006). Characterization of the functional properties of soy milk cake fermented by *Bacillus* sp. *Food Science and Biotechnology*, *15*(5), 704.

Oh, S.-M., Jang, E.-K., Seo, J.-H., Ryu, M.-J., & Lee, S.-P. (2007). Characterization of γ-polyglutamic acid produced from the solid-state fermentation of soybean milk cake using *Bacillus* sp. *Food Science and Biotechnology*, *16*(4), 509–514.

Peñas, E., Préstamo, G., & Gomez, R. (2004). High pressure and the enzymatic hydrolysis of soybean whey proteins. *Food Chemistry*, *85*(4), 641–648.

Peñas, E., Préstamo, G., Polo, F., & Gomez, R. (2006). Enzymatic proteolysis, under high pressure of soybean whey: Analysis of peptides and the allergen Gly m 1 in the hydrolysates. *Food Chemistry*, *99*(3), 569–573.

Pérez-López, E., Cela, D., Costabile, A., Mateos-Aparicio, I., & Rupérez, P. (2016). *In vitro* fermentability and prebiotic potential of soyabean Okara by human faecal microbiota. *British Journal of Nutrition*, *116*(6), 1116–1124.

Radočaj, O., & Dimić, E. (2013). Valorization of wet okara, a value-added functional ingredient, in a coconut-based baked snack. *Cereal Chemistry*, *90*, 256–262. doi:10.1094/CCHEM-11-12-0145-R

Rashad, M. M., Mahmoud, A. E., Abou, H. M., & Nooman, M. U. (2011). Improvement of nutritional quality and antioxidant activities of yeast fermented soybean curd residue. *African Journal of Biotechnology*, *10*(28), 5504–5513.

Redondo-Cuenca, A., Villanueva-Suárez, M. J., & Mateos-Aparicio, I. (2008). Soybean seeds and its by-product okara as sources of dietary fibre. Measurement by AOAC and Englyst methods. *Food Chemistry*, *108*(3), 1099–1105.

Rusydi, R., & Yakupitiyage, A. (2019). Potential of tofu wastewater as a medium in Cyanobacteria nostocmuscorum culture for its biomass and lipid production. *IOP Conference Series: Earth and Environmental Science*, *348*, 1–7.

Satyantini, W. H., Pratiwi, R. M., Sahidu, A. M., & Nindarwi, D. D. (2019). Growth of *Bacillus* sp. and *Flavobacterium* sp. in culture media with the addition of liquid whey tofu waste. *IOP Conference Series: Earth and Environmental Science*, *236*(012092), 1–8.

Shi, M., Yang, Y., Li, Y., Wang, Y., & Zhang, Z. (2011). Optimum condition of ecologic feed fermentation by Pleurotus ostreatus using soybean curd residue as raw materials. *International Journal of Biology*, *3*(4), 2.

Singh, A., & Banerjee, R. (2013). Peptide enriched functional food adjunct from soy whey: A statistical optimization study. *Food Science and Biotechnology*, *22*(SUPPL. 1), 65–71.

Singh, A., Meena, M., Kumar, D., Dubey, A. K., & Hassan, M. I. (2015). Structural and functional analysis of various globulin proteins from soy seed. *Critical Reviews in Food Science and Nutrition*, *55*(11), 1491–1502.

Stanojevic, S. P., Barac, M. B., Pesic, M. B., Jankovic, V. S., & Vucelic-Radovic, B. V. (2013). Bioactive proteins and energy value of okara as a byproduct in hydrothermal processing of soy milk. *Journal of Agricultural and Food Chemistry*, *61*(38), 9210–9219.

Suryanti, V., Handayani, D. S., Marliyana, S. D., & Suratmi, S. (2017). Physicochemical properties of biosurfactant produced by *Pseudomonas fluorescens* grown on whey tofu. *IOP Conference Series: Material Science and Engineering*, *176*(012003), 1–6.

Thi, L. N., Champagne, C. P., Lee, B. H., & Goulet, J. (2003). Growth of *Lactobacillus paracasei* ssp. *paracasei* on tofu whey. *International Journal of Food Microbiology*, *89*(1), 67–75.

Tu, C., Azi, F., Huang, J., Xu, X., Xing, G., & Dong, M. (2019). Quality and metagenomic evaluation of a novel functional beverage produced from soy whey using water kefir grains. *LWT - Food Science and Technology*, *113*(June), 108258.

Tu, C., Tang, S., Azi, F., Hu, W., & Dong, M. (2019). Use of kombucha consortium to transform soy whey into a novel functional beverage. *Journal of Functional Foods*, *52*(November 2018), 81–89.

Tu, Z., Chen, L., Wang, H., Ruan, C., Zhang, L., & Kou, Y. (2014). Effect of fermentation and dynamic high pressure microfluidization on dietary fibre of soybean residue. *Journal of Food Science and Technology*, *51*(11), 3285–3292.

Turhan, S., Temiz, H., & Sagir, I. (2007). Utilization of wet okara in low-fat beef patties. *Journal of Muscle Foods*, *18*, 226–235. doi:10.1111/j.1745-4573.2007.00081.x

Vahvaselkä, M., & Laakso, S. (2010). Production of *cis*-9, *trans*-11-conjugated linoleic acid in camelina meal and okara by an oat-assisted microbial process. *Journal of Agricultural and Food Chemistry*, *58*(4), 2479–2482.

van der Riet, W. B., Wight, A. W., Cilliers, J. J. L., & Datel, J. M. (1989). Food chemical investigation of tofu and its byproduct okara. *Food Chemistry*, *34*(3), 193–202.

Villanueva, M. J., Yokoyama, W. H., Hong, Y. J., Barttley, G. E., & Rupérez, P. (2011). Effect of high-fat diets supplemented with okara soybean by-product on lipid profiles of plasma, liver and faeces in Syrian hamsters. *Food Chemistry*, *124*(1), 72–79.

Villanueva-Suarez, M.-J., Perez-Cozar, M.-L., Mateos-Aparicio, I., & Redondo-Cuenca, A. (2016). Potential fat-lowering and prebiotic effects of enzymatically treated okara in high-cholesterol-fed Wistar rats. *International Journal of Food Science and Nutrition*, *67*(7), 828–833.

Villanueva-Suárez, M. J., Pérez-Cózar, M. L., & Redondo-Cuenca, A. (2013). Sequential extraction of poly-saccharides from enzymatically hydrolyzed okara byproduct: Physicochemical properties and *in vitro* fermentability. *Food Chemistry*, *141*(2), 1114–1119.

Vong, W. C., Hua, X. Y., & Liu, S. (2018). Solid-state fermentation with *Rhizopus oligosporus* and *Yarrowia lipolytica* improved nutritional and flavour properties of okara. *LWT - Food Science and Technology*, *90*, 316–322.

Vong, W. C., Lim, X. Y., & Liu, S.-Q. (2017). Biotransformation with cellulase, hemicellulase and *Yarrowia lipolytica* boosts health benefits of okara. *Applied Microbiology and Biotechnology*, *101*(19), 7129–7140.

Vong, W. C., & Liu, S.-Q. (2017). Changes in volatile profile of soybean residue (okara) upon solid-state fermentation by yeasts. *Journal of the Science of Food and Agriculture*, *97*(1), 135–143.

Vong, W. C., & Liu, S.-Q. (2018). Bioconversion of green volatiles in okara (soybean residue) into esters by coupling enzyme catalysis and yeast (*Lindnera saturnus*) fermentation. *Applied Microbiology and Biotechnology*, *102*(23), 10017–10026.

Vong, W. C., & Liu, S. Q. (2019). The effects of carbohydrase, probiotic *Lactobacillus paracasei* and yeast *Lindnera saturnus* on the composition of a novel okara (soybean residue) functional beverage. *LWT - Food Science and Technology*, *100*(October 2018), 196–204.

Vong, W. C., Yang, K. L. C. A., & Shao–Quan, L. I. U. (2016). Okara (soybean residue) biotransformation by yeast Yarrowia lipolytica. *International journal of food microbiology*, *235*, 1–9.

Wang, H.-J., & Murphy, P. A. (1996). Mass balance study of isoflavones during soybean processing. *Journal of Agricultural and Food Chemistry*, *44*(8), 2377–2383.

Wang, L., Wu, Z., Zhao, B., Liu, W., & Gao, Y. (2013). Enhancing the adsorption of the proteins in the soy whey wastewater using foam separation column fitted with internal baffles. *Journal of Food Engineering*, *119*(2), 377–384.

Wang, S., Chang, T., Wang, C., Shi, L., Wang, W., Yang, H., & Cui, M. (2015). Effect of okara size on the property of pork meat gel. *Journal of Food Quality*, *38*, 248–255. doi:10.1111/jfq.12144

Wang, S., Wang, X., Tian, Y., & Cui, Y. (2020). Nutrient recovery from tofu whey wastewater for the economi-cal production of docosahexaenoic acid by *Schizochytrium* sp. S31. *Science of the Total Environment*, *710*, 1–7.

Wang, S.-K., Wang, X., Miao, J., & Tian, Y.-T. (2018). Tofu whey wastewater is a promising basal medium for microalgae culture. *Bioresource Technology*, *253*, 79–84.

Wang, T., Qin, G.-X., Sun, Z.-W., & Zhao, Y. (2014). Advances of research on glycinin and β-conglycinin: A review of two major soybean allergenic proteins. *Critical Reviews in Food Science and Nutrition*, *54*(7), 850–862.

Wang, Y., & Serventi, L. (2019). Sustainability of dairy and soy processing: A review on wastewater recycling. *Journal of Cleaner Production*, *237*, 1–8.

Wongkhalaung, C., Leelawatcharamas, V., & Japakaset, J. (2009). Utilization of soybean residue to produce monacolin K-cholesterol lowering agent. *Songklanakarin Journal of Science and Technology*, *31*(1), 35–39.

Xiao, Y., Wang, L., Rui, X., Li, W., Chen, X., Jiang, M., & Dong, M. (2015). Enhancement of the antioxidant capacity of soy whey by fermentation with *Lactobacillus plantarum* B1-6. *Journal of Functional Foods*, *12*, 33–44.

Xu, X., Liu, H., & Zhou, Y. (2012). Study on the meitauza production from okara by *Actinomucor elegans* and *Zymomonas mobilis*. In E. Zhu & S. Sambath (Eds.), *Information Technology and Agricultural Engineering* (Vol. 134, pp. 329–336). Springer.

Xu, Y., Ye, Q., Zhang, H., Yu, Y., Li, X., Zhang, Z., & Zhang, L. (2019). Naturally fermented acid slurry of soy whey: High-throughput sequencing-based characterization of microbial flora and mechanism of tofu coagulation. *Frontiers in Microbiology, 10*(May), 1–12.

Yogo, T., Ohashi, Y., Terakado, K., Nezu, Y., Hara, Y., Tagawa, M., Kageyama, H., & Fujisawa, T. (2011). Influence of dried okara-tempeh on the composition and metabolites of fecal microbiota in dogs. *International Journal of Applied Research in Veterinary Medicine, 9*(2), 181–188.

Yokomizo, A., Takenaka, Y., & Takenaka, T. (2002). Antioxidative activity of peptides prepared from okara protein. *Food Science and Technology Research, 8*(4), 357–359.

Yokota, T., Hattori, T., Ohami, H., Ohishi, H., & Watanabe, K. (1996). Repression of acute gastric mucosal lesions by antioxidant-containing fraction from fermented products of okara (bean-curd residue). *Journal of Nutritional Science and Vitaminology, 42*(2), 167–172.

Yokota, T., Hattori, T., Ohishi, H., Ohami, H., & Watanabe, K. (1996). Effect of oral administration of crude antioxidant preparation from fermented products of okara (bean curd residue) on experimentally induced inflammation. *LWT - Food Science and Technology, 29*(4), 304–309.

Zhang, Y., Zhou, J., Pan, L., Dai, Z., Liu, C., Wang, J., & Zhou, H. (2019). Production of bacteriocin-like substances by *Bacillus* spp. JY-1 in soy whey. *Advances in Biochemistry, 7*(3), 65–70.

Zhang, Z., Lv, G., Pan, H., Pandey, A., He, W., & Fan, L. (2012). Antioxidant and hepatoprotective potential of endo-polysaccharides from *Hericium erinaceus* grown on tofu whey. *International Journal of Biological Macromolecules, 51*(5), 1140–1146.

Zhu, D., Sun, H., Li, S., Hu, X., Yuan, X., Han, C., & Zhang, Z. (2015). Influence of drying methods on antioxidant activities and immunomodulatory of aqueous extract from soybean curd residue fermentated by *Grifola frondosa*. *International Journal of Biology, 7*(1), 82.

Zhu, H., Suzuki, T., Tsygankov, A. A., Asada, Y., & Miyake, J. (1999). Hydrogen production from tofu wastewater by *Rhodobacter sphaeroides* immobilized in agar gels. *International Journal of Hydrogen Energy, 24*(4), 305–310.

Zhu, Y., Wang, Z., & Zhang, L. (2019). Optimization of lactic acid fermentation conditions for fermented tofu whey beverage with high-isoflavone aglycones. *LWT - Food Science and Technology, 111*(May), 211–217.

Zhu, Y.-P., Yamaki, K., Yoshihashi, T., Ohnishi Kameyama, M., Li, X.-T., Cheng, Y.-Q., Mori, Y., & Li, L.-T. (2010). Purification and identification of 1-deoxynojirimycin (DNJ) in okara fermented by *Bacillus subtilis* B2 from Chinese traditional food (meitaoza). *Journal of Agricultural and Food Chemistry, 58*(7), 4097–4103.

Zhu, Y. P., Fan, J. F., Cheng, Y. Q., & Li, L. T. (2008). Improvement of the antioxidant activity of Chinese traditional fermented okara (meitauza) using *Bacillus subtilis* B2. *Food Control, 19*(7), 654–661.

Zu, X., Zhang, Z., Che, H., Zhang, G., Yang, Y., & Li, J. (2010). Nattokinase's extraction from *Bacillus subtilis* fermented soybean curd residue and wet corn distillers' grain and fibrinolytic activities. *International Journal of Biology, 2*(2), 120.

11 Okara
A Soybean By-Product with Interesting Properties in Nutrition and Health

Alejandra García-Alonso, Inmaculada Mateos-Aparicio,
María Dolores Tenorio-Sanz, María José Villanueva-Suarez,
and Araceli Redondo-Cuenca

CONTENTS

11.1   INTRODUCTION

Okara is the residue of soybean milling after extraction of the aqueous fraction used for producing tofu and soy drink. In this process, about 1.2 kg of fresh okara results from processing 1 kg soybean to yield tofu (Mateos-Aparicio et al., 2010a; Vong & Liu, 2016). Thus, large quantities of okara are produced every year worldwide and sometimes they are related to a significant disposal problem. The environmentally friendly disposal of these residues is a major challenge for the industry (Seibel, 2018). Thus, the search for alternative uses and value aggregation has gained attention in recent years (Scialabba, 2014).

Okara has been known and used for years in Asian countries due to their high consumption of soy and derived products, and it has gradually spread throughout the planet and garnered more and more interest. It has been commonly utilized for animal feed or discarded, due to its high

DOI: 10.1201/9781003030294-11

perishability. However, okara still has significant nutritious components: 40–60% carbohydrates, 20–30% protein, and 10–20% lipids (all on a dry basis) (Li et al., 2012). This makes okara a suitable substrate for biotransformation and there is great interest in adding value to this by-product and finding new alternatives for its use.

If we do a review of the literature related to okara, according to Web of Knowledge (WoK), on Web of Science 800 articles have been published to date. The first article dates from 1957 (Donomae et al., 1957), with another publication in 1960 and 1976 and two in 1979. Since then, we can find an increase in the number of articles in each decade, with an average of 5 articles a year in the 80s, 10 in the 90s, and 21 in the year 2000, passing in the last decade to an average production of 38 articles a year, with a growing evolution as can be appreciated in Figure 11.1.

Most of the articles are aimed at a better understanding of this by-product with a high interest in its nutritional composition, the presence of bioactive compounds, and its use as a potential new ingredient. In recent years, given the large amount of waste generated by agri-food technological processes and their environmental impact and economical losses, the need to integrate this waste back into the food chain for reuse and revaluation, within the so-called circular economy, has become evident. Circular economy (CE) is a conceptual model which has been used for better use of resources and minimization of waste in a closed-loop approach which could be appropriate for waste management. CE is a much-discussed pathway toward sustainability. In the interest of achieving a CE, alternative, value-added uses of okara have been investigated across the world.

In this sense, the studies of recent years on okara give very favorable and encouraging results for its possible use and the reduction of the waste generated, giving it a new utility of interest to develop new functional foods enriched with compounds obtained from okara or with the okara itself properly modified, as will be seen in the following sections of this chapter. This is in accordance with the fact that consumers are increasingly demanding beneficial health foods and much more care about the environment.

Therefore, the objective of this chapter is to review the bibliography that allows us to make the okara better known. To do this, we will start with a description of okara, its production, its composition, and the definition of the bioactive compounds present in it, then we will present the positive effects on health that derive from the use of this by-product directly or from its components, and, finally, we will show the current ways of using okara and the lines of future interest in its development.

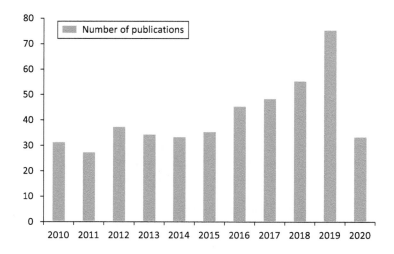

FIGURE 11.1 Evolution in the number of publications about okara in the last ten years.

11.2 OKARA: DEFINITION, WORLD PRODUCTION, OBTAINMENT, AND CHEMICAL COMPOSITION

11.2.1 DEFINITION AND WORLD PRODUCTION

Okara is the by-product remaining after the aqueous fraction is extracted and filtered in soy milk and tofu production. It is a white-yellowish material consisting of the insoluble parts of soybean seed. It is called *dòuzhā* in Chinese, *okara* in Japanese, and *biji* or *kongbiji* in Korean. This by-product has been used in the traditional gastronomy of these countries for thousands of years. In Japan, it is used in a side dish called *unohana* which consists of okara cooked with soy sauce, mirin, sliced carrots, burdock root, and shiitake mushrooms. Okara can be used to prepare Indonesian tempeh fermented by *Rhizopus oligosporus* (Shurtleff & Aoyagi, 1979). Okara is eaten in the Shandong cuisine of eastern China by steaming a wet mixture of okara that has been formed into blocks of *zha doufu* (KeShun, 1999). Currently, several researchers have reported in the literature the healthy effect of these recipes with okara that have been used for millennia in Asian countries. *Meitauza* is a traditional food in China based on okara previously steamed and covered with rice below 20° C for 8–14 days to produce natural fermentation by *Actinomucor elegans* and *Zymomonas mobilis*, which were isolated by Xu et al. (2012). The fermentation of the okara-tempeh by *Rhizopus oligosporus* produced increases in the prebiotic properties (Yogo et al., 2011). Okara-koji fermented with *Aspergillus oryzae* was added to biscuits and cupcakes (Matsuo, 1999) and also used as a flour substitute to extend the shelf-life of high-fat bakery products. *Meju* is a traditional Korean cake that is fermented by *A. oryzae* and dried naturally. Lee et al. (2013) replaced the soy ingredient with okara in this recipe to get okara-meju, which provided additional health and organoleptic benefits. The biovalorization of okara as a useful substrate for fungi, bacteria, and yeast fermentation is therefore very important for the production of functional ingredients as will be explained in section 11.4.

The statistical data on production worldwide of okara is not easily available. The production worldwide of okara is estimated to be more than four million tons, most of which is produced by Asian countries including Japan, Korea, China, and Singapore. The manufacture of tofu by the soy product industry produces around 800,000 tons of okara in Japan, about 310,000 tons in Korea, and around 2,800,000 tons in China annually. Moreover, the production and consumption in non-Asian countries have increased and are continuously accompanied by the accumulation of the okara by-product (Kamble & Rani, 2020).

The manufacture of soy products generates annually around 1.1–1.2 kg of fresh okara for every kilogram of soy processed into soy milk or tofu products. These high volumes of by-products pose a disposal problem because the high moisture content (70–80%) makes okara very perishable and susceptible to decomposition, which means economic losses and socio-environmental problems. Okara is commonly used as compost and animal feed, and excess okara is disposed of as waste. However, as previously mentioned, okara still contains many nutrients, minerals, and phytochemicals, making this by-product suitable to search for alternatives for its use as a source of ingredients with high added value and functionality (Colletti et al., 2020; Vong & Liu, 2016).

11.2.2 OKARA OBTAINMENT

Okara, as mentioned before, is the byproduct remaining after soy milk and tofu production. The sequence of soy milk processing steps can be carried out using the Japanese or Chinese method (Figure 11.2). In the Japanese way of soy milk manufacture, soaked whole soybeans are first cooked before grinding and filtering; in the Chinese way, the raw soybeans are first ground and then extracted with water, filtered, and then heated (Vong & Liu, 2016).

In general, in the process soybeans are first soaked in water to allow complete hydration of the protein and other components. The amount of water that is added, the time, and the temperature

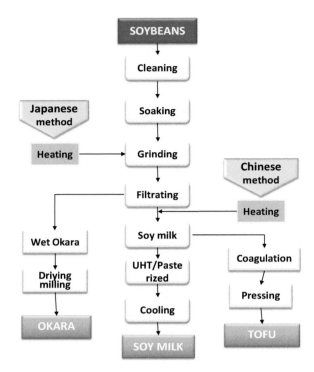

FIGURE 11.2   Okara-obtaining procedure.

of this step will depend on soybean variety which depends on the climate, soil composition, and agricultural methods, among other factors. After this mixture is ground into slurry, filtration or centrifugation is conducted to separate the liquid and solid parts; these are soy milk and okara, respectively. During the step of grinding soybeans, soy lipoxygenase and hydroperoxide lyase can react with the unsaturated fatty acids, primarily linoleic acid, leading to the formation of aroma compounds that explain the off-flavors in raw soy milk. As these enzymes are usually denatured at temperatures above 80° C, the Chinese method of soy milk manufacture, where soybeans are ground before the filtrate is boiled, is likely to produce okara that has more "beany" and "green" characteristics (Yuan & Chang, 2007).

The thermal treatment to which soybeans are submitted has the purpose of destroying the antinutritional factors and all microorganisms including the spores and enzyme activity in the case of sterilization, increasing the product shelf-life. The homogenization during this treatment helps to minimize the size of the particles and advantage their distribution in soy milk, improving the mouth-feel of the drink and making the particles stable in suspension during storage.

Filtration separates the residue rich in fiber and insoluble proteins, which is the okara, from the protein-oil dispersion-emulsion which is known as soy milk. After this, the process can go ahead to produce tofu. Milk proteins are coagulated using a solution of calcium and magnesium salts for 10–30 min at a temperature between 70–85° C, forming a gel-like structure that expels the serum toward the outside and the tofu is finally obtained.

11.2.3   CHEMICAL COMPOSITION

The composition of okara mainly depends on the cultivar of soybean, the incidence of environmental factors, and the processing conditions used during soy milk production (Colletti et al., 2020), and also the different methods applied in the analysis of each component.

The general composition of okara has been analyzed by various authors; a summary of the results is shown in Table 11.1. On the other hand, values for minerals, vitamins, and amino acids are reported in Tables 11.2 and 11.3, respectively. Okara has also a high amount of phytochemical compounds that are described in Section 11.3 of this chapter, providing information on the interesting possibilities of this by-product as a potential healthy ingredient.

The proximate composition (expressed as g/100 g dry matter) shows protein (18–35%), fat (3–20%), dietary fiber (12–76%), and ash (3.4–4.2%) (Redondo-Cuenca et al., 2008; Mateos-Aparicio et al.,

TABLE 11.1
Proximate Composition (g/100 g Dry Matter)

Moisture*	Protein	Carbohydrates	Fat	Fiber	Ash	References
	28.5	2.6	9.8	55.5	3.6	Redondo-Cuenca et al., 2008
	33.4	4.1	8.5	54.3	3.7	Mateos-Aparicio et al., 2010a
	34.7	37.9	3.2		4.2	Vishwanathan et al., 2011
	25.5	32.6	12	12.2	3.9	Rashad et al., 2011
1.38	34.6	21.1	19.1	20.37	3.44	Ostermann-Porcel et al., 2017
	26		10	53,8		Hu et al., 2019
4.9	18.1		3.6	76.4	3.4	Li et al., 2019
81.6	3.52	12,2	1.73		0.88	*Colletti et al., 2020
85.0	3.65		1.38	9.2		*Zhu et al., 2008

* Data expressed as wet matter.

TABLE 11.2
Minerals and Vitamins (mg/100 g)

*Minerals	dm	% PRI	wm		
Potassium (mg)	1350	2.5	213		
Sodium (mg)	30	3.7	9		
Calcium (mg)	320	8.0	80		
Magnesium (mg)	130	10.5	26		
Iron (mg)	0.62	2.3	1.3		
Zinc (mg)	0.29	0.9	0.56		
Manganese	0.21	3.2	0.4		
Cupper	0.1	3.2	0.2		
**Vitamins	Soy	% PRI	Okara	% PRI	
Ascorbic acid (mg)	6	30	0		
Thiamine (mg)	0.87	30	0.02	0.5	
Riboflavin (mg)	0.87	24	0.02	0.5	
Niacin (mg)	1.62	30	0.1	0.2	
Pantothenic acid (mg)	0.79	30	0.09	0.54	
Pyridoxine (mg)	0.38	87	0.12	2.1	
Folate (µg)	375	30	26	2	

PRI = Population Reference Intake.
*Mateos-Aparicio et al., 2010a; dm = dry matter.
*Colletti et al., 2020; wm = wet matter.
** Colletti et al., 2020.

TABLE 11.3

Amino Acid Content

Amino Acids	*mg/g Protein	*mg/g Protein	**g/100 g Okara
Aspartic acid	117	101	3.13
Threonine	41	37	1.15
Serine	50	51	1.58
Glutamic acid	195	160	4.97
Glycine	46	39	1.22
Alanine	46	43	1.34
Cysteine + methionine	26	17	0.52
Valine	51	47	1.45
Isoleucine	51	70	2.17
Leucine	81	108	3.37
Tyrosine + phenylalanine	95	34 + 97	1.07 + 3.03
Lysine	65	54	1.68
Histidine	28	27	0.84
Arginine	75	60	1.86
Proline	36	41	1.26

*Chan & Ma, 1999.
**Kumar et al., 2016.

2010a; Rashad et al., 2011; Vishwanathan et al., 2011; Li et al., 2019). Data could be expressed as wet okara, just after processing, then the moisture is around 80%, and the concentration of nutrients is much lower (Colletti et al., 2020; Zhu et al., 2008).

Okara is an important source of protein, with a good essential amino acid profile (Table 11.3). The most important concentration of essential amino acids is, in decreasing order, leucine, followed by phenyalanine, isoleucine, lysine, and threonine, all with similar content, and methionine with a lower amount. Among non-essential amino acids are glutamic acid and aspartic acid (Table 11.3). Okara protein presents low water solubility, hindering its incorporation as an ingredient into the food system in the elaboration of food products. The improvement of nutritional quality and functional properties by mild acid treatment will enhance the utilization of okara protein as a food ingredient by increasing the *in vitro* digestibility and the available lysine content of okara protein. The ratio of essential amino acids to total amino acids in okara is similar to soy milk and tofu (Chan & Ma, 1999; Vishwanathan et al., 2011). The protein composition can be separated through its sedimentation properties in four fractions: 2S, 7S, 11S, and 15S globulins. The composition of the two main fractions, 7S (β-conglycinin) and 11S (glycinin), was correlated with the physicochemical properties of okara protein. β-conglycinin is relatively lower in sulfur-containing amino acids than glycinin and thus is nutritionally lower in score (Kaviani & Kharabian, 2008; Singh et al., 2015).

β-conglycinin is the major protein in the 7S globulin fraction. Two more fractions, basic 7S globulin (Bg7S) and γ-conglycinin, have been identified, which together account for only a small percentage of 7S. The Bg7S is a glycoprotein suitable for its high content of cysteine residues. Glycinin is a heterogeneous oligomer and could be separated into two different fractions constituted by acidic polypeptides relatively rich in methionine and basic polypeptide chains relatively poor in methionine and with two and three cysteine residues that could be linked to the acidic and basic chains (Singh et al., 2015; Tao et al., 2019). Nevertheless, most of the water-soluble proteins are extracted in soy milk, therefore the protein of okara is predominantly non-extractable (de Figueiredo et al., 2018). Usually, these okara proteins are tightly linked to non-starch polysaccharides such as cellulose, hemicellulose, etc. Also, the extraction of these proteins is difficult because they are located inside the intact plant cells and it is necessary to apply a strong treatment capable of disrupting the cell wall (Preece et al., 2015). Tao et al. (2019) obtained okara proteins by an alkaline extraction with different pretreatments (homogenization, ultrasonic, and steam-cooking treatments) and observed significant changes in the protein structure affecting their functional properties, improving the foaming ability and stability of emulsion, and increasing solubility, water holding capacity, and viscosity.

Trypsin inhibitors make up about 5.19–14.4% of the okara protein, although they can be inactivated with adequate heat treatment (Stanojevic et al., 2013).

The amount of dietary fiber (DF) is high but the fraction distribution is not equilibrated. The total dietary fiber (TDF) ranges between 76.4 and 54.3%, insoluble dietary fiber (IDF) from 75.1 to 50.1% constituted by cellulose and hemicelluloses, and soluble dietary fiber (SDF) with the lowest percentage 1.3–4.7% (Redondo-Cuenca et al., 2008; Li et al., 2019; Hu et al., 2019). The significant differences between the percentages in the fractions are an excellent alternative to look for a treatment that improves this relationship, increasing SDF fraction. Okara fiber, mainly insoluble, has low water retention, swelling, and oil-holding capacities because these properties are more related to the SDF. The indigestible fraction shows good suitability for fermentation by *Bifidobacteria*, indicating that it could be used as a prebiotic ingredient. The polysaccharides in the cell wall of okara are mainly composed of the monomers glucose and galactose, higher than uronic acid, arabinose, and xilose in similar proportions (Redondo-Cuenca et al., 2008).

Regarding the fat content, okara's fatty acids profile presents polyunsaturated fatty acids such as linoleic acid (C18:2), representing around 54% of the total fatty acids content (Mateos-Aparicio et al., 2010a) and linolenic acid (C18:3). It contains also considerable amounts of oleic acid (C18:1) and a moderate quantity of saturated acids such as palmitic and stearic acids (C16:0 and C18:0) (Mateos-Aparicio et al., 2010a; Kumar et al., 2016).

Okara also contains minerals and vitamins (Table 11.2), although during the different stages of the okara obtainment process from soybean seeds, significant losses of each micronutrient can occur, except for calcium.

The comparison between soybean and the generated okara (results showed the same moisture) shows a significant reduction in all minerals except in the case of calcium which increases from 16 to 43%. Calcium interacts with denatured soy proteins (glycinin and β-conglycinin) by heat, and its presence promotes the formation of complexes with soybean protein hydrolyzates (Speroni et al., 2010). Regarding vitamins, being water-soluble, there is a large reduction except in the case of pyridoxine. The percentage of losses for these vitamins was greater than 65%, except for folic acid, which was lower (37%).

EFSA (2017) defines the concept of PRI (Population Reference Intake) used in Europe as the adequate level of nutrient intake for practically all (97–98%) of a population. If we apply this concept to okara micronutrients to evaluate their contribution to nutrient intake, the results are those explained below.

The minerals most represented in okara were potassium, calcium, magnesium, and iron. Regarding the three first minerals, they contribute to PRI with ranges of 9–12%, 10–13%, and 12%, respectively. Iron provides to daily recommendations a significant percentage of the PRI by assuming 14%.

In relation to vitamins, pyridoxine, niacin, pantothenic acid, and folic acid still are in appreciable amounts in okara, being the intakes of folic acid and pyridoxine which have greater contributions to the PRI (12% and 11% respectively).

11.3   BIOACTIVE COMPOUNDS AND HEALTH IMPLICATIONS

In this section, we are going to describe the main bioactive compounds that are found in okara as well as their health-positive implications. The composition of okara differs depending on the cultivar of soybean, the method of soy milk processing, and the number of water-soluble components extracted from the ground soybeans (Vong & Liu, 2016). Okara has an interesting nutritional value, containing phenolic compounds and dietary fiber (DF) (Santos et al., 2019; Guimarães et al., 2020). Several studies have suggested that agri-food industry by-products could be a remarkable source of a wide range of bioactive compounds (Gullón et al., 2020). Bioactive ingredients of okara have been reported to safeguard against hypolipidemic, hypocholesterolemic, and type 2 diabetes syndrome by reducing glycemic index (GI) ingestion (Kamble et al., 2019).

11.3.1 Dietary Fiber and Oligosaccharides

DF is an essential component of a healthy diet and its positive relationship with human health has been established by the scientific community (Kaczmarczyk et al., 2012). More than 50% of functional foods available in the market have DF as an active component (Giuntini & Menezes, 2011; Macagnan, da Silva, & Hecktheuer, 2016). The importance of food fiber has promoted the growth of a large potential market for fiber-rich products and ingredients, as well as the research of new sources of DF that can be used as ingredients in the food industry, such as agronomic by-products that have traditionally been undervalued (Redondo-Cuenca et al., 2015).

The definition for DF adopted by the Codex Alimentarius in 2009 includes carbohydrate polymers that are not hydrolyzed by the endogenous enzymes in the small intestine of humans and thus includes resistant starch (RS). This definition also includes oligosaccharides of degrees of polymerization (DP) 3–9, but the decision on whether to include these oligosaccharides in the dietary fiber value was left to the discretion of national authorities (Mccleary, 2019).

In Europe, EFSA (2010) issued a document that describes DF as the

> non-digestible carbohydrates plus lignin, including non-starch polysaccharides (NSP) – celluloses, hemicelluloses, pectins, hydrocolloids (i.e., gums, mucilages, β-glucans), resistant oligosaccharides – fructo-oligosaccharides (FOS), galacto-oligosaccharides (GOS), other resistant oligosaccharides, resistant starches – consisting of physically enclosed starch, some types of raw starch granules, retrograded amylose, chemically and/or physically modified starches, and lignin.

The characteristics of DF make it an important ingredient in formulating diverse functional foods due to its beneficial effects (Gray, 2006; Redondo-Cuenca et al., 2006).

The beneficial effects and effectiveness of DF depend not only on fiber intake but also on fiber composition, its organizational structure and physicochemical characteristics, and associated bioactive compounds, which are directly related to its plant source and preparation methods (Elleuch et al., 2011; Macagnan et al., 2016).

Over the last few decades, much information has been discovered about the different compounds present in DF, such as prebiotics (due to the profile of fermentability of specific substances and their interaction with colonic microflora) (Giuntini & Menezes, 2011) and associated bioactive compounds (which can confer antioxidant status to the body) (Macagnan et al., 2016).

DF is the primary component and bioactive constituent of okara. Based on the differences in solubility, total dietary fiber (TDF) can be divided into insoluble dietary fiber (IDF) and soluble dietary fiber (SDF). The study of the content, type, and the main monomers (Redondo-Cuenca et al., 2008; Mateos-Aparicio et al., 2010c; Mateos-Aparicio et al., 2010d), the characterization of cell wall polysaccharides and *in vitro* physicochemical properties (Mateos-Aparicio et al., 2010c), and the *in vitro* and animal studies of health properties (Préstamo et al., 2007; Jiménez-Escrig et al., 2008; Villanueva et al., 2011) can predict the potential physiological effects *in vivo* of the DF of this by-product.

Okara is rich in DF (TDF 54.3, IDF 50.1, SDF 4.2 g/100 g dry matter) (Mateos-Aparicio et al., 2010a) and mainly consists of cellulose, hemicelluloses, and pectins (Li et al., 2019). The portion of SDF in okara is very low. This affects the development and utilization values of okara. Therefore, increasing the SDF content is an important way to improve the quality and enhance the utilization of okara (Li et al., 2019).

TABLE 11.4

Monomeric Composition of Okara Dietary Fiber (g/100 g Dry Matter)

	IDF	SDF
Rhamnose	0.47	0.37
Fucose	0.37	0.07
Arabinose	5.17	0.73
Xylose	4.23	0.09
Mannose	0.91	0.28
Galactose	8.31	2.11
Glucose	13.07	0.27
Uronic acids	3.10	1.60
Total	35.63	5.51

Source: Redondo-Cuenca et al., 2008.

Redondo-Cuenca et al. (2008) studied the monomeric composition of the fiber of okara (Table 11.4) and indicate that IDF is characterized by high glucose content, followed by galactose and arabinose, as well as considerable amounts of uronic acids and xylose. In the case of SDF, the most representative monomers are galactose, uronic acids, and arabinose.

The study of Mateos-Aparicio et al. (2010c) contributed to the knowledge of cell wall polysaccharides of okara and physicochemical properties. Cellulose is the main polymer, in the hemicellulosic fractions a large proportion of xyloglucans exists, and slightly methylated rhamnogalacturonan regions vary, substituted by arabinans, galactans, and/or arabinogalactans in the pectin-rich fractions. The results of swelling capacity (SC) (9.44 mL/g) and water retention capacity (WRC) (8.33 g/g) are similar to those of citrus pulp, pea and soybean cotyledons, and Spanish edible seaweeds (Guillon & Champ, 2002; Robertson et al., 2000; Rupérez & Saura-Calixto, 2001) and greater than apple pulp and pea hull (Robertson et al., 2000). The knowledge of okara's polysaccharides and their swelling property and water retention capacity could be useful for its development as a potential source of functional ingredients. From a technological point of view, fibers are used for texturing and to increase volume, above all in manufacturing low-calorie foods, essentially due to their capacity to absorb water (Oakenfull, 2001; Mateos-Aparicio et al., 2010c).

Soybeans are an important source of oligosaccharides such as raffinose and stachyose. These oligosaccharides are also called α-galactosides, α-galactooligosaccharides, or oligosaccharides of the family of the raffinose. *In vitro* studies have shown that human digestive enzymes cannot hydrolyze this type of oligosaccharides, due to lacking α-galactosidase, so they can reach the colon intact where they are fermented by bifidobacteria (Corzo et al., 2015). The highest amounts of oligosaccharides of okara are α-galactosides (stachyose + raffinose 1.4 g/100 g dm). Moreover, inulin was detected in okara (0.4 g/100 g dm) (Mateos-Aparicio et al., 2010a). As a result of soybean processing for soy drink or tofu preparation, okara residue is mainly an insoluble and indigestible by-product (Espinosa-Martos & Rupérez, 2009), which contains low soluble carbohydrates (CHO).

DF intake is generally accepted as part of a positive diet for improving health. The physiological and physicochemical properties of DF are dependent on the individual components present, particularly the soluble and insoluble fractions and other associated compounds (Redondo-Cuenca et al., 2015). Okara is high in DF but is mainly insoluble. Most of the studies are directed to increasing the soluble fraction. Thus, research on the physiological effects of okara is limited, especially on unmodified by-products.

Okara has many bioactivities such as preventing hypoglycemia or diabetes, hyperlipidemia or hypercholesterolemia, cardiovascular diseases, obesity, antioxidant properties, and promoting the

TABLE 11.5

Health Improvement Due to Physiological Effects of Okara

Physiological Effect	Reference
Prevention of obesity	Matsumoto et al., 2007; Villanueva-Suárez et al., 2016
Prevention of diabetes	Hosokawa et al., 2016; Ismaiel, Yang, & Cui, 2017; Xu et al., 2000; Duru et al., 2018; Lecerf et al., 2020
Prevention of hyperlipidemia or hypercholesterolemia	Lemes et al., 2014; Nishibori et al., 2018; Villanueva et al., 2011; Wang & Li, 1996; Bedani et al., 2015; Préstamo et al., 2007; Skudelska & Nogowski, 2007; Kitawaki et al., 2007
Antioxidant properties	Amin & Mukhrizah, 2006; Croge et al., 2018; Mateos-Aparicio et al., 2010b; Yokomizo et al., 2002; Zhu et al., 2008a; Jiménez-Escrig et al., 2008; Jimenez-Escrig et al., 2010; Mateos-Aparicio et al., 2010a; Ruffer & Kulling, 2006; Voss et al., 2019
Prebiotic effect, promoting the growth of probiotics	Pérez-López et al., 2018; Pérez-López et al., 2016c; Vieira et al., 2017; Jiménez-Escrig et al., 2008
Prevention of cancer	Ahmad et al., 2013; Guo et al., 2020; Pavese et al., 2014; Zhang et al., 2019; Tuli et al., 2019; Zhao et al., 2019
Menopausal symptoms and menopausal bone loss reduction	Hirose et al., 2016; Irrera et al., 2017; Liu et al., 2009; Zheng et al, 2016

growth of probiotics (Li et al., 2019). Therefore, okara can be useful as a functional ingredient with health-promoting attributes. Table 11.5 shows the most important physiological effects of okara.

Préstamo et al. (2007) examined the effects of okara in rats, concluding that it could be useful as a weight-loss dietary supplement. Other authors indicate that due to the high DF content, okara supplementation shows body-weight loss, beneficial properties on lipid metabolism, and potential prebiotic effect, as reported earlier in rats and Syrian hamsters (Espinosa-Martos & Rupérez, 2009; Locke et al., 2015; Huang et al., 2016; Jiménez-Escrig et al., 2008; Villanueva et al., 2011).

In the study of Villanueva et al. (2011), male golden Syrian hamsters were fed high-fat diets supplemented with okara for three weeks. The okara soybean by-product showed hypocholesterolemic and hypolipidemic effects when hamsters were fed high-fat diets supplemented with 20% okara fiber over three weeks. Okara could improve the nutritional quality of other foods and is a promising functional ingredient for use in the development of fiber-rich foods (Redondo-Cuenca et al., 2015).

In vitro experiments have indicated that okara is a potential source of antioxidant components (Amin & Mukhrizah, 2006). The antioxidant activities of the soluble fractions from okara would seem to confirm the strong role of pectins in the reductive power and free radical scavenging activity of cell wall polysaccharides from edible plants reported by other authors, although a contribution from residual proteins cannot be ruled out (Mateos-Aparicio et al., 2010d).

Préstamo et al. (2007) showed that *in vivo* colonic fermentation of okara resulted in a lower pH and higher total short-chain fatty acid production. Jiménez-Escrig et al. (2008) observed that DF from okara protects the gut environment in terms of antioxidant status and prebiotic effect. Due to its prebiotic capacity, okara can enhance short-chain fatty acid release and improve calcium and magnesium absorption in the organism. This may be associated with restoring several beneficial groups in gut microbiota and preventing the bacterial drop caused by a high-fat diet after okara consumption (Pérez-López et al., 2018; Guimarães et al., 2018).

11.3.2 PHENOLIC COMPOUNDS AND ISOFLAVONES

Soybeans contain many phenolic compounds and are recognized as a major dietary source of isoflavones that have been categorized as the most important group of soluble phenolic compounds due to their biological activities (Zaheer & Aktar, 2017). Isoflavones are a subgroup of naturally occurring flavonoids consisting of heterocyclic phenols, non-nutritive substances, which have been linked with potential health benefits. These compounds are described as phytoestrogen compounds since they have a chemical structure similar to estradiol and exhibit estrogenic activity (Ludueña et al., 2007; Mateos-Aparicio et al., 2013; Cederroth & Nef, 2009). Twelve isoflavone components with different chemical structures and bioavailability have been isolated from soybeans: Three aglycones (daidzein, genistein, and glycitein) and their respective nine glucosidic conjugates of three types: β-glucosides (daidzin, genistin, and glycitin), acetyl-glucosides (acetyl-daidzin, acetyl-genistin, and acetyl-glycitin) and malonyl-glucosides (malonyl-daidzin, malonyl-genistin, and malonyl-glycitin) (Jackson et al., 2002; Genovese et al., 2006). The total isoflavones content is represented mostly by conjugated forms, mainly β-glucosides (genistin, daidzin, and, to a lesser extent, glycetin) and malonyl-glucosides, and only 2–3% are in the form of isoflavone aglycones (Achouri et al., 2005; Cederroth & Nef, 2009). Soybean has an isoflavone content ranging from 20 to > 400 mg/100 g depending on diverse factors such as variety, environmental growing conditions, harvests, geographic location, and storage conditions (Liu, 2004; AESAN, 2007; USDA, 2008; Jankowiak, 2014a; Mateos-Aparicio et al., 2008; Szymczak et al., 2017).

Although lower than in other soy-based foods, the isoflavone concentration in okara is significant and this by-product could be considered as a way of incorporating isoflavones in the diet as an ingredient of several foods (Jankowiak et al., 2014c). The total isoflavone content of okara depends on the soybean variety used, with the concentration ranging between 0.02 and 0.12 g/100 g solids (Jackson et al., 2002). Approximately 1/3 of the isoflavones present in soy are transferred to the by-product in the process of obtaining the aqueous extract. However, the percentage varies between 12 and 40% of the raw soybean isoflavones depending on the processing conditions (Jackson et al., 2002; Giri & Mangaraj, 2012). Okara can contain the same 12 isoflavones of the soybeans, although with a different profile compared to the raw material due to the major processing operations: Soaking, grinding, and heating steps involved in soy milk and okara production.

Since the predominant isoflavones in soybean are the genistein series, the wet okara presents a higher content of this type of isoflavone (Genovese et al., 2006). Ludueña et al. (2007) reported differences in the isoflavone profile according to the origin of the transgenic or non-transgenic culture. Okara of non-transgenic varieties has a content of genistein higher than that of daidzein, while this relationship changes in genetically modified varieties.

In the okara, isoflavones are mainly in the β-glucoside form, but there is a greater percentage of aglycones in the okara when compared with soy beverage (Wang & Murphy, 1996; Villares et al., 2011). Therefore, the distribution of the individual isoflavones in okara is shifted towards the β-glucoside and aglycon forms at the expense of a decrease in the malonyl-b-glucoside content during the soaking and heating steps in the manufacturing of soy milk (Jankoviak et al., 2014b). Jackson et al. (2002) reported the instability of the malonyl conjugates when exposed to heat. The thermal treatment causes the decarboxylation of malonyl-glucosides that yield acetyl derivatives. Furthermore, they would be transformed into glucosides via ester hydrolysis. Depending on the conditions, disruption of the glycosidic bond could also occur, resulting in aglycones release. Numerous researchers reported the absence or low concentration of acetyl-glucosides in okara (Jankoviak, 2014a; Jackson et al., 2002; Rinaldi et al., 2000). Regarding the method of soy milk production, cold-grinding instead of hot-grinding of the slurry leads to an increased amount of aglycones due to prolonged β-glucosidase activity (Prabhakaran & Perera, 2006). Different authors attributed the higher concentration of aglycones in okara to their lower polarity compared with other forms, making them remain in the residue during soy milk production (Murphy et al., 2002; Genovese et al., 2006). It should be noted that the greater proportion of aglycones in the soy by-product is associated with better biological activity in humans. This is explained by the fact that conjugated glycosides have to be hydrolyzed prior to absorption and therefore have lower bioavailability than the corresponding aglycones with low molecular weight which facilitates diffusion (Justus et al., 2019).

Due to its high moisture (75–80 g/100 g), okara deteriorates rapidly. In order to extend its shelf-life, okara is submitted to a drying process. Processing conditions, time, and temperature affect the kinetics of isoflavone degradation and also could promote the hydrolysis of conjugated glycosides and favor the formation of aglycones, with greater bioavailability and thus, greater biological activity (Muliterno et al., 2017).

During the drying process, interconversion and thermal degradation of isoflavone forms can occur. The interconversion between different forms of isoflavones by thermal effect favors the decrease in malonyl- and acetyl-glucosides and the increase in glycosides and aglycones. Resistance to thermal degradation was reported by Lee and Lee (2009) due to isoflavone–protein interaction that produces thermal stability. When the protein is denatured, isoflavones are exposed, reducing their stability, and high drying temperatures ($\geq 70°$ C) produce significant isoflavone loss. Muliterno et al. (2017) recommended temperature and time conditions (T = $50°$ C; t < 6 h) to obtain a dried okara with high isoflavone content, mainly in the aglycone form.

Since non-conjugated isoflavones, particularly genistein and daidzein, display greater bioavailability (Izumi et al., 2000; Sirotkin & Harrath, 2014; Nkurunziza et al., 2019; Villares et al., 2011), many studies have focused on processing technologies aimed at increasing conversion reactions to enhance the formation of these compounds considering a possible application in food or nutraceutical industry. Although the popular *Aspergillus niger* fermentation has been the standard technology in commercial β-glucosidase production for hydrolysis, there are emerging approaches utilizing various microorganisms, different than the filamentous fungus cited above, as potential mechanisms for the biotransformation of β-glucoside isoflavones into isoflavone aglycones in okara (Chan Vong and Liu, 2019; Doan et al., 2019; Santos et al., 2018). Alternatively, the transformation of glycosides to aglycones is reported by different authors during okara enzymatic hydrolysis using added peptidases to produce protein hydrolysis. Since isoflavones are bound to the hydrophobic interior of globular soybean proteins, an additional effect of isoflavone release and conversion to aglycones is produced (Garcia Pereira et al., 2019; Justus et al., 2019).

Many studies have suggested the association of isoflavone from soy-derived foods intake and the lower incidence of several pathologies such as cancer, cardiovascular disease, diabetes mellitus, osteoporosis, or menopausal symptoms (Table 11.5). Isoflavones are reported to act as a phytoestrogen in humans, binding to both α and β estrogen receptors, and causing an estrogenic or anti-estrogenic effect according to the type of estrogen receptor. Health-promoting effects of soybean isoflavones related to risk reduction in breast cancer, prostate cancer, menopausal symptoms, or osteoporosis have been attributed to their estrogenic-like activity and great interest has been observed in the past decades on the role of phytoestrogens as an alternative to hormone replacement therapy. These compounds also possess high antioxidant activity and metal-ion chelating properties and there is scientific evidence that the reactive oxygen species and free radicals generated during cellular metabolism or peroxidation of lipids play important roles in the pathogenesis of certain chronic pathologies such as coronary heart disease, diabetes mellitus, or certain types of cancer (Ibidapo et al., 2019).

Although numerous studies associate isoflavone consumption with health benefits, some results are contradictory or inconclusive and there is some controversy about it. The evidence provided is insufficient to establish a cause and effect relationship to substantiate health benefit claims by EFSA and greater standardization and documentation of clinical trial data of soy isoflavones are needed (EFSA, 2011; EFSA, 2012).

In addition to isoflavones, there are other phenolic compounds present in okara that are potentially effective in human health thanks to their antioxidant activities (Tyug et al., 2010). Pelaes Vital et al. (2018) reported a total phenolic content in this by-product of 106.7 mg gallic acid equivalents (GAE)/100 g and flavonoid content of 32.7 mg/100 g quercetin equivalents. The phenolic compounds other than isoflavones are mostly phenolic acids. Kim et al. (2006) reported that soybean contains syringic, chlorogenic, gallic, and ferulic acids. Okara, originating from soybean, might also have a significant amount of these phenolic compounds. Phenolic compounds play an important role in stabilizing lipid peroxidation and inhibiting various types of oxidizing enzymes and, thus, provide protection against oxidative damage from free radicals and contribute to antioxidant activities (Craft et al., 2012). However, the most important polyphenols of the okara by-product in relation to biological activities are the associated isoflavones and they have been the most widely investigated.

11.3.3 PROTEINS AND BIOACTIVE PEPTIDES

Soybean and its products are widely recognized as low-cost protein sources due to their excellent nutritional and functional properties. Okara, the major by-product of the soy milk industry, which is rich in proteins, as mentioned in the previous sections, could possibly release under physiological conditions potential bioactive peptides.

Biologically active peptides are food-derived peptides that can exhibit diverse activities, including opiate-like, mineral binding, immunomodulatory, antimicrobial, antioxidant, anticancer, antidiabetic, antithrombotic, antihypertensive, hypocholesterolemic, and blood-pressure-lowering actions (Erdmann et al., 2008; Puchalska et al., 2014; Nongonierma & FitzGerald, 2017). Bioactive peptides have been detected in different animal and vegetable protein sources, milk peptides being by far the most commonly known source (Pihlanto et al., 2008). They are extracted from non-antioxidant precursor proteins from different origins by the activity of either proteolytic microorganisms or isolated enzymes.

The importance of antioxidants is recognized by the World Health Organization which has been arguing in favor of increasing worldwide consumption of dietary sources of antioxidants, food intake being the main form of acquiring these compounds (WHO, 1990). Peptides can exert antioxidant activity by the same mechanisms as those observed for other antioxidants.

In general, the antioxidant activity of proteins has been related to their amino acid composition. Therefore, through enzymatic hydrolysis, it is possible to increase this capacity, since the protein cleavage favors the exposure of antioxidant amino acids in proteins (Elias et al., 2008).

Biologically active peptides can be hydrolyzed during different stages of gastrointestinal (GI) digestion before being transferred and absorbed into the intestinal epithelium (Picariello et al., 2010). The GI tract is known to be one of the major barriers in the human body. The conditions in the GI tract, such as digestive enzymes and pH values in the stomach, might influence the structures and functions of the peptides (Ao & Li, 2013). It is necessary that bioactive peptides remain intact during food processing and GI digestion before reaching their target sites to accomplish physiological effects.

Okara proteins are resistant to complete digestion by gastrointestinal enzymes and some essential amino acids' low bioavailability is due to their poor solubility (Chan & Ma, 1999). Through soy proteins' enzymatic hydrolysis, we can obtain bioactive peptides and free amino acids. Studies performed with okara have shown some bioactivities, especially in protein hydrolyzates (Sbroggio et al., 2016). These peptides are specific protein fragments that possess antioxidant activity (Zhang et al., 2018; Coscueta et al., 2016), i.e. can control oxidative processes in food and humans, and also may have antihypertensive (Coscueta et al., 2016; Guan et al., 2017) and antidiabetic activities (Singh et al., 2014) among other potential physiological activities.

Several studies have analyzed okara peptides for their antioxidant properties (Yokomizo et al., 2002) and health-promoting effects of bioactive peptides (Jimenez-Escrig et al., 2010) derived from okara with promising results both *in vitro* and in animal models.

As mentioned above, studies have been carried out on okara protein showing that it resists complete digestion by the gastrointestinal enzymes pepsin and pancreatin. The low-molecular-weight fraction of these digestion-resistant peptides (less than 1 kDa) is highly efficient in inhibiting angiotensin-converting enzyme (ACE) and shows the most remarkable antioxidant activity, possibly due to the high presence of hydrophobic amino acids (Jimenez-Escrig et al., 2010). This research evidenced that the consumption of okara protein may result in health benefits based on the bioavailability and bioactivity of the obtained peptide.

In the same line, Voss et al. (2019) carried out a study on dried okara to test the effect of enzymatic hydrolysis on the antioxidant capacity and antihypertensive activity of the final hydrolysates. They analyzed the peptides profile and respective sequence, as well as free amino acids composition, concluding that it is possible to obtain valuable extracts from okara with different biological properties depending on the enzyme used to promote the hydrolysis.

In general, based on the literature, it can be assumed that okara protein and resulting bioactive peptides make this by-product useful either as a functional ingredient or as a food supplement for blood-pressure-lowering or antioxidant applications.

11.3.4   HEALTH IMPLICATIONS

Consumers are increasingly demanding beneficial health foods and okara can be a useful by-product to be used to improve the quality of the diet. Studies have indicated that this by-product is a source of bioactive compounds and presents prebiotic and antioxidant activity. Taking into account all the bioactive compounds that we can find in okara, it can be accepted, based on several studies both *in vivo* or *in vitro*, that its use in food, as an ingredient or functional compound, has important implications in human health. Some of them are summarized in Table 11.5.

11.4   OKARA VALORIZATION FOR FOOD USES

As was commented in the previous sections, okara presents an interesting nutritional and bioactive profile. However, it is wasted or at most used as animal feed. Thus, considerable attention has been recently paid to its reuse (Yang et al., 2020). To use this by-product, it is necessary to find the most suitable procedure, from the point of view of cost-effectiveness and yield, to obtain its bioactive compounds and/or the way to re-introduce it in the food chain again. This section includes a review of the approaches to valorize okara.

11.4.1   DIRECT USES

Even though it is not easy to introduce okara in the food industry due to its own characteristics, such as moisture content, some products have been developed using this by-product. Okara has been used as a direct ingredient because of its nutritional composition in China and Japan for years to provide dietary fiber and protein easily and cheaply (Colletti et al., 2020). Moreover, okara has been processed to obtain a soy-based snack (Katayama & Wilson, 2008), gluten-free biscuits (Ostermann-Porcel et al., 2017) bread and pancakes (Aplevic & Demiate, 2007; Bowles & Demiate, 2006; Suda et al., 2007; Wickramarathna & Arampath, 2003), dried noodles (Sun & Yang, 2010), rice noodles (Kang et al., 2018), functional pasta enriched with okara (Kamble et al., 2019), and corn tortillas (Waliszewski et al., 2002). In these types of products, okara can replace partially the wheat flour to decrease the energy value and increase the content of fiber, protein, and unsaturated fats, or can be used, as in the case of corn tortillas, to improve the amino acid profile of the product, or to reduce starch digestion, and thus glycemic index, as in the studies of rice noodles and functional pasta mentioned above.

Besides, other non-based cereal food can be functionalized by adding okara. The nutritional and functional properties of soy yogurt were increased with the addition of okara (Bedani et al., 2014). Genta et al. (2002) developed a soy candy with higher protein availability by introducing okara in the formulation. Other studies were performed on the development of healthy drinks (Li et al., 2005) or more functional sausages (Huang et al., 2004).

In some of the food products developed using okara (Genta et al., 2002; Waliszewski et al., 2002) an undesirable flavor, called a "beany" flavor, was a significant problem for the development of an okara-based or soy-based food product. This "beany" flavor or odor is a result of the oxidation of unsaturated fatty acids by lipoxygenase enzymes during the processing of soy protein products (Wilson, 1995).

Pelaes-Vital et al. (2018) have enriched a dairy product, milk with omega-3 fatty acids, with okara as a source of antioxidants with good results, avoiding lipid oxidation, and an increase in the antioxidant capacity of the final product due to the high content of phenolic compounds.

However, the previously mentioned uses are done after a drying process that can reduce or change the chemical composition of this by-product. The possible use of wet okara is scarcely considered in the revised literature, although it has been included in low-fat beef burgers (Su et al., 2013; Turhan et al., 2007), in a coconut-based baked snack (Radocaj & Dimic, 2013), and in freshwater products (Rinaldi et al., 2000). Some approaches have been made to improve the use of fresh okra as an ingredient. Park et al. (2015) developed okara cookies using fresh okara; the cookies were baked using only okara or okara with additives: Starch, soy flour, and hydroxypropyl methylcellulose. Based on their results, the additives held the water content of fresh okara during the production of okara cookies; further, the additives allowed the formulation of healthier snack foods. Recently, Guimarães et al. (2018) used fresh okara to enrich a vegetable paste with a final product with low energetic value and high protein, total dietary fiber and isoflavones, and a good antioxidant activity. More research has to be done on this topic.

In general, we have to point out that okara is important not only from a bioactive point of view but also because of its digestibility and structural properties, and a good understanding of these aspects of okara is important before it is used in the formulation of food products for human consumption. Several works on the structural characteristics and digestibility of okara showed that the incorporation of okara into various food products contributes to a good *in vitro* digestibility of protein and starch as well as an altered structural strength (Kamble et al., 2019; Kang et al., 2018; Li et al., 2013; Voss et al., 2018).

11.4.2 TREATMENTS FOR OKARA VALORIZATION

Okara consists of about 80% water and this high moisture content makes it difficult to handle and too expensive to dry by conventional means; therefore it is a very perishable residue from the soybean industry. It is usually treated as waste and is either used as feed, fertilizer, or is landfilled; indeed, particularly in Japan, most is burned to generate carbon dioxide (Colletti et al., 2020). Thus, discarding okara as waste is a potential environmental problem because it is highly susceptible to putrefaction, and finding a way to reincorporate it into the food chain is more than necessary (Pérez-López et al., 2016b).

Okara, like other vegetable by-products, is abundant and represents an inexpensive material, and because of its chemical composition is a promising source of bioactive compounds. Trying the use of this material should be focused on finding an adequate treatment to extract the mentioned bioactive substances and/or to stabilize the whole by-product, always considering the cost-effectiveness and yield.

As was commented, the high moisture is a handicap for using this by-product, and usually, a pre-treatment stage to reduce the water content is necessary. The most used techniques for the adjustment of the water content consist of 1) concentration with thermal and/or vacuum conditions; 2) thermal drying; 3) freeze drying; 4) de-watering with mechanical pressing; and 5) centrifugation and/or filtration. Despite these treatments' ability to condition the product and stabilize it, they result in rising costs of use (Galanakis, 2012; Mateos-Aparicio & Matias, 2019).

Other pre-treatments are more focused on the yield and decreasing costs, such as ultrasonication, steam explosion, microwave, etc. In fact, they are used to assist chemical extraction with better results than extraction by itself. Nevertheless, they use in most occasions organic solvents generating a new residue with negative environmental impact. The mentioned operation should be part of the valorization process, but not the only method. There are procedures that allow an entire valorization of the whole material, or at least ones that can be used in the residues after extraction trying to get a holistic valorization (Mateos-Aparicio & Matias, 2019).

The procedures used for okara valorization could be divided as follows: 1) bioprocessing, 2) chemical treatments, and 3) physical treatments.

11.4.2.1   Bioprocessing

Bioprocessing is commonly referred to as fermentation, although the use of enzymes is considered a bioprocess too. The use of enzymes and the fermentation of okara have been studied on several occasions. Specifically, an enzymatic treatment based on the use of the commercial food-grade Ultraflo® enzyme has been applied to hydrolyze the dietary fiber polysaccharides from okara obtaining a soluble dietary-fiber-enriched product that presented more healthy properties (Pérez-López et al., 2016a). The enzymatic methods are usually fast and do not consume large volumes of solvents, but they are considered expensive and the industrial scale-up could be complicated (Puri et al., 2012). Fermentation, however, seems to have more advantages in okara valorization. This process can be used to improve the nutritional value and the techno-functional properties of okara, increasing the contents of proteins, isoflavones, soluble dietary fiber, and decomposing phytic (Colletti et al., 2020). In fact, fungal fermentation has been studied to isolate bioactives, such as 8-hydroxydaidzein (Fujita et al., 2004), monacolin K (Wongkhalaung et al., 2009), low-molecular-weight carbohydrates, and antioxidant peptides (Li et al., 2016). Moreover, fungal fermentation has been used to obtain a polysaccharide with antioxidant and immunomodulatory activities (Zhu et al., 2015).

On the other hand, okara can be prepared as a fermented food for direct consumption using fungi (Colletti et al., 2020). This is the case of *meitauza* from China, which is a traditional food made with okara (Nagai et al., 2014); however, other foods can be prepared using okara although it is not the regular way, like tempeh (Yogo et al., 2011), okara-koji, an ingredient which can be used in baked foods such as cupcakes and biscuits (Matsuo, 1999), miso-like fermented seasonings (Matsuo & Takeuchi, 2003) and *meju*, a traditional Korean cake (Lee et al., 2013). The mentioned studies about the use of fungi-fermented okara in different traditional foods remark on the additional health benefits such as an increase in antioxidant activity, improved lipid profile, and also presentation of antimutagenic and prebiotic effects.

Okara can be fermented also by bacteria involving the *Bacillus* species. For instance, okara fermented by *B. subtilis* shows an increase in antioxidant activity (Zhu et al., 2008). The use of bacterial fermentation of this by-product is mainly used for the production of bioactive compounds such as 1-deoxyrimycin, an inhibitor of α-glucosidase useful in diabetes management (Zhu et al., 2010), and 9-Cis-11-trans conjugated linoleic acid, in this last case obtained from the fermentation of okara with *Propionibacterium freudenreichii* ssp. *Shermanii* (Vahvaselka & Laakso, 2010).

Yeast fermentation has been used too, using mainly solid-state fermentation (Colletti et al., 2020). This is the case of yeast-fermented okara food with an improved nutritional profile getting a protein and ash increase (Rashad et al., 2011). Furthermore, okara can be biotransformed with solid-state fermentation to modify its aroma and improve the organoleptic quality (Vong & Liu, 2017).

11.4.2.2   Chemical Treatments

Chemical treatments consisting mainly of solvent-based extraction of bioactives are widespread because they are easy to perform. However, very often they result in low yields, accumulation of used solvents, and the majority are very environmentally hazardous (Mateos-Aparicio & Matias, 2019).

Ethanol is the most utilized solvent due to its cheap price and GRAS (generally recognized as safe) consideration according to the Food and Drug Administration (FDA) (Galanakis, 2012). Moreover, organic solvents such as hexane and dichloromethane are able to extract lipophilic compounds and phenols, and mixtures of acetone and water are frequently used for antioxidant extraction. One of the most valuable compounds from okara is isoflavones. They can be extracted with a mixture of alcohol-water, the most effective being between 50% and 80% ethanol. The mixture minimizes the extraction of sugars and peptides that the use of only water produces (Jankowiak et al., 2015).

Furthermore, the hydrolysis of biopolymers, particularly polysaccharides, using acidic and/or alkaline conditions are chemical treatments that can be used. In the case of okara, the cell wall polysaccharides can be hydrolyzed, which allows the separation of pectic polysaccharides, hemicelluloses, and cellulose (Mateos-Aparicio et al., 2010a, 2010b).

Regarding the mentioned disadvantages of the use of solvents, mainly organic ones, the tendency today is the assisted extraction, such as microwave-assisted extraction (MAE) or ultrasound-assisted extraction (Mateos-Aparicio & Matias, 2019). MAE with water has been used in okara to solubilize neutral carbohydrates, primarily arabinose and galactose, and polyphenols (Tsubaki et al., 2009).

11.4.2.3 Physical Treatments

The physical treatments used in the valorization of by-products can be divided into thermal and non-thermal procedures. The thermal ones, i.e. pasteurization, sterilization, hot air-drying, and freeze drying, extend shelf-life. They can be used directly over the whole by-product and this retains characteristics to be considered a functional ingredient. Nevertheless, these procedures can affect the nutritional profiles (i.e. vitamin degradation) and sensorial properties (i.e. taste, color, etc.) (Mateos-Aparicio & Matias, 2019).

Hydrothermal processes are the most interesting thermal processes for valorization, and especially supercritical fluid extraction, which involves the use of a substance above its critical temperature and pressure exhibiting liquid and gas properties (Mateos-Aparicio & Matias, 2019). Okara has been extracted using ethanol-modified supercritical carbon dioxide to obtain oil components, mainly fatty acids, phytosterols, and polyphenolic compounds because of the use of ethanol (Quitain et al., 2006). Moreover, subcritical water, which is when temperatures are greater than $100°$ C and below the critical point ($374°$ C), can extract isoflavone aglycones (Nkurunziza et al., 2019).

Among non-thermal treatments, high hydrostatic pressure (HHP) processing is an environmentally friendly process used to extend products' shelf-life. It consists of subjecting the product to a high level of isostatic pressure (300–600 MPa) transmitted by water (Figure 11.3). The highlight of HHP treatment is that can maintain the sensory (taste, flavor) and nutritional properties (Arshadi et al., 2016).

HHP technique has been used to solubilize dietary fiber from the soybean by-product okara. Indeed, soluble dietary fiber increases because of the contents of arabinose, mannose, and uronic acids at the expense of the corresponding monosaccharides of insoluble dietary fiber, resulting in an improvement of hydration properties and a better oil holding capacity (Mateos-Aparicio et al., 2010b).

HHP assisted by enzymes (Ultraflo® or Viscozyme®) is more effective at releasing soluble carbohydrates than HHP as a unique treatment (Pérez et al., 2016a, 2016b, 2017). Indeed, HHP assisted by food-grade enzymes is capable of producing a nutritionally improved okara ingredient, with a

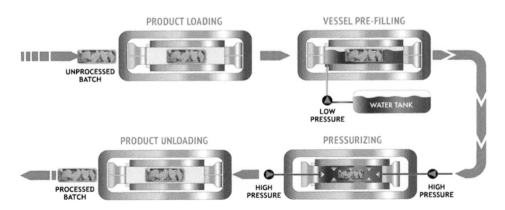

FIGURE 11.3   Diagram of operation of a unit of high hydrostatic pressure. Source: www.hiperbaric.com/en /high-pressure.

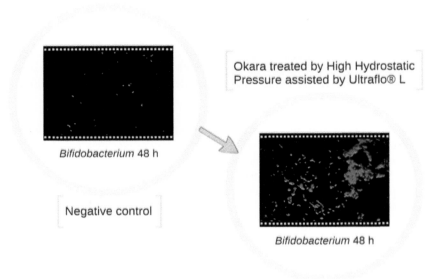

FIGURE 11.4 Bifidobacteria increase after 48 h when okara treated by high hydrostatic pressure assisted by Ultraflo®L enzyme was fermented.

more balanced soluble to insoluble fiber content. Raw okara and okara treated by HHP assisted by Ultraflo® present potential prebiotic effects because they have the capacity to promote beneficial bacteria growth such as *bifidobacterial* and *lactobacilli* which are major in the treated samples (Pérez-López et al., 2016c) (Figure 11.4).

Furthermore, the okara treated by the synergistic treatment of HHP and enzymes reduce plasma triglycerides and can help to reduce the microbiota imbalance produced by fat-high diets (Pérez-López et al., 2018; Mateos-Aparicio et al., 2019).

11.5 SUMMARY

Okara is a by-product of the process of obtaining soya milk. Although it has been known in Asian countries for many years and has been used in animal feed, its use in human food, worldwide, is relatively recent and for the moment somewhat limited considering the wide possibilities that this by-product has. This chapter aims to help in the knowledge of okara, on the one hand from the point of view of its obtainment and composition, and on the other hand from the point of view of its interesting beneficial properties in the control of numerous processes related to an improvement of human health due to the presence of bioactive compounds. This last aspect has made a section of the chapter focus on indicating how okara can be used directly and the processes of biovalorization aimed at enhancing its use in food, as an ingredient or a component that gives added functionality to food products made with it.

REFERENCES

Achouri, A., Boye, J.I., Belanger, D. (2005). Soybean isoflavones: Efficacy of extraction conditions and effect of food type on extractability. *Food Research International*, 38(10): 1199–1204. doi: 10.1016/j.foodres.2005.05.005.

AESAN (2007). Informe del Comité Científico de la Agencia Española de Seguridad Alimentaria y Nutrición (AESAN) en relación con las consecuencias asociadas al consumo de isoflavonas. *Revista del Comité Científico*, 5: 77–94.

Ahmad, A., Biersack, B., Li, Y., Bao, B., Kong, D., Ali, S., et al. (2013). Perspectives on the role of isoflavones in prostate cancer. *American Association of Pharmaceutical Scientists Journal*, 15(4): 991–1000.

Amin, I., Mukhrizah, O. (2006). Antioxidant capacity of methanolic and water extracts prepared from food-processing by-products. *Journal of the Science of Food and Agriculture*, 86(5): 778–784. doi: 10.1002/jsfa.2414.

Ao, J., Li, B. (2013). Stability and antioxidative activities of casein peptide fractions during simulated gastrointestinal digestion in vitro: Charge properties of peptides affect digestive stability. *Food Research International*, 52(1): 334–341.

Aplevic, K.S., Demiate, I.M. (2007). Physicochemical analyses of commercial samples of cheese bread premix and production of cheese breads with addition of okara. *Ciencia Agrotec*, 31(5): 1416–1422.

Arshadi, M., Attard, T.M., Lukasik, R.M., Brncic, M., da Costa Lopes, A.M., Finell, M., et al. (2016). Pretreatment and extraction techniques for recovery of added value compounds from wastes throughout the agri-food chain. *Green Chemistry*, 18(23): 6160–6204.

Bau, T.R., Ida, E.I. (2015). Soymilk processing with higher isoflavone aglycone content. *Food Chemistry*, 183: 161–168.

Bedani, R., Campos, M.M., Castro, I.A., Rossi, E.A., Saad, S.M. (2014). Incorporation of soybean by-product okara and inulin in a probiotic soy yoghurt: Texture profile and sensory acceptance. *Journal of the Science of Food and Agriculture*, 94(1): 119–125.

Bedani, R., Rossi, E.A., Cavallini, D.C.U., Pinto, R.A., Vendramini, R.C., Augusto, E.M., et al. (2015). Influence of daily consumption of symbiotic soy-based product supplemented with okara soybean by-product on risk factors for cardiovascular diseases. *Food Research International*, 73: 142–148. doi: 10.1016/j.foodres.2014.11.006.

Bowles, S., Demiate, I.M. (2006). Caracterização físico-química de okara e aplicação em pães do tipo francês. *Food Science and Technology*, 26(3): 652–659.

Cederroth, C.R., Nef, S. (2009). Soy, phytoestrogens and metabolism: A review. *Molecular and Cellular Endocrinology*, 304(1–2): 30–42. doi: 10.1016/j.mce.2009.02.027.

Chan Vong, W., Liu, S.Q. (2019). Biovalorisation of okara (soybean residue) for food and nutrition. *Trends in Food Science and Technology*, 52: 139–147.

Chan, W.M., Ma, C.Y. (1999). Acid modification of proteins from soymilk residue (okara). *Food Research International*, 32(2): 119–127.

Colletti, A., Attrovio, A., Boffa, L., Mantegna, S., Cravotto, G. (2020). Valorisation of by-products from soybean (Glycine max (L.) Merr.) processing. *Molecules*, 25(9): 2129.

Commission, C.A. (2009). ALINORM 09/32/3. Of, R., Session, T. H. E. S., the, O. F., committee, E., The, O. F., & Alimentarius, C, June: 23–26.

Corzo, N., Alonso, J.L., Azpiroz, F., Calvo, M.A., Cirici, M., Leis, R., et al. (2015). Prebiotics: Concept, properties and beneficial effects. *Nutricion Hospitalaria*, 31 (suppl 1). doi: 10.3305/nh.2015.31.sup1.8715.

Coscueta, E.R., Amorim, M.M., Voss, G.B., Nerli, B.B., Picó, G.A., Pintado, M.E. (2016). Bioactive properties of peptides obtained from Argentinian defatted soy flour protein by Corolase PP hydrolysis. *Food Chemistry*, 198: 36–44.

Craft, B.D., Kerrihard, A.L., Amarowicz, R., Pegg, R.B. (2012). Phenol-based antioxidants and the in vitro methods used for their assessment. *Comprehensive Reviews in Food Science and Food Safety*, 11(2): 148–173.

Croge, C., De Felix, D.S., Araújo, P., Gallina, M., Gallina, M.Z., Matumoto-Pintro, P.T. (2018). Okara residue as source of antioxidants against lipid oxidation in milk enriched with omega-3 and bioavailability of bioactive compounds after in vitro gastrointestinal digestion. *Journal of Food Science and Technology*, 55(4): 1518–1524.

de Figueiredo, V.R.G.D., Yamashita, F., Vanzela, A.L.L., Ida, E.I., Kurozawa, L.E. (2018). Action of multi-enzyme complex on protein extraction to obtain a proteinconcentrate from okara. *Journal of Food Science and Technology*, 55(4): 1508–1517.

Doan, D.T., Luu, D.P., Nguyen, T.D., Thi, B.H., Thi, H.M., Do, H.N., et al. (2019). Isolation of Penicillium citrinum from roots of Clerodendron Cyrtophyllum and application in biosynthesis of aglycone isoflavones from soybean waste fermentation. *Foods*, 8(11): 554. doi: 10.3390/foods8110554.

Donomae, I., Matsumoto, Y., Kokubu, T., Koide, R. (1957). Production of coronary heart disease in the rabbit by lanolin feeding. *Circulation Research*, V: 645–649.

Duru, K.C., Kovaleva, E.G., Danilova, I.G., van der Bijl, P., Belousova, A.V. (2018). The potential beneficial role of isoflavones in type 2 diabetes mellitus. *Nutrition Research*, 59: 1–15.

EFSA (2010). Scientific Opinion on Dietary Reference Values for carbohydrates and dietary fiber. *EFSA Journal*, 8(3): 1462.

EFSA (2011). Scientific Opinion on the substantiation of health claims related to soy isoflavones and protection of DNA, proteins and lipids from oxidative damage (ID 1286, 4245), maintenance of normal blood LDL-cholesterol concentrations (ID 1135, 1704a, 3093a), reduction of vasomotor symptoms associated with menopause (ID 1654, 1704b, 2140, 3093b, 3154, 3590), maintenance of normal skin tonicity. *EFSA Journal*, 9: 7, 2264: (ID 1704a), contribution to normal hair growth (ID 1704a, 4254), "cardiovascular health" (ID 3587), treatment of prostate cancer (ID 3588) and "upper respiratory tract" (ID 3589) pursuant to Article 13(1) of Regulation (EC) No 1924/20061.

EFSA (2012). Scientific Opinion on the substantiation of health claims related to soy isoflavones and maintenance of bone mineral density. *EFSA Journal*, 10: 8, 2847: (ID 1655) and reduction of vasomotor symptoms associated with menopause (ID 1654, 1704, 2140, 3093, 3154, 3590).

EFSA (2017). Dietary Reference Values for nutrients Summary report. https://efsa.onlinelibrary.wiley.com/doi/epdf/10.2903/sp.efsa.2017.e15121.

Elias, R.J., Kellerby, S.S., Decker, E.A. (2008). Antioxidant activity of proteins and peptides. *Critical Reviews in Food Science and Nutrition*, 48(5): 430–441. doi: 10.1080/10408390701425615.

Elleuch, M., Bedigian, D., Roiseux, O., Besbes, S., Blecker, C., Attia, H. (2011). Dietary fibre and fibre-rich by-products of food processing: Characterisation, technological functionality and commercial applications: A review. *Food Chemistry*, 124(2): 411–421.

Erdmann, K., Cheung, B.W.Y., Schro¨der, H. (2008). The possible roles of food-derived bioactive peptides in reducing the risk of cardiovascular disease. *Journal of Nutrional Biochemistry*, 19(10): 643–654 -2.

Espinosa-Martos, I., Rupérez, P. (2009). Indigestible fraction of okara from soybean: Composition, physicochemical properties and in vitro fermentability by pure cultures of Lactobacillus acidophilus and Bifidobacterium bifidum. *European Food Research and Technology*, 228(5): 685–693.

Fujita, T., Funako, T., Hayashi, H. (2004). 8-Hydroxydaidzein, an aldose reductase inhibitor from okara fermented with Aspergillus sp. HK-388. *Bioscience,. Biotechnolology and Biochemistry*, 68(7): 1588–1590.

Galanakis, C.M. (2012) Recovery of high added-value components from food wastes: Conventional, emerging technologies and commercialized applications. *Trends in Food Science & Technology*. 26: 68–87.

Garcia Pereira, D., Justus, A., Gabriel Falcão, E., de Souza Rocha, T., Iouko Ida, E., Kurozawa, L.E. (2019). Enzymatic hydrolysis of okara protein concentrate by mixture of endo and exopeptidase. *Journal of Food Processing and Preservation*, 43(10): e14134. doi: 10.1111/jfpp.14134.

Genovese, M.I., Davila, J., Lajolo, F.M. (2006). Isoflavones in processed soybean products from Ecuador. *Brazilian Archives of Biology and Technology*, 49(5): 853–859.

Genta, H.D., Genta, M.L., Alvarez, N.V., Santana, M.S. (2002). Production and acceptance of a soy candy. *Journal of Food Engineering*, 53(2): 199–202.

Giri, S.K., Mangaraj, S. (2012). Processing influences on composition and quality attributes of soymilk and its powder. *Food Engineering Reviews*, 4(3): 149–164. doi: 10.1007/s12393-012-9053-0.

Giuntini, E.B., Menezes, E.W. (2011). Fibra alimentar. Série de Publicações ILSI Brasil. *Funções Plenamente Reconhecidas de Nutrientes*, Vol. 18. ILSI: São Paulo, 23.

Gray, J. (2006). Dietary fibre. Definition and analysis. *Physiology. Health*. ILSI Europe. Concise Monograph Series.

Guan, H., Diaoa, X., Jianga, F., Hana, J., Konga, B. (2017). The enzymatic hydrolysis of soy protein isolate by Corolase PP under high hydrostatic pressure and its effect on bioactivity and characteristics of hydrolysates. *Food Chemistry*, 245: 89–96.

Guillon, F., Champ, M.M.-J. (2002). Carbohydrate fractions of legumes: Uses in human nutrition and potential for health. *British Journal of Nutrition*, 88 (suppl3): 293–306.

Guimarães, R.M., Ida, E.I., Falcão, H.G., de Rezende, T.A.M., Silva, J.de S., Alves, C.C.F., et al. (2020). Evaluating technological quality of okara flours obtained by different drying processes. *LWT, Food Science and Technology*, 123. doi: 10.1016/j.lwt.2020.109062.

Guimarães, R.M., Silva, T.E., Lemes, A.C., Boldrin, M.C.F., Pereira, M.A., Silva, F.G., et al. (2018). Okara: A soybean by-product as an alternative to enrich vegetable paste. *LWT- Food Science and Technology*, 92: 593–599.

Gullón, P., Gullón, B., Romaní, A., Rocchetti, G., Lorenzo, J.M. (2020). Smart advanced solvents for bioactive compounds recovery from agri-food by-products: A review. *Trends in Food Science and Technology*, 101: 182–197.

Guo, S., Wang, Y., Li, Y., Lid, Y., Feng, C., Li, Z. (2020). Daidzein-rich isoflavones aglycone inhibits lung cancer growth through inhibition of NF-κB signaling pathway. *Immunology Letters*, 222: 67–72.

Ibidapo, O., Henshaw, F., Shittu, T., Afolabi, W., Afolabi, W. (2019). Bioactive components of malted millet (Pennisetum glaucum), Soy Residue "okara" and wheat flour and their antioxidant properties. *International Journal of Food Properties*, 22(1): 1886–1898.

Hirose, A., Terauchi, M., Akiyoshi, M., Owa, K., Kato, K., Kubota, T. (2016). Low-dose isoflavone aglycone alleviates psychological symptoms of menopause in Japanese women: A randomized, double-blind, placebo-controlled study. *Archives of Gynecology and Obstetrics*, 293(3): 609–615.

Hosokawa, M., Katsukawa, M., Tanaka, H., Fukuda, H., Okuno, S., Tsuda, K., et al. (2016). Okara ameliorates glucose tolerance in GK rats. *Journal of Clinical Biochemistry and Nutrition*, 58(3): 216–222.

Hu, Y., Piao, Chunhong, Chena, Yue, Zhou, Yanan, Wang, Dan, Hansong, Yu, et al. (2019). Soybean residue (okara) fermentation with the yeast Kluyveromyces Marxianus. *Food Bioscience*, 31. 100439.

Huang, H., Krishnan, H.B., Pham, Q., Yu, L.L., Wang, T.T.Y. (2016). Soy and gut microbiota: Interaction and implication for human health. *Journal of Agricultural and Food Chemistry*, 64(46): 8695–8709.

Huang, W., Cao, L., Ma, Y., Wang, J., Jiang, Y. (2004). Preparation of nutritional sausage with soybean fiber. *Meat Industry*, 9: 11–13.

Irrera, N., Pizzino, G., D'Anna, R., Vaccaro, M., Arcoraci, V., Squadrito, F., Altavilla, D., Bitto, A. (2017). Dietary management of skin health: The role of genistein. *Nutrients*, 9(6): 622.

Ismaiel, M., Yang, H., Cui, M. (2017). Evaluation of high fibers okara and soybean bran as functional supplements for mice with experimentally induced Type 2 diabetes. *Polish Journal of Food and Nutrition Sciences*, 67(4): 327–337.

Izumi, T., Piskula, M.K., Osawa, S., Obata, A., Tobe, K., Saito, M., et al. (2000). Soy isoflavone aglycones are absorbed faster and in higher amounts than their glucosides in humans. *Journal of Nutrition*, 130(7): 1695–1699.

Jackson, C.J.C., Dini, J.P., Lavandier, C., Rupasinghe, H.P.V., Faulkner, H., Poysa, V., et al. (2002). Effects of processing on the content and composition of isoflavones during manufacturing of soy beverage and tofu. *Process Biochemistry*, 37(10): 1117–1123.

Jankoviak, L. (2014a). Separation of Isoflavones from Okara: Process Mechanisms & Synthesis. PhD thesis. Wageningen University: Wageningen, NL. ISBN: 978-94-6257-064-1. http://edepot.wur.nl/312325 Accessed: 10 February 2020.

Jankowiak, L., Kantzas, N., Boom, R., van der Goot, A. (2014b). Isoflavone extraction from okara using water as extractant. *Food Chemistry*, 160: 371–378. doi: 10.1016/j.foodchem.2014.03.082.

Jankowiak, L., Mendez Sevillano, D., Boom, R.M., Ottens, M., Zondervan, E., Van der Goot, A.J. (2015). A process synthesis approach for isolation of isoflavones from okara. *Industrial and Engineering Chemistry Research*, 54(2): 691–699.

Jankowiak, L., Trifunovic, O., Boom, R., van der Goot, A.J. (2014c). The potential of crude okara for isoflavone production. *Journal of Food Engineering*, 124: 166–172.

Jimenez-Escrig, A., Alaiz, M., Vioque, J., Ruperez, P. (2010). Health-promoting activities of ultra-filtered okara protein hydrolysates released by in vitro gastrointestinal digestion: Identification of active peptide from soybean lipoxygenase. *European Food Research and Technology*, 230(4): 655–663.

Jiménez-Escrig, A., Tenorio, M.D., Espinosa-Martos, I., Rupérez, P. (2008). Health-promoting effects of a dietary fiber concentrate from the soybean byproduct okara in rats. *Journal of Agricultural and Food Chemistry*, 56(16): 7495–7501.

Justus, A., Garcia Pereira, D., Idaa, E.J., Kurozawaa, L.E. (2019). Combined uses of an endo- and exopeptidase in okara improve the hydrolysates via formation of aglycone isoflavones and antioxidant capacity. *LWT-Food Science and Technology*, 115: 108467.

Kaczmarczyk, M.M., Miller, M.J., Freund, G.G. (2012). The health benefits of dietary fiber: Beyond the usual suspects of type 2 diabetes mellitus, cardiovascular disease and colon cancer. *Metabolism: Clinical and Experimental*, 61(8): 1058–1066.

Kamble, D.B., Rani, S. (2020). Bioactive components, in vitro digestibility, microstructure and application of soybean residue (okara): A review. *Legume Science*, 2(1): e32.

Kamble, D.B., Singh, R., Rani, S., Pratap, D. (2019). Physicochemical properties, in vitro digestibility and structural attributes of okara enriched functional pasta. *Journal of Food Processing and Preservation*, 43(12): 1–9.

Kang, M.J., Bae, I.Y., Lee, H.G. (2018). Rice noodle enriched with okara: Cooking property, texture, and in vitro starch digestibility. *Food Bioscience*, 22: 178–183.

Katayama, M., Wilson, L.A. (2008). Utilization of okara, a byproduct from soymilk production, through the development of soy-based snack food. *Journal of Food Science*, 73(3): 152–157.

Kaviani, B., Kharabian, A. (2008). Improvement of the nutritional value of soybean [Glycine max (L.) Merr.] seed with alteration in protein subunit of glycinin (11S globulin) and beta-conglycinin (7S globulin). *Turkey Journal of Biology*, 32: 91–97.7.

KeShun, L. (1999). Chapter 6. Oriental soyfoods. In Y.W. Ang Catharina et al. (Eds.), *Asian Foods: Science and Technology*. CRC Press. ISBN 978-1566767361.

Kim, J.A., Jung, W. S., Chun, S. C., Yu, C. Y., Ma, K. H., Gwag, J. G., Chung, I. M. (2006). A correlation between the level of phenolic compounds and the antioxidant capacity in cooked-with-rice and vegetable soybean (Glycine max L.) varieties. *European Food Research Technology*, 224: 259–270. doi: 10.1007/s00217-006-0377-

Kitawaki, R., Takagi, N., Iwasaki, M., Asao, H., Okada, S., Fukuda, M. (2007). Plasmacholesterol-lowering effects of soymilk and okara treated by lactic acid fermentation in rats. *Journal of the Japanese Society for Food Science and Technology (Japan)*, 54(8): 379–382.

Kumar, V., Rani, A., Husain, L. (2016). Investigations of amino acids profile, fatty acids composition, isoflavones content and antioxidative properties in soy okara. *Asian Journal of Chemistry*, 28(4): 903–906.

Lecerf, J.M., Arnoldi, A., Rowland, I., Trabald, J., Widhalme, K., Aiking, H., Messina, M. (2020). Soyfoods, glycemic control and diabetes. *Nutrition Clinique et Metabolisme* 34(2): 141–148. ISSN 09850562.

Lee, S., Lee, J. (2009). Effects of oven-drying, roasting, and explosive puffing process on isoflavone distribution in soybeans. *Food Chemistry*, 112(2): 316–320.

Lee, S.-I., Lee, Y.-K., Kim, S.-D., Lee, J.-E., Choi, J., Bak, J.-P., et al. (2013). Effect of fermented soybean curd residue (FSCR, SCR-meju) by Aspergillus oryzae on the anti-obesity and lipids improvement. *Journal of Journal of Nutrition and Health*, 46(6): 493–502.

Lemes, S.F., Lima, F.M., De Almeida, A.P.C., Ramalho, A.D.F.S., De Lima, R.S.R., Michelotto, L.F., et al. (2014). Nutritional recovery with okara diet prevented hypercholesterolemia, hepatic steatosis and glucose intolerance. *International Journal of Food Sciences and Nutrition*, 65(6): 745–753.

Li, B., Lu, F., Nan, H., Liu, Y. (2012). Isolation and structural characterisation of okara polysaccharides. *Molecules*, 17(1): 753–761.

Li, B., Yang, Wei, Nie, Yuanyang, Kang, Fangfang, Goff, H. Douglas, Cui, Steve W. (2019). Effect of steam explosion on dietary fiber, polysaccharide, protein and physicochemical properties of okara. *Food Hydrocolloids*, 94: 48–56.

Li, J., Wang, Q., He, M. (2005). Development of a soybean residue fiber and vitamin drink. *Science and Technology of Food Industry*, 26: 111–113.

Li, S., Chen, Y., Li, K., Lei, Z., Zhang, Z. (2016). Characterization of physicochemical properties of fermented soybean curd residue by *Morchella esculenta*. *International Biodeterioration and Biodegradation*, 109: 113–118.

Li, S., Zhu, D., Li, K., Yang, Y., Lei, Z., Zhang, Z. (2013). Soybean curd residue: Composition, utilization, and related limiting factors. *ISRN Industrial Engineering*, 8.

Liu, J., Ho, S.C., Su, Yi-xiang, Zhang, C-x, Chen, Y-m (2009). Effect of long- term intervention of soy isoflavones on bone mineral density in women: A meta-analysis of randomized controlled trials. *Bone*, 44(5): 948–953.

Liu, K. (2004). Soy isoflavones: Chemistry, processing effects, health benefits, and commercial production. In K. Liu (Ed.), *Soybeans as Functional Foods and Ingredients*. AOCS Press: Champaign, IL, 52–72.

Locke, A.E., Kahali, B., Berndt, S.I., Justice, A.E., Pers, T.H., Day, F.R. et al. (2015). Genetic studies of body mass index yield new insights for obesity biology. *Nature*.

Ludueña, B., Chichizola, C., Franconi, C. (2007). *Isoflavonas en Soja, Contenido de Daizceina y Genisteina y Su Importancia Biológica*. Bioquímica y Patología Clínica, 71, Vol. 1, 54–66.

Macagnan, F.T., da Silva, L.P., Hecktheuer, L.H. (2016). Dietary fibre: The scientific search for an ideal definition and methodology of analysis, and its physiological importance as a carrier of bioactive compounds. *Food Research International*, 85: 144–154.

Mateos-Aparicio, I., Mateos-Peinado, C., Jiménez-Escrig, A., Rupérez, P. (2010a). Multifunctional antioxidant activity of polysaccharide fractions from the soybean byproduct okara. *Carbohydrate Polymers*, 82(2): 245–250.

Mateos-Aparicio, I., Mateos-Peinado, C., Rupérez, P. (2010b). High hydrostatic pressure improves the functionality of dietary fibre in okara by-product from soybean. *Innovative Food Science and Emerging Technologies*, 11(3): 445–450.

Mateos-Aparicio, I., Matias, A. (2019). Food industry processing by-products in food. In: Galanakis (Ed.) *The Role of Alternative and Innovative Food Ingredients and Products in Consumer Wellness*. Elsevier, 239–283pp.

Mateos-Aparicio, I., Pérez-López, E., Rupérez, P. (2019). Valorisation approach for the soyban by-product okara using high hydrostatic pressure. *Current Nutrition and Food Science*, 15(6): 548–550.

Mateos-Aparicio, I., Redondo Cuenca, A., Villanueva Suárez, M.J., Tenorio Sanz, M.D. (2013) Chapter 4. Soybean: Overview of the nutritional profile and the implications for the nutrition and health effects. In H. Satou, R. Nakamura (Eds.), *Legumes: Types, Nutritional Composition and Health Benefits*. Nova Publishers: New York, 125–158.

Mateos-Aparicio, I., Redondo Cuenca, A., Villanueva-Suárez, M.J., Zapata-Revilla, A. (2008). Soybean, a promising health source. *Nutrición Hospitalaria*, 23(4): 305–312.

Mateos-Aparicio, I., Redondo-Cuenca, A., Villanueva-Suarez, M.J. (2010c). Isolation and characterisation of cell wall polysaccharides from legume by-products: Okara (soymilk residue), pea pod and broad bean pod. *Food Chemistry*, 122(1): 339–345.

Mateos-Aparicio, I., Redondo-Cuenca, A., Villanueva-Suarez, M.-J., Zapata-Revilla, M.-A., Tenorio-Sanz, M.-D. (2010d). Pea pod, broad bean pod and okara, potential sources of functional compounds. *LWT – Food Science and Technology*, 43(9): 1467–1470.

Matsumoto, K., Watanabe, Y., Yokoyama, S.I. (2007). Okara, soybean residue, prevents obesity in a diet-induced murine obesity model. *Bioscience, Biotechnology and Biochemistry*, 71(3): 720–727.

Matsuo, M. (1999). Application of okara koji, okara fermented by Aspergillus oryzae, for cookies and cupcakes. *Journal of Home Economics of Japan*, 50: 1029–1034.

Matsuo, M., Takeuchi, T. (2003). Preparation of low salt miso-like fermented seasonings using soy-oncom and okara-oncom (fermented soybeans and okara with Neurospora intermedia) and their antioxidant activity and antimutagenicity. *Food Science and Technology Research*, 9(3): 237–241.

Mccleary, B.V. (2019). Total dietary fiber (codex definition) in foods and food ingredients by a rapid enzymatic-gravimetric method and liquid chromatography: Collaborative study, first Action 2017.16. *Journal of AOAC International*, 102(1): 196–207.

Muliterno, M.M., Rodrigues, D., Sanches de Lima, F., Ida, E.I., Kurozawa, L.E. (2017). Conversion/degradation of isoflavones and color alterations during the drying of okara. *LWT - Food Science and Technology*, 75: 512–519.

Murphy, P.A., Barua, K., Hauck, C.C. (2002). Solvent extraction selection in the determination of isoflavones in soy foods. *Journal of Chromatography. Part B*, 777(1–2): 129–138.

Nagai, T., Li, L.T., Ma, Y.L., Sarkar, P.K., Nout, R., Park, K.Y., et al. (2014). Diversity of plant-based food products involving alkaline fermentation. In P.K. Sarkar, M.R. Nout(Eds.), *Handbook of Indigenous Foods Involving Alkaline Fermentation*. CRC Press: Boca Raton, FL, 78–87.

Nishibori, N., Kishibuchi, R., Morita, K. (2018). Suppressive effect of okara on intestinal lipid digestion and absorption in mice ingesting high-fat diet. *International Journal of Food Sciences and Nutrition*, 69(6): 690–669.

Nkurunziza, D., Pendletonb, P., Chuna, B.S. (2019). Optimization and kinetics modeling of okara isoflavones extraction using subcritical water. *Food Chemistry*, 295: 613–621.

Nongonierma, A.B., FitzGerald, R.J. (2017). Strategies for the discovery and identification of food protein-derived biologically active peptides. *Trends in Food Science and Technology*, 69: 289–305.

Oakenfull, D. (2001). Physicochemical properties of dietary fiber: Overview. In S.S. Cho, M.L. Dreher (Eds.), *Handbook of Dietary Fiber*. Marcel Dekker Editions: New York, 195-206.

Ostermann-Porcel, M.V., Quiroga-Panelo, N., Rinaldoni, A.N., Campderrós, M.E. (2017). Incorporation of okara into gluten-free cookies with high quality and nutritional value. *Journal of Food Quality*: 1–8.

Park, J., Choi, I., Kim, Y. (2015). Cookies formulated from fresh okara using starch, soy flour and hydroxy-propyl methylcellulose have high quality and nutritional value. *Lebensmittel-Wissenschaft und --Technologie- Food Science and Technology*, 63(1): 660–666.

Pavese, J.M., Krishna, S.N., Bergan, R.C. (2014). Genistein inhibits human prostate cancer cell detachment, invasion, and metastasis. *American Journal of Clinical Nutrition*, 100 (suppl): 431S–46S.

Pelaes Vital, A.C., Croge, C., da Silva, D.F., Araújo, P.J., Gallina, M.Z., Matumoto-Pintro, P.T. (2018). Okara residue as source of antioxidants against lipid oxidation in milk enriched with omega-3 and bioavailability of bioactive compounds after in vitro gastrointestinal digestión. *Journal of Food Science and Technology*, 55(4): 1518–1524. doi: 10.1007/s13197-018-3069-2.

Pérez-López, E., Cela, D., Costabile, A., Mateos-Aparicio, I., Rupérez, P. (2016c). In vitro fermentability and prebiotic potential of soyabean Okara by human faecal microbiota. *British Journal of Nutrition*, 116(6): 1116–1124.

Pérez-López, E., Mateos-Aparicio, I., Rupérez, P. (2016a). Okara treated with high hydrostatic pressure assisted by Ultraflo® L: Effect on solubility of dietary fibre. *Innovative Food Science and Emerging Technologies*, 33: 32–37.

Pérez-López, E., Mateos-Aparicio, I., Rupérez, P. (2016b). Low molecular weight carbohydrates released from Okara by enzymatic treatment under high hydrostatic pressure. *Innovative Food Science and Emerging Technologies*, 38: 76–82.

Pérez-López, E., Mateos-Aparicio, I., Ruperez, P. (2017). High hydrostatic pressure aided by food-grade enzymes as a novel approach for Okara valorization. *Innovative Food Science and Emerging Technologies*, 42: 197–203.

Pérez-López, E., Veses, A.M., Redondo, N., Tenorio-Sanz, M.D., Villanueva, M.J., Redondo-Cuenca, A., Marcos, E., Mateos-Aparicio, I., Rupérez, P., Rupérez, P. (2018). Soybean Okara modulates gut microbiota in rats fed a high-fat diet. *Bioactive Carbohydrates and Dietary Fibre*, 16: 100–107.

Picariello, G., Ferranti, P., Fierro, O., Mamone, G., Caira, S., Di Luccia, A., et al. (2010). Peptides surviving the simulated gastrointestinal digestion of milk proteins: Biological and toxicological implications. *Journal of Chromatography. Part B*, 878(3–4): 295–308.

Pihlanto, A., Akkanen, S., Korhonen, H.J. (2008). ACE—Inhibitory and antioxidant properties of potato (Solanum tuberosum). *Food Chemistry*, 109(1): 104–112.

Prabhakaran, M.P., Perera, C.O. (2006). Effect of extraction methods and UHT treatment conditions on the level of isoflavones during soymilk manufacture. *Food Chemistry*, 99(2): 231–237.

Preece, K.E., Drost, E., Hooshyar, N., Krijgsman, A., Cox, P.W., Zuidam, N.J. (2015). Confocal imaging to reveal the microstructure of soybean processing materials. *Journal of Food Engineering*, 147(2): 8–13.

Préstamo, G., Rupérez, P., Espinosa-Martos, I., Villanueva, M.J., Lasunción, M.A. (2007). The effects of okara on rat growth, cecal fermentation, and serum lipids. *European Food Research and Technology*, 225(5–6): 925–928.

Puchalska, P.M., García, M.C., Marina, M.L. (2014). Identification of native angiotensin-I converting enzyme inhibitory peptides in commercial soybean based infant formulas using HPLC-Q-ToF-MS. *Food Chemistry*, 157: 62–69.

Puri, M., Sharma, D., Barrow, C.J. (2012). Enzyme-assisted extraction ofbioactives from plants. *Trends in Biotechnology January*, 30(139): 37–44.

Quitain, A.T., Oro, K., Katoh, S., Moriyoshi, T. (2006). Recovery of oil components of okara by ethanol-modified supercritical carbon dioxide extraction. *Bioresource Technology*: 1509–1514.

Radocaj, O., Dimic, E. (2013). Valorization of wet okara, a value-added functional ingredient, in a coconut-based baked snack. *Cereal Chemistry Journal*, 90(3): 256–262.

Rashad, M.M., Mahmoud, A.E., Abou, H.M., Nooman, M.U. (2011). Improvement of nutritional quality and antioxidant activities of yeast fermented soybean curd residue. *African Journal of Biotechnology*, 10: 5504–5513.

Redondo-Cuenca, A., Villanueva-Suárez, M.J., Goñi, I. (2015). Healthy dietary fibers from plant food by-products. In Y.H. Hui, E.Ö. Evranuz(Eds.), *Handbook of Vegetable Preservation and Processing*, 2nd ed. *Food Science and Technology*, 25–54. CRC Press. ISBN: 97814822-12297.

Redondo-Cuenca, A., Villanueva-Suarez, M.J., Mateos-Aparicio, I. (2008). Soybean seeds and its by-product okara as sources of dietary fibre. Measurement by AOAC and Englyst methods. *Food Chemistry*, 108(3): 1099–1105.

Redondo-Cuenca, A., Villanueva-Suárez, M.J., Rodríguez-Sevilla, M.D., Mateos-Aparicio, I. (2006). Chemical composition and dietary fibre of yellow and green commercial soybeans (Glycine max). *Food Chemistry*, 101(3).

Rinaldi, V.E.A., Ng, P.K.W., Bennink, M.R. (2000). Effects of extrusion on dietary fiber and isoflavone contents of wheat extrudates enriched with wet okara. *Cereal Chemistry Journal*, 77(2): 237–240.

Robertson, J.A., De Monredon, F.D., Dysseler, P., Guillon, F., Amadò, R., Thibault, J.F. (2000). Hydration properties of dietary fibre and resistant starch: A European collaborative study. *LWT - Food Science and Technology*, 33(2): 72–79.

Ruffer, C., Kulling, S.E. (2006). Antioxidant activity of isoflavones and their major metabolites using different in vitro assays. *Journal of Food Agriculture Chemistry*, 54(8): 2926–2931.

Rupérez, P., Saura-Calixto, F. (2001). Dietary fibre and physicochemical properties of edible Spanish seaweeds. *European Food Research and Technology*, 212(3): 349–354.

Santos, D.C., Oliveira Filho, J.G., Silva, J.S., Sousa, M.F., Vilela, M.S., Silva, M.A.P., et al. (2019). Okara flour: Its physicochemical, microscopical and functional properties. *Nutrition and Food Science*, 49(6): 1252–1264.

Santos, V., Nascimento, C., Schimidt, C., Mantovani, D., Dekker, R., Alves da Cuna, M.A. (2018). Solid-state fermentation of soybean okara: Isoflavones biotransformation, antioxidant activity and enhancement of nutritional quality. *LWT-Food Science and Technology*, 92: 509–515.

Sbroggio, M.F., Montilha, M.S., Ribeiro, V., De Figueiredo, G., Georgetti, S.R., Kurozawa, L.E. (2016). Influence of the degree of hydrolysis and type of enzyme on antioxidant activity of okara protein hydrolysates. *Food Science and Technology*, 36(2): 375–381.

Scialabba, N.E. (2014). Food wastage footprint: Full-cost accounting. http://www.fao.org/3/a-i3991e.pdf. Accessed 18/07/16.

Seibel, N.F. (2018). Chapter 4. *Soja: Cultivo, Benefícios e Processamento*, 1st ed. CRV: Curitiba.

Shurtleff, W., Aoyagi, A. (1979). Tofu and soymilk production, Volume 2. *The Book of Tofu*. ISBN 1928914047.

Singh, A., Meena, M., Kumar, D., Dubey, A.K., Hassan, M.I. (2015). Structural and functional analysis of various globulin proteins from soy seed. *Critical Reviews in Food Science and Nutrition*, 55(11): 1491–1502.

Singh, B. P., Vij, S., Hati, S. (2014). Functional significance of bioactive peptides derived from soybean. *Peptides*, 54: 171–179.

Sirotkin, A.V., Harrath, A.H. (2014). Phytoestrogens and their effects. *European Journal of Pharmacology*, 741: 230–236. doi: 10.1016/j.ejphar.2014.07.057.

Skudelska, K., Nogowski, L. (2007). Genistein: A dietary compound inducing hormonal and metabolic changes. *Journal of Steroid Biochemistry and Molecular Biology*, 105(1–5): 37–45.

Speroni, F., Jung, S., de Lamballerie, M. (2010). Effects of calcium and pressure Treatmenton thermal gelation of soybean protein. *Journal of Food Science*, 75: 1.

Stanojevic, S.P., Barac, M.B., Pesic, M.B., Jankovic, V.S., Vucelic-Radovic, B.V. (2013). Bioactive proteins and energy value of okara as a byproduct in hydrothermalprocessing of soy milk. *Journal of Agricultural and Food Chemistry*, 61(38): 9210–9219.

Su, S.I.T., Pedroso Yoshida, C.M., Contreras-Castillo, C.J., Quiñones, E.M., Venturini, A.C. (2013). Okara, a soymilk industry by-product, as a non-meat protein source in reduced fat beef burgers. *Food Science and Technology*, 33: 52–56.

Suda, T., Kido, Y., Tsutsui, S., Tsutsui, D., Fujita, M., Nakaya, Y. (2007). Nutritional evaluation of the new okara powder for food processing material. *Foods Food Ingredients Journal of Japan*, 212: 320.

Sun, X., Yang, Y. (2010). Study on cooking quality of noodle of okara fiber. *Grain Processing*, 1: 57–59.

Szymczaka, G.Z., Wójciak-Kosiorb, M., Sowab, I., Zapałab, K., Strzemskib, M., Kocjanb, R. (2017). Evaluation of isoflavone content and antioxidant activity of selected soy taxa. *Journal of Food Composition and Analysis*, 57: 40–48.

Tao, X., Cai, Y., Liu, T., Long, Z., Huang, L., Deng, X., et al. (2019). Effects of pretreatments on the structure and functional properties of okara protein. *Food Hydrocolloids*, 90: 394–402.

Tsubaki, S., Nakauchi, M., Ozaki, Y., Azuma, J.-I. (2009). Microwave heating for solubilization of polysaccharide and polyphenol from soybean residue (Okara). *Food Science and Technology Research*, 15(3): 307–314.

Tuli, H.S., Tuorkey, M.J., Thakral, F., Sak, K., Kumar, M., Sharma, A.K., et al. (2019). Molecular mechanisms of action of genistein in cancer: Recent advances. *Frontiers in Pharmacology*, 10(1336): 1–16.

Turhan, S., Temiz, H., Sagir, I. (2007). Utilization of wet okara in low-fat beef patties. *Journal of Muscle Foods*, 18(2): 226–235.

Tyug, T.S., Prasad, N., Ismail, A. (2010). Antioxidant capacity, phenolics and isoflavones in soybean by-products. *Food Chemistry*, 123(3): 583–589. doi: 10.1016/j.foodchem.2010.04.074.

USDA (2008). *Database for the Isoflavone Content of Selected Foods, Release 2.0*. Nutrient Data Laboratory. Beltsville Human Nutrition Research Center: Beltsville, MD.

Vahvaselka, M., Laakso, S. (2010). Production of cis-9, trans-11-conjugated linoleic acid in Camelina meal and okara by an oat-assisted microbial process. *Journal of Agriculture and Food Chemistry*, 58(4): 2479–2482.

Vieira, A.D.S., Bedani, R., Albuquerque, M.A.C., Biscola, V., Saad, S.M.I. (2017). The impact of fruit and soybean by-products and amaranth on the growth of probiotic and starter microorganisms. *Food Research International*, 97: 356–363.

Villanueva, M.J., Yokoyama, W.H., Hong, Y.J., Barttley, G.E., Rupérez, P. (2011). Effect of high-fat diets supplemented with okara soybean by-product on lipid profiles of plasma, liver and faeces in Syrian hamsters. *Food Chemistry*, 124(1): 72–79.

Villanueva-Suárez, M.-J., Pérez-Cózar, M.-L., Mateos-Aparicio, I., Redondo-Cuenca, A. (2016). Potential fat-lowering and prebiotic effects of enzymatically treated okara in high-cholesterol-fed Wistar rats. *International Journal of Food Sciences and Nutrition*, 67(7): 828–833.

Villares, A., Rostagno, M.A., García-Lafuente, A., Guillamón, E., Martínez, J.A. (2011). Content and profile of isoflavones in soy-based foods as a function of the production process. *Food and Bioprocess Technology*, 4(1): 27–38.

Vishwanathan, K., Singh, V., Subramanian, R. (2011). Influence of particle size on protein extractability from soybean and okara. *Journal of Food Engineering*, 102(3): 240–246.

Vong, W.C., Liu, S.Q. (2016). Biovalorisation of okara (soybean residue) for food and nutrition. *Trends in Food Science and Technology*, 52: 139–147.

Vong, W.C., Liu, S.Q. (2017). Changes in volatile profile of soybean residue (okara) upon solid-state fermentation by yeasts. *Journal of the Science of Food and Agriculture*, 97(1): 135–143.

Voss, G.B., Osorio, H., Valente, L.M.P., Pintado, M.M. (2019). Impact of thermal treatment and hydrolysis by alcalase and Cynara cardunculus enzymes on the functional and nutritional value of Okara. *Process Biochemistry*, 83: 137–147.

Voss, G.B., Rodríguez-Alcalá, L.M., Valente, L.M.P., Pintado, M.M. (2018). Impact of different thermal treatments and storage conditions on the stability of soybean byproduct (okara). *Journal of Food Measurement and Characterization*, 12(3): 1981–1996.

Waliszewski, K.N., Pardio, V., Carreon, E. (2002). Physicochemical and sensory properties of corn tortillas made from nixtamalized corn flour fortified with spent soymilk residue (okara). *Journal of Food Science*, 67(8): 3194–3197.

Wang, C., Li, S. (1996). Influence of okara fiber on lipid metabolism and hemorheology of rats. *Acta Nutrimenta Sinica*, 18(2): 168–174.

Wang, H.J., Murphy, P.A. (1996). Mass balance study of isoflavones during soybean processing. *Journal of Agricultural and Food Chemistry*, 44(8): 2377–2383.

WHO (1990). Diet, nutrition, and the prevention of chronic diseases. Geneva: WHO http://apps.who.int/iris/bitstream/10665/42665/1/ WHO_TRS_916.pdf.

Wickramarathna, G.L., Arampath, P.C. (2003). Utilization of okara in bread making. *The Ceylon Journal of Science (Biological Science)*, 31: 29–33.

Wilson, L.A. (1995). Soy foods. In D.R. Erickson (Ed.), *Practical Handbook of Soybean Processing and Utilization*. AOCS Press and the United Soybean Board: Champaign, IL, 428–459.

Wongkhalaung, C., Leelawatcharamas, V., Japakaset, J. (2009). Utilisation of soybean residue to produce monacolin K-cholesterol lowering agent. *Songklanakarin Journal of Science and Technology*, 31: 35–39.

Xu, H., Wang, Y., Liu, H., Zheng, J., Xin, Y. (2000). Influence of soybean fibers on blood sugar and blood lipid metabolism and hepatic-nephritic histomorphology of mich with stz-induced diabetes. *Acta Nutrimenta Sinica*, 22(2): 171–174.

Xu, X., Liu, H., Zhou, Y. (2012). Study on the meitauza production from okara by Actinomucor elegans and Zymomonas mobilis. In: E. Zhu, S. Sambath (Eds.), *Information Technology and Agricultural Engineering*. Springer: Berlin Heidelberg.

Yang, L-C, Zu, T-J, Yang, F-C. (2020). Biovalorization of soybean residue (okara) via fermentation with Ganoderma lucidum and Lentinus edodes to attain products with high anti-osteoporotic effects. *Journal of Biosciences and Bioengieneering*, 129(4):514–518.

Yogo, T., Ohashi, Y., Terakado, K., Nezu, Y., Hara, Y., Tagawa, M., Fujisawa, T. (2011). Influence of dried okara-tempeh on the composition and metabolites of faecal microbiota in dogs. *International Journal of Applied Research in Veterinary Medicine*, 9: 181–188.

Yokomizo, A., Takenaka, Y., Takenaka, T. (2002) Antioxidative activity of peptides prepared from okara protein, Takenaka, Takenaka, T. *Food Science and Technology Research*, 8(4): 357–359.

Yuan, S., Chang, S.K. (2007). Selected odor compounds in soymilk as affected by chemical composition and lipoxygenases in five soybean materials. *Journal of Agricultural and Food Chemistry*, 55(2): 426–431.

Zaheer, K., Akhtar, M.H. (2017). An updated review of dietary isoflavones: Nutrition, processing, bioavailability and impacts on human health. *Critical Reviews in Food Science and Nutrition*, 57(6): 1280–1293.

Zhang, H., Gordon, R., Li, W., Yang, X., Pattanayak, A., Fowler, G., et al. (2019). Genistein treatment duration effects biomarkers of cell motility in human prostate. *PLoS One*, 14(3): e0214078. doi: 10.1371/journal.pone.0214078.

Zhang, Q., Tong, X., Qi, B., Wang, Z., Li, Y., Sui, X., et al. (2018). Changes in antioxidant activity of alcalase-hydrolyzed soybean hydrolysate under simulated gastrointestinal digestion and transepithelial transport. *Journal of Functional Foods*, 42: 298–305.

Zhao, T.T., Jin, F., Li, J.-F., Xu, Y.-Y., Dong, H.T., Liu, Q., et al. (2019). Dietary isoflavones or isoflavone-rich food intake and breast cancer risk: A meta-analysis of prospective cohort studies. *Clinical Nutrition*, 38(1): 136–145.

Zheng, X., Lee, S.K., Chun, O.K. (2016). Soy isoflavones and osteoporotic bone loss: A review with an emphasis on modulation of bone remodeling. *Journal of Medicine and Food*, 19(1): 1–14.

Zhu, D., Sun, H., Li, S., Hu, X., Yuan, X., Han, C., et al. (2015). Influence of drying methods on antioxidant activities and immunomodulatory of aqueous extract from soybean curd residue fermented by *Grifola frondosa*. *International Journal of Biology*, 7(1): 82.

Zhu, Y.P., Cheng, Y.Q., Wang, L.J., Fan, J.F., Li, L.T. (2008). Enhanced antioxidative activity of Chinese traditionally fermented okara (meitauza) prepared with various microorganism. *International Journal of Food Properties*, 11(3): 519–529.

Zhu, Y.P., Fan, J.F., Cheng, Y.Q., Li, L.T. (2008a). Improvement of the antioxidant activity of Chinese traditional fermented okara (Meitauza) using Bacillus subtilis B2. *Food Control*, 19(7): 654–661.

Zhu, Y.P., Yamaki, K., Yoshihashi, T., Ohnishi Kameyama, M., Li, X.T., Cheng, Y.Q., et al. (2010). Purification and identification of 1-deoxynojirimycin (DNJ) in okara fermented by Bacillus subtilis B2 from Chinese traditional food (meitaoza). *Journal of Agricultural and Food Chemistry*, 58(7): 4097–4103.

12 Polysaccharides from Soybean Hulls and Their Functional Activities

He Liu, Shengnan Wang, Lina Yang,
Hong Song, and Dayu Zhou

CONTENTS

12.1 SOYBEAN HULLS

12.1.1 SOYBEAN HULLS' POLYSACCHARIDES

12.1.1.1 Composition of Polysaccharides

Soy hulls (Figure 12.1) are the seed coat of soybeans which comprise approximately 8% of the soybean. They are one of the main by-products generated during the initial cracking process, which is the process commonly used in soybean oil and protein production. However, huge amounts of soy hulls are not well utilized and are discarded as waste now (Kim et al., 2015). Recently, soy hulls have been considered a valuable source of novel polysaccharides since they contain about 30% pectin, 20% cellulose, and 50% hemicellulose (Gnanasambandam & Proctor, 1999). Moreover, the potential health benefits of soy-based foods have created a demand for more studies on the soy hull polysaccharide as food a ingredient.

The soybean hull polysaccharides consisted of galactose, xylose, galacturonic acid, arabinose, glucose, and rhamnose with a molar ratio of 4.7:2.5:1.5:3.8:3.2:3.0 (Liu et al., 2013; Wang et al., 2019a). The molecular weight of soybean hull polysaccharides is around 500 kDa. However, their molecular weights fractioned by using ethanol at 20%, 40%, and 60% (v/v), are 381.83, 285.30, and 124.21 kDa, respectively. It has been reported that mannose, galacturonic acid, and galactose were the major monosaccharide components in soybean hull polysaccharides. Their fractions have different monosaccharide building units as shown in Table 12.1, and all of them are in a class of heteropolysaccharides. Protein and uronic acid were detected in all soy hull soluble polysaccharide (SHSP) fractions, which indicated that all soybean hull polysaccharide fractions belonged to the class of acid glycoprotein compounds (Wang et al., 2019a). Similar results were obtained in other soy polysaccharides (Furuta & Maeda, 1999) which contain about 5% protein.

DOI: 10.1201/9781003030294-12

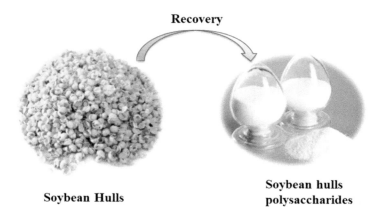

FIGURE 12.1 Pictures of soybean hulls and soybean hull polysaccharides.

TABLE 12.1
Monosaccharide Composition of Different SHSP Fractions

Monosaccharide Composition (mol %)	Molar Composition (mol %)		
	SHSP20	SHSP40	SHSP60
Rhamnose	3.69	2.28	12.79
Arabinose	6.47	1.38	6.55
Galactose	12.97	19.13	29.39
Glucose	0.86	0.33	nd
Xylose	0.95	nd	0.85
Mannose	27.38	61.36	38.43
Galacturonic acid	47.67	15.52	10.22
Glucuronic acid	nd	nd	1.77

nd = not determined.

Dietary fiber (DF) is a kind of polysaccharide and cannot be decomposed by human digestive tract enzymes (Mudgil & Barak, 2013). In 2009, the Codex Alimentarius Commission defined DF as composed of carbohydrate polymers with ten or more monomer units that cannot be hydrolyzed by endogenous enzymes in the human small intestine, including edible carbohydrate polymers that exist in food, and they are obtained or synthesized from food raw materials by physical, enzymatic, or chemical methods. Carbohydrate polymers are good for health (Mudgil & Barak, 2013). Soybean hull DF can be divided into insoluble DF (IDF) and soluble DF (SDF) (Chen et al., 2014). IDF is mainly composed of the cell wall, such as cellulose, lignin, and some hemicellulose, and could promote intestinal peristalsis and reduce the retention time of food in the intestine (Dai & Chau, 2017). SDF is composed of hemicellulose polysaccharides, such as non-digestible oligosaccharides, arabinoxylans, β-glucan, some hemicellulose, pectin, and inulin (Dai & Chau, 2017). SDF is generally hydrophilic amorphous fiber that relates to the metabolism of carbohydrates and lipids and is easy to be moistened by intestinal fluid. Therefore, it can be used by intestinal microbiota.

12.1.1.2 Chemical and Molecular Properties
The SHSP can be fractioned with 20% ethanol (SHSP 20), 40% ethanol (SHSP 40), and 60% ethanol (SHSP60), the results indicated that the apparent viscosity of all SHSP solutions was related to the concentration and shear rate, and it presented a positive change to concentration and a negative

effect on the shear rate. The shear-thinning was more pronounced in the lower shear rate (0.1 to 1 s-1) than higher (1 to 100 s-1). The SHSP40 has the highest viscosity (0.37 Pa·S), compared with the SHSP20 (0.32 Pa·S) and the SHSP60 (0.22 Pa·S) at 0.04 mg mL^{-1}. The significant properties of soy hull soluble polysaccharides are gel formation and binding with divalent metal ions (Liu et al., 2013). The gelation of polysaccharides is related to the molecular weight, monosaccharide composition, concentration, temperature, pH, and the presence of cross-linking agents. All of those factors affect the gelation process (Ni et al., 2018). Moreover, gelation is very sensitive to salt and can be controlled by the presence of specific cations such as Ca^{2+}, Mg^{2+}, K^+, Cs^+, Na^+, and Rb^+, since they increase the mediate helix-helix aggregation and double helix stability (Almeida et al., 2017;). The SHSP20 and SHSP40 were able to form a gel with Mg^{2+}, K^+, and Na^+ which can resist structural damage upon heating. The gelation ability of the SHSPs induced by Mg^{2+}, K^+, and Na^+ leads to a decrease in the order of SHSP20, SHSP40, and SHSP60.

The conformations of SHSP fractions showed significant changes in size and shape at the concentration of 5 µg/ mL on the mica surface (Figure 12.2). The SHSP20 is composed of localized networks, tenuous wormlike strands, and a small proportion of non-uniform spherical structures. The lengths (130–710 nm), widths (39–143 nm), and height (1.8–4.9 nm) of these fibers varied, suggesting that these fibers may involve some degree of side-by-side aggregation of helices. Meanwhile, the SHSP40 possessed fragmented structures and a small proportion of non-uniform spherical structures. The SHSP40 comprised mostly compact and spherical particles with heights of 0.4 nm–3.1 nm, widths of 19–121 nm, and lengths of 200 nm, 142 nm, 68 nm, etc. The SHSP60 molecules were aggregated into flakes with heights from 1.2–5.6 nm and lengths of 381 nm, 458 nm, and 791 nm (Wang et al., 2019a).

The water retention capacity (WRC) and water swelling capacity (WSC) of the soybean hull polysaccharide (such as DF) refer to its ability to retain water within its matrix. The WRC of insoluble dietary fibers (IDF) was 4.32 g/g, and for soluble dietary fibers (SDF) was 7.95 g/g. The IDF and SDF had WSC values of 4.39 mL/g and 9.94 mL/g, respectively. Recent studies have suggested that the higher the SDF contents, the higher the WSC value (Navarro-González et al., 2011). The oil adsorption capacity (OAC) assesses dietary fiber's (DF's) ability to absorb fat. The OAC of IDF and SDF were 1.48 g/g and 3.74 g/g, respectively. Previous studies have reported that the hydration ability of DF was also closely associated with the surface structure (Zhu et al., 2018; Chu et al., 2019). The SDF has a large surface area because of its porous structure as observed in the SEM micrograph. Its surface tension strength stabilized moisture in the capillary structures, thus potentially explaining its higher WRC, WSC, and OAC values. The glucose adsorption capacity (GAC) can be considered an *in vitro* index of the effect of DF on the delayed absorption of glucose in the gastrointestinal tract, which aids in the evaluation of postprandial blood glucose levels (Chu et al., 2019). The GAC of IDF was 28.20–168.38%, and for SDF it was 24.34–147.11%. The insoluble fiber particles pose physical obstacles to glucose molecules, thus resulting in the entrapment of glucose within the fiber-forming network (Chi-Fai et al., 2004). The bile acid retardation index (BRI) predicts the effects of DF in the gastrointestinal tract on delaying the absorption of bile acid (Maet al.,, 2015). Taurocholic acid was synthesized from taurine and bile acid through an esterification reaction. The bile salts in the terminal ileum or large intestine were metabolized by fermentation in the colon and/or excreted into the feces (Grundy et al., 2016). The process may be related to changes in the gut microbiota. The values of BRI increased with increasing dialysis time. The BRI of IDF was 3.29–17.24% and for SDF it was 3.86–5.39%.

12.1.1.3 Physiological Functions of Polysaccharides

12.1.1.3.1 Antidiabetes

Diabetes is a chronic systemic disease related to genetic factors and a variety of environmental factors. Diabetes currently affects over 350 million people worldwide and another 1 billion people in the world are pre-diabetic and may eventually end up with full-blown diabetes (Lotfy et al., 2016).

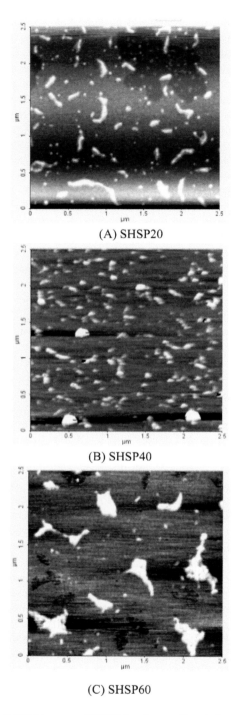

(A) SHSP20

(B) SHSP40

(C) SHSP60

FIGURE 12.2   Comparison of AFM images of SHSP fractions at 5 μg /mL.

About 10% of diabetic patients are type 1 diabetes (T1D) (Bakay & Hakonarson, 2012). The patients with T1D require insulin injections because of the absolute deficiency of insulin secretion. However, most patients have not lost their insulin secretion completely, and their insulin is relatively deficient. This type of diabetes is called type 2 diabetes (T2D), with an incidence rate of about 90% (Bakay

& Hakonarson, 2012). It costs around 1.2trillion USD to diagnose, treat, and care for both T1D and T2D patients globally. Soybean hull polysaccharides play a very significant role in regulating the balance of microbial nutrition in the gastrointestinal tract, and cholesterol-like metabolism, which can effectively improve T2D.

Recently extensive studies have confirmed the gut microbiota is related to the occurrence and development of diabetes and other diseases. The number of probiotics in stool decreased in diabetic patients, and probiotics introduced into the intestine could improve the symptoms of T2D (Larsen et al., 2010; Yadav et al., 2007). Soy hull polysaccharides promote the growth of *Lactobacillus* in the intestinal tract of diabetic rats, and the total cholesterol, low-density lipoprotein, and blood sugar in the plasma were all reduced (Figures 12.3, 12.4) (Yang et al., 2020, 2020c). Meanwhile, body weight and insulin resistance are also found to be improved. Soy hull polysaccharides promote the growth of beneficial intestinal bacteria, inhibit the growth of harmful bacteria, significantly increase the content of short-chain fatty acids, and reduce the intestinal pH value. Therefore, soy hull polysaccharides play a healthy role in the body (Lin et al., 2022; Yang et al., 2020, 2020b). Polysaccharides reduce blood sugar by changing the intestinal microbiota structure of insulin resistance, making the intestinal microbiota structure of T2D patients similar to that of normal people.

Because the mechanism of T2D is very complicated and is often induced by various environmental factors on the basis of complex genetic background. Among them, insulin signal transduction disorder is one of the common important links that many factors lead to the formation of diabetes (Quinn, 2002). T2D can occur in three levels: The pre-receptor level, receptor level, and post-receptor level. At present, there are many insulin-signaling pathways, including insulin receptor (InsR), insulin receptor substrate (IRS), and phosphatidylinositol-3-kinase (PI3K), as well as many downstream signal factor proteins. There are three main signaling pathways to reduce blood sugar, AMPK (AMP-activated protein kinase)/PI3K mediated regulation of glucose uptake, IKK mediated insulin resistance, and PPARγ/NF-κB to control the oxidative stress and inflammation (Nandipati et al., 2017). Therefore, taking the receptors and enzymes related to the pathogenesis of diabetes as the targets, the correlation between the anti-T2D activity of soybean hull polysaccharides and the signal pathway of insulin membrane receptors has broad prospects at the cellular and molecular levels.

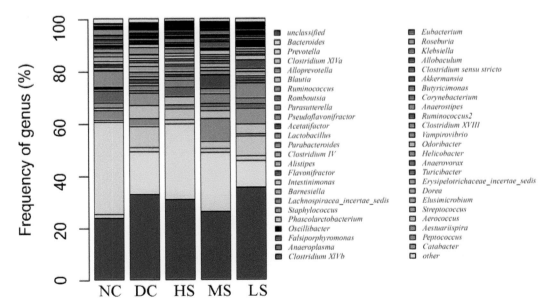

FIGURE 12.3 Response of the gut microbiota to soybean hull polysaccharide treatment at the genus level. Source: Yang et al. (2020).

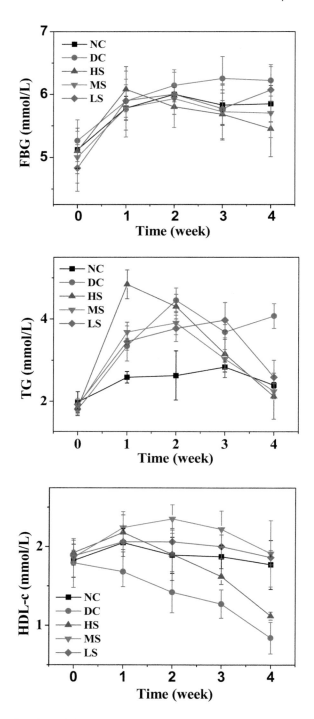

FIGURE 12.4 Effect of soybean hull polysaccharides on the serum FBG levels, serum TG levels, and the ratio of HDL-c in each group of rats fed a high-fat-high-sucrose diet. Source: Lin et al. (2022).

12.1.1.3.2   Lowering Blood Pressure and Cholesterol

Cardiovascular disease is the leading cause of death and disability worldwide. Hypertension is a major risk factor for cardiovascular morbidity and mortality. Endothelial dysfunction is one of the important initiators of vascular pathogenesis in hypertension. Impaired endothelial function can be initiated as a consequence of either reduced production/release of endothelial-dependent relaxation factors such as nitric oxide (NO) and prostacyclin (PGI2), and overproduction of endothelial-dependent contraction factors such as reactive oxygen species, or these two aspects combined (Wong et al., 2010). Wang et al. (2019) demonstrated that white mulberry fruit polysaccharides induced endothelium-dependent relaxation in rat mesenteric artery and stimulated NO production in endothelial cells. The results of the study showed that *in vivo* treatment with white mulberry fruit polysaccharides reduced mean arterial pressure significantly in both normotensive and spontaneously hypertensive rats. The polysaccharides obtained from *Cymodocea nodosa* have displayed an important antihypertensive activity (IC50 = 0.43 mg/ml) in a dose-dependent manner (Kolsi et al., 2016). These novel findings provided the foundational knowledge for further mechanistic studies examining the effects of polysaccharides from plants, particularly against hypertension.

Some randomized controlled trials have suggested that increasing fiber intake can reduce systolic and diastolic blood pressure; however, the reduction is small (Streppel et al., 2005). Threapleton and colleagues reported their meta-analysis of randomized clinical trials in which they tried to differentiate the effects of soluble and insoluble dietary fibers. The results confirmed that higher consumption of fiber, insoluble fiber, and cereal-vegetable fiber could reduce the risk of cardiovascular and coronary heart disease (Threapleton et al., 2013; Osborn & Sinn, 2007).

It is well known that increased serum total cholesterol (TC) and low-density lipoprotein (LDL-C) are the major risk factors for cardiovascular diseases. The high level of serum high-density lipoprotein (HDL-C) protects against cardiovascular diseases. Because of the side effects of synthetic drugs, there is an increasing interest in developing functional foods to treat hypercholesterolemia. Among these functional foods, many plant polysaccharides were proved to have remarkable hypocholesterolemic effects, such as polysaccharides from soybean, white kidney bean, red kidney bean, small black soybean, field bean, and lentil (Bai et al., 2020). Carboxymethylation of polysaccharides from *Morchella angusticeps* Peck has shown the ability to reduce cholesterol via up-regulation of hepatic protein expression of CYP7A1 and LDL-R and down-regulation of HMG-CoA in rats (Li et al., 2017). The study was conducted on mice fed a high-fat diet using polar extracts from *Cyclocarya paliurus* leaves; the results indicated that cholesterol level was lowered considerably. It was attributed to the cholesterol-lowering effect of ChE and conversion of cholesterol to bile acids by a molecular mechanism that involved an increase in the expression level of CYP7A1 and inhibition of HMG-CoA reductase, thus resulting in suppression of biosynthesis of cholesterol (Jiang et al., 2015).

An abundance of results suggests that the incorporation of dietary fiber residue in fat- and cholesterol-rich diets could eliminate the increases of total cholesterol, triglyceride, low-density lipoprotein cholesterol, and high-density lipoprotein cholesterol in the serum (Wu et al., 2020; Chen et al., 2014; O'Keefe, 2019). Those effects potentially arise because certain fibrous constituents essentially bind sterolic derivatives and cholesterol, thereby preventing their absorption in the body (Kreuzer et al., 2002). In contrast, high-density lipoprotein cholesterol (HDL-C) plays a fundamental role in hindering cholesterol transport, protecting against vascular disease, exerting anti-inflammatory effects, activating nitric oxide synthase, inciting prostacyclin release, and promoting endothelial repair (Esa et al., 2011; Tardif, 2010). Therefore, the presence of SDF can significantly promote the concentration of HDL-C while reducing the concentration of TC, TG, LDL-C, and AI in the human body. The effects can be attributed to the presence of carboxymethyl cellulose (CMC), a major component of natural fiber from numerous sources that can entrap cholesterol crystals (Chen et al., 2014).

12.1.1.3.3 Anti-Oxidation

Oxidation is a common reaction and an essential biological process to many organisms for the production of energy. However, abundant reactive oxygen species are uncontrolled producing some oxidative reactions *in vivo*. These oxygen species can react with macromolecules, leading to epidemic diseases such as cancer, rheumatoid arthritis, and atherosclerosis (Finkel & Holbrook, 2000; Moskovitz et al., 2002). Synthetic antioxidants such as butylated hydroxyanisole and butylated hydroxytoluene have been suspected to be responsible for liver damage and carcinogenesis (Qi et al., 2005). Therefore, it is necessary to develop and utilize some naturally derived antioxidants so that they can defend the human body against free radicals and retard the progress of many chronic diseases (Kinsella et al., 1993). Some potent natural polysaccharides with low cytotoxicity from plants are explored as novel potential antioxidants in applications of the food industry and pharmaceutics (Lai et al., 2010; Grace et al., 2016).

Soybean hull polysaccharides (SSCPs) have been implicated as bioactive antioxidants. Three water-soluble polysaccharides, named A-SSCP, S-SSCP, and H-SSCP were extracted from soybean hulls by microwave-assisted ammonium oxalate, and microwave-assisted sodium citrate and hot-water extraction, respectively. The effects of different extractions on the physicochemical characterization *in vitro* of antioxidant soybean hull polysaccharides were investigated. The results indicated that soybean hull polysaccharidess have the same monosaccharide compositions, but significant differences in their contents. However, they all had typical characteristics of polysaccharides. An ascending dose-dependent manner was found for the *in vitro* antioxidant of three polysaccharides (e.g. against DPPH, ABTs radicals, and reducing power). A-SSCP showed obvious better antioxidant activities than S-SSCP and H-SSCP (Table 12.2), which were attributed to the higher contents of arabinose, galacturonic acid, protein, sulfate radicals, and lower molecular weight (Mw) (Mirzadeh et al., 2020).

12.1.1.3.4 Anticancer

In recent years, polysaccharides have been proven to play an important role in different stages of cancer development, including gastrointestinal cancer (colon cancer, esophageal cancer, and pancreatic cancer), reproductive system cancer (breast cancer, cervical cancer, and prostate cancer), and

TABLE 12.2

SSCPs' Antioxidant Activities and Matrix for Correlation Analysis

	DPPH Radicals	ABTS Radicals	Total Reducing Ability
A-SSCP	76.9%	85.3%	1.03
S-SSCP	63.6%	68.7%	0.95
H-SSCP	63.1%	50.9%	0.35
Mannose	−0.8**	−0.9**	−0.8**
Galactose	−0.9**	−0.9**	−0.7*
Rhamnose	0.6	0.8**	0.9**
Arabinose	0.9**	0.9**	0.7*
Glucose	0.4	0.06	−0.2
Galacturonic acid	0.8**	0.9**	0.8**
Mw	−0.5	−0.8**	−0.8**
Protein	0.8**	0.9**	0.9**
Sulfate radicals	0.9**	0.8**	0.6

** Correlation is significant at $p < 0.01$ level.

* Correlation is significant at $p < 0.05$ level.

lung cancer (Yu et al., 2018). The anticancer mechanism of polysaccharides mainly includes three aspects:

1) Promoting the release of cholesterol and toxin, and regulating intestinal microbiota. Polysaccharides can increase the volume of intestinal contents and reduce the concentration of carcinogens (Jacobs, 1986). They can increase excretion by combining cholesterol and bile acid, and also regulate the structure of intestinal microbiota and increase the content of SCFA (Yang et al., 2020). Those aspects inhibit cancer.
2) Regulating cancer cell cycle and apoptosis. Polysaccharides can block the cancer cell cycle and activate the caspase apoptosis signal pathway. They also up-regulate Bax and DR5 expression and down-regulate Bcl-xL expression to promote cancer cell apoptosis (Vanessa et al., 2012).
3) Activating tumor suppressor pathways. Polysaccharides mainly inhibit MAPK, Wnt, and PKC signaling pathways to inhibit cancer.

12.1.1.3.5   Anti-Obesity

Obesity has become a serious health problem in industrialized countries. As one of the modern "civilized diseases", obesity has become a global common health issue. The levels of blood lipids, blood cholesterol, triglycerides, and low-density lipoprotein in obese patients are generally high, whereas the level of high-density lipoprotein is relatively low. Diet is the main factor affecting the composition and function of intestinal microbiota, and the role of intestinal microbiota in obesity seems to be important. Soy hull polysaccharides improve intestinal microbiota by supplying microbial substrates. Both *Lactobacillus* and *Bifidobacteria* can reduce blood cholesterol, enhance the early dissociation of bile, and remove extracellular cholesterol through absorption and precipitation. The soy hull polysaccharides are fermented to produce SCFA and promote local pH reduction (Yang et al., 2020c). The pH reduction may regulate the composition of intestinal microbiota and inhibit the growth of *Enterobacteriaceae* which may be one of the sources for LPS. For overweight individuals, soy hull polysaccharides intake can significantly prevent their weight gain (Yang et al., 2020b). Therefore, polysaccharides play an important role in maintaining intestinal function, maintaining the intestinal barrier, and alleviating inflammatory reactions in obese patients (Yang et al., 2020, 2020b, 2020c).

12.2   RECOVERY OF SOYBEAN POLYSACCHARIDES

The studies of the extraction methods and characteristics of the polysaccharide from soy hulls are relatively limited. In previous studies, soy hull polysaccharide has been commonly extracted by hot acid solution, precipitated, membrane filtered, and purified with alcohols to remove impurities (Kim et al., 2015). Acid and alkali hydrolysis have been applied to obtain the nanofiber from lignocellulosic wastes with an increase of insoluble fiber contents (Alemdar & Sain, 2008). The acidic pectin extraction and continuous alkali treatments for the residual insoluble fraction will result in effective recovery of both pectin and insoluble fiber from soy hulls and contribute to positive impacts on improving functional properties (Alemdar & Sain, 2008).

Liu et al. (2013) studied the extraction of soybean hull polysaccharides by hot-compressed water in a batch system. The results showed that an optimum temperature of 160° C and a short extraction time of 60 min were suitable for the preparation of soybean hull polysaccharides. In the sugar composition of the polysaccharide products, arabinose constituted 35.6–46.9%. Wang et al. (2019a, 2019b) reported that dry soybean hulls were ground into powder (to pass a 60-meshes sieve), and were extracted by ammonium oxalate assisted by the microwave radiation method. The fractionation procedure is shown in Figure 12.5. In detail, soybean hull powder was dispersed in water at 85° C containing 0.6 (%, wt) ammonium oxalate, then treated in a microwave oven at 450 W, maintained at 95° C for 10 min. The dispersions were passed through a 200-mesh filter cloth, and

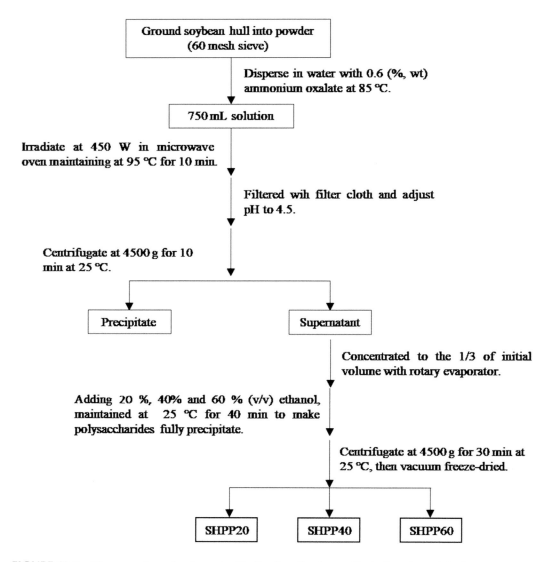

FIGURE 12.5 The procedure of polysaccharide fractionation by gradient ethanol precipitation.

the pH of filtrates reached 4.5, centrifuged at 4,500 g for 10 min. After that, the supernatant was concentrated by a vacuum rotary evaporator. Then ethanol was added (95%, v/v) to the concentrated solution, until final ethanol contents were 20%, 40%, and 60% (v/v) respectively. After 40 min, the precipitate was centrifuged at 4,500 g for 30 min at 25° C and dried in a vacuum freeze-drier, resulting in SHSP20, SHSP40, and SHSP60 fractions. In addition, the membrane separation method can also be used to separate polysaccharides from soybean hulls.

12.3   SOYBEAN-POLYSACCHARIDES-RICH FUNCTIONAL FOODS

Soy hull polysaccharides have emulsifying properties, thickening properties, and anti-aging properties. For example, instant rice (Figure 12.6A) with soybean polysaccharides has low GI properties, while resisting starch aging and increasing the palatability and shelf-life of the product. Adding soybean polysaccharides in multigrain beverages (Figure 12.6B) could reduce the aggregation,

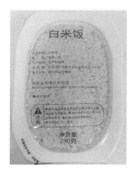

(A) Low glycemic index rice with soybean hull polysaccharides

(B) Soy pumpkin seed drink stabilized by soybean hull polysaccharides

(C) The emulsions stabilized by soybean hull polysaccharides at concentration of 0.50 wt%.

FIGURE 12.6 Food ingredient products of soybean hull polysaccharides.

coalescence, stratification, phase separation, and other phenomena of droplets in the beverage, consequently enhancing the palatability and stability of the beverage and greatly extending the shelf-life of the product.

Food oil-in-water (O/W) emulsion is a thermodynamically unstable multiphase system, which is prone to generate aggregation, flocculation, and other unstable phenomena. Therefore, stabilizing agents (such as amphiphilic polymers and solid particles) are generally added to avoid the phase separation between the oil and aqueous phases (Berton-Carabin et al., 2014; Charoen et al., 2012; Salvia-Trujillo et al., 2016). Soy protein isolate (SPI) can be used to stabilize O/W emulsions due to the high emulsifying activity and surface properties. However, the stabilizing behavior of SPI is easily affected by temperature, pH, ionic strength, and other factors, resulting in stability reduction of emulsions (Wang et al., 2020a). Tran et al. (2013) found that the addition of soy soluble polysaccharides above 0.25 wt% can improve the aggregation stability and inhibit phase separation of emulsion that is stabilized by using 0.50 wt% SPI under the isoelectric point of SPI due to the steric repulsion. The soybean hull polysaccharides (SHP) concentrations had a significant effect on the emulsifying stability of SPI/SHP. The addition of 0.05–0.15 wt% SHP delayed the movement of droplets because of the higher zeta-potential between droplets, decreased the particle size of droplets, and significantly improved the stability of emulsions (Figure 12.6C) (Wang et al., 2020a, 2020b).

In recent years, various methods have been proposed to fabricate bioactive delivery systems, e.g. emulsions, liposomes, polymersomes, micelles, hydrogels, and polymer conjugates (McClements, 2017). Meanwhile, microparticles with core-shell structures have been applied to bioactive delivery systems (Duan et al., 2016). The potential appeal of core-shell hydrogel beads was reflected in the following aspect as increasing bioactive efficiency and absorption, and maintaining bioactive required release time without reducing the frequency and quantity of dosing until they reach the desired site in the body after ingestion (Kumar et al., 2015). In addition, bioactivity can also be protected by gel beads against a harsh pH, gastrointestinal (GI) enzymes via the oral route, so that it could be absorbed and utilized by the human body to the utmost extent (Chai et al., 2018; Darwish et al., 2010). Recently, the gel bead carriers with core-shell structures have also been increasing in functionalization (Hu et al., 2019). The novel generation of a pH-responsive carrier can release the loaded bioactive by protonation (or deprotonation) of ionizable groups based on the difference in acid and alkali microenvironment during GI. Soybean hull polysaccharides as natural, biodegradable, biocompatible polymers have been used in food, cosmetics, and pharmaceutical industries as thickeners, stabilizers, and gelling agents (Bao et al., 2018; Leung et al., 2018; Wang et al., 2020c). They are successfully constructed a core-shell bioactive delivery system using natural food-grade materials based on soybean hull polysaccharides which is a degradable natural plant polysaccharide. Using the ionotropic gelation method, soybean hull polysaccharide beads with mesh-like structures were fabricated by mixing hydrophobic soy isoflavone with hydrophilic soybean hull polysaccharides. The bead exhibited an easily controlled *in vitro* release characteristic under pH regimes that corresponded to dissolution media. Furthermore, ion-SHP@ZIF-8 (Zeolitic imidazolate framework-8) microspheres were acquired through a gentle mixing method, and the shaping of ZIF-8 crystals was successfully achieved via mixing with metal cation-induced SHP gels and K^+, Na^+, Ca^{2+}, and Mg^{2+} could endow such hybrids with a three-dimensional structure without breaking the structure of the internal ZIF-8 crystals (Wang et al., 2020d).

12.4 SUMMARY

Soybean hull polysaccharides can be extracted by using a hot acid solution, precipitated, membrane filtered, and purified with alcohols to remove impurities. They consist of arabinose, galactose, xylose, glucose, and rhamnose, and the molecular weight is 500 kDa. Soybean hull polysaccharides are able to form gels with Mg^{2+}, K^+, and Na^+ which can resist structural damage upon heating. Soybean hull polysaccharides play a significant role in regulating the balance of microbial nutrition in the gastrointestinal tract, as cholesterol-like metabolism to effectively improve T2D, and as bioactive antioxidants. Soybean hull dietary fiber is mainly composed of the cell wall, such as cellulose, lignin, and some hemicellulose, which can promote intestinal peristalsis and reduce the retention time of food in the intestine. The physical and chemical properties of soybean hull dietary fiber indicated that they significantly influence the nutritive values and physiological functions of soybean foods. It can play an important role in maintaining intestinal function, maintaining the intestinal barrier, and alleviating inflammatory reactions in obese patients, and regulatinh the intestinal microbiota, increasing SCFA contents, and decreasing the blood glucose of T2D patients. Based on the active function of soybean polysaccharides, it can support functional food development, such as emulsions, micelles, and hydrogels to fabricate bioactive delivery systems. In conclusion, soybean hull polysaccharides, a new type of polysaccharide, have good physicochemical and functional properties and can be used as a food emulsifier, thickener, and nutritional supplement in food, with broad application prospects.

REFERENCES

Alemdar, A., Sain, M. (2008). Isolation and characterization of nanofibers from agricultural residues – Wheat straw and soy hulls. *Bioresource Technology*, 99(6): 1664–1671.

Almeida, N., Rakesh, L., Mueller, A., Hirschi, S., Zhang, Y. (2017). Viscoelastic properties of konjac glucomannan in the presence of salts. *Journal of Thermal Analysis and Calorimetry*, 131(3): 2547–2553.

Bai, Z., Meng, J., Huang, X., Wu, G., Zuo, S., Nie, S. (2020). Comparative study on antidiabetic function of six legume crude polysaccharides. *International Journal of Biological Macromolecules*, 154: 25–30.

Bakay, M., Hakonarson, H.J.D.M. (2012). What have genome-wide association studies contributed to the understanding of the pathogenesis and future management of type 1 diabetes? *Diabetes Management*, 2(2): 77–80.

Bao, H., Zhou, R., You, S., Wu, S., Wang, Q., Cui, S.W. (2018). Gelation mechanism of polysaccharides from Auricularia auricula–judae. *Food Hydrocolloids*, 76: 35–41.

Berton–Carabin, C.C., Ropers, M.H., Genot, C. (2014). Lipid oxidation in oil –in –water emulsions: Involvement of the interfacial layer. *Comprehensive Reviews in Food Science and Food Safety*, 13(5): 945–977.

Chai, J., Jiang, P., Wang, P., Jiang, Y., Li, D., Bao, W., Liu, B., Liu, B., Zhao, L., Norde, W., Yuan, Q., Ren, F., Li, Y. (2018). The intelligent delivery systems for bioactive compounds in foods: Physicochemical and physiological conditions, absorption mechanisms, obstacles and responsive strategies. *Trends in Food Science and Technology*, 78: 144–154.

Charoen, R., Jangchud, A., Jangchud, K., Harnsilawat, T., Decker, E.A., Mcclements, D.J. (2012). Influence of interfacial composition on oxidative stability of oil–in–water emulsions stabilized by biopolymer emulsifiers. *Food Chemistry*, 131(4): 1340–1346.

Chen, Y., Ye, R., Yin, L., Zhang, N. (2014). Novel blasting extrusion processing improved the physicochemical properties of soluble dietary fiber from soybean residue and in vivo evaluation. *Journal of Food Engineering*, 120: 1–8.

Chi-Fai, C., Chen, C.H., Lin, C.Y. (2004). Insoluble fiber-rich fractions derived from Averrhoa carambola: Hypoglycemic effects determined by in vitro methods. *LWT - Food Science and Technology*, 37(3): 331–335.

Chu, J., Zhao, Z., Lu, F., Bie, X., Zhang, C. (2019). Improved physicochemical and functional properties of dietary fiber from millet bran fermented by *Bacillus natto*. *Food Chemistry*, 294: 79–86.

Dai, F.J., Chau, C.F. (2017). Classification and regulatory perspectives of dietary fiber. *Journal of Food and Drug Analysis*, 25(1): 37–42.

Darwish, M.S.A., Peuker, U., Kunz, U., Turek, T. (2010). Bi-layered polymer–magnetite core/shell particles: Synthesis and characterization. *Journal of Materials Science*, 46(7): 2123–2134.

Duan, H., Lü, S., Gao, C., Bai, X., Qin, H., Wei, Y., Liu, M. (2016). Mucoadhesive microparticulates based on polysaccharide for target dual drug delivery of 5-aminosalicylic acid and curcumin to inflamed colon. *Colloids and Surfaces, Part B: Biointerfaces*, 145: 510–519.

Esa, N.M., Kadir, K.K.A., Amom, Z., Azlan, A. (2011). Improving the lipid profile in hypercholesterolemia-induced rabbit by supplementation of germinated brown rice. *Journal of Agricultural and Food Chemistry*, 59(14): 7985–7991.

Finkel, T., Holbrook, N.J. (2000). Oxidants, oxidative stress and the biology of ageing. *Nature*, 408(6809): 239–247.

Furuta, H., Maeda, H. (1999). Rheological properties of water-soluble soybean polysaccharides extracted under weak acidic condition. *Food Hydrocolloids*, 13(3): 267–274.

Gnanasambandam, R., Proctor, A. (1999). Preparation of soy hull pectin. *Food Chemistry*, 65(4): 461–467.

Grace, M.H., Esposito, D., Timmers, M.A., Xiong, J., Yousef, G., Komarnytsky, S., Lila, M.A. (2016). Chemical composition, antioxidant and anti-inflammatory properties of pistachio hull extracts. *Food Chemistry*, 210: 85–95.

Grundy, M.M.L., Edwards, C.H., Mackie, A.R., Gidley, M.J., Butterworth, P.J., Ellis, P.R. (2016). Reevaluation of the mechanisms of dietary fibre and implications for macronutrient bioaccessibility, digestion and postprandial metabolism. *The British Journal of Nutrition*, 116(5): 816–833.

Hu, B., Han, L., Ma, R., Phillips, G.O., Nishinari, K., Fang, Y. (2019). All–natural food–grade hydrophilic–hydrophobic core–shell microparticles: Facile fabrication based on gel–network–restricted antisolvent method. *ACS Applied Materials and Interfaces*, 11(12): 11936–11946.

Jacobs, L.R. (1986). Modification of experimental colon carcinogenesis by dietary fibers. *Advances in Experimental Medicine and Biology*, 206: 105–118.

Jiang, C., Wang, Q., Wei, Y., Yao, N., Wu, Z., Ma, Y., Lin, Z., Zhao, M., Che, C., Yao, X., Zhang, J., Yin, Z. (2015). Cholesterol–lowering effects and potential mechanisms of different polar extracts from Cyclocarya Paliurus leave in hyperlipidemic mice. *Journal of Ethnopharmacology*, 176: 17–26.

Kim, H.W., Lee, Y.J., Kim, Y.H.B. (2015). Efficacy of pectin and insoluble fiber extracted from soy hulls as a functional non–meat ingredient. *LWT – Food Science and Technology*, 64(2): 1071–1077.

Kinsella, J., Frankel, E., German, J., Kanner, J. (1993). Possible mechanisms for the protective role of antioxidants in wine and plant foods. *Food Technology*, 47: 85–89.

Kolsi, R.B.A., Fakhfakh, J., Krichen, F., Jribi, I., Chiarore, A., Patti, F.P., Blecker, C., Allouche, N., Belghith, H., Belghith, K. (2016). Structural characterization and functional properties of antihypertensive *Cymodocea nodosa* sulfated polysaccharide. *Carbohydrate Polymers*, 151: 511–522.

Kreuzer, M., Hanneken, H., Wittmann, M., Gerdemann, M.M., Machmuller, A. (2002). Effects of different fibre sources and fat addition on cholesterol and cholesterolrelated lipids in blood serum, bile and body tissues of growing pigs. *Journal of Animal Physiology and Animal Nutrition*, 86(3–4): 57–73.

Kumar, V.A., Shi, S., Wang, B.K., Li, I.C., Jalan, A.A., Sarkar, B., Wickremasinghe, N.C., Hartgerink, J.D. (2015). Drug–triggered and cross–linked self–assembling nanofibrous hydrogels. *Journal of the American Chemical Society*, 137(14): 4823–4830.

Lai, F., Wen, Q., Li, L., Wu, H., Li, X. (2010). Antioxidant activities of water–soluble polysaccharide extracted from mung bean (*Vigna radiata* L.) hull with ultrasonic assisted treatment. *Carbohydrate Polymers*, 81(2): 323–329.

Larsen, N., Vogensen, F.K., Berg, F., Nielsen, D.S., Andreasen, A.S., Pedersen, B.K., Al–Soud, W.A., Sørensen, S.J., Hansen, L.H., Jakobsen, M. (2010). Gut microbiota in human adults with type 2 diabetes mellitus differs from non-diabetic adults. *PLOS ONE*, 5: 9085.

Leung, H., Arrazola, A., Torrey, S., Kiarie, E. (2018). Utilization of soy hulls, oat hulls, and flax meal fiber in adult broiler breeder hens. *Poultry Science*, 97(4): 1368–1372.

Li, Y., Yuan, Y., Lei, L., Li, F., Zhang, Y., Chen, J., Zhao, G., Wu, S., Yin, R., Ming, J. (2017). Carboxymethylation of polysaccharide from Morchella angusticepes peck enhances its cholesterol–lowering activity in rats. *Carbohydrate Polymers*, 172: 85–92.

Lin, Q., Yang, L.N., Han, L., Wang, Z.Y., Luo, M.S., Zhu, D.S., Liu, H., Li, X., Feng, Y. (2022). Effects of soy hull polysaccharide on the gut microbiota in rats fed a high–fat–high–sucrose diet. *Food Science and Human Wellness*, 11(1): 49–57.

Liu, H., Guo, X., Li, J., Zhu, D., Li, J. (2013). The effects of MgSO4, d–glucono –δ–lactone (GDL), sucrose, and urea on gelation properties of pectic polysaccharide from soy hull. *Food Hydrocolloids*, 31(2): 137–145.

Lotfy, M., Adeghate, J., Kalasz, H., Singh, J., Adeghate, E.J.C.D.R. (2016). Chronic complications of diabetes mellitus: A mini review. *Current Diabetes Reviews*, 13(1): 3–10.

Ma, M.M., Mu, T.H., Sun, H., Zhang, M., Chen, J., Yan, Z. (2015). Optimization of extraction efficiency by shear emulsifying assisted enzymatic hydrolysis and functional properties of dietary fiber from deoiled cumin (*Cuminum cyminum* L.). *Food Chemistry*, 179: 270–227.

McClements, D.J. (2017). Recent progress in hydrogel delivery systems for improving nutraceutical bioavailability. *Food Hydrocolloids*, 68: 238–245.

Mirzadeh, M., Arianejad, M.R., Khedmat, L. (2020). Antioxidant, antiradical, and antimicrobial activities of polysaccharides obtained by microwave-assisted extraction method: A review. *Carbohydrate Polymers*, 229: 115421.

Moskovitz, J., Yim, M.B., Chock, P.B. (2002). Free radicals and disease. *Archives of Biochemistry and Biophysics*, 397(2): 354–359.

Mudgil, D., Barak, S. (2013). Composition, properties and health benefits of indigestible carbohydrate polymers as dietary fiber: A review. *International Journal of Biological Macromolecules*, 61: 1–6.

Nandipati, K.C., Subramanian, S., Agrawal, D.K. (2017). Protein kinases: Mechanisms and downstream targets in inflammation mediated obesity and insulin resistance. *Molecular and Cellular Biochemistry*, 426(1–2): 27–45.

Navarro-González, I., García-Valverde, V., García-Alonso, J., Periago, J. (2011). Chemical profile, functional and antioxidant properties of tomato peel fiber. *Food Research International*, 44(5): 1528–1535.

Ni, X., Wang, K., Wu, K., Corke, H., Nishinari, K., Jiang, F. (2018). Stability, microstructure and rheological behavior of konjac glucomannan–zein mixed systems. *Carbohydrate Polymers*, 188: 260–267.

O'Keefe, S.J. (2019). The association between dietary fibre deficiency and high-income lifestyle-associated diseases: Burkitt's hypothesis revisited. *The Lancet Gastroenterology and Hepatology*, 4(12): 984–996.

Osborn, D., Sinn, J. (2007). Probiotics in infants for prevention of allergic disease and food hypersensitivity. *Cochrane Database of Systematic Reviews*, 4(4): 6475.

Qi, H., Zhang, Q., Zhao, T., Chen, R., Zhang, H., Niu, X., Li, Z. (2005). Antioxidant activity of different sulfate content derivatives of polysaccharide extracted from *Ulva pertusa* (Chlorophyta) in vitro. *International Journal of Biological Macromolecules*, 37(4): 195–199.

Quinn, L. (2002). Mechanisms in the development of type 2 diabetes mellitus mellitus. *The Journal of Cardiovascular Nursing*, 16(2): 1–16.

Salvia–Trujillo, L., Decker, E.A., Mcclements, D.J. (2016). Influence of an anionic polysaccharide on the physical and oxidative stability of omega-3 nanoemulsions: Antioxidant effects of alginate. *Food Hydrocolloids*, 52: 690–698.

Streppel, M., Arends, L., Veer, P., Grobbee, D., Geleijnse, J. (2005). Dietary fber and blood pressure: A meta-analysis of randomized placebo-controlled trials. *Archives of Internal Medicine*, 165(2): 150–156.

Tardif, J.C. (2010). Emerging high-density lipoprotein infusion therapies: Fulfilling the promise of epidemiology? *Journal of Clinical Lipidology*, 4(5): 399–404.

Threapleton, D., Greenwood, D., Evans, C. (2013). Dietary fibre intake and risk of cardiovascular disease: Systematic review and meta-analysis. *British Medical Journal*, 347: 6879.

Tran, T., Rousseau, D. (2013). Stabilization of acidic soy protein –based dispersions and emulsions by soy soluble polysaccharides. *Food Hydrocolloids*, 30(1): 382–392.

Vanessa, R.S., Anna, Giros, Rosa, M.X., Lourdes, F., Mike, G., Anna, A., Xavier, L. (2012). Stool-fermented Plantago ovata husk induces apoptosis in colorectal cancer cells independently of molecular phenotype. *British Journal of Nutrition*, 107(11): 1591–1602.

Wang, C., Cheng, W., Bai, S., Ye, L., Du, J., Zhong, M., Liu, J., Zhao, R., Shen, B. (2019). White mulberry fruit polysaccharides enhance endothelial nitric oxide production to relax arteries in vitro and reduce blood pressure in vivo. *Biomedicine and Pharmacotherapy*, 116: 109022.

Wang, S., Shao, G., Yang, J., Liu, J., Wang, J., Zhao, H., Yang, L., Liu, H., Zhu, D., Li, Y., Jiang, L. (2020c). The production of gel beads of soybean hull polysaccharides loaded with soy isoflavone and their pH-dependent release. *Food Chemistry*, 313: 126095.

Wang, S., Yang, J., Shao, G., Qu, D., Zhao, H., Yang, L., Zhu, L., He, Y., Liu, H., Zhu, D. (2020a). Soy protein isolated–soy hull polysaccharides stabilized O/W emulsion: Effect of polysaccharides concentration on the storage stability and interfacial rheological properties. *Food Hydrocolloids*, 101: 105490.

Wang, S., Yang, J., Shao, G., Qu, D., Zhao, H., Zhu, L., Yang, L., Li, R., Li, J., Liu, H., Zhu, D. (2020b). Dilatational rheological and nuclear magnetic resonance characterization of oil–water interface: Impact of pH on interaction of soy protein isolated and soy hull polysaccharides. *Food Hydrocolloids*, 99: 105366.

Wang, S., Zhao, H., Shao, G., Yang, L., Zhu, L., Li, J., Zhou, D., Song, H., Liu, H., Zhu, D., He, Y. (2020d). Microstructural analysis of ZIF–8 particles using soy hull polysaccharide gel as a coating induced by different metal cations. *Microporous and Mesoporous Materials*, 306: 110408.

Wang, S., Zhao, L., Li, Q., Liu, C., Han, J., Zhu, L., Zhu, D., He, Y., Liu, H. (2019a). Rheological properties and chain conformation of soy hull water–soluble polysaccharide fractions obtained by gradient alcohol precipitation. *Food Hydrocolloids*, 91: 34–39.

Wang, S., Zhao, L., Li, Q., Liu, C., Han, J., Zhu, L., Zhu, D., He, Y., Liu, H. (2019b). Impact of Mg^{2+}, K^+, and Na^+ on rheological properties and chain conformation of soy hull soluble polysaccharide. *Food Hydrocolloids*, 92: 218–227.

Wong, W.T., Wong, S.L., Tian, X.Y., Huang, Y. (2010). Endothelial dysfunction: The common consequence in diabetes and hypertension. *Journal of Cardiovascular Pharmacology and Therapeutics*, 55(4): 300–307.

Wu, W., Hu, J., Gao, H., Chen, H., Fang, X., Mu, H., Han, Y., Liu, R. (2020). The potential cholesterol-lowering and prebiotic effects of bamboo shoot dietary fibers and their structural characteristics. *Food Chemistry*, 332: 127372.

Yadav, H., Jain, S., Sinha, P.R. (2007). Antidiabetic effect of probiotic dahi containing *Lactobacillus acidophilus* and *Lactobacillus casei* in high fructose fed rats. *Nutrition*, 23(1): 62–68.

Yang, L.N., Huang, J.H., Luo, M.S., Wang, Z.Y., Zhu, L.J., Wang, S.N., Zhu, D.S., Liu, H. (2020a). The influence of gut microbiota on the rheological characterization of soy hull polysaccharide and mucin interactions. *RSC Advances*, 10(5): 2830–2840.

Yang, L.N., Zhao, Y.F., Huang, J.H., Zhang, H.Y., Lin, Q., Han, L., Liu, J., Wang, J., Liu, H. (2020b). Insoluble dietary fiber from soy hulls regulates the gut microbiota in vitro and increases the abundance of Bifidobacteriales and Lactobacillales. *Journal of Food Science and Technology*, 57(1): 152–162.

Yu, Y., Shen, M., Song, Q., Xie, J. (2018). Biological activities and pharmaceutical applications of polysaccharide from natural resources: A review. *Carbohydrate Polymers*, 183: 91–101.

Zhu, Y., Chu, J., Lu, Z., Lv, F., Bie, X., Zhang, C., Zhao, H. (2018). Physicochemical and functional properties of dietary fiber from foxtail millet (*Setaria italic*) bran. *Journal of Cereal Science*, 79: 456–461.

13 Effects of Novel Processing Methods on Structure, Functional Properties, and Health Benefits of Soy Protein

Hao Hu, Ahmed Taha, and Ibrahim Khalifa

CONTENTS

DOI: 10.1201/9781003030294-13

13.1   INTRODUCTION

Soybean products play a significant nutritional role as one of the primary protein sources for millions of people worldwide. Mainly soybean traditional foods such as tofu, miso, and tempeh have been consumed by Asian people for more than 2,000 years (Fukushima, 1991). Recently, soybean proteins as a by-product of soy oil production have been commonly used in many food products due to their high nutritional quantity and reasonable price (Endres et al., 2001). Moreover, studies showed that consuming soy proteins could reduce blood cholesterol and cardiovascular diseases risk (Ramdath et al., 2017). Furthermore, soy proteins' functional properties, such as foaming, gelling, and emulsifying properties, increase their potential to be introduced to the food industries (Nishinari et al., 2014).

As a mixture of several subunits, soy protein was classified into 2S, 7S, 11S, and 15S based on their sedimentation coefficients. The 7S refers to β-conglycinin, while the 11S refers to glycinin, and they represent 80% of the total protein content. The 7S is a glycoprotein with a molecular mass of 150–200 kDa and comprises three subunits of α, ά, and β with molecular weights 68, 72, and 52 kDa, respectively (Nishinari et al., 2014). These three subunits have similar amino acid sequences (Fukushima, 1991). The 11S is a hexamer with a molecular weight of around 300–380 kDa and each subunit consists of basic (~20 kDa) and acidic (~35 kDa) polypeptides. The acidic and basic polypeptides are linked by a disulfide bond (Peng & Nielsen, 1986). Among all amino acids, glutamate and aspartate contents are considerably higher in both globulins. The amount of sulfur-containing amino acids and tryptophan in 11S is remarkably higher than that of 7S (Young, 1991). Each 11S molecule contains two sulfhydryl groups (-SH) and 20 S-S, while 7S has no -SH radicals and two S-S bonds per molecule (Fukushima, 1991). The contents of SH groups and S-S bonds could influence the functional properties of soy proteins.

Soy proteins have been successfully applied in many food products due to their functional properties such as solubility, texture, mouthfeel, gelling, foaming, and emulsifying properties (Nishinari et al., 2014). The functional performance of proteins can be controlled by their molecular weight, charge distribution, primary structure, quaternary structure, conformation state, and the environment (Kinsella, 1979b). Soy protein has unique characteristics such as weak electrostatic repulsions, large molecular weight, ease of aggregate, and low solubility that could negatively influence their functional properties (Taha et al., 2018). Therefore, several processing techniques such as high pressure, ultrasound, microwave, irradiation, ultraviolet, and pulsed electric field have been used to improve the techno-functional properties of soy proteins. In this chapter, the effects of novel processing technologies on the structural and functional properties as well as the bioactivity and health benefits of soy proteins will be discussed.

13.2   STRUCTURE, FUNCTIONAL PROPERTIES, AND HEALTH BENEFITS OF SOY PROTEIN

On the structure and physicochemical properties of proteins, the techno-functional properties of proteins can be categorized into surface properties, protein–protein interactions, and hydration. The solubility, viscosity, dispersibility, adhesion, swelling, and wettability are related to the hydration properties. The gelation and precipitation are linked to the protein–protein interaction, while the surface tension, foaming, and emulsifying properties are related to the surface properties of proteins (Yang & Powers, 2016). The functional properties of soy proteins are essential in controlling the quality of the final soy-based products. Therefore, understanding the relations between the three-dimensional structures and techno-functional properties of soy protein molecules could enhance the functional properties of these molecules. During the heat treatment of soy proteins, the polypeptide chains are unfolded, and the buried amino acid side residues are exposed to the surface. The protein molecules are combined through the exposed groups such as -SH and S-S, or hydrophobic bonding (Fukushima, 2011; Kinsella, 1979). The active groups (-SH and hydrophobic groups) in this case must be located at an accessible location at protein molecules' surfaces. The glycinin gel is more turbid and more rigid than that of β-conglycinin due to the larger numbers of SH groups in glycinin. However, β-conglycinin has much stronger emulsifying properties than glycinin because of the easily unfolded structure and higher hydrophobicity of β-conglycinin compared to glycinin (Tang, 2017). The protein denaturation and aggregation are practically measured by determining the solubility. Hence, the solubility of a protein can be considered a good indicator of its functionality. Due to their large molecular weight, soy proteins exhibited lower solubility, which limited their applications (Kinsella, 1979; Nishinari et al., 2014). Therefore, some processing technologies such as ultrasound and high pressure have been used to improve the solubility of soy proteins (Hu et al., 2013c). Recently, soy proteins have become popular in Western countries due to their health benefits, such as reducing the risk of cardiovascular diseases (Fukushima, 2011). Therefore, this chapter will also discuss the effects of processing technologies on the bioactivity and health properties of soy proteins.

13.3 EFFECTS OF ULTRASOUND PROCESSING ON THE STRUCTURE, FUNCTIONAL PROPERTIES, AND HEALTH BENEFITS OF SOY PROTEIN

Ultrasound is acoustic waves with frequencies of more than 20 kHz (Awad et al., 2012). Ultrasound waves are classified into high-intensity low-frequency (HIU, 10–1000 W cm^{-2}), and low-intensity high-frequency (< 1 W cm^{-2}) ultrasound (Ashokkumar, 2015). HIU has attracted much attention in many food processing applications, particularly for improving the functional properties of food proteins (Hu et al., 2013c, 2015a, 2015b; Jambrak et al., 2008, 2009, 2014; O'Sullivan et al., 2016a, 2017; Zheng et al., 2019b; Zhou et al., 2016a). Ultrasound devices generate acoustic cavitation resulting in the generation and the collapse of air bubbles in the system. The collapse of gas bubbles enables the release of reactive radicals to promote chemical reactions (Leong et al., 2017). Moreover, several physical effects such as high temperature, shock waves, turbulence, high-speed liquid jet, pressures, and extreme physical shearing could be generated as a result of the oscillation and collapse of acoustic cavitation bubbles (Ashokkumar, 2011). These chemical and physical effects can influence the structure and functional properties of soy proteins.

13.3.1 EFFECTS OF HIU ON THE STRUCTURE OF SOY PROTEINS

Several studies have been conducted to investigate the effects of HIU on the structural changes of soy proteins. For instance, O'Sullivan et al. (2016) found that ultrasound treatment (~34 W cm^{-2} for 2 min) did not significantly reduce the molecular weight profile of soy protein isolate (SPI) solutions (1 wt %) (O'Sullivan et al., 2016b). Similarly, Hu et al. (2013c) noticed that HIU did not alter the protein profile of SPI. However, in the same study, Hu et al. concluded that HIU could alter the secondary structure and increase the free sulfhydryl content of SPI (Hu et al., 2013c). But in another study on the effects of HIU on the structural properties of soy protein fractions (7S and 11S), the authors revealed that HIU could change the tertiary and quaternary structures of 7S and 11S (Hu et al., 2015b). These structural changes could be attributed to the fact that HIU could break the intermolecular disulfide bonds. Therefore, the HIU effects on the structure of soy protein depend on the HIU and solution conditions.

13.3.2 EFFECTS OF HIU ON THE FUNCTIONAL PROPERTIES OF SOY PROTEINS

Ultrasound treatment showed promising improvements in the techno-functional properties of soy proteins. It was concluded that HIU significantly improved the solubility of soy proteins. The solubility of commercial SPI in distilled water increased considerably with increasing ultrasonic intensity and time (Hu et al., 2013c). The ultrasound treatment changed the conformational changes so that more hydrophilic amino acid residues were exposed to water (Jambrak et al., 2009). Moreover, HIU can decrease the particle size of protein molecules and increase protein–water interactions, consequently increasing the protein solubility (Arzeni et al., 2012) (Figure 13.1A). The effects of ultrasound on soy glycinin (11S) at different ionic strengths were studied (Figure 13.1B). 11S had a higher solubility at an ionic strength of 0.6 compared to those at 0.06 and 0.2 after 5 min of sonication. It was also found, at an ionic strength of 0.6, that the solubility of 11S significantly increased after 5 min of sonication (at 20 kHz, 80 W cm^{-2}). However, after sonication for more than 5 min, the solubility gradually decreased with increasing the sonication time (Zhou et al., 2016a), while the solubility of 11S at ionic strengths of 0.06 and 0.2 gradually increased with increasing the sonication time. These results were consistent with particle size findings, where the solubility increased when the particle size decreased (Zhou et al., 2016a).

The foaming properties of proteins are crucial for many food applications. Ultrasound treatment significantly enhanced the foaming capacity of soy proteins (Jambrak et al., 2009; Morales et al., 2015). The enhanced solubility and surface hydrophobicity of soy proteins after ultrasound treatment could be the reason for the foaming capacity improvement. The protein unfolding and exposing the hydrophobic groups to the surface of protein molecules, induced by ultrasound treatment, could facilitate the adsorption of protein molecules at the air-water interface to encapsulate air bubbles and thereby enhance the foaming capacity (Ren et al., 2020). Furthermore, the foaming capacity of soy proteins could be improved by reduction in particle size. The smaller the particle size, the faster protein molecules could be adsorbed to the air-water interface. It was also concluded that there were synergistic effects between ultrasound treatment and temperature on the foaming capacity of soy proteins. In other words, the foaming capacity significantly improved when the ultrasound treatment was applied together with heat treatment (Morales et al., 2015). Nevertheless, Ren et al. (2020) found that the foaming stability of SPI decreased after ultrasound treatment. This could be due to the fact that HIU reduced the viscosity and destroyed the protein–protein interaction. These changes could result in a weak protein network and less stable protein film at the interface and therefore reduce the foaming stability of soy proteins (Morales et al., 2015; Ren et al., 2020).

Soy proteins have been used as natural emulsifiers due to their amphiphilic structure and biodegradability, as well as high nutritional values. However, the poor solubility of native soy proteins reduces the stability and shelf-life of soy protein emulsions. Therefore, several researchers hypothesized that ultrasound treatment could improve the emulsifying properties of soy proteins. It was proved that HIU can enhance the emulsifying activity index (EAI) and emulsifying stability index

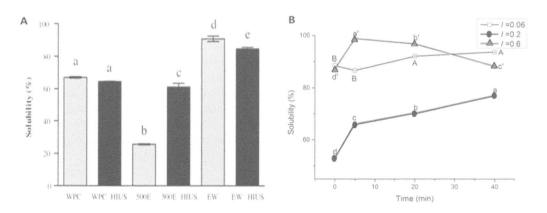

FIGURE 13.1 (A) Water solubility (%) of control and ultrasound (HIUS)-treated proteins, including whey protein concentrate (WPC), soy protein isolate (500E), and egg white protein (EW) solubility. (B) Effect of HIU (20 kHz at 80 W cm^{-2} for 5, 20, or 40 min) on protein solubility of glycinin at different ionic strengths. Sources: (A) Arzeni et al. (2012); (B) Zhou et al. (2016a).

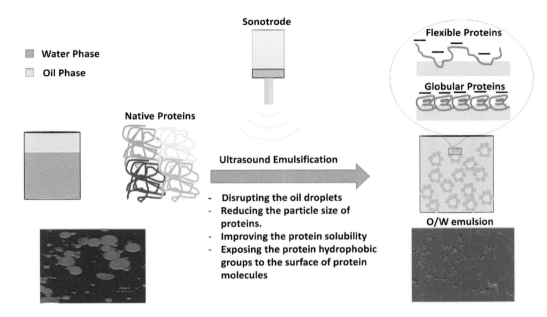

FIGURE 13.2 A schematic diagram of the possible mechanism of ultrasound emulsification of protein-stabilized emulsions. Source: Taha et al. (2020).

(ESI) of SPI (Jambrak et al., 2009; Ren et al., 2020; Yang et al., 2018), soy glycinin (11S, at ionic strengths of 0.2 or 0.6) (Zhou et al., 2016b), and soy β-conglycinin (7S) (Hu et al., 2015a). As shown in Figure 13.2 (Taha et al., 2020), during the ultrasound emulsification process, the acoustic cavitation effects not only disrupt the oil droplets but also reduce the particle size of protein molecules, improve the protein solubility, and expose the hydrophobic groups to the interface. Those changes could facilitate the formation of strong protein film at the water/oil interface, and reduce the interfacial tension, thus improving the emulsifying properties of soy proteins (Taha et al., 2020).

Gelling property of soy proteins is vital in many foodstuff applications such as tofu and tempeh in Asian countries. Several coagulants such as glucono-δ-lactone (GDL), $CaSO_4$, and $CaCl_2$ were used to prepare tofu by curdling soy milk with these coagulants (Hu et al., 2013b). The gelation properties are related to water holding capacity (WHC), swelling ability, and viscosity of rehydrated soy proteins (Hua et al., 2005). HIU showed positive effects on the gelling properties of soy proteins. For instance, ultrasound pretreatment improved the WHC and gel strength of $CaSO_4$-induced SPI gels (Hu et al., 2013b) (Figure 13.3A) as well as acid (GDL)-induced SPI gels (Hu et al., 2013a). Moreover, the gel strength of soy proteins produced by different denaturation methods was improved significantly after sonication for 25 min (Figure 13.3B) (Zheng et al., 2019a). As a result of HIU, the particle size of soy protein molecules decreased while the free SH-groups and surface hydrophobicity increased. These changes could facilitate intermolecular and SS bonds formation, resulting in a uniform and dense microstructure of $CaSO_4$-induced SPI gels. The changes in the network and SS bonds could reduce the solubility of proteins and improve the WHC and gel strengths of SPI gels (Hu et al., 2013b).

13.3.3 Effects of HIU on the Health Benefits of Soy Proteins

The physical effects of ultrasound could degrade the protein structure and open up the hydrophilic groups. The opening up of structure can improve the protein solubility and allows the protease to bind more easily with the protein substrate. Chen et al. (2011) found that the ultrasound pretreatment significantly improved the accessibility of 7S and 11S subunits to papain hydrolysis. Xu et al. (2015)

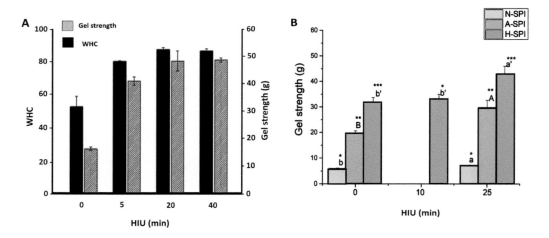

FIGURE 13.3 (A) Effect of HUS pretreatment time (0, 5, 20, and 40 min) on the WHC and gel strength of $CaSO_4$-induced SPI gels (Hu et al., 2013b). (B) Effect of HIU on cold gel strength of native SPI (N-SPI), alcohol denatured SPI (A-SPI), and heat moisture denatured SPI (H-SPI) (Zheng et al., 2019a).

concluded that HIU pretreatment enhanced the antioxidant activity of highly denatured soy protein meal hydrolysate prepared using neutrase. Cui et al. (2019) reported that the maximum emulsification and emulsifying activities of SPI-WPI mixtures of 76.46 m^2/g and 22.83 min, respectively, were obtained with 300 W ultrasonic treatment. A slight redshift in the UV-Vis absorption spectrum was observed after this treatment, indicating exposure of internal groups outside of protein molecules and protein structure changes. This study demonstrated that ultrasonic treatment was effective in improving the functional properties of the SPI-WPI mixture.

13.4 EFFECTS OF HIGH-PRESSURE PROCESSING ON THE STRUCTURE, FUNCTIONAL PROPERTIES, AND HEALTH BENEFITS OF SOY PROTEIN

The commercial benefits of high pressure became available to the food industries in the late 1980s. High-pressure processing (HPP), particularly at pressures higher than 100 MPa, has met consumers' rising demand, introducing new food products with a fresh appearance and natural flavor to the market (Rastogi et al., 2007). The HPP has been mainly introduced as non-thermal processing technology for sufficient microbial inactivation with minimal effects on the nutritional and sensorial properties of foods (Murchie et al., 2005). Recently, HPP has been extensively utilized to alter the structural and functional properties of food systems (Cadesky et al., 2017; He et al., 2016). During HPP, pressure and temperature can facilitate the protein denaturation or conformational changes and therefore influence the functional properties of food proteins (Messens et al., 1997).

13.4.1 EFFECTS OF HPP ON THE STRUCTURE OF SOY PROTEINS

Many studies have been conducted to investigate the effects of HPP on the structural and conformational changes of soy proteins. It was concluded that HPP could lead to the formation or rearrangement of S-S bonds, hydrogen bonds, hydrophobic or electrostatic interactions (Yang & Powers, 2016). For instance, HPP unfolded the structure of SPI and its subunits (7S and 11S) that probably attributed to the exposure of hydrophobic groups, the cleavage of weak van der Waals forces, and hydrogen bonds (Puppo et al., 2004). Zhang et al. (2005) concluded that the denaturation of soy proteins in soy milk occurred at 400 MPa for 11S and 300 MPa for 7S (Zhang et al., 2005). In another study, it was found that HPP can dissociate soy glycinin to subunits at pressures ˃ 300 MPa. With HPP at 500 MPa for 10 min, the ordered α-helix and β-sheet structures of glycinin are destroyed and converted to the random coil. The results also showed that HPP could alter the tertiary and quaternary structure of soy glycinin (Savadkoohi et al., 2016; Zhang et al., 2003). These structural changes could be due to the fact that HPP treatment can disturb S-S bonding that links the subunits of glycinin and dissociate glycinin into subunits.

13.4.2 EFFECTS OF HPP ON THE FUNCTIONAL PROPERTIES AND HEALTH BENEFITS OF SOY PROTEINS

Solubility is a critical factor that influences the gelling, foaming, and emulsifying properties of proteins. The solubility of a protein is related to the structures of native, intermediate, and denatured proteins. The protein–protein interactions and aggregation can be promoted when hydrophobic groups (from nonpolar amino acid residues) are abundant in protein molecules (Yang & Powers, 2016). Torrezan et al. (2007) found that pH and protein concentration had more pronounced effects on the solubility of SPI compared to the effect of pressure. They concluded that HPP (at pH 6.84, 2% SPI concentration, at a pressure of 450 MPa) slightly increased the solubility of SPI (Torrezan et al., 2007). Moreover, Speroni et al. (2009) observed that HPP enhanced the solubility of SPI, 7S, and 11S in water. The increase in protein solubility was probably because the HPP treatment could increase the ability of the protein to interact with water. Another reason could be the formation of nanometric thermodynamic stable aggregates due to HPP treatment (Speroni et al., 2009). On the other hand, Wang et al. (2008) found that HPP treatment (200–600 MPa) slightly decreased the solubility of SPI (at a protein concentration of 1%). The unfolding of globulins and the exposure of SH and hydrophobic groups during pressurization could result in aggregation and loss of solubility (Wang et al., 2008). The results' inconsistency might be because some researchers used freeze-dried soy proteins for their experiments while others used freshly prepared soy proteins from soybean meal flour. This could result in different aggregation states of the used protein samples causing different solubility patterns.

The foaming and emulsifying properties are related to the surface properties of proteins. In terms of foaming properties, the foaming capacity of soy proteins was enhanced at a pressure range of 200–300 MPa for 5–15 min and then decreased at higher pressure levels (Li et al., 2011). Moreover, the foaming stability gradually reduced with increasing pressure levels (Figure 13.4). The partial denaturation and improved solubility of soy proteins after pressurization at 200–300 MPa could promote the adsorption of protein particles during the bubbling and thereby decrease the air–water surface tension. While increasing the pressure above 300 MPa could cause an extensive aggregation,

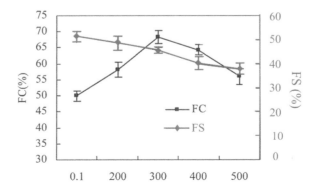

FIGURE 13.4 The foaming capacity (FC) and stability (FS) (%) of SPI dispersions at different pressure levels for 15 min. Source: adapted from Li et al. (2011).

leading to the unbalance between the rigidity and flexibility of protein molecules (Li et al., 2011). These changes could negatively influence the foaming properties of soy proteins. The influences of HPP on the emulsifying properties of soy proteins were studied. Molina et al. (2001) found that the EAI of 7S reached the highest value at a pressure of 400 MPa then decreased at higher pressure levels while the EAI value of 11S peaked at 200 MPa (Molina et al., 2001). At 400 MPa, the extensive denaturation of 11S could lead to aggregation because of the oxidation of SH groups and the formation of S-S bonds. These changes can reduce the solubility of 11S and thereby reduce its emulsifying properties. On the other hand, the pressure of 400 MPa could dissociate 7S into partially denatured subunits with improved surface activity. Similar results were reported by Wang et al. (2008), who found that the EAI of SPI reached the peak value at 400 MPa (Wang et al., 2008). The partial denaturation of HHP treated SPI at 400 MPa increased the surface hydrophobicity due to the exposure of buried hydrophobic groups and therefore improved the EAI of SPI (Puppo et al., 2004).

During the gelation process, a three-dimensional (3D) network is formed from denatured or partially denatured protein molecules. The well-ordered gel matrix can entrap significant amounts of water through protein–solvent and protein–protein interactions (Van Camp et al., 1997). Heat treatment has been used as a conventional processing method to denature soy proteins during the tofu-making process. However, heat treatment could mainly influence the hydrogen-bonded networks, but HPP more effectively disrupts specific electrostatic and hydrophobic interactions (Zhang et al., 2005). It was found that soy gel cannot be formed at a pressure level of less than 300 MPa, and gel strengths of soy gels significantly increased when the pressure increased from 400 to 600 MPa (Zhang et al., 2005). Cruz et al. (2009) studied the effects of HPP combined with heat treatment on the gelling properties of soy milk. The firmness of soy yogurt reached the highest value when soy milk was treated at a pressure of 300 MPa and 40° C (Cruz et al., 2009). In another study, compared to conventional heat treatment (95° C, 15 min), HPP (300 MPa and 50° C) of soy milk produced soy yogurt with improved WHC and firmness (Ferragut et al., 2009). Increasing the number of physicochemical bonds within the soy protein gel could improve its ability to hold water within the 3D network. The extensive denaturation derived from heat treatment and the mechanical effects of HPP could expose the SH groups to the surface, facilitating the formation of S-S bonds. Moreover, the smaller particle sizes of treated soy proteins could form a strong and homogeneous network (Cruz et al., 2009). Chen et al. (2019) studied the interactions between soy proteins and tea polyphenols under high-pressure processing treatment. HPP-treated soy proteins showed a more marked protective effect on tea polyphenols. Moreover, the addition of tea polyphenols also altered the molecular weight profile of soy proteins, probably due to the hydrogen bonding and hydrophobic interaction (Chen et al., 2019). The effects of HPP on the contents and profile of isoflavones in soy milk were investigated (Jung et al., 2008). It was found that the contents of isoflavone in the pressurized soy milk did not change significantly due to the pressure alone. However, the pressure associated with temperature during HPP may have significantly modified the isoflavones of soy milk (Jung et al., 2008). HPH treatments did not change the primary structure of proteins but caused slight modifications of secondary structure that were detected. Interestingly, free-SH groups in BSA increased with increasing pressure due to partial unfolding, while decreasing in WPI, possibly due to disaggregation and compaction of native WPI aggregates. HPH pretreatment allowed an enhancement of the BSA hydrolysis reaction rate, while the compaction of WPI aggregates caused a reduced hydrolysis extent. To conclude, HPH caused positive changes in the protein conformation and affected the degree of enzymatic hydrolysis (Carullo et al., 2020).

13.5 EFFECTS OF IRRADIATION ON THE STRUCTURE, FUNCTIONAL PROPERTIES, AND HEALTH BENEFITS OF SOY PROTEIN

Irradiation is an emerging non-thermal and eco-friendly technology used in many food applications. During the irradiation treatment, the food samples are exposed to ionizing radiation (x-rays, electron beams, or γ-rays) mainly to destroy pathogenic bacteria with less effect on the appearance and texture of samples. When food materials are exposed to ionizing radiation passes, they adsorb the energy, leading to the ionization of the atoms and molecules of these food materials (Kuan et al., 2013). The US Food and Drug Administration (FDA) and the World Health Organization (WHO) recommended that irradiated foods up to 10 kGy are considered safe for humans (FDA, 2004). When the processing is performed using ionizing radiation, the results suggested that high energy density per atomic transition could cleave molecules and produce ionization (Paramita & Rekha, 2009). Besides its applications in food preservation, irradiation has recently been used to alter the structure and functional properties of food proteins. Gamma (γ) irradiation of proteins is performed to change the structure of food proteins in order to obtain desired functional properties and develop new food products for consumers.

13.5.1 Effects of Irradiation on the Structure of Soy Proteins

Byun et al. (1994) studied the effects of different γ-irradiation dose levels on the structural changes of soy proteins. They found that γ-irradiation did not change the molecular weight and subunit patterns of soy proteins, meaning that the primary structure of soy proteins was not influenced by γ-irradiation. However, when soy proteins were irradiated at 20 kilograys (kGy), the secondary structure of 7S and 11S components changed, while 11S was more sensitive to irradiation (20 kGy). The results concluded that γ-irradiation partially denatured soy proteins and changed their secondary structure and conformational state (Byun et al., 1994). Similar findings were observed with sunflower protein isolate (Malik et al., 2017). Irradiation caused cleavage of S-S bonds in food proteins, disrupted the ordered structure, and thus induced denaturation or polymerization. The SDS-PAGE profile of SPI film showed that γ-irradiation, at dose levels higher than 16 kGy, damaged the protein structure and caused fragmentation and aggregation (Lee et al., 2005). Similar SDS-PAGE profiles were observed by Meinlschmidt et al. (2016), who found that γ-irradiation at dose levels higher than 25 kGy caused protein aggregations. Higher molecular weight aggregates of irradiated SPI samples could be formed due to the formation of protein–protein cross-linking interactions, electrostatic and hydrophobic interactions, and the formation of S-S bonds (Lee et al., 2005; Meinlschmidt et al., 2016).

13.5.2 Effects of Irradiation on the Functional Properties and Health Benefits of Soy Protein

The solubility of irradiated proteins is influenced by the molecular weight profile of food proteins. The reduction in the molecular weight at a low dose of radiation could improve the solubility of food proteins. However, the aggregation and fragmentation of irradiated proteins at a high dose of irradiation can reduce the solubility of these proteins (Malik et al., 2018). Early in 1988, Afify and Shousha concluded that a low dose of irradiation did not significantly affect the solubility of soy proteins (Afify & Shousha, 1988). The effects of irradiation (0–7.5 kGy) on the solubility of soy proteins were observed. The solubility of soy proteins was gradually reduced and reached the lowest value at 7.5 kGy (Afify et al., 2011). Radiation can also influence the viscosity of soy protein solutions. For example, γ-irradiation decreased the viscosity of SPI solutions due to the fact that the oxygen radicals generated by water radiolysis during irradiation can cause conformational changes of protein molecules (Lee et al., 2005). Similarly, the viscosity of soy protein isolate and glycerol solution gradually decreased with increasing irradiation dose from 0 to 25 kGy, probably due to the structural breakdown in proteins (Sabato & Lacroix, 2002). The emulsifying and foaming properties of food proteins could be influenced by irradiation treatment, depending on the structure of proteins and the dose of irradiation. The EAI and ESI of grass carp myofibrillar protein decreased significantly after irradiation at 6 kGy (Shi et al., 2015). After irradiation at 7.5 kGy, the emulsifying and foaming properties of SPI-maltose complexes reached the maximum value (Wang, Zhang, Wang, Wang, et al., 2020). In another study, the freeze-thaw stability of SPI-maltose complexes' emulsions improved when irradiated at 7.5 and 12.5 kGy (Wang et al., 2020). The irradiation treatment could change the conformational state of protein molecules, exposing the hydrophobic groups to the surface proteins and thereby improving the emulsifying and foaming properties of soy proteins. Moreover, the denaturation of protein molecules could increase the protein–protein interactions, forming a strong protein film at the oil/water or water/air interface, improving the emulsifying and foaming properties of these proteins.

13.6 EFFECTS OF MICROWAVE PROCESSING ON THE STRUCTURE, FUNCTIONAL PROPERTIES, AND HEALTH BENEFITS OF SOY PROTEIN

Microwave heating (MH) has been exclusively used as a reaction method for many years. The mechanism behind the MH effects is that the electromagnetic energy interacts with the exposed materials at the molecular level. Then, the energy is converted to heat through the motion of molecules during MH (Aslan et al., 2008). Moreover, using MH could reduce the reaction time and improve the yield in chemical synthesis compared to traditional heating methods (Cai et al., 2004). The effects of MH on the chemical reactions are induced by both thermal and non-thermal effects (de la Hoz et al., 2005). Thermal effects of MH are a result of the inverted heat transfer, the heterogeneous microwave field throughout the exposed materials, and the selective absorption of the radiation by polar substances. Those effects would significantly improve processes, change selectivity, and perform reactions that did not happen under normal conditions. The thermal effects are dominant in most microwave reactions and only superheating contributes to the acceleration of reaction. For example, Obermayer et al. (2009) employed silicon carbide reaction vials in a microwave reactor to confirm these effects in most cases. Guan et al. (2011) suggested that improvement of MH-assisted soy protein isolate-saccharide graft reactions was mainly due to MH-effect. To improve the techno-functional properties of soybean proteins, MH-assisted phosphorylation was applied (Wang et al., 2012). It was found that, after microwave treatment at 600 W for 3 min, the phosphorylation level of soybean protein isolates reached 35.72 mg g^{-1} and emulsifying activity and stability indexes increased significantly. The protein solubility was increased by 26% but a 13.5% decrease was observed in the apparent viscosity. The MH treatment caused a significant increase in the surface hydrophobicity, charge density, and content of sulfhydryl groups. Soybean curds pretreated with MH heating to 65, 80, and 95° C have a shelf-life of 16, 21, and 27 days, respectively, compared to seven days of control (Wu et al., 1977). The percentages of true digestibility were found to be 73, 84, 87, and 81 when the soybeans were MH for 0, 9, 12, and 15 min, respectively (Hafez et al., 1985). MH also affected the functional properties of soybean proteins (Armstrong et al., 1979), but more studies are needed.

13.7 EFFECTS OF OHMIC HEATING ON THE STRUCTURE, FUNCTIONAL PROPERTIES, AND HEALTH BENEFITS OF SOY PROTEIN

Ohmic heating (OH) is considered an emerging heating technology that is used for thermal processing. In this method, the food samples are located between two electrodes serving as a resistor and an alternating electric current (AC) passed through the circuit. Then, the heat is generated because of the electrical resistance. The electric energy is immediately transformed into heat, leading to an increase in the temperature. The OH devices simply consist of an electrical circuit, which is comprised of resistance and a source of voltage and current. The food sample is made part of an electrical circuit. The OH is considered as an alternate processing technique for thermal treatment of food samples, bypassing conventional heating systems rapidly. The difference between OH and MH is that in MH processing, the electric energy is converted to microwaves through the magnetron before heating the food samples (Sun, 2012). Milk pasteurization was the earliest application of OH technology; however, the high costs involved, and the corrosion of the electrodes led to the abandonment of the technology. Research has currently allowed the application of OH to the processing of fruits, vegetables, meat products, and other foods, allowing them to manufacture products with higher quality and better nutritional value than those by existing techniques (Pereira et al., 2015). Therefore, the OH offers several advantages such as high energy efficiency, the absence of hot surfaces, uniform and rapid heating of solids and liquids, simple, environmentally friendly, and cost-effective technology.

The OH also allows high-speed heating, overcoming heat transfer limitations that naturally occur during conventional thermal treatments (Pereira et al., 2015). Recent investigations found that the effects of electric fields on globular proteins caused conformational changes in protein structure, which results in protein films and gels with distinctive properties (Rodrigues et al., 2015). OH processing promoted structural alterations which facilitated the production of gels with less water holding capacity. The recommendation on applying electrical fields at sterilization temperatures could change the conformational state of proteins, and thus bring new technological and functional properties to the produced protein gels. It was also found that lipid-protein film with higher quality was obtained by OH, and the film formation rate, as well as the rehydration capacity of protein-lipid film, was improved significantly by OH (Lei et al., 2007). The changes in physical and chemical characteristics of soybean proteins during OH and conventional heating were also studied. The OH imbibed water, decreased the surface hydrophobicity, desaturated all 7S soy protein, and damaged soybean proteins' structure (Cha, 2011). Li et al. (2018) showed that the number of free amino groups increased by 14%, total sulfhydryl, surface hydrophobicity, functional properties, and foam stability decreased, and protein structure insignificantly changed after treating with OH. Thus, it can be assumed that OH is conducive to the improvement of functional properties, digestion, and absorption of soybean milk protein. The OH was also used, for example, to enhance the diffusion of soy milk from soybeans (Kim et al., 1995). Moreover, the OH was used as an efficient and convenient heating measure in soy proteins–based tofu making (Lien et al., 2014; Li et al., 2018).

13.8 EFFECTS OF PULSED ELECTRIC FIELD PROCESSING ON THE STRUCTURE, FUNCTIONAL PROPERTIES, AND HEALTH BENEFITS OF SOY PROTEIN

Pulsed electric field (PEF) technology has attracted much attention in the last decade due to its high energy efficiency and environmentally friendly properties. It has been used as a non-thermal process to destroy microorganisms while retaining nutritional quality and food flavor (Marco-Molés et al., 2011). During PEF treatment, a number of pulses of high voltage electric field are applied for a short time (from nanoseconds to milliseconds) to the food material between two electrodes. PEF is an ideal alternative treatment for the processing of high-protein foods such as milk and dairy ingredients or products containing heat-sensitive components because of its reduced heating effects. The main mechanism behind the effects of PEF on protein structure is polar groups of proteins that can absorb energy, and this energy can cause protein aggregation or generate free radicals. The generated free radicals can disrupt various protein–protein interactions such as hydrogen bonds, electrostatic and hydrophobic interactions, Van der Waals forces, and disulfide bonds. Therefore, protein structure and functional properties can be changed (Wang et al., 2008). The demand for natural and healthy products has triggered the need to develop several non-thermal approaches for food processing, of which PEF has proven to be a very valuable alternative food process. Several available studies found that PEF treatment can cause conformational changes in protein structure via the formation of free radicals. This could damage the protein structure and thus influence the techno-functional properties of proteins. However, it is worth mentioning that assessing the effects of PEF is complicated due to the existence of many factors that could influence this process such as temperature, pH, ionic strength of the solvent, and concentration. Therefore, further investigations on the mechanism of PEF effects on food proteins are recommended. Moreover, most of the available studies mainly focus on plant proteins and lack information on the applications of PEF on different animal proteins that can be an area for future research. In addition, the production of harmful substances generated by PEF needs to be studied before transferring the technology to the food industry. The nutritional and digestive properties of protein treated by electric fields are also topics worthy of future study. PEF enhanced the apparent viscosity of soy milk, which increased with the increase in electric field intensity and the consistency index (Xiang et al., 2011). PEF also altered the secondary structure of soy proteins, implying that the PEF treatment technique may be a novel method for preparing protein nanotubes (Liu et al., 2011).

13.9   COLD PLASMA PROCESSING ON THE STRUCTURE, FUNCTIONAL PROPERTIES, AND HEALTH BENEFITS OF SOY PROTEIN

Atmospheric cold plasma (ACP) processing is a fast, non-thermal, flexibly operated, and environmentally friendly technique. The mechanism of ACP involves using several high-energy radicals, such as hydroxyl radicals, superoxide, atomic oxygen, and nitric oxide, to break the covalent bonds and initiate numerous chemical reactions. Several studies used ACP to modify the surface properties of biomacromolecules, including different proteins (Dong et al., 2017; Ekezie et al., 2019). The above studies suggested that ACP could change the secondary and tertiary structure of proteins, and thus modify their physicochemical and techno-functional properties. In another study, it was found that protein glycosylation was associated with the degree of changes in the protein tertiary structure (Huang et al., 2013). It could be hypothesized that using high-energy radicals during ACP processing to disrupt the protein structure might activate specific reaction sites (Barranco et al., 2016), therefore accelerating the Maillard-type glycation reaction. Studies also have stated that the grafting of plasma-activated surfaces could improve their functional properties (Sun et al., 2006). ACP was previously suggested to modulate the functionality of dry bulk materials in the food sector (Bußler et al., 2015). ACP was used as a novel method for modifying protein film properties (Moosavi et al., 2020). Recently, Yepez et al. (2016) investigated ACP as a method to produce trans-free POHs. In this study, it was found that the direct exposure of soy oil to ACP can decrease the iodine value from 131 to 92 after 12 h and increase saturated fatty acids by 32.3%. They also concluded that ACP could reduce polyunsaturated fatty acids without the formation of trans-fats. Specifically, α-linolenic and linoleic acids were reduced from 9.5 to 3.7% and from 48.2 to 36%, respectively. Conversely, to reduce iodine value from 130 to 90, soybean oil will have to be hydrogenated for 1–4 h under pressure levels of 1–3 atm at a temperature range of 150–235° C, in the presence of 0.01–0.08% nickel catalyst (Patterson, 1994). Meanwhile, it was found that ACP can be used to produce partially hydrogenated soybean oil without any trans-fatty acid. Direct and remote ACP were used to treat the soy allergens (β-conglycinin and glycinin) (Meinlschmidt et al., 2016). SDS-PAGE analysis showed a new band was formed at 50 kDa. Moreover, protein bands of these two proteins became less visible after ACP treatment. Additionally, the studies on the milk allergens (casein, β-lactoglobulin, and α-lactalbumin) revealed that no obvious changes in the IgE binding values and gel band intensities were observed after the exposure to a remote plasma source. A significant improvement in solubility and emulsion stability was also shown following ACP (Ji et al., 2018). ACP also decreased the immunoreactivity of soy proteins (Meinlschmidt et al., 2016), showing the benefits of ACP to reduce food allergens.

13.10   EFFECTS OF ULTRAVIOLET PROCESSING ON THE STRUCTURE, FUNCTIONAL PROPERTIES, AND HEALTH BENEFITS OF SOY PROTEIN

The effects of ultraviolet (UV) radiation on food materials have recently been investigated. UV processing has attracted attention because the absorbance of electromagnetic radiation can form free radicals in amino acids, leading to the formation of intermolecular covalent bonds (Díaz et al., 2016). The influences of UV radiation on several protein films, including soy proteins, egg albumin, gluten, fish gelatin, corn zein, peanut protein, and sodium caseinate films, have been studied (Liu et al., 2004; Otoni et al., 2012; Rhim et al., 2000). These studies found that UV radiation could change the physicochemical properties and color of films. However, the results were variable, possibly because of the differences in the nature of the used protein materials and the treatment conditions. In these works, UV was used as a pretreatment of films prepared from heat-treated protein solutions. The nature and turbidity of treated samples can influence the absorbed energy and transmittance of UV light. At the highest dose, UV treatment was able to affect most mechanical properties and solubility of proteins when applied to the film-forming solution. These films showed significantly lower solubility, higher puncture deformation, puncture strength, and tensile strength than untreated films (Díaz et al., 2016). Films treated with the highest dose of UV radiation in solution showed tensile strength like heat-treated films. The UV radiation caused color changes in the films and made them greener and darker compared to the untreated films. The effect of UV treatment on color was more pronounced when applied to the film-forming solution. The UV radiation of solutions at high doses could increase the concentration of free sulfhydryl groups and induce aggregate formation compared to the untreated films, but the changes were lower than those observed in heat-treated films. Likewise, it was confirmed that UV-C irradiation could alter the solubility and conformations of proteins unveiling the potential of UV-C radiation for flour modification (Kumar et al., 2020). UV modifies the properties of the films in different ways depending on the pH of the solutions (Díaz et al., 2017). Gennadios et al. (1998) studied the effects of conventional UV light on the physicochemical properties of soy protein films. It was found that UV irradiation could increase the mechanical strength of soy protein films. Moreover, UV irradiation increased the sharpness of immobile bands in electrophoretic patterns, meaning that covalent cross-links in UV-treated films were further developed (Gennadios et al., 1998). The residual immunoreactivity of UV-treated soy protein samples was determined using the Sandwich ELISA technique. The results concluded that UV treatment altered immunoreactivity. This could be attributed to the photothermal, photochemical, and photophysical effects induced by UV treatment that could cause protein aggregation and crosslinking or change the allergen conformation and therefore diminish the antibody-binding ability of the modified epitopes (Meinlschmidt et al., 2016).

13.11 CONCLUSION

Soy proteins have been extensively applied in many food industries in recent decades because of their good functional properties and health benefits. Several promising processing technologies have been used to change the structure of soy protein in order to improve the functional properties and health benefits of soy proteins. Moreover, the effects of the novel technologies of soy protein depend on soy protein's conformational and aggregation states as well as solution conditions. While some of these technologies, such as ultrasound and high-pressure processing, have been previously applied on a large scale, others such as irradiation and cold plasma processing still need further research to develop effective large-scale devices.

REFERENCES

Afify, A. E. M. M. R., Rashed, M. M., Mahmoud, E. A., & El-Beltagi, H. S. (2011). Effect of gamma radiation on protein profile, protein fraction and solubility's of three oil seeds: Soybean, peanut and sesame. *Notulae Botanicae Horti Agrobotanici Cluj-Napoca*, 39(2), 90–98.

Afify, A. E. M. M. R., & Shousha, M. A. (1988). Effect of low-dose irradiation on soybean protein solubility, trypsin inhibitor activity, and protein patterns separated by polyacrylamide gel electrophoresis. *Journal of Agricultural and Food Chemistry*, 36(4), 810–813.

Armstrong, D. L., Stanley, D. W., & Maurice, T. J. (1979). *Functional Properties of Microwave-Heated Soybean Proteins, Functionality and Protein Structure*, ACS Symposium Series (92), 147–172.

Arzeni, C., Martínez, K., Zema, P., Arias, A., Pérez, O. E., & Pilosof, A. M. R. (2012). Comparative study of high intensity ultrasound effects on food proteins functionality. *Journal of Food Engineering, 108*(3), 463–472.

Ashokkumar, M. (2011). The characterization of acoustic cavitation bubbles - An overview. *Ultrasonics Sonochemistry, 18*(4), 864–872.

Ashokkumar, M. (2015). Applications of ultrasound in food and bioprocessing. *Ultrasonics Sonochemistry, 25*(1), 17–23.

Aslan, K., & Geddes, C. D. (2008). A review of an ultrafast and sensitive bioassay platform technology: Microwave-accelerated metal-enhanced fluorescence. *Plasmonics, 3*(2), 89–101.

Awad, T. S., Moharram, H. A., Shaltout, O. E., Asker, D., & Youssef, M. M. (2012). Applications of ultrasound in analysis, processing and quality control of food: A review. *Food Research International, 48*(2), 410–427.

Barranco, A., Borras, A., Gonzalez-Elipe, A. R., & Palmero, A. (2016). Perspectives on oblique angle deposition of thin films: From fundamentals to devices. *Progress in Materials Science, 76*, 59–153.

Bußler, S., Steins, V., Ehlbeck, J., & Schlüter, O. (2015). Impact of thermal treatment versus cold atmospheric plasma processing on the techno-functional protein properties from Pisum sativum 'Salamanca'. *Journal of Food Engineering, 167*, 166–174.

Byun, M. -W., Kang, I., Hayashi, Y., Matsumura, Y., & Mori, T. -J. (1994). Effect of γ-irradiation on soya bean proteins. *Journal of the Science of Food and Agriculture, 66*(1), 55–60.

Cadesky, L., Walkling-Ribeiro, M., Kriner, K. T., Karwe, M. V., & Moraru, C. I. (2017). Structural changes induced by high-pressure processing in micellar casein and milk protein concentrates. *Journal of Dairy Science, 100*(9), 7055–7070.

Cai, L., Liu, X., Tao, X., & Shen, D. (2004). Efficient microwave-assisted cyanation of aryl bromide. *Synthetic Communications, 34*(7), 1215–1221.

Carullo, D., Donsì, F., & Ferrari, G. (2020). Influence of high-pressure homogenization on structural properties and enzymatic hydrolysis of milk proteins. *LWT, 130*, 109657.

Cha, Y. H. (2011). Effect of ohmic heating on characteristics of heating denaturation of soybean protein. *The Korean Journal of Food and Nutrition, 24*(4), 740–745.

Chen, G., Wang, S., Feng, B., Jiang, B., & Miao, M. (2019). Interaction between soybean protein and tea polyphenols under high pressure. *Food Chemistry, 277*, 632–638.

Chen, L., Chen, J., Ren, J., & Zhao, M. (2011). Effects of ultrasound pretreatment on the enzymatic hydrolysis of soy protein isolates and on the emulsifying properties of hydrolysates. *Journal of Agricultural and Food Chemistry, 59*(6), 2600–2609.

Cruz, N. S., Capellas, M., Jaramillo, D. P., Trujillo, A. J., Guamis, B., & Ferragut, V. (2009). Soymilk treated by ultra high-pressure homogenization: Acid coagulation properties and characteristics of a soy-yogurt product. *Food Hydrocolloids, 23*(2), 490–496.

Cui, Q., Wang, L., Zhou, G., Dong, Y., Wang, X., & Jiang, L. (2019). Effect of ultrasonic treatment on functional properties of soy protein isolate-whey protein isolate mixture. *Shipin Kexue/Food Science, 40*(23), 111–116.

de la Hoz, A., Diaz-Ortiz, A., & Moreno, A. (2005). Microwaves in organic synthesis. Thermal and non-thermal microwave effects. *Chemical Society Reviews, 34*(2), 164–178.

Díaz, O., Candia, D., & Cobos, Á. (2016). Effects of ultraviolet radiation on properties of films from whey protein concentrate treated before or after film formation. *Food Hydrocolloids, 55*, 189–199.

Díaz, O., Candia, D., & Cobos, Á. (2017). Whey protein film properties as affected by ultraviolet treatment under alkaline conditions. *International Dairy Journal, 73*, 84–91.

Dong, S., Gao, A., Xu, H., & Chen, Y. (2017). Effects of dielectric barrier discharges (DBD) cold plasma treatment on physicochemical and structural properties of zein powders. *Food and Bioprocess Technology, 10*(3), 434–444.

Ekezie, F. G. C., Cheng, J. H., & Sun, D. W. (2019). Effects of atmospheric pressure plasma jet on the conformation and physicochemical properties of myofibrillar proteins from king prawn (litopenaeus vannamei). *Food Chemistry, 276*, 147–156.

Endres, J. G. (2001). *Soy Protein Products: Characteristics, Nutritional Aspects, and Utilization.* The American Oil Chemists Society.

FDA (2004). Overview of irradiation of food and packaging. FDA. https://www.fda.gov/food/irradiation-food-packaging/overview-irradiation-food-and-packaging.

Ferragut, V., Cruz, N. S., Trujillo, A., Guamis, B., & Capellas, M. (2009). Physical characteristics during storage of soy yogurt made from ultra-high pressure homogenized soymilk. *Journal of Food Engineering, 92*(1), 63–69.

Fukushima, D. (2011). Soy proteins. *Handbook of Food Proteins*. Woodhead Publishing, 210–232.

Fukushima, Danji (1991). Recent progress of soybean protein foods: Chemistry, technology, and nutrition. *Food Reviews International*, *7*(3), 323–351.

Gennadios, A., Rhim, J. W., Handa, A., Weller, C. L., & Hanna, M. A. (1998). Ultraviolet radiation affects physical and molecular properties of soy protein films. *Journal of Food Science*, *63*(2), 225–228.

Guan, J. J., Zhang, T. B., Hui, M., Yin, H. C., Qiu, A. Y., & Liu, X. Y. (2011). Mechanism of microwave-accelerated soy protein isolate–saccharide graft reactions. *Food Research International*, *44*(9), 2647–2654.

Hafez, Y. S., Mohamed, A. I., Hewedy, F. M., & Singh, G. (1985). Effects of microwave heating on solubility, digestibility and metabolism of soy protein. *Journal of Food Science*, *50*(2), 415–417.

He, X., Mao, L., Gao, Y., & Yuan, F. (2016). Effects of high pressure processing on the structural and functional properties of bovine lactoferrin. *Innovative Food Science and Emerging Technologies*, *38*, 221–230.

Hu, H., Cheung, I. W. Y., Pan, S., & Li-Chan, E. C. Y. (2015a). Effect of high intensity ultrasound on physicochemical and functional properties of aggregated soybean β-conglycinin and glycinin. *Food Hydrocolloids*, *45*, 102–110.

Hu, H., Fan, X., Zhou, Z., Xu, X., Fan, G., Wang, L., Huang, X., Pan, S., & Zhu, L. (2013a). Acid-induced gelation behavior of soybean protein isolate with high intensity ultrasonic pretreatments. *Ultrasonics Sonochemistry*, *20*(1), 187–195.

Hu, H., Li-Chan, E. C. Y., Wan, L., Tian, M., & Pan, S. (2013a). The effect of high intensity ultrasonic pretreatment on the properties of soybean protein isolate gel induced by calcium sulfate. *Food Hydrocolloids*, *32*(2), 303–311.

Hu, H., Wu, J., Li-Chan, E. C. Y., Zhu, L., Zhang, F., Xu, X., Fan, G., Wang, L., Huang, X., & Pan, S. (2013c). Effects of ultrasound on structural and physical properties of soy protein isolate (SPI) dispersions. *Food Hydrocolloids*, *30*(2), 647–655.

Hu, H., Zhu, X., Hu, T., Cheung, I. W. Y., Pan, S., & Li-Chan, E. C. Y. (2015b). Effect of ultrasound pretreatment on formation of transglutaminase-catalysed soy protein hydrogel as a riboflavin vehicle for functional foods. *Journal of Functional Foods*, *19*, 182–193.

Hua, Y., Cui, S. W., Wang, Q., Mine, Y., & Poysa, V. (2005). Heat induced gelling properties of soy protein isolates prepared from different defatted soybean flours. *Food Research International*, *38*(4), 377–385.

Huang, X., Tu, Z., Wang, H., Zhang, Q., Shi, Y., & Xiao, H. (2013). Increase of ovalbumin glycation by the Maillard reaction after disruption of the disulfide bridge evaluated by liquid chromatography and high resolution mass spectrometry. *Journal of Agricultural and Food Chemistry*, *61*(9), 2253–2262.

Jambrak, A. R., Lelas, V., Mason, T. J., Krešić, G., & Badanjak, M. (2009). Physical properties of ultrasound treated soy proteins. *Journal of Food Engineering*, *93*(4), 386–393.

Jambrak, A. R., Mason, T. J., Lelas, V., Herceg, Z., & Herceg, I. L. (2008). Effect of ultrasound treatment on solubility and foaming properties of whey protein suspensions. *Journal of Food Engineering*, *86*(2), 281–287.

Jambrak, A. R., Mason, T. J., Lelas, V., Paniwnyk, L., & Herceg, Z. (2014). Effect of ultrasound treatment on particle size and molecular weight of whey proteins. *Journal of Food Engineering*, *121*(1), 15–23.

Ji, H., Dong, S., Han, F., Li, Y., Chen, G., Li, L., & Chen, Y. (2018). Effects of dielectric barrier discharge (DBD) cold plasma treatment on physicochemical and functional properties of peanut protein. *Food and Bioprocess Technology*, *11*(2), 344–354.

Jung, S., Murphy, P. A., & Sala, I. (2008). Isoflavone profiles of soymilk as affected by high-pressure treatments of soymilk and soybeans. *Food Chemistry*, *111*(3), 592–598.

Kalla, A. M., & Devaraju, R. (2017). Microwave energy and its application in food industry: A reveiw. *Asian Journal of Dairy and Food Research*, *36*(1), 37–44.

Kim, J., & Pyun, Y. (1995, July). Extraction of soy milk using ohmic heating. In 9th Congress of Food Sci. Tech.

Kinsella, J. E. (1979a). Functional properties of soy proteins. *Journal of the American Oil Chemists' Society*, *56*(3), 242–258.

Kuan, Y. H., Bhat, R., Patras, A., & Karim, A. A. (2013). Radiation processing of food proteins - A review on the recent developments. *Trends in Food Science and Technology*, *30*(2), 105–120.

Kumar, A., Nayak, R., Purohit, S. R., & Rao, P. S. (2020). Impact of UV-C irradiation on solubility of Osborne protein fractions in wheat flour. *Food Hydrocolloids*, *110*, 105845.

Lee, M., Lee, S., & Song, K. B. (2005). Effect of γ-irradiation on the physicochemical properties of soy protein isolate films. *Radiation Physics and Chemistry*, *72*(1), 35–40.

Lei, L., Zhi, H., Xiujin, Z., Takasuke, I., & Zaigui, L. (2007). Effects of different heating methods on the production of protein–lipid film. *Journal of Food Engineering*, *82*(3), 292–297.

Leong, T. S. H., Martin, G. J. O., & Ashokkumar, M. (2017). Ultrasonic encapsulation – A review. *Ultrasonics Sonochemistry*, *35*(B), 605–614.

Li, H., Zhu, K., Zhou, H., & Peng, W. (2011). Effects of high hydrostatic pressure on some functional and nutritional properties of soy protein isolate for infant formula. *Journal of Agricultural and Food Chemistry*, *59*(22), 12028–12036.

Li, X., Ye, C., Tian, Y., Pan, S., & Wang, L. (2018). Effect of ohmic heating on fundamental properties of protein in soybean milk. *Journal of Food Process Engineering*, *41*(3), e12660.

Lien, C. C., Shen, Y. C., & Ting, C. H. (2014). Ohmic heating for tofu making—A pilot study. *Journal of Agricultural Chemistry and Environment*, *3*(02), 7–13.

Liu, C. C., Tellez-Garay, A. M., & Castell-Perez, M. E. (2004). Physical and mechanical properties of peanut protein films. *LWT - Food Science and Technology*, *37*(7), 731–738.

Liu, Y. Y., Zeng, X. A., Deng, Z., Yu, S. J., & Yamasaki, S. (2011). Effect of pulsed electric field on the secondary structure and thermal properties of soy protein isolate. *European Food Research and Technology*, *233*(5), 841.

Malik, M. A., Sharma, H. K., & Saini, C. S. (2017). Effect of gamma irradiation on structural, molecular, thermal and rheological properties of sunflower protein isolate. *Food Hydrocolloids*, *72*, 312–322.

Malik, M. A., Sharma, H. K., Saini, C. S., Sharma, H. K., & Saini, C. S. (2018). Gamma irradiation of food proteins: Recent developments in modification of proteins to improve their functionality. In *Technologies in Food Processing*. Apple Academic Press, 129–150.

Marco-Molés, R., Rojas-Graü, M. A., Hernando, I., Pérez-Munuera, I., Soliva-Fortuny, R., & Martín-Belloso, O. (2011). Physical and structural changes in liquid whole egg treated with high-intensity pulsed electric fields. *Journal of Food Science*, *76*(2), C257–C264.

Meinlschmidt, P., Ueberham, E., Lehmann, J., Reineke, K., Schlüter, O., Schweiggert-Weisz, U., & Eisner, P. (2016). The effects of pulsed ultraviolet light, cold atmospheric pressure plasma, and gamma-irradiation on the immunoreactivity of soy protein isolate. *Innovative Food Science and Emerging Technologies*, *38*, 374–383.

Messens, W., Van Camp, J., & Huyghebaert, A. (1997). The use of high pressure to modify the functionality of food proteins. *Trends in Food Science and Technology*, *8*(4), 107–112.

Molina, E., Papadopoulou, A., & Ledward, D. A. (2001). Emulsifying properties of high pressure treated soy protein isolate and 7S and 11S globulins. *Food Hydrocolloids*, *15*(3), 263–269.

Moosavi, M. H., Khani, M. R., Shokri, B., Hosseini, S. M., Shojaee-Aliabadi, S., & Mirmoghtadaie, L. (2020). Modifications of protein-based films using cold plasma. *International Journal of Biological Macromolecules*, *142*, 769–777.

Morales, R., Martínez, K. D., Pizones Ruiz-Henestrosa, V. M., & Pilosof, A. M. R. (2015). Modification of foaming properties of soy protein isolate by high ultrasound intensity: Particle size effect. *Ultrasonics Sonochemistry*, *26*, 48–55.

Murchie, L. W., Cruz-Romero, M., Kerry, J. P., Linton, M., Patterson, M. F., Smiddy, M., & Kelly, A. L. (2005). High pressure processing of shellfish: A review of microbiological and other quality aspects. *Innovative Food Science and Emerging Technologies*, *6*(3), 257–270.

Nishinari, K., Fang, Y., Guo, S., & Phillips, G. O. (2014). Soy proteins: A review on composition, aggregation and emulsification. *Food Hydrocolloids*, *39*, 301–318.

O'Sullivan, J. J., Park, M., Beevers, J., Greenwood, R. W., & Norton, I. T. (2017). Applications of ultrasound for the functional modification of proteins and nanoemulsion formation: A review. *Food Hydrocolloids*, *71*, 299–310.

O'Sullivan, J., Murray, B., Flynn, C., & Norton, I. (2016a). The effect of ultrasound treatment on the structural, physical and emulsifying properties of animal and vegetable proteins. *Food Hydrocolloids*, *53*, 141–154.

O'Sullivan, J., Park, M., & Beevers, J. (2016b). The effect of ultrasound upon the physicochemical and emulsifying properties of wheat and soy protein isolates. *Journal of Cereal Science*, *69*, 77–84.

Obermayer, D., Gutmann, B., & Kappe, C. O. (2009). Microwave chemistry in silicon carbide reaction vials: Separating thermal from non-thermal effects. *Angewandte Chemie International Edition*, *48*(44), 8321–8324.

Otoni, C. G., Avena-Bustillos, R. J., Chiou, B. S., Bilbao-Sainz, C., Bechtel, P. J., & McHugh, T. H. (2012). Ultraviolet-B radiation induced cross-linking improves physical properties of cold-and warm-water fish gelatin gels and films. *Journal of Food Science*, *77*(9), E215–E223.

Paramita, B., & Rekha, S. (2009). Effect of irradiation on food texture and rheology. In *Novel Food Processing: Effects on Rheological and Functional Properties*, (103–122).

Patterson, H. B. W. (1994). *Hydrogenation of Fats and Oils: Theory and Practice*. AOCS Press.

Peng, I. C., & Nielsen, S. S. (1986). Protein-protein interactions Between soybean beta-conglycinin (B1-B6) and myosin. *Journal of Food Science, 51*(3), 588–590.

Pereira, R., Rodrigues, R. M., Teixeira, J. A., & Vicente, A. A. (2015). *Aquecimento óhmico: Uma ferramenta ao serviço da biotecnologia.* University of Minho, Portugal, (April 2015).

Puppo, C., Chapleau, N., Speroni, F., De Lamballerie-Anton, M., Michel, F., Añón, C., & Anton, M. (2004). Physicochemical modifications of high-pressure-treated soybean protein isolates. *Journal of Agricultural and Food Chemistry, 52*(6), 1564–1571.

Ramdath, D., Padhi, E., Sarfaraz, S., Renwick, S., & Duncan, A. (2017). Beyond the cholesterol-lowering effect of soy protein: A review of the effects of dietary soy and its constituents on risk factors for cardiovascular disease. *Nutrients, 9*(4), 324.

Rastogi, N. K., Raghavarao, K. S. M. S., Balasubramaniam, V. M., Niranjan, K., & Knorr, D. (2007). Opportunities and challenges in high pressure processing of foods. *Critical Reviews in Food Science and Nutrition, 47*(1), 69–112.

Ren, X., Li, C., Yang, F., Huang, Y., Huang, C., Zhang, K., & Yan, L. (2020). Comparison of hydrodynamic and ultrasonic cavitation effects on soy protein isolate functionality. *Journal of Food Engineering, 265,* 109697.

Rhim, J. W., Gennadios, A., Handa, A., Weller, C. L., & Hanna, M. A. (2000). Solubility, tensile, and color properties of modified soy protein isolate films. *Journal of Agricultural and Food Chemistry, 48*(10), 4937–4941.

Rodrigues, R. M., Martins, A. J., Ramos, O. L., Malcata, F. X., Teixeira, J. A., Vicente, A. A., & Pereira, R. N. (2015). Influence of moderate electric fields on gelation of whey protein isolate. *Food Hydrocolloids, 43,* 329–339.

Sabato, S. F., & Lacroix, M. (2002). Radiation effects on viscosimetry of protein based solutions. *Radiation Physics and Chemistry, 63*(3–6), 357–359.

Savadkoohi, S., Bannikova, A., Mantri, N., & Kasapis, S. (2016). Structural modification in condensed soy glycinin systems following application of high pressure. *Food Hydrocolloids, 53,* 115–124.

Shi, Y., Li, R. Y., Tu, Z. C., Ma, D., Wang, H., Huang, X. Q., & He, N. (2015). Effect of γ-irradiation on the physicochemical properties and structure of fish myofibrillar proteins. *Radiation Physics and Chemistry, 109,* 70–72.

Speroni, F., Beaumal, V., de Lamballerie, M., Anton, M., Añón, M. C., & Puppo, M. C. (2009). Gelation of soybean proteins induced by sequential high-pressure and thermal treatments. *Food Hydrocolloids, 23*(5), 1433–1442.

Sun, D. W. (Ed.) (2012). *Thermal Food Processing: New Technologies and Quality Issues.* CRC Press.

Sun, H. X., Zhang, L., Chai, H., & Chen, H. L. (2006). Surface modification of poly (tetrafluoroethylene) films via plasma treatment and graft copolymerization of acrylic acid. *Desalination, 192*(1–3), 271–279.

Taha, A., Hu, T., Zhang, Z., Bakry, A. M., Khalifa, I., Pan, S., & Hu, H. (2018). Effect of different oils and ultrasound emulsification conditions on the physicochemical properties of emulsions stabilized by soy protein isolate. *Ultrasonics Sonochemistry, 49,* 283–293.

Taha, A., Ahmed, E., Ismaiel, A., Ashokkumar, M., Xu, X., Pan, S., & Hu, H. (2020). Ultrasonic emulsification: An overview on the preparation of different emulsifiers-stabilized emulsions. *Trends in Food Science and Technology, 105,* 363–377.

Tang, C.-H. (2017). Emulsifying properties of soy proteins: A critical review with emphasis on the role of conformational flexibility. *Critical Reviews in Food Science and Nutrition, 57*(12), 2636–2679.

Torrezan, R., Tham, W. P., Bell, A. E., Frazier, R. A., & Cristianini, M. (2007). Effects of high pressure on functional properties of soy protein. *Food Chemistry, 104*(1), 140–147.

Van Camp, J., Messens, W., Clément, J., & Huyghebaert, A. (1997). Influence of ph and sodium chloride on the high pressure-induced gel formation of a whey protein concentrate. *Food Chemistry, 60*(3), 417–424.

Wang, X. B., & Chi, Y. J. (2012). Microwave-assisted phosphorylation of soybean protein isolates and their physicochemical properties. *Czech Journal of Food Sciences, 30*(2), 99–107.

Wang, X. S., Tang, C. H., Li, B. S., Yang, X. Q., Li, L., & Ma, C. Y. (2008). Effects of high-pressure treatment on some physicochemical and functional properties of soy protein isolates. *Food Hydrocolloids, 22*(4), 560–567.

Wang, Y., Zhang, A., Wang, X., Xu, N., & Jiang, L. (2020). The radiation assisted-Maillard reaction comprehensively improves the freeze-thaw stability of soy protein-stabilized oil-in-water emulsions. *Food Hydrocolloids, 103,* 105684.

Wang, Y., Zhang, A., Wang, Y., Wang, X., Xu, N., & Jiang, L. (2020). Effects of irradiation on the structure and properties of glycosylated soybean proteins. *Food and Function, 11*(2), 1635–1646.

Wu, M. T., & Salunkhe, D. K. (1977). Extending shelf-life of fresh soybean curds by in-package microwave treatments. *Journal of Food Science, 42*(6), 1448–1450.

Xiang, B. Y., Simpson, M. V., Ngadi, M. O., & Simpson, B. K. (2011). Effect of pulsed electric field on the rheological and colour properties of soy milk. *International Journal of Food Sciences and Nutrition, 62*(8), 787–793.

Xu, J., Zhao, Q., Qu, Y., & Ye, F. (2015). Antioxidant activity and anti-exercise-fatigue effect of highly denatured soybean meal hydrolysate prepared using neutrase. *Journal of Food Science and Technology, 52*(4), 1982–1992.

Yang, F., Liu, X., Huang, Y., Huang, C., & Zhang, K. (2018). Swirling cavitation improves the emulsifying properties of commercial soy protein isolate. *Ultrasonics Sonochemistry, 42*, 471–481.

Yang, J., & Powers, J. R. (2016). Effects of high pressure on food proteins. In Food *Engineering Series*. Springer, (353–389).

Yepez, X. V., & Keener, K. M. (2016). High-voltage atmospheric cold plasma (HVACP) hydrogenation of soybean oil without trans-fatty acids. *Innovative Food Science and Emerging Technologies, 38*, 169–174.

Young, V. R. (1991). Soy protein in relation to human protein and amino acid nutrition. *Journal of the American Dietetic Association, 91*(7), 828–835.

Zhang, H., Li, L., Tatsumi, E., & Isobe, S. (2005). High-pressure treatment effects on proteins in soy milk. *LWT - Food Science and Technology, 38*(1), 7–14.

Zhang, H., Li, L., Tatsumi, E., & Kotwal, S. (2003). Influence of high pressure on conformational changes of soybean glycinin. *Innovative Food Science and Emerging Technologies, 4*(3), 269–275.

Zheng, T., Li, X., Taha, A., Wei, Y., & Hu, T. Fatamorgana, P. B., Zhang, Z., Liu, F., Xu, X., Pan, S., & Hu, H. (2019). Effect of high intensity ultrasound on the structure and physicochemical properties of soy protein isolates produced by different denaturation methods. *Food Hydrocolloids, 97*(July), 105216.

Zhou, M., Liu, J., Zhou, Y., Huang, X., Liu, F., Pan, S., & Hu, H. (2016). Effect of high intensity ultrasound on physicochemical and functional properties of soybean glycinin at different ionic strengths. *Innovative Food Science and Emerging Technologies, 34*, 205–213.

14 Fermentation of Soybeans – Technology, Nutritional Properties, and Effects

Runni Mukherjee, Runu Chakraborty, and Abhishek Dutta

CONTENTS

14.1 INTRODUCTION

Soybean, which is a native oilseed plant of Asia, is widely used worldwide as a source of oil. However, oil is not the only product that is procured from soybean seeds. A large variety of other products and by-products are also obtained from the seeds either before or after oil is extracted. Soybean and its products are particularly favored all over the world for their high protein concentration as well as for being rich in other bioactive compounds which have been reported to impart various health benefits upon consumption (Messina et al., 1994; Mukherjee et al., 2016).

Interest in the composition of soy products has exponentially grown because numerous potential anticarcinogens and other therapeutic agents have been reported to be found in soybeans and related products (Messina et al., 1994). Soy foods are said to provide a protective effect on the breast, intestine, liver, bladder, prostate, skin, and stomach from cancer development (Messina et al., 1994). The initial observation that Asian women with a diet high in soybean have a lower incidence of breast cancer (Messina et al., 1994) has been corroborated by epidemiological studies (Wu et al., 1998).Growing evidence has also shown that consumption of soybeans may reduce the risk of other chronic and lifestyle diseases like osteoporosis (Barnes et al., 1994), chronic renal disease (Fico et al., 2000; Ranich et al., 2001), plasmacholesterol (Hoet al., 2000), and coronary heart disease (Lucas et al., 2001).

Soybean seeds are grown primarily to produce oil which has a unique fatty acid profile; it predominantly contains unsaturated fatty acids, including monounsaturated fatty acids and polyunsaturated

DOI: 10.1201/9781003030294-14

fatty acids. Linoleic acid, and to a much lesser extent, α-linolenic acid, are the most abundant fatty acids of soybean oil. Other than oil, soymilk is also a popular and commonly used product of soybean which is a highly nutritious dairy replacement for lactose-intolerant people. High protein soy flour, soy chunks, etc. are also available from soybean (Figure 14.1).

Various important and valuable by-products are also obtained during and after oil extraction or other processing of soybean seeds (Figure 14.1). Some of these by-products are highly nutritious, like deoiled meal. This by-product is very rich in proteins with a balanced amino acid profile and various bioactive compounds like the seeds but also contains some antinutritional factors. This meal is commonly used as animal feed but has its limitations due to the presence of some undesirable compounds. The recent focus of soybean meal research is on the use of fermentation to eliminate the undesirable compounds and thereby increase the bioavailability of the nutritional components (Mukherjee et al., 2016).

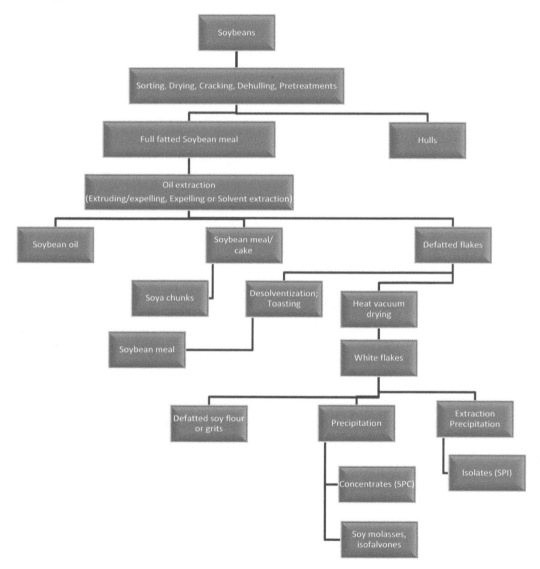

FIGURE 14.1 Flowchart for soybean seed processing and production stages of various products during processing.

Other than soybean meal, some other important by-products are hulls, lecithin, as well as soybean protein concentrates or isolates.

Even with so many varied products and high nutritional potentials, the presence of certain antinutritional factors and allergen proteins limit the wide-scale acceptability of soybean and its associated products. The presence of these compounds makes raw soybean seeds inedible for ruminants as well as non-ruminants (Li et al., 1990) but various processing methods including fermentation are regularly being used to overcome the possible adverse effects of consuming soybeans (Mukherjee et al., 2016).

Fermentation is a traditional method that has been reported to not only improve nutritional content but also decrease undesirable compounds, thereby recreating a better product with improved nutritional value (Frias et al., 2008). A wide variety of microorganisms have been investigated for their capabilities to ferment soybeans and the resulting products have been shown to have better digestibility and reduced immunoreactivity among other benefits (Frias et al., 2008; Song et al., 2008). Soybean seeds as well as deoiled meal are now regularly fermented to investigate the various beneficial effects of fermentation on their physicochemical as well as nutritional properties. Both fields call for further research and development.

Various analytical and empirical models are also being tried and used for the development and optimization of processes for the fermentation of soybean products and subsequent product developments (Mukherjee et al., 2019).

14.1.1 Products and By-Products

Soybeans are valued for their versatilities in producing a varied amount of valuable products and very promising by-products. Some of the most common products and by-products are discussed in the following sections.

14.1.1.1 Products

Oils: Soybean oil is second only to palm oil in terms of production scale and provides about 22% of the total world fats and oils. It is highly rich in unsaturated oils, the unsaponifiables containing sterols, hydrocarbons, and tocopherols (Hammond et al., 2004). Soybean oil is used for a variety of purposes in food either with or without processing. These include cooking and salad oils, spreads and shortenings, mayonnaise, and salad dressings. Some 12–13% of soybean oil currently has non-food uses (Woodfield & Harwood, 2017).

Soy milk: Vegan alternatives to cow's milk, cheese, and yogurt are produced from soy. Soy is also used to produce infant formulas for those infants who are intolerant to lactose or cow's milk protein. Soy milk is made from whole soybeans or full-fat soy flour (Fehily, 2004).

Soy flour: Soy flour is finely ground soybean meal and can be used directly or subjected to further processing. For gluten-free flour combinations, soybean flour is an ideal choice as the contributor of proteins. The addition of soy flour to bread flour thus far has been the most satisfactory method of increasing the protein concentration in bread as well as overcoming the lysine limitation of wheat flour (Penfield & Campbell, 1990).

Soy protein concentrates/isolates *(SPC, SPI)*: Soy protein has been shown to support health needs across the lifespan as a source of lean, cholesterol-free, and lactose-free protein. Soy protein is the only vegetable protein that is a complete protein as it meets all essential amino acid requirements to support the normal growth and development of infants and children (Paulsen, 2009). SPC, SPI, and textured vegetable proteins are also used in petfoods (Mukherjee et al., 2016).

14.1.1.2 By-Products

Hull: The soybean hull, also known as the seed coat, is obtained as a by-product. On a dry weight basis, hulls constitute about 8% of the total seed, depending on the variety and the seed size (Gnanasambandam & Proctor, 1999). Sometimes separated hulls are toasted and ground before being blended back into defatted soy meal to make a meal containing 44% protein. As of today, there are no industrial markets for soybean hulls; all hulls are sold for roughage in feeding livestock.

Soybean meal (SBM): Soybean meal is one of the largest volumes of both plant protein meals and animal feed ingredient resources available in the world (Mukherjee et al., 2016). The most common forms are solvent extracted (44% crude protein, as-fed basis) and dehulled, solvent extracted (48% crude protein, as-fed basis) (Mukherjee et al., 2016). Soybean meal contains large quantities of essential amino acids (approximately 45% of total amino acids) and in terms of essential amino acid balance, methionine is more limiting than lysine (Hong et al., 2004; Song et al., 2008). Soybean meal is extensively used in feed for swine, poultry, aquaculture, horses, and other ruminants as well as non-ruminants. However, newly weaned animals, as well as older animals, cannot digest soy proteins, and thus soybean meal must be used only in limited amounts in combination with other feed protein sources for younger and old animals (Mukherjee et al., 2016).

14.2   FERMENTATION

Fermentation is one of the oldest forms of food processing and preservation (Ross et al., 2002), and also has the capacity to improve the nutritional and functional properties of the original product (Frias et al., 2008). Fermentation has been widely used to increase the bioavailability of nutrients and reduce the levels of antinutritional factors of soybean (Hotz & Gibson, 2007). Several studies (Frias et al., 2008; Song et al., 2008) have also confirmed the ability of the fermentation process in degrading antinutritive and allergenic compounds of soybean, thereby increasing the possibilities of utilization of various processed products. A wide variety of microorganisms have been used to ferment soybeans for nutritional enhancement. The fermentation process is facilitated by the use of a mold or a bacterium. The fermentation conditions and nutritional quality of the fermented products thus produced can vary depending on the type of microorganism used. *Aspergillus* is the most popular species for fermentation due to its capacity to produce enzymes such as hemicellulases, hydrolases, pectinases, protease, amylase, lipases, and tannases (Mathivanan et al., 2006). In the case of bacterial fermentation, various *Lactobacillus* species and *Bacillus subtilis* are preferred (Han et al., 2001; Amadou et al., 2011).

14.2.1   BACTERIAL FERMENTATION

Traditionally, *Bacillus* spp. has been used to produce fermented soy-based foods (Han et al., 2001). Various types of bacterially fermented soybean foods are consumed in various Asian countries, including China, Indonesia, Japan, Korea, and Vietnam. Some of the most popular ones are:

a) Natto: It is a Japanese fermented food made from soybeans. Boiled rice straw was originally used as the source of fermentation bacteria. After putting steamed soybeans into the semi-sterilized straws they are closed tightly and kept at 40° C for one day. Spore-forming *Bacillus* species, especially *Bacillus subtilis*, naturally present on the surface of rice straws, act as a starter culture for the fermentation. These days, a pure culture of *B. subtilis* is used as a starter (Ruiz-Larrea et al., 1997).

b) Cheonggukjang: It is a fermented soybean paste used in Korean cuisine. It is predominantly fermented with *B. subtilis* for short periods (approximately two days) without salt or other seasonings. It has a paste-like texture, and also includes some whole, uncrushed soybeans (Ruiz-Larrea et al., 1997; Isanga & Zhang, 2008).

c) Doenjang or soybean paste: It is a type of fermented bean paste made entirely of soybean and brine. It is also a by-product of soup soy sauce production. It is sometimes used as a relish. Doenjang can be eaten as a condiment in raw-paste form with vegetables, as a flavored seasoning, or even as a dipping condiment. However, it is more commonly mixed with garlic and sesame oil (Ruiz-Larrea et al., 1997; Isanga & Zhang, 2008).

14.2.2 Fungal Fermentation

Several species of the *Aspergillus* genus have been used to ferment soybeans like *A. oryzae*, *A. usamii*, *A. awamori*, and *A. niger* to name a few. Zamora and Veum (1979) worked on fermentation of dehulled soybeans with *A. oryzae* and *Rhizopus oligoporous*, established the importance of fermentation in nutritional quality improvement, and also created the path for future research work. Some of the most popular fungal fermented soybean foods are:

a) Tempeh: It is a traditional Indonesian soy product. It is made by a natural culturing and controlled fermentation process that binds soybeans into a cake form. A special fungus *Rhizopus oligosporus*, usually marketed under the name tempeh starter, is used for the fermentation. It is especially popular on the island of Java, where it is a staple source of protein. Like tofu, tempeh is made from soybeans, but it is a whole soybean product with different nutritional characteristics and textural qualities. Tempeh's fermentation process and its retention of the whole bean give it a higher content of protein, dietary fiber, and vitamins. It has a firm texture and an earthy flavor, which becomes more pronounced as it ages (Song et al., 2008).

b) *Sufu*: It is pickled tofu that is made by fermenting soybean with *Actinomucor elegans*, *Mucor racemosus*, or *Rhizopus* spp.

c) Soy nugget: For making nuggets, large whole soybeans are soaked and steamed and then mixed with roasted wheat or glutinousrice flour before fermenting with the koji mold *Aspergillus oryzae*. After several days of incubation, the resultant soybean koji is packed in kegs with saltwater and various spices, seeds, and/or root ginger slivers (and occasionally rice wine), then aged for several months. The soy nuggets are then sun-dried (Song et al., 2008).

d) *Miso*: It is possibly the most important fermented food in Japanese cuisine. It is basically fermented soybean paste, but rice and several cereal grains and even other seeds can be combined into an extremely wide variety of miso that differ according to the combination of ingredients. The basic production process is essentially the same for all recipes: Rice, barley, or soybeans are steamed, cooled, and inoculated with the koji mold *Aspergillus oryzae*. When the koji has become established it is added, as a seed culture, to a mixture of washed, cooked, cooled, and crushed soybeans, water, and salt. This is placed in vats and allowed to ferment for 12 to 15 months to allow the proteins in the mixture to be broken down slowly and naturally, forming a paste that is flavorsome and nutritious (Song et al., 2008).

14.2.3 Mixed Culture Fermentation

Sometimes more than one type of microorganisms is used in combination to complete the fermentation. The best example of such a method is the production of soy sauce, which has been used in China for more than 2,500 years, so it is one of the world's oldest condiments (Gao et al., 2011). It is a seasoning agent with a salty taste and a distinctly meaty aroma, although it is made by a complex fermentation process of a combination of soybeans and wheat in water and salt, in which carbohydrates are fermented to alcohol and lactic acid and proteins are broken down to peptides and amino acids. This traditional brewing or fermentation method can take six months or more to produce the finest soy sauce, which should be a dark brown, transparent, salty liquid with a balanced flavor and aroma. The brown color is a result of sugar caramelization during the six- to eight-month maturation process. Three major groups of microorganisms are involved in soy sauce fermentation: *Aspergillus oryzae* and *A. sojaei*are involved in the koji production, and communities of halotolerant bacteria (*Pediococcus halophilus*) and yeasts (*Saccharomyces rouxii*) are responsible for the moromi fermentation. Enzymes produced by all these various microbes hydrolyze the raw materials during the complex soy sauce fermentation process (Gao et al., 2011).

14.3   NUTRITIONAL IMPROVEMENTS DURING FERMENTATION

Fermentation with fungal or bacterial strains results in a lot of changes in the physicochemical as well as nutritional properties of soybeans. Although major changes are similar irrespective of the microorganisms used, certain minor changes vary, and depending on them the final product characterizations also vary (Mukherjee et al., 2019).

Crude protein and crude fat contents in soybeans increase after fungal and bacterial fermentation. However, there are no changes in calcium and phosphorus after fermentation (Chen et al., 2010). The increase in protein and fat contents may be due in part to the decreased carbohydrate content after fermentation. Dry matter disappearance during fermentation can be accounted for by the utilization of nutrients by the microorganism. Carbohydrates in the soybeans and meals may be used as substrates for energy and the synthesis of fatty acids. Soy proteins may be used for building structural proteins and nucleic acids (Mukherjee et al., 2016).

Fermentation affects the characteristics of proteins in soybeans. When the sizes of peptides are grouped as large (60 kDa), medium (20–60 kDa), and small (20 kDa), unfermented soybeans contain greater amounts of large- and medium-sized peptides than fermented soybeans (Cordle, 2004), whereas the amount of small-sized peptide in fermented soybeans is greater (Teng et al., 2012). Likewise, soybean meals contain greater amounts of large- and medium-sized peptides than fermented soybean meals, whereas the amount of small-sized peptides in fermented soybean meals is greater (Li et al., 1990; Teng et al., 2012). Increased amounts of small-sized peptides in these fermented soybeans and fermented soybean meals is due to partial digestion of large-sized peptides in soybeans and soybean meals by proteases secreted by *A. oryzae* during fermentation. It is well documented that *A. oryzae*s ecretes proteases, including aminopeptidases, serine endopeptidases, aspartic endopeptidases, etc. (Teng et al., 2012). Much like fungal fermentation, bacterial fermentation also decreases the protein size, which can be attributed to the enzymes of the organisms and to the fermentation process itself (Hong et al., 2004). *In vitro* trypsin digestibility also increases after fermentation, thereby improving nutritional and functional properties compared to the unfermented soybeans (Frias et al., 2008).

Since crude protein increases after fermentation, the amino acid content of fermented soybean also increases. However, fungal fermentation of soybeans does not affect the contents of most essential amino acids including lysine, tryptophan, methionine, arginine, histidine, and phenylalanine, whereas the contents of glycine, aspartic acid, threonine, valine, and glutamic acid increase after fermentation. Leucine, isoleucine, and serine in soybeans also increase after fungal fermentation. The total amount of amino acids however does not change significantly after fermentation (Teng et al., 2012). This may indicate that the increase in protein content is due to an increase in selected amino acids rather than being due to an increase in all amino acids, which may be attributable to the preferential utilization of some selected amino acids by *A. oryzae*. Fermentation with lactic acid bacteria like *L. plantarum* results in protein hydrolysis and increased liberation of free amino acids, thus the resulting product has a significantly higher total free amino acid content as compared to raw soybean. However, histidine, threonine, methionine andphenylalanine contents do not change whereas leucine, isoleucine, aspartic acid, and proline increase after fermentation (Hong et al., 2004). When soybean is fermented with *B. subtilis*, the concentration of small-sized proteins increases along with the contents of arginine, serine, threonine, aspartic acid, alanine, and glycine, but proline content decreases (Teng et al., 2012). Different proteinase profiles, secretion abilities, and fermentation temperatures of the two organisms are possibly the reasons behind the differences in the amino acid profiles.

Trypsin inhibitor (TI) content in soybeans and soybean meals is reduced by 91% after fungal fermentation. Similar to fungal fermentation, bacterial strains also degrade various antinutritional factors of soybeans including TI (Teng et al., 2012).

Fermentation with bacterial strains results in higher antioxidant activity, too. The increased concentrations of certain amino acids such as histidine, serine, valine, and lysine after fermentation are thought to have a relation with the increased antioxidant property (Amadou et al., 2011).

TABLE 14.1

Comparison of Antioxidative Properties of Unfermented and Fermented (Kinema) Soybeans[1]

Antioxidative Properties	Unfermented Soybean	Fermented Soybean (Kinema)
Total phenolic content	3.3 mg/g (dw)	8.05 mg/g (dw)
Fe^{3+} reducing power	3.8 mgAAE/g (dw)	9.3 mgAAE/g (dw)
Fe^{2+} chelating activity	22%	64%
Prevention of linoleic acid oxidation	36%	44%

[1]Moktan et al. (2008).

Concentrations of phenolic compounds also increase after fermentation, thereby increasing both antioxidant and metal chelating activity (Moktan et al., 2008). Table 14.1 lists comparisons of antioxidative properties of fermented and unfermented soybeans. Fermentation using *B. subtilis* results in higher crude protein and lower TI content (Teng et al., 2012). These findings support the fact that fermentation of soybean is indeed a complex processsthat is strongly controlled by the enzymes involved and thus the type of organism selected for fermentation is one of the major factors in determining the levels of various nutritional components in the final products.

Several research groups (Frias et al., 2008; Song et al., 2008; Teng et al., 2012) have tried to compare the nutritional quality of products produced by both fungal and bacterial fermentation. A fermentation study (Teng et al., 2012) involving *A. oryzae* and *B. subtilis* showed a consistent increase in crude protein during fermentation which occurs due to microbial growth. However, the proportion of soluble protein increased by19.4% and 63.11%, respectively, which may occur due to stronger hydrolysis by *B. subtilis* fermentation. The percentages of degradation of TI for both fermentations were of comparable values, increase in antioxidant activity was also of equivalent levels. But in the case of *in vitro* digestibility and degradation of antigenic proteins, *B. subtilis* was found to be more efficient.

Studies done with yeast (*Saccharomyces cereviseae*) and bacterial strains (*L. plantarum* and *Bifidobacterium lactis*) by Song et al. (2008) indicated that both fermentations significantly reduce immunoreactivity, which is a major benefit in terms of acceptability and health issues. Both fermentations resulted in the breakdown of larger antigenic proteins, thereby considerably increasing the amount of smaller-sized peptides (15 kD) and reducing the immunoreactivity. These findings are in partial agreement with another study (Frias et al., 2008) which found that fermentation with *A. oryzae* or *R. oryzae* resulted in a much lower reduction in immunoreactivity compared to fermentation with *B. subtilis* or *L. plantarum*, thus making the bacterial strains a better choice for reducing the immunoreactivity. This may be attributed to the fact that slower-growing fungi result in less viable microorganisms, generating lower epitomes of alteration and thus higher immunoreactivity compared to the bacterial fermented product.

While comparing the amino acid contents of products from yeast and bacterial fermentation (Song et al., 2008), it is found that among the essential amino acids, isoleucine and methionine amounts did not change significantly after fermentation (both bacterial and yeast fermentation), whereas during bacterial fermentation cysteine decreases. Taking into consideration the original concentrations of essential amino acids in soybeans, fermentation with *S. cerevisae* can be recommended due to a sharp increase in cysteine, an essential amino acid that is present in lower concentrations in soybeans (Song et al., 2008). In the case of fermentation with fungal and bacterial strains, the changes in the amino acid profile are comparable with an increase in the majority of the essential and non-essential amino acid contents (Frias et al., 2008; Teng et al., 2012). The variations in the amino acid profiles may arise due to the differences in proteinase profiles and secretion abilities of the organisms. The dissimilarity in the final products can be attributed to the different metabolic activities and enzymes involved with the two species utilized for fermentation, although in both typesof fermentation, the antinutrients decrease and nutritional quality improves.

These findings also indicate that the choice of microorganism for fermentation should be done according to the nutritional requirement of the final fermented product.

14.4   HEALTH BENEFITS OF FERMENTED SOYBEAN

14.4.1   Bioactive Compounds

Research in the past few decades has shown that soybean seeds are highly rich in proteins, peptides, fatty acids, isoflavones, and other secondary metabolites with potential bioactivities (Messina et al., 1994; Gibbs et al., 2004). These compounds exhibit antioxidant and anticarcinogenic properties along with potential benefits against cardiovascular disease, osteoporosis, and menopausal symptoms, among others (Gibbs et al., 2004). Table 14.2 lists the most important bioactive compounds found in fermented soybeans as well as their beneficial and/or adverse effects on human health upon consumption. Effects of fermentation on these compounds are varied.

Kunitz-Trypsin inhibitor (KTI), Bowman-Birk inhibitor (BBI), and lunasin are three major and best-characterized bioactive proteins/peptides of soybean seeds. KTI and BBI are serine protease inhibitors with molecular weights of 20.1 and 8 kDa respectively (Wu et al., 1998). Both of these proteins have been shown to exhibit anticarcinogenic and/or anti-invasive/metastatic activities (Gibbs et al., 2004). However, fermentation results in almost complete elimination of these inhibitors (Teng et al., 2012).

Lunasin is a 2S-albumin derived peptide and is not a protease inhibitor. It consists of 43 amino acids with a molecular weight of 5.5 kDa and exhibits several beneficial health effects, similar to BBI and KTI (Wu et al., 1998). There is compelling evidence gathered over the years which

TABLE 14.2

List of Various Bioactive Compounds Present in Fermented Soybean and Their Potential Health Benefits as Well as Possible Adverse Effects upon Consumption

Bioactive Compounds	Health Benefits	Adverse Effects	References
Isoflavones	Estrogen-like activities, anticarcinogenic activities, cardiovascular disease prevention, antiatherogenic activities, hypolipidemic agent, antidiabetic activities, antimicrobial activities	Production of thymic and immune abnormalities, potential estrogenic or immunotoxic effects, may induce apoptosis or cytotoxicity at high dose	Ruiz-Larrea et al., 1997; Isanga & Zhang, 2008
Saponins	Antiviral activity against HIV, antioxidative activity, anti-tumor-promotion and growth inhibition of tumors, anticarcinogenic activity, cholesterol-lowering properties, anti-osteoporosis activity, anti-inflammatory activity	May induceapoptosis or necrosis or cytotoxicity at high dose	Konoshima et al., 1992; Isanga & Zhang, 2008
Phytosterols	Anticancer effects, prostatic hyperplasia-lowering effects, stimulation of plasminogen activating factor, reduction in serum cholesterol		Moghadasian & Frohlich, 1999; Isanga & Zhang, 2008
Peptides (lunasin)	Antioxidative, anti-inflammatory, anticarcinogenic, and cholesterol-lowering properties		Lule et al., 2005

endorses the antioxidative, anti-inflammatory, anticancerous, and cholesterol-lowering properties of this particular peptide (Arjmandi et al., 1998). Fermentation, especially with LAB, is a potential method for the enhancement of this bioactive peptide (Rizzello et al., 2012).

Soybean isoflavones are present as glucosides (daidzin, genistin, and glycitin), malonylglucosides (malonyldadzin, malonylgenistin, and malonylglycitin), acetylglucosides (acetyldaidzin, acetylgenistin, and acetylglycitin), and aglycones (daidzein, genistein, and glycitein) (Barnes et al., 1994). The aglycones are structurally more favorable for bioabsorption and thus fermented products which contain the aglycones have enhanced bioavailability and antioxidant properties (Ruiz-Larrea et al., 1997). The core structure of the isoflavones resembles that of 17-β-estradiol and thus these are able to bind with the estrogen receptors, exhibiting numerous beneficial health effects. A growing body of evidence suggests isoflavones can inhibit hormones associated with breast and prostate cancers because of their structure which is similar to estrogen. Moreover, isoflavones were also shown to possess antioxidant and anti-inflammatory properties (Barnes et al., 1994; Fico et al., 2000).

Soyasaponins, on the other hand, are triterpenoid glycosides, containing oleanane-type aglyconeand polysaccharide chains (Ranich et al., 2001). In soybean, saponins are present in two major forms as group A and group B soyasaponins. Together, these soyasaponins possess anti-inflammatory, antioxidative, antiglycemic, anticarcinogenic, hepatoprotective, and cholesterol-lowering properties (Ranich et al., 2001). Soyasapogenols are the aglycone forms of soyasaponins and are formed during the fermentation of soy foods. It was reported that these soyasapogenols are better absorbed in the blood as compared to soyasaponins and therefore, these are supposed to be more potent than their corresponding soyasaponins (Ranich et al., 2001).

In addition to these, fermented soybean contains several other bioactive compounds such as carotenoids, tocopherol, phenolic acids, and flavonoids with potential health benefits. Therefore, soybean seeds can be included in the daily diet to improve overall health with these non-toxic and inexpensive bioactive compounds.

14.4.2 HEALTH BENEFITS OF TRADITIONAL FERMENTED SOY FOODS

Fermented foods offer a natural source of bioavailable saprophytic organisms, reflecting the external ecologic terrain as well as the internal terrain of the organism (Figure 14.2). Thus, fermented

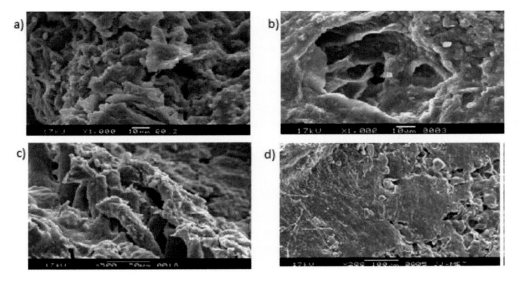

FIGURE 14.2 SEM micrographs showing changes in ultrastructures and absence/presence of microorganisms before and after fermentation. Source: Mukherjee et al. (2018).

foods are best chosen that are produced locally with soil-based organisms (Han et al., 2001). The other advantage is the natural presence of prebiotics. Microorganisms involved in fermented foods are mostly lactic acid bacteria (LAB), bacilli, other Gram-positive and a few Gram-negative bacteria, yeasts, and filamentous molds. Ethnic fermented foods have been prepared and consumed for centuries for nutritional supplements, stability, taste, aroma, and flavor, and some have been used for therapeutic purposes. The microflorae of fermented foods impart health-promoting benefits to the consumers (Han et al., 2001; Fehily, 2004).

Consumption of natto has been associated with beneficial stimulation of the immune system (Ruiz-Larrea et al., 1997). In modern times, endospore formulations have been used to treat diarrhea, allergies, and arthritis in humans and to enhance growth in domestic animals and aquaculture. Research indicates significant daily human consumption of soybean isoflavones can increase bone density, thereby lowering the risk of osteoporosis; help to prevent prostate enlargement; and reduce the risk of certain types of cancers (e.g. breast cancer, colon cancer, lung cancer, prostate cancer, and uterine cancer); they also exhibit antioxidant properties (Ruiz-Larrea et al., 1997; Isanga & Zhang, 2008). Because soyfoods are a good source of dietary potassium, an increase in soyfood consumption would help reduce potassium deficiencies. Differences in the level of consumption of traditionally prepared soy foods (miso, tofu, tempeh, natto) are believed to contribute to the large difference in prostate cancer incidence and mortality betweenAsian and Western males. A large-scale epidemiologic study by Hebert et al. (1998) found that soy-derived products offered highly significant protection against prostate cancer. Animal studies reveal that soy isoflavones, particularly genistein, inhibit prostate cancer growth in cell cultures (Ruiz-Larrea et al., 1997). In rat models, genistein has been found to offer significant chemopreventive activity against advanced prostate cancer. Possible mechanisms include estrogenic properties and inhibition of 5-α reductase. Soy foods contain protease inhibitors, saponins, and phytates, all of which have putative anticarcinogenic effects (Isanga & Zhang, 2008).

Genistein and biochanin A, but not daidzein, isolated from soyfood, increased the tumor latency period, decreased the promotion of tumors, and/or decreased tumor multiplicity in rodents with chemically induced mammary carcinogenesis. Chemoprotection was observed when exposure to genistein and soy products occurred prepubertally but was most effective when exposure occurred both prepubertally and during adulthood (Isanga & Zhang, 2008). Data from investigations into isoflavones support their usefulness in a chemoprevention program, but more research is needed to determine effective dosage in supplement form. Most studies showed that a diet containing 5–50% fermented soypaste, in the form of cheonggukjang, doenjang, and meju, leads to a reduction in body weight gain or adiposity in high-fat-induced obese or diabetic rodents (Ruiz-Larrea et al., 1997; Isanga & Zhang, 2008).

A number of cardioprotective benefits have been attributed to dietary isoflavones including a reduction in LDL cholesterol, inhibition of pro-inflammatory cytokines and cell adhesion proteins, a potential reduction in the susceptibility of the LDL particle to oxidation, inhibition of platelet aggregation, and an improvement in vascular reactivity (Isanga & Zhang, 2008). Therefore, it is suggested that improvements in cholesterol levels may be helpful for people with hypercholesterolemia.

14.5 SOYBEAN MEAL FERMENTATION

Soybean meal, the by-product of soybean processing, is also being subjected to fermentation in order to widen its applicability. Soybean meal is predominantly used in animal feed and gradually it is being replaced with fermented meals. Soybean meal is most commonly fermented with *A. oryzae* and *L. plantarum* although *Bacillus* species are also used frequently (Mukherjee et al., 2016). The fermentation process can be a solid-state fermentation or a submerged fermentation and the meal can be subjected to both processes depending on its initial state, i.e. crude without any alteration of moisture content or texture or dried and ground flour to make fine particles thatare readily

TABLE 14.3

Comparison of Nutritional Quality of Unfermented, Fungi-Fermented, and Bacteria-Fermented Soybean Meal[1]

Nutritional Components Affected by Fermentation	Unfermented Soybean Meal	Types of Organism Used for Fermentation	
		Fermented Soybean Meal	
		Fungi	Bacteria
Crude protein content (%)	34.5	37.4	37.5
Soluble protein content (%)	20	24	33
In vitro digestibility (pepsin)	60.5	67.4	76
Antioxidant activity (%)	8	27	38
Small-sized peptides (< 15 kD) (%)	5	35	63

[1]Hong et al. (2004); Teng et al. (2012); Mukherjee et al. (2016).

dissolvable in water. The former approach, being an alternative to the latter approach, has numerous advantages including productivity and cost due to the possibility of using agro-industrial residues and/or by-products as nutrient sources, as well as support for microorganism development (Rigo et al., 2010).

Much like soybeans, fermentation brings about various physicochemical changes in soybean meal too (Table 14.3). And the benefits of feeding fermented meal to animals are well documented (Mukherjee et al., 2016). Fermented meal thus has emerged as an appropriate alternative to raw meal for use as animal feed. Several beneficial effects, including increased average daily gain, improved growth performance, better protein digestibility, decreased immunological reactivity, and undesirable morphological changes are observed when fermented meal is fed to non-ruminants.

14.6 FUTURE SCOPE

Conversion of low-cost by-products into value-added products has always been the main focus of the fermentation industry, and currently, its scope is being expanded to include the use of probiotics to ferment various by-products apart from dairy products. Soybeans and soybean meals are promising products that have not yet reached their full potential use and should be explored further to widen their range of applications. Fermenting SBM with well-known probiotics has not only the possibility of improving the nutritional values of SBM, but can also contribute to the improvement of intestinal health, thereby converting a by-product into a value-added product (Mukherjee et al., 2019). The purpose to develop optimized fermentation processes needs a suitable mathematical model that has the ability to mimic the complex nonlinearity of mixed-culture fermentations without compromising its prediction capabilities. In biotechnological applications, the prediction of a suitable model becomes difficult due to the complex nature of biocatalysts and microorganisms and also due to the lack of reliable information about their interactions with the environment as well as physical phenomena like nutrient solubility, oxygen transfer, and water availability. Artificial Neural Network (ANN) has the advantage that it can mimic almost every form of nonlinearity and thus can make accurate forecasts even when the process behavior is complex and nonlinear and data are unstructured, making it a useful tool for complex biotechnological processes like fermentation (Baughman & Liu, 1995). Genetic algorithm (GA), an artificial intelligence-based stochastic nonlinear optimization method, can be used to optimize thein put space of a trained ANN model (Mukherjee et al., 2019).

14.7 SUMMARY

Soybean is considered the treasure trove of nature as it not only contains a high concentration of proteins but also has several other bioactive compounds as well as a balanced amino acid profile. These bioactive compounds impart several health benefits to consumers and thus soybeans are very popular among many ethnic groups as a staple food. Fermented soybean is the preferred method of consumption, as fermentation not only modifies texture, flavor, and taste, but also increases the bioavailabilities of its various bioactive compounds by increasing their concentrations as well as by decreasing the antinutritive factors present in it which interfere with bioabsorption. Several traditional foods of Asian origin contain fermented soybeans like natto, soy sauce, miso, soy nuggets, etc. and they have been well documented to have beneficial effects against cancers, cardiovascular diseases, renal issues, tumors, osteoporosis, and other chronic diseases. All the products that are produced from soybeans like oil, textured vegetable proteins, soy milk, etc., as well as by-products like soybean meal, have beneficial effects on consumers and are thus highly sought after. Fermentation of soybean meal is an emerging trend and newer applications of the fermented product are emerging every day.

REFERENCES

Amadou, I., Le, G. W., Shi, Y. H., Jin, S. (2011). Reducing, radical scavenging, and chelation properties of fermented soy protein meal hydrolysate by *Lactobacillus plantarum* Lp6. *International Journal of Food Properties*, 14(3): 654–665.

Arjmandi, B. H., Getlinger, M. J., Goyal, N. V., Alekel, L., Hasler, C. M., Juma, S., Drum, M. L., Hollis, B. W., Kukreja, S. C. (1998). Role of soy protein with normal or reduced isoflavone content in reversing bone loss induced by ovarian hormone deficiency in rats. *American Journal of Clinical Nutrition*, 68(6) Supplement: 1358S–1363S.

Barnes, S., Kirk, M., Coward, L. (1994). Isoflavones and their conjugates in soy foods: Extraction conditions and analysis by HPLC-mass spectrometry. *Journal of Agricultural and Food Chemistry*, 42(11): 2466–2474.

Baughman, D. R., Liu, Y. A. (1995). *Neural Networks in Bioprocessing and Chemical Engineering*. Cambridge, MA: Academic Press.

Chen, C. C., Shih, Y. C., Chiou, P. W. S., Yu, B. (2010). Evaluating nutritional quality of single stage- and two stage-fermented soybean meal. *Asian-Australasian Journal of Animal Sciences*, 23(5): 598–606.

Cordle, C. T. (2004). Soy protein allergy: Incidence and relative severity. *Journal of Nutrition*, 134(5): 1213S–1219S.

Fehily, A. M. (2004). NUTRITION | Soy-based foods. In ColinWrigley (Ed.), *Encyclopedia of Grain Science* (pp. 348–354). Amsterdam: Elsevier.

Fico, G., Braca, A., Bilia, A. R., Tome, F., Morelli, I. (2000). Flavonol glycosides from the flowers of *Aconitum paniculatum*. *Journal of Natural Products*, 63(11): 1563–1565.

Frias, J., Song, Y. S., Martínez-Villaluenga, C., De Mejia, E. G., Vidal-Valverde, C. (2008). Immunoreactivity and amino acid content of fermented soybean products. *Journal of Agricultural and Food Chemistry*, 56(1): 99–105.

Gao, X., Cui, C., Ren, J., Zhao, H., Zhao, Q., Zhao, M. (2011). Changes in the chemical composition of traditional Chinese-type soy sauce at different stages of manufacture and its relation to taste. *International Journal of Food Science and Technology*, 46(2): 243–249.

Gibbs, B. F., Zougman, A., Masse, R., Mulligan, C. (2004). Production and characterization of bioactive peptides from soy hydrolysate and soy-fermented food. *Food Research International*, 37(2): 123–131.

Gnanasambandam, R., Proctor, A. (1999). Preparation of soy hull pectin. *Food Chemistry*, 65(4): 461–467.

Hammond, E. G., Johnson, L. A., Murphy, P. A. (2004). SOYBEAN | Grading and marketing. In C.Wrigley (Ed.), *Encyclopedia of Grain Science* (pp. 155–159). Amsterdam: Elsevier.

Han, B. Z., Rombouts, F. M., Nout, M. J. R. (2001). A Chinese fermented soybean food. *International Journal of Food Microbiology*, 65(1–2): 1–10.

Hebert, J. R., Hurley, T. G., Olendzki, B. C., Teas, J., Ma, Y., Hamp, J. S. (1998). Nutritional and socioeconomic factors in relation to prostate cancer mortality: A cross-national study. *Journal of the National Cancer Institute*, 90(21): 1637–1647.

Ho, S. C., Woo, J. L. F., Leung, S. S. F., Sham, A. L. K., Lam, T. H., Janus, E. D. (2000). Intake of soy products is associated with better plasma lipid profiles in the Hong Kong Chinese population. *Journal of Nutrition*, 130(10): 2590–2593.

Hong, K. J., Lee, C. H., Kim, S. W. (2004). *Aspergillus oryzae* 3.042GB-107 fermentation improves nutritional quality offood soybeans and feed soybean meals. *Journal of Medicinal Food*, 7(4): 430–434.

Hotz, C., Gibson, R. S. (2007). Traditional food-processing and preparation practices to enhancing the bioavailability of micronutrients in plant-based diets. *Journal of Nutrition*, 137(4): 1097–1100.

Isanga, J., Zhang, G. N. (2008). Soybean bioactive components and their implications to health—A review. *Food Reviews International*, 24(2): 252–276.

Konoshima, T., Kokumai, M., Kozuka, M., Tokuda, H., Nishino, H., Iwashima, A. (1992). Anti-tumor-promoting activities of afromosin and soya saponin I isolated from *WistariaBrachybotrys*. *Journal of Natural Products*, 55(12): 1776–1778.

Li, D. F., Nelssen, J. L., Reddy, P. G., Blecha, F., Hancock, J. D., Allee, G., Goodband, R. D., Klemm, R. D. (1990). Transient hypersensitivity to soybean meal in the early-weaned pig. *Journal of Animal Sciences*, 68(6): 1790–1799.

Lucas, E. A., Khalil, D. A., Daggy, B. P., Arjmandi, B. H. (2001). Ethanol-extracted soy protein isolate does not modulate serum cholesterol in golden Syrian hamsters: A model of postmenopausal hypercholesterolemia. *Journal of Nutrition*, 131(2): 211–214.

Lule, V. K., Garg, S., Pophaly, S. D. (2005). Potential health benefits of lunasin A multifaceted soy-derived bioactive peptide. *Journal of Food Science*, 80: R485–494.

Mathivanan, R., Selvaraj, P., Nanjappan, K. (2006). Feeding of fermented soybean meal on broiler performance. *International Journal of Poultry Science*, 5: 868–872.

Messina, M. J., Persky, V., Setchell, D. R., Barnes, S. (1994). Soy intake and cancer risk: A review of the *in vitro* and *in vivo* data. *Nutrition and Cancer*, 21(2): 113–131.

Moghadasian, M. H., Frohlich, J. J. (1999). Effects of dietary phytosterols on cholesterol metabolism and atherosclerosis clinical and experimental evidence. *American Journal of Medicine*, 107(6): 588–594.

Moktan, B., Saha, J., Sarkar, P. K. (2008). Antioxidant activities of soybean as affected by *Bacillus*-fermentation to kinema. *Food Research International*, 41(6): 586–593.

Mukherjee, R., Chakraborty, R., Dutta, A. (2016). Role of fermentation in improving nutritional quality of soybean meal – A review. *Asian Australasian Journal of Animal Sciences*, 29(11): 1523–1529.

Mukherjee, R., Chakraborty, R., Dutta, A. (2018). Evaluation of the effect of fermentation on protein and phenolic profiles of soybean meal. In *Biospectrum 2018, 2nd International Conference on Biotechnology and Biological Sciences*, Kolkata, India.

Mukherjee, R., Chakraborty, R., Dutta, A. (2019). Comparison of optimization approaches (RSM and ANN-GA) for a novel mixed culture approach in soybean meal fermentation. *Journal of Food Process Engineering*, 42(5): e13124.

Paulsen, P. V. (2009). Isolated soy protein usage in beverages. In P.Paquin (Ed.), *Woodhead Publishing Series in Food Science, Technology and Nutrition, Functional and Speciality Beverage Technology*, 13 (pp. 318–345). Sawston, UK: Woodhead Publishing.

Penfield, M. P., Campbell, A. M. (1990). CHAPTER 19 - YEAST BREADS. In M. P.Penfield, A. M.Campbell (Eds.), *Food Science and Technology, Experimental Food Science* (3rd edition) (pp. 418–441). Cambridge, MA: Academic Press.

Ranich, T., Bhathena, S. J., Velasquez, M. T. (2001). Protective effects of dietary phytoestrogens in chronic renal disease. *Journal of Renal Nutrition*, 11(4): 183–193.

Rigo, E., Ninow, J. L., Di Luccio, M., Oliveira, J. V., Polloni, A. E., Remonatto, D., Arbter, F., Vardanega, R., de Oliveira, D., Treichel, H. (2010). Lipase production by solid fermentation of soybean meal with different supplemens. *LWT - Food Science and Technology*, 43(7): 1132–1137.

Rizzello, C. G., Nionelli, L., Coda, R., Gobbetti, M. (2012). Synthesis of the cancer preventive peptide lunasin by lactic acid bacteria during sourdough fermentation. *Nutrition and Cancer*, 64(1): 111–120.

Ross, P. R., Morgan, S., Hill, C. (2002). Preservation and fermentation: Past, present and future. *International Journal of Food Microbiology*, 79(1–2): 3–16.

Ruiz-Larrea, M. B., Mohan, A. R., Paganga, G., Miller, N. J., Bolwell, G. P., Rice-Evans, C. A. (1997). Antioxidant activity of phytoestrogenic isoflavones. *Free Radical Research*, 26(1): 63–70.

Song, Y. S., Frias, J., Martinez-Villaluenga, C., Vidal-Valdeverde, C., de Mejia, E. G. (2008). Immunoreactivity reduction of soybean meal by fermentation, effect on amino acid composition and antigenicity of commercial soy products. *Food Chemistry*, 108(2): 571–581.

Teng, D., Gao, M., Yang, Y., Liu, B., Tian, Z., Wang, J. (2012). Bio-modification of soybean meal with *Bacillus subtilis* or *Aspergillus oryzae*. *Biocatalysis and Agricultural Biotechnology*, 1(1): 32–38.

Woodfield, H. K., Harwood, J. L. (2017). Oilseed crops: Linseed, rapeseed, soybean, and sunflower. In B.Thomas, B. G.Murray, D. J.Murphy (Eds.), *Encyclopedia of Applied Plant Sciences* (2nd edition) (pp. 34–38). Cambridge, MA: Academic Press.

Wu, A. H., Ziegler, R. G., Nomura, A. M. Y., West, D. W., Kolonel, L. N., Horn-Ross, P. L., Hoover, R. N., Pike, M. C. (1998). Soy intake and risk of breast cancer in Asians and Asian Americans. *American Journal of Clinical Nutrition*, 68(6) Supplement: 1437S–1443S.

Zamora, R. G., Veum, T. L. (1979). Whole soybeans fermented with *Aspergillus oryzae* and *Rhizopus oligosporus* for growing pigs. *Journal of Animal Sciences*, 48(1): 63–68.

15 Soybean Processing By-Products and Potential Health Benefits

Philip Davy and Quan V. Vuong

CONTENTS

15.1 INTRODUCTION

Soybeans are considered a globally important commodity, owing to their vast utilization around the world, and are prized for their nutritional and economic benefits. Global estimates have indicated that 100 million hectares of soybeans are cultivated every year (Iqbal et al., 2019) and it is a predominant plant crop, with over 50% of land used for growing legumes devoted to soybean cultivation (Sulieman et al., 2015). Soybeans are used in industrial products such as lubricants, biodiesel, and hydraulic fluids. Soybeans have high nutritional content and typically contain 40% proteins, 35% carbohydrates, and 20% oil and are used as both animal fodder and a food source for human consumption (Easwar Rao & Viswanatha Chaitanya, 2020). Most soybeans intended for human consumption are used to produce soybean oil and soy milk.

DOI: 10.1201/9781003030294-15

The largest producers of soybean oil around the world are China, the USA, Brazil, and Argentina (Alvim & Fochezatto, 2020), with China reporting $145 billion of revenue in 2019 (Ibisworld, 2019). Soy milk is the most consumed alternative milk in the world. China is also a leading producer in the global market, with approximately $12 billion in gross revenue. It has been reported that for every one ton of soybean oil produced, this results in 4.5 tons of soybean by-products, whereas for every one ton of dried soybeans used in the production of soy milk, this results in 1.1 tons of soybean by-products to be produced, accounting for a large volume of waste material (Khare et al., 1995).

These by-products are rich sources of proteins, fiber, fats, simple carbohydrates (CHO), potassium, sodium, calcium, iron, magnesium, trace vitamins and minerals, isoflavones, and soyasaponins. These by-products are a potential source of functional ingredients for food and pharmaceutical products. However, the nutritional composition for both oil and milk waste was found to be influenced by processing conditions, post-processing treatments, growing conditions, and cultivars (O'Toole, 1999; Radočaj & Dimić, 2013; Van der Riet et al., 1989; Voss et al., 2018). This chapter describes sources of soybean by-products, outlines their nutrient and phytochemical content, discusses their potential health benefits and factors affecting their potential health benefits, and microbial treatment for improving their health benefits.

15.2 SOURCES OF SOYBEAN PROCESSING BY-PRODUCTS

15.2.1 SOYBEAN PRESS CAKE AND SOYBEAN MEAL DERIVED FROM SOYBEAN OIL PRODUCTION

Soybean oil can be extracted either by mechanical means or by organic solvents. Hexane is the most common solvent being used in industrial settings due to cheaper production costs and high extraction yields, with heptane, butane, and ethanol used to a lesser degree (Cheng & Rosentrater, 2017a). Solvent extraction accounts for the majority of oil production with over 99% of endogenous oil being recovered (Bargale et al., 1999), compared to 70% yield from a mechanical, expeller/extruder process (Nelson et al., 1987). As the mechanical processing does not require the use of chemical reagents, there is less environmental and safety issues associated with this method, which is a major concern for the solvent methods (Cheng et al., 2018).

Further processing stages are required to remove excess solvent from both the liquid products and solid waste materials resulting in higher production costs, however, this is countered by an increase in oil recovery. When comparing the associated cost of each method it has been shown that on set-up and maintenance of production, that an output of over 12 million kilograms of soybean oil annually, profitability can be achieved for the mechanical extrusion/expeller processing. While hexane solvent extraction requires an annual minimum output of 173 million kg of oil. To produce crude oil, processing typically involves, cleaning, cracking, dehulling, flaking, extraction, solvent recovery, and separation of solid waste. Crude oil can then be treated further to produce refined oils for use as edible oil and margarine or for industrial applications. These processes include degumming, neutralization, bleaching, and deodorization (Cheng & Rosentrater, 2017b).

Degumming is a process to remove phosphatides, which are also known as gums and lecithin. Water, steam, food-grade acids, and caustic substances can be added to the crude oil to aid in the removal of these phosphatides. This process is undertaken to stabilize the crude oil for storage and transportation and to prepare it for other stages of refining. The gums and lecithin are considered as manufacturing by-products. Lecithin is a valuable commodity and is used in many food lines as an emulsifier. However, global production of lecithin currently outweighs its usage, resulting in an excess, which is disposed of with the other soybean oil processing wastes. Acids and caustic substances will affect the usability of lecithin extracted under these conditions (Erickson, 1995).

Neutralization is a process where free fatty acids are saponified by the addition of a caustic solution, with sodium hydroxide being the predominant reagent. This soapstock is removed by washing the oil with hot water. Removal of the free fatty acids is undertaken to improve taste and odor as well as increase stability. Neutralization can be conducted in conjunction with degumming, however, this will convert the lecithin to soapstock and it will be removed as waste during this process (Kuleasan & Tekin, 2008). Other methods that eliminate free fatty acids include distillation, esterification, and solvent extraction (Bhosle & Subramanian, 2005).

Bleaching is a process to remove peroxides from the oil and is a major contributor to stability and flavor (Tai & Lin, 2007; Wiedermann, 1981). The other purpose of this stage is to remove traces of minerals, gums, phosphatides, and fatty acids that may be present and to adjust the color of the final oil. The color of the oil can be affected by the processing stages or can be from the soybean itself (Ghorbanpour, 2018). The oil is filtered through acidified activated clays, carbon, or silicas. The most common industrial method uses bentonite clay as the bleaching medium (Liu et al., 2008).

Deodorization is the final stage in the production of edible oil and is the removal of unwanted odorous compounds and any free fatty acids that remain. This practice results in a neutral flavor and aroma which can be classified as acceptable. This is accomplished by injecting steam into the oil to break down and carry away the volatile compounds responsible for odor (Zehnder, 1995). This process also removes tocopherols that are present in high quantities in crude soybean oil. Tocopherols are strong antioxidants and will help to prevent oxidative damage to unsaturated fatty acids (Shimada et al., 2000).

Both the mechanical and solvent extraction processes that produce crude oil create large volumes of solid waste materials. These are the remaining insoluble fractions of the soybean cotyledon and are often referred to by different names. The mechanical processing waste is commonly referred to as soybean press cake (SPC), while the solvent by-product is known as soybean meal (SBM), however, it is not uncommon to find these terms being used interchangeably. Due to the high proportion of proteins that remain, these by-products are highly prized as a feed stock for many commercial animal producers. Figures 15.1 and 15.2 show a basic production schematic of the process for both methods, where these by-products are removed in the processing lines and the end products of the process.

Refining processes also yield many by-products, and include fatty acids, phosphatides, soapstock, tocopherols, and acidified clays from the decoloration stage. Phosphatides and soapstock are

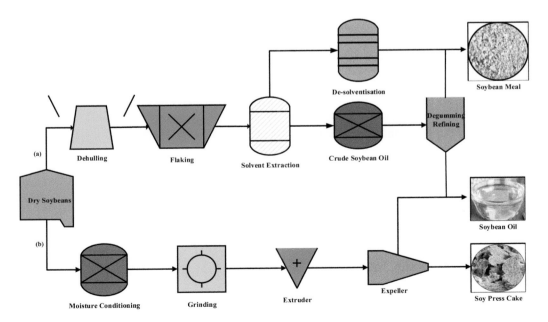

FIGURE 15.1 Production diagram of soybean oil from (A) solvent extraction technique and (B) extrusion technique.

FIGURE 15.2   Soybean oil end products. Soybean meal (left); soybean oil (center); soybean press cake (right).

typically added to the SBM. Tocopherols that are removed during the deodorization stage can be reused to produce vitamin E for use in food products and supplements (Shimada et al., 2000).

15.2.2   SOYBEAN BY-PRODUCT (OKARA) DERIVED FROM SOY MILK PRODUCTION

The manufacturing of soy milk typically starts with soaking of the dried beans for up to 24 hrs, before grinding with water at a ratio of between 8 and 10:1 (v/w) water to bean. Filtration or mechanical separators are used to remove the insoluble residue, followed by thermal processing to extend shelf-life and to denature antinutrients such as lipoxygenase and trypsin inhibitors, then packaging for sale (Liener, 1994). Figures 15.3 and 15.4 show a basic production schematic and images of soy milk and by-products.

The flavor of soy milk and its subsequent residue is determined by the variety of bean, heating, filtration, and grinding processes employed. In the Japanese traditional method, the soaked beans and water are typically heated before being ground and separated, while in the Chinese method, only the soy milk is heated after completion of the grinding and filtration processes. The soy milk by-product (SMB) is then removed at this stage. These traditional production methods, which only differ when heat is applied, generally result in a stronger flavor, which is typically described as "beany" and is generally not favored by most consumers (Vong & Liu, 2016). This flavor has been attributed to the action of lipoxygenase, which catalyzes the degradation of polyunsaturated fatty acids (Liu, 1997).

Modern processes have been developed to improve yield while reducing the effect of lipoxygenase and other antinutritive compounds to produce a more desirable flavor. The insoluble fraction of the soybean cotyledon, or SMB, generated from this fluid milk production is typically sold as cattle fodder or dumped in landfills (Rinaldi et al., 2000) and is also commonly known by its Japanese name of okara.

Due to the high water and nutrient contents that remain, lipid oxidation and growth of microorganisms result in rapid degradation of SMB, which can lead to environmental issues when it is being disposed of, and can hamper further utilization of these by-products. To mitigate these issues, extra processing needs to be applied which generally relies on the removal of water using heat treatments or freeze-drying, however, these processing techniques are often expensive and require specialized equipment, thus adding more financial cost to the manufacturing process (Vong & Liu, 2016).

15.3   NUTRIENT AND PHYTOCHEMICAL CONTENT OF SOYBEAN PROCESSING BY-PRODUCTS

15.3.1   NUTRIENT CONTENT OF SOYBEAN PRESS CAKE AND SOYBEAN MEAL

The approximate nutritional composition (Table 15.1) of soybean oil wastes, from the solvent extraction and extruder/expeller, has been shown to differ, with major differences in composition due to the variation in manufacturing processes which has been previously discussed. The total carbohydrate

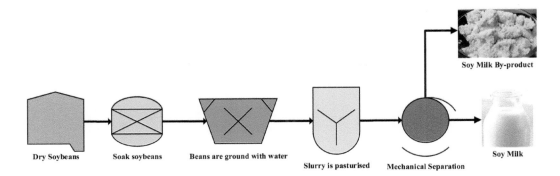

Soy Milk By-product

Dry Soybeans   Soak soybeans   Beans are ground with water   Slurry is pasturised   Mechanical Separation   Soy Milk

FIGURE 15.3   Production diagram to produce soymilk and its by-product (SMB).

FIGURE 15.4   Soy milk (left) and soy milk by-products (right).

content of SBM has been shown to be 34.8–37.2% and 38.2% for SPC. This was determined by the difference method and includes both simple and complex carbohydrate sources. It is a simple approximation and determined as the difference (in weight) between the dry matter and proteins, fats, and ash (BeMiller, 2017).

Proteins are the major constituents in both SBM and SPC, of which SBM contains approximately 44.4–54.6% of proteins, whereas SPC has about 45.1–52.6% of proteins. Due to its abundance of proteins, this waste material is attractive for its utilization as a feed stock. With the building blocks of proteins being amino acids (AA), the content of both waste streams has been well reported, showing a total of 18 AA for each containing ten essential/conditionally essential and eight non-essential (Grieshop et al., 2003; Jaworski & Stein, 2017).

The dry matter content of amino acid in SBM includes arginine (3.5–4.7%), histidine (1.2–1.7%), isoleucine (2.3–2.8%), leucine (3.7–4.9%), lysine (3.0–4.1%), methionine (0.6–0.9%), phenylalanine (2.4–3.3%), threonine (1.7–2.6%), valine (2.4–3.0%), tryptophan (0.74%), alanine (2.0–3.8%), aspartame (5.2–6.9%), cysteine (0.6–0.7%), glutamate (8.5–11.5%), glycine (2.1–2.7%), proline (2.4–3.7%), serine (1.9–3.3%) and tyrosine (1.6–1.9%). While the levels of these AAs have been shown to be similar in SPC with concentrations of arginine (3.1–3.8%), histidine (1.2–1.4%), isoleucine (2.1–2.4%), leucine (3.7–4.9%), lysine (3.0–4.1%), methionine (0.6–0.9%), phenylalanine (2.4–3.3%), threonine (1.7–2.6%), valine (2.4–3.0%), tryptophan (0.74%), alanine (2.0–3.8%), aspartame (5.2–6.9%), cysteine (0.6–0.7%), glutamate (8.5–11.5%), glycine (2.1–2.7%), proline (2.4–3.7%), serine (1.9–3.3%), and tyrosine (1.6–1.9%) (Karr-Lilienthal et al., 2006).

TABLE 15.1

Summary of the Nutritive and Non-Nutritive Composition of Soybean Oil and Soy Milk By-Products.

Nutritional Component	SBM	SPC	SMB	Method	Reference
Carbohydrate (g/100g)	34.8–37.2	38.2	3.8–5.3	SBM/SPC difference method SMB sum of starch, mono, and oligosaccharides	(Grieshop et al., 2003; Van der Riet et al., 1989)
Protein (g/100g)	44.4–54.6	45.1–52.6	24.5–37.5	SBM/SPC combustion analyzer SMB Kjeldahl	(Grieshop et al., 2003; Karr-Lilienthal et al., 2006; O'Toole, 1999)
Fat (g/100g)	2.18–4.8	4.9–11.3	9.3–22.3	Soxhlet	(Banaszkiewicz, 2011; Grieshop et al., 2003; Karr-Lilienthal et al., 2006; O'Toole, 1999)
Total dietary fiber (g/100g)	17.0–20.5	18.8–40.2	52.8–58.1	Enzymatic gravimetric method AOAC 991.43	(Behera et al., 2013; Grieshop et al., 2003; Jaworski & Stein, 2017; Van der Riet et al., 1989)
Insoluble fiber (g/100g)	17.0	34.6	0.2–43.6	Enzymatic gravimetric method AOAC 991.43	(Behera et al., 2013; Jaworski & Stein, 2017; Van der Riet et al., 1989)
Soluble fiber (g/100g)	1.6	5.6	12.6–14.6	Enzymatic gravimetric method AOAC 991.43	(Behera et al., 2013; Jaworski & Stein, 2017; Van der Riet et al., 1989)
Ash (g/100g)	6.5–8.1	6.6–7.4	3.0–6.36	AOAC method 942.05	(Behera et al., 2013; Grieshop et al., 2003; Jaworski & Stein, 2017; Kamble & Rani, 2020; Karr-Lilienthal et al., 2006; Van der Riet et al., 1989)
Isoflavones (mg/100g)	257.7–306.0	167–501	101.8–260.3	Solvent extraction/HPLC	(Eldridge, 1982; Jankowiak et al., 2014; Kao & Chen, 2006; Kao et al., 2008; Méndez Sevillano et al., 2014; Zuo et al., 2008)
Saponins (mg/100g)	565	127	103	Solvent extraction/HPLC	(Jacobsen et al., 2018; Kao et al., 2007b; Lee et al., 2005)
TPC (mg/100g)	340–1,680	29.3–104.2	130–360	Folin-Coicalteu method	(Guimarães et al., 2020; Sharifi et al., 2013; Vital et al., 2018; Yang et al., 2019; Zamindar et al., 2017)
Antioxidant capacity (mg/100g)	1.03–1.86 (mM)		24.5–37.8	DPPH decoloration	(Silva & Perrone, 2015; Vital et al., 2018)

As this waste has been derived from oil extraction processes, the content of fat in these waste materials is reduced. In comparison with SBM, which contains 2.2–4.8% fat, SPC has a greater content of fat at 4.9–11.3% on a dry basis; this is due to the lower efficiency of this mechanical process to extract oil (Banaszkiewicz, 2011; Grieshop et al., 2003; Karr-Lilienthal et al., 2006). The fatty acid (FA) content of oils in both SBM and SPC have high concentrations of palmitic (C16:0), stearic (C18:0), oleic (C18:1), linoleic (C18:2n6), and linolenic (18:3n3). The general trend of these components is 14% saturated FA, 25% monounsaturated FA, and 61% polyunsaturated FA, with the dominant FA being linoleic at 55% alone (Kralik et al., 2008; Waldroup & Waldroup, 2005; Wiseman & Salvador, 1991).

The dietary fiber content of any food can be divided into soluble fiber (SF) and insoluble fiber (IF). The SBM has a reported total dietary fiber (TDF) concentration of 17–20.5% (Grieshop et al., 2003). A study has shown that samples of SBM had a mean TDF content of 18.8%, which consisted of 16.97% insoluble fiber (IF) and 1.83% soluble fiber (SF). The TDF in SPC has been reported to be higher than that of SBM with ranges in content by 18.8–40.2%. At the highest reported level of TDF (40.2%), the SPC had 5.6% SF and 34.6% IF (Behera et al., 2013; Jaworski & Stein, 2017).

Ash is the gross determination of mineral content and is determined by incineration of samples in a muffle furnace, typically at temperatures of 500–550° C, until all carbon residues are removed. Ash is the difference between the initial and final weight of samples at the end of the ashing cycle. The ash content of both oil waste products is similar with contents of 6.5–8.1% for SBM and 6.6–7.4% for SPC (Behera et al., 2013; Grieshop et al., 2003; Jaworski & Stein, 2017; Karr-Lilienthal et al., 2006).

15.3.2 Nutrient Content of Soy Milk By-Product

Levels of nutritional components have been reported for SMB (Table 15.1), with some variation in their concentrations. Ash content ranges from 3.0–6.36% of dried weight (Kamble & Rani, 2020), while the total CHO content in SMB is 3.8–5.3%. The dominant macronutrient in SMB is TDF, which accounts for 52.8 to 58.1%, of those 40.2–43.6% is IF and 12.6–14.6% is SF (Van der Riet et al., 1989). These polymer chains consist of arabinose, galactose, xylose, and galacturonic acid with lower concentrations of rhamnose, glucose, fructose, and mannose (Redondo-Cuenca et al., 2008).

Proteins are a prominent component of SMB, however, the levels are considerably lower than that of the waste of the oil extractions, with content ranging from 24.5% to 37.5%. The amino acid composition of SMB has been reported by multiple sources with different results. Eight essential amino acids and the same amount of non-essential amino acids were detected in SMB (Kumar et al., 2016). The SMB contains arginine (1.86%), histidine (0.84%), isoleucine (2.7%), leucine (3.7%), lysine (1.7%), methionine (0.52%), phenylalanine (3.0%), threonine (1.15%), valine (1.5%), alanine (1.3%), aspartame (3.13%), glutamate (5.0%), glycine (1.2%), proline (1.3%), serine (1.6%), and tyrosine (1.1%). Interestingly, tryptophan and cysteine were not detected due to hydrolysis from the acidic test conditions, while another study found that SMB contains a total of 17 amino acids, which include ten essential and conditional amino acids and seven non-essential amino acids including cysteine (2.03 g/16g of nitrogen) (Puechkamut & Panyathitipong, 2012).

As soy milk is an aqueous extraction, it stands to reason that high levels of oil still remain with this by-product, ranging from 9.3% to 22.3% (O'Toole, 1999). The linoleic acid concentration was significantly higher ($p < 0.05$) than all other components at 49.76% of the oil fraction. Compared to soybean oil, the occurrence of alpha-linolenic acids is low (5.14%). It was assumed that since this polyunsaturated fatty acid is easily oxidized, it has been degraded during the manufacturing of soy milk and the drying process required to stabilize SMB. The fatty acids reported were palmitic (C16:0), stearic (C18:0), oleic (C18:1), linoleic (C18:2n6), and alpha-linolenic (18:3n3) (Kumar et al., 2016). Consuming oils that have a high concentration of unsaturated fatty acids can reduce low-density lipoproteins (LDL) and increase high-density lipoproteins (HDL), which reduces the occurrence of cardio vascular diseases (Mishra & Manchanda, 2012).

15.3.3   Phytochemical Content of Soybean Processing By-Products

The major phytochemicals found in soybean processing by-products (SPB) are phenolic compounds and soyasaponins, with both being linked with potential health benefits while exhibiting great power as antioxidants (Wiboonsirikul et al., 2013). The total phenolic content (TPC) of SBM has been reported to range from 340 to 1,680 mg gallic acid equivalent (GAE)/100 g, while the content of SPC and SMB have been shown to be at levels of 29.3–104.2 mg GAE/100 g and 130–360 mg GAE/100 g, respectively (Guimarães et al., 2020; Sharifi et al., 2013; Vital et al., 2018; Yang et al., 2019; Zamindar et al., 2017). Isoflavones are a major phenolic group and have gathered significant attention. Concentrations of isoflavones in SBM have been shown from 257.7 to 306.0 mg/100 g (Eldridge, 1982; Zuo et al., 2008), with SPC having a greater range in content from 167 to 501.0 mg/100 g (Kao & Chen, 2006; Kao et al., 2008), and SMB at levels of 101.8–260.3 mg/100 g (Jankowiak et al., 2014; Méndez Sevillano et al., 2014). The concentrations of saponins are 565 mg/100 g in SBM, 127 mg/100 g in SPC, while SMB has saponin content of 103 mg/100 g (Jacobsen et al., 2018; Kao et al., 2007b; Lee et al., 2005). In addition, soybean processing by-products also possess potent antioxidant power. The 2,2-diphenyl-1-(2,4,6-trinitrophenyl) hydrazyl (DPPH) free radical scavenging capacity of SBM has been reported as 1.03–1.86 mM Trolox equivalent (TE)/100 g, while 24.5–37.8 mg TE/100 g for SMB (Mareček et al., 2017; Silva & Perrone, 2015; Vital et al., 2018).

15.4   POTENTIAL HEALTH BENEFITS OF SOYBEAN PROCESSING BY-PRODUCTS

As mentioned earlier, soybean processing by-products (SBP) are a rich source of proteins, soluble (SF) and insoluble fiber (IF), isoflavones, and saponins. These components play an important role in nutrition and are known to support good human health (Lu et al., 2013; Rinaldi et al., 2000; Singh et al., 2017a; Zhong-Hua et al., 2015). The incorporation of a specific class, or group of compounds, can be used to target a dietary shortfall or help remedy health issues. Their incorporation into a diet can be in the form of an extract which concentrates a specific class of nutrient (i.e. proteins and fiber), incorporation of crude by-products into existing dietary staples, or the formulation of a nutraceutical from the non-nutritive fractions (i.e. isoflavones). Table 15.2 provides a summary of the health benefits of soybean meal, press cake, and milk by-products.

15.4.1   Soybean Processing By-Products as a Rich Source of Fiber for Health Benefits

Dietary fibers are known to be fermented by the gut microbiota, releasing nutrients that feed endogenous bacteria and provide minor nutrients to the body (Dhingra et al., 2012; Trowell et al., 1985). The benefits to human health from dietary fiber include reduction of risk for coronary heart disease, diabetes, obesity, hypolipidemia, and some forms of cancer, while improving satiety and the function of gut microbiota alongside providing bulk and reducing transit time of stools (Mann & Cummings, 2009; McDougall et al., 1996). Due to its health potential, fiber has been incorporated into foods, however, the addition of fiber can influence the organoleptic characteristics. For example, incorporation of fiber can improve the water- and oil-holding capacity, gelation, and emulsification properties and reduce syneresis (Elleuch et al., 2011).

SMB has been incorporated into bread and resulted in a lower glycaemic index (GI) due to the high fiber content of the SMB (Lu et al., 2013). Similarly, SMB has been added to rice noodles and pasta, and the results show that GI in these fortified foods is much lower than the control (Kamble et al., 2019; Kang et al., 2018b), revealing that fiber derived from SPB can help reduce the GI of dietary staples.

The major fiber fraction of soybean by-products is IF with lesser amounts of SF or the water-soluble fraction (Redondo-Cuenca et al., 2008). The monomers of these polymer chains consist of arabinose, galactose, xylose, and galacturonic acid with lower concentrations of rhamnose, glucose,

TABLE 15.2

Summary of Health Benefits of SBP

Nutritional Component	Benefits to Human Health	Reference
Fiber	Fiber has been incorporated into bread, pasta, and rice noodles and shown to reduce glycaemic index which slows glucose incorporation. Diet containing 10% SMB dietary fiber can reduce weight gain and lower serum cholesterol while increasing antioxidants and short-chain fatty acids in the cecum. Fiber has benefits for people with type 2 diabetes.	(Lu et al., 2013; Kamble et al., 2019; Kang et al., 2018; Jiménez-Escrig et al., 2008; Simpson et al., 1981)
Protein	Inactivation of trypsin inhibitors and lectins by thermal treatments allows incorporation and digestion of proteins. Inactivated trypsin inhibitors increase the cysteine content and improve the amino acid profile. Heat also increases the accumulation of Bg7S subunits which contain lysine and sulfur amino acids. High sulfur-containing amino acids improve use as a complementary protein, as many plant-based proteins lack these components.	(Stanojevic et al., 2012, 2013)
Fats	SMB and SPC contain mostly polyunsaturated fatty acids and monounsaturated fatty acids, which can be classed as a source of good fats. Consuming dietary fats with higher levels of these fatty acids aids in reducing LDL and can improve heart health.	(Kumar et al., 2016) (Mishra & Manchanda, 2012)
Isoflavones	Isoflavones can have a positive effect on breast and prostate cancers, reduce menopausal symptoms, and reduce loss of calcium which can lead to osteoporosis and osteopenia. Isoflavones can improve the risk of CVD and insulin sensitivity in postmenopausal women. Isoflavones were coupled with flavan-3-ols in fortified foods. Isoflavones can act as an anti-inflammatory agent, retard hepatoma and prostate cancer cell growth, improve photoaging protection from UVB light, and reduce dermatitis when applied as a topical cream.	(Atmaca et al., 2008; Kang et al., 2010; Curtis et al., 2012) (Kao et al., 2007a, 2007b; Wang et al., 2009; Chiang et al., 2007; Chu et al., 2020)
Saponins	Soyasaponins can exhibit antiviral action when used as an adjuvant on herpes virus. Soyasaponins can also inhibit cytomegalovirus, influenza, and HIV type 1. Consuming a diet rich in saponins can help to lower total cholesterol, plasma leptins, and triglycerides, while also improving blood glucose levels and hepatic metabolism. Saponin can also have a positive effect as an antidiabetic, anticoagulant, and antithrombotic, while lowering lipids and having hypocholesterolemic effects.	(Hayashi et al., 1997; Kim et al., 2009; Park et al., 2005, 2009; Messina & Barnes, 1991; Zhong-Hua et al., 2015)
Phenolics and antioxidant properties	All phenolic compounds can influence the antioxidant capacity of soybean processing by-products, with saponins and isoflavones each being shown to be major contributors to this action. Antioxidants can scavenge free radicals in the body and play a role in preventing atherogenesis and oxidation of LDL and cellular DNA. Phytosterols have been selectively purified from the deodorizer distillate produced during the purification of crude soybean oil. These phytosterols have been shown to lower the absorption of cholesterol. Phytosterols have been shown to reduce hypercholesterolemia and help to prevent CVD.	(Wiboonsirikul et al., 2013) (Pizzino et al.; Rice-Evans & Miller, 1996; Rice-Evans et al., 1997) (Cabral & Klein, 2017; Singh et al., 2017b; Torres et al., 2009)

fructose, and mannose (Mitsuoka, 2002). It has been shown that a diet rich in these types of fibers can improve control of type 2 diabetes or impaired glucose metabolism (Simpson et al., 1981). A study reported that a diet consisting of 10% SMB dietary fiber can be linked with a lowered weight gain and lowered total serum cholesterol while also showing an increase in cecum antioxidants and short-chain fatty acid (SCFA) production (Jiménez-Escrig et al., 2008).

15.4.2 SOYBEAN PROCESSING BY-PRODUCTS AS A RICH SOURCE OF PROTEINS FOR HEALTH BENEFITS

The Food and Agriculture Organization of the United Nations (FAO/UN) has highlighted the importance of proteins as vital support to the health and wellbeing of the global population (FAO/UN, 2013). Soy proteins are used in many areas of food production for their oil and water binding qualities, contribution to texture, viscosity, gelation, cohesiveness, emulsion stabilities, and foaming capabilities, amongst other things (Jooyandeh, 2011). Raw soybeans possess trypsin inhibitors (Kunitz and Bowmen-Burke) and lectins, which negatively affect the bioavailability of proteins, resulting in low digestibility and assimilation. Thermal treatment of soy has been shown to inactivate these proteinaceous compounds, resulting in improved recovery of proteins and their digestibility. Other factors also affect the degree of inactivation including, pH, temperature, pressure, time, and cultivars (Stanojevic et al., 2013).

Urease is an endogenous enzyme present in raw soybeans and its activity can be used as an indirect analytical method for determining the extent of trypsin inhibitors present. As urease and trypsin inhibitors are both affected by heat in the same way, this relationship is used to determine the activity of both. Urease activity releases ammonia which can be measured by a change in pH (Ibáñez et al., 2020). A change in pH between 0.05 and 0.20 will indicate that both urease and trypsin have been sufficiently inactivated (Wachiraphansakul & Devahastin, 2007). The urease activity of SBM, SPC, and SMB have been reported showing a difference in pH values of 0.03–0.10, 0.01–0.1, and 0.04–0.08 respectively. These results indicate that during processing that the trypsin inhibitors are sufficiently inactivated. The presence of inactivated trypsin inhibitors also improves the nutritional quality of these by-products, as they are cysteine-rich proteins and can add these to the amino acid composition (Grieshop et al., 2003; Karr-Lilienthal et al., 2006; Stanojevic et al., 2013).

Due to the high level of proteins reported, utilization of SPBs as a nutritional source for applications in food products has great potential. The proteins found in SPC have greater solubility and digestibility than those of SBM and have the added advantage of not being produced with an organic solvent, affording SPC more interest as a food ingredient (Endres, 2001). With a lower content of proteins than its oil counterparts, SMB has been studied for its potential as a dietary source of protein; it has been found that SMB showed positive attributes, has great potential to be used as a plant-based protein, and does not share the same restraints as raw soybeans. As part of this the antinutrient lectin, which is responsible for lower intestinal absorption of nutrients, was shown to be low in SMB, as it is removed with soy milk (Stanojevic et al., 2013). One of the major proteins extracted after processing at high temperatures is Bg7S. These proteins have elevated levels of the amino acid, cysteine, which correlated to a higher concentration in SMB. Therefore, SMB has elevated levels of sulfur-containing amino acids and lysine, showing it as a good complimentary source of proteins, as many plants lack these amino acids (Stanojevic et al., 2012).

15.4.3 SOYBEAN PROCESSING BY-PRODUCTS AS A RICH SOURCE OF ISOFLAVONES FOR HEALTH BENEFITS

Isoflavones are biologically active metabolites that are found in high concentrations in soybeans and their derivatives. In their natural state, they are mainly found in the biological inactive, glycosidic form, but can be hydrolyzed by glucosidase in the gut to an aglycone. These aglycones have a higher biological activity and can be consumed in this form, increasing bioavailability. There has been a total of 12 isoflavones isolated from soy and their concentration in SPB varies. The aglycone versions are daidzein, genistein, and glycitein, with the glycosides being, daidzin, genistin, and glycitin. The glycosidic isoflavones can also alternatively have either an acetyl or malonyl functional group as well. In SPB the composition of these forms will vary and depend upon the extent and type of processing applied.

Isoflavones are also known as phytoestrogens due to their ability to act on estrogen receptors in the body because of their structural similarity to estradiol and have been shown to have mild modulatory effects (Zaheer & Humayoun Akhtar, 2017). It has been proposed that isoflavones can have positive effects on breast and prostate cancer, menopausal symptoms, osteoporosis, and osteopenia while reducing the occurrence of cardiovascular diseases (CVD) and diabetes (Atmaca et al., 2008; Curtis et al., 2012; Kang et al., 2010). Debate exists as to whether isoflavones are truly as advantageous as first thought. There are studies that have reported the deleterious effects of isoflavones on breast and prostate cancers, male and female fertility, and hypothyroidism. As the majority of studies conducted in these areas have been on various rodent models, conjecture still remains with many authors suggesting further research is required to achieve true consensus (Cederroth & Nef, 2009; D'Adamo & Sahin, 2014).

Isoflavones powders, which comprise glycosides, malonyl-glycosides, acetyl-glycosides, aglycones, and two mixtures of the 12 isoflavones (ISO-1 and ISO-2), were developed. ISO-1 is simply the isoflavone extracted from crude SPC containing all 12 isoflavones and ISO-2 is a formulation of equal weights of the four isoflavone extracts adjusted for total isoflavone concentration (Kao et al., 2005). The anti-inflammatory effects of these powders were tested on mice, who were fed a solution containing a total isoflavone content of 50 mg/kg body weight for four weeks, and found that isoflavone powders could reduce inflammatory markers tested (Kao et al., 2007b). In addition, the aglycone and acetyl-glucosides fractions extracted from SPC showed strong inhibition on antiproliferation hepatoma cells (HepGH2) at a dose of 50 µg/mL (Kao et al., 2007a). ISO-1 and a mixture of aglycones and acetyl-glycoside fractions have also been shown to influence prostate cell lines. Although not all markers tested were observed to be affected, it was concluded that these fractions had a significant ability to retard prostate cancer cells (Wang et al., 2009).

The ISO-1 extract has also been shown to have a positive effect as a topical treatment in several *in vitro* and *in vivo* mouse studies. Photoaging and UVB protective effects of ISO-1 have been determined in both keratinocytes (Chiang et al., 2007) and in mice (Chiu et al., 2009) treated topically with this extract. The results from both studies have shown that there is a protective nature and it is able to impede the destruction of UVB-induced cell death and was able to reduce epidermal thickening and the expression of cell damage markers. Both studies have concluded that ISO-1 extract has the potential to be incorporated into cosmetics due to its photo-aging protection of skin. The ISO-1 extract can also mitigate the inflammatory effects of keratinocytes on induced dermatitis in mice, showing a potential benefit of incorporation in topical applications (Chu et al., 2020).

15.4.4 SOYBEAN PROCESSING BY-PRODUCTS AS A RICH SOURCE OF SAPONINS FOR HEALTH BENEFITS

Soybeans are a well-documented source of the secondary metabolite classes known as saponins. These triterpenoid compounds have been shown to exhibit fungicidal, insecticidal, and antimicrobial characteristics, playing a direct role in plant defenses. These same compounds have also been shown to benefit human health in aiding in the inhibition of multiple diseases and conditions (Tantry & Khan, 2013). The actions of soyasaponins are varied, with investigations reporting positive effects as a cancer adjuvant with antitumor, antidiabetic, anticoagulant, antithrombotic, lipid-lowering, and hypocholesterolemic effects (Messina & Barnes, 1991; Zhong-Hua et al., 2015). An *in vitro* study on soyasaponins has shown a defined effect on specific viral strains. Soyasaponins have a reduced effect on the herpes simplex virus 1 and exhibited antiviral effects when coupled with acyclovir. Further to this, an inhibitory effect was observed on the human cytomegalovirus, influenza virus, and HIV type 1 (Hayashi et al., 1997).

A study that fed mice a high fat diet, has shown that crude saponins from SPC at 0.5%, 1.0%, and 1.5% (total feed weight) found that the saponin rich diet significantly ($p < 0.05$) lowered plasma leptin, triglycerides, and total cholesterol, while showing higher exertion of triglycerides. The high-fat diet also induced glucose intolerance in the control group, however, soy saponins diets were able to improve the postprandial glucose levels, while also increasing hepatic liver metabolism (Kim et al., 2009). Saponins in soy might modulate the endogenous globulin proteins resulting in a lower affinity for chymotrypsin (Shimoyamada et al., 1998). Crude saponin extracts from SPC have also been found to reduce markers tested on selected cancer cells (Park et al., 2005) and human colon cancer cells (Park et al., 2009). It should be noted that during cooking 75% of saponins from defatted soybean flour can be lost, thus processing conditions need to be considered when incorporating saponins into foods.

15.4.5 OTHER FUNCTIONAL COMPONENTS FROM SOYBEAN PROCESSING BY-PRODUCTS FOR HEALTH BENEFITS

Another bioactive compound with a significant effect on cardiovascular disease is phytosterols, named due to their structural similarity and cellular functionality to cholesterols. Research has shown that plant sterols and stanols can reduce uptake of dietary cholesterols, lowering LDL cholesterol levels, while showing anti-inflammatory and anti-atherogenicity activity. They may also play a role as an antioxidant and an anticancer agent. Clinical studies have been able to show an intake of 2 g of phytosterols a day can reduce LDLs without significantly affecting the concentration of HDL-cholesterol and triglycerides. Phytosterols from soy have been shown to lower the absorption of cholesterol, leading to lower serum levels (Cabral & Klein, 2017; Quilez et al., 2003; Singh et al., 2017b).

During deodorization of crude soybean oil, a distillate deodorizer by-product is produced which contains phytosterols, free fatty acids, tocopherols, and squalene. A two-step enzymatic process has been developed, which was able to obtain sterol esters, tocopherols, and fatty acid esters (Torres et al., 2007). In a further study, the phytosterols and tocopherols were selectively separated and phytosterols purified by supercritical carbon dioxide (Torres et al., 2009). Tocopherols have also been selectively purified from soybean oil deodorizing distillate using lipase and have multiple improvements compare to the common industrial procedures (Shimada et al., 2000). Tocopherols are commonly used as food ingredients for their ability to act as an antioxidant. In soybean waste, the presence of high concentrations of isoflavones, saponins, flavonoids, and other phenolic components contribute to the overall antioxidant capacity. These compounds have been highlighted for their ability to act as antioxidants, by either donating hydrogen or scavenging oxygen radicals and play a role in preventing atherogenic, oxidation of LDL and cellular DNA (Pizzino et al., 2017; Rice-Evans & Miller, 1996; Rice-Evans et al., 1997).

15.5 FACTORS AFFECTING THE POTENTIAL HEALTH BENEFITS OF SOYBEAN PROCESSING BY-PRODUCTS

Variations in the nutritional content and quality of SPB can be affected by multiple agricultural and manufacturing factors, such as regionality, growing conditions, cultivars, harvesting, storage, and processing practices. For example, the SBM from Brazil has higher proteins, neutral detergent fiber, iron, and raffinose than SBM from the USA and Argentina, whereas the SBM from the USA and Argentina has higher lysine, methionine, cysteine, and threonine as compared to that in SBM from Brazil and India (Ibáñez et al., 2020). Variations within countries and between processing plants have been studied as well. Furthermore, types of soybean waste including SBM and SPC are different in crude proteins, fat content, total dietary fiber, dry matter, and organic matter(Grieshop et al., 2003; Karr-Lilienthal et al., 2006). Different cultivars also lead to different compositions of proteins and phytochemicals (Grizotto et al., 2008; Van der Riet et al., 1989; Yu et al., 2018).

Post-manufacturing processes have been found to affect the composition of SMB. Therefore, SMB typically requires post-manufacturing treatment once it is separated from the soy milk because rapid spoilage can occurs due to microbial activities (Grizotto & Aguirre, 2011). Treatment after processing often relies on the removal of water by rapid drying to minimize spoilage and typically employs techniques that require large amounts of energy, which is a major cost to producers. Other treatments that have been applied generally employ controlled fermentation or biotransformation to yield a more stable product by bacteria, yeast, and enzymatic treatment. Traditional styles of drying such as forced air tray dryers and drum dryers have been used but the results are not favorable due to degradation of proteins and elevated costs (de Aguirre et al., 1981; Travaglini et al., 1980). Vacuum tray drying has been found more effective for the removal of water (Sengupta et al., 2012).

Novel post-processing treatments have been developed which focus on improving specific health qualities and are summarized in Table 15.3. Increases in dietary SF can improve specific health concerns such as early stages of irritable bowel syndrome and can contribute to appetite regulation (Bijkerk et al., 2009; Delargy et al., 1997). The application of high hydrostatic pressure of between 200–600 MPa and temperatures of 30–60° C have been applied to SMB resulting in an increase of both SF and IF (Mateos-Aparicio et al., 2010). Both bacteria and yeasts have also been applied to improve the concentration of SF. For example, *Saccharomyces cerevisiae* (baker's yeast) has been applied as a substrate to convert fiber to a more digestible form. This process also allowed the hydrolysis of isoflavones to their biologically more active aglycones and a higher concentration of proteins, polyphenolics, and antioxidant potential (Santos et al., 2018). Treatment with commercial enzyme mixtures such as Ultraflo®-L and Viscozyme®-L coupled with high hydrostatic pressure has been found to result in a significant increase in TDF, soluble polysaccharides, and low-molecular-weight mono, di, and oligo CHO (Pérez-López et al., 2016a, 2016b, 2017). These mixed enzymes can also increase the total extractable proteins by hydrolyzing cell wall CHO (de Figueiredo et al., 2018). In addition, treatment of SBM with chemicals such as dilute sulphuric acid at concentrations of 0.9–1.9% is found to increase crude proteins from 48% to 58% (Luján-Rhenals et al., 2014), while the extraction of oil by supercritical fluid extract improves recovery yields, with the main components being linoleic acids, phytosterols, and phytostanols (Quitain et al., 2006).

15.6 UTILIZATION OF SOYBEAN PROCESSING BY-PRODUCTS FOR FUNCTIONAL FOODS

Functional foods are generally recognized by the incorporation of a specific active ingredient/s, which may improve a specific health condition or increase the nutritional value of a specific food item (Varzakas et al., 2016). They aim to provide higher nutritional value beyond what the original food contains. This can be accomplished by recognizing dietary shortfalls and offering alternatives to the norm, which are similar in organoleptic characteristics. The following demonstrates that in combining waste utilization with novel food product development functional foods can be created with the potential to be included in dietary staples.

15.6.1 INCORPORATION OF DRIED POWDERS AS AN INGREDIENT IN COMMON FOOD ITEMS

Utilization of bulk amounts of these by-products to reduce waste while adding to health benefits needs to be further investigated. Because of the surplus created, the cost is low which highlights SPBs as a potential ingredient to produce novel functional food products. Research into this area is continual, but progress is reduced due to limitations already highlighted. A significant amount of research has been undertaken on specific foods that incorporate these dry powders, which aim to retain good consumer acceptability and include composite wheat flours, gluten-free flour, breads, biscuits and cookies, soy crackers, cakes, sausages, frankfurters, beef patties, noodles, cheese analog, and various other food items.

Studies have been conducted to develop composite wheat flours and tested on bread and bakery products, which has shown good functionality in organoleptic testing. The development of quality

TABLE 15.3

Post-Manufacturing and Biological Treatments to Improve Nutritional Qualities of Soybean Processing By-Products

Nutrient	Treatment and Results	Reference
Fiber	High hydrostatic pressure under optimum conditions of 400 MPa and 60° C can increase both SF and IF. Solid-state fermentation with *S. cerevisiae* increases in SF, isoflavone aglycones, proteins, and polyphenols. Enzyme mixture Ultraflo®-L and Viscozyme®-L coupled with hydrostatic pressure significantly increases in TDF, soluble polysaccharides, and low molecular weight mono, di, and oligo CHO.	(Mateos-Aparicio et al., 2010; Santos et al., 2018; Pérez-López et al., 2016a, 2016b, 2017)
Protein	Sulphuric acid treatment at the optimal concentration of 0.9–1.9% could increase the crude proteins by 10% while also increasing fermentable sugars. Multi-enzyme complex could increase extractable proteins. Hydrothermal processing under high pressure showed increases in extractable protein and high concentration of Bg7S glycoproteins.	(Luján-Rhenals et al., 2014; de Figueiredo et al., 2018; Stanojevic et al., 2012)
Lipid	Supercritical fluid extraction with CO_2/EtOH as co-solvent could improve recovery lipid yield. Fermentation with *Lactobacillus plantarum* could improve the fatty acid profile while retaining the probiotic bacteria during drying, leading to better application as a functional food.	(Quitain et al., 2006; Quintana et al., 2017)
Isoflavones	Solid-state fermentation with *S. cerevisiae* or Mucor could increase the conversion of isoflavones to their more digestible aglycone form. Mucor also increased the total recoverable isoflavones. Fermentation of soy milk mixed with SMB by *L. acidophilus* LA3 could also increase the conversion of aglycone isoflavones and was used to produce a probiotic sauce.	(Santos et al., 2018; Jing, 2009; de Moraes Filho et al., 2016, 2018)
Biotransformation	SMB has been utilized as a bacterial substrate which resulted in a significant increase in folate. Fermentation followed by fortification with inulin as a prebiotic has yielded a synbiotic supplement that can lower LDL and LDL/HDL ratio, having a positive effect on CVD risk. Fermentation with koji mold can increase the protein content by 10% while also decreasing the peptide size and inactivating trypsin inhibitors, resulting in better solubility. Whereas SPC fermented with *S. cerevisiae* can increase ash, total CHO and dietary fiber, and conversion of aglycone isoflavones.	(de Albuquerque et al., 2016; Bedani et al., 2015; Hong et al., 2004; Silva et al., 2018)

gluten-free flour allows sufferers of gluten intolerance or celiac disease to consume foods typically made from wheat flour. Freeze-dried SMB has been suggested as a good drying method to produce a gluten-free flour, compared to microwave drying and rotary drum dryer, when tested for rheological properties. The dried SBM was also mixed with manioc flour to produce a dough that performed well when tested with a texture analyzer. Samples of the flour were also treated with lipase which increased solubility and swelling capacity which was regarded as a favorable increase. (Ostermann-Porcel et al., 2017).

SMB has been used in a gluten-free flour formula, using commercial improvers that are used with existing composite flours mixtures. This flour was tested in baked goods with American biscuit and cookie dough, as well as muffins and pancake batters. When SMB flour was used to make the American biscuits and cookies, texture and volume were negatively affected and did not produce an acceptable product. It was hypothesized that the absence of gluten and the increased water holding capacity of SMB flour modify organoleptic qualities. However, when this flour mixture was used to make both pancakes and muffins it out-performed commercial gluten-free flour in both physical and sensory panel tests (Aguado, 2010).

Gluten-free cookies which incorporate a significant level of SMB have been reported in the literature. The substitution of SMB (0–50%) into a cookie dough made with manioc flour was developed and at a replacement of 30%. The SMB cookies had acceptable qualities when assessed by a panel of 70 untrained judges for color, odor, taste, and texture. A negative effect on physical properties was observed as the content of SMB increased when compared to control. Volume, spread ratio, color, and hardness were affected by inclusion, proposed to be due to the high fiber content. The nutritional and energy value was also noted to have increased with substitution (Ostermann-Porcel et al., 2017). A composite flour mixture of SMB, broken rice, and rice bran, all industry by-products, has been developed for biscuits. The flour mixture was optimized to produce a final product that performed like a commercial standard for physical properties of volume, color, spread, weight, and texture. Samples that met specific requirements were stored for ten months to assess any changes. During storage lipids, carbohydrates and energy decreased, while the concentration of moisture and protein was significantly ($p < 0.05$) increased. Textural properties were also observed to have changed during extended storage. It was proposed that syneresis was attributed to the changes in moisture and texture (Tavares et al., 2016).

Bread and bread dough products are typically made with white wheat flour and are devoid of many nutrients, especially fiber. Unless fortification with fiber is included, the concentration of this essential nutritional component in white bread is minimal. This is due to the removal of the outer bran layers and germ of the wheat grain, which contain the majority of dietary fibers, during the milling of wheat flour (Taylor et al., 2005). Several sources have reported that SMB can be incorporated into wheat flour. A substitution of 10–15% dried SMB and wheat flour mixture has been suggested to produce a bread with similar organoleptic characteristics compared to standard white bread. From sensory testing, studies have suggested that at 10% substitution acceptability was high and not significantly different ($p > 0.05$) from a white bread control product while it was noted that an increase ($p < 0.05$) in textural qualities, crust color, total dietary fiber, protein, and fats was observed (Lu et al., 2011; Silva et al., 2009; Wickramarathna & Arampath, 2003; Xiao & Xiangyang, 2010; Yang et al., 2012). Figure 15.5 shows the effect of SMB flour on bread volume when incorporated into a basic bread formulation. As the amount of SMB is increased there is an effect on the volume of the loaf produced. Reduced formation of gluten due to the presence of fiber and removal of gluten-forming proteins will contribute to the loss of volume.

Steamed breads and noodles which are a staple part of the diet for many Asian cultures have the same nutritional limitations as breads as they are generally made with white flour as well. When

FIGURE 15.5 Bread produced with SMB composite flour mix. Left to right: Control (0%), 5% SMB, 10% SMB, 15% SMB.

SMB was added to steamed bread, it was found that 15% of the wheat flour could be replaced, while still achieving a desirable product. Once gluten was added to the dough, the final product had high similarities to the control. When SMB replaces wheat flour in noodles with substitution up to 25% and fortified with gluten, it results in very minor character changes compared to the control (Fei et al., 2013).

Other studies have also focused on incorporating SMB into noodles and pasta. A recent study tested the incorporation of SMB flour in rice noodles between 5% and 15% which resulted in negative textural and cooking properties (Kang et al., 2018a). To improve the formulation, alginate was incorporated in the rice noodle formulae before cooking, followed by a 30 min soaking in a calcium chloride bath before the noodles were dried for storage. This increased the textural and cooking properties to a level comparable ($p > 0.05$) to commercial noodles. The *in vitro* GI was tested, which showed that when SMB was substituted at 10%, the release of glucose was significantly reduced, demonstrating a hypoglycaemic effect. Wheat noodles and macaroni formulated with SMB have been tested for acceptability against the nine-point hedonic scale. Noodles containing 30% SMB flour were rated at 7.90 ("like very much") on the hedonic scale, while macaroni, with 20% SMB flour in the same study was reported an acceptability rating of 7.77 (between "like moderately" and "like very much") (Ahlawat & Punia, 2012).

SMB dried in a solar tunnel has been tested on butter cookies, sponge cake rusks (dried cake fingers popular in India), and doughnuts substituted at 20% for wheat flour. During hedonic testing, panel members classed all of these products as "like very much" for the categories of color, appearance, aroma, texture, taste, and overall acceptability (Ahlawat & Punia, 2012). Further to these products, a cake mixture at a substitution level of 5%, 10%, 15%, and 20% SMB wheat flour has been developed. A tasting panel has determined that no difference between control and 5% SMB flour ($p > 0.05$) was detected, and overall likeness was achieved for the cake up to 15% (Li et al., 2014). Other research has focused on incorporating fresh SMB (wet) into baked products, bypassing the drying stage and its related costs. A cookie has been formulated with the incorporation of wheat starch, soy flour, and hydroxypropyl methylcellulose (HMC), at ratios of 1:25 with fresh SMB on the organoleptic qualities. The authors have determined that the highest acceptability was achieved by the incorporation of both soy flour and HMC which showed improved water-holding capacity, better shelf life, and a crispier firmer texture compared to the control (100% SMB). With the inclusion of high proportions of SMB, significant functional benefits for health were provided, as high concentrations of TDF and protein were reported (Park et al., 2015). Further to this, SBM has been used to produce low-carb biscuits, which was high in proteins, fiber, and isoflavone aglycones formulated with both fermented and non-fermented meal. Solid-state fermentation was carried out with *S. cerevisiae*. When tested by a sensory panel the biscuits formulated without fermented meal had greater acceptability while the fermented samples, however, had a greater accumulation of aglycones (de Oliveira Silva et al., 2018). A further study was able to show that fermentation improved the metabolism of isoflavones *in vivo* when these biscuits were consumed (de Oliveira Silva et al., 2020). A coconut baked snack has also been developed by replacing dried coconut with wet SMB at 10–50% with good results. The researchers determined that when coconut was substituted at both 30% and 40%, a trained panel rated this snack had an overall score 8.4/9 on a sensory scale. All samples that contained SMB had sensory scores above the control (100% coconut) while also having a significant increase ($p < 0.05$) in the fiber and a reduction of total fats (Radočaj & Dimić, 2013).

Partially dried SMB has been used to make a "clone" of a Japanese-style okara snack product for the American market. The new snacks, which were baked in oil, were tested by a trained sensory panel and rated higher in acceptability scores for the descriptors of oily aroma, beany aroma, cinnamon aroma, saltiness, sweetness, beany flavour, oily flavour, cinnamon flavor, color, and crunchiness. It was reported that 82% of panelists preferred the new SMB product over the original product. The proximate nutritive content of this new snack showed that both fiber and protein were increased (Katayama & Wilson, 2008).

With a significant concentration of proteins remaining in SMB, utilization as an additive in meat products is beneficial, while the presence of fiber and protein can act as an emulsifier. This has been applied to frankfurter-style sausages which were compared in a sensory test against a commercial product, the results of which showed that incorporation at the highest level tested (4% SMB flour) acceptability for color, odor, taste, texture, was not affected, with participants indicating they would "probably buy" or "certainly buy" this product (Grizotto et al., 2012). Similar results have been reported on pork meat gel cooked in a casing, which included 5% of SMB (Chang et al., 2014). Beef sausages have also been developed with significant rates of substitution of 20% being reported while still retaining good sensory attributes. It was reported that as SMB increased so too did the texture and water-holding capacity of the final product with a decrease in firmness, cohesion, springiness, chewiness, and color. When assessing the proximate nutrient components, carbohydrates, ash, and fiber levels rose with the incorporation of SMB, while lipids and protein decreased (Noriham et al., 2016).

Hamburgers are quite often regarded as energy-dense fast food. Novel research has the potential to deliver a product that has a higher nutritional density. Soy waste has been utilized to formulate burger patties, replacing textured soy protein with SMB and bacon. This study showed that SMB performed well when compared to textured soy protein and could be used as a replacement when added at a concentration of 8%. The burger patties had comparable moisture retention, water absorption, and holding capacity, with incorporation of both soy derivatives with minimal difference in results. As the concentration of the soy protein components was increased, so too were the water- and moisture-holding capabilities. It was hypothesized that the concentration of amino acids with polar side chains on the surface of the proteins increased these effects (Falcão et al., 2015). Bacon performed poorly in these tests and when comparing textured soy protein, SMB reduced the oil holding capabilities of the burger patties, while having lower sensory scores. As fat absorption is directly related to perceived flavor, this could account for a reduction in this characteristic in the sensory trial (El Nasri & El Tinay, 2007). Others have reported similar results when incorporating wet SMB into a low-fat patty, with concentrations of 20% and 22.5%. Sensory testing revealed that SMB increased the hardness of the cooked beef patties while reducing the chewiness, springiness, and cohesion. These studies reported that juiciness, appearance, tenderness, and overall acceptability were not statistically ($p > 0.05$) different from the control, while moisture loss was conserved in the test patties. As an effect of this, the perceived juiciness was reported by each study to be higher and has demonstrated that fats can be removed from beef patties and replaced by wet SMB (Su et al., 2013; Turhan et al., 2007).

Due to the high concentration of plant-based proteins in SPB, it is an excellent source of this essential nutrient to include in foods designed for people following either vegetarian or vegan diets as competent sources in these diets can be challenging. However, when good dietary planning is undertaken, the benefits of a plant-based diet are achieved (Melina et al., 2016). A vegetable-based pâté-style paste was formulated using fresh SMB at concentrations of 33–50% and was mixed with carrot, mayonnaise, and seasonings. Sensory and preference testing by a panel of 50 untrained judges with a nine-point hedonic scale was employed to assess acceptability of these pâtés, and the sample containing 33% fresh SMB had the greatest acceptability with a mean of 8 or "like very much", while showing a significant difference ($p < 0.05$) in preference from the other samples. Nutritional analysis has shown these samples to have concentrations of total dietary fiber at 5.76 g/100 g, a protein content of 3.07 g/100 g, β-carotene at 0.411 mg/100 mL, and total isoflavones of 0.15 μmol/g (wet weight), while having a low caloric value of 375 kJ/100 g (Guimarães et al., 2018). While not considered a targeted vegetarian product, SMB has been included in the formulation of peanut butter. This common vegetable-based spread could be considered energy-dense due to high concentrations of macronutrients, especially oils. The incorporation of SMB into peanut butter with a partial substitution for peanuts (0–25%) has been undertaken for physiochemical properties and sensory acceptability (Nasution et al., 2012). When analyzing the texture of SMB samples, the force required to penetrate and exit the sample by probe significantly increased ($p < 0.05$). With the addition of 2% glycerol monostearate, the spreadability of the butter was increased and was used in sensory testing compared to commercial peanut butter. The results of sensory trials determined that there was not a statistically significant ($p > 0.05$) difference when substituted with up to 15% SMB for color, odor, mouthfeel, taste, and overall acceptability, but it was significantly different ($p < 0.05$) for samples above this. The total fat did not significantly ($p > 0.05$) change with incorporation, however, saturated fats were reduced from 29.7% (control) to 1.12% for the 15% SMB peanut butter.

This section has highlighted that SPBs can be used as an ingredient to formulate functional foods and would also include different methods of drying or preservation of waste streams that maintain nutritional qualities. The selection of the preservation technique involves the consideration of equipment availability and costs, the time required to dry SMB, characteristics of the final product, and energy cost. Figure 15.6 proposes a process schematic to produce functional foods from these waste materials.

15.6.2   UTILIZATION AS SUBSTRATES FOR IMPROVEMENT OF HEALTH BENEFITS

As SMB is a rich source of nutrients, it can be applied as a substrate to produce probiotic and synbiotic foods. SMB has been used as a growth medium for *Lactobacillus plantarum* which increased both the unsaturated/saturated fatty acid and polyunsaturated/monounsaturated fatty acid ratios. It was reported that this application retained the probiotic strain during drying, thus providing potential functional food ingredients for further applications (Quintana et al., 2017). In addition, solid-state fermentation of SMB can hydrolyze isoflavones to their more readily available form, with a significant increase of the aglycone, while fermentation with Mucor can significantly increase the total recoverable isoflavone and aglycone concentrations (Jing, 2009; Santos et al., 2018). Similarly, fermentation of soy milk mixed with SMB has shown a good conversion of isoflavone aglycones and the fermented mixture can be used to further develop a probiotic creamy sauce from the soy milk fermented with *Lactobacillus acidophilus* LA3 (de Moraes Filho et al., 2016, 2018).

SMB can be potentially used as a substrate for *Lactobacillus rhamnosus* LGG, *Lactobacillus rueteri* RC-14, *Bifidobacterium longum* subsp. *infantis* BB-02, and *Bifidobacterium longum* BB-46 to increase folate content (de Albuquerque et al., 2016). In addition, SMB contains a high concentration of prebiotic fiber such as inulin, with an ability to support the colonization of probiotic bacteria, thus it has the potential to be incorporated into a synbiotic product. A double-blind clinical trial has been conducted to test a synbiotic supplement produced from SMB fortified with inulin as a

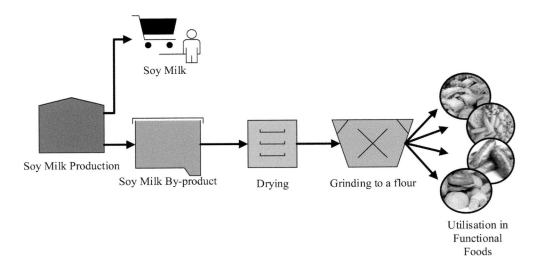

FIGURE 15.6   Proposed process schematic for the utilization of SPB in functional foods.

prebiotic and fermented with *Lactobacillus acidophilus* LA-5 and *Bifidobacterium animalis* subsp. *lactis* BB-12, and the starter *Streptococcus thermophilus*. The results showed that a synbiotic supplement formulated with SMB has a potential effect on lowering lipids *in vivo* (Bedani et al., 2015).

Fermentation by fungi can modify a broad range of nutritional components which can lead to an increase in bioavailability of proteins, CHO, and isoflavones while improving the antioxidant capacity of SMB. For example, fermentation using *Aspergillus oryzae* (koji) GB-107 on the proteins in SBM can improve the proteins concentration by 10%, while reducing all peptides to a small size (< 20 kDa) and eliminating the majority of trypsin inhibitors (Hong et al., 2004). In addition, fermentation using *S. cerevisiae* on SPC can increase ash, total CHO, and dietary fiber all significantly (p < 0.05), while decreasing lipids and simple sugars. More isoflavones can also be converted to their aglycone during fermentation (de Oliveira Silva et al., 2018).

15.7 CONCLUSION

Soybean by-products generated from soybean oil and soy milk production are available in large quantities and have been considered as waste due to limited applications. Due to the high water and nutrient content, there is rapid deterioration, and this leads to environmental issues. These by-products are a rich source of proteins, soluble and insoluble fiber, isoflavones, and saponins, thus they can be potentially used as functional food ingredients and are linked with health benefits. Soybean by-products can reduce the GI and lower weight gain and total serum cholesterol. Soybean by-products can be a potential source of plant-based proteins for the human diet. Isoflavones in soybean by-products such as daidzin, genistin, and glycitin can be linked with the prevention of breast and prostate cancer, menopausal symptoms, osteoporosis and osteopenia, cardiovascular diseases, and diabetes. Soyasaponins have been associated with the prevention of cancer and antidiabetic, anticoagulant, antithrombotic, lipid-lowering, and hypocholesterolemic effects.

To optimize the preservation or increase the bioavailability of beneficial compounds found in these by-products, further treatments can be employed. Leveraging the existing cohort of knowledge of post-processing methods, drying techniques and biotransformation can aid in the utilization as a functional food ingredient or the production of nutraceuticals. It has been well established that conversion of isoflavone aglycones can be achieved through many different physical and biochemical pathways.

As these by-products are abundant with low value, further studies are needed to investigate their use as food ingredients to add more value for the soybean industry and provide more health potentials for consumers. Existing research has shown that common dietary food items can have these industrial by-products incorporated in their formulations to produce functional foods which provide health benefits to consumers. These processing by-products are capable of being integrated into everyday food items to form a range of diets, such as the incorporation of SMB into white wheat flour for commercial bread, pasta, or noodle production. These foods are part of the everyday diets in many countries around the world, they can have the greatest global benefits to aid human health while reducing environmental impacts.

As the production of soybean products worldwide is increasing, the environmental and financial impacts are growing too. Real-world strategies that utilize large quantities of SPB need to be implemented to develop sellable products. As consumer interest in healthy food options is now common in mainstream markets, utilization of these waste materials as a nutritional ingredient has high market appeal.

REFERENCES

Aguado, A. C. (2010). *Development of Okara Powder as a Gluten Free Alternative to All Purpose Flour for Value Added Use in Baked Goods* (Master Of Science Master Thesis), University of Maryland, College Park, MD.

Ahlawat, D., & Punia, D. (2012). Effect of solar tunnel and freeze drying techniques on the organoleptic acceptability of products prepared incorporating okara (soy by-product). *International Journal of Food and Sciences*, *3*(3), 10–13.

Alvim, A., & Fochezatto, A. (2020). The impacts of soy-based biodiesel on the main soy producers in the international market. *Journal of Agricultural Studies*, *8*(2), 498.

Atmaca, A., Kleerekoper, M., Bayraktar, M., & Kucuk, O. (2008). Soy isoflavones in the management of post-menopausal osteoporosis. *Menopause*, *15*(4), 748–757.

Banaszkiewicz, T. (2011). Nutritional value of soybean meal. *Soybean and Nutrition*, 1–20.

Bargale, P., Ford, R., Sosulski, F., Wulfsohn, D., & Irudayaraj, J. (1999). Mechanical oil expression from extruded soybean samples. *JAOCS, Journal of the American Oil Chemists' Society*, *76*(2), 223–229.

Bedani, R., Rossi, E. A., Cavallini, D. C. U., Pinto, R. A., Vendramini, R. C., Augusto, E. M., ... Saad, S. M. I. (2015). Influence of daily consumption of synbiotic soy-based product supplemented with okara soybean by-product on risk factors for cardiovascular diseases. *Food Research International*, *73*, 142–148.

Behera, S., Indumathi, K., Mahadevamma, S., & Sudha, M. (2013). Oil cakes–a by-product of agriculture industry as a fortificant in bakery products. *International Journal of Food Sciences and Nutrition*, *64*(7), 806–814.

BeMiller, J. N. (2017). Carbohydrate analysis. In S. S. Nielsen (Ed.), *Food Analysis* (pp. 333–360). Cham: Springer International Publishing.

Bhosle, B., & Subramanian, R. (2005). New approaches in deacidification of edible oils––A review. *Journal of Food Engineering*, *69*(4), 481–494.

Bijkerk, C., De Wit, N., Muris, J., Whorwell, P., Knottnerus, J., & Hoes, A. (2009). Soluble or insoluble fibre in irritable bowel syndrome in primary care? Randomised placebo controlled trial. *BMJ: British Medical Journal*, *339*, b3154.

Cabral, C. E., & Klein, M. R. S. T. (2017). Phytosterols in the treatment of hypercholesterolemia and prevention of cardiovascular diseases. *Arquivos brasileiros de cardiologia*, *109*(5), 475–482.

Cederroth, C. R., & Nef, S. (2009). Soy, phytoestrogens and metabolism: A review. *Molecular and Cellular Endocrinology*, *304*(1–2), 30–42.

Chang, T., Wang, C., Wang, S., Shi, L., Yang, H., & Cui, M. (2014). Effect of okara on textural, color and rheological properties of pork meat gels. *Journal of Food Quality*, *37*(5), 339–348.

Cheng, M.-H., & Rosentrater, K. A. (2017a). Economic feasibility analysis of soybean oil production by hexane extraction. *Industrial Crops and Products*, *108*, 775–785.

Cheng, M.-H., & Rosentrater, K. A. (2017b). Profitability analysis of soybean oil processes. *Bioengineering*, *4*(4), 83.

Cheng, M.-H., Sekhon, J. J. K., Rosentrater, K. A., Wang, T., Jung, S., & Johnson, L. A. (2018). Environmental impact assessment of soybean oil production: Extruding-expelling process, hexane extraction and aqueous extraction. *Food and Bioproducts Processing*, *108*, 58–68.

Chiang, H.-S., Wu, W.-B., Fang, J.-Y., Chen, B.-H., Kao, T.-H., Chen, Y.-T., ... Hung, C.-F. (2007). UVB-protective effects of isoflavone extracts from soybean cake in human keratinocytes. *International Journal of Molecular Sciences*, *8*(7), 651–661.

Chiu, T.-M., Huang, C.-C., Lin, T.-J., Fang, J.-Y., Wu, N.-L., & Hung, C.-F. (2009). In vitro and in vivo anti-photoaging effects of an isoflavone extract from soybean cake. *Journal of Ethnopharmacology*, *126*(1), 108–113.

Chu, T., Wu, N.-L., Hsiao, C.-Y., Li, H.-J., Lin, T.-Y., Ku, C.-H., & Hung, C.-F. (2020). An isoflavone extract from soybean cake suppresses 2, 4-dinitrochlorobenzene-induced contact dermatitis. *Journal of Ethnopharmacology* 263, 113037.

Curtis, P. J., Sampson, M., Potter, J., Dhatariya, K., Kroon, P. A., & Cassidy, A. (2012). Chronic ingestion of flavan-3-ols and isoflavones improves insulin sensitivity and lipoprotein status and attenuates estimated 10-year CVD risk in medicated postmenopausal women with type 2 diabetes: A 1-year, double-blind, randomized, controlled trial. *Diabetes Care*, *35*(2), 226–232.

de Aguirre, J.M., Travaglini, D.A., Cabral, A.C.D., Travaglini, M.M.E., Silveira, E.T.F., Sales, A.M., ... Ferreira, V.L.P. (1981). The drying and storage of the residue from the water extraction soymilk process [Brazil]. *Boletim Do Instituto de Tecnologia de Alimentos*.

de Albuquerque, M. A. C., Bedani, R., Vieira, A. D. S., LeBlanc, J. G., & Saad, S. M. I. (2016). Supplementation with fruit and okara soybean by-products and amaranth flour increases the folate production by starter and probiotic cultures. *International Journal of Food Microbiology*, *236*, 26–32.

de Figueiredo, V. R. G., Yamashita, F., Vanzela, A. L. L., Ida, E. I., & Kurozawa, L. E. (2018). Action of multi-enzyme complex on protein extraction to obtain a protein concentrate from okara. *Journal of Food Science and Technology*, *55*(4), 1508–1517.

de Moraes Filho, M. L., Busanello, M., & Garcia, S. (2016). Optimization of the fermentation parameters for the growth of Lactobacillus in soymilk with okara flour. *LWT - Food Science and Technology, 74*, 456–464.

de Moraes Filho, M. L., Busanello, M., Prudencio, S. H., & Garcia, S. (2018). Soymilk with okara flour fermented by Lactobacillus acidophilus: Simplex-centroid mixture design applied in the elaboration of probiotic creamy sauce and storage stability. *LWT, 93*, 339–345.

de Oliveira Silva, F., Lemos, T. C., Sandôra, D., Monteiro, M., & Perrone, D. (2020). Fermentation of soybean meal improves isoflavone metabolism after soy biscuit consumption by adults. *Journal of the Science of Food and Agriculture, 100*(7), 2991–2998.

de Oliveira Silva, F., Miranda, T. G., Justo, T., da Silva Frasão, B., Conte-Junior, C. A., Monteiro, M., & Perrone, D. (2018). Soybean meal and fermented soybean meal as functional ingredients for the production of low-carb, high-protein, high-fiber and high isoflavones biscuits. *LWT, 90*, 224–231.

Delargy, H., O'sullivan, K., Fletcher, R., & Blundell, J. (1997). Effects of amount and type of dietary fibre (soluble and insoluble) on short-term control of appetite. *International Journal of Food Sciences and Nutrition, 48*(1), 67–77.

Dhingra, D., Michael, M., Rajput, H., & Patil, R. T. (2012). Dietary fibre in foods: A review. *Journal of Food Science and Technology, 49*(3), 255–266.

Easwar Rao, D., & Viswanatha Chaitanya, K. (2020). Changes in the antioxidant intensities of seven different soybean (Glycine max (L.) Merr.) cultivars during drought. *Journal of Food Biochemistry, 44*(2), e13118.

El Nasri, N. A., & El Tinay, A. (2007). Functional properties of fenugreek (Trigonella foenum graecum) protein concentrate (Trigonella foenum Graecum) protein concentrate. *Food Chemistry, 103*(2), 582–589.

Eldridge, A. C. (1982). Determination of isoflavones in soybean flours, protein concentrates, and isolates. *Journal of Agricultural and Food Chemistry, 30*(2), 353–355.

Elleuch, M., Bedigian, D., Roiseux, O., Besbes, S., Blecker, C., & Attia, H. (2011). Dietary fibre and fibre-rich by-products of food processing: Characterisation, technological functionality and commercial applications: A review. *Food Chemistry, 124*(2), 411–421.

Endres, J. G. (2001). *Soy Protein Products: Characteristics, Nutritional Aspects, and Utilization.* The American Oil Chemists Society.

Erickson, D. R. (1995). Degumming and lecithin processing and utilization. In D. R. Erickson (Ed.). *Practical Handbook of Soybean Processing and Utilization* (pp. 174–183). AOCS Press.

Falcão, H. G., Seibel, N. F., & Yamaguchi, M. M. (2015). Optimization of beef patties formulation with textured soy protein, okara and bacon using a simplex-centroid mixture design. *International Journal of Latest Research in Science and Technolog, 4*(6), 104–109.

FAO/UN (2013). *Dietary Protein Quality Evaluation in Human Nutrition: Report of an FAO Expert Consultation, 31 March–2 April, 2011.* Auckland, New Zealand: Food and Agriculture Organization of the United Nations.

Fei, L., Zhenkun, C., Yang, L., & Bo, L. (2013). The Effect of okara on the qualities of noodle and steamed bread. *Advance Journal of Food Science and Technology, 5*(7), 960–968.

Ghorbanpour, M. (2018). Soybean oil bleaching by adsorption onto bentonite/iron oxide nanocomposites. *Journal of Physical Science, 29*(2), 113–119.

Grieshop, C. M., Kadzere, C. T., Clapper, G. M., Flickinger, E. A., Bauer, L. L., Frazier, R. L., & Fahey, G. C. (2003). Chemical and nutritional characteristics of United States soybeans and soybean meals. *Journal of Agricultural and Food Chemistry, 51*(26), 7684–7691.

Grizotto, R. K., & Aguirre, J. M. d. (2011). Study of the flash drying of the residue from soymilk processing-" okara". *Food Science and Technology (Campinas), 31*(3), 645–653.

Grizotto, R. K., Aguirre, J. M. d., Bruns, R. E., Bombonati, A. Y., & Claus, M. L. (2008). *Physical and Chemical Characteristics of Soymilk Residue ("Okara") Obtained from Brazilian Soybean Cultivars.* International Commission of Agricultural Engineering (CIGR), Institut fur Landtechnik.

Grizotto, R. K., Andrade, J. C. d., Miyagusku, L., & Yamada, E. A. (2012). Physical, chemical, technological and sensory characteristics of Frankfurter type sausage containing okara flour. *Food Science and Technology, 32*(3), 538–546.

Guimarães, R. M., Ida, E. I., Falcão, H. G., Rezende, T. A. M. d., Silva, J. d. S., Alves, C. C. F., ... Egea, M. B. (2020). Evaluating technological quality of okara flours obtained by different drying processes. *LWT, 123*, 109062.

Guimarães, R. M., Silva, T. E., Lemes, A. C., Boldrin, M. C. F., da Silva, M. A. P., Silva, F. G., & Egea, M. B. (2018). Okara: A soybean by-product as an alternative to enrich vegetable paste. *LWT, 92*, 593–599.

Hayashi, K., Hayashi, H., Hiraoka, N., & Ikeshiro, Y. (1997). Inhibitory activity of soyasaponin II on virus replication in vitro. *Planta Medica*, *63*(2), 102–105.

Hong, K.-J., Lee, C.-H., & Kim, S. W. (2004). Aspergillus oryzae GB-107 fermentation improves nutritional quality of food soybeans and feed soybean meals. *Journal of Medicinal Food*, *7*(4), 430–435.

Ibáñez, M., de Blas, C., Cámara, L., & Mateos, G. (2020). Chemical composition, protein quality and nutritive value of commercial soybean meals produced from beans from different countries: A meta-analytical study. *Animal Feed Science and Technology*, *267*, 114531.

Ibisworld (2019). Cooking oil production in China. Retrieved from http://ezproxy.newcastle.edu.au/login?url=http://search.ebscohost.com/login.aspx?direct=true&db=edsibw&AN=edsibw.41A505A5&site=eds-live.

Iqbal, N., Hussain, S., Raza, M. A., Yang, C., Safdar, M. E., Brestic, M., … Yang, W. (2019). Drought tolerance of soybean (Glycine max L. Merr.) by improved photosynthetic characteristics and an efficient antioxidant enzyme system under a split-root system. *Frontiers in Physiology*, *10*, 786.

Jacobsen, H. J., Kousoulaki, K., Sandberg, A.-S., Carlsson, N.-G., Ahlstrøm, Ø., & Oterhals, Å. (2018). Enzyme pre-treatment of soybean meal: Effects on non-starch carbohydrates, protein, phytic acid, and saponin biotransformation and digestibility in mink (Neovison vison). *Animal Feed Science and Technology*, *236*, 1–13.

Jankowiak, L., Kantzas, N., Boom, R., & van der Goot, A. J. (2014). Isoflavone extraction from okara using water as extractant. *Food Chemistry*, *160*, 371–378.

Jaworski, N., & Stein, H. (2017). Disappearance of nutrients and energy in the stomach and small intestine, cecum, and colon of pigs fed corn-soybean meal diets containing distillers dried grains with solubles, wheat middlings, or soybean hulls. *Journal of Animal Science*, *95*(2), 727–739.

Jiménez-Escrig, A., Tenorio, M. D., Espinosa-Martos, I., & Rupérez, P. (2008). Health-promoting effects of a dietary fiber concentrate from the soybean byproduct okara in rats. *Journal of Agricultural and Food Chemistry*, *56*(16), 7495–7501.

Jing, X. (2009). Change of content and configuration of isoflavones in process of soybean residue fermentation with Mucor. *China Brewing*, *5*, 033.

Jooyandeh, H. (2011). Soy products as healthy and functional foods. *Middle-East Journal of Scientific Research*, *7*(1), 71–80.

Kamble, D. B., & Rani, S. (2020). Bioactive components, in vitro digestibility, microstructure and application of soybean residue (okara): A review. *Legume Science*, *2*(1), e32.

Kamble, D. B., Singh, R., Rani, S., & Pratap, D. (2019). Physicochemical properties, in vitro digestibility and structural attributes of okara-enriched functional pasta. *Journal of Food Processing and Preservation*, *43*(12), e14232.

Kang, M., Bae, I., & Lee, H. (2018). Rice noodle enriched with okara: Cooking property, texture, and in vitro starch digestibility. *Food Bioscience*, *22*, 178–183.

Kang, X., Zhang, Q., Wang, S., Huang, X., & Jin, S. (2010). Effect of soy isoflavones on breast cancer recurrence and death for patients receiving adjuvant endocrine therapy. *Canadian Medical Association Journal*, *182*(17), 1857–1862.

Kao, T.-H., & Chen, B.-H. (2006). Functional components in soybean cake and their effects on antioxidant activity. *Journal of Agricultural and Food Chemistry*, *54*(20), 7544–7555.

Kao, T.-H., Chien, J.-T., & Chen, B.-H. (2008). Extraction yield of isoflavones from soybean cake as affected by solvent and supercritical carbon dioxide. *Food Chemistry*, *107*(4), 1728–1736.

Kao, T.-H., Huang, R.-F. S., & Chen, B.-H. (2007a). Antiproliferation of hepatoma cell and progression of cell cycle as affected by isoflavone extracts from soybean cake. *International Journal of Molecular Sciences*, *8*(11), 1095–1110.

Kao, T., Lu, Y., & Chen, B. (2005). Preparative column chromatography of four groups of isoflavones from soybean cake. *European Food Research and Technology*, *221*(3–4), 459–465.

Kao, T., Wu, W., Hung, C., Wu, W., & Chen, B. (2007b). Anti-inflammatory effects of isoflavone powder produced from soybean cake. *Journal of Agricultural and Food Chemistry*, *55*(26), 11068–11079.

Karr-Lilienthal, L. K., Bauer, L. L., Utterback, P. L., Zinn, K. E., Frazier, R. L., Parsons, C. M., & Fahey, G. C. (2006). Chemical composition and nutritional quality of soybean meals prepared by extruder/expeller processing for use in poultry diets. *Journal of Agricultural and Food Chemistry*, *54*(21), 8108–8114.

Katayama, M., & Wilson, L. (2008). Utilization of okara, a byproduct from soymilk production, through the development of soy-based snack food. *Journal of Food Science*, *73*(3), S152–S157.

Khare, S. K., Jha, K., & Gandhi, A. P. (1995). Citric acid production from Okara (soy-residue) by solid-state fermentation. *Bioresource Technology*, *54*(3), 323–325.

Kim, S.-M., Seo, K.-I., Park, K.-W., Jeong, Y.-K., Cho, Y.-S., Kim, M.-J., … Lee, M.-K. (2009). Effect of crude saponins from soybean cake on body weight and glucose tolerance in high-fat diet induced obese mice. *Journal of the Korean Society of Food Science and Nutrition, 38*(1), 39–46.

Kralik, G., Gajčević, Z., & Škrtić, F. (2008). The effect of different oil supplementations on laying performance and fatty acid composition of egg yolk. *Italian Journal of Animal Science, 7*(2), 173–183.

Kuleasan, S., & Tekin, A. (2008). Alkaline neutralization of crude soybean oil by various adsorbents. *European Journal of Lipid Science and Technology, 110*(3), 261–265.

Kumar, V., Rani, A., & Husain, L. (2016). Investigations of amino acids profile, fatty acids composition, isoflavones content and antioxidative properties in soy okara. *Asian Journal of Chemistry, 28*(4), 903.

Lee, Y.-B., Lee, H.-J., Kim, C.-H., Lee, S.-B., & Sohn, H.-S. (2005). Soy isoflavones and soyasaponins: Characteristics and physiological functions. *Journal of Applied Biological Chemistry, 48*(2), 49–57.

Li, W., Chenjie, W., Tong, C., Liu, S., Hong, Y., & Min, C. (2014). Effect of okara on the sensory quality of cake. *Research in Health and Nutrition, 2*, 1–4.

Liener, I. E. (1994). Implications of antinutritional components in soybean foods. *Critical Reviews in Food Science and Nutrition, 34*(1), 31–67.

Liu, K. (1997). *Soybeans : Chemistry, Technology, and Utilization*. New York: Chapman & Hall.

Liu, Y., Huang, J., & Wang, X. (2008). Adsorption isotherms for bleaching soybean oil with activated attapulgite. *Journal of the American Oil Chemists' Society, 85*(10), 979–984.

Lu, F., Li, B., Zhang, Y., & Zhang, Z. (2011). Application of bean curd residue in bread. *Science and Technology of Food Industry, 9*, 085.

Lu, F., Liu, Y., & Li, B. (2013). Okara dietary fiber and hypoglycemic effect of okara foods. *Bioactive Carbohydrates and Dietary Fibre, 2*(2), 126–132.

Luján-Rhenals, D., Morawicki, R. O., Mendez-Montealvo, G., & Wang, Y. J. (2014). Production of a high-protein meal and fermentable sugars from defatted soybean meal, a co-product of the soybean oil industry. *International Journal of Food Science and Technology, 49*(3), 904–910.

Mann, J., & Cummings, J. (2009). Possible implications for health of the different definitions of dietary fibre. *Nutrition, Metabolism, and Cardiovascular Diseases, 19*(3), 226–229.

Mareček, V., Mikyška, A., Hampel, D., Čejka, P., Neuwirthová, J., Malachová, A., & Cerkal, R. (2017). ABTS and DPPH methods as a tool for studying antioxidant capacity of spring barley and malt. *Journal of Cereal Science, 73*, 40–45.

Mateos-Aparicio, I., Mateos-Peinado, C., & Rupérez, P. (2010). High hydrostatic pressure improves the functionality of dietary fibre in okara by-product from soybean. *Innovative Food Science and Emerging Technologies, 11*(3), 445–450.

McDougall, G. J., Morrison, I. M., Stewart, D., & Hillman, J. R. (1996). Plant cell walls as dietary fibre: Range, structure, processing and function. *Journal of the Science of Food and Agriculture, 70*(2), 133–150.

Melina, V., Craig, W., & Levin, S. (2016). From the academy: Position of the Academy of Nutrition and Dietetics: Vegetarian diets. *Journal of the Academy of Nutrition and Dietetics, 116*(12), 1970–1980.

Méndez Sevillano, D., Jankowiak, L., van Gaalen, T. L. T., van der Wielen, L. A. M., Hooshyar, N., van der Goot, A.-J., & Ottens, M. (2014). Mechanism of isoflavone adsorption from okara extracts onto food-grade resins. *Industrial and Engineering Chemistry Research, 53*(39), 15245–15252.

Messina, M., & Barnes, S. (1991). The role of soy products in reducing risk of cancer. *Journal of the National Cancer Institute, 83*(8), 541–546.

Mishra, S., & Manchanda, S. (2012). Cooking oils for heart health. *J Pre Cardio, 1*(3), 123–131.

Mitsuoka, T. (2002). Prebiotics and intestinal flora. *Bioscience and Microflora, 21*(1), 3–12.

Nasution, Z., Tung, M., & Faridah, Y. (2012). Okara: An alternative ingredient for partial substitution of peanuts in peanut butter. *Peanut Science, 13*(1), 18–20.

Nelson, A. I., Wijeratne, W. B., Yeh, S. W., Wei, T., & Wei, L. (1987). Dry extrusion as an aid to mechanical expelling of oil from soybeans. *Journal of the American Oil Chemist Society, 64*(9), 1341–1347.

Noriham, A., Ariffaizuddin, R. M., Noorlaila, A., & Zakry, A. F. (2016). Potential use of okara as meat replacer in beef sausage. *Jurnal Teknologi, 78*(6–6), 13–18.

O'Toole, D. K. (1999). Characteristics and use of okara, the soybean residue from soy milk production – A review. *Journal of Agricultural and Food Chemistry, 47*(2), 363–371.

Ostermann-Porcel, M. V., Quiroga-Panelo, N., Rinaldoni, A. N., & Campderrós, M. E. (2017). Incorporation of okara into gluten-free cookies with high quality and nutritional value. *Journal of Food Quality, 2017*.

Ostermann-Porcel, M. V., Rinaldoni, A. N., Rodriguez-Furlán, L. T., & Campderrós, M. E. (2017). Quality assessment of dried okara as a source of production of gluten-free flour. *Journal of the Science of Food and Agriculture, 97*(9), 2934–2941.

Park, J., Choi, I., & Kim, Y. (2015). Cookies formulated from fresh okara using starch, soy flour and hydroxypropyl methylcellulose have high quality and nutritional value. *LWT - Food Science and Technology*, *63*(1), 660–666.

Park, K.-U., Kim, J.-Y., & Seo, K.-I. (2009). Antioxidative and cytotoxicity activities against human colon cancer cells exhibited by edible crude saponins from soybean cake. *Korean Journal of Food Preservation*, *16*(5), 754–758.

Park, K.-U., Wee, J.-J., Kim, J.-Y., Jeong, C.-H., Kang, K.-S., Choi, Y.-S., & Seo, K.-I. (2005). Anticancer and immuno-activities of edible crude saponin from soybean cake. *Journal of the Korean Society of Food Science and Nutrition*, *34*(10), 1509–1513.

Pérez-López, E., Mateos-Aparicio, I., & Rupérez, P. (2016a). Low molecular weight carbohydrates released from Okara by enzymatic treatment under high hydrostatic pressure. *Innovative Food Science and Emerging Technologies*, *38*, 76–82.

Pérez-López, E., Mateos-Aparicio, I., & Rupérez, P. (2016b). Okara treated with high hydrostatic pressure assisted by Ultraflo® L: effect on solubility of dietary fibre. *Innovative Food Science and Emerging Technologies*, *33*, 32–37.

Pérez-López, E., Mateos-Aparicio, I., & Rupérez, P. (2017). High hydrostatic pressure aided by food-grade enzymes as a novel approach for Okara valorization. *Innovative Food Science and Emerging Technologies*, *42*, 197–203.

Pizzino, G., Irrera, N., Cucinotta, M., Pallio, G., Mannino, F., Arcoraci, V., … Bitto, A. (2017). Oxidative stress: Harms and benefits for human health. *Oxidative Medicine and Cellular Longevity*, 169.

Puechkamut, Y., & Panyathitipong, W. (2012). Characteristics of proteins from fresh and dried residues of soy milk production. *Kasetsart Journal of Natural Science*, *46*(5), 804–811.

Quilez, J., Garcia-Lorda, P., & Salas-Salvado, J. (2003). Potential uses and benefits of phytosterols in diet: Present situation and future directions. *Clinical Nutrition*, *22*(4), 343–351.

Quintana, G., Gerbino, E., & Gómez-Zavaglia, A. (2017). Okara: A nutritionally valuable by-product able to stabilize Lactobacillus plantarum during freeze-drying, spray-drying, and storage. *Frontiers in Microbiology*, *8*, 641.

Quitain, A. T., Oro, K., Katoh, S., & Moriyoshi, T. (2006). Recovery of oil components of okara by ethanol-modified supercritical carbon dioxide extraction. *Bioresource Technology*, *97*(13), 1509–1514.

Radočaj, O., & Dimić, E. (2013). Valorization of wet okara, a value-added functional ingredient, in a coconut-based baked snack. *Cereal Chemistry Journal*, *90*(3), 256–262.

Redondo-Cuenca, A., Villanueva-Suárez, M. J., & Mateos-Aparicio, I. (2008). Soybean seeds and its by-product okara as sources of dietary fibre. Measurement by AOAC and Englyst methods. *Food Chemistry*, *108*(3), 1099–1105.

Rice-Evans, C., & Miller, N. (1996). Antioxidant activities of flavonoids as bioactive components of food. *Biochemical Society Transactions*, *24*(3), 790–795.

Rice-Evans, C., Miller, N., & Paganga, G. (1997). Antioxidant properties of phenolic compounds. *Trends in Plant Science*, *2*(4), 152–159.

Rinaldi, V., Ng, P., & Bennink, M. (2000). Effects of extrusion on dietary fiber and isoflavone contents of wheat extrudates enriched with wet okara. *Cereal Chemistry Journal*, *77*(2), 237–240.

D'Adamo, C.R. & Sahin, A. (2014). Soy foods and supplementation: A review of commonly perceived health benefits and risks. *Alternative Therapies in Health and Medicine*, *20*(Supplement 1), 39.

Santos, V. A. Q., Nascimento, C. G., Schimidt, C. A., Mantovani, D., Dekker, R. F., & da Cunha, M. A. A. (2018). Solid-state fermentation of soybean okara: Isoflavones biotransformation, antioxidant activity and enhancement of nutritional quality. *LWT*, *92*, 509–515.

Sengupta, S., Chakraborty, M., Bhowal, J., & Bhattacharya, D. (2012). Study on the effects of drying process on the composition and quality of wet okara. *International Journal of Science. Environmental and Technology*, *1*(4), 319–330.

Sharifi, M., Naserian, A. A., & Khorasani, H. (2013). *Effect of Tannin Extract from Pistachio by Product on In Vitro Gas Production*.

Shimada, Y., Nakai, S., Suenaga, M., Sugihara, A., Kitano, M., & Tominaga, Y. (2000). Facile purification of tocopherols from soybean oil deodorizer distillate in high yield using lipase. *Journal of the American Oil Chemists' Society*, *77*(10), 1009–1013.

Shimoyamada, M., Ikedo, S., Ootsubo, R., & Watanabe, K. (1998). Effects of soybean saponins on chymotryptic hydrolyses of soybean proteins. *Journal of Agricultural and Food Chemistry*, *46*(12), 4793–4797.

Silva, F. d. O., & Perrone, D. (2015). Characterization and stability of bioactive compounds from soybean meal. *LWT - Food Science and Technology*, *63*(2), 992–1000.

Silva, L. H. d., Paucar-Menacho, L. M., Vicente, C. A., Salles, A. S., Steel, C. J., & Chang, Y. (2009). Development of loaf bread with the addition of "okara" flour. *Brazilian Journal of Food Technology, 12*(1/4), 315–322.

Simpson, H. C. R., Lousley, S., Geekie, M., Simpson, R. W., Carter, R. D., Hockaday, T. D. R., & Mann, J. I. (1981). A high carbohydrate leguminous fibre diet improves all aspects of diabetic control. *The Lancet, 317*(8210), 1–5.

Singh, B., Singh, J. P., Singh, N., & Kaur, A. (2017a). Saponins in pulses and their health promoting activities: A review. *Food Chemistry, 233*, 540–549.

Singh, B., Yadav, D., & Vij, S. (2017b). Soybean bioactive molecules: Current trend and future prospective. In *Bioactive Molecules in Food* (pp. 1–29). Springer International Publishing.

Stanojevic, S. P., Barac, M. B., Pesic, M. B., Jankovic, V. S., & Vucelic-Radovic, B. V. (2013). Bioactive proteins and energy value of okara as a byproduct in hydrothermal processing of soy milk. *Journal of Agricultural and Food Chemistry, 61*(38), 9210–9219.

Stanojevic, S. P., Barac, M. B., Pesic, M. B., & Vucelic-Radovic, B. V. (2012). Composition of proteins in okara as a byproduct in hydrothermal processing of soy milk. *Journal of Agricultural and Food Chemistry, 60*(36), 9221–9228.

Su, S. I. T., Yoshida, C. M. P., Contreras-Castillo, C. J., Quiñones, E. M., & Venturini, A. C. (2013). Okara, a soymilk industry by-product, as a non-meat protein source in reduced fat beef burgers. *Food Science and Technology (Campinas), 33*, 52–56.

Sulieman, S., Ha, C. V., Nasr Esfahani, M., Watanabe, Y., Nishiyama, R., Pham, C. T. B., … Tran, L.-S. P. (2015). DT2008: A promising new genetic resource for improved drought tolerance in soybean when solely dependent on symbiotic N2 fixation. *BioMed Research International, 2015*.

Tai, Y.-H., & Lin, C.-I. (2007). Variation of peroxide value in water-degummed and alkali-refined soy oil during bleaching under vacuum. *Separation and Purification Technology, 56*(3), 257–264.

Tantry, M. A., & Khan, I. A. (2013). Saponins from Glycine max Merrill (soybean). *Fitoterapia, 87*, 49–56.

Tavares, B. O., Silva, E. P. d., Silva, V. S. N. d., Soares Junior, M. S., Ida, E. I., & Damiani, C. (2016). Stability of gluten free sweet biscuit elaborated with rice bran, broken rice and okara. *Food Science and Technology (Campinas), 36*(2), 296–303.

Taylor, M. R., Brester, G. W., & Boland, M. A. (2005). Hard white wheat and gold medal flour: General Mills' contracting program. *Review of Agricultural Economics, 27*(1), 117–129.

Torres, C. F., Fornari, T., Torrelo, G., Señoráns, F. J., & Reglero, G. (2009). Production of phytosterol esters from soybean oil deodorizer distillates. *European Journal of Lipid Science and Technology, 111*(5), 459–463.

Torres, C. F., Torrelo, G., Senorans, F. J., & Reglero, G. (2007). A two steps enzymatic procedure to obtain sterol esters, tocopherols and fatty acid ethyl esters from soybean oil deodorizer distillate. *Process Biochemistry, 42*(9), 1335–1341.

Travaglini, D., Silveira, E., Travaglini, M., Vitti, P., Pereira, L., de Aguirre, J., … Figueiredo, I. (1980). The processing of soy milk residue mixed with corn grits. *Boletim do instituto de tecnologia de alimentos (Brazil), 17*(3), 275–296.

Trowell, H., Burkitt, D., & Heaton, K. (1985). *Dietary Fibre, Fibre-Depleted Foods and Disease*. Academic Press, Orlando, Florida.

Turhan, S., Temiz, H., & Sagir, I. (2007). Utilization of wet okara in low-fat beef patties. *Journal of Muscle Foods, 18*(2), 226–235.

Van der Riet, W., Wight, A., Cilliers, J., & Datel, J. (1989). Food chemical investigation of tofu and its byproduct okara. *Food Chemistry, 34*(3), 193–202.

Varzakas, T., Zakynthinos, G., & Verpoort, F. (2016). Plant food residues as a source of nutraceuticals and functional foods. *Foods, 5*(4), 88.

Vital, A. C. P., Croge, C., da Silva, D. F., Araújo, P. J., Gallina, M. Z., & Matumoto-Pintro, P. T. (2018). Okara residue as source of antioxidants against lipid oxidation in milk enriched with omega-3 and bioavailability of bioactive compounds after in vitro gastrointestinal digestion. *Journal of Food Science and Technology, 55*(4), 1518–1524.

Vong, W. C., & Liu, S.-Q. (2016). Biovalorisation of okara (soybean residue) for food and nutrition. *Trends in Food Science and Technology, 52*, 139–147.

Voss, G., Rodríguez-Alcalá, L., Valente, L., & Pintado, M. (2018). Impact of different thermal treatments and storage conditions on the stability of soybean byproduct (okara). *Journal of Food Measurement and Characterization, 12*(3), 1981–1996.

Wachiraphansakul, S., & Devahastin, S. (2007). Drying kinetics and quality of okara dried in a jet spouted bed of sorbent particles. *LWT - Food Science and Technology, 40*(2), 207–219.

Waldroup, P., & Waldroup, A. (2005). Fatty acid effect on carcass: The influence of various blends of dietary fats added to corn-soybean meal based diets on the fatty acid composition of broilers. *International Journal of Poultry Science, 4*(3), 123–132.

Wang, B. F., Wang, J. S., Lu, J. F., Kao, T. H., & Chen, B. H. (2009). Antiproliferation effect and mechanism of prostate cancer cell lines as affected by isoflavones from soybean cake. *Journal of Agricultural and Food Chemistry, 57*(6), 2221–2232.

Wiboonsirikul, J., Mori, M., Khuwijitjaru, P., & Adachi, S. (2013). Properties of extract from okara by its subcritical water treatment. *International Journal of Food Properties, 16*(5), 974–982.

Wickramarathna, G., & Arampath, P. (2003). Utilization of okara in bread making. *Journal of Bio-Science, 31*, 29–33.

Wiedermann, L. (1981). Degumming, refining and bleaching soybean oil. *Journal of the American Oil Chemists' Society, 58*(3Part1), 159–166.

Wiseman, J., & Salvador, F. (1991). The influence of free fatty acid content and degree of saturation on the apparent metabolizable energy value of fats fed to broilers. *Poultry Science, 70*(3), 573–582.

Xiao, L. C. D. H. L., & Xiangyang, W. Z. L. (2010). Effects of extruded soybean residue on dough characteristics and bread quality of flour. *Journal of the Chinese Cereals and Oils Association, 12*, 005.

Yang, J., Wu, X.-b., Chen, H.-l., Sun-waterhouse, D., Zhong, H.-b., & Cui, C. (2019). A value-added approach to improve the nutritional quality of soybean meal byproduct: Enhancing its antioxidant activity through fermentation by Bacillus amyloliquefaciens SWJS22. *Food Chemistry, 272*, 396–403.

Yang, L., Bo, L., Fei, L., & Mingshan, S. (2012). Effect of okara powder on textural properties of bread dough. *Journal of Henan Institute of Science and Technology (Natural Sciences Edition), 3*, 015.

Yu, H., Liu, R., Hu, Y., & Xu, B. (2018). Flavor profiles of soymilk processed with four different processing technologies and 26 soybean cultivars grown in China. *International Journal of Food Properties, 20*(Sup 3), 1–12.

Zaheer, K., & Humayoun Akhtar, M. (2017). An updated review of dietary isoflavones: Nutrition, processing, bioavailability and impacts on human health. *Critical Reviews in Food Science and Nutrition, 57*(6), 1280–1293.

Zamindar, N., Bashash, M., Khorshidi, F., Serjouie, A., Shirvani, M. A., Abbasi, H., Sedaghatdoost, A. (2017). Antioxidant efficacy of soybean cake extracts in soy oil protection. *Journal of Food Science and Technology, 54*(7), 2077–2084.

Zehnder, C. T. (1995). Deodorization. In *Practical Handbook of Soybean Processing and Utilization* (pp. 239–257). Elsevier.

Zhong-Hua, L., Hong-Lian, G., Rui-Ling, L., & Jin-Hui, Z. (2015). Extraction and antioxidant activity of soybean saponins from low-temperature soybean meal by MTEH. *Open Biotechnology Journal, 9*(1), 178–184.

Zuo, Y. B., Zeng, A. W., Yuan, X. G., & Yu, K. T. (2008). Extraction of soybean isoflavones from soybean meal with aqueous methanol modified supercritical carbon dioxide. *Journal of Food Engineering, 89*(4), 384–389.

16 Korean Traditional Fermented Soybean Foods and Their Functionalities

Dong Hwa Shin and Su Jin Jung

CONTENTS

DOI: 10.1201/9781003030294-16

16.1   INTRODUCTION

Soybeans have contributed to the human race not only as an important food resource but also as the most important protein source besides animal proteins in Asian countries. Southeast Asian countries including Korea, China, and Japan have utilized various soybean fermented foods for a long time, including soybeans themselves. In recent years, consumption of fermented soybean products has significantly increased due to some functionalities, such as the creation of lots of bioactive substances through the fermentation process.

Until some years ago, soybean products in some countries like the USA, Canada, and South American countries were mainly used as a source for obtaining oils and animal feed, but recently soybeans have been used as food resources due to the functionalities of some components in soybeans. In particular, it has been highlighted that soybeans represent new tastes and flavors in addition to presenting bioactive effects by creating new functional substances through the fermentation process. Although soybean proteins are insoluble, they can be decomposed to soluble amino acids or peptides through fermentation and that significantly affects their tastes and flavors. In particular, it improves their functionalities by decomposing flavonoid substances through the action of enzymes, which are generated by microorganisms during the fermentation process, and that has been stressed as an additional merit in fermented soybean foods.

Functionalities in fermented foods have been well known throughout the world and many countries try to present the excellence of their own foods by combining it with their dietary cultures. Advantages of fermented foods are that they create new tastes and flavors based on microorganisms, and the microorganisms' functionalities have also been emphasized. From a different viewpoint, it is considered that the probiotic role in microorganisms contributes to the fermentation processes. Studies on the role of microorganisms applied to the guaranteed safety of fermented foods have been actively conducted and may introduce different positive roles in future.

Special fermented soybean foods in Korea are one of the typical traditional foods and have shown great popularity in their dietary culture. There are plenty of research works related to typical processing methods and functionalities related those products.

In this chapter we will present unique process methods and functionalities of each product including cultural background and microorganisms related to fermentation based on various research works until now.

16.2 OVERALL REVIEW OF KOREAN FERMENTED SOYBEAN PRODUCTS

16.2.1 CULTURAL BACKGROUND

In the Korean dietary culture, fermented products are an important base of the ordinary daily diet and go back more than a century. The main resources of eating habits are grains, including rice and barley, etc. Most of the grains are boiled and eaten with some other side dishes including fermented soybean products for giving a salty and umami taste. According to the above conditions, various different fermented foods have been introduced utilizing various agricultural products, such grains, vegetables, and beans, before the Christian era, and that becomes an opportunity for presenting differences and originalities in Korean-style foods from generation to generation. In particular, different alcoholic beverages, rice cakes, meats, and fish are offered in memorial services based on the harvest ceremony in agricultural cultures, and that allows the establishment of producing techniques of various alcoholic beverages and special food products. Fermented foods have been produced based on natural phenomena at the beginning of the world and have largely been developed by various fermentation techniques through improving their quality and diversifying their products. Fermented foods in Korea have been presented in a coexistent manner together with traditional and industrial ways. These foods still have been produced as different products and have been distributed in different ways. Representative traditional soybean fermented foods in Korea are simply summarized in Table 16.1. The detailed production methods of each product and their characteristics are to be introduced in the following sections.

Fermented soybean products mostly use soybeans as a major matrix and are based on the functions of microorganisms. There are only a few fermented foods based on grains or vegetables in Western dietary habits, and soybean products are also very limited. In the East, however, there are different soybean products and fermented soybean products have been used as the most important source of seasoning in their dishes. The history of producing such fermented soybean foods goes back to before the Christian era. Fermented soybean products have been used to provide some salty and savory flavors in cooked vegetables and meats for Korean main diets of boiled rice and dishes. It has been largely used in most Korean ethnic foods as a seasoning. Also, Korean foods represent different tastes and flavors depending on the fermented soybean products used compared to other global foods.

The fermented soybean products produced in Korea are largely divided into four different types, namely *ganjang*, *doenjang*, *gochujang*, and *cheonggukjang*. These products show different main and extra ingredients in their fermentation processes and represent different seasoning methods in

---

TABLE 16.1

Ethnic Fermented Soybean Foods and Their Characteristics in Korea

Raw Material	Food Name	Characteristics
Soybean	*Ganjang*	Soy sauce. Brown colored liquid.
	Doenjang	Brown colored paste made by *meju*.
	Gochujang	Red pepper used. Fermented spicy condiment.
	Cheonggukjang	Solely soybean used and fermented by *Bacillus*. High-temperature fermentation (40–43 °C)

processing foods (Shin et al., 2011). The most important thing in producing such fermented soybean products is the main raw material, soybeans. Soybeans are stored and distributed as dried products after harvest and represent quality differences between varieties. In addition to the quality of the main material, the fermentation process and condition determined by microorganisms significantly affect the quality and properties of final products.

Functionalities of fermented soybean products are scientifically known well. Various functional ingredients are already included in soybeans. In particular, a large amount of isoflavones (genistin, genestein, daidzin, and daidzein), which are a type of phytochemical, represent some functionalities in themselves (Liu, 1997). In addition, phenolic compounds (syringic, vanillic, chlorogenic, ferulic, and cinnamic acid), lignan, carotenoids (lutein and α- and β-carotene), fatty acids like linoleic acid, linolenic acid, and lecithin, choline, and saponin are included in soybean (Park, 2009). As these ingredients in soybeans represent their own particular physiological activities, the value of soybean is reevaluated by the people who have no concern with using soybean as foods. In addition, according to some related studies on such issues, it has been known that fermented soybean products create new functional substances by action of various microbes.

16.2.2   THE ORIGIN OF SOYBEAN

It is hard to say when we began to use soybean in the human diet. According to archeological records, the first cultivation year of soybean is estimated to be about 4,000 years ago. Globally, wild soybeans were found in Korea as the living base of ancient Dongyi tribes, along the coast of the Yangtze River in China, Manchuria, and Siberia (Nawaz et al., 2020). In general, in the estimation of the origin of crops, wild variety is a very important index to predict the origin of the cultivation region. In the Korean Peninsula, carbonized soybeans were found in some areas in the Bronze Age and that means evidence for the cultivation of soybeans (Kwon et al., 2006). It is clear that soybean first emerged as a domesticated crop in northern China, Manchuria, and the Korean Peninsula. Table 16.2 shows the historic sites of soybean that were found in the Korean Peninsula according to their era (Lee, 1992).

It has been known that the cultivation of soybean originated in the era of the middle period in the agricultural age of the New Stone Age to the Bronze Age (around 1500 BC). Some pieces of earthenware studded with soybean in the Bronze Age were found at the Paldang submerged area in the suburb of Seoul. Choi (2009) proposed that the first people who cultivated soybean as a food in the history of mankind lived around 4000–2000 BC. Also, it is assumed that the first cultivation of soybeans was started in arable fields at the hub of Baekdu Mountain. It is considered that soybean has contributed to the table of Dongyi tribes, ancestors of the Korean people, as an important protein source in such primitive ages that showed insufficient nutrients. Thus, it can be seen that

TABLE 16.2
Unearthed Remains of Soybean in Korean Peninsula

Remains	Kind	State	Period (Earthenware)
Ohdong, Heryong County, Hambuk	Each grain of soybean, red bean as carbonization	Bottom of habitat site, Bronze Age	Bronze/Iron Age (minimum pottery)
Honam, Samseok section, Pyongyang	Carbonization grain of foxtail millet, proso millet, sorghum bicolor, soybean	Habitat site of No. 36	Bronze Age (top type pottery)
Submerged area, Gyeonggi Paldang	Soybean, red beans		Bronze Age
Buwon-dong, Gimhae County, Gyeongnam	Rice, wheat, and husk of soybeans (3 each)	A district, 11th floor	

soybean plays an important role in forming the early countries in Northeast Asia due to the increase in military powers on the basis of improving their nutrients (Kwon et al., 2005a).

16.2.3 Raw Materials

For making fermented soybean products, the major raw materials are soybean, rice, red pepper, salt, and some other grains including barley.

16.2.3.1 Soybean

Soybean species cultured in Korea can be classified into beans used for fermented condiments, vegetable salads, cooking with rice, and unripe green beans. About 43 different species are cultivated for producing various soybean products; the Jangyup soybean was first bred in 1978 and Jungmo 3003 was bred in 2009 (Rural Development Administration, National Institute of Crop Science, 2014). Some species are chosen for fermented products. Continuous studies are required to develop ideal species for each target usage. The main components of soybeans are protein (30–50%) and fat (14–24%), and soybean contains the greatest amount of protein among vegetables. Most of the protein (63–90%) is glycinin, such as globulin and phaseolin (17%), and legumelin is present as well. These are insoluble proteins that do not possess any specific taste based on their high polymer characteristics (Snyder & Kwon, 1987; Liu, 1997). Soybean lacks essential sulfuric amino acids but has plenty of lysine, which is lacking in grains, hence eating soybeans with rice complements nutrients.

Soybean proteins do not have any taste themselves, but the taste is acquired when they are broken down into amino acids and peptides and can be used as various condiments. The fermentation process is the most important biological reaction of soybean to produce taste materials by mold and bacteria, and taste is produced by proteases induced from microorganisms breaking down the proteins into amino acids and peptides as a soluble state.

16.2.3.2 Rice

The origin of rice cultivation goes back to Assam of Kanjis in India and spread to Burma, Thailand, and China about 11,500 years ago. Afterwards, rice cultivation may have started at the Yangchow River (Chi & Hung, 2010). In Korea, a surprisal rice husk which is 12,500 years old was found at the Sorori Chungbuk (Kim et al., 2013b), Korea. It means the history of rice cultivation goes back a long time in this peninsula. At that time cultivation methods may not have been the same as at present. There are two most common popular species, namely non-glutinous and glutinous short rice in Korea. To produce soybean products, two varieties are used for each product. For making *gochujang meju*, mostly non-glutinous rice is used, but the base for *gochujang* uses glutinous rice after malt hydrolysis. There are many varieties of rice and any of them could be used for making fermented products. The annual production of rice reaches approximately four million tons a year in Korea. The rice used for fermented soybean products is not an important factor for the quality of the products but it provides a sweet taste by malt hydrolysis and viscous properties.

16.2.3.3 Red Pepper

16.2.3.3.1 Varieties of Red Peppers and Their Characteristics

Capsicum, which originates from the Greek word "kapto" meaning stimulating, is included in cayenne, chili, jalapeno, green bell, paprika, and tabasco. Varieties of chili (red) peppers are available in various forms. They are perennial in tropical regions but annual in temperate regions and their shapes are lineal, cone, oviform, and globoid. When they are immature, they are green or violet but when they become mature, the color changes to red, orange, yellow, or purple, and their spice strength also changes (Kim & Ku, 2001). The red pepper consumed in Korea is the perennial *Capsicum annuum* L. belonging to *Solanaceae*, and about 500 different species and varieties are being cultivated. According to the Nonghyup organization (Korea), about 160 species are currently being cultivated and 21 species comprise about 65%. The most popular varieties are Manida, super Manida, and Dogyachungchung, but new varieties are being developed every year. They can be supplied for green groceries or for dried powdered products and can be classified as "spicy" and "not spicy". Dried red peppers contain 11–14% water, 15–18% crude fat, 12–16% protein, 0.3–0.8% essential oil, and carbohydrates such as pentosane and galactose. Vitamins A, C, and E are plentiful in fresh chilis and chili pepper leaves (Kim & Ku, 2001). The main components of the spicy taste are capsaicinoids and capsaicin. Dihydrocapsaicin and nordihydrocapsaicin also belong to this group. Red peppers are the most important spice in Korean food culture. Koreans consume 1.9 g of red pepper powder and 4.9 g of red pepper paste every day, indicating high consumption of red pepper (Korea National Health and Nutrition Examination Survey, 2007). Thai people consume about 5 g of chili pepper daily, which is the highest consumption in the world, and consumption by Koreans is similar to this. Some foreign cooks have stated that red peppers have brought innovative changes to food culture compared to the use of fire in cooking. As a result, highly consumed red peppers have a huge impact on physiological functions, and a number of studies regarding this matter have been conducted in Korea and in other countries.

Many studies have been conducted about capsaicin, which is the main capsaicinoid in red pepper. Studies about physiological functions have been conducted mainly with animals and those with humans are rarely seen. According to these studies, capsaicin affects lipid metabolism, and capsaicin-injected animals show significantly lower serum triglyceride levels and tend to have less fatty tissue. Moreover, capsaicin has an impact on the contraction and relaxation of cardiac muscle and is associated with calcium release in cardiac muscle and the delay of muscle movement. Furthermore, impacts on gastric function, activation of renal function, anti-inflammatory, anticancer, and antioxidant effects have been demonstrated (Chun & Suh, 1980). Apart from these, studies regarding accelerating digestion and impacts on neurons have also been published. The pharmacological dose of capsaicin ranges from 3 mg to 120 mg, although 30 mg to 1.2 g has also been suggested. Research results regarding the physiological impacts of chili pepper are controversial. Some studies have reported that chili pepper increases the onset of gastric cancer, whereas others have shown that chilis decrease the onset of cancer and the rate of mutations (Dietrich, 1972). However, it is widely accepted that chili pepper has positive physiological effects according to the administered dose (Mozsik, 2018). Although there are also results stating that chilis have positive influences on the immunological response and suppress tumors (Popescu et al., 2021), they also hinder nitroso compound generation and increase insulin secretion. A number of studies have suggested the possibility of obesity control and prevention of fat accumulation by consuming red peppers. Thus, red peppers have a huge impact on physiological function, and consuming a large amount of red pepper results in more positive effects than negative effects. A number of foods are made with red peppers and the following are examples from China, Japan, and other Asian countries.

16.2.3.4 Salt

Salt affects the taste of all foods. Fermented foods are no exception.

For making fermented soybean foods, sun-dried salt is always used and is a vital ingredient for taste and preservation by inhibiting putrefactive bacteria. Sun-dried salt is highly nutritive because various inorganic substances are included as well as NaCl. Contents can differ depending on the production area, production method, and production period even if it is made from seawater containing similar components (Lee et al., 2007a). Various types of salts are prepared, including rock salt which is solidified, sun-dried salt, which is made by evaporating seawater, purified salt, which is made by crystallizing melted salt, refined salt, which is made by separating NaCl, lake salt, which is made when a lake evaporates, normal salt, which is made when subterranean saltwater evaporates, and natural salt, which is made by boiling seawater (Kim et al., 2009). Sun-dried salt for traditional fermented foods is usually kept for six months to one year to remove the impure ingredients including magnesium chloride and calcium chloride, and it is used when salt feels sufficiently dry. Condiments can taste bitter due to magnesium and calcium salt when the brine is not sufficiently removed.

Refined salt can be used as a condiment but sun-dried salt is currently used the most, as sun-dried salt is permitted by the Korean Food Sanitation Act (January 4, 2008). In recent years, the use of salt made from deep seawater is drawing attention.

16.2.3.5 Meju

Meju is one of the most important starters to produce all different types of fermented soybean products except *cheonggukjang* which uses solely cooked soybean and is fermented by *Bacillus subtilis* simultaneously. The most common preparation method is as follows. Soybeans are boiled with water for three to four hours after washing and steeping in water. The soaked soybeans should be fully cooked and that becomes an important factor for deciding the quality of *meju*. The cooked soybeans are mashed by pounding and shaped like a block at a proper size (usually 15 × 15 × 20 cm) by hand. The shaped soybeans lump is *meju*, and is to be fermented for four to five days at room temperature until its surface is dry. The fermentation is to be carried out at 25–30 °C for one to two months. In the fermentation process, the drying of *meju* is continued and then bacteria and molds are naturally generated during drying. Then, fermentation and maturation proceed at the same time. During the drying of *meju*, there are some apertures on *meju*. The fermentation and maturation are continued through the winter. There are two types of *meju*. One is *meju* for making *ganjang* and *doenjang*, which uses solely soybean, but *meju* for making *gochujang* uses soybean with non-glutinous rice at a ratio of 6:4. It is called *gochujang meju*. This *meju* contains a starch source as rice and natural microorganisms concerned produce enzymes related to the hydrolysis of starch and produce a sweet taste in *gochujang* during fermentation. Finished *meju* then continues to dry in sunlight and is stored. Dried *meju* is used as the basic material of *ganjang* (soybean sauce) and *doenjang* (soybean paste).

For producing *meju*, the general procedure is as seen in Figure 16.1, Figure 16.2 and Figure 16.3. All microorganisms concerned broadly in *meju* are seen in Table 16.3.

Aspergillus and *Bacillus* genus are the dominant microorganisms concerned in fermented soybean products. In general, the two strains are mainly involved in fermented soybean products, but ethnic fermented soybean products are connected to various natural strains and species. During the traditional production of *doenjang*, *meju* (Hong, Kim, & Samson, 2015) is a major ingredient. The microorganisms present in fermented *meju* are the molds *Mucor* and *Aspergillus*; the bacteria *Bacillus*, *Lactobacillus*, and *Streptococcus*; and the yeasts *Saccharomyces*, *Zygosaccharomyces*, *Pichia*, *Hansenula*, and *Devryomyecs* (Park, 2009).

16.2.3.6 Biological Changes during Soybean Fermentation

Many different microorganisms (mostly selected species) are concerned with the cooked soybean during the first steps of fermentation and produce typical compounds slowly that express special tastes and flavors. All biological conversion in *meju* by microorganisms concerned is shown in Figure 16.4 (Cho et al., 2020).

As shown in Figure 16.4, many different biological changes proceed during fermentation and produce various active compounds for functional actions and taste.

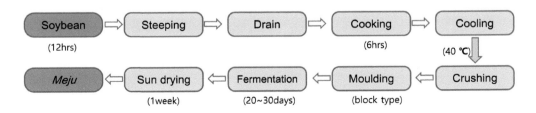

FIGURE 16.1 Procedure of *meju* preparation.

FIGURE 16.2   Traditional fermented *meju*, block types.

FIGURE 16.3   Mass production system of *meju*.

TABLE 16.3

Microorganisms Isolated from *Meju*

Mold	*Aspergillus flavus, Asp. fumigatus, Asp. niger, Asp. oryzae, Asp. retricus, Asp. spinosa, Asp. terreus, Asp. wentii*
	Botrytis cinerea
	Mucor adundans, Mucor circinelloides, Mucor griseocyanus, Mucor jasseni, Mucor hiemalis, Mucor racemosus
	Penicillium citrinum, Pen. griseopurpureum, Pen. griesotula, Pen. kaupscinskii, Pen. lanosum, Pen. thomii, Pen. turalense
	Rhizopus chinencis, Rhi. nigricans, Rhi. oryzae, Rhi. sotronifer
Yeast	*Candida edax, Candida incommenis, Candida utilis*
	Hansenula anomala, Han. senula capsulata, Han. senula holstii
	Rhodotorula flaca, Rhodotorula glutinis
	Saccharomyces sp., *Sac. exiguus, Sac. cerevisiae, Sac. kluyveri*
	Zygosaccharomyces japoninus, Zygo. saccharomyces rouxii
Bacteria	*Bacillus citreus, Bac. circulans, Bac. licheniformis, Bac. megaterium, Bac. mesentricus, Bac. subtilis, Bac. pumilis*
	Lactobacillus sp.
	Pedicocus sp., *Pedicoccus acidilactici*

Source: Choi et al. (1995).

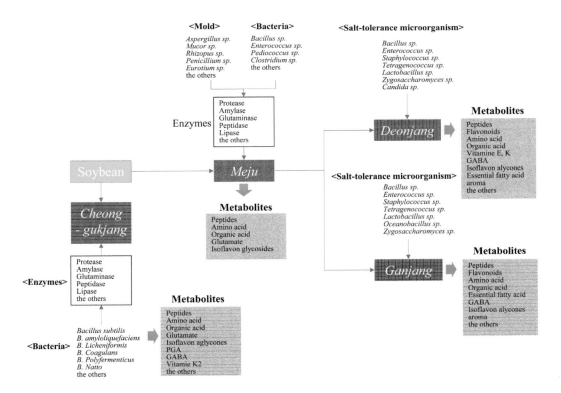

FIGURE 16.4 Flow diagram of biological changes of fermented soybean products during fermentation.

16.3 DOENJANG

16.3.1 INTRODUCTION

Doenjang is one of the traditional fermented soybean products and shows a paste-type product with a yellowish-brown color. Also, it has the longest history of all traditional seasonings in fermented soybean products (Shin, 2011). *Doenjang* has usually been used as a source of seasonings. In addition, it has largely been cooked as a pot stew and a sauce with vegetables. *Doenjang* shows a unique fermentation flavor and a savory taste due to nucleic and amino acids which are produced during the fermentation period.

16.3.2 PREPARATION METHOD

There is a wide category of two types of *doenjang* which are classified into traditional and factory-made products. Producing traditional *doenjang* deeply depends upon natural fermentation including temperature, humidity, and some other conditions, but factory-made products, microorganisms, and fermentation conditions are fully controlled. Factory-made *doenjang* preparation methods are simply presented in Figure 16.5.

After making *meju*, *doenjang*, and *ganjang* are produced by the process presented in Figure 16.6. In this process, some clay jars are used in a traditional process and some plastics, stainless steels, cement tanks, etc. are also used according to purpose.

As shown in Figure 16.6, solid and liquid materials are separated after aging. The liquid is processed into *ganjang* and the solid produces *doenjang* respectively in a traditional method. *Doenjang* is ingested as soup at Korean tables almost every day and has been known as an excellent protein source with various functional ingredients (Kwon et al., 2011). For seasoning *doenjang* soup, some garlic, onion, and red pepper powder are added to the soup according to personal tastes.

16.3.3 MICROORGANISMS CONCERNED

Aspergillus and *Bacillus* genus are mainly related to commercial fermented soybean products, in which they manage the taste and flavor of *doenjang*. Ethnic fermented soybean products are

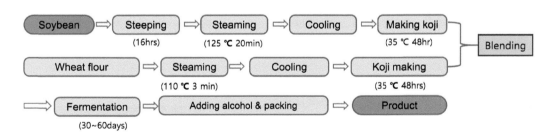

FIGURE 16.5   Common process of factory-made *doenjang* making.

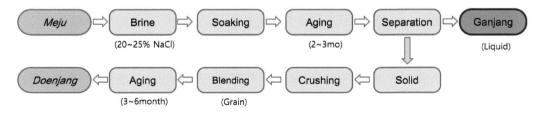

FIGURE 16.6   Procedure of *doenjang* and *ganjang* preparation.

TABLE 16.4

Daidzein and Genistein Contents of Soybean, *Meju*, and *Doenjang* (mg/kg Dry Basis)

	Soybeans	*Meju*	*Doenjang*
Daidzein, free	106 ± 7^{a1}	269 ± 38^{b}	578 ± 70^{d}
Daidzein, total	406 ± 29^{a}	433 ± 41^{bc}	538 ± 59^{c}
Daidzein, aglycones (%)	26.03 ± 0.96^{a}	61.96 ± 3.34^{b}	107.68 ± 10.26^{a}
Genistein, free	95 ± 20^{a}	137 ± 16^{b}	455 ± 10^{d}
Genistein, total	486 ± 86^{a}	200 ± 7^{b}	538 ± 57^{b}
Genistein, aglycones (%)	19.49 ± 1.1^{a3}	68.52 ± 6.62^{b}	85.26 ± 8.72^{d}
D/G ratio[2]	0.85 ± 0.09^{ab}	2.16 ± 0.17^{c}	1.00 ± 0.10^{b}

[1] In each column, different alphabets in superscript show statistically significant difference ($p < 0.05$).
[2] Total daidzein contents/total genistein contents ratio.

connected to various natural strains and species that originated in *meju*. For the traditional production of *doenjang*, *meju* is a major ingredient. The microorganisms present in fermented *meju* also appear in *doenjang* (Table 16.3). *Mucor* and *Aspergillus* are the main mold groups, the bacteria are *Bacillus*, *Lactobacillus*, and *Streptococcus*, and the yeasts are *Saccharomyces*, *Zygosaccharomyces*, *Pichia*, *Hansenula*, and *Devryomyecs* genus (Park, 2009).

16.3.4 NUTRITIONAL COMPOSITION

The major ingredient of *doenjang* is soybeans, which include lipids and protein at about 20% and 40% respectively, and 12 different isomers including daidzein, genistein, and glycitein, known as isoflavones and phytoestrogens (Shin, 2011). It has been known that isoflavones are the glycoside of phenolic compounds and represent effects of preventing breast cancer and prostate diseases (Pagliacci et al., 1994; Severson et al., 1989). Also, various functionalities of soybeans presented in Korean *doenjang* have been recognized (Kim et al., 1999a). Based on this study, it can be seen that the ingredients of isoflavones are newly generated during the fermentation process of soybean (Table 16.4).

As Table 16.4 shows, isoflavones can change into more bioactive compounds and that show healthy effects in humans. The chemical composition of *doenjang* in general compared with other fermented soybean foods is in Table 16.5 (RDA, 2016).

Doenjang produces water-soluble flavor substances by dissolving protein, fat, and carbohydrates based on different enzymes produced by microorganisms during its fermentation process.

16.3.5 FUNCTIONALITY AND HEALTH BENEFITS

Doenjang represents various physiological functions through the substances newly produced by its fermentation process in addition to various functionalities in soybeans. The major physiological active substances in *doenjang* are not only trypsin inhibitor, vitamin E, and unsaturated fatty acids (linoleic acid), but also phytoestrogen and isoflavones, which are the category of 12 types of isomers such as daidzein, genistein, and glycitein (Banaszkiewicz, 2011; Shin, 2011). The health functionalities of *doenjang* are known for their anticancer, antihypertensive, anti-atherogenic, antithrombosis, anti-obesity, immunomodulatory, and antidementia effects.

16.3.5.1 Anticancer Effect

The anticancer activity in *doenjang* increases significantly in traditional *doenjang* with a long fermentation period (Shin, Jung, & Chae, 2015). It has been reported that *doenjang* can also help detoxify the liver and suppress tumors (Choi et al., 1998). *Doenjang* intake has been confirmed to

TABLE 16.5
Chemical Composition of the Fermented Soybean Products

	Proximate Compositions			Dietary Fiber		Minerals					Vitamins				
											A		B₁	B₂	C
	Water	Protein	Fat	Ash	CHO	Ca	P	Fe	K	NA	Valent	β-Carotene	Thiamin	Riboflavin	Ascorbic acid
Foods	kcal	g	g	g	g			mg			RE	μg		mg	
Doenjang	47.8 −193	13.7	7.3	12.3	18.9	84	208	2.5	647	3,748	0	0	0.04	0.12	0
Gochujang	44.6 −134	4.9	1.1	8.2	43.8	40	90	2.2	822	3,164	408	2445	0.17	0.52	5
Cheonggukjang	70.7 −108	10.2	0.8	3.4	14.9	96	177	3.8	602	961	13	80	0.15	0.29	0
Ganjang (*Soy sauce*)	60 −83	3.7	0.01	19.3	17	38	155	0.9	390	7,157	0	0	0.02	0.08	0

TABLE 16.6

Antitumor Activity of *Deonjang*

Samples	Tumor Weight (g)	Inhibition Rate (%)
Sarcoma 180 cancer cell (A) + PBS	5.8 ± 0.3[a]	–
(A) + fermented *doenjang* (3 months)	5.4 ± 0.2[a]	7
(A) + fermented *doenjang* (6 months)	4.7 ± 0.3[b]	19
(A) + fermented *doenjang* (24 months)	3.6 ± 0.2[c]	38

[a–c] Significant differences based on the Duncan's multiple range test ($p < 0.05$).

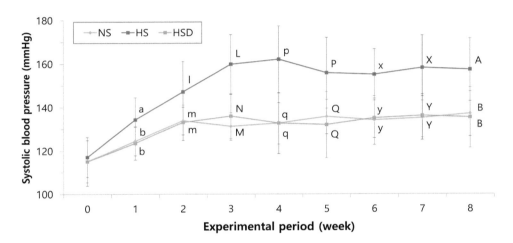

FIGURE 16.7 Changes in systolic blood pressure of SD rats for eight weeks; values are shown as the means ± SD. Values with different letters within a column are significantly different from each other at $p < 0.05$ by Duncan's multiple range test. NS = normal-salt diet; HS = high-salt diet; HSD = high-salt with *doenjang* diet.

significantly increase the activity of glutathione S-transferase, the elimination of cancer cells, and the activity of natural killer cells. In particular, as the fermentation period of *doenjang* increased, the weight loss of tumors and the inhibitory effect of tumor generation were excellent (Table 16.6). This effect was more beneficial in anticancer activity by increasing the amount of aglycones such as genistein and daidzein, which are physiologically active substances produced during the fermentation process of *doenjang* through changing it to a form that is easy to digest.

16.3.5.2 Antihypertensive Effect

In general, there is a high risk of blood pressure rising due to the high salinity contained in *doenjang*. However, according to a preclinical study by Mun, Park, and Cha (2019), blood pressure was significantly reduced in the group that consumed *doenjang* with high-fat diets compared to the group that consumed the same amount of table salts considering the metabolism of sodium (Figure 16.7). This blood pressure reducing effect represented a more positive result by the substance of arginine-proline, a dipeptide that indicates ACE inhibition activity in *doenjang* with a long fermentation period (Kim, Lee, & Kwon, 1999b). That is, the effect and relevance of the increased risk of developing high blood pressure on the intake of salts included in traditional fermented foods (*doenjang*) are minimal.

In addition, a substance that shows a function of ACE inhibition activity in *doenjang* has been recognized as arginine-proline (Kim et al., 1999b), and the ACE inhibition capability is increased in the fermentation process of *doenjang*.

16.3.5.3   Body Weight Control and Anti-Atherogenic Effects

Doenjang can lower risk factors for arteriosclerosis and cardiovascular diseases. *Doenjang* improves lipid metabolism by mainly inhibiting the reduction and synthesis of the concentrations of VLDL-C and LCL-C and inhibits TC level reduction and LDL-C oxidation of blood (Kim & Lee, 2002). Fermented *doenjang* burned body fat and was effective in suppressing visceral fat and in weight loss by facilitating the β-oxidation process (Kwak, Park, & Song, 2012). The results of the abdominal CT after 12 weeks of ingestion of *doenjang* in overweight and obese adults with variants of obesity genes (Cha et al., 2012) showed that the area of visceral fat was reduced by 8.6 cm² which was more beneficial to the effect of abdominal obesity as it reduced the visceral fat by about eight times more than the placebo group (Table 16.7). In particular, in the case of mutant alleles of obese PPAR-γ2, the effects of body fat reduction, weight loss, and antioxidation played a greater role (Cha et al., 2014).

In an experiment on animals, a *doenjang* diet positively affected lipid metabolism by inhibiting the synthesis of VLDL-C and LCL-C instead of synthesizing HDL-C and represented an effect of decreasing blood flow TC and of improving high cholesterol (Kim & Lee, 2002). It has been determined that *doenjang* shows an anticoagulation effect due to a thrombin inhibitor (Jang, In, & Chae, 2004). Consumption of *doenjang* contributes to fat burning and weight loss by promoting a β-oxidation process in fatty acids (Kwak et al., 2012).

16.3.5.4   Boosting Immune System

The surfactin secreted by *Bacillus subtilis*, a dominant species in *doenjang*, has an antiviral effect (Johnson et al., 2019). Boiled soybeans do not contain any immunomodulation substances (KFSP) but they have been identified in *doenjang* (Kim et al., 2014). The KFSP substances in traditional *doenjang* have four times higher immune-enhancing activity than factory-made *doenjang* and ten times higher KFSP content than Japanese soybean paste (miso) (Kim et al., 2014; Jang, Kim, & Kang, 2009).

TABLE 16.7

Change of Body Weight and Composition and Abdominal Fat Area Measurements at Zero Weeks and 12 Weeks of Study

Parameters	*Doenjang* Group (n = 26)			Placebo Group (n = 25)			
	0 Weeks	12 Weeks	Change[2]	0 Weeks	12 Weeks	Change	P Value[3]
Body weight (kg)	70.2 ± 1.8 1[1]	69.4 ± 1.8	−0.8**	67.8 ± 1.7	67.6 ± 1.8	−0.3	<0.001
Body fat mass (kg)	22.1 ± 1.0	21.4 ± 0.9	−0.7**	22.1 ± 0.9	21.8 ± 0.9	−0.4*	<0.001
Body fat (%)	31.7 ± 1.2	31.1 ± 1.6	−0.6**	32.6 ± 0.9	32.2 ± 0.9	−0.4	0.007
Total fat (cm²)	333.1 ± 3.5	248.8 ± 10.5	−84.3***	327.9 ± 38.8	253.6 ± 12.3	−74.4*	0.788
Visceral fat (cm²)	74.4 ± 4.8	65.9 ± 3.6	−8.6***	65.8 ± 4.0	65.3 ± 3.4	−0.60	0.041
Subcutaneous fat (cm²)	258.6 ± 23.7	182.9 ± 9.7	−75.7**	262.1 ± 39.3	188.3 ± 10.8	−73.8*	0.960
VSR	0.35 ± 0.03	0.39 ± 0.03	0.04 *	0.31 ± 0.03	0.37 ± 0.02	0.06***	0.447

[1] Values are expressed as means ± SE. WHR = waist to hip ratio; VSR = visceral to subcutaneous ratio.

[2] Values in this column represent the difference between the mean changes scores of the *doenjang* group and those of the placebo group.

[3] P-values derived from repeated measures analysis (per protocol) after adjusting for age, gender, and BMI.

*p < 0.05, **p < 0.01, ***p < 0.001. P-values indicate significant differences in the variables between zero weeks and 12 weeks, which were evaluated by paired t-test.

16.3.5.5 Antidementia Effect

Recently, the brain nerve protection effect was confirmed by the intake of *doenjang*. The administration of *doenjang* showed a decrease in β-amyloid peptide (Aβ) levels by controlling the gene expression involved in the production and decomposition of amyloid-beta in the brain. In the case of *doenjang*, in particular, the inhibition of brain neurodegeneration was greater than that of steamed soybeans (Ko et al., 2019). Thus, it suggested that the physiological activation compound produced during the fermentation and ripening process of *doenjang* may be involved in increasing brain nerve protection.

16.3.6 ETHNIC AND SOCIO-ECONOMIC VALUE

Doenjang is one of the essential side dishes on Korean tables almost every day and has been positioned as an important plant protein source. *Doenjang* is usually ingested as soup as a side dish to boiled rice. Koreans intake an average of 8.8 g *doenjang* every day and the annual gross domestic product and export are 90,370 and 4,214 tons (MDFS, 2019) and 6.15 million dollars (MDFS, 2019) respectively. Korean traditional foods have been propagated throughout the world due to a recent Korean wave and preferences for *doenjang* in foreign consumers have also been increased. Both the factory-produced and traditional products are released to the market and some consumers prefer the traditional fermented soybean products even though there are some gaps between the factory-produced and traditional products. Figure 16.8 shows pictures of *doenjang* and *ganjang*.

16.4 GOCHUJANG

16.4.1 INTRODUCTION

For producing fermented hot pepper pastes (*gochujang*), the mixture of soybeans and grains (rice and others) is cooked and mashed. Then, the mashed mixture is shaped and fermented by molds and bacteria introduced to make *gochujang meju*. Then, the mixture of *gochujang meju*, hot pepper flours, and a rice syrup digested by malt, which is produced using some barley malts, is to be fermented for a long time (Shin et al., 2001). *Gochujang* products are divided into two categories: The traditional one produced by using traditional fermentation methods, and the factory-made for mass production. Traditional *gochujang* is usually produced by natural fermentation, and factory *gochujang* represents different ways of managing microbes and fermentation conditions. Recently,

FIGURE 16.8   Aging of *doenjang* and *ganjang*.

gochujang products have been made by housewives in average households and produced by middle and large factories based on traditional methods. Factory *gochujang* is produced in a mass production system and covers almost 80–90% of *gochujang* markets in Korea. Traditional *gochujang* is usually sold to old-aged consumers, and factory-made *gochujang* is largely consumed because its major markets are young people and foodservice businesses.

Gochujang uses very hot-flavored red pepper as an extra ingredient different from other fermented soybean products, and some grains, such as rice and barley, are applied to *gochujang* for supplementing a sweet taste through saccharifying it. For producing *gochujang*, a specific type of *meju* that is made of mixing rice and soybeans with a ratio of 6:4 is required. Then, the mixture is steamed, formed, and fermented. The major extra ingredient, red pepper, is originated from South America and was introduced in Korea in the sixteenth century or thereabouts (Kwon et al., 2011). *Gochujang* is a unique fermented spice only in Korea and has been deeply rooted in the Korean diet.

16.4.2 PREPARATION METHOD

For producing *gochujang*, it requires *gochujang meju* different from other fermented soybean products. *Gochujang meju* is a mixture of soybeans and starch and produces enzymes naturally proliferated by microorganisms according to the compositions of matrices (Shin, 2011).

The general preparation procedure for *gochujang meju* is seen in Figure 16.9.

Gochujang has been produced by both traditional and industrial methods and the *gochujang meju* is used in the traditional method. In the industrial method, koji, which is proliferated by pure strains, is used.

The *gochujang* produced by the process presented in Figure 16.10 is used as a spice source for various foods and a large amount of *gochujang* is used to seasoning foods compared to other countries' spices.

Gochujang represents a paste-type product and can be used in different types of foods like dressings, seasoned bar rice cakes, stews, and so on. *Gochujang* has a balance in foods based on its own hot-flavored taste as well as sweet and savory tastes.

16.4.3 MICROORGANISMS CONCERNED

Gochujang uses various ingredients compared to those of *doenjang*. The species and genus of microorganisms in the traditional *gochujang*, which includes a large amount of red pepper powder, malt, grain, *ganjang*, etc., including soybeans, are largely distributed (Kwon et al., 2009). As the factory-made *gochujang* is processed using koji, which is made of pure strains, the strain is relatively simple and the predominant strain is the *Bacillus* species. The microbial load in *gochujang* is about 10^7–10^8/g and the frequency of presenting the microorganisms is determined as *Bacillus velegensis* > *B. amyloliquefaciens* > *B. subtilis*, and the *Oceanobacillus* is also detected in addition to the halophilic microorganisms determined as *B. ligueformis* > *B. subtilis* > *B. velezensis* > *B.*

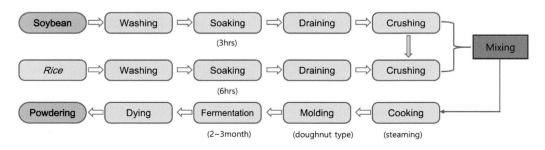

FIGURE 16.9   Illustration of the preparation method of *gochujang meju*.

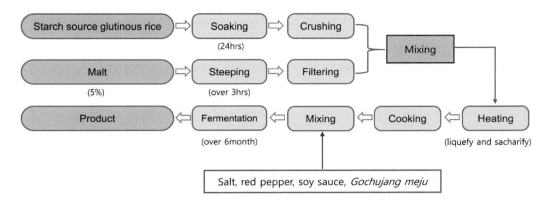

FIGURE 16.10 General procedure for preparing *gochujang*.

amyloliqueformis (Nam, Park, & Lim, 2012). Yeasts revealed in *gochujang* are *Zygosaccharomyces* and *Candida lactis* (Park, 2009). Also, *Z. rouxii* shows the highest detection rate (Park, 2009). *Aspergillus* is the predominant strain and *Penicillium* and *Rhizopus* are also detected (Kim et al., 2013a). These would usually proliferate on the surface of *gochujang* instead of inside it (Kim et al., 2013a).

16.4.4 NUTRITIONAL COMPOSITION

The capsaicin, which represents the hot-flavored taste from red pepper and some sugar induced from grains, significantly affects the taste and nutrition. The general ingredients of *gochujang* consist of 44.6% moisture, 4.9% protein, 1.1% fat, 8.2% ash, 43.8% carbohydrate, 40 mg of calcium, 90 mg of phosphorus, 2.2mg of iron, and 822 mg of potssium (RDA, 2016). Vitamins A, B_1, B_2, C, and niacin are also included, with a particularly high content of β-carotene, 244μg (RDA,2016). The content of saccharides in the nutritional composition of *gochujang* is highly present and the total amount of saccharides is about 43–59%, of which the contents of glucose and maltose represent the highest rates (Shin et al., 2011). Various free amino acids are detected and the high rates of glutamic acid and proline affect the taste of *gochujang*. The major ingredient of nucleic acid is CMP (Shin et al., 2011). The capsaicin that is the characteristic taste of *gochujang*, a hot-flavored taste, is included as about 200–300 mg.

16.4.5 FUNCTIONALITY AND HEALTH BENEFITS

Gochujang is a fermentation product of various substances that have been created or converted into new substances through a fermentation process of mixed substances such as capsaicin, which represents the spicy taste contained in red pepper powder, *meju*, and starch. In the various physiological activation functions of *gochujang*, the fermented traditional *gochujang* has been significantly improved compared to the non-fermented *gochujang* and red pepper powder (capsaicin). Also, the physiological activity was different depending on the fermentation period of *gochujang* and the use of ingredients.

 Gochujang represents different functionalities caused by both the capsaicin generated from red pepper and some other fermentation products.

16.4.5.1 Anticancer Effect

The anticancer and tumor metastasis inhibition effects of *gochujang* were different depending on whether it was fermented and how it was made (traditional *gochujang* > factory-made *gochujang*). As the fermentation period of traditional *gochujang* is longer than that of factory-made *gochujang*, the anticancer activity function is improved significantly (Song, Kim, & Lee, 2008). In traditional *gochujang* and factory-made fermented *gochujang*, the tumor cell inhibition rate for the transition of cancer cells to the lungs were 45% and 23% respectively, while non-fermented *gochujang* was 17%; the tumor control effect was the most effective in the traditional *gochujang* fermented for six months (Park et al., 2001; Cui et al., 2002) (Table 16.8).

16.4.5.2   Anti-Obesity Effect

The anti-obesity effect of the six-month fermented traditional *gochujang* was more significant than that of the *gochujang* and red pepper powder (capsaicin) without fermentation (Choo, 1999; Rhee et al., 2003). This effect means that the synergy of the non-glycoside isoflavone in *meju*, the capsaicin of red pepper, and the metabolites produced during its fermentation had a greater effect (Kim et al., 2004). Although the main ingredients of red pepper powder, capsaicin, and dietary fiber, may have some effect on body fat accumulation and fat decomposition partially, the secondary metabolites of *gochujang* may have a positive effect on energy and glucose metabolism. Recently, the anti-obesity effect showed significant differences depending on the type of ingredients used in *gochujang* (Son et al., 2020; Son et al., 2018). Some reports have shown that capsaicin in red pepper stimulates the spinal cord to increase metabolism by stimulating the secretion of adrenaline hormones in the adrenal gland, and it uses the glycogen and fat cells stored in the liver as energy to promote body fat decomposition (Choo, 1999; Koo et al., 2008). In particular, the capsaicin of red pepper can contribute to the reduction of body fat by increasing the β-adrenergic activity, which is a brown fat tissue (Koh, 2006) (Table 16.9).

In the results of the abdominal CT after consuming 32 g of *gochujang* tablets daily in overweight and obese adults for 12 weeks, it showed that the ratio of internal fat and subcutaneous fat decreased

---

TABLE 16.8

Inhibition Effects of the Methanol Extract in *Gochujang* and Red Pepper Powder on Generating Tumors in Sarcoma-180 Cancer Cell Implanted Balb/c Mouse Models

Samples	Tumor Weight (g)	Inhibition Rate (%)
Sarcoma-180 cancer cell (A) + PBS (control group)	6.0 ± 0.1^a	–
(A) + traditional *gochujang* (0 days fermentation)	50 ± 0.9^{ab}	17
(A) + traditional *gochujang* (6 months fermentation)	6.0 ± 0.1^c	45
(A) + factory made *gochujang*	4.5 ± 0.1^{bc}	23
(A) + red pepper powder	4.7 ± 0.3^b	22

Source: Park et al. (2001).
[a-c] Significant differences in a–c Duncan's multiple range tests ($p < 0.05$).

TABLE 16.9

Effect of *Gochujang* in High Fat Dietary Mouse Model (for 21 Days) on the Body Weight

	Normal	High Fat	*Gochujang* (Before)	*Gochujang* (6 Months Fermentation)	Red Pepper
Initial weight (g)	199.9 ± 7.5	198.0 ± 6.4	200.7 ± 9.6	199.9 ± 9.7	199.9 ± 2.7
Final weight (g)	333.8 ± 0.5^b	382.8 ± 1.4^a	362.6 ± 2.5	354.5 ± 0.1^c	376.5 ± 5.1^a
Increment (g/day)	4.4 ± 0.2^d	6.4 ± 0.1^a	5.8 ± 0.4^b	5.6 ± 0.5^c	6.2 ± 0.3^{ab}

Source: Rhee et al. (2003).
[a-d] Means significant difference by Duncan's multiple range test ($p < 0.05$).

significantly compared to the placebo group, resulting in a significant reduction in abdominal obesity (Cha et al., 2013).

16.4.5.3 Blood Glucose Control Effect

The administration of fermented *gochujang* (5% *gochujang* powder) in insulin-resistant diabetic animal models showed improvements in the sensitivity of insulin and the resistance of glucose in the blood (Kwon et al., 2009). Also, the 12-week intake of *gochujang* for overweight and obese adults with variations in obesity-related peroxisome genes, peroxisome proliferator activator receptor-γ (PPARγ2), in clinical trials represents an improvement in the insulin sensitivity through controlling blood glucose, suggesting it could help prevent and improve cardiovascular disease in the future (Lee et al., 2017). Therefore, the administration of *gochujang* showed that personalized prescriptions were needed in the future as metabolic activity reactions in the body varied depending on the polymorphism of the obesity gene PPARγ2.

16.4.5.4 Atherosclerosis and Cardiovascular Effect

The effect of lipid profiles improvement of *gochujang* showed significant decreases in blood TG and TC levels in the fermented traditional *gochujang* (Choo, 1999). It is estimated that the capsaicin of *gochujang* increases the secretion of catecholamine in adrenal medullas by activating the sympathetic nervous system, which positively affects energy and lipid metabolism (Diepvens, Westerterp, & Westerterp-Plantenga, 2007). In clinical studies conducted with random assignment and double-blind methods, the intake of *gochujang* for overweight and obese adults (Kim et al., 2010) and hyperlipidemic adults (Lim et al., 2015) is expected to significantly reduce blood TC and LDL-C levels compared to the placebo group. In the future, long-term consumption of *gochujang* is expected to help prevent and manage cardiovascular diseases.

16.4.5.5 Stress Control and Mitigation (Autonomic Nervous System Stabilization)

The autonomic nervous system acts as a computer that automatically controls our bodies, divided into sympathetic and parasympathetic nervous systems. The sympathetic nervous system is involved when feeling stress and fear during exercise, but the parasympathetic nervous system plays a more important role when sleeping. Normally, the two nervous systems maintain the homeostasis of our bodies by balancing the promotion and inhibition. As the functions of the autonomic nervous system are degraded, the responsiveness and adaptability of the body to external stimuli are significantly reduced, which can be seen as a state of decrease in heart rate variation. The test of autonomic nervous functions evaluates sympathetic and parasympathetic nervous functions by measuring the changes in blood pressures for the Valsalva maneuver, changes in postures, changes in the heart rate in repetitive inspiration and expiration, standing, and applying a grip (Nguewa et al., 2011). An increase in the pain of the sympathetic nervous system caused by stress increases the oxidative stress hormone catecholamine, which contributes to the growth of tumor cells (Thaker et al., 2006). Consumption of *gochujang* for 12 weeks in hyperlipidemia patients has been shown to help improve or stabilize autonomic nervous functions by controlling and alleviating stress through reducing breathing (difference in the heart rate between repetitive inspiration and expiration). In future, continuous consumption of *gochujang* can help control stress, prevent coronary artery diseases, and reduce risks (Im, 2013) (Table 16.10).

16.4.6 Ethnic and Socio-Economic Value

Gochujang has been introduced as the latest fermented soybean product but has remarkably increased in its use as a seasoning source that corresponds with the Korean taste preferring a hot-flavored spice of *gochujang*. The personal daily intake of *gochujang* is about 6.4 g and the gross domestic product is 134,191 tons (MDFS, 2019). Also, it has been exported to Japan, America, Europe, and other countries and the amount of exports is about 15,216 tons (MDFS, 2019). The

TABLE 16.10

Autonomic Nervous Functions (Deep Breathing) in *Gochujang* and Placebo Groups for Zero and 12 Weeks in the Study

	Gochujang Group (n = 13)			Placebo Group (n = 13)			
	0 wk	12 wk	P-Value[1]	0 wk	12 wk	P-Value[1]	P-Value[2]
Breathing	0.2 ± 0.3	0.0 ± 0.0	0.008**	0.1 ± 0.2	0.2 ± 0.4	0.387	0.027*
ECG	0.4 ± 0.4	0.3 ± 0.5	0.613	0.4 ± 0.4	0.5 ± 0.5	0.613	0.470
Valsalva	0.3 ± 0.4	0.4 ± 0.4	0.570	0.3 ± 0.4	0.6 ± 0.5	0.022*	0.340
Upright	0.2 ± 0.2	0.1 ± 0.2	0.337	0.0 ± 0.1	0.1 ± 0.2	0.337	0.192
Handgrip	0.0 ± 0.0	0.0 ± 0.0	–	0.0 ± 0.0	0.0 ± 0.0	–	–
T-score		0.8 ± 0.7	0.279	0.8 ± 0.6	1.4 ± 1.0	0.073	0.035*

Source: Im (2013).

Abnormal = 1, borderline = 0.5, normal = 0. All values are presented as mean ± SD.

[1] Analyzed by paired t-test.

[2] Independent t-test, $p < 0.05$, **$p < 0.01$.

FIGURE 16.11   Pictures of *gochujang meju* and a final product.

exports and regions have recently been expanded. In addition, the production has been changed from one-man-based businesses to enterprise scales according to changes in social conditions. Recently its commercialization rates are recorded at about 80–90%. The rates are increasing from now on and it is expected that the amount of homemade traditional *gochujang* will be largely decreased (Ministry of Food and Drug Safety, 2019). Figure 16.11 shows *gochujang meju* and a final product.

16.5   *CHEONGGUKJANG*

16.5.1   INTRODUCTION

Cheonggukjang is a traditional fermented soybean food that is fully fermented within 2–3 days. *Cheonggukjang* represents a particular quality characteristic caused by some microorganisms that shows a palatable taste and a unique smell due to the protein generated by protease, which is produced from the proliferation of *Bacillus* sp. In addition, it produces a sticky viscous material. Although there is a similar product in Japan, natto, it shows different fermentation bacteria and dietary usages to *cheonggukjang*. Korean *cheonggukjang* is usually used in stews mixed with other foods rather than eating alone. However, natto is eaten on its own. Various physiological activities in *cheonggukjang* have been largely known, and *cheonggukjang* has been produced using traditional methods in average households. Also, it can easily be produced in households due to its easy production and it has been used as a side dish for boiled rice. *Cheonggukjang* shows the shortest fermentation period and the highest fermentation temperature (40–43° C) compared to other fermented soybean products. It is a fermented soybean product without applying any extra ingredients compared to other fermented soybean products and has a unique flavor and a peculiar taste. Although it is similar to Japanese natto and Indian *kinema*, it shows differences in production methods, applied microorganisms, and ways of intake (Lee et al., 2011; Tamang, 2015).

16.5.2 Preparation Method

In producing *cheonggukjang*, well-matured soybeans are prepared and washed. Then, the soybeans are fully steeped in water at 10–16 °C and are steamed and dehydrated in order to undergo fermentation (Shin, 2011). Cooked soybeans are to be fermented in a jar. Traditionally some rice straws are inserted into the soybeans or used at the bottom and upper sections of the jar for inoculating bacteria. It has been known that rice straws contain bacteria that have high protease activities (Lee, Lee, & Chung, 1971).

The use of rice straws represents a certain way of natural inoculation in which *Bacillus subtilis* specifically contributes to the fermentation process (Santos et al., 2007). Also, the fermentation takes 30–35 hours at 40–43 °C. The fermentation time is largely varied according to temperature and applied bacteria. Also, a humidity condition is important to the fermentation. Tastes and odors are different according to the fermentation period. Recently a two-stage fermentation method that uses two fermentation stages at 42–43 °C and 50–53 °C is used to remove a disgusting smell. After completing the fermentation of cooked soybeans, some sticky and viscous materials are presented at the surface of soybeans and a unique smell is also generated. The process is simply presented in Figure 16.12.

The fermentation of *cheonggukjang* is performed at around 42–43 °C and *Bacillus* strains are selectively proliferated. As the fermentation is completed within a short period of time, one to two days, it represents lots of adhesive and sticky substances. After completing the fermentation of *cheonggukjang*, it can be finished by adding some salts and spices including garlic. Recently *cheonggukjang* has been distributed as a frozen product with a specific amount and shape (Shin, 2011).

16.5.3 Microorganism Concerned

Cheonggukjang has a long historic background of being produced through natural fermentation and is still produced in the home or on small-business scales based on natural fermentation processes (Jang et al., 2006). In Japan, natto strains were separated 100 years ago and named *Bacillus natto*

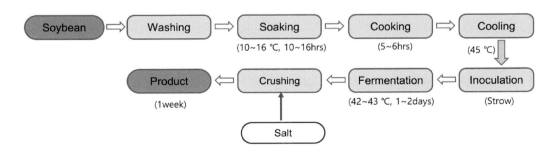

FIGURE 16.12 *Cheonggukjang* preparation method.

Sawamura (Shurtleff & Aoyagi, 2012). Then, it was identified that *B. subtilis* was the predominant strain (Shurtleff & Aoyagi, 2012). It is still known that the predominant strain in *cheonggukjang* is *Bacillus subtilis* (Shin, 2011). Recently new strains that can reduce the unique odor in *cheong-gukjang* significantly have developed for business purposes (Kim et al., 2003). The fermentation strains related to *cheonggukjang* are identified as a carbon source and use sucrose, fructose, and so on. Also, they contribute to creating adhesive and sticky substances in *cheonggukjang* (Baek et al., 2010).

Although all vegetative cells are inactivated through the cooking process in producing *cheong-gukjang*, spore formers survive and lead to the fermentation of *cheonggukjang* (Kwon et al., 2006). The amount of the biogenic amine, which is a harmful component produced during the fermenta-tion process, can be reduced by selecting a specific strain (Han et al., 2006).

16.5.4 NUTRITIONAL COMPOSITION

Although *cheonggukjang* is initiated from soybeans, it creates its own flavor and taste with various substances due to the completion of its fermentation process. Table 16.11 represents the comparison of the general composition between soybeans and fermented *cheonggukjang* (Kim & Hahm, 2002).

Amino acids that remarkably affect flavors through the completion of a fermentation process are largely created by the protease, which is generated by involved strains; in particular, there exist huge increases in amino acids and sticky substances (Baek et al., 2008). The largest contents of amino acids are glutamic acids and alanines (Baek et al., 2008). Specifically, 24 h after the fermentation, *cheonggukjang* shows an increase in pH levels during its fermentation and represents a pH over 8 (Ju & Oh, 2009; Joo, 1971). In addition, soluble proteins are largely increased as much as 40–80 kDa and differ from those of natto (Santos et al., 2007). The viscous substance that is a particular material in *cheonggukjang* represents 61% of crude proteins. Also, the glutamic acid in amino acids shows the highest content, 32%, and a fibrinolytic activity is presented (Lee, Kim, & Kim, 1991). In the fermentation process of *cheonggukjang*, the viscous substance is different according to applied bacteria (Baek et al., 2008), and the molecular weight of the substance is about 15,000–65,000. The viscous substance usually consists of glutamate and fructose, and the composition of polymers is varied according to their proliferation conditions (Lee et al., 1992). In addition, there are some reports that the odors, compositions, and threshold values in *cheonggukjang* are changed according to bacteria applied to the fermentation of *cheonggukjang* (Kim et al., 2003).

16.5.5 FUNCTIONALITY AND HEALTH BENEFITS

Cheonggukjang has a decomposition process in which the fibers and carbohydrates that make up the soybean peel or cell membrane are decomposed into sugars by β-amylase during fermentation, and the soybean protein is decomposed into amino acids by *Bacillus subtilis*. The protease produced by microbial strains that have a significant effect on flavor after the fermentation process of *cheong-gukjang* increases the production of amino acids and sticky substances (Baek et al., 2008). Soybean protein increases the production of several amino acids, such as peptone and polypeptide by the protease, and the total amino acid content of *cheonggukjang* is about 12%. During the fermentation

TABLE 16.11

Proximate Composition of *Cheonggukjang*

Product	Protein(g)	Lipid (g)	Carbohydrate (g)	Fiber (g)	Ca (mg)	Fe (mg)	K (mg)	B_1 (mg)	B_2 (mg)
Steamed soybean	16.0	9.0	7.6	2.1	70	2.0	570	0.22	0.09
Cheonggukjang	16.5	10.0	9.8	2.3	90	3.3	660	0.70	0.56

process of soybeans, the vitamin B_2 increases by 5–10 times compared to boiled soybeans, and new micronutrients are created in addition to vitamin K and water-soluble essential amino acids. In addition to the nutritional properties of soybeans, *cheonggukjang* is rich in dietary fiber and isoflavones, which are important for health functionality, and contains a large amount of important physiological active substances, especially polyglutamic acid (gamma-PGA), phenolic acid, phospholipids, saponin, trypsin inhibitor, and phytic acid (Sung et al., 2005). These substances show various physiological activities such as improving digestion and intestinal functions, anti-obesity, blood glucose control, antiatherogenic effect, fibrinolysis effect, blood pressure control, lipid improvement, anticancer effect, and immune regulation.

16.5.5.1 Enhance Digestive Absorption and Nutrients

The sticky substance of *cheonggukjang* is a type of levan, fructan, and polyglutamic acid, which is polymerized with fructose and glutamic acid. Although there is little vitamin K_2 and vitamin B_{12} in regular beans, these are newly produced by microbial bacteria during the fermentation process of *cheonggukjang*, and their vitamin and mineral contents and digestive absorption rates are further increased. The absorption rate of protein in the small intestine is only 65% when consuming soybeans, but it rises to more than 95% when consuming *cheonggukjang*. The amount of isoflavone in soybeans increases by about 21 times when it becomes *cheonggukjang* compared to boiled soybeans (Lee et al., 2007b). The isoflavone of boiled soybeans shows a decrease in absorption rate as combined with sugars, but is changed to aglyconic forms, genistein and daidzein, by enzymes that remove sugars according to the fermentation of the soybeans. This increases the absorption rate of digestion in the body, which further enhances the function of physiological activation (Table 16.12).

16.5.5.2 Antihypertensive Effect

Cheonggukjang has a better effect on antihypertension than boiled soybeans. This effect positively affects antihypertension by inhibiting ACE enzyme activity by amino acids such as valine and tyrosine produced during the fermentation process. According to a pre-clinical study that studied the antihypertensive effects of *cheonggukjang* (Yang, Lee, & Song, 2003), it reported that the both control group (casein) and steamed soybeans group showed an increase in blood pressure, while the administration of *cheonggukjang* showed no increase in blood pressure.

16.5.5.3 Atherosclerosis and Cardiovascular Effect

The protease, which is a protein decomposition enzyme included in *cheonggukjang*, can be expected to reduce HMG-Co reductase activity in the body to prevent the synthesis of cholesterol and to have

TABLE 16.12

Bioactive Compound Content in Cooked Soybean and *Cheonggukjang*

Phytochemical			Steamed Soybean	*Cheonggukjang*
Isoflavone	Glucoside	Daidzin	15–57	79–93
		Genistin	36–86	87–91
		Glycitin	2–6	10–12
	Aglycone	Daidzein	0.3–5	4–7
		Genistein	0.2–5	3–4
		Glycitein	0.1–0.6	11–13
Gamma-PGA			–	ç
Ammonia			–	ç
Protein absorption (%)			65	95

an antithrombogenic effect (Koh, 2006). It helps the thrombus dissolution and the inhibition of thrombus generation by the fibrinolytic system related to blood circulation included in *cheonggukjang*. Also, *cheonggukjang* is rich in dietary fibers and indigestible carbohydrate, which can help prevent arteriosclerosis by suppressing the absorption of neutral fat and cholesterol in the intestines and increasing lipid excretion through excrements (Lae, 2005). The anti-arteriosclerosis effect of *cheonggukjang* was verified in a clinical study with random assignment and double-blind methods for overweight and obese adults (Back et al., 2011). In particular, the *cheonggukjang* intake group showed a significant decrease in the arteriosclerosis index, Apolipoprotein B, compared to the placebo group (Figure 16.13).

16.5.5.4 Antioxidant Activity

The major antioxidants of *cheonggukjang* are known as isoflavones, phenolic acids, chlorogenic acids, isochlorogenic acids, tocopherol, amino acids, peptides, and saponins (Lee et al., 2001). The isoflavone content in *cheonggukjang* is twice as high as that of steamed soybeans (Lee et al., 2007b), and the antioxidant activity is proportional to the contents of polyphenols and isoflavones (Devi et al., 2009). In particular, the antioxidant effect of *cheonggukjang* is higher than that of steamed soybeans because free amino acids, peptides, and phenolic acids produced in its fermentation process are involved in the antioxidant activity. The genistein contained in *cheonggukjang* shows antioxidant effects by inhibiting the generation of superoxide anions and removing a tumor promotion factor, hydrogen peroxide (Ryu et al., 2007).

16.5.5.5 Antidiabetic Activities

Cheonggukjang is rich in dietary fiber and has a lower blood glucose index, which makes it more advantageous for controlling blood glucose. According to a study on blood glucose in the animal models with type 2 diabetes (Shin, 2011), *cheonggukjang* significantly decreased the concentration of fasting blood glucose and plasma insulin compared to boiled soybeans and the control groups (casein intake). This was beneficial for blood glucose control by promoting the secretion of insulin in the pancreas by a substance called trypsin inhibitor contained in *cheonggukjang*, while also increasing insulin sensitivity. An eight-week intake of *cheonggukjang* in adults with impaired fasting blood glucose showed a significant decrease in the level of the fasting blood glucose (FBG) along with decreases in blood TC and LDL-C concentrations, helping to improve blood glucose and

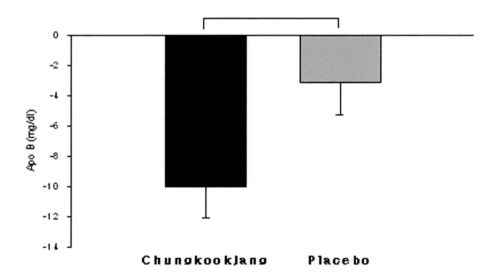

FIGURE 16.13   Changes in Apo B during the intervention period $p < 0.05$.

lipid metabolism (Shin et al., 2011). This effect is shown to inhibit the activity of the carbohydrate digestive enzyme (a-glucosidase) in the small intestine through the intake of *cheonggukjang*, while soluble fiber improves insulin resistance by decreasing the movement rate of the digestive canal of the diet, contributing to the blood glucose control effect (Choi et al., 2016).

16.5.5.6 Anti-Obesity Activity

Proper exercise and calorie intake are important to prevent and treat obesity. In particular, the intake of *cheonggukjang*, which is rich in dietary fiber, may help with weight control. In studies with hyperlipidemia and obesity-induced animal models, the administration of *cheonggukjang* was excellent for anti-obesity effects by inhibiting body fat accumulation, reducing weight and body fat, and improving serum lipid composition (Koh, 2006; Lee et al., 2001; Choi et al., 2016; Byun et al., 2016b). The intake of *cheonggukjang* pills (about 35 g/day) for 12 weeks for overweight and obese adults has had a beneficial effect on improving obesity and lipid metabolism compared to placebo groups (Byun et al., 2016a). In particular, the lipid metabolism index, Apo A1, increased in men, and Apo B decreased significantly in women, indicating an improvement in the antiatherogenic index (Byun et al., 2016a). In addition to the *in vitro* absorption promotion factors of isoflavone contained in *cheonggukjang,* dietary fiber and indigestible sugars are believed to have had a more beneficial effect in preventing obesity by inhibiting intestinal lipid absorption and increasing lipid excretion.

16.5.5.7 Improvement of Intestinal Functions

Although constipation may be caused by lack of exercise or abuse of constipation drugs, weak abdominal muscle strength, and sensitive nerves, the most important reason is lack of fiber intake. Fiber has 40 times more moisture than its own weight. *Cheonggukjang* is rich in insoluble fiber (cellulose), hemicellulose, and soluble fiber (pectin), which increase the viscosity of intestinal contents and slow hunger (Lee & Hwang, 1997). In addition, *Bacillus* in *cheonggukjang* have excellent intestinal promotion effects and insoluble dietary fiber stimulates the colon wall to shorten the passage time and help improve beneficial intestinal microbial growth and intestinal functions (Lae, 2005; Park, Lee, & Kang, 1988). There are about 100 billion bacteria per 100 g of *cheonggukjang*, and the galactooligosaccharides, stachyose and raffinose, included in *cheonggukjang* also promote the synthesis of intestinal vitamins along with dietary fiber. A beneficial bacterium that promotes digestive absorption by promoting intestinal peristalsis, *Bifidobacteria*, inhibits the proliferation of harmful bacteria and external infestation bacteria in the intestines and helps to suppress and reduce the production of ammonia, amine, and indole produced by putrefying bacteria (Park et al., 1988).

16.5.5.8 Immunity Boosting and Anti-Allergic Activity

Cheonggukjang can help improve immune activation in biological defense and control sensitive immune functions (hypersensitive reactions). *Cheonggukjang* contains a large amount of gamma-polyglutamic acid (γ-polyglutamic acid or γ-PGA), a polymer produced by the fermented *Lactobacillus subtilis*. The ingredient of *cheonggukjang* that helps to control the immune system is γ-PGA, a polymer nucleic acid. This substance stimulates the immune cell of TLR4 (toll-like receptor 4) in the mucous membrane of the small intestine by promoting the secretion of IFN-γ in the Th1 lymphocyte. Then, it controls the immune system by removing viruses through increasing NK-cell activity using the mucous membrane immune system (Jang et al., 2009; Hahm et al., 2004; Chang et al., 2005). In addition, *cheonggukjang* also improves allergies (atopic dermatitis) caused by hypersensitive immune reactions (Cho et al., 2019; Baek et al., 2015). As adults who have hypersensitive immune responses (allergic skin) in clinics have been given 35 g of *cheonggukjang* tablets (raw *cheonggukjang*, 27 g/d) per day for 12 weeks, the amount of swelling caused by histamine has decreased significantly, helping to alleviate symptoms of skin hypersensitivity (Baek et al., 2015) (Figure 16.14).

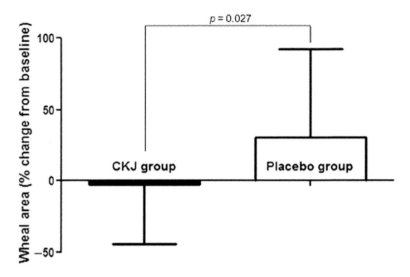

FIGURE 16.14 Percentage change from baseline in histamine-induced wheal area. Values are presented as mean ± standard deviation. Analyzed with independent t-test and p-values compared to placebo.

FIGURE 16.15 Pictures of *cheonggukjang* and cuisine.

16.5.6 Ethnic and Socio-Economic Value

Cheonggukjang is one of the fermented soybean products with a unique flavor and is usually enjoyed in the winter season as a pot stew together with boiled rice. Also, it has savory tastes that are preferred by Koreans. In recent years, *cheonggukjang* has been eaten throughout the four seasons (Shin, 2011). Although there is little consumption in the younger generation due to its pungent odor, there are steady demands from the elders. According to developments of production technologies, there are new products that almost remove such odor and that leads to increased consumption. The annual gross production of *cheonggukjang* is 14,319 tons and some products are exported (MDFS, 2019). Although the recent major export consumers are ethnic Koreans, there is a possibility of increasing the demand of native populations through increasing consumers and improving its flavors. Figure 16.15 shows *cheonggukjang*.

16.6 GANJANG

16.6.1 INTRODUCTION

Ganjang has largely been used in East Asia countries, such as Korea, Japan, and China, including some Southeast Asian countries. It is a fermented soybean product and has been known throughout the world as soy sauce. Traditional *ganjang* in Korea is made by separating the liquid part in which the mixture of *meju* and brine is aged for three to four months. The liquid part is used as an undiluted solution (Figure 16.6). In the case of commercially produced *ganjang*, it uses koji, which is made of soybeans and flour, as the major ingredient. Then, it is steeped in brine and is fermented and aged. The fermented product is used as an undiluted solution of *ganjang* after filtering the fermented product. *Doenjang* is not produced in this process but solely *ganjang*.

The traditional *ganjang* has a savory taste because the soybean protein is dissolved into amino acids by the enzymes produced by different microorganisms in the fermentation process of *meju*. In the case of factory-made *ganjang*, however, its taste is created by dissolving the soybeans and flour starch during the fermentation process of the enzyme, which is usually created by *Aspergillus* and *Bacillus subtilis* (Liu et al., 2015). Because of using flour, the taste of the final product is sweet compared with traditional products.

Recently, the demand for factory-made products largely exceeds that for traditional products. The traditional products have limitedly been used for specific consumers and purposes. Thus, the level of demand is also limited. Although there are some studies on the functionality of *ganjang*, an effort to decrease salt has been made at a national level in order to avoid its high salt content (Song, Jeong, & Baik, 2015). *Ganjang* is classified into traditional and factory-made *ganjang*. These two types of *ganjang* show some different production processes. For producing traditional *ganjang*, it is necessary to produce *meju*. Traditional *meju* has been made by housewives in average households and handed down to the present day. For factory-scale production, koji is prepared. *Ganjang* is prepared by itself, not together with *doenjang*.

After preparing *meju* following the process mentioned above, *meju* is washed using fresh water and dried again. Then, dried *meju* is soaked in brine (18–19 Be'). The ratio of *meju* to brine is 1:3 or more brine depending upon the quality of *ganjang*. For maintaining a uniform quality, the concentration of brine and the ratio of *meju* to brine are to be specifically determined. Also, *meju* should be immersed under the brine because it floats on the surface of the brine which is the main source of contamination of *meju* by unwanted microorganisms.

After soaking *meju* in brine, some red peppers and charcoal are applied to the brine in order to prevent bad smells and germs. It is not scientifically confirmed but traditionally believed. After *meju*

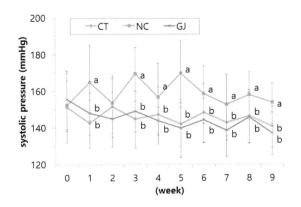
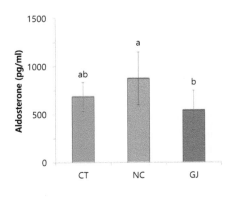

FIGURE 16.16 Systolic blood pressure and aldosterone for experimental period. Points are mean ± SD (n = 8). Values with different letters indicate statistical difference ($p < 0.05$) by Duncan's multiple range test (a > b). CT = control; NC = NaCl group; GJ = *ganjang* group.

is soaked in brine it is left to mature for about 60 days or longer in a sunny place and occasionally left open to sunlight scorching. After soaking *meju* in brine for three to four months, *meju* is to be separated from brine using a filtering patch or mesh. The separated solution is used as raw *ganjang*, and the separated solid matter is used to produce *doenjang* after mixing with new *meju* or cooked rice, barley, or other grains.

16.6.2 FUNCTIONALITY AND HEALTH BENEFITS

The health functionalities of *ganjang* have been consistently reported in studies on anti-hypertension (Mun et al., 2017), anticancer (Song et al., 2018; Hur et al., 2020), anti-inflammation, and antioxidant effects (Kim et al., 2020). In the results of comparing the effects on blood pressure by administering a traditional fermented soybean sauce (*ganjang*) and table salts to animals at the same salinity, the blood pressure in the traditional fermented soybean sauce was significantly lower than the salt (Figure 16.16). That is, despite the high Na content of *ganjang*, the reason for the blood pressure decreasing effect is assumed to have been the increase in the excretion of Na. This means that there are many limitations to constrain soybean fermented foods simply with sodium intake. Recently, it has been confirmed that there is a significant difference in the effects of anticancer activation depending on the type of *ganjang* (Song et al., 2018) and the fermentation period (Hur et al., 2020). The fermented *ganjang* with sesame seeds greatly increased the expression of colonic p53 compared to other *ganjang*, resulting in a strong anticancer effect (Song et al., 2018).

In particular, the fermented *ganjang* and sesame-added *ganjang* showed significantly lower concentrations of serum tumor necrosis factor-α, interferon-γ, IL-6, and IL-17-α than the acid hydrolysis *ganjang*, helping to inhibit the occurrence of colorectal cancer. In the anticancer effect of *ganjang*, the *ganjang* fermented for five to ten years showed significant increases in the activity of the splenocytes and the natural killer cells compared to the *ganjang* fermented for less than 15 years and more than 15 years, resulting in significant anticancer effects (Hur et al., 2020). In addition, the polysaccharide separated from traditional Korean *ganjang* is found to be very excellent in physiological activities for antioxidant and anti-inflammatory effects as it is dose-dependent compared to the polysaccharide separated from Japanese soy sauce (Kim et al., 2020).

16.7 SUMMARY

The origin of soybeans has been located in Manchuria, East Asia. Then, soybeans were propagated to various countries of the world. In some Southeast Asian countries, in particular Korea, China, and Japan, soybeans have been variously used in dietary life and contributed to improving nutrients as protein sources. The history of using soybeans goes back thousands of years. On the other hand, in Western countries, soybeans have been largely used as animal feed and oil sources. However, recently these countries have also been using soybeans as food materials and producing various soybean processing products. Soybeans hold an important position in Korean dietary life as one of the five major grains and have been taken by themselves or with boiled rice. Also, soybeans have been variously used by processing them or by growing them as vegetables. Then, soybeans have been taken as vegetables, powders, and other various types of foods different from other countries. In particular, fermentation methods using molds, bacteria, and yeasts have been developed using cooked soybeans and that leads to producing *doenjang* (fermented soybean paste), *gochujang* (fermented hot pepper soybean paste), *cheonggukjang*, and *ganjang* (soybean sauce). These products have been used in Korean dietary life as the primary seasonings and side dishes. Also, these fermented soybean products play important roles in differentiating Korean foods from other foods.

Recently, phytochemicals including isoflavones in soybeans have been known as functional ingredients. Also, it is verified that soybean proteins and their hydrolysates have physiological activities from the results of various studies. Therefore, people around the world represent interests in the use of soybeans and their fermented products as foods. In addition, lots of processed soybean products have been produced and distributed due to their effects on preventing various chronic diseases through simple pharmacological treatments only.

In Korea, regarding the characteristics of using soybeans, various soybean products have been developed using specific fermentation processes. It has been scientifically proven that various functionalities are newly created during several fermentation processes. Also, studies on this issue have been continuously conducted. Regarding the specific functionalities of soybean products, they contribute to anticancer and antitumor effects, improving blood circulation, and preventing chronic diseases. In the case of *gochujang*, it is known that *gochujang* represents significant effects on constraining obesities.

Although Korean fermented foods contain a large amount of salt known to harm health, many prior studies have shown that the intake of fermented soybean products has beneficial effects such as relieving metabolic syndromes, boosting immunity, nerve protection, antihypertension, anticancer, anti-inflammatory, and anti-obesity effects.

Some benefits of fermented soybean products can be summarized as follows.

It is possible to significantly increase the nutritional aspects and digestibility of soybeans by converting insoluble components to soluble ones by fermentation. Fermented soybean products show an increase in amino acids, vitamins, and minerals due to the decomposition of soybean components through the fermentation process compared to the intake of soybean foods.

In most cases, the intestinal movement can be improved due to increases in fiber and beneficial intestinal microorganisms included in the fermented soybean products. As the soybean protein is decomposed into peptides and amino acids through a fermentation process, it can help people sleep well due to increased secretion of sleeping substances such as tryptophan and melatonin.

The substances contained in fermented soybean foods can help prevent and improve cardiovascular diseases by facilitating blood circulation. The consumption of fermented soybean products can reduce cholesterol by reducing the activity of 3-hydroxyl 3-methyl glutaryl CoA (HMG-CoA) reductase, which is a control enzyme for cholesterol composition rate in liver cells and increasing bile acid excretion through feces. In addition, it can reduce the levels of the blood lipid concentration, which is the atherogenic index, and the apolipoprotein B concentration to prevent arteriosclerosis, and can enable thrombosis to help relax blood vessels and control blood pressure. Also, the fermented soybean products have enough fiber to reduce the movement rate of food in the digestive tract and improve insulin resistance to help blood glucose management, as well as having anti-obesity effects.

Poly-gamma-glutamic acid (poly-γ-glutamic acid, γ-PGA), a sticky mucous component of *cheonggukjang*, promotes the secretion of IFN-γ in Th1 lymphocytes, and *doenjang* promotes NK-cells to activate immune cells in the body. This can help improve immunity, relieve hypersensitive immune reactions, and prevent cancer.

REFERENCES

Back, H. I., S. R. Kim, J. A. Yang, M. G. Kim, S. W. Chae & Y.-S. Cha (2011) Effects of chungkookjang supplementation on obesity and atherosclerotic indices in overweight/obese subjects: A 12-week, randomized, double-blind, placebo-controlled clinical trial. *Journal of Medicinal Food*, 14(5), 532–537.

Baek, H. I., S. Y. Jung, K.-C. Ha, H.-M. Kim, E.-K. Choi, S.-J. Jung, E.-O. Park, S.-W. Shin, M.-G. Kim, S.-K. Yun, D. Kwon, H. Yang, M. Kim, H. Kang, J. Kim, D. Jeong, S. Jo, B. Cho & S. Chae (2015) Effect of Chongkukjang on histamine-induced skin wheal response: A randomized, double-blind, placebo-controlled trial. *Journal of Ethnic Foods*, 2(2), 52–57.

Baek, J. G., S. M. Shim, D. Y. Kwon, H.-K. Choi, C. H. Lee & Y.-S. Kim (2010) Metabolite profiling of Cheonggukjang, a fermented soybean paste, inoculated with various Bacillus strains during fermentation. *Bioscience, Biotechnology, and Biochemistry*, 122(4), 1313–1319. doi: 1008022094-1008022094.

Baek, L. M., L.-Y. Park, K.-S. Park & S.-H. Lee (2008) Effect of starter cultures on the fermentative. *Korean Journal of Food Science and Technology*, 40, 400–405.

Banaszkiewicz, T. (2011) *Nutritional Value of Soybean Meal, Soybean and Nutrition.* Hany El-Shemy (Ed.), *ISBN*, 978-953.

Byun, M., O. Yu, Y. Cha & T. Park (2016a) Korean traditional chungkookjang improves body composition, lipid profiles and atherogenic indices in overweight/obese subjects: A double-blind, randomized, crossover, placebo-controlled clinical trial. *European Journal of Clinical Nutrition*, 70(10), 1116–1122.

Byun, M. S., O. K. Yu, H. H. Baek & Y. S. Cha (2016b) Fermented soybean products chungkookjang and Nattδ inhibits intracellular lipid accumulation in adipocyte and prevents weight gain in high-fat diet induced obese mice. *The FASEB Journal*, 30, 1176.30–1176.30.

Cha, Y.-S., S.-R. Kim, J.-A. Yang, H.-I. Back, M.-G. Kim, S.-J. Jung, W. O. Song & S.-W. Chae (2013) Kochujang, fermented soybean-based red pepper paste, decreases visceral fat and improves blood lipid profiles in overweight adults. *Nutrition and Metabolism*, 10(1), 24.

Cha, Y.-S., Y. Park, M. Lee, S.-W. Chae, K. Park, Y. Kim & H.-S. Lee (2014) Doenjang, a Korean fermented soy food, exerts antiobesity and antioxidative activities in overweight subjects with the PPAR-γ2 C1431T polymorphism: 12-week, double-blind randomized clinical trial. *Journal of Medicinal Food*, 17(1), 119–127.

Cha, Y.-S., J. Yang, H.-I. Back, S.-R. Kim, M.-G. Kim, S.-J. Jung, W. O. Song & S.-W. Chae (2012) Visceral fat and body weight are reduced in overweight adults by the supplementation of Doenjang, a fermented soybean paste. *Nutrition Research and Practice*, 6(6), 520–526.

Chang, J.-H., Y.-Y. Shim, S.-H. Kim, K.-M. Chee & S.-K. Cha (2005) Fibrinolytic and immunostimulating activities of *Bacillus* spp. strains isolated from Chungkuk-jang. *Korean Journal of Food Science and Technology*, 37, 255–260.

Chi, Z. & H.-c. Hung (2010) The emergence of agriculture in southern China. *Antiquity*, 84(323), 11–25.

Cho, B. O., J. Y. Shin, J.-s. Kim, D. N. Che, H. J. Kang, D.-Y. Jeong & S. I. Jang (2019) Soybean fermented with *Bacillus amyloliquefaciens* (cheonggukjang) ameliorates atopic dermatitis-like skin lesion in mice by suppressing infiltration of mast cells and production of IL-31 cytokine. *Journal of Microbiology and Biotechnology*, 29(5), 827–837.

Cho, S. H., S. Y. Kim, S. H. Ryu, H. I. Lim, G. S. Kang & G. S. Kim (2020) *Traditional Jangryu Safety Management Manual*. Ministry of Agriculture. Ministry of Agriculture, Food and Rual Affairs, Wanju, Korea, 10.

Choi, D. K. (2009) Origin of soybean, Jang, Shi and Tofu, and reevaluation of distribution: From traditional reference and nuderearth pots. *The Journal of Korean Historical Folklife*, 30, 363–427.

Choi, J.-H., P. B. Tirupathi Pichiah, M.-J. Kim & Y.-S. Cha (2016) Cheonggukjang, a soybean paste fermented with *B. licheniformis*-67 prevents weight gain and improves glycemic control in high fat diet induced obese mice. *Journal of Clinical Biochemistry and Nutrition*, 59, 31–38.

Choi, S.-Y., M.-K. Lee, K.-S. Choi, Y.-J. Koo & W.-S. Park (1998) Changes of fermentation characteristics and sensory evaluation of kimchi on different storage temperature. *Korean Journal of Food Science and Technology*, 30, 644–649.

Choi, S. H., M. H. Lee, S. K. Lee & M. J. Oh (1995) Microflora and enzyme activity of conventional Meju, and isolation of useful mould. *Korean Journal of Agricultural Science*, 22, 188–196.

Choo, J. (1999) Body-fat suppressive effects of capsaicin through β-adrenergic stimulation in rats fed a high-fat diet. *Korean Journal of Nutrition*, 32, 533–539.

Chun, J.-K. & C.-S. Suh (1980) The relationship between the storage humidity and the sorption rate of red-pepper powder. *Applied Biological Chemistry*, 23, 1–6.

Cui, C.-B., S.-W. Oh, D.-S. Lee & S. Ham (2002) Effects of the biological activities of ethanol extract from Korean traditional Kochujang added with sea tangle (*Laminaria longissima*). *Korean Journal of Food Preservation*, 9, 1–7.

Devi, M. A., M. Gondi, G. Sakthivelu, P. Giridhar, T. Rajasekaran & G. Ravishankar (2009) Functional attributes of soybean seeds and products, with reference to isoflavone content and antioxidant activity. *Food Chemistry*, 114(3), 771–776.

Diepvens, K., K. R. Westerterp & M. S. Westerterp-Plantenga (2007) Obesity and thermogenesis related to the consumption of caffeine, ephedrine, capsaicin, and green tea. *American Journal of Physiology. Regulatory Integrative and Comparative Physiology*, 292(1), 77–85.

Dietrich, W. (1972) *Prophetie und Geschichte: Eine redaktionsgeschichtliche Untersuchung zum deuteronomistischen Geschichtswerk*. Vandenhoeck & Ruprecht, Gottingen, Germany.

Hahm, J., T. Lee, J. Lee, C. Park, M. Sung & H. Poo (2004) Antitumor effect of poly-γ-glutamic acid by modulating cytokine production and NK cell activity. Abstract No. H003 presented at *2004 International Meeting of the Federation of Korean Microbiological Societies*, Seoul, Korea. October, 21–22.

Han, G.-H., K.-N. Bahn, Y.-W. Son, M.-R. Jang, C.-H. Lee, S.-H. Kim, D.-B. Kim, S.-B. Kim & T.-Y. Cho (2006) Evaluation of biogenic amines in Korean commercial fermented foods. *Korean Journal of Food Science and Technology*, 38, 730–737.

Hong, S., D. Kim & R. Samson (2015) Aspergillus associated with Meju, a fermented soybean starting material for traditional soy sauce and soybean paste in Korea. *Mycobiology*, 43(3), 218–224.

Hur, J., M. J. Kim, S. P. Hong & H. J. Yang (2020) Anticancer effects of Ganjang with different aging periods. *Journal of the Korean Society of Food Culture*, 35, 215–223.

Im, J. (2013) *The Effect of Kochujang Pills on Blood Lipids Profiles in Hyperlipidemia Subjects: A 12 Weeks, Randomized, Double-Blind, Placebo-Controlled Clinical Trial.* Master's thesis, Chonbuk National University, Jeonju.

Jang, C.-H., J.-K. Lim, J.-H. Kim, C.-S. Park, D.-Y. Kwon, Y.-S. Kim, D.-H. Shin & J.-S. Kim (2006) Change of isoflavone content during manufacturing of Cheonggukjang, a traditional Korean fermented soyfood. *Food Science and Biotechnology*, 15, 643–646.

Jang, I.-H., M.-J. In & H.-J. Chae (2004) Manufacturing method for traditional doenjang and screening of high fibrin clotting inhibitory samples. *Applied Biological Chemistry*, 47, 149–153.

Jang, S.-N., K.-L. Kim & S.-M. Kang (2009) Effects of PGA-LM on CD4+ CD25+ foxp3+ Treg cell activation in isolated CD4+ T cells in NC/Nga mice. *Microbiology and Biotechnology Letters*, 37, 160–169.

Johnson, B. A., A. Hage, B. Kalveram, M. Mears, J. A. Plante, S. E. Rodriguez, Z. Ding, X. Luo, D. Bente, S. S. Bradrick, A. N. Freiberg, V. Popov, R. Rajsbaum, S. Rossi, W. K. Russell & V. D. Menachery (2019) Peptidoglycan-associated cyclic lipopeptide disrupts viral infectivity. *Journal of Virology*, 93(22), 1–15.

Joo, H.-K. (1971) Studies on the manufacturing of Chungkukjang. *Korean Journal of Food Science and Technology*, 3, 64–67.

Ju, K.-E. & N.-S. Oh (2009) Effect of the mixed culture of *Bacillus subtilis* and *Lactobacillus plantarum* on the quality of Cheonggukjang. *Korean Journal of Food Science and Technology*, 41, 399–404.

Kim, D.-H., S.-H. Kim, S.-W. Kwon, J.-K. Lee & S.-B. Hong (2013a) Fungal diversity of rice straw for meju fermentation. *Journal of Microbiology and Biotechnology*, 23(12), 1654–1663.

Kim, H., J. Park, J. Jung & D. Hwang (2020) Anti-oxidative and anti-inflammatory Activities of polysaccharide isolated from Korean-style soy sauce. *Biomedical Science Letters*, 26(1), 51–56.

Kim, H.-L., T.-S. Lee, B.-S. Noh & J.-S. Park (1999a) Characteristics of the stored samjangs with different doenjangs. *Korean Journal of Food Science and Technology*, 31, 36–44.

Kim, J.-G. & I.-K. Lee (2002) Effects of dietary supplementation of Korean soybean paste (Done-jang) on the lipid metabolism in rats fed a high fat and/or a high cholesterol diet. In: *Proceedings of the Korean Society of Food Hygiene and Safety Conference*, 170–171. The Korean Society of Food Hygiene and Safety, Seoul, Korea.

Kim, J. H., Y. Jia, J. G. Lee, B. Nam, J. H. Lee, K. S. Shin, B. S. Hurh, Y. H. Choi & S. J. Lee (2014) Hypolipidemic and antiinflammation activities of fermented soybean fibers from meju in C57BL/6 J mice. *Phytotherapy Research*, 28(9), 1335–1341.

Kim, J. H., D. H. Lee, S. H. Lee, S. Y. Choi & J. S. Lee (2004) Effect of *Ganoderma lucidum* on the quality and functionality of Korean traditional rice wine, yakju. *Journal of Bioscience and Bioengineering*, 97(1), 24–28.

Kim, J.-M., J.-H. Yoon, K.-S. Ham, I.-C. Kim & H.-L. Kim (2009) Hazards for the sea salt production procedures and its improvement. *Safe Food*, 4, 8–13.

Kim, K. & Y. Hahm (2002) Recent studies about physiological functions of Chungkkokjang and functional enhancement with genetic engineering. *Institute of Molecular Biology and Genetics*, 16, 1–18.

Kim, K. J., Y.-J. Lee, J.-Y. Woo & A. T. Jull (2013b) Radiocarbon ages of Sorori ancient rice of Korea. *Nuclear Instruments and Methods in Physics Research, Section B*, 294, 675–679.

Kim, S.-H., Y.-J. Lee & D.-Y. Kwon (1999b) Isolation of angiotensin converting enzyme inhibitor from Doenjang. *Korean Journal of Food Science and Technology*, 31, 848–854.

Kim, W.-J. & K.-H. Ku (2001) *Sensory Evaluation of Food.* Hyoilbook, Seoul, 60–120.

Kim, Y., Y.-J. Park, S.-O. Yang, S.-H. Kim, S.-H. Hyun, S. Cho, Y.-S. Kim, D. Y. Kwon, Y.-S. Cha, S. Chae & H. K. Choi (2010) Hypoxanthine levels in human urine serve as a screening indicator for the plasma total cholesterol and low-density lipoprotein modulation activities of fermented red pepper paste. *Nutrition Research*, 30(7), 455–461.

Kim, Y.-S., H.-J. Jung, Y.-S. Park & T.-S. Yu. (2003) Characteristics of flavor and functionality of *Bacillus subtilis* K-20 Chunggukjang. *Korean Journal of Food Science and Technology*, 35, 475–478.

Ko, J. W., Y.-S. Chung, C. S. Kwak & Y. H. Kwon (2019) Doenjang, a Korean traditional fermented soybean paste, ameliorates neuroinflammation and neurodegeneration in mice fed a high-fat diet. *Nutrients*, 11(8), 1702.

Koh, J.-B. (2006) Effects of cheonggukjang added *Phellinus linteus* on lipid metabolism in hyperlipidemic rats. *Journal of the Korean Society of Food Science and Nutrition*, 35(4), 410–415.

Koo, B., S.-H. Seong, D. Y. Kown, H. S. Sohn & Y.-S. Cha (2008) Fermented Kochujang supplement shows anti-obesity effects by controlling lipid metabolism in C57BL/6J mice fed high fat diet. *Food Science and Biotechnology*, 17, 336–342.

Kwak, C. S., S. C. Park & K. Y. Song (2012) Doenjang, a fermented soybean paste, decreased visceral fat accumulation and adipocyte size in rats fed with high fat diet more effectively than nonfermented soybeans. *Journal of Medicinal Food*, 15(1), 1–9.

Kwon, D. Y., S. M. Hong, I. S. Ahn, M. J. Kim, H. J. Yang & S. Park (2011) Isoflavonoids and peptides from meju, long-term fermented soybeans, increase insulin sensitivity and exert insulinotropic effects in vitro. *Nutrition*, 27(2), 244–252.

Kwon, D. Y., S. M. Hong, I. S. Ahn, Y. S. Kim, D. W. Shin & S. Park (2009) Kochujang, a Korean fermented red pepper plus soybean paste, improves glucose homeostasis in 90% pancreatectomized diabetic rats. *Nutrition*, 25(7–8), 790–799.

Kwon, D. Y., J. S. Jang, J. E. Lee, Y. S. Kim, D. H. Shin & S. Park (2006) The isoflavonoid aglycone-rich fractions of chungkookjang, fermented unsalted soybeans, enhance insulin signaling and peroxisome proliferator-activated receptor-γ activity in vitro. *Biofactors*, 26(4), 245–258.

Lae, P. (2005) *The Effect of Fermented Soybean Powder on Improvement of Constipating Patients Receiving Maintenance Hemo Dialysis.* MS thesis, Kyungpook National University, Daegu (Korea).

Lee, B.-Y., D.-M. Kim & K.-H. Kim (1991) Physico-chemical properties of viscous substance extracted from Chungkook-jang. *Korean Journal of Food Science and Technology*, 23, 599–604.

Lee, H. & E. Hwang (1997) Effects of alginic acid, cellulose and pectin level on bowel function in rats. *Korean Journal of Nutrition*, 30, 465–477.

Lee, J.-J., C.-H. Cho, J.-Y. Kim, D. Kee & H.-B. Kim (2001) Antioxidant activity of substances extracted by alcohol from chungkookjang powder. *Korean Journal of Microbiology*, 37, 177–181.

Lee, K., J. Park, C. Choi, H. Song, S. Yoon, H. Yang & K. Ham (2007a) Salinity and heavy metal contents of solar salts produced in Jeollanamdo Province of Korea. *Journal of the Korean Society of Food Science and Nutrition*, 36(6), 753–758.

Lee, K.-H., H.-J. Lee & M.-K. Chung (1971) Studies on Chung-Kook-Jang (Part I)-On the changes of soy-bean protein in manufacturing Chung-Kook-Jang. *Applied Biological Chemistry*, 14, 191–200.

Lee, S., H.-S. Eom, M. Yoo, Y. Cho & D. Shin (2011) Determination of biogenic amines in Cheonggukjang using ultra high pressure liquid chromatography coupled with mass spectrometry. *Food Science and Biotechnology*, 20(1), 123–129.

Lee, S. W. (1992) Historical study of old Korean dietary life in the East Asian countries. *Hyangmun-SA*, Seoul, Koea, 1–20.

Lee, Y., Y.-S. Cha, Y. Park & M. Lee (2017) PPARγ2 C1431T polymorphism interacts with the antiobesogenic effects of Kochujang, a Korean fermented, soybean-based red pepper paste, in overweight/obese subjects: A 12-week, double-blind randomized clinical trial. *Journal of Medicinal Food*, 20(6), 610–617.

Lee, Y.-L., S.-H. Kim, N.-H. Choung & M.-H. Yim (1992) A study on the production of viscous substance during the chungkookjang fermentation. *Applied Biological Chemistry*, 35, 202–209.

Lee, Y.-W., J.-D. Kim, J. Zheng & K. H. Row (2007b) Comparisons of isoflavones from Korean and Chinese soybean and processed products. *Biochemical Engineering Journal*, 36(1), 49–53.

Lim, J.-H., E.-S. Jung, E.-K. Choi, D.-Y. Jeong, S.-W. Jo, J.-H. Jin, J.-M. Lee, B.-H. Park & S.-W. Chae (2015) Supplementation with *Aspergillus oryzae*-fermented kochujang lowers serum cholesterol in subjects with hyperlipidemia. *Clinical Nutrition*, 34(3), 383–387.

Liu, K. (1997) Chemistry and nutritional value of soybean components. In: *Soybeans*. Springer, Boston, MA, 25–113.

Liu, X., J. Y. Lee, S.-J. Jeong, K. M. Cho, G. M. Kim, J.-H. Shin, J.-S. Kim & J. H. Kim (2015) Properties of a bacteriocin produced by *Bacillus subtilis* EMD4 isolated from ganjang (soy sauce). *Journal Microbiology Biotechnology Microbiology Biotechnology*, 25(9), 1493–1501.

MDFS (2019) *Food Production Performance in the Korea (Fermented Soy Products)*. Ministry of Food and Drug Safety(A Statistical Publication), Osong, Korea, 85.

Mozsik, G. (2018) Introductory chapter: The general problems of human clinical nutritional and pharmacological observations of capsaicin in human beings and patients. In: *Capsaicin and Its Human Therapeutic Development*. IntechOpen, London, United Kingdom.

Mun, E.-G., J. E. Park & Y.-S. Cha (2019) Effects of doenjang, a traditional Korean soybean paste, with high-salt diet on blood pressure in Sprague–Dawley rats. *Nutrients*, 11(11), 2745.

Mun, E.-G., H.-S. Sohn, M.-S. Kim & Y.-S. Cha (2017) Antihypertensive effect of Ganjang (traditional Korean soy sauce) on Sprague-Dawley Rats. *Nutrition Research and Practice*, 11(5), 388–395.

Nam, Y. D., S. l. Park & S. I. Lim (2012) Microbial composition of the Korean traditional food "kochujang" analyzed by a massive sequencing technique. *Journal of Food Science*, 77(4), M250–M256.

Nawaz, M. A., X. Lin, T.-F. Chan, J. Ham, T.-S. Shin, S. Ercisli, K. S. Golokhvast, H.-M. Lam & G. Chung (2020) Korean wild soybeans (Glycine soja Sieb & Zucc.): Geographic distribution and germplasm conservation. *Agronomy*, 10(2), 214.

Nguewa, J., E. Sobngwi, E. Wawo, M. Azabji-Kenfack, M. Dehayem, E. Ngassam & J. Mbanya (2011) Cardiac autonomic neuropathy: Characteristics and associated factors in a group of patients with type 2 diabetes. In: *Diabetes and Metabolism*. Masson Editeur, MASSON EDITEUR, MOULINEAUX CEDEX 9 FRANCE–.

Pagliacci, M., M. Smacchia, G. Migliorati, F. Grignani, C. Riccardi & I. Nicoletti (1994) Growth-inhibitory effects of the natural phyto-oestrogen genistein in MCF-7 human breast cancer cells. *European Journal of Cancer*, 30(11), 1675–1682.

Park, K. (2009) *Science and Functionality of Fermented Soybean*. Seoul: Korean Jang Cooperative, 38–44.

Park, K.-Y., K.-R. Kong, K.-O. Jung & S.-H. Rhee (2001) Inhibitory effects of Kochujang extracts on the tumor formation and lung metastasis in mice. *Preventive Nutrition and Food Science*, 6, 187–191.

Park, S., H. Lee & K. Kang (1988) A study on the effect of oligosaccharides on growth of intestinal bacteria *Korean Journal of Dairy Science*, 6(3),187–191.

Popescu, G. D. A., C. Scheau, I. A. Badarau, M.-D. Dumitrache, A. Caruntu, A.-E. Scheau, D. O. Costache, R. S. Costache, C. Constantin, M. Neagu & C. Caruntu (2021) The effects of capsaicin on gastrointestinal cancers. *Molecules*, 26(1), 94.

RDA (2016) *Korean Food Composition Table (9th Revision)*, 18. Rural Development Administration, Wanju, Korea, 418–428.

Rhee, S., K. Kong, K. Jung & K. Park (2003) Decreasing effect of kochujang on body weight and lipid levels of adipose tissues and serum in rats fed a high-fat diet. *Journal of the Korean Soceity of Food Science and Nutrition*, 32(6), 882–886.

Ryu, B.-M., K. Sugiyama, J.-S. Kim, M.-H. Park & G.-S. Moon (2007) Studies on physiological and functional properties of Susijang, fermented soybean paste. *Journal of the Korean Society of Food Science and Nutrition*, 36(2), 137–142.

Santos, I., I.-Y. Sohn, H.-S. Choi, S.-M. Park, S.-H. Ryu, D.-Y. Kwon, C.-S. Park, J.-H. Kim, J.-S. Kim & J.-K. Lim (2007) Changes of protein profiles in cheonggukjang during the fermentation period. *Korean Journal of Food Science and Technology*, 39, 438–446.

Severson, R. K., A. M. Nomura, J. S. Grove & G. N. Stemmermann (1989) A prospective study of demographics, diet, and prostate cancer among men of Japanese ancestry in Hawaii. *Cancer Research*, 49(7), 1857–1860.

Shin, D. H. (2011) Utilization of soybean as food stuffs in Korea. In: *Soybean and Nutrition*. Ed. H. A. El-Shemy, Giza, Egypt, 81–110.

Shin, D.-H., E.-Y. Ahn, Y.-S. Kim & J.-Y. Oh (2001) Changes in physicochemical characteristics of kochujang prepared with different koji during fermentation. *Korean Journal of Food Science and Technology*, 33, 256–263.

Shin, D.-H., S.-J. Jung & S.-W. Chae (2015) Health benefits of Korean fermented soybean products. In: *Health Benefits of Fermented Foods*. Ed. J. P. Tamang. New York: CRC Press, 395–431.

Shin, S.-K., J.-H. Kwon, Y.-J. Jeong, S.-M. Jeon, J.-Y. Choi & M.-S. Choi (2011) Supplementation of cheonggukjang and red ginseng cheonggukjang can improve plasma lipid profile and fasting blood glucose concentration in subjects with impaired fasting glucose. *Journal of Medicinal Food*, 14(1–2), 108–113.

Shurtleff, W. & A. Aoyagi (2012) *History of Natto and Its Relatives (1405–2012)*. Soyinfo Center, Lafayette, Indiana.

Snyder, H. E. & T. Kwon (1987) *Soybean Utilization*. AVI. Van Nostrand Reinhold, Newyork, USA.

Son, H. K., H. W. Shin, E. S. Jang, B. S. Moon, C. H. Lee & J. J. Lee (2020) Gochujang prepared using rice and wheat koji partially alleviates high-fat diet-induced obesity in rats. *Food Science and Nutrition*, 8(3), 1562–1574.

Son, H.-K., H.-W. Shin, E.-S. Jang, B.-S. Moon, C.-H. Lee & J.-J. Lee (2018) Comparison of antiobesity effects between gochujangs produced using different koji products and tabasco hot sauce in rats fed a high-fat diet. *Journal of Medicinal Food*, 21(3), 233–243.

Song, H.-S., Y.-M. Kim & K.-T. Lee (2008) Antioxidant and anticancer activities of traditional Kochujang added with garlic porridge. *Journal of Life Science*, 18(8), 1140–1146.

Song, J.-L., J.-H. Choi, J.-H. Seo & K.-Y. Park (2018) Fermented Ganjangs (soy sauce and sesame sauce) attenuates colonic carcinogenesis in azoxymethane/dextran sodium sulfate-treated C57BL/6J mice. *Journal of Medicinal Food*, 21(9), 905–914.

Song, Y.-R., D.-Y. Jeong & S.-H. Baik (2015) Effects of indigenous yeasts on physicochemical and microbial properties of Korean soy sauce prepared by low-salt fermentation. *Food Microbiology*, 51, 171–178.

Sung, M. H., C. Park, C. J. Kim, H. Poo, K. Soda & M. Ashiuchi (2005) Natural and edible biopolymer poly-γ-glutamic acid: Synthesis, production, and applications. *Chemical Record*, 5(6), 352–366.

Tamang, J. P. (2015) Naturally fermented ethnic soybean foods of India. *Journal of Ethnic Foods*, 2(1), 8–17.

Thaker, P. H., L. Y. Han, A. A. Kamat, J. M. Arevalo, R. Takahashi, C. Lu, N. B. Jennings, G. Armaiz-Pena, J. A. Bankson, M. Ravoori, W. M. Merritt, Y. G. Lin, L. S. Mangala, T. J. Kim, R. L. Coleman, C. N. Landen, Y. Li, E. Felix, A. M. Sanguino, R. A. Newman, M. Lloyd, D. M. Gershenson, V. Kundra, G. Lopez-Berestein, S. K. Lutgendorf, S. W. Cole & A. K. Sood (2006) Chronic stress promotes tumor growth and angiogenesis in a mouse model of ovarian carcinoma. *Nature Medicine*, 12(8), 939–944.

Yang, J., S. Lee & Y. Song (2003) Improving effect of powders of cooked soybean and chongkukjang on blood pressure and lipid metabolism in spontaneously hypertensive rats. *Journal of the Korean Society of Food Science and Nutrition*, 32(6), 899–905.

17 Traditional Chinese Fermented Soybean Products and Their Physiological Functions

Yali Qiao, Zhen Feng, Xuejing Fan, Gefei Liu,
Shuang Zhai, Yang Li, and Baokun Qi

CONTENTS

17.1 INTRODUCTION

Soybean is a basic food ingredient in traditional Asian countries and is consumed as an important protein source. Soybean foods are generally used as non-meat or non-milk alternative food products in Western countries due to their high protein content and multifold healthful function. Soybean is rich in proteins (40–50%), lipids (20–30%), and carbohydrates (26–30%) on a dry basis (Vagadia et al., 2017). Soybean foods contain essential nutrients and some other functional ingredients, especially in fermented soybean foods, such as isoflavones, bioactive peptides, saponins, and phytosterols that are regarded as powerhouses of phytochemicals. These components also are used as dietary supplements due to their beneficial effects on health (Rizzo, 2020). A number of epidemiological studies have demonstrated the related advantages of soybean products in lowering risks for menopausal symptoms, cardiovascular diseases, and several cancers including breast, prostate, and colon (He & Chen, 2013).

DOI: 10.1201/9781003030294-17

Fermented soybean products in China are primarily salty and savory, e.g. soybean paste, soy sauce, douchi, sufu, etc. They are commonly used as seasoning agents or condiments to flavor foods such as stir-fries, stews, and soups. The fermentation process changes the physicochemical properties and enriches sensory properties of soybean foods, including color, flavor, and texture (Xiang et al., 2019). The differences in the color, flavor, and texture of fermented soybean products are dependent on the substrate, the microbial strains, and environmental factors. Fermented soybean foods have been consumed in most Asian centuries and are an important part of the Asian diet, especially in Eastern Asia. Microbes should have the ability to hydrolyze organic matter, ultimately transforming the chemical constituents, improving the bio-availability of nutrients, and generating bioactive components in the fermentation process of soybean products (Lokuruka, 2011). Therefore, the traditional fermented soybean products in Asia are considered to possess more health-promoting benefits than the super-processed soy products that are consumed in the West (Peterson et al., 2012). It is increasingly recognized that Asian populations consuming a moderate amount of fermented soybean foods are at lower risk of overall diseases. The fermentation process removes the trypsin inhibitors on the coating of soybeans. These inhibitors interfere with the absorption of nutritive compounds. The fermentation process is also thought to convert iron and copper to their most beneficial forms and enhance their bioavailability (Nkhata et al., 2018). Fermenting or culturing soybean has also been shown to enhance the bio-availability of isoflavone and render isoflavone precursors genistin and daidzin to their active isoflavone forms, genistein and daidzein. Fermented soybean products, such as douchi, sufu, and soy sauce, are staples of many Chinese diets, with many regions boasting milieu-specific products. Over the last few years, there has been interested in fermented soybean foods in Western countries because of their beneficial effects on health (Melini et al., 2019). Moreover, consumption of fermented soybeans has been reported to serve multifold benefits in health including improving bone health, reducing cancer risk, and preventing the progression of hypertension and cardiovascular diseases. The scientific evidence for beneficial effects and the underlying health mechanisms of these fermented soybean foods remains to be elucidated (Cao et al., 2019).

The purpose of this chapter is to introduce the basic production processes of douchi, sufu, soy sauce, soybean paste, natto, and tempeh. The chapter provides an overall summary of the studies on the investigation of the health-promoting components formed upon soybean fermentation and the compositional changes, action mechanisms, and bioavailability of these components in the abovementioned soybean fermented products.

17.2 DOUCHI

Douchi is a solid fermented soybean product with unique flavor and taste. It is widely consumed in China. Douchi is made from soybeans or black soybeans and the processing is composed mainly of fungal solid-state fermentation (pre-fermentation), followed by salting and maturation (post-fermentation). *Aspergillus*, *Mucor*, *Rhizopus*, or bacteria can be used for the fermentation process. Apart from being consumed as an important protein source and as a flavoring condiment for cooking, douchi can be used for pharmaceutical purposes (Zhang et al., 2006).

The first mention of douchi is in *Si Ji*, which was written by Si Maqian (about 104 BC). The production of douchi had been greatly developed in the Han dynasty (206 BC). Zhang Zhongjing in the Han dynasty recorded that soup cooked with cape jasmine and douchi was used to treat colds and anorexia in *Talking About Typhoid* (Chinese medical literature). In *Ben Cao Gang Mu* (Chinese medical literature) written by Li Shizhen in the Ming dynasty (AD 1368–1644), the curative effects of douchi included appetite enhancement, digestion promotion, and asthma prevention.

Based on the microorganisms used, douchi products can be classified into *Aspergillus*-type (Liuyang douchi and Yangjiang douchi), *Mucor*-type (Yongchuan douchi), and bacterial type (Qianxi douchi and Babao douchi). The types of starter culture depend on the local climate and environment, as well as the manufacturer's preference. Among them, *Aspergillus*-type douchi is the earliest and most widespread product in the world (Zhang et al., 2007). Additionally, douchi can be sorted into salt douchi and Dan douchi depending on whether it is salt-fermented. Most of the douchi produced in China is salt douchi which has a strong taste and is primarily used as a seasoning agent due to its excellent flavor. The realization of the negative effects of high salt content on health leads to the production of Dan douchi which has a lower salt content.

The disadvantages of traditional manufacturing process of douchi are complicated processing technology and long production cycle. Utilization of natural inoculation during the processing makes the mass production of douchi impossible.

17.2.1 Processing Technology of Douchi

It is well-known that douchi making consists of three major processes: Preparation of soybean (soaking and cooking), pre-fermentation (preparation of douchi *qu*), and post-fermentation (Figure 17.1) (Zhang et al., 2007; He et al., 2016). Pre-fermentation is an aerobic fermentation process using various microorganisms as a starter culture. As mentioned above, *Aspergillus*, *Mucor*, or bacteria could be inoculated into pretreated soybeans to make douchi *qu*. Post-fermentation is an anaerobic fermentation process, which is crucial for developing the flavor and nutrients of douchi.

In China, for the traditional method of producing douchi, black soybeans are generally selected as the raw material of douchi, which promotes its bright black color. The soybeans are soaked in water for 3–4 h and then steamed or boiled until soft. After cooling, the soybeans are spread on mats and fermented naturally by bacteria adhering to the mats or suspended in the atmosphere and fermented at room temperature for 3–4 days to harvest matured *qu*. The predominant microorganisms depend on incubation conditions. The water, salt, a little sugar, spices, and flavor such as capsicum paste are mixed with *qu*. The mixture is packed into bottles for fermentation. Fermentation normally continues for several months under natural conditions. In China, bacterial-type douchi is also known as water douchi due to its high moisture content during fermentation.

For the modern production method, the inoculation of douchi *qu* generally utilizes the pure starter culture. Beans are inoculated with *Aspergillus* (0.3%) or *Mucor* strain spores (0.5%) at 26–30° C for 5–6 days or 5–10° C for 15–20 days, respectively. For *Aspergillus*-type douchi, after the douchi *qu* is mature, abundant mycelium and spores appear. It is washed with water in order to remove the spores, mycelium, and a part of the enzymes to avoid a mold odor and bitter taste in the final product. After washing and mixing with condiments and spices, the mixture is encased in a container, sealed, and fermented. *Aspergillus*-type douchi is fermented at 30–35° C for 7–40 days whereas *Mucor*-type douchi is fermented at about 20° C for 10–12 months. The taste and characteristic flavor of final products are closely associated with the microbial community diversity in douchi during post-fermentation.

17.2.2 The Development of Functional Components of Douchi

17.2.2.1 Anti-Oxidative Components

Reactive oxygen or free radicals are associated with the progression of lifestyle-related diseases, including diabetes mellitus, neurodegenerative disorders, cataract development, and rheumatoid arthritis (Phaniendra et al., 2015). Compared with unfermented soybean, douchi enhances antioxidant activity. The extracts of douchi exhibit antioxidant activity by α,α-diphenyl-β-pricrylhydrazyl (DPPH) scavenging activity and reducing ability of ferrous ions (2.65 to 23.05 lmol TE/g and 1.03 mmol FE/100 g on average, respectively). The lowest DPPH and FRAP values were detected in the raw soybeans (1.52 lmol TE/g and 0.39 mmol FE/100 g, respectively) (Xu et al., 2015). As reported by Wang et al. (2008), the activities of superoxide dismutase, catalase, and glutathione peroxidase in the liver and kidney of douchi-fed rats elevated significantly. In the douchi-fed group, lipid peroxidation in the liver and kidney was also repressed.

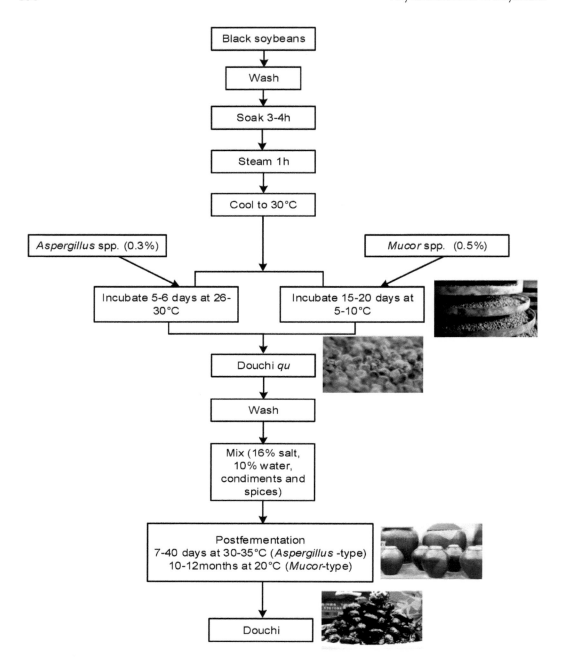

FIGURE 17.1   Schematic diagram of the douchi production process.

Isoflavones from soybeans are mainly categorized into daidzin groups, genistin groups, and glycitin groups, each of which exists in four forms: aglycones (daidzein, genistein, and glycitein), β-glucoside (genistin, daidzin, and glycitin), acetyl glucoside (6"-O-acetyldaidzin, 6"-O-acetylgenistin, and 6"-O-acetylglycitin), and malonyl glucoside (6"-O-malonyldaidzin, 6"-O-malonylgenistin, and 6"-O-malonylglycitin) (Collison, 2008) (Figure 17.2).

Their antioxidant capacities are associated with the number of hydroxyl groups in their structure. The bioformation of isoflavone aglycones from glucosides is catalyzed by β-glucosidase. Due

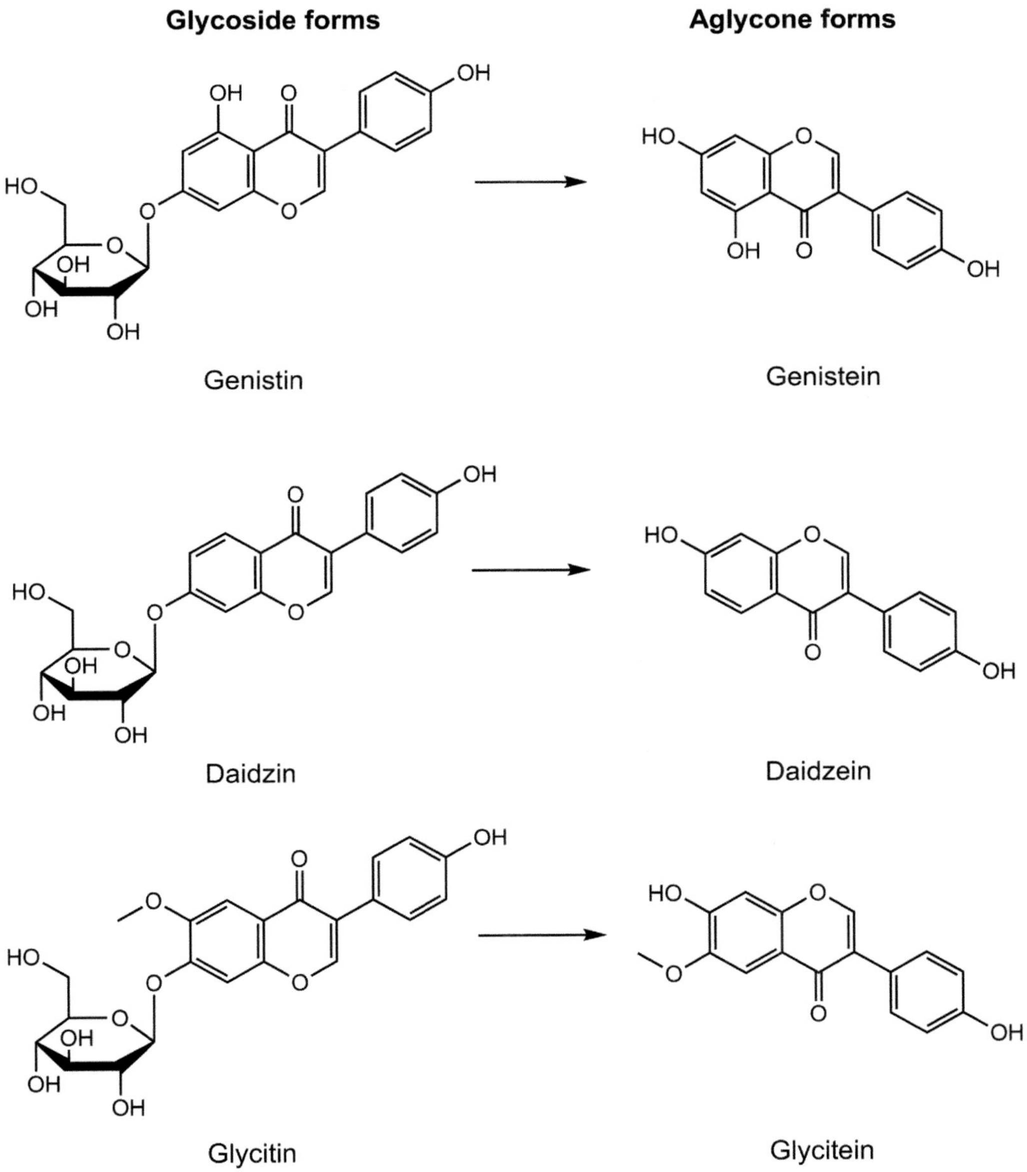

FIGURE 17.2 Molecular structure of isoflavones. Isoflavones are present in a biologically inactive glycoside form (genistin, daidzin, and glycitin) in soybean. β-glucosidases cleave the glucosyl residue and generate the biologically active aglycones form (genistein, daidzein, and glycitein) of isoflavones during digestion.

to its superior activity for hydrolyzing the β-glucosidic linkage of acetylglycoside and malonyl-glycoside, β-glucosidase is considered to be the key enzyme for the bioconversion of isoflavone isomers (Yuksekdag et al., 2017). However, isoflavone aglycones found in soybean are in trace quantities. In fact, the manufacturing procedures of douchi, such as soaking, cooking, and fermentation, cause significant differences in the redistribution of isoflavone isomers. Although pre-fermentation decreased total isoflavone levels in douchi, biotransformation of isoflavonoid aglycones, especially daidzein, genistein, and glycitein, was observed. Additionally, more than 90% of the isoflavones in douchi are in the form of aglycones after post-fermentation (Wang et al., 2007). Isoflavonoid aglycones have been suggested to possess greater antioxidant activity than their glycosides (Chiang et al., 2016; Huang et al., 2016). Glycone isoflavones have a higher antioxidant capability and bio-availability than glucoside isoflavones (Pyo et al., 2005; Kao & Chen, 2006). Izumi et al. (2000) found aglycone soy isoflavones were absorbed faster and in higher amounts than their glucosides in a human study. To convert glucoside isoflavones to aglycone isoflavones in a soy product could increase its antioxidant activity and bioavailability. Thereby, increased levels of isoflavonoid agly-cone are considered to be beneficial in douchi's anti-oxidant ability. Furthermore, the β-glucosidase activity was inhibited by a high NaCl supplementation during douchi processing, leading to prevent-ing isoflavone glucosides from being converted into aglycones (Wang et al., 2007).

In bacterial-type douchi, low-molecular-weight peptides rather than isoflavones play an impor-tant role in variations in antioxidant activity. As reported by Fan et al. (2009), the bioconversion of isoflavones had marginal effects on enhancing radical scavenging activities of douchi fermented by *Bacillus subtilis* B1. The activity of β-glucosidase was lower in bacterial-type douchi than fungi-type douchi, which was related to the microorganisms involved in fermentation. Proteases from *B. subtilis* B1 showed high activity, causing the accumulation of low-molecular-weight peptides. The low-molecular-weight peptides, especially consisting of hydrophobic, aromatic, acidic, and basic amino acids had a marked association with antioxidant activity. Wu et al. (2017) have evaluated the antioxidant activity of douchi protein hydrolysate, and the fraction of molecular weight 10–50 kDa exhibits higher antioxidant activity than other fractions.

Phenolic acid compounds generated during fermentation also contribute to the anti-oxidative activity of douchi. The term "phenolic acid" includes hydroxy and other functional derivatives of benzoic acid (C6■C1) and cinnamic acid (C6■C3) (Rice-Evans & Packer, 2003; Kumar & Pandey, 2013). Figure 17.3 gives the structures of the basic representatives of these acids.

Three phenolic acid components isolated from douchi, vanillic acid, syringic acid, and ferulic acid, showed strong DPPH radical scavenging activity, of which IC_{50} values were found to be 65, 20, and 58 μM, respectively (Chen et al., 2005). Further, vanillic acid and ferulic acid effectively inhibited the oxidation of soybean oil, as detected by a decrease in peroxide values and p-anisidine values and relatively low reduction in iodine values (Naz et al., 2005).

Phenolic components are considered to be crucial bioactive components, since they are condu-cive to the overall antioxidant profiles in soy foods. Molecular structures, particularly the number

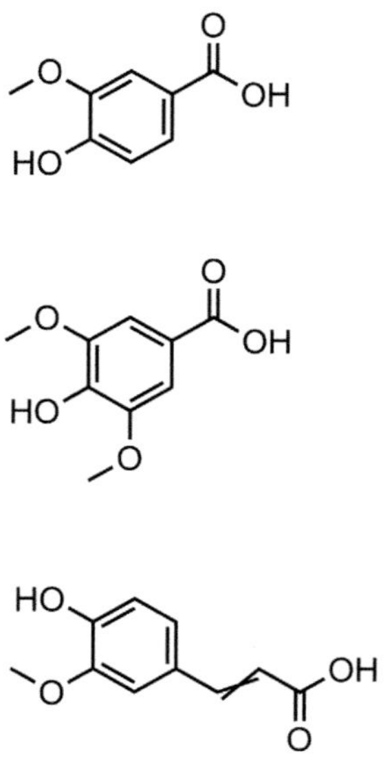

FIGURE 17.3   Chemical formulas of vanillic acid, syringic acid, and ferulic acid.

and positions of the hydroxyl groups and the nature of substitutions on the aromatic rings, confer to phenolic compounds the capacity of inactivating free radicals that is referred to as the structure-activity relationship (SAR) (Minatel et al., 2017).

Compared with other fermented soybean products, douchi showed relatively higher total phenolic content levels (10.06 mg GAE/g). Further, the total phenolic content levels in different brands of douchi are distinct, probably due to the inhibition of microorganism activity caused by the variance of ingredients ratio, microflora, and particular manufacturing processes (Xu et al., 2015). The increase in total phenolic content during fermentation is attributed to the mobilization of phenolic compounds from their bound form to a free state through enzymes produced during fermentation (Salar et al., 2016). During fermentation, phenolic compounds are metabolized and modified by fermenting organisms into other conjugates, glucosides, and/or related forms. Such a metabolism of phenolic compounds during fermentation has been reported to increase their bioavailability (Gänzle et al., 2014) and lead to the generation of compounds that impact flavor (Czerny & Schieberle, 2002; Katina et al., 2012).

Further, the Maillard reaction takes place between soybean proteins, its hydrolyzed peptides, and reducing sugar. The product of the Maillard reaction, melanoidin, possesses strong antioxidative activity. The melanoidin skeleton contains a peptide structure, which consists of aspartic acid, glutamic acid, arginine, lysine, and proline (Qin & Ding, 2006).

17.2.2.2 Anti-Hypertensive Components

Hypertension, known as a major public health problem, is the most serious risk factor for cardiovascular and cerebrovascular diseases. The renin-angiotensin system is a humoral mechanism within the body that is indispensable for the regulation of blood pressure（Drenjančević et al., 2011）. Angiotensin I-converting enzyme (ACE; EC 3.4.15.1) is a central component of the renin-angiotensin system. It is a dipeptidyl carboxypeptidase and catalyzes the conversion of the inactive decapeptide angiotensin I into the octapeptide angiotensin II. Octapeptide angiotensin II is an effective vasoconstrictor agent which catalyzes the degradation of bradykinin, leading to the elevation of blood pressure (He et al., 2013). Therefore, it has been established that inhibition of ACE activity is a potent way to treat hypertension.

It has been reported that douchi exhibits ACE inhibitory activities, indicating its enormous potential as a source of antihypertensive substances (Li et al., 2010; Zhang et al., 2006). At present, the ACE inhibitory activities are principally associated with the hydrolysis of soybean protein and the generation of bioactive peptides (Li-Chan et al., 2015). The ACE inhibitory peptide with the highest inhibited activity is composed of phenylalanine, isoleucine, and glycine with a ratio of 1:2:5 in *Aspergillus Egyptiacus*–fermented douchi (Zhang et al., 2006).

Li et al. investigated the stability of ACE inhibitory activities of douchi under various treatments of processing and simulated digestion conditions. The results showed that the ACE inhibitory activities of douchi were, to a large degree, steadily treated with high temperature and different pH levels and able to withstand the simulated gastrointestinal digestion by pepsin, trypsin, and chymotrypsin (Wang et al., 2015). The stability of ACE inhibitory activities under treatments at high temperatures might be involved in the heat tolerance of ACE inhibitory peptides and other bioactive components (Shuangquan et al., 2008). Protease related to the simulated gastrointestinal digestion, especially pepsin with lower specificity, catalyzes the hydrolyzation of protein to generate more oligopeptides and bioactive peptides. Furthermore, pepsin catalyzes the releases of peptides with C-terminal hydrophobic and aromatic AA residues through cleaving related peptide bonds. These peptides have been demonstrated to be beneficial for exerting ACE inhibitory activity (Nelson et al., 2000). The ACE inhibitory activities also rely on the type of starter cultures. ACE inhibitory activity reduced significantly in the ripening process of bacterial-type douchi; however, it increased in fungi-type douchi. *Aspergillus oryzae*–fermented douchi exhibited the highest inhibitory activity after maturing for 30 days (Li et al., 2010). Proteases originated from various starter cultures are capable of acting on distinctive catalytic sites of soybean protein and producing peptides with different structures

(Nora et al., 2006). The structure of peptides considerably affects the ACE inhibitory activities (Li et al., 2004).

17.2.2.3 Anti-Inflammatory Components

As mentioned above, douchi is a typical herbal medicine for treating typhoid fever in traditional medicine. For heat-involved diseases, such as heartburn, inflammation, and the common cold, douchi is applied as a supplementary and alternative medicine (Aum et al., 2016). Douchi has curative roles in inflammatory allergic diseases that are related to T_H2 immunomodulatory activity, such as atopic dermatitis (AD) (Jung et al., 2016).

A deficiency of ceramides (represented by PKC level) in the stratum corneum is an essential etiologic aspect in the early events of AD (Hogan et al., 2012). Another main characteristic of AD is the imbalance of T_H1/T_H2 immune response, being caused by inducing T helper (Th) 2 cytokines, for instance, interleukin (IL)-4 in skin lesions (Georas et al., 2006). The levels of PKC and IL-4 were significantly decreased in the douchi-treated group. The mast cell degranulation induced by IL-4 promotion increases the symptoms of AD by releasing inflammatory mediators such as substance P or MMP-9 （Baram et al., 2001）. Administering douchi was observed to alleviate skin damage and decrease histological changes of AD-like skin symptoms in mice. The decreases of IL-4 and PKC caused the decrease of inflammatory factors, for instance, substance P, inducible nitric oxide synthase (iNOS), and matrix metallopeptidase 9 (MMP-9) (Jung et al., 2016). Aum et al. (2016) observed that douchi extracts treated with hataedock exhibited anti-inflammatory activities on AD-induced NC/Nga mice. Hataedock treatment with douchi extract alleviated AD symptoms by reducing the release of inflammatory cytokines, such as Fc ε receptor, substance P, and NF-κB, at the early stage of AD. Moreover, douchi treated with hataedock promoted skin lipid barrier formation via stimulating the differentiation of keratinocytes (Kim et al., 2017).

γ-linolenic acid (GLA 18:3, △6,9,12) is an important n-6 polyunsaturated fatty acid (PUFA) and is indispensable for the structure of membrane lipids (Zhang et al., 2017). The metabolism of GLA leads to the production of anti-inflammatory eicosanoids, such as leukotrienes (Wang et al., 2012). GLA possesses many healthy and medicinal roles, for instance, selective tumoricidal and anti-inflammatory activity (Itoh et al., 2010; Kim, 2012).

It has been demonstrated that GLA selectively kills tumor cells without harming the normal cells. After GLA treatment on human hepatocellular cell lines Huh7, cell proliferation was attenuated and the generation of reactive oxygen species and apoptosis were promoted. Further, the translational levels of antioxidant proteins, including heme oxygenase-1 (HO-1), aldo-keto reductase 1 family C1, C4, and thioredoxin, increased. Among them, the HO-1 protein levels were overexpressed, contributing to the antioxidant protection against oxidative stress caused by GLA in Huh7 cells. GLA treatment is an effective therapeutic modality for patients with advanced hepatocellular carcinoma (Shinji et al., 2010).

The primary inflammatory cells observed in diabetic kidneys are monocytes/macrophages. The monocytes/macrophages infiltrate and are attracted to the target tissue leading to renal injury. The process is mediated by various chemokines and adhesion molecules such as MCP-1 and ICAM-1 (Sara et al., 2010). Emerging evidence has shown that GLA treatment has a positive effect on renoprotection due to its anti-inflammatory effects. The attenuation of inflammation under diabetic conditions is partially mediated by inhibiting elevated MCP-1 and ICAM-1 expression (Kim et al., 2012).

Mucor racemosus–fermented douchi contains large amounts of GLA and the content of GLA is partly dependent on the fermentation conditions, including inoculum form, inoculum amounts, fermentation temperature, and fermentation time. Under the optimal fermentation conditions (inoculum was 5.30×10^7 spores/10 g, 26° C), the GLA production of *M. racemosus* was elevated (Lu et al., 2011). It also has been observed that *R. oryzae* from douchi could produce large amounts of GLA. Additionally, RoD6D, high homology to fungal △6-fatty acid desaturase genes, was isolated from *R. oryzae*. The encoded product of RoD6D displayed △6-fatty acid desaturase activity, promoting the accumulation of γ-linolenic acid (Lu & Zhu, 2015).

17.2.2.4 Components with Anti-α-Glucosidase Activity

α-glucosidase inhibitor commonly acts as a therapeutic method for type II diabetes by integrating with intestine α-glucosidase and restricting the level of postprandial blood glucose (Hossain et al., 2020).

Aqueous douchi samples were collected from various areas of China and exerted various degrees of anti-α-glucosidase activity against rat intestinal α-glucosidase (Chen et al., 2007). Furthermore, the activity of anti-α-glucosidase was associated with the carbohydrate levels in douchi samples, suggesting that the α-glucosidase inhibitor probably possesses carbohydrate structure (Jing et al., 2007). Moreover, the anti-α-glucosidase activities of douchi were affected by the fermentation conditions, microorganisms, and additives (Chen et al., 2007). For example, Song et al. (2001) observed that the anti-α-glucosidase activity of douchi *qu* fermented with *A. oryzae* was higher than that of *A. elegans* and *R. arrhizus*. However, the levels of anti-α-glucosidase activities in douchi *qu* were extremely low during the first 60 h of pre-fermentation. Comparatively, high activity of α-glucosidase inhibition was observed in the douchi *qu* produced with *Bacillus subtilis* B2 (Zhu et al., 2011). The results suggested that the anti-α-glucosidase activities in douchi *qu* partly relied on the types of microorganisms. The highest anti-α-glucosidase activities were observed in douchi *qu* fermented with *A. orzyzae* at 5.0% and 7.5% salt levels; this variation most probably contributed to the ability of *A. orzyzae* to adapt to habitats with low water activity (Song et al., 2001). A sharp increase of anti-α-glucosidase activity in douchi *qu* was observed during the first 48 h of fermentation. Additionally, the fermentation humidity also is considered as an influencing factor of anti-α-glucosidase activity. The lowest anti-α-glucosidase activities of douchi *qu* were at 70% relative humidity and the highest were at 90% (Zhu et al., 2011).

Deoxynojirimycin (DNJ), a representative natural aza-sugar, is a D-glucose analog with an NH-group substituting for the oxygen atom of the pyranose ring. It has been illustrated that DNJ and some of its derivatives are potent glucosidase inhibitors. They bind at the active site substituting the substrates and serve as a competitive inhibitor blocking the reaction catalyzed by these enzymes (Romaniouk et al., 2004). DNJ is detected in douchi *qu* fermented with *B. subtilis* B2 and it obviously plays a crucial role in the anti-α-glucosidase activity of douchi (Zhu et al., 2011).

Additionally, genistein, an aglycon form of isoflavones found in soybean and existing in douchi (Chen et al., 2015), was considered as a candidate for an α-glucosidase inhibitor of yeast (Lee & Lee, 2001). It was demonstrated to be a reversible, slow-binding, and non-competitive inhibitor of α-glucosidase.

17.3 SUFU

Sufu, also known as to-fu-ru, is a Chinese traditional fermented food that has spread since the Han dynasty (Han et al., 2001). Because of its strong umami taste and unique aroma, sufu is widely consumed as an appetizer or as a side dish mainly with rice or steamed bread by Chinese and Southeast Asians.

Sufu can be categorized into red sufu, white sufu, grey sufu, and other types based on the product characteristics in color and flavor. Red sufu is the most traditional sufu. Its surface is a natural red, and the cut surface is a yellowish white with a mellow taste. In addition to soybeans, the raw material for making red sufu is taro. Generally, high-quality liquor is added to make sufu savory and mellow before sealing. Guilin Huaqiao white sufu is one of the most famous mold-fermented white sufus with an annual production of 17,000,000 jars in Guilin, China. The production of Guilin sufu is an entire process from refining to filtering, shaping, pressing, drying, and mildewing. The selection of materials is excellent. The sufu curd is small, smooth, and soft with an orange and transparent surface (He et al., 2020). Grey sufu, also known as stinky tofu, is ripened with a special dressing mixture with an unpleasant odor. Wanzhihe sufu in Beijing is an indigenous celebrated grey sufu brand (Liu et al., 2012).

On the basis of the starter culture used, sufu can be grouped into three types, which are mold-fermented tofu, naturally fermented tofu, and bacteria-fermented tofu (Wang et al., 2009). *Mucor* is one of the most suitable organisms for preparing mold-type sufu. The preparation process of *Mucor* is similar to that of *Rhizopus*; the difference is that *Rhizopus* is more suitable to a higher temperature (37° C) (Li & Tan, 2009). The manufacturing technique of bacteria-type sufu is much different than for other types of sufu. Kedong sufu is a typical bacteria-fermented sufu in China. The ripening stage of Kedong sufu takes more than six months including pre-fermentation and post-fermentation (Feng et al., 2013).

To date, pure starters isolated from the autochthonous microbiota of traditional sufu have been applied in the industrial production of sufu. More information about sufu microecology contributes to the development of specific starters and/or adjunct cultures to obtain a consistent flavor and texture, a standard color, and a shorter ripening time.

17.3.1   Processing Technology of Sufu

There are various methods for producing different types of sufu in China. The processes for the production of sufu consist of four major steps, namely, 1) preparation of tofu; 2) *pehtze* fermentation; 3) salting or brining; and 4) ripening (Figure 17.4).

Soybeans are soaked in water and ground on a stone mortar into a slurry. After filtration, the slurry is diluted and pressed to obtain soymilk. It is then coagulated by adding calcium salts. Excess water has been removed, and through pressing the precipitate forms tofu, which can be cut into different shapes and sizes. The tofu curds are placed on a bamboo tray and set in the air for two days at room temperature. The trays are surrounded by straw for natural inoculation and fermentation. The traditional sufu preparation is a contaminative process. Tofu curds need to be exposed to the sun for several hours before inoculation. Utilizing sun to dehydrate the tofu cubes could prevent them from bacterial spoilage.

The modern method to prepare sufu is based on the traditional method with modification. The tofu cubes are sprayed with a suspension of spores of *Actinomucor*, *Mucor*, or *Rhizopus*. The loaded trays are transferred into incubators. After 36–40 h at around 25° C, the trays are cooled by aeration. When a slightly yellowish-white color covers the tofu curd, that means the formation of fresh *pehtze* is finished (Li & Tan, 2009). *Pehtze* is then placed in a container for salting by means of sprinkling salt evenly on each layer of *pehtze*. *Pehtze* can also be soaked in a saturated salt solution.

Based on the different types of sufu, salt, sugar, alcoholic beverage, and spices are added into salted *pehtze* during the ripening process. An alcoholic beverage can act as a preservative against microbial growth and confer a unique flavor and aroma. There are many types of spices, such as pepper, star anise, cinnamon, ginger, chili, and so on. In addition to adjusting the flavor of fermented sufu, spices can make sufu less susceptible to bacterial spoilage. Ang-kak is added into the saltwater serving as a colorant for red sufu. Ang-kak is red yeast rice made by the mycelium of *Monascus* (Chaisrisook, 2002). The mixture composed of salted *pehtze* and auxiliary ingredients with a ratio of 2:1 is impacted into jars and sealed with clay and sheath leaves of bamboo. The ripening stage of traditional sufu preparation is continued for a longer period of six months; however, modern processes take about two to three months. Reducing the salt content (80 kg^{-1}) can shorten the ripening time to 40 days, which is of benefit to manufacturers (Han et al., 2003).

17.3.2   The Development of Functional Components of Sufu

17.3.2.1   Anti-Hypertensive Components

It has been established that peptides and peptide fractions with ACE inhibitory activity present in sufu products could be beneficial for hypertension reduction. For example, sufu extract mainly composed of peptides with weights less than 10 kDa inhibited ACE activity that indicated potential anti-hypertensive activity (Wang et al., 2003). A similar observation was reported by Hang and Zhao (2012) who observed ACE inhibitory activity *in vitro* of the extract of sufu fermented by

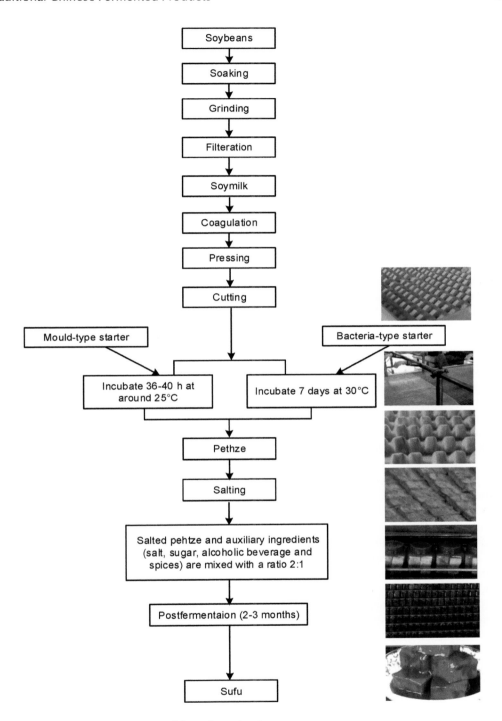

FIGURE 17.4 Schematic diagram of the sufu production process.

Mucor spp. Furthermore, spontaneously hypertensive rats in the sufu-group showed a more significant anti-hypertensive effect compared to the control group. The ACE activity in kidneys was significantly lower, which could be associated with the antihypertensive effect of sufu (Kuba et al., 2004). Moreover, the total cholesterol in serum was reduced by feeding with sufu. Patients with hypercholesterolemia could develop atherosclerosis, resulting in decreased arterial elasticity, which may aggravate hypertension (Kuba et al., 2004; van Rooy & Pretorius, 2014).

Amino acid composition analysis showed that the amino acid sequences of the ACE inhibitors in sufu were Ile-Phe-Leu (IC50, 44.8 mM) and Trp-Leu (IC50, 29.9 mM) (Kuba et al., 2003). Ile-Phe-Leu was observed in the primary structures of the α- and β-subunits of β-conglycinin, while Trp-Leu was in the B-, B1A-, and BX-subunits of glycinin. The release of Ile-Phe-Leu from β-conglycinin could be catalyzed by *M. purpureus* and/or *A. oryzae* used in the fermentation process. Although the basic subunit in glycinin was not readily degraded by these enzymes, Trp-Leu might be released from the basic subunit during long-term fermentation. According to Hang and Zhao (2012), the extract of sufu with a higher ACE inhibitory effect contained more total hydrophobic amino acids or proline. Some ACE inhibitory peptides possess hydrophobic amino acids, especially Pro, such as Val-Leu-Ile-Val-Pro and Leu-Ala-Ile-Pro-Val-Asn-Lys-Pro from soybean proteins (Mallikarjun et al., 2006)

NaCl content is the principal influence factor of ACE inhibitory activities. As observed by Ma et al. (2013), the salting process caused a decrease of ACE inhibitory activities in sufu. The absorption of NaCl inhibited mold growth and its protease activity, thereby causing the reduction of peptides generation and the decrease of ACE inhibitory activity. In addition, the disappearance of water-soluble matter and the changes in dry matter might be another reason. The ACE inhibitory activity is also significantly affected by manufacturing procedures. A slight reduction of ACE inhibitory activity was observed in low-salt sufu, caused by heat treatment, which suggests the thermostability of low-salt sufu. And the ACE inhibitory activity of low-salt sufu increased under acidic conditions (Ma et al., 2014). The variation of process conditions impacts the hydrolysis extent of proteins or peptides during fermentation, which could affect ACE inhibitory activity.

Aside from ACE inhibitory peptides, γ-aminobutyric acid (GABA) has been demonstrated to be a potential hypotensive agent. GABA (γ-aminobutyric acid) is a biologically active four-carbon non-protein amino acid widely distributed in nature among microorganisms, plants, and animals (Ramos-Ruiz et al., 2018). GABA can depress the elevation of blood pressure in animals and human subjects. The administered GABA ranged from 0.3 to 300 mg/kg and caused a decrease in spontaneously hypertensive rats (Kimura et al., 2002). A single oral dose of GABA is effective in lowering blood pressure in spontaneously hypertensive rats (Hayakawa et al., 2004). GABA could inhibit noradrenaline liberates from sympathetic nerves in the mesenteric arterial bed through presynaptic $GABA_B$ receptors, resulting in the antihypertensive effect of GABA (Hayakawa et al., 2002). The biosynthesis of GABA is generally a one-step reaction of decarboxylating glutamate to GABA, catalyzed by glutamate decarboxylase (GAD) (Figure 17.5) (Dhakal et al., 2012).

Lv et al. (2020) observed two GABA synthesis pathways, the putrescine utilization (Puu) pathway and the acetylation pathway (Figure 17.6) of *Kocuria kristinae*, and one of the dominant strains in Kedong sufu during fermentation. Additionally, it has been reported that *Lactococcus* isolated from sufu displayed a capacity to synthesize GABA, such as *L. brevis*, *L. casei*, and *L. curieae* (Huang et al., 2018; Bao et al., 2020; Han et al., 2001; Wang et al., 2015). *Bacillus licheniformis* in sufu is favorable for the accumulation of GABA (Xie et al., 2018).

Glutamic acid is abundant in soy protein, and fermentation is considered an effective means for GABA enrichment, particularly in an anaerobic environment. Sufu ripening is commonly limited to

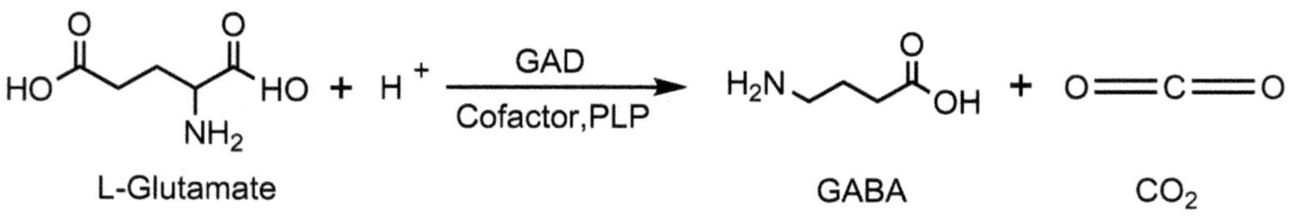

FIGURE 17.5 Decarboxylation of L-glutamate to GABA by glutamate decarboxylase (GAD). PLP = pyridoxal-5'-phosphate.

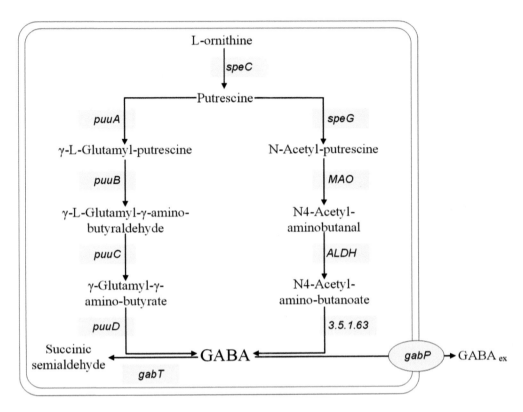

FIGURE 17.6 GABA synthesis pathways, the putrescine utilization (Puu) pathway, and the acetylation pathway observed in *Kocuria kristinae*.

strict anaerobic conditions, contributing to GABA accumulation. During fermentation, the GABA content is increased from 32.64 to 133.13 mg/100 g dry matter (Ma et al., 2013). Even though the GABA level is elevated during fermentation, the salting process sustained the reduction of the GABA level. This reduction was probably associated with water loss during salting due to the solubility of GABA (Ma et al., 2013). Compared with other types of fermented soybean products, sufu samples presented relatively higher GABA levels. Red sufu had relatively lower GABA content as compared with other sufu samples; the differences in the sources of soybean materials and different fermentation technologies may have led to the variation in GABA content (Xu et al., 2017).

17.3.2.2 Anti-Oxidative Components

Numerous studies have reported the antioxidant activity of sufu. The methanol extract of sufu exerts potent antioxidant activity, such as DPPH radical scavenging activity, Fe^{2+}-chelating ability, and reducing power, compared with non-fermented tofu extract (Cai et al., 2016). Wang et al. (2003) suggested that peptides could be the predominant compounds that are in charge of the antioxidant effect in mold-fermented sufu extract. A similar conclusion was obtained by Cai et al. (2016). The antioxidant activity of ABTS radical reducing power and ferric reducing power was correlated with short-chain peptides (especially for 500–1,000 Da fraction) in white sufu. Proteolysis exposed hydrophilic amino residue side-chain groups for small peptides, resulting in more opportunity to trap the water-soluble ABTS radical cation than large molecular peptides with less hydrophilic residue side chains (Gülçin, 2010).

It has been known that the isoflavone composition was changed during sufu processing. The contents of aglycones were elevated, while the corresponding contents of glucosides were reduced.

The former exhibited higher antioxidative activity than the latter. The alternation of isoflavone isomers depended on the activity of β-glucosidase, which was affected by the NaCl level (Li-Jun et al., 2004).

Cai et al. (2016) investigated the alternation in the content and distribution of glucosides (daidzin and genistin) and their corresponding aglycones during the manufacturing process of white sufu. The results indicated that the aglycone content (daidzein and genistein) increased 0.26- and 0.91-fold, respectively, during fermentation. During the ripening stage, almost all glucosides transform into aglycones and daidzein increased to 0.11 g 100 g^{-1} DM, and genistein increased to 0.054 g 100 g^{-1} DM. But the contents of both were steady and briefly decreased for the last 15 days of ripening, probably due to the further conversion of aglycones to other chemical compounds.

As reported by Xie et al. (2018), the contents of daidzin (1.05 mg/100 g DM) and genistein (0.68 mg 100 g^{-1} DM) were the highest in white sufu. Red sufu, jiang sufu, and cabbage sufu contained the lowest levels of daidzin (0.04 mg/100 g), genistein (0.21 mg/100 g), and daidzein (0.18 mg/100 g), respectively. These differences were associated with the raw materials, fermentation time, manufacturing process, and isoflavone extraction parameters. For example, the daidzin contents of soybean cultivars were significantly affected by location and cultivar. Daidzin contents of soybean cultivars grown in northern China were 35.2 mg/100 g, which was higher than that in southern China (27.6 mg/100 g) (Xie et al., 2018). The antioxidant effects of the sufu were related to the ripening temperature and the duration of the ripening period. The effect of sufu for DPPH radical-scavenging effect and Fe^{2+} iron-chelating ability increased as the ripening period was extended. The samples at 16 days exhibited the highest reducing power followed by that from sufu ripened for 12, eight, four, and zero days in descending order (Huang et al., 2011). Fermentation temperature was capable of transforming the isoflavone distribution in sufu fermented with *A. elegans*. The contents of isoflavone aglycones were higher when fermented at 26° C than at 32° C; the abundant isoflavone aglycones were favorable for improving physiological properties. Further, 6'-O-malonyl-glucosides were detected when fermented at 32° C, while they did not exist at 26° C (Yin et al., 2005). Han et al. (2003) also reported that the generation of enzyme or enzyme activity by *A. elegans* was not enough to prepare a good quality of sufu when ripened at over 30° C or below 25° C. However, the extract of sufu fermented with *A. oryzae* showed a higher DPPH radical scavenging effect and Fe^{2+} iron-chelating ability when ripened at 45° C than those samples ripened at 37° C and 25° C (Huang et al., 2011). The growth of *Mucor flavus* and the capacity of protease activity to degrade soybean protein was prominent when ripened at below the 15° C that were suitable for sufu production (Cheng et al., 2009). As reported by Cheng et al. (2011), the alternation of isoflavone constituents could be observed when producing sufu fermented with *M. flavus* under a lower temperature (13° C). Sufu fermentation resulted in a significant increase in isoflavone aglycones contents, accompanied by a decrease in isoflavone glucosides. The former accounted for 99.4% of total isoflavones in sufu fermented with *M. flavus* under 15% NaCl level at 13° C (Cheng et al., 2011).

17.3.2.3 Anti-Inflammatory Components

Butyric acid is a short-chain fatty acid with the formula CH3-CH2-CH2-COOH, generated by the colon – in particular, by the proximal colonic microbiota. Colonocytes cells obtained the most effective energy from butyric acid. Butyric acid promotes proliferation and differentiation of normal enterocytes when it depresses cell proliferation in a colon carcinoma cell line *in vitro* due to its inhibitory effect on histone deacetylases (Sossai, 2012). Butyric acid's anti-inflammatory action has been the subject of much attention. Butyric acid causes the down-regulation of the expressions of cytokine genes, such as IL-6, IL-8, TNF-α, and TGF-β, through inhibiting NF-κB (Meijer et al., 2010). Butyrate enemas are utilized as a treatment for inflammatory bowel disease and other colonic inflammatory diseases. It has been suggested that the inhibitory effects of butyric acid in these diseases are primarily due to the regulation of energy availability in colonocytes. However, the cellular signaling mechanism that mediates the defensive roles of butyric acid in inflammatory bowel disease has not been completely elaborated (Mishiro et al., 2013).

Butyric acid was observed to be the principal volatile organic acid detected in grey sufu, which was much higher than that of other foods known (Ma et al., 2013). The accumulation of butyric acid could be caused by some butyric acid-producing bacteria, such as *C. butyricum* (Takahashi et al., 2018). Further, lactic acid could convert to butyric acid in *C. butyricum* and contribute to the elevation of butyric acid content (Detman et al., 2019). However, the evidence regarding whether grey sufu contains butyric acid-producing bacteria or not has not been shown. The undesirable flavor of butyric acid at high concentrations restricts its food application. Grey sufu is a Chinese traditional mold-fermented soybean product with a strong and offensive odor, but is attractive to certain consumers. Thus, grey sufu rich in butyric acid is beneficial for its application promotion due to its potential efficacy in several diseases.

17.4 SOY SAUCE

Soy sauce is a fermented soybean liquid, serving as a condiment and seasoning in many Asian countries with a salty taste and distinct fragrance. The production of soy sauce started 2,000 years ago in China during the Han dynasty. The first historical record mentions that tamari-type soy sauce was introduced to Japan before AD 700. The soy sauce for *chiang-yu* (Chinese) and *shoyu* (Japanese) appeared around the sixteenth century, and especially *shoyu* is widely consumed by the general population. Around the seventeenth century, *shoyu* was industrially produced in different regions of Japan.

Soybeans are the only material for producing Chinese soy sauce; no cereals or only a small amount of cereals are added. According to fermentation processes, Chinese soy sauce is divided into two types: Low-salt solid-state fermentation soy sauce and high-salt liquid-state fermentation soy sauce. The production process of the first type is a rapid fermentation process and its fermentation period is just about 30 days. The ripening of the second type of soy sauce takes about six months, and the high-salt liquid-state fermentation soy sauce has a better taste and fragrance than those of low-salt solid-state fermentation (Zhao et al., 2013). In China's National Standard of brewing soy sauce, the quality grade of soy sauce is based on the content of amino acid nitrogen. The contents of amino acid nitrogen for special grade, first grade, second grade, and third grade are $\geq$ 0.80, $\geq$ 0.70, $\geq$ 0.55 (or 0.60), $\geq$ 0.40 g / 100 mL, respectively.

17.4.1 Processing Technology of Soy Sauce

The manufacturing processes of soybean paste basically consist of three stages: *koji* production, brine fermentation, and refining (Figure 17.7).

The substrates of *koji* are soybean and wheat. The soybeans are washed and soaked in water overnight, then steamed in a batch or continuous cooker. The wheat is roasted and crushed. After cooling down the soybean and wheat to 25–30° C, the soy-wheat mixture is inoculated with *A. oryzae*, *A. sojae*, or *A. tamarii* for *koji* making. The proteins, starch, and other macromolecules in the soybean-wheat mixture are decomposed by proteinases, peptidases, cellulases, and amylases from the mash microflora to be used as nutrients.

A brine containing a high concentration of sodium salt (17.5–20% w/v) is added to make *moromi*. The brine fermentation is facilitated by involving cultures ordinarily containing *Tetragenococcus halophilus*, *Pediococcus acidilactici*, *Lactobacillus delbrueckii*, other lactic acid bacteria, and yeast strains of *Zygosaccharomyces rouxii* and *Candida versitalis*. During the initial stage of brine fermentation, the *Aspergillus* species are salt-sensitive and quickly lyse and die. *L. delbrueckii* and *T. halophilus* are grown and produce lactic acid to decrease the pH. Yeast strains are dominant strains during the rest of the fermentation. Salt-tolerant yeasts, *Z. rouxii* and *C. versitalis*, are more salt- and acid-tolerant than the lactic acid bacteria. *Z. rouxii* can synthesize ethanol and higher alcohols, such as butyl alcohol, isoamyl methanol, and amyl methanol. 4-hydroxy-2 (or 5) -ethyl-5 (or 2) -methyl-3 (2H) -furanone (HEMF) can also be generated by *Z. rouxii*, which plays a vital role in the aroma

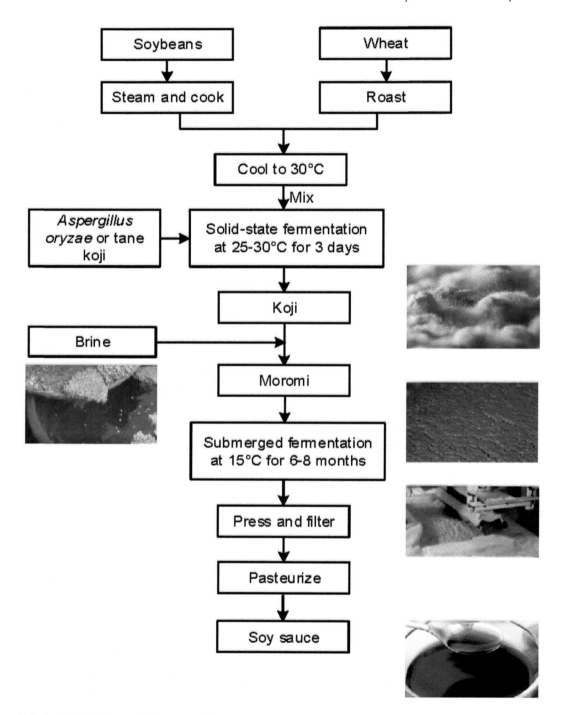

FIGURE 17.7   Schematic diagram of the soy sauce production process.

formation of soy sauce (Kenji et al., 2014). *C. versatilis* can produce 4-ethyl-guaiacol (4-EG), which is one of the characteristic aroma components in soy sauce. The adding time of salt-tolerant microorganisms is critical for brine fermentation since flavor components in soy sauce would be affected. It was reported that the flavor of soy sauce adding *Torulopsis* and salt-tolerant lactic acid bacteria on the fifteenth day and adding *Sacchromyces* on the forty-fifth day was richer than the other samples (Wan et al., 2011). The brine fermentation generally takes six to eight months, therefore, shortening this period is critical and new processes for soy sauce brewing are desirable.

After fermentation, the *moromi* is pressed and filtrated to make the soluble soy sauce separated from the solid residue. The refined raw soy sauce is generally equivalent to heat at 80° C in order to inactivate a large number of microbial and enzymatic reactions. Many flavor compounds such as pyrazine, organic acids, phenols, aldehydes, mercaptans, and their inter-reaction compounds would appear in the results of non-enzymatic reaction.

17.4.2   THE DEVELOPMENT OF FUNCTIONAL COMPONENTS OF SOY SAUCE

17.4.2.1   Anti-Oxidative Components

Soy sauce has been reported to possess strong antioxidant activities, indicating that soy sauce could serve as a superior source of dietary antioxidants (Li et al., 2010; Zhao et al., 2013).

As a potent anti-oxidant, soy sauce could effectively scavenge free radicals, such as ABTS and DPPH (Aoshima & Ooshima, 2009; Long et al., 2000). Dark soy sauce also was confirmed to be a potent antioxidant by scavenging free radical ABTS *in vitro*. Dark soy sauce also exhibited a rapid antioxidant effect against lipid peroxidation *in vivo* (Lee et al., 2006). The addition of soy sauce decreased the concentrations of the 2-thiobarbituric acid, peroxide, and conjugated diene ($P < 0.05$) during the storage period in raw beef patties. Soy sauce significantly suppressed the formation of primary and secondary metabolites of lipid oxidation (Kim et al., 2013). Daidzein (isoflavone), 4-ethylguaiacol, catechol, 4-ethylphenol, and Maillard reaction products are regarded as the key antioxidants in soy sauce (Li et al., 2017; Li, Zhao, Zhao, & Cui, 2011).

It has been investigated that the products of Maillard reaction have antioxidant characteristics in soy sauce. Among them, melanoidins were the major substances of brown-colored products with strong antioxidant activities (Wang et al., 2007). The structural characteristics of melanoidins have been elucidated in several studies. Carbohydrate residues could be a part of the melanoidin back-bone, observed by NMR and MS analyses. Further, amino acids could be associated with the generation of the chromophore unit, observed by the carbonyl resonance and non-saturated proton signals analyses (Cammerer et al., 2002). The antioxidant activities of melanoidins have the metal chelating capacity of these compounds. Melanoidins are capable of chelating transition metals due to their anionic nature. It also has put forward that the nitrogen atoms in melanoidin are in charge of the chelation of copper ions. Besides, the radical scavenging activity of melanoidin could be the other principal mechanism for its antioxidant activity (Wang et al., 2011). *Staphylococcus* sp. SSB48 and *B. amyloliquefaciens* SSB6 isolated from fermented *moromi* have the ability to degrade melanoidin without interfering with aromatic complexity (Det-udom et al., 2019). Furanones and pyranones, such as 4-hydroxy-2,5-dimethyl-3(2H)-furanone (HDMF), 4-hydroxy-2 (or 5) -ethyl-5 (or 2) -methyl-3(2H)-furanone(HEMF), 4-hydroxy-5-methyl-3(2H)-furanone (HMF), 3-hydroxy-2-methyl-4-pyranone (maltol), and 5-hydroxy-2-hydroxymethyl-1,4-pyranone (kojic acid) are predominant contributors to the flavour of soy sauce (Li et al., 2017) (Figure 17.8).

HMF, HDMF, and HEMF were proved to possess antioxidative properties, and the order of potency was as follows: HEMF > HDMF > HMF (Takigawa & Shibuya, 2012). Especially, HMF and HDMF (as well as HEMF) could reduce the concentration of hydrogen peroxide in human polymorphonuclear leucocytes aroused by arachidonic acid or 12-o-tetradecanoylphor-bol-13-acetate. In addition, HDMF, HMF, and HEMF could reduce benzo[a]pyrene-initiated forestomach neoplasia of mice by reacting at the post-initiation stage (Kataoka, 2005).

HEMF was produced by cultivating halotolerant yeast, such as *Z. rouxii*, including amino-carbonyl reaction products based on ribose and amino acids (Sugawara et al., 2007; Hayashida et al., 2000). Ohata et al. (2007) demonstrated that the five-member ring and the methyl group on the HEMF side chain were formed from a five-carbon chemical compound generated by the

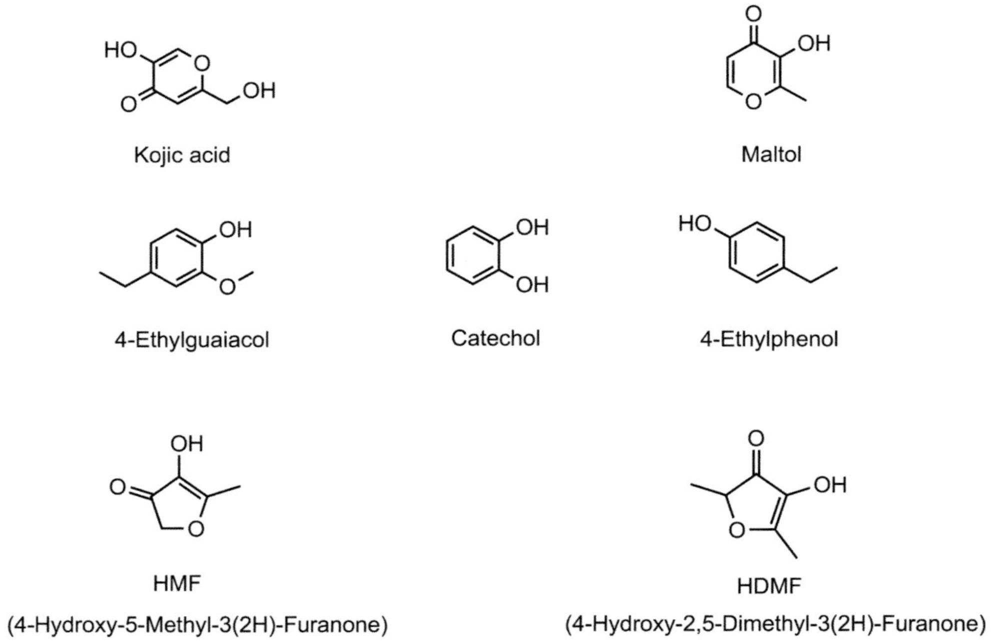

FIGURE 17.8   Chemical formulas of antioxidative compounds in soy sauce (kojic acid, maltol, 4-ethylguaia-col, catechol, 4-ethylphenol, HMF, and HDMF).

amino-carbonyl reaction of ribose and glycine (the C5 precursor) and that the ethyl group on the HEMF side chain was formed from a two-carbon chemical compound generated by the glucose metabolism of yeast (the C2 precursor). Ohata et al. (2007) also assumed that acetaldehyde is effective as the C2 precursor and that the C5 and C2 precursors were most likely enzymatically combined by yeast.

Overmuch increasing the heating time or temperature suppressed the formation of HEMF in soy sauce "moromi mash" containing Maillard reaction products (Hayashida et al., 2000). Besides, more HEMF was produced with higher NaCl content, total nitrogen content, and pH of the medium (Yaginuma et al., 2002). Nunomura and Sasaki also proposed a HEMF-enhancing effect of NaCl. HEMF was primarily generated by halophilic "soy sauce yeasts", namely, *Zygosaccharomyces rouxii*, which could explain the phenomenon mentioned above (Nunomura, 2006).

Maltol and kojic acid in soy sauce were demonstrated to have strong radical scavenging ability. Kojic acid is a chelating organic acid freely soluble in water, ethanol, and acetone. Kojic acid is biologically generated via the direct fermentation of glucose by *A. oryzae* and *A. flavus* to propose the biosynthesis pathway of kojic acid. Glucose transformed into glucose catalyzed by glucose dehydrogenase leads to d-gluconic acid transformed into d-gluconic acid. However, direct evidence for the involvement of these enzymes in kojic acid biosynthesis was insufficient (Goldberg et al., 2009). Maltol was one of several antioxidative substances identified in an ethyl acetate extract of dark soy sauce (Wang et al., 2007), which was produced by a selected strain of *Z. rouxii* (Wang et al., 2007).

The phenolic compounds such as 4-ethylguaiacol, catechol, and 4-ethylphenol were identified as crucial small molecule antioxidant compounds in soy sauce. The contribution rates of these phenolic compounds to the antioxidative activity of soy sauce were in the order 4-ethylguaiacol > catechol > 4-ethylphenol (Li et al., 2017). Among them, 4-ethylguaiacol could be produced from lignin pyrolysis by yeast in fermentation and is an important odor-active component in soy sauce (Lee et al., 2006). Candida species are vital for the development of aroma in soy sauce via generating phenolic compounds such as 4-ethylguaiacol (Suezawa et al., 2007). Vinylphenols (4-vinylphenol,

4-vinylcatechol, and 4-vinylguaiacol) could be reduced into ethylphenols (including 4-ethylphenol, 4-ethylcatechol, and 4-ethylguaiacol) through the catalysis of vinylphenol reductase. The corresponding genes were currently detected in *Lactobacillus plantarum* (Santamaría et al., 2018). Catechol came from the raw materials of soybean and wheat or the fungal metabolites of *Aspergillus oryzae* from tryptophan in soy sauce (Li, Zhao, & Parkin, 2011).

17.4.2.2 Immunomodulatory Components

The immunomodulatory and antiallergic effects of soy sauce polysaccharides (SPS) have been demonstrated in both *in vitro* and *in vivo* studies.

During the fermentation of soy sauce, the proteins of the raw materials are entirely degraded into peptides and free amino acids through the catalysis of microbial proteolytic enzymes. However, polysaccharides derived from the cell wall in soy possessed anti-enzymatic properties, therefore existing after fermentation. The cell wall polysaccharides of soybeans in soy sauce with a great amount of galacturonic acid could be hydrolyzed to a slight extent during the koji and *moromi* stages. These polysaccharides constitute approximately 1% (w/v) of the soy sauce (Imamura & Matsushima, 2013). Mast cells are essential for inflammatory and immediate allergic reactions, such as type I hypersensitivity. Mast cells are capable of releasing chemical mediators, such as histamine, proteases, chemotactic factors, cytokines, and metabolites of arachidonic acid. The release could be induced by IgE-dependent immune mechanisms. IgE-dependent mast cell degranulation occurred in response to certain antigens (allergens). IgE-mast cells could blanket the plasma membranes of these immune cells (Amin, 2012).

The antiallergic activities of SPS were demonstrated by evaluating the release of histamine from antigen-induced basophilic leukemia (RBL-2H3) cells in rats (Kobayashi et al., 2004). RBL-2H3 cells are associated with type I hypersensitivity via releasing histamine, which is similar to the function of mast cells. SPS exhibited an inhibitory effect in the liberation of histamine from RBL-2H3 cells in a concentration-dependent manner. Moreover, the cytotoxicity of SPS is much lower than that of ketotifen, and results suggested that SPS is a potentially safe condiment for antiallergic foods (Kobayashi et al., 2004). Moreover, orally administered SPS exhibited a significant suppressive effect on the passive cutaneous reaction elicited in a mouse ear model of type I hypersensitivity (Kobayashi et al., 2004). Oral supplementation of SPS is a potent treatment for sufferers with perennial allergic rhinitis and seasonal allergic rhinitis in two double-blind placebo-controlled clinical studies. In a four-week randomized, double-blind, placebo-controlled parallel-group study, patients with perennial allergic rhinitis interfered with 600 mg of SPS alleviated the symptom such as runny nose, sore throat, and eye itching (Kobayashi et al., 2004). In an eight-week randomized, double-blind, placebo-controlled parallel-group study, patients with seasonal allergic rhinitis because of Japanese cedar pollen treated with 600 mg of SPS significantly alleviated symptoms such as sneezing, nasal stuffiness, and hindrance of daily life (Kobayashi et al., 2005).

SPS possessed immunomodulatory effects in systemic immunity through activating macrophages and regulating the balance of Th1/Th2 cell responses *in vitro* and *in vivo*. Macrophages play diverse roles in the defense line of the innate immune system against invasive insults (Wynn & Levy, 2010). SPS enhanced the consumption of glucose by peritoneal macrophages *in vitro* and oral administration of SPS to mice increased the capacity of peritoneal macrophages to consume glucose *in vivo* (Matsushita et al., 2006). Th1/Th2 cell response shifted to a predominantly Th2 cell response, one of the characteristics in allergic diseases, causes the elevation of IgE by B cells (Deo et al., 2010). SPS significantly suppressed the generation of IL-4 and the elevation of IFN-g, thus regulating the balance of Th1/Th2 cell responses through shifting toward predominantly Th1 cell responses (Matsushita et al., 2006). Therefore, SPS may play a role in the suppression of IgE production by B cells through adaptation of the balance of Th1/Th2.

Furthermore, SPS increased the production of immunoglobulin A (IgA) originated from Peyer's patch cells *in vitro* and oral supplementation with 1.5 mg per day of SPS significantly enhanced the concentration of IgA in the intestine in BALB/c mice. SPS in human intestinal epithelial cell line

Caco-2 could be transported across cell monolayers and the uronic acid was absorbed in intestines in a time-dependent manner (Kobayashi & Makio, 2008).

Tetragenococcus halophilus Th221 isolated from soy sauce was confirmed to possess an immunomodulatory activity that promotes T helper type 1 (Th1) immunity. *T. halophilus* Th221 prompted strong interleukin (IL)-12 production by mouse peritoneal macrophages *in vitro*. Oral ingestion of *T. halophilus* Th221 promoted Th1-dependant contact sensitivity and suppressed Th2 immunity by inhibiting the production of IgE. It is beneficial for mitigating allergic symptoms (Masuda et al., 2008). In a randomized, double-blind, and placebo-controlled study conducted over eight weeks, patients with perennial allergic rhinitis were treated by oral administration of Th221, alleviating symptoms significantly and decreasing serum IgE concentrations (Nishimura et al., 2009).

17.5   FUTURE TRENDS

For thousands of years, a variety of traditional soybean fermented foods which represent different ethnic, social, and cultural histories are developed by humans on every continent. The development of the international food trade and the extensive urbanization processes significantly influenced the traditional soybean fermented foods in many parts of the world. In fact, these traditional foods in every country have undergone changes driven by technical innovations and consumer demands. For some of these foods, especially those produced in the developing regions of the world, the general manufacturing methods and scale of manufacture have changed relatively little.

One of the main challenges of fermented soybean foods will be how to manage their large-scale production without losing the unique flavors, textures, and other characteristics of the traditional products. Through applying modern biotechnology, such as excellent strains selection, metabolic engineering technology, modern fermentation technology, optimizing the industrial production fermentation process. The optimization of the fermentation process could contribute to obtaining the optimal composition of raw materials, reduction of production costs, improvement of product quality, and achieving independent innovation of products.

Another challenge of soybean fermented foods facing scientists will be research and development on functional soybean fermented foods. During the last few decades, scientific studies on fermentation-derived bioactive compounds and the physiological function of fermented soybean foods have gradually increased. Currently, fermented soybean foods consumed not only have nutritional values, wholesomeness, or palatability, but have beneficial health functions. However, only a few reliable and specific functional products targeting subgroups such as patients with diabetes or hypertension have commercialized and reached the market. Advanced and precise technologies for accurate identification and isolation of bioactive components and molecular studies on the molecular mechanisms of how the effective components exert their functions are required for further developing functional fermented soybean foods.

REFERENCES

Amin, K. (2012). The role of mast cells in allergic inflammation. *Respiratory Medicine*, 106(1), 9–14.

Aoshima, H., and Ooshima, S. (2009). Anti-hydrogen peroxide activity of fish and soy sauce. *Food Chemistry*, 112(2), 339–343.

Aum, S. H., Ahn, S. H., Park, S. Y., Cheon, J. H., and Kim, K. B. (2016). The anti-inflammatory effects of Hataedock taken Douchi extracts on atopic dermatitis-like skin lesion of NC/Nga mouse. *The Journal of Pediatrics of Korean Medicine*, 30(2), 1–9.

Bao, W., Huang, X., Liu, J., Han, B., and Chen, J. (2020). Influence of Lactobacillus brevis on metabolite changes in bacteria-fermented Sufu. *Journal of Food Science*, 85(1), 165–172.

Baram, D., Vaday, G. G., Salamon, P., Drucker, I., Hershkoviz, R., and Mekori, Y. A. (2001). Human mast cells release metalloproteinase-9 on contact with activated T cells: Juxtacrine regulation by TNF-alpha. *Journal of Immunology*, 167(7), 4008–4016.

Cai, R. C., Li, L., Yang, M., Cheung, H. Y., and Fu, L. (2016). Changes in bioactive compounds and their relationship to antioxidant activity in white Sufu during manufacturing. *International Journal of Food Science and Technology*, 51(7), 1721–1730.

Cammerer, B., Jalyschko, W., and Kroh, L. W. (2002). Intact carbohydrate structures as part of the melanoidin skeleton. *Journal of Agricultural and Food Chemistry*, 50(7), 2083–2087.

Cao, Z. H., Green-Johnson, J. M., Buckley, N. D., and Lin, Q. Y. (2019). Bioactivity of soy-based fermented foods: A review. *Biotechnology Advances*, 37(1), 223–238.

Chaisrisook, C. (2002). Mycelial reactions and mycelial compatibility groups of red rice mould (monascus purpureus). *Mycological Research*, 106(3), 298–304.

Chen, J., Cheng, Y. Q., Yamaki, K., and Li, L. T. (2007). Anti-α-glucosidase activity of Chinese traditionally fermented soybean (douchi). *Food Chemistry*, 103(4), 1091–1096.

Chen, J., and Rosenthal, A. (2015). Modifying food texture. In Guo, J., and Yang, X. Q. (eds.) *Texture Modification of Soy-Based Products*. Woodhead Publishing, 237–255.

Chen, Y. C., Sugiyama, Y., Abe, N., Kuruto-Niwa, R., Nozawa, R., and Hirota, A. (2005). DPPH radical-scavenging compounds from dou-chi, a soybean fermented food. *Bioscience, Biotechnology and Biochemistry*, 69(5), 999–1006.

Cheng, Y. Q., Hu, Q., Li, L. T., Saito, M., and Yin, L. J. (2009). Production of Sufu, a traditional Chinese fermented soybean food, by fermentation with. Mucor flavus at Low temperature. *Food Science and Technology Research*, 15(4), 347–352.

Cheng, Y. Q., Zhu, Y. P., Hu, Q., Li, L. T., Saito, M., Zhang, S. X., and Yin, L. J. (2011). Transformation of isoflavones during Sufu (a traditional Chinese fermented soybean curd) production by fermentation with Mucor flavus at low temperature. *International Journal of Food Properties*, 14(3), 629–639.

Chiang, C. M., Wang, D. S., and Chang, T. S. (2016). Improving free radical scavenging activity of soy isoflavone glycosides daidzin and genistin by 3'-hydroxylation using recombinant Escherichia coli. *Molecules*, 21(12), E1723.

Chou, C. C., and Hwan, C. H. (1994). Effect of ethanol on the hydrolysis of protein and lipid during the ageing of a Chinese fermentation soy bean curd—Sufu. *Journal of the Science of Food and Agriculture*, 66(3), 393–398.

Collison, M. W. (2008). Determination of total soy isoflavones in dietary supplements, supplement ingredients, and soy foods by high-performance liquid chromatography with ultraviolet detection: Collaborative study. *Journal of AOAC International*, 91(3), 489–500.

Czerny, M., and Schieberle, P. (2002). Important aroma compounds in freshly ground wholemeal and white wheat flour-identification and quantitative changes during sourdough fermentation. *Journal of Agricultural and Food Chemistry*, 50(23), 6835–6840.

Deo, S., Mistry, K., Kakade, A., and Niphadkar, P. (2010). Role played by th2 type cytokines in ige mediated allergy and asthma. *Lung India: Official Organ of Indian Chest Society*, 27(2), 66–71.

Detman, A., Mielecki, D., Chojnacka, A., Salamon, A., Błaszczyk, M. K., and Sikora, A. (2019). Cell factories converting lactate and acetate to butyrate: Clostridium butyricum and microbial communities from dark fermentation bioreactors. *Microbial Cell Factories*, 18(1), 36.

Det-udom, R., Prakitchaiwattana, C., and Mahawanich, T. (2019). Autochthonous microbes and their key properties in browning reduction during soy sauce fermentation. *LWT-Food Science and Technology*, 111, 378–386.

Dhakal, R., Bajpai, V. K., and Baek, K. H. (2012). Production of GABA (γ-aminobutyric acid) by micro-organisms: A review. *Brazilian Journal of Microbiology: [Publication of the Brazilian Society for Microbiology]*, 43(4), 1230–1241.

Drenjančević-Perić, I., Jelaković, B., Lombard, J. H., Kunert, M. P., Kibel, A., and Gros, M. (2011). High-salt diet and hypertension: Focus on the renin-angiotensin system. *Kidney and Blood Pressure Research*, 34(1), 1–11.

Fan, J., Zhang, Y., Chang, X., Saito, M., and Li, Z. (2009). Changes in the radical scavenging activity of bacterial-type douchi, a traditional fermented soybean product, during the primary fermentation process. *Bioscience, Biotechnology and Biochemistry*, 73(12), 2749–2753.

Feng, Z., Gao, W., Ren, D., Chen, X., and Li, J. J. (2013). Evaluation of bacterial flora during the ripening of kedong Sufu, a typical Chinese traditional bacteria-fermented soybean product. *Journal of the Science of Food and Agriculture*, 93(6), 1471–1478.

Gänzle, M. G. (2014). Enzymatic and bacterial conversions during sourdough fermentation. *Food Microbiology*, 37, 2–10.

Georas, S. N., Guo, J., De Fanis, U., and Casolaro, V. (2006). T-helper cell type-2 regulation in allergic disease. *European Respiratory Journal*, 26(6), 1119–1137.

Giunti, S., Barutta, F., Cavallo Perin, P., and Gruden, G. (2010). Targeting the mcp-1/ccr2 system in diabetic kidney disease. *Current Vascular Pharmacology*, 8(6), 849–860.

Goldberg, I., and Rokem, J. S. (2009). Organic and Fatty Acid Production, Microbial. *Encyclopedia of Microbiology* (3rd ed.), 421–442. Academic Press.

Gülçin, İ. (2010). Antioxidant properties of resveratrol: A structure-activity insight. *Innovative Food Science and Emerging Technologies*, 11(1), 210–218.

Han, B. Z., Rombouts, F. M., and Nout, M. (2001). A Chinese fermented soybean food. *International Journal of Food Microbiology*, 65(1–2), 1–10.

Han, B. Z., Wang, J. H., Rombouts, F. M., Frans, M., and Nout, M. J. R. (2003). Effect of nacl on textural changes and protein and lipid degradation during the ripening stage of Sufu, a Chinese fermented soybean food. *Journal of the Science of Food and Agriculture*, 83(9), 899–904.

Hang, M., and Zhao, X. H. (2012). Fermentation time and ethanol/water-based solvent system impacted in vitro ace-inhibitory activity of the extract of mao-tofu fermented by mucor spp.*CyTA-Journal of Food*, 10(2), 1–7.

Hayakawa, K., Kimura, M., and Kamata, K. (2002). Mechanism underlying γ-aminobutyric acid-induced antihypertensive effect in spontaneously hypertensive rats. *European Journal of Pharmacology*, 438(1–2), 107–113.

Hayakawa, K., Kimura, M., Kasaha, K., Matsumoto, K., Sansawa, H., and Yamori, Y. (2004). Effect of a γ-aminobutyric acid-enriched dairy product on the blood pressure of spontaneously hypertensive and normotensive Wistar–Kyoto rats. *British Journal of Nutrition*, 92(3), 411–417.

Hayashida, Y., Eto, K., Tamura, Y., Kakimoto, M., Tominaga, K., Tanaka, T., Nishimura, K., Kuriyama, H., Ohba, R., and Slaughter, J. C. (2000). 4-hydroxy-3(2H)-furanones HDMF and HEMF production in shoyu mash. *Journal of Soy Sauce Research and Technology*, 26, 123–127.

He, F. J., and Chen, J. Q. (2013). Consumption of soybean, soy foods, soy isoflavones and breast cancer incidence: Differences between Chinese women and women in Western countries and possible mechanisms. *Food Science and Human Wellness*, 2(3–4), 146–161.

He, G., Huang, J., Liang, R., Wu, C., and Zhou, R. (2016). Comparing the differences of characteristic flavour between natural maturation and starter culture for Mucor-type Douchi. *International Journal of Food Science and Technology*, 51(5), 1252–1259.

He, H. L., Liu, D., and Ma, C. B. (2013). Review on the angiotensin-I-converting enzyme (ACE) inhibitor peptides from marine proteins. *Applied Biochemistry and Biotechnology*, 169(3), 738–749.

He, R. Q., Wan, P., Liu, J., and Chen, D. W. (2020). Characterisation of aroma-active compounds in Guilin Huaqiao white Sufu and their influence on umami aftertaste and palatability of umami solution. *Food Chemistry*, 321, 126739.

Hogan, M. B., Peele, K., and Wilson, N. W. (2012). Skin barrier function and its importance at the start of the atopic march. *Journal of Allergy*, 2012, 901940.

Hossain, U., Das, A. K., Ghosh, S., and Sil, P. C. (2020). An overview on the role of bioactive α-glucosidase inhibitors in ameliorating diabetic complications. *Food and Chemical Toxicology: An International Journal Published for the British Industrial Biological Research Association*, 145, 111738.

Huang, S., Su, S., Chang, J., Lin, H., Wu, W., Deng, J., and Huang, G. J. (2016). Antioxidants, anti-inflammatory, and antidiabetic effects of the aqueous extracts from Glycine species and its bioactive compounds. *Botanical Studies*, 57(1), 38.

Huang, X., Yu, S., Han, B., and Chen, J. (2018). Bacterial community succession and metabolite changes during Sufu fermentation. *LWT-Food Science and Technology*, 97, 537–545.

Huang, Y. H., Lai, Y. J., and Chou, C. C. (2011). Fermentation temperature affects the antioxidant activity of the enzyme-ripened Sufu, an oriental traditional fermented product of soybean. *Journal of Bioscience and Bioengineering*, 112(1), 49–53.

Imamura, M., and Matsushima, K. (2013). Suppression of umami aftertaste by polysaccharides in soy sauce. *Journal of Food Science*, 78(8), C1136–C1143.

Itoh, S., Taketomi, A., Harimoto, N., Tsujita, E., Rikimaru, T., Shirabe, K., Shimada, M., and Maehara, Y. (2010). Antineoplastic effects of gamma linolenic acid on hepatocellular carcinoma cell lines. *Journal of Clinical Biochemistry and Nutrition*, 47(1).

Izumi, T., Piskula, M. K., Osawa, S., Obata, A., Tobe, K., Saito, M., Kataoka, K., Kubota, Y., and Kikuchi, M. (2000). Soy isoflavone aglycones are absorbed faster and in higher amounts than their glucosides in humans. *Journal of Nutrition*, 130(7), 1695–1699.

Jing, C., Yong-Qiang, C., Xiao-Qing, L., Hai-Di, X. U., Jian, S., Li-Te, L. I., and Yamaki, K. (2007). Study on in vitro anti-α-glucosidase activity of Chinese fermented douchi water-extract. *Food Science*.

Jung, A. R., Ahn, S. H., Park, I. S., Park, S. Y., Jeong, S. I., Cheon, J. H., and Kim, K. (2016). Douchi (fermented Glycine max Merr.) alleviates atopic dermatitis-like skin lesions in NC/Nga mice by regulation of PKC and IL-4. *BMC Complementary and Alternative Medicine*, 16(1), 1–14.

Kao, T., and Chen, B. (2006). Functional components in soybean cake and their effects on antioxidant activity. *Journal of Agricultural and Food Chemistry*, 54(20), 7544–7555.

Kataoka, S. (2005). Functional effects of Japanese style fermented soy sauce (shoyu) and its components. *Journal of Bioscience and Bioengineering*, 100(3), 227–234.

Katina, K., Juvonen, R., Laitila, A., Flander, L., Nordlund, E., Kariluoto, S., Piironen, V., and Poutanen, K. (2012). Fermented wheat bran as a functional ingredient in baking. *Cereal Chemistry Journal*, 89(2), 126–134.

Kenji, U., Jun, W., Takeshi, A., Daisuke, W., Yoshinobu, M., and Hitoshil, S. (2015). Screening of high-level 4-hydroxy-2 (or 5)-ethyl-5 (or 2)-methyl-3(2h)-furanone-producing strains from a collection of gene deletion mutants of saccharomyces cerevisiae. *Applied and Environmental Microbiology*, 81(1), 453–460.

Kim, D. H., Yoo, T. H., Lee, S. H., Kang, H. Y., Nam, B. Y., Kwak, S. J., Kim, J. K., Park, J. T., Han, S. H., and Kwak, S. J. (2012). Gamma linolenic acid exerts anti-inflammatory and anti-fibrotic effects in diabetic nephropathy. *Yonsei Medical Journal*, 53(6).

Kim, H. W., Choi, Y. S., Choi, J. H., Kim, H. Y., Hwang, K. E., Song, D. H., Lee, S. Y., Lee, M. A., and Kim, C. J. (2013). Antioxidant effects of soy sauce on color stability and lipid oxidation of raw beef patties during cold storage. *Meat Science*, 95(3), 641–646.

Kim, H. Y., Ahn, S. H., Yang, I., and Kim, K. (2017). Effect of skin lipid barrier formation on hataedock treatment with douchi. *Journal of Korean Medicine*, 38(2), 41–52.

Kimura, M., Hayakawa, K., and Sansawa, H. (2002). Involvement of γ-aminobutyric acid (GABA) B receptors in the hypotensive effect of systemically administered GABA in spontaneously hypertensive rats. *Japanese Journal of Pharmacology*, 89(4), 388–394.

Kobayashi, M. (2005). Immunological functions of soy sauce: Hypoallergenicity and antiallergic activity of soy sauce. *Journal of Bioscience and Bioengineering*, 100(2), 144–151.

Kobayashi, M., Matsushita, H., Tsukiyama, R., Saito, M., and Sugita, T. (2005). Shoyu polysaccharides from soy sauce improve quality of life for patients with seasonal allergic rhinitis: A double-blind placebo-controlled clinical study. *International Journal of Molecular Medicine*, 15(3), 463.

Kobayashi, M., Matsushita, H., Yoshida, K., Tsukiyama, R., Sugimura, T., and Yamamoto, K. (2004). In vitro and in vivo anti-allergic activity of soy sauce. *International Journal of Molecular Medicine*, 14(5), 879–884.

Matsushita, H., Kobayashi, M., Tsukiyama, R-I., Fujimoto, M., Suzuki, M., Tsuji, K., and Yamamoto, K. (2008). Stimulatory effect of shoyu polysaccharides from soy sauce on the intestinal immune system. *International Journal of Molecular Medicine*, 22, 243–247.

Kuba, M., Tanaka, K., Tawata, S., Takeda, Y., and Yasuda, M. (2003). Angiotensin I-converting enzyme inhibitory peptides isolated from tofuyo fermented soybean food. *Bioscience Biotechnology Biochemistry*, 67, 1278–1283.

Kuba, M., Shinjo, S., and Yasuda, M. (2004). Antihypertensive and hypocholesterolemic effects of tofuyo in spontaneously hypertensive rats. *Journal of Health Science*, 50(6), 670–673.

Kumar, S., and Pandey, K. A. (2013). Chemistry and biological activities of flavonoids: An overview. *The Scientific World Journal*, 11–12, 1–16.

Lee, C. Y., Isaac, H. B., Wang, H., Huang, S. H., Long, L. H., Jenner, A. M., Kelly, R. P., and Halliwell, B. (2006). Cautions in the use of biomarkers of oxidative damage; the vascular and antioxidant effects of dark soy sauce in humans. *Biochemical and Biophysical Research Communications*, 344(3), 906–911.

Lee, D. S., and Lee, S. H. (2001). Genistein, a soy isoflavone, is a potent alpha-glucosidase inhibitor. *FEBS Letters*, 501(1), 84–86.

Li, F. J., Yin, L. J., Lu, X., and Li, L. T. (2010). Changes in angiotensin I-converting enzyme inhibitory activities during the ripening of douchi (a Chinese traditional soybean product) fermented by various starter cultures. *International Journal of Food Properties*, 13(3), 512–524.

Li, G. H., Le, G. W., Shi, Y. H., and Shrestha, S. (2004). Angiotensin I-converting enzyme inhibitory peptides derived from food proteins and their physiological and pharmacological effects. *Nutrition Research*, 24(7), 469–486.

Li, H., Lin, L., Feng, Y., Zhao, M., Li, X., Zhu, Q., and Xiao, Z. (2017). Enrichment of antioxidants from soy sauce using macroporous resin and identification of 4-ethylguaiacol, catechol, daidzein, and 4-ethylphenol as key small molecule antioxidants in soy sauce. *Food Chemistry*, 240, 885–892.

Li, Y., Zhao, H., Zhao, M., and Cui, C. (2010). Relationships between antioxidant activity and quality indices of soy sauce: An application of multivariate analysis. *International Journal of Food Science and Technology*, 45(1), 133–139.

Li, Y., Zhao, M., and Parkin, K. L. (2011). β-Carboline derivatives and diphenols from soy sauce are in vitro quinone reductase (QR) inducers. *Journal of Agricultural and Food Chemistry*, 59(6), 2332–2340.

Li, Z. G., and Tan, H. Z. (2009). *Traditional Chinese Foods: Production and Research Progress*. New York: Nova Science Publishers.

Li-Chan, E. C. (2015). Bioactive peptides and protein hydrolysates: Research trends and challenges for application as nutraceuticals and functional food ingredients. *Current Opinion in Food Science*, 1, 28–37.

Li-Jun, Y., Li-Te, L., Zai-Gui, L., Tatsumi, E., and Saito, M. (2004). Changes in isoflavone contents and composition of Sufu (fermented tofu) during manufacturing. *Food Chemistry*, 87(4), 587–592.

Liu, Y. P., Miao, Z. W., Guan, W., and Sun, B. G. (2012). Analysis of organic volatile flavor compounds in fermented stinky tofu using spme with different fiber coatings. *Molecules*, 17(4).

Lokuruka, M. N. (2011). Effects of processing on soybean nutrients and potential impact on consumer health: An overview. *African Journal of Food, Agriculture, Nutrition and Development*, 11(4), 5000–5017.

Long, L. H., Chua, D., Kwee, T., and Halliwell, B. (2000). The antioxidant activities of seasonings used in Asian cooking. Powerful antioxidant activity of dark soy sauce revealed using the ABTS assay. *Free Radical Research Communications*, 32(2), 181–186.

Lu, H., Zhang, B. B., and Wu, Z. H. (2011). Studies on Mucor racemosus fermentation to manufacture gamma-linolenic acid functional food douchi. *Food Science and Technology Research*, 16(6), 543–548.

Lu, H., and Zhu, Y. (2015). Screening and molecular identification of overproducing γ-linolenic acid fungi and cloning the delta 6-desaturase gene. *Biotechnology and Applied Biochemistry*, 62(3), 316–322.

Ma, Y. L., Wang, J. H., Cheng, Y. Q., Yin, L. J., and Li, L. T. (2013). Some biochemical and physical changes during manufacturing of grey Sufu, a traditional Chinese fermented soybean curd. *International Journal of Food Engineering*, 9(1), 45–54.

Ma, Y. L., Wang, J. H., Cheng, Y. Q., Yin, L. J., Liu, X. N., and Li, L. T. (2014). Selected quality properties and angiotensin I-converting enzyme inhibitory activity of low-salt Sufu, a new type of Chinese fermented tofu. *International Journal of Food Properties*, 17(7–10), 2025–2038.

Mallikarjun Gouda, K. G., Gowda, L. R., Rao, A. A., and Prakash, V. (2006). Angiotensin I-converting enzyme inhibitory peptide derived from glycinin, the 11S globulin of soybean (Glycine max). *Journal of Agricultural and Food Chemistry*, 54(13), 4568–4573.

Matsushita, H., Kobayashi, M., Tsukiyama, R., and Yamamoto, K. (2006). In vitro and in vivo immunomodulating activities of shoyu polysaccharides from soy sauce. *International Journal of Molecular Medicine*, 17(5), 905–909.

Matsushita, H., Kobayashi, M., Tsukiyama, R.I., Fujimoto, M., Suzuki, M., Tsuji, K., and Yamamoto, K. (2008). Stimulatory effect of shoyu polysaccharides from soy sauce on the intestinal immune system. *International Journal of Molecular Medicine*, 22(2), 243–247.

Meijer, K., de Vos, P., and Priebe, M. G. (2010). Butyrate and other short-chain fatty acids as modulators of immunity: What relevance for health?. *Current Opinion in Clinical Nutrition and metabolic Carer*, 13(6), 715–721.

Melini, F., Melini, V., Luziatelli, F., Ficca, A. G., and Ruzzi, M. (2019). Health-promoting components in fermented foods: An up-to-date systematic review. *Nutrients*, 11(5), 1189.

Minatel, I. O., Borges, C. V., Ferreira, M. I., Gomez, H. A. G., and Lima, G. P. P. (2017). Phenolic compounds: Functional properties, impact of processing and bioavailability. *Phenolic Compounds - Biological Activity*, 8, 1–24.

Mishiro, T., Kusunoki, R., Otani, A., Ansary, M. M. U., Tongu, M., Harashima, N., Yamada, T., Sato, S., Amano, Y., Itoh, K., Ishihara, S., and Kinoshita, Y. (2013). Butyric acid attenuates intestinal inflammation in murine dss-induced colitis model via milk fat globule-egf factor 8. *Laboratory Investigation; A Journal of Technical Methods and Pathology*, 93(7), 834–843.

Naz, S., Sheikh, H., Siddiqi, R., and Sayeed, S. A. (2005). Oxidative stability of olive, corn and soybean oil under different conditions. *Food Chemistry*, 88(2), 253–259.

Nelson, D. L., and Cox, M. M. (2000). Amino acids, peptides, and proteins. In Nelson, D. L., and Cox, M. M. (eds.), *Lehninger Principles of Biochemistry*. New York, 115–158.

Nishimura, I., Igarashi, T., Enomoto, T., Dake, Y., Okuno, Y., and Obata, A. (2009). Clinical efficacy of halophilic lactic acid bacterium Tetragenococcus halophilus Th221 from soy sauce moromi for perennial allergic rhinitis. *Allergology International: Official Journal of the Japanese Society of Allergology*, 58(2), 179–185.

Nkhata, S. G., Ayua, E., Kamau, E. H., and Shingiro, J. B. (2018). Fermentation and germination improve nutritional value of cereals and legumes through activation of endogenous enzymes. *food Science and Nutrition*, 6(8), 2446–2558.

Nora, N. T., Esther, S. D., and Wisdom, K. A. (2006). The comparative ability of four isolates of Bacillus subtilis to ferment soybeans into dawadawa. *International Journal of Food Microbiology*, 106(2), 145–152.

Nunomura, N. (2006). Flavor components of soy sauce, HEMF. *Journal of the Brewing Society of Japan*, 101(3), 151–160.

Ohata, M., Kohama, K., Morimitsu, Y., Kubota, K., and Sugawara, E. (2007). The formation mechanism by yeast of 4-hydroxy-2(or 5)-ethyl-5(or 2)-methyl-3(2H)-furanone in Miso. *Bioscience, Biotechnology and Biochemistry*, 71(2), 407–413.

Phaniendra, A., Jestadi, D. B., and Periyasamy, L. (2015). Free radicals: properties, sources, targets, and their implication in various diseases. *Indian journal of clinical biochemistry*, 30(1), 11–26.

Peterson, J. J., Dwyer, J. T., Jacques, P. F., and McCullough, M. L. (2012). Associations between flavonoids and cardiovascular disease incidence or mortality in European and US populations. *Nutrition Reviews*, 9(9), 491–508.

Pyo, Y., Lee, T., and Lee, Y. (2005). Effect of lactic acid fermentation on enrichment of antioxidant properties and bioactive isoflavones in soybean. *Journal of Food Sciencei*, 70(3), s215–s220.

Qin, L. K., and Ding, X. L. (2006). Extractions of melanoidins in the long- ripenned douchiba(dcb) and analyses of amino acid compositions in its peptide Skeltons. *Journal of Food Science*, 27(1), 125–129.

Ramos-Ruiz, R., Poirot, E., and Flores-Mosquera, M. (2018). GABA, a non-protein amino acid ubiquitous in food mliberateatrices. *Cogent Food and Agriculture*, 4(1), 1534323.

Rice-Evans, C. A., and Packer, L. (2003). *Flavonoids in Health and Disease*. Boca Raton: CRC Press, 1–43.

Rizzo, G. (2020). The antioxidant role of soy and soy foods in human health. *Antioxidants*, 9(7), 635.

Romaniouk, A. V., Anne, S., Jie, F., and Vijay, I. K. (2004). Synthesis of a novel photoaffinity derivative of 1-deoxynojirimycin for active site-directed labeling of glucosidase I. *Glycobiology*, 4, 301–310.

Salar, R. J., Purewal, S. S., and Bhatti, M. S. (2016). Optimization of extraction conditions and enhancement of phenolic content and antioxidant activity of pearl millet fermented with Aspergillus awamori MTCC-54. *Resource-Efficient Technologies*, 2(3), 148–157.

Santamaría, L., Reverón, I., López de Felipe, F., Blanca, D. L. R., and Muñoz, R. (2018). Unravelling the reduction pathway as alternative metabolic route to hydroxycinnamate decarboxylation in lactobacillus plantarum. *Applied and Environmental Microbiology*, 01123–01118.

Quan, S., Tsuda, H., and Miyamoto, T. (2008). Angiotensin I-converting enzyme inhibitory peptides in skim milk fermented with Lactobacillus helveticus130B4 from camel milk in Inner Mongolia, China. *Journal of the Science of Food and Agriculture*, 88(15), 2688–2692.

Song, M. H., Nah, J. Y., Han, Y. S., Han, D. M., and Chae, K. (2001). Promotion of conidial head formation in Aspergillus oryzae by a salt. *Biotechnology Letters*, 23(9), 689–691.

Sossai, P. (2012). Butyric acid: What is the future for this old substance? *Swiss Medical Weekly*, 142(2324).

Suezawa, Y., and Suzuki, M. (2007). Bioconversion of ferulic acid to 4-vinylguaiacol and 4-ethylguaiacol and of 4-vinylguaiacol to 4-ethylguaiacol by halotolerant yeasts belonging to the genus candida. *Bioscience, Biotechnology and Biochemistry*, 71(4), 1058–1062.

Sugawara, E., Ohata, M., Kanazawa, T., Kubota, K., and Sakurai, Y. (2007). Effects of the amino-carbonyl reaction of ribose and glycine on the formation of the 2(or 5)-ethyl-5(or 2)-methyl-4-hydroxy-3(2H)-furanone aroma component specific to miso by halo-tolerant yeast. *Bioscience Biotechnology and Biochemistry*, 71(7), 1761–1763.

Takahashi, M., McCartney, E., Knox, A., Francesch, M., Oka, K., Wada, K., Ideno, M., Uno, K., KozOwski, K., Jankowski, J., Gracia, M. I., Morales, J., Kritas, S. K., Esteve-Garcia, E., and Kamiya, S. (2018). Effects of the butyric acid-producing strain Clostridium butyricum MIYAIRI 588 on broiler and piglet zootechnical performance and prevention of necrotic enteritis. *Animal Science Journal = Nihon chikusan Gakkaiho*, 89(6), 895–905.

Takigawa, H., and Shibuya, Y. (2012). The metabolites of food microorganisms. *Drug Discovery: Research in Pharmacognosy*, 227.

Vagadia, B. H., Vanga, S. K., and Raghavan, V. (2017). Inactivation methods of soybean trypsin inhibitor—A review. *Trends in Food Science and Technology*, 115–125.

van Rooy, M. J., and Pretorius, E. (2014). Obesity, hypertension and hypercholesterolemia as risk factors for atherosclerosis leading to ischemic events. *Current Medicinal Chemistry*, 21(19), 2121–2129.

Wan, S. P., Wang, C. L., Hou, L. H., and Cao, X. L. (2011). Effect of adding salt-tolerant microorganisms on the flavor of soy-sauce mash, International Conference on Remote Sensing, Environment and Transportation. *Engineering, Nanjing*, 2011, 7500–7502.

Wang, H. Y., Qian, H., and Yao, W. R. (2011). Melanoidins produced by the Maillard reaction: Structure and biological activity. *Food Chemistry*, 128(3), 573–584.

Wang, H., Jenner, A. M., Lee, C. Y., Shui, G., Tang, S. Y., Whiteman, M., Wenk, M. R., and Halliwell, B. (2007). The identification of antioxidants in dark soy sauce. *Free Radical Research*, 41(4), 479–488.

Wang, L. J., Yin, L. J., Li, D., Zou, L., Saito, M., Tatsumi, E., and Li, L. (2007). Influences of processing and NaCl supplementation on isoflavone contents and composition during douchi manufacturing. *Food Chemistry*, 101(3), 1247–1253.

Wang, L. J., Yin, L. J., Li, D., Zou, L., Saito, M., Tatsumi, E., and Li, L. (2007). Influences of processing and NaCl supplementation on isoflavone contents and composition during douchi manufacturing. *Food Chemistry*, 101(3), 1247–1253.

Wang, L., Saito, M., Tatsumi, E., and Li, L. (2003). Antioxidative and angiotensin I-converting enzyme inhibitory activities of Sufu (fermented tofu) extracts. *Japan Agricultural Research Quarterly: JARQ*, 37(2), 129–132.

Wang, D., Wang, L. J., Zhu, F. X., Zhu, J. Y., Chen, X. D., Zou, L., and Saito, M. (2008). In vitro and in vivo studies on the antioxidant activities of the aqueous extracts of Douchi (a traditional Chinese salt-fermented soybean food). *Food Chemistry*, 107(4), 1421–1428.

Wang, R. Z., Wei, X. Y., Shen, Z. H., Wu, Z. G., Huang, D. P., and Qu, G. Y. (2009). The characteristic Sufu in China. In Tu, R. L., and Li, G. G. (eds.) *The Production of Sufu in China*. Beijing: Light Industry Press, 230–232.

Wang, X., Lin, H., and Gu, Y. (2012). Multiple roles of dihomo-γ-linolenic acid against proliferation diseases. *Lipids in Health and Disease*, 11(1), 25.

Wang, Y., Li, F., Chen, M., Li, Z., Liu, W., and Wang, C. (2015). Angiotensin I-converting enzyme inhibitory activities of Chinese traditional soy-fermented douchi and Soypaste: Effects of processing and simulated gastrointestinal digestion. *International Journal of Food Properties*, 18(4), 934–944.

Wu, L., Jiang, A., Jing, Y., Zheng, Y., and Yan, Y. (2017). Antioxidant properties of protein hydrolysate from Douchi by membrane ultrafiltration. *International Journal of Food Properties*, 20(5), 997–1006.

Wynn, J. L., and Levy, O. (2010). Role of innate host defenses in susceptibility to early-onset neonatal sepsis. *Clinics in Perinatology*, 37(2), 307–337.

Xiang, H., Sun-Waterhouse, D., Waterhouse, G. I. N., Cui, C., and Ruan, Z. (2019). Fermentation-enabled wellness foods: A fresh perspective. *Food Science and Human Wellness*, 8(3), 203–243.

Xie, C., Zeng, H., Li, J., and Qin, L. (2018). Comprehensive explorations of nutritional, functional and potential tasty components of various types of Sufu, a Chinese fermented soybean appetizer. *Food Science and Technology*.

Xu, L., Cai, W. X., and Xu, B. J. (2017). A Systematic assesment on vitamins (B2, B12) and GABA profiles in fermented soy products marketed in China. *Journal of Food Processing and Preservation*, 41(5), e13126.

Xu, L., Du, B., and Xu, B. (2015). A systematic, comparative study on the beneficial health components and antioxidant activities of commercially fermented soy products marketed in china. *Food Chemistry*, 174, 202–213.

Yaginuma, A., and Nunomura, N. (2002). Formation and environmental factors of a flavor component (HEMF) by soy sauce yeast. *Journal of the Brewing Society of Japan*, 97(9), 608–614.

Yin, L. J., Li, L. T., Liu, H., Saito, M., and Tatsumi, E. (2005). Effects of fermentation temperature on the content and composition of isoflavones and β-glucosidase activity in Sufu. *Bioscience, Biotechnology and Biochemistry*, 69(2), 267–272.

Yuksekdag, Z., Cinar Acar, B., Aslim, B., and Tukenmez, U. (2017). β-glucosidase activity and bioconversion of isoflavone glycosides to aglycones by potential probiotic bacteria. *International Journal of Food Properties*, 20(sup3), S2878–S2886.

Zhang, J. H., Tatsumi, E., Ding, C. H., and Li, L. T. (2006). Angiotensin I-converting enzyme inhibitory peptides in douchi, a Chinese traditional fermented soybean product. *Food Chemistry*, 98(3), 551–557.

Zhang, J. H., Tatsumi, E., Fan, J. F., and Li, L. T. (2007). Chemical components of Aspergillus-type Douchi, a Chinese traditional fermented soybean product, change during the fermentation process. *International Journal of Food Science and Technology*, 42(3), 263–268.

Zhang, Y., Luan, X., Zhang, H., Garre, V., Song, Y., and Ratledge, C. (2017). Improved γ-linolenic acid production in Mucor circinelloides by homologous overexpressing of delta-12 and delta-6 desaturases. *Microbiol Cell Factories*, 16(1), 1–9.

Zhao, G., Yao, Y., Wang, X., Hou, L., Wang, C., and Cao, X. (2013). Functional properties of soy sauce and metabolism genes of strains for fermentation. *International Journal of Food Science and Technologyl*, 48(5), 903–909.

Zhu, Y. P., Li, X. T., Huang, Z. G., Li, L., and Su, Y. C. (2011). Improving anti-α-glucosidase activity of douchi koji using a newly isolated strain of bacillus subtilis b2. *International Journal of Food Engineering*, 7, 1.

18 Value-Added Processing and Function of Okara

Sainan Wang, Mohammed Sharif Swallah,
Jiaxin Li, and Hansong Yu

CONTENTS

18.1 INTRODUCTION

Soybean curd (tofu) and soya milk are traditional Asian soy food products and are now consumed worldwide. In soybean production to obtain protein isolates and other end-products, including but not limited to soy milk and soybean curd (tofu), a bulk amount of fiber-rich residue called okara is generated, thus, it is the by-product/residue produced during the processing of soy foods. It is popularly known by its Japanese name okara, tofuzha, or duozha (Chinese), and biji/bejee (Korean), respectively. Okara is obtained after the extraction of the aqueous fraction, i.e. the water-insoluble by-product generated when soybeans are macerated/soaked, crushed, and pressed in the soy milk production, which is reported to reflect a significant problem of disposal (Swallah et al., 2021a; Kamble & Rani, 2020). The soy milk production process comprises an aqueous extraction process of soybeans: Soaking the bean in water, the mechanical grinding of the beans, the filtration of the insoluble portion, as well as the pasteurization of the milk solution. This milk is evidenced to contain varied nutritional and antinutritional components such as proteins, vitamins, fats, minerals, carbohydrates, and phytochemicals such as soy saponins and isoflavones (Davy & Vuong, 2020). In this process, 1 kg soybean yields 1.2 kg fresh okara after the addition of water in the maceration stage. Hence, it is justifiable that large amounts (1.4 billion tons) are produced annually worldwide, and yet are less utilized (Nagai et al., 2014, Guimarães et al., 2018). The insoluble fraction, i.e. the cotyledon of the soybean or the soybean residue formed from the milk production, is typically sold as animal fodder, used as fertilizers, dumped as landfill, or discarded as waste, which can consequently lead to an environmental burden as stated earlier (Rinaldi, Ng, & Bennink, 2000). As Asia is the largest manufacturer of soy milk and tofu, it has been reported that the amount of okara generated from tofu production is approximately 2.8 million tons in China, 800,000 tons in Japan, and 310,000 tons in Korea (Fu et al., 2017). However, the effective utilization of okara has not yet been achieved owing to its high perishability and undesirable flavor and texture, which are all triggered by its high moisture content (Vong, Hua, & Liu, 2018; Swallah et al., 2021a). Owing to the high moisture and nutrient content possessed in okara, lipid oxidation as well as the growth of microorganisms yield to rapid degradation and can lead to environmental problems. This action has the potential to pollute the surrounding land when wastes are dumped off or landfilled, which can cause runoff during times of rain, obstructing the surrounding lands and water system (Vong & Liu, 2016). To amend these issues, additional processing techniques need to be adopted, which generally depend on the removal of water via freeze drying or heat treatments; however, these processing methods are often cost-intensive and again demand specialized equipment, i.e. incurring extra financial cost to the manufacturing process (Davy & Vuong, 2020). Although incineration is an option to curb the disposal of this soybean-derived waste, it is evidenced to be accompanied by a great environmental issue owing to the large sum of carbon dioxide exposed to the atmosphere (Li et al., 2013). Hence, the valorization of okara will be essential to help utilize the untapped precious nutrients and as well serve to minimize the socio-environmental and the economic burdens caused by this waste disposal (Hu et al., 2019; Kang, Bae, & Lee, 2018; Swallah et al., 2021a). This chapter seeks to summarize the nutritional and functional components of okara as well as discuss its value-added processing and the development of new health-promoting foods.

18.2 NUTRITIONAL AND PHYTOCHEMICAL COMPONENTS OF OKARA

In recent years, considerable preference has been given to the use of agricultural food wastes as they possess several groups of substances that are beneficial for the development of functional foods (Das et al., 2021). The main components of okara are the beans' coating and the broken cotyledon cells (Colletti et al., 2020), which are comprised of crude fiber, soluble dietary fiber, and insoluble dietary fiber, and are suggested in varied reports to aid in numerous biological processes as well as play vital roles in the fight against syndromes of different origins (Swallah et al., 2021a). Hence, okara is regarded as a significant source of dietary fiber due to its high composition and low cost. Nevertheless, the chemical composition is impacted by the processing and extraction method, and also by the fraction of water-soluble components obtained from the ground beans and whether further extractable constituents have been extracted from the residue or not, and the cultivar of soybean used. Different cultivars differ in fatty acids compositions, crude protein, and lipid contents, as well as lipoxygenase activities (Vong & Liu, 2016; Kamble & Rani, 2020; Swallah et al., 2021a). The processing sequence and procedures for the beans are essential as they determine the fate of all water-soluble extracts in the beans. For example, there is variation in the methods by which the Japanese and Chinese process their soy milk and soybean curd. With the Chinese method, the soaked beans are rinsed, and the beans are then crushed and the residue/okara is then filtered off with water, and afterward, the extract is heated; in the Japanese way, the soaked beans are first cooked before grinding and filtering (O'Toole, 1999; Li et al., 2019; Vong & Liu, 2016; Swallah et al., 2021a). Figure 18.1 presents the varied schematic illustration for the steps involved in processing soy milk and the generation of the soybean residue/okara (Swallah et al., 2021a; Vong & Liu, 2016; Kamble & Rani, 2020; Guimarães et al., 2020), and the general composition and physicochemical properties of okara are shown in Table 18.1.

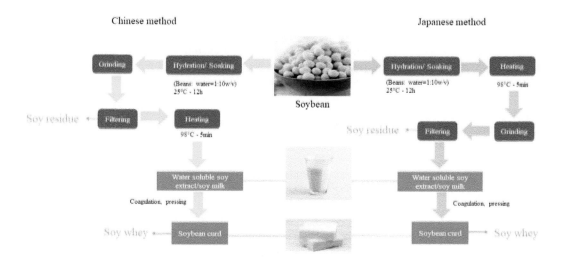

FIGURE 18.1 Schematic illustration for production of soy milk and generation of soybean residue/okara used by Chinese and Japanese.

TABLE 18.1

Nutritional Content and Physicochemical Properties of Okara

Items	Nutrients	Content	Physicochemical Properties	References
Macronutrients	Dietary fiber (%)	52.80–58.10	Water retention, adsorption, viscosity, ion exchange, and fermentation ability	(Vong & Liu 2016; Gupta et al. 2018; Shin & Jeong 2015)
	Protein (%)	25.40–33.40	Solubility, emulsifying, gelling, and foaming properties	
	Fat (%)	9.30–10.90	Emulsifying properties	
Micronutrients	**Minerals (‰)**	936.0–1,350.0		(Stanojevic et al. 2014)
	K	260.0–428.0	Water retention and adsorption	
	Ca	130.0–165.0		
	Mg	16.0–96.0		
	Na	0.60–11.0		
	Fe	0.30–3.50		
	Zn	0.20–3.10		
	Mn	0.10–1.20		
	Cu	0.82–1.04		
	B3	0.48–0.59		
	Vitamins (‰)			
	B1	0.03–0.04	Water-soluble	(Vong & Liu 2016)
	B2	36.33–37.03	Water-insoluble	(Laurenz et al. 2017)
	Isoflavones (‰)			

18.2.1 DIETARY FIBER

Dietary fiber (DF) content in okara accounts for 52.80–58.10% approximately (Vong & Liu, 2016; Gupta, Lee, & Chen, 2018). The DF comprises non-digestible polysaccharides that prevent the onset of several metabolic diseases through physicochemical properties and intestinal fermentation. DF can be divided into soluble dietary fiber (SDF) and insoluble dietary fiber (IDF). Elleuch et al.

reported that solubility is related to the structure of the polysaccharides, which can be regular (insoluble) or irregular (soluble) on the backbone or as side chains. The presence of a substitution group such as -COOH or SO42- increases the solubility (Elleuch et al., 2011). The content of SDF in okara is relatively low in contrast with the content of IDF. Chemical and enzymatic treatment, microorganism fermentation, high-pressure treatment, and micronization technology have been used to increase the SDF content of okara. Regarding enzymatic treatment, both acid and alkali treatments are suggested to aid the increase of the SDF content of okara. However, the water holding capacity and expansibility of the fiber were decreased while darkness in fiber color was increased by acid and alkali treatments. During homogenization, okara fiber was subjected to strong shearing. Glycosidic linkages of polysaccharides were broken by micronization and cavitation. During high-temperature cooking, crystalline areas in hemicelluloses and cellulose were forced by water molecules, resulting in the disruption of combined molecular chains and the release of some soluble hemicelluloses. DF presents a three-dimensional and multiphase network structure, in which there are amorphous and crystalline regions, as well as hydrophilic and hydrophobic surfaces. The maintenance of the network structure depends on chemical bonds of different strengths and physical effects. It exhibits many physical and chemical properties related to physiology, biochemistry, and nutrition, including hydration properties, viscosity, adsorption capacity, fermentation properties, and ion exchange. The water holding capacity of DF depends on the type, preparation method, and particle size, which is generally 1.5 to 25 times its weight. Moreover, the IDF has good water retention and adsorption due to the loss of porous structure and active groups on the surface. Also, the IDF goes beyond only aiding in the absorption of macromolecules such as cholesterol, bile acids, and heavy metals, with functions like the promotion of intestinal peristalsis, thereby expelling toxins. The viscosity of DF is mainly determined by the chemical structure, temperature, solvent, and concentration. Among them, agar, pectin, guar gum, etc. have good viscosity and gelling properties. These viscous polysaccharides directly interact with water molecules at low concentrations to increase the viscosity of the solution. At high concentrations, the polysaccharide molecules interact and twist into a network structure, which greatly increases the viscosity. Among them, the viscosity of SDF is greater than that of IDF (Wen et al., 2017). In the intestine, the SDF can stably be dispersed in water and entangled with glucose to form a mixture with a certain viscosity. The mixture binds water molecules, absorbs mineral cations, and serves as a fermentation agent for intestinal microorganisms (Dong et al., 2019). Besides, the SDF has a higher capacity to form gels and act as an emulsifier, enabling it to be readily incorporated into food products (Yan, Ye, & Chen, 2015). DF can be fermented and degraded by the microbial flora in the colon to produce short-chain fatty acids. The short-chain fatty acids produced by fermentation in the colon can participate in the metabolism of different organs in the body, which play an irreplaceable role. The degree of degradation of DFs is affected by solubility, particle size, and intake mode. The structure of DFs contains active groups such as carboxyl and hydroxyl groups, which are weakly acidic cations and can be reversibly exchanged with organic cations. It can also exchange with cations such as calcium, zinc, lead, etc., and can preferentially exchange harmful ions like lead, which are further excreted out of the body via feces. These physicochemical properties give the DF, a key nutritive role in the healthy diet framework of the human body, outstanding physiological functions, including the prevention of cardiovascular disease, oxidation and cancer, control of obesity, and improvement of diabetes (Dahl & Stewart, 2015; Sawicki and Livingston, 2017).

18.2.2 PROTEIN

The protein content of okara is approximately 25.40–33.40% (Shin & Jeong, 2015; Gupta et al., 2018). Soy protein isolates can be divided into 2S (α-glycinin), 7S (β-conglycinin), 11S (glycinin), and 15S, of which 7S and 11S account for more than 80%. The functional properties of protein refer to the physicochemical properties of processing, including solubility, emulsifying, foaming, and gelling properties. Solubility is a prerequisite for the functional properties of proteins, which is related to temperature, pH, and ionic concentration. The association and dissociation reaction of 11S glycinin plays an important role in these functions (Huang et al., 2017). At low pH and ion concentration, 11S glycinin can not only dissociate into the 2S glycinin component to increase solubility but also can polymerize with increasing pH. Singh et al. showed that when the pH is below 6.5, the solubility of the 11S glycinin is lower than that of the 7S glycinin (Singh et al., 2015). The 7S glycinin has good emulsifying and foaming properties due to the high content of tryptophan, methionine, cysteine, and low lysine content. The 11S glycinin has a high content of sulfur-containing amino acids, which makes it a tight structure stabilized by disulfide bonds (Aoki, Taneyama, & Inami, 1980). The ratio of 7S/11S is approximately 0.5–1.3, and its ratio is an indicator of the functional and nutritional value of soy protein (Aoki et al., 1980; Zhang et al., 2015). Due to the different functional properties of 7S glycinin and 11S glycinin, they have different physiological functions, like prevention of cardiovascular disease, oxidation, and cancer (Xu, Mukherjee, & Chang, 2018).

18.2.3 FATS

Okara contains approximately 9.30–10.90% fats, which are made of glycerin and fatty acids (Vong & Liu, 2016; Gupta et al., 2018). Among them, fatty acids account for more than 80%, which consists of linoleic acid (54.10% of the total fatty acids), oleic acid (20.40%), palmitic acid (12.30%), linolenic acid (8.80%), and stearic acid (4.70%) (Mateos-Aparicio et al., 2010). Fats are considered one of the most elemental nutrients for humans. Linoleic acid and linolenic acid are polyunsaturated fatty acids. They have important effects on human health, such as reducing blood triglyceride levels, moisturizing skin, preventing atherosclerosis, and cholesterol deposition in blood vessels (Alejandre et al., 2017; Timilsena et al., 2017). In addition, lecithin, cephalin, and inositol phospholipids are substances in okara essential for the human brain and liver, which have good effects in protecting cell membranes, reducing blood fat, delaying aging, and preventing fatty liver.

18.2.4 MINERALS

Okara also contains a considerable amount of minerals, mainly potassium (936–1,350%), calcium (260–428%), and magnesium (130–165%). Minerals in okara have water retention and adsorption properties, which leads to promoting human health. Potassium can maintain the normal function of nerves and muscles. Calcium is an important component of bones and teeth, which can stabilize the conformation of protein and enzymes. Magnesium maintains the activity of many enzymes in normal muscle. The proper amount of minerals in the okara is used as a dietary supplement in capsules or tablets to meet the needs of the human body. At the same time, minerals have a very high antioxidant capacity that contributes to the nutritional value of the raw okara (Stanojevic et al., 2014).

18.2.5 VITAMINS

Okara contains specific amounts of B vitamins, such as vitamin B1 (0.82–1.04%), vitamin B2 (0.48–0.59%), and vitamin B3 (0.03–0.04%). B vitamins are soluble in water and are more conducive to human digestion and absorption (Vong & Liu, 2016). Studies have shown that vitamin B1 is essential for the normal function of the heart, muscles, and nerves. Also, it is beneficial for the treatment of certain metabolic disorders. Yilmaz et al. found that vitamin B1 has protective effects on alcoholic liver injury in rats (Yilmaz et al., 2015). Vitamin B2 plays an important role in energy metabolism and prevents radiation-related esophagitis and cancer significantly. Liu et al. conducted a meta-analysis and found that vitamin B2 intake is negatively correlated with the risk of colon cancer (Liu et al., 2015). In addition, vitamin B3 plays a key role in energy metabolism. More and more, studies have shown that vitamin B3 can provide therapeutic benefits in various inflammation-based diseases (Zhang et al., 2012; Salman & Naseem, 2015). Ma et al. reported that vitamin B3 can reduce blood cholesterol and low-density lipoprotein (LDL) (Ma et al., 2014).

18.2.6 ISOFLAVONES

Soy foods and products are shown in several reports to contain a relatively high and varied group of phenolic compounds, including flavonoids, phenolic acids, and non-flavonoids (Swallah et al., 2021b). Their important role in our diet as bioactive compounds has been widely studied, with growing evidence revealing their role in reducing chronic disease risks, such as cardiovascular diseases, immune dysfunction, diabetes, cancers, and age-related eye problems, which are all linked with the antioxidative properties of these phenolic compounds (Swallah et al., 2021b; Jiang, Cai, & Xu, 2013).

Soybean possesses up to 12 varied categories of isoflavones, which can be classified into three main subgroups (i.e., glycitein, genistein, and daidzein), all of which can take four varied forms: β-glucosidase, malonylglucosides, acetyl-glucosides, and aglycones, which constitute the main phenolic components and have been ascribed to performing many health-encouraging functions (Villares et al., 2011; Li et al., 2013). A recent study on okara's composition was reviewed, and the authors documented that the total isoflavones content is 355 mg/g (dry weight basis). The concentration of malonyl glucosides, aglycones, acetyl glucosides, and isoflavone glucosides in the residue was found to be 196.8, 54.1, 3.2, and 103.2 mg/g, respectively (Li et al., 2013). As discussed earlier, okara is also expected to contain the same 12 isoflavones, however, the processing method and conditions are evidenced to influence the content and the original profiles of isoflavones in okara (Villares et al., 2011). The hot grinding process resulted in higher extraction of isoflavones into soy milk than cold grinding. However, direct or indirect heating during ultra-high-temperature processing depicted no significant influence on isoflavones concentration. The major metabolic pathways of isoflavones are glucuronidation and sulfation. Most isoflavones in okara are hydrolyzed by β-glucosidase to aglycones, thereby improving the physiological activity (Islam et al., 2015; Wang et al., 2017; Raman & Doble, 2015). De Toledo et al. studied the effect of gamma irradiation on total phenolics, trypsin inhibitors, and tannins in both cooked and raw soybeans. The authors reported that after exposure to gamma radiation, both the cooked and raw beans displayed an increased phenolic content and decreased tannins and trypsin inhibitors (De Toledo et al., 2007). Tannins have been reported to exert an antinutrient impact on protein digestibility. In contrast with other reported techniques, the rate of trypsin inhibition was reported to be less (Davy & Vuong, 2020). In addition, since flavonoid aglycon is barely soluble in water, the hydroxyl group of the isoflavones can increase its solubility by glycosidation. Isoflavones are excellent natural antioxidants that can bind to excess free radicals and inhibit the peroxidation of fats (Laurenz et al., 2017). Some authors have demonstrated the positive effects of isoflavones on the prevention of cardiovascular disease and inhibition of prostate cancer (Stanojevic et al., 2014). Isoflavones are weak estrogens and have been proposed as hormone substitutes for postmenopausal women. The health merits of isoflavones include but are not limited to anticancer and anti-inflammatory properties, cardiovascular defenses, and the enzyme-repressive roles of isoflavones are mainly associated with their antioxidant capacity, which is comparable to or much better than that of other polyphenols (Swallah et al., 2020a, 2020b). Consequently, the isoflavones extraction is of great interest in the production of dietary supplements. However, debate still exists as to whether isoflavones are really beneficial as first thought. There are reported studies with harmful functions of isoflavones on prostate and breast cancers, male and female fertility, as well as hypothyroidism. Since the majority of studies examined in these areas have been on various animal models, speculation still remains, with many researchers recommending that further research is essential to achieve consensus (Cederroth & Nef, 2009; Kitagawa et al., 1984).

18.3 OKARA PROCESSING IN THE FOOD INDUSTRY

The food industry is on a quest to develop new food products with superior health-enhancing characteristics. China is the largest market for soy foods in the world, mainly soy milk and tofu, generating a large volume of waste as discussed earlier. Owing to its rich nutrition, it is an elevating trend to add okara into foods either as a whole or portion replacement (Godfrey, 2002). The high solvent-binding properties of okara make it a perfect and low-cost ingredient with which to increase yield in bakery and meat products (Mateos-Aparicio et al., 2010). Specifically, the fortification of soybean-derived ingredients into a variety of food products with the aim of providing beneficial functions to the body has been on the rise over the years and has stimulated much interest from the food industry. These are named "functional foods" (Genovese & Lajolo, 2002). Okara has been used as a food supplement for human consumption and animal nutrition for several years, primarily in China and Japan, both in processed and raw forms, to more easily offer a fair intake of the nutritional claim for fiber and protein. Okara can be a fractional substitute for wheat flour and soy flour, as well as other food-producing components to aid in the enhancement of the protein and fiber content (Li, Qiao, & Lu, 2012). The high content of carbohydrates, proteins, and other forms of nutrients rooted in okara makes it an ideal substrate for microbial fermentation. Yeasts, bacteria, and fungal fermentation of okara are suggested to decrease the content of raw fiber, increase the content of soluble fiber, isoflavones, proteins, and amino acids, as well as decomposing phytic acid, yielding an improvement in the processing properties and nutritional value (Vong & Liu, 2017). The use of okara in varied food formulation such as bread, drinks, pancakes, sausages, cake, biscuits, candies, and nutritional flour has been studied and demonstrated earlier in numerous reports (Kang et al., 2018; Pan, Liu, & Shiau, 2018; Kamble et al., 2019; Wu, 2003; Zhao & Kong, 2009).

18.3.1 Processing Technology and Processes

The nutritional composition of okara has been widely explored and was found to be affected by post-processing treatments, processing conditions, and the specific cultivar of soybean. Macronutrients levels differed in different experiments with proteins ranging from 24.5 to 37.5 g/100 g, carbohydrates 3.8 to 5.3 g/100 g, total dietary fiber 14.5 to 58.1 g/100 g, and lipids 9.3 to 22.3 g/100 g of dry weight (O'Toole, 1999; Voss et al., 2018; Radočaj & Dimić, 2013). To increase the potential application of okara as a starting material for producing food ingredients or for animal feed formulation and as a supplement in existing dietary staples, it is essential to recognize the production process of soy milk, the by-product it produces, as well as further processing treatments and its qualities, to enable easy amendment for the associated issues. The flavor of soy milk and its by-products is dictated by the type/variety of the bean, filtration, heating, and grinding processes adapted as discussed earlier (Swallah et al., 2021a; Davy & Vuong, 2020).

The traditional soy milk production process generally yields a stronger flavor, typically termed as "beany" and mostly not the favorite of most consumers (Vong & Liu, 2016). This unpleasant flavor has been linked to the action of lipoxygenase, which catalyzes the degradation of polyunsaturated fatty acids (Liu, 1997). Modern processes have been industrialized to improve yield while reducing the influence of lipoxygenase and other antinutritive compounds to generate a more acceptable flavor. It is evidenced that the application of heat has a major effect on the final flavor profiles and characteristics of soy milk owing to the thermal degradation of trypsin inhibitors and lipoxygenase. The grinding of soaked soybeans at 3° C and 80° C yielded a low level of flavor-forming compounds, whereas 3° C alone exerted a higher protein concentration attributable to higher solubility (Mizutani & Hashimoto, 2004). Approximately 5.19–14.4% of the okara protein content is made of trypsin inhibitors and is reported to be inactivated after sufficient heat treatment (Stanojevic et al., 2013), which may be employed in industrial settings. The antinutrient effects of okara have been reduced following the use of other novel techniques, including but not limited to infrared treatment (Yalcin & Basman, 2015), industrialized dielectric heat treatments (Berlinet et al., 2006), and irradiation with gamma rays (De Toledo et al., 2007). The use of infrared treatment, gamma irradiation, and microwave-assisted extraction was applied in soybean processing. An increase in protein content, viscosity, protein solubility, and digestibility (Varghese & Pare,

2019), inactivation of trypsin inhibitors and lipoxygenase (Yalcin & Basman, 2015), increase in total phenolic content, and decrease in tannins and trypsin inhibitors (De Toledo et al., 2007) were recorded. An improved protein content is reported to be linked with reduced soybean particle size during grinding (Vishwanathan, Singh, & Subramanian, 2011). It has again been reported that the available protein level is reduced following bean blanching at high temperatures before grinding, a known technique that aids to reduce the effect of lipoxygenase. High-pressure and -temperature pasteurization at 115° C has been applied to soy milk following the removal of the residue (okara), exerting higher aggregation of proteins in contrast with unpressurized pasteurization, while maintaining high acceptability during sensory observation (Zuo et al., 2016). Recently, Xu et al. compared the beany flavor of wet and dry soybean after heat treatment. The authors reported that dry beans with heat treatment of 100° C before grinding with water could decrease the formation of negative/bad flavors from lipid oxidation. The effect was further corrected via boiling the soybean slurry before filtration and removal of the by-products. This study again compared the impacts of different cultivars grown in different regions of China and discovered that cultivars deficient in lipoxygenase yield the lowest beany flavor (Yu et al., 2017). Again, Chauhan and Chauhan's study confirmed that when seeds are blanched for 30 mins in a 0.5% solution of sodium bicarbonate, phytic acid, trypsin inhibitors, as well as saponins were all markedly ($p < 0.05$) decreased by this treatment (Chauhan & Chauhan, 2007). Furthermore, other treatment techniques, which include phytochemical, chemical, and enzyme treatments have as well been adopted to reduce the antinutrient effects (Davy & Vuong, 2020). Microbial bioconversion of okara proteins may present few benefits. Thus, its bioconversion into smaller proteins may uplift its solubility and hence produce bioactive peptides and/or amino acids (Swallah et al., 2021a). A recent study recommends that it is essential to put into consideration all possible effects of fermentation on the molecular weights of amino acids profile, peptides, and the inhibitory activity of trypsin since they exert a role in impacting the overall functional characteristics such as solubility and foaming properties, and the bioactivity of soybean residual content (Colletti et al., 2020; Vong & Liu, 2016). The current study to increase okara utilization via drying processes and production of new food ingredients is summarized in Figure 18.2.

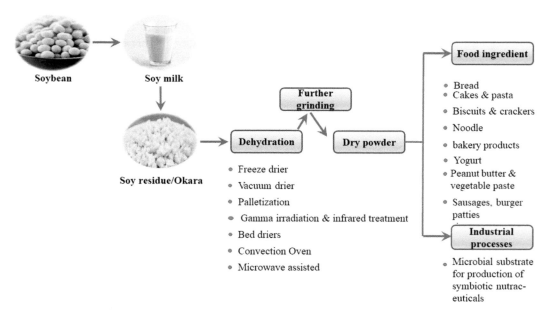

FIGURE 18.2   Okara utilization via drying processes and production of new food ingredients.

18.4 NEW OKARA PRODUCTS

Okara's rich profile of protein, lipid, dietary fiber, and phytochemicals, such as isoflavone, phytosterol, and lignan (Zhong-Hua et al., 2015; Li et al., 2012; Jankowiak et al., 2014), and its view as a useful health-elevating functional ingredient of food products, has resulted in the application of okara in bakery foods, beverages, dairy products, meat food, and other industrial products. Its addition adds a significant improvement to food technology and quality. It maintains the inherent flavor and texture of food and plays a role to produce healthy food. In addition, it also improves the water retention capacity of food, improves sensory quality, and extends shelf-life. It also meets an individual's demand for a low-fat and low-calorie healthy diet. The effect of okara supplementation on food function and properties are summarized in Table 18.2.

18.4.1 BAKED FOODS

The DFs, protein, and fat in the okara go beyond increasing the nutritional content of the baked goods, with additional effects such as improving the texture and flavor of the finished product (Park, Choi, & Kim, 2015). Since there is no special requirement for supplementing gluten during the baking of bread and biscuits, okara can be used instead of wheat flour. The bread and biscuits fortified with okara have good sensory qualities. Mudgil et al. improved the sensory quality of biscuits by adding SDF to the biscuits (Mudgil, Barak, & Khatkar, 2017). Belghith et al. used DFs instead of wheat flour to prepare the bread. This bread has better sensory quality and physicochemical properties than ordinary bread (Fendri et al., 2016). Besides, the foaming properties of the okara protein did not only enhanced the volume and the fluffiness of the bread but also yielded a better texture and appearance of the bread. The fortification of okara in the noodle-making process is suggested to improve the quality of noodles owing to the strong water-holding capacity of okara DFs. A study by Sereewat et al. confirmed that the addition of okara instead of wheat flour can increase the protein content as well as improve the quality of the cooked noodles (Sereewat et al., 2015).

18.4.2 BEVERAGES AND DAIRY PRODUCTS

Adding 0.5–1.5% of DFs to the beverage can significantly improve brew ability, stability, and dispersibility of the beverage. Moreover, DFs have adsorptive properties, which can reduce cholesterol levels by promoting the excretion of bile acids. Lee et al. evaluated the nutritional compositions as well as the antioxidant capacities on radicals and oxidative stress in okara yogurt through fermentation. The results showed that yogurt has good quality and nutritional value (Lee et al., 2018).

18.4.3 FRIED PUFFED FOODS

Soybean puffed food has opened up a new way for the utilization of okara. Lu et al. used okara as raw materials to produce soybean puffed food. Processed products did not only portray crispy characteristics and attractive flavors but also showed improvement over the disadvantages of coarse tofu and bean flavor (Lu et al., 2015).

18.4.4 MEAT PRODUCTS

Okara protein is among the most used sources of vegetable proteins in the meat industry, due to its interesting technological characteristics, such as its emulsifying properties, gelling capability, texture-improving capacity, and water-binding capacity (Soares et al., 2014). The addition of okara changes the processing characteristics of meat products, such as canned ham, lunch, and pork, while increasing the fiber content. The polydextrose in DFs is a low-molecular-weight polymer, which is not easily digested and absorbed. It is added to meat products to give the product a rich and smooth taste, and sensory characteristics of fats.

TABLE 18.2

Effect of Okara Supplementation on Food Function and Properties

Experimental Model	Dietary Formulations	Effect on Food Properties and Function	Conclusion Remarks	Reference
Okara and vital gluten on physicochemical properties of noodle	Added portion of okara (0%, 5%, 10%)	Increased total phenolics, flavonoids, and radical-scavenging activity. 10–15% okara reduced optimum cooking time, extensibility tensile strength, and elasticity of noodle.	5% or 10% dried okara powder plus 6% vital gluten might be best in making noodles with increased phytochemicals and consumer's sensory satisfaction.	(Pan et al. 2018)
Application of okara to enrich vegetable paste	High moisture (80.77–81.42%) Low lipid (5.62–7.62%) Low calorie (95.14–108.14 kcal)	Increased β-carotene (0.411 mg/100 mL). Elevated antioxidant activity. Increased isoflavones (0.15 μmol/gFM).	The sample with the lowest content of okara (34 g/100 g) presented the highest average of 8.0 in the acceptance test and was also considered the tasters' favorite one.	(Guimarães et al. 2018)
Starch digestibility of steamed rice bread fortified with okara	Added portion of okara (0%, 7%, 14%, 21%)	Improved elasticity and viscidity. Decreased hardness, cohesiveness, and chewiness. Increased amylose content, slowly digestible starch, resistant starch. Reduced predicted glycemic index (pGI) from 79.14 to 74.17–68.91	Okara can potentially modify the texture and starch digestibility of steamed rice bread.	(Tang et al. 2020)
Gastrointestinal stress in synbiotic soy yogurt with okara during storage for 28 days	Soy yogurt + okara Soy yogurt + guava pulp Soy yogurt + mango pulp	Increased survival rates (%) of *L. acidophilus* La-5 and *B. animalis* Bb-12, ranging from 8 to 9 log cfu/g after simulated gastrointestinal conditions. Improved probiotic strains functionality.	In this study, okara endorsed probiotic functionality in modulated intestinal conditions, however, the addition of fruit pulps might lead to a reduction.	(Bedani et al. 20144)
Digestibility of rice noodle enriched with okara	Added portion of okara (0%, 5%, 10%, 20%)	Improved cooking loss, adhesiveness, and hardness with increasing level of okara. Reduced water absorption, cohesiveness, and swelling index. 0%, 5%, and 10% okara decreased in vitro starch digestibility. 10% okara reduced predicted glycemic index.	10% okara can be used to produce health-beneficial rice noodles with reduced in vitro starch digestibility and improve cooking quality.	(Kang et al. 2018)
Digestibility and structural attributes of okara-enriched functional pasta	Added okara contents (10–50%)	No structural changes, decreased glycemic index (27.41 ± 0.05–12.38 ± 0.01). 50% okara encouraged total phenolic content and antioxidant activity (158.37 ± 0.40 to 232.90 ± 0.85 mg GAE/100 g and 10.87 ± 0.10%–56.21 ± 0.05%).	The study showed that pasta enriched with okara has the potential to be commercialized on the industrial level to develop nutritional enriched functional pasta.	(Kamble et al. 2019)

18.5 HEALTH-ENCOURAGING FUNCTIONS OF PREBIOTIC OKARA

As earlier mentioned, okara contains significantly high levels of dietary fiber and proteins, and substantial amounts of mineral elements and isoflavones, which give it a high nutritional value as well as a potential prebiotic function. It is therefore potentially useful as a functional ingredient with health-promoting properties (Jiménez-Escrig et al., 2008). Prebiotics are non-digestible parts of food termed carbohydrates that act as fibers. They reach the colon unaltered and are used by the intestinal microorganisms, serving as food for "good" intestinal bacteria, as well as boosting their growth, colonization, and sustainability in the digestive tract (Swallah et al., 2021a; Hijová, Bertková, & Štofilová, 2019). The most commonly used prebiotics in human research include Galactooligosaccharide, Arabinoxylan-oligosaccharides, Fructooligosaccharide, Xylo-oligosaccharide, and Soybean oligosaccharides (Hijová et al., 2019; Sawicki and Livingston, 2017). Okara as a prebiotic has been explored in varied research (i.e. *in vivo*/*in vitro*) using *Lactobacillus acidophilus* and *Bifidobacterium bifidum* (Swallah et al., 2021a; Villanueva-Suárez et al., 2016; Bedani, Rossi, & Saad, 2013; Jiménez-Escrig et al., 2008). When okara is consumed, it selectively stimulates the activity and/or growth of certain gastrointestinal microbes that can confer health functions to the host (Saad et al., 2013). The conversion of okara insoluble fibers into soluble fibers was also recorded when *Streptococcus thermophilus* and *Lactobacillus delbrueckii* subspecies *bulgaricus* were used (Tu et al., 2014). Okara can serve as a surface for bacteria-cell adhesion, that is, facilitating substrate absorption as well as cell growth. Treatment with β-glucanase (Ultraflo L®) improved the content of okara-soluble dietary fibers and consequently enhanced fermentation by *B. bifidum* (Villanueva-Suárez, Pérez-Cózar, & Redondo-Cuenca, 2013). Okara is reported to have played an essential role in averting hyperlipidemia and may also be proficient as a weight-loss dietary supplement with a potential prebiotic function (Colletti et al., 2020). Villanueva et al. studied the effect of high-fat diets fortified with okara on lipid profiles of the plasma, feces, and liver of male Syrian hamsters following three weeks of feeding. The plasma levels of triglyceride (TG), very low-density lipoprotein (VLDL) plus low-density lipoprotein (LDL) cholesterol, and total cholesterol in the Syrian hamsters with okara supplemented diet decreased significantly. TG, total lipid, and esterified cholesterol concentrations in the liver were decreased. All okara-supplemented hamster groups recorded an increased fecal excretion of total lipids, free cholesterol, TGs, and total nitrogen. The results recommend that the main components of okara, protein and dietary fiber, may be linked to cholesterol and total-lipid reduction in plasma and the liver, and increased fecal production in the group fed high-fat diet (Villanueva et al., 2011). In another study, mice were fed either a high-fat diet supplemented with dry okara (10%, 20%, or 40%) or a high-fat diet (14% crude fat) for ten weeks (Matsumoto, Watanabe, & Yokoyama, 2007). The intake dose of okara hindered the development of body weight as well as white epididymal adipose tissue in a dependent manner, and again prevented an increase in plasma lipids, including LDL cholesterol, total cholesterol, and unesterified fatty acids. Okara consumption also inhibited liver steatosis. The RT-PCR (real-time reverse-polymerase transcriptase chain reaction) revealed that okara intake yielded to the down-regulation of the fatty-acid synthesis gene in the liver and overexpression of the 7α-hydroxylase cholesterol gene.

In addition, researchers believe that a large intake of fiber-rich foods (i.e. mainly soluble fiber) can help regulate blood sugar levels after meals since it can reduce the rate of carbohydrate uptake in the intestines. Okara is reported to possess about 50% fiber and 25% protein, hence making it an ideal food supplement for diabetics (Colletti et al., 2020). Xu et al. studied the effect of okara on lipids, blood sugar levels, and hepatic-nephritic histomorphology of rats with streptozotocin-induced diabetes. The total serum cholesterol, TGs, and glycemia of diabetic rats fed okara-containing food for five weeks reduced significantly, whereas high-density lipoprotein (HDL) cholesterol significantly increased. The authors concluded that okara can significantly decrease the plasma levels of lipids and sugar, enhance blood sugar and lipid metabolism, as well as protect the liver and kidneys of diabetic rats/patients (Xu et al., 2000). In fact, the health-promoting functions of okara have been studied (i.e., *in vivo*, *in vitro*, and human study) for many years in Asia and other parts of the globe in treating chronic disease and are summarized in Table 18.3.

TABLE 18.3
Health Encouraging Functions of Okara (In Vivo/In Vitro)

Experimental Model	Outcome	Author Suggestion	Reference
In vivo (Wistar rats fed high-fat diet), fed okara (20%), for four weeks	Reduced body weight and triglycerides, increased SCFA production, amino acid metabolism, mineral absorption, and microbial protection; no effect on *Firmicutes:Bacteroidetes* ratio, *Bacteroides* and *B. coccoides-E. rectale* groups in control group, increased *C. leptum* and *Bacteroides* population in feces, *Enterobacteriaceae* (cecal content) and *Enterococcus* (fecal and cecal content) groups.	Okara exerts health-promoting attributes in vivo and could be further used as prebiotic and functional ingredient in foods	(Pérez-López et al. 2018)
In vivo (Female Wistar rats fed a standard rat diet) fed dietary rich okara (10%), for four weeks	Reduced body weight gain, total cholesterol, elevated antioxidant status and butyrogenic effect in the cecum, increased apparent absorption and true retention of calcium.	The development of an innovative soybean by-product rich in dietary fiber could be valuable as a functional ingredient with health-promoting effects	(Jiménez-Escrig et al. 2008)
In vivo (High-cholesterol-fed Wistar rats) for four weeks	Reduced liver and serum triglyceride levels, pH of fecal contents, increased total lipids, triglycerides and bile acids in feces, increased SCFA production.	Enzymatically treated okara fiber can promote intestinal transit by increasing fecal bulk	(Villanueva-Suárez et al. 2016)
In vitro (Water jet [WJ] treated okara and water jet treated microcrystalline cellulose [MCC]), effect on α-amylase inhibition and butyrate production using *Roseburia intestinalis*	Improved inhibition of α-amylase activities by WJ-treated okara than WJ-treated MCC, increased butyrate production by *Roseburia intestinalis* in WJ-treated okara.	These results depict that WJ system can be used on okara to improve repressed α-amylase activities and butyrate production by gut microbiota	(Nagano et al. 2020)
In vivo (High fat-fed Syrian hamsters), fed 13% or 20% okara fiber for three weeks	Increased fecal excretion of total lipids, triglycerides, free cholesterol, and total nitrogen. No changes in feed intake and body weight gain. 20% okara group: Reduced plasma triglycerides, VLDL- plus LDL cholesterol and total cholesterol, decreased liver total lipids, triglycerides, and total esterified cholesterol concentrations.	Okara might aid in the inhibition of hyperlipidemia and could be used as a natural ingredient for functional food preparation	(Villanueva et al. 2011)

(Continued)

TABLE 18.3 (CONTINUED)

Health Encouraging Functions of Okara (In Vivo/In Vitro)

Experimental Model	Outcome	Author Suggestion	Reference
In vitro (Fermentability and prebiotic potential of okara using human fecal slurries), using 16S rRNA-based fluorescence in situ hybridization and HPLC	Promoted SCFA plus lactic acid, increased beneficial bacteria (*bifidobacteria* and *lactobacilli*), reduced potentially harmful bacterial groups (*Clostridia* and *Bacteroides*).	The differences observed between fructo-oligosaccharides and okara substrates could be accredited to the great complexity of okara's cell wall, which needs longer times to be fermented than other easily digested molecules. Hence, allowing a prolonged potential prebiotic effect. These findings support an *in vitro* potential prebiotic effect of okara	(Pérez-López et al. 2016)
In vivo (Wistar Hannover female rats), control group (fed standard rat chow) and treated group (fed a mixture of the standard rat chow plus okara) for four weeks	No changes in food intake, reduced growth rate, and feeding efficiency, increased fecal weight and moisture, reduced lower pH, improved cecal weight, increased total SCFA production in okara-fed group compared to control group. No changes in albumin, uric acid, protein, bilirubin, or glucose content in rat serum for both groups.	Okara is a rich source of low-cost dietary fiber and protein and might be effective as a dietary weight-loss supplement with prebiotic effect potentials	(Préstamo et al. 2007)
In vivo (Senescence-accelerated mouse prone 8 [SAMP8] mice), fed standard diet, or a diet containing (7.5% or 15%, w/w) okara, for 26 weeks	15% okara-fed group: Decreased body weight, increased fecal weight, and altered cecal microbiota composition compared with the control group, no changes in serum lactic acid and butyric acid levels. 7.5% okara-fed group: Increased NeuN intensity in the hippocampus than control mice, reduced inflammatory cytokine TNF-α, increased brain-derived neurotrophic factor, improved acetylcholine synthesizing enzyme. Increased acetylcholine level in the brain.	Oral administration of okara could delay cognitive decline without drastically changing gut microbiota	(Corpuz et al. 2019)
In vivo (High fat-fed C57BL/6J male mice) for 12 weeks	Reduced body weight and epididymal fat weight. Decreased serum and hepatic lipid profiles. Increased fecal triacylglycerol and total cholesterol levels. Improved PPAR-α expression, ↓ PPAR-γ and FAS levels.	Okara intake appears to protect mice against diet-induced obesity and metabolic dysregulation associated with obesity	(Kim et al. 2016)
In vivo (High lard fed Goto-Kakizaki [GK] type 2 diabetes male rats) for two weeks	No changes in body weight gain or food intake, reduced plasma glucose levels, enhanced mRNA expression levels of PPARγ, adiponectin, and GLUT4.	The study suggested that okara can play significant role in treating type 2 diabetes	(Hosokawa et al. 2016)
In vivo (Human type 2 diabetes mellitus outpatients) fed okara for two weeks	Elevated food intake (fiber 6.9 to 12.6 g), reduced fasting blood glucose (6.3 to 5.4 mmol/L), reduced fructosamine (319 to 301 μmol/L).	Okara increased fiber intake and subsequently improved blood glucose in DM patients	(Nguyen et al. 2019)

18.6 CONCLUSION

This current chapter reports on the nutritional and functional components of okara, and its value-added processing as well as the development of new health-promoting foods. From the literature survey presented, okara has a high nutritional value and is capable of being integrated into the food sector or our everyday food items for a range of diets such as to partially replace traditional flour, with the added potential merit of exerting functional components. It could be concluded that the fortification of okara into white flour for commercial bread, cake, pasta, and/or noodle production has the uppermost potential as saleable food products. The desirable functionality of okara protein and starch and its high nutritional and biological value make it commercially suitable. As these foods are incorporated into the diets of many countries across the world, thus they can have great benefits for human health and decrease environmental impacts. The hypoglycemic function of okara shows it can be used to make functional food that has the potential to provide varied health benefits.

As the flavor compounds, nutrients, and antinutritional components of soy and soy-derived products are dependent on the processing practices of manufacturers, it would be apposite to tailor the innovation and development of new foods to match already existing industrial practices. In addition, channeling research into novel treatments and drying techniques, or the adoption of existing pre- and post-manufacturing practices, is likely to increase the storage potential, functionality, and utilization of okara.

ACKNOWLEDGMENTS

This work was funded and supported by China Agriculture Research System of MOF and MARA (Project No. CARS-04).

REFERENCES

Alejandre, M., D. Passarini, I. Astiasarán & D. Ansorena (2017) The effect of low-fat beef patties formulated with a low-energy fat analogue enriched in long-chain polyunsaturated fatty acids on lipid oxidation and sensory attributes. *Meat Science*, 134, 7–13.

Aoki, H., O. Taneyama & M. Inami (1980) Emulsifying properties of soy protein: Characteristics of 7S and IIS proteins. *Journal of Food Science*, 45(3), 534–538.

Bedani, R., M. M. Campos, I. A. Castro, E. A. Rossi & S. M. Saad (2014) Incorporation of soybean by-product okara and inulin in a probiotic soy yoghurt: Texture profile and sensory acceptance. *Journal of the Science of Food and Agriculture*, 94(1), 119–125.

Bedani, R., E. A. Rossi & S. M. I. Saad (2013) Impact of inulin and okara on Lactobacillus acidophilus La-5 and Bifidobacterium animalis Bb-12 viability in a fermented soy product and probiotic survival under in vitro simulated gastrointestinal conditions. *Food Microbiology*, 34(2), 382–389.

Berlinet, C., P. Brat, J. M. Brillouet & V. Ducruet (2006) Ascorbic acid, aroma compounds and browning of orange juices related to PET packaging materials and pH. *Journal of the Science of Food and Agriculture*, 86(13), 2206–2212.

Cederroth, C. R. & S. Nef (2009) Soy, phytoestrogens and metabolism: A review. *Molecular and Cellular Endocrinology*, 304(1–2), 30–42.

Chauhan, O. & G. Chauhan (2007) Anti-nutrients in soybeans at different stages of soy milk production. *Journal of Food Science and Technology-Mysore*, 44, 378–380.

Colletti, A., A. Attrovio, L. Boffa, S. Mantegna & G. Cravotto (2020) Valorisation of by-products from soybean (Glycine max (L.) Merr.) processing. *Molecules*, 25(9), 2129.

Corpuz, H. M., M. Arimura, S. Chawalitpong, K. Miyazaki, M. Sawaguchi, S. Nakamura & S. Katayama (2019) Oral administration of okara soybean by-product attenuates cognitive impairment in a mouse model of accelerated aging. *Nutrients*, 11(12), 2939.

Dahl, W. J. & M. L. Stewart (2015) Position of the Academy of Nutrition and Dietetics: Health implications of dietary fiber. *Journal of the Academy of Nutrition and Dietetics*, 115(11), 1861–1870.

Das, A. K., P. K. Nanda, N. R. Chowdhury, P. Dandapat, M. Gagaoua, P. Chauhan, M. Pateiro & J. M. Lorenzo (2021) Application of pomegranate by-products in muscle foods: Oxidative indices, colour stability, shelf life and health benefits. *Molecules*, 26(2), 467.

Davy, P. & Q. V. Vuong (2020) Soy milk by-product: Its composition and utilisation. *Food Reviews International*, 1–23.

De Toledo, T., S. Canniatti-Brazaca, V. Arthur & S. Piedade (2007) Effects of gamma radiation on total phenolics, trypsin and tannin inhibitors in soybean grains. *Radiation Physics and Chemistry*, 76(10), 1653–1656.

Dong, J. l., L. Wang, J. Lü, Y. y. Zhu & R. l. Shen (2019) Structural, antioxidant and adsorption properties of dietary fiber from foxtail millet (Setaria italica) bran. *Journal of the Science of Food and Agriculture*, 99(8), 3886–3894.

Elleuch, M., D. Bedigian, O. Roiseux, S. Besbes, C. Blecker & H. Attia (2011) Dietary fibre and fibre-rich by-products of food processing: Characterisation, technological functionality and commercial applications: A review. *Food Chemistry*, 124(2), 411–421.

Fendri, L. B., F. Chaari, M. Maaloul, F. Kallel, L. Abdelkafi, S. E. Chaabouni & D. Ghribi-Aydi (2016) Wheat bread enrichment by pea and broad bean pods fibers: Effect on dough rheology and bread quality. *LWT*, 73, 584–591.

Fu, Z. q., M. Wu, X. y. Han & L. Xu (2017) Effect of okara dietary fiber on the properties of starch-based films. *Starch-Stärke*, 69(11–12), 1700053.

Genovese, M. I. & F. M. Lajolo (2002) Isoflavones in soy-based foods consumed in Brazil: Levels, distribution, and estimated intake. *Journal of Agricultural and Food Chemistry*, 50(21), 5987–5993.

Godfrey, P. (2002) Soy Products as Ingredients-Farm to the Table. *Innovations in Food Technology*, 14, 1–3.

Guimarães, R. M., E. I. Ida, H. G. Falcão, T. A. M. de Rezende, J. de Santana Silva, C. C. F. Alves, M. A. P. da Silva & M. B. Egea (2020) Evaluating technological quality of okara flours obtained by different drying processes. *LWT*, 123, 109062.

Guimarães, R. M., T. E. Silva, A. C. Lemes, M. C. F. Boldrin, M. A. P. da Silva, F. G. Silva & M. B. Egea (2018) Okara: A soybean by-product as an alternative to enrich vegetable paste. *LWT*, 92, 593–599.

Gupta, S., J. J. Lee & W. N. Chen (2018) Analysis of improved nutritional composition of potential functional food (Okara) after probiotic solid-state fermentation. *Journal of Agricultural and Food Chemistry*, 66(21), 5373–5381.

Hijová, E., I. Bertková & J. Štofilová (2019) Dietary fibre as prebiotics in nutrition. *Central European Journal of Public Health*, 27(3), 251–255.

Hosokawa, M., M. Katsukawa, H. Tanaka, H. Fukuda, S. Okuno, K. Tsuda & N. Iritani (2016) Okara ameliorates glucose tolerance in GK rats. *Journal of Clinical Biochemistry and Nutrition*, 58(3), 216–222.

Hu, Y., C. Piao, Y. Chen, Y. Zhou, D. Wang, H. Yu. & B. Xu (2019) Soybean residue (okara) fermentation with the yeast Kluyveromyces marxianus. *Food Bioscience*, 31, 100439.

Huang, L., X. Ding, C. Dai & H. Ma (2017) Changes in the structure and dissociation of soybean protein isolate induced by ultrasound-assisted acid pretreatment. *Food Chemistry*, 232, 727–732.

Islam, M. A., R. Bekele, J. H. vanden Berg, Y. Kuswanti, O. Thapa, S. Soltani, F. R. van Leeuwen, I. M. Rietjens & A. J. Murk (2015) Deconjugation of soy isoflavone glucuronides needed for estrogenic activity. *Toxicology in Vitro : An International Journal Published in Association with BIBRA*, 29(4), 706–715.

Jankowiak, L., O. Trifunovic, R. M. Boom & A. J. van der Goot (2014) The potential of crude okara for isoflavone production. *Journal of Food Engineering*, 124, 166–172.

Jiang, S., W. Cai & B. Xu (2013) Food quality improvement of soy milk made from short-time germinated soybeans. *Foods*, 2(2), 198–212.

Jiménez-Escrig, A., M. D. Tenorio, I. Espinosa-Martos & P. Rupérez (2008) Health-promoting effects of a dietary fiber concentrate from the soybean byproduct okara in rats. *Journal of Agricultural and Food Chemistry*, 56(16), 7495–7501.

Kamble, D. B. & S. Rani (2020) Bioactive components, in vitro digestibility, microstructure and application of soybean residue (okara): A review. *Legume Science*, 2(1), e32.

Kamble, D. B., R. Singh, S. Rani & D. Pratap (2019) Physicochemical properties, in vitro digestibility and structural attributes of okara-enriched functional pasta. *Journal of Food Processing and Preservation*, 43(12), e14232.

Kang, M. J., I. Y. Bae & H. G. Lee (2018) Rice noodle enriched with okara: Cooking property, texture, and in vitro starch digestibility. *Food Bioscience*, 22, 178–183.

Kim, H.-S., O.-K. Yu, M.-S. Byun & Y.-S. Cha (2016) Okara, a soybean by-product, prevents high fat diet-induced obesity and improves serum lipid profiles in C57BL/6J mice. *Food Science and Biotechnology*, 25(2), 607–613.

Kitagawa, I., M. Yoshikawa, T. Hayashi & T. Taniyama (1984) Quantitative determination of soyasaponins in soybeans of various origins and soybean products by means of high performance liquid chromatography. *Yakugaku Zasshi: Journal of the Pharmaceutical Society of Japan*, 104(3), 275–279.

Laurenz, R., P. Tumbalam, S. Naeve & K. D. Thelen (2017) Determination of isoflavone (genistein and daidzein) concentration of soybean seed as affected by environment and management inputs. *Journal of the Science of Food and Agriculture*, 97(10), 3342–3347.

Lee, J. H., C. E. Hwang, E. J. Cho, Y. H. Song, S. C. Kim & K. M. Cho (2018) Improvement of nutritional components and in vitro antioxidative properties of soy-powder yogurts using Lactobacillus plantarum. *Journal of Food and Drug Analysis*, 26(3), 1054–1065.

Li, B., M. Qiao & F. Lu (2012) Composition, nutrition, and utilization of okara (soybean residue). *Food Reviews International*, 28(3), 231–252.

Li, Q., S. Yang, Y. Li, Y. Huang & J. Zhang (2019) Antioxidant activity of free and hydrolyzed phenolic compounds in soluble and insoluble dietary fibres derived from hulless barley. *LWT*, 111, 534–540.

Li, S., D. Zhu, K. Li, Y. Yang, Z. Lei & Z. Zhang (2013) Soybean curd residue: Composition, utilization, and related limiting factors. *ISRN Industrial Engineering*, 2013, 1–8.

Liu, K. (1997) Chemistry and nutritional value of soybean components. In: *Soybeans*, 25–113. Springer.

Liu, Y., Q.-Y. Yu, Z.-L. Zhu, P.-Y. Tang & K. Li (2015) Vitamin B2 intake and the risk of colorectal cancer: A meta-analysis of observational studies. *Asian Pacific Journal of Cancer Prevention: APJCP*, 16(3), 909–913.

Sawicki, C. M., K. A. Livingston, N. M. McKeown, M. Obin, S. B. Roberts & M. Chung (2017) Dietary fiber and the human gut microbiota: Application of evidence mapping methodology. *Nutrients*, 9(2), 125.

Lu, F., X.-d. Chen, B. Li & S.-l. Li (2015) *Preparation of Okara Puffed Food by Twin Screw Extrusion Soybean Science*, 34(2), 306–309.

Ma, L., B. H. Lee, R. Mao, A. Cai, Y. Jia, H. Clifton, S. Schaefer, L. Xu & J. Zheng (2014) Nicotinic acid activates the capsaicin receptor TRPV1: Potential mechanism for cutaneous flushing. *Arteriosclerosis, Thrombosis, and Vascular Biology*, 34(6), 1272–1280.

Mateos-Aparicio, I., A. Redondo-Cuenca, M.-J. Villanueva-Suárez, M.-A. Zapata-Revilla & M.-D. Tenorio-Sanz (2010) Pea pod, broad bean pod and okara, potential sources of functional compounds. *LWT - Food Science and Technology*, 43(9), 1467–1470.

Matsumoto, K., Y. Watanabe & S.-i. Yokoyama (2007) Okara, soybean residue, prevents obesity in a diet-induced murine obesity model. *Bioscience, Biotechnology, and Biochemistry*, 0702080286–0702080286.

Mizutani, T. & H. Hashimoto (2004) Effect of grinding temperature on hydroperoxide and off-flavor contents during soymilk manufacturing process. *Journal of Food Science*, 69(3), SNQ112–SNQ116.

Mudgil, D., S. Barak & B. Khatkar (2017) Cookie texture, spread ratio and sensory acceptability of cookies as a function of soluble dietary fiber, baking time and different water levels. *LWT*, 80, 537–542.

Nagai, T., L. Te Li, Y. L. Ma, P. K. Sarkar, R. Nout, K. Y. Park, J. K. Jeong, J. E. Lee, G. Im. Lee & C. H. Lee (2014) Diversity of plant-based food products involving alkaline fermentation. In *Handbook of Indigenous Foods Involving Alkaline Fermentation*, 7–187. CRC Press.

Nagano, T., R. Hirano, S. Kurihara & K. Nishinari (2020) Improved effects of okara atomized by a water jet system on α-amylase inhibition and butyrate production by Roseburia intestinalis. *Bioscience, Biotechnology, and Biochemistry*, 84(7), 1–8.

Nguyen, L. T., T. H. Nguyen, L. T. Nguyen, S. Kamoshita, T. P. Tran, H. T. Le, F. Shimura & S. Yamamoto (2019) Okara improved blood glucose level in Vietnamese with type 2 diabetes mellitus. *Journal of Nutritional Science and Vitaminology*, 65(1), 60–65.

O'Toole, D. K. (1999) Characteristics and use of okara, the soybean residue from soy milk production a review. *Journal of Agricultural and Food Chemistry*, 47(2), 363–371.

Pan, W.-C., Y.-M. Liu & S.-Y. Shiau (2018) Effect of okara and vital gluten on physico-chemical properties of noodle. *Czech Journal of Food Sciences*, 36(4), 301–306.

Park, J., I. Choi & Y. Kim (2015) Cookies formulated from fresh okara using starch, soy flour and hydroxypropyl methylcellulose have high quality and nutritional value. *LWT - Food Science and Technology*, 63(1), 660–666.

Pérez-López, E., D. Cela, A. Costabile, I. Mateos-Aparicio & P. Rupérez (2016) In vitro fermentability and prebiotic potential of soyabean Okara by human faecal microbiota. *British Journal of Nutrition*, 116(6), 1116–1124.

Pérez-López, E., A. Veses, N. Redondo, M. Tenorio-Sanz, M. Villanueva, A. Redondo-Cuenca, A. Marcos, E. Nova, I. Mateos-Aparicio & P. Rupérez (2018) Soybean Okara modulates gut microbiota in rats fed a high-fat diet. *Bioactive Carbohydrates and Dietary Fibre*, 16, 100–107.

Préstamo, G., P. Rupérez, I. Espinosa-Martos, M. J. Villanueva & M. A. Lasunción (2007) The effects of okara on rat growth, cecal fermentation, and serum lipids. *European Food Research and Technology*, 225(5–6), 925–928.

Radočaj, O. & E. Dimić (2013) Valorization of wet okara, a value-added functional ingredient, in a coconut-based baked snack. *Cereal Chemistry Journal*, 90(3), 256–262.

Raman, M. & M. Doble (2015) κ-carrageenan from marine red algae, Kappaphycus alvarezii–A functional food to prevent colon carcinogenesis. *Journal of Functional Foods*, 15, 354–364.

Rinaldi, V., P. Ng & M. Bennink (2000) Effects of extrusion on dietary fiber and isoflavone contents of wheat extrudates enriched with wet okara. *Cereal Chemistry Journal*, 77(2), 237–240.

Saad, N., C. Delattre, M. Urdaci, J.-M. Schmitter & P. Bressollier (2013) An overview of the last advances in probiotic and prebiotic field. *LWT - Food Science and Technology*, 50(1), 1–16.

Salman, M. & I. Naseem (2015) Riboflavin as adjuvant with cisplatin: Study in mouse skin cancer model. *Frontiers in Bioscience: Elite*, 7, 278–291.

Sereewat, P., C. Suthipinittham, S. Sumathaluk, C. Puttanlek, D. Uttapap & V. Rungsardthong (2015) Cooking properties and sensory acceptability of spaghetti made from rice flour and defatted soy flour. *LWT - Food Science and Technology*, 60(2), 1061–1067.

Shin, D. & D. Jeong (2015) Korean traditional fermented soybean products: Jang. *Journal of Ethnic Foods*, 2(1), 2–7.

Singh, A., M. Meena, D. Kumar, A. K. Dubey & M. I. Hassan (2015) Structural and functional analysis of various globulin proteins from soy seed. *Critical Reviews in Food Science and Nutrition*, 55(11), 1491–1502.

Soares, S., J. S. Amaral, M. B. P. Oliveira & I. Mafra (2014) Quantitative detection of soybean in meat products by a TaqMan Real-time PCR assay. *Meat Science*, 98(1), 41–46.

Stanojevic, S. P., M. B. Barac, M. B. Pesic, V. S. Jankovic & B. V. Vucelic-Radovic (2013) Bioactive proteins and energy value of okara as a byproduct in hydrothermal processing of soy milk. *Journal of Agricultural and Food Chemistry*, 61(38), 9210–9219.

Stanojevic, S. P., M. B. Barac, M. B. Pesic, S. M. Zilic, M. M. Kresovic & B. V. Vucelic-Radovic (2014) Mineral elements, lipoxygenase activity, and antioxidant capacity of okara as a byproduct in hydrothermal processing of soy milk. *Journal of Agricultural and Food Chemistry*, 62(36), 9017–9023.

Swallah, M. S., H. Fan, S. Wang, H. Yu & C. Piao (2021a) Prebiotic impacts of soybean residue (Okara) on Eubiosis/Dysbiosis condition of the gut and the possible effects on liver and kidney functions. *Molecules*, 26(2).

Swallah, M. S., H. Fu, H. Sun, R. Affoh & H. Yu (2020a) The impact of polyphenol on general nutrient metabolism in the monogastric gastrointestinal tract. *Journal of Food Quality*, 2020.

Swallah, M. S., H. Sun, R. Affoh, H. Fu & H. Yu (2020b) Antioxidant potential overviews of secondary metabolites (polyphenols) in fruits. *International Journal of Food Science*, 2020.

Swallah, M. S., H. Yu, C. Piao, H. Fu, Z. Yakubu & F. L. Sossah (2021b) Synergistic two-way interactions of dietary polyphenols and dietary components on the gut microbial composition: Is there a positive, negative, or neutralizing effect in the prevention and management of metabolic diseases? *Current Protein and Peptide Science*, 22, 1–10.

Tang, Z., J. Fan, Z. Zhang, W. Zhang, J. Yang, L. Liu, Z. Yang & X. Zeng (2020) Insights into the structural characteristics and in vitro starch digestibility on steamed rice bread as affected by the addition of okara. *Food Hydrocolloids*, 113, 106533.

Timilsena, Y. P., B. Wang, R. Adhikari & B. Adhikari (2017) Advances in microencapsulation of polyunsaturated fatty acids (PUFAs)-rich plant oils using complex coacervation: A review. *Food Hydrocolloids*, 69, 369–381.

Tu, Z., L. Chen, H. Wang, C. Ruan, L. Zhang & Y. Kou (2014) Effect of fermentation and dynamic high pressure microfluidization on dietary fibre of soybean residue. *Journal of Food Science and Technology*, 51(11), 3285–3292.

Varghese, T. & A. Pare (2019) Effect of microwave assisted extraction on yield and protein characteristics of soymilk. *Journal of Food Engineering*, 262, 92–99.

Villanueva-Suárez, M.-J., M.-L. Pérez-Cózar, I. Mateos-Aparicio & A. Redondo-Cuenca (2016) Potential fat-lowering and prebiotic effects of enzymatically treated okara in high-cholesterol-fed Wistar rats. *International Journal of Food Sciences and Nutrition*, 67(7), 828–833.

Villanueva-Suárez, M. J., M. L. Pérez-Cózar & A. Redondo-Cuenca (2013) Sequential extraction of polysaccharides from enzymatically hydrolyzed okara byproduct: Physicochemical properties and in vitro fermentability. *Food Chemistry*, 141(2), 1114–1119.

Villanueva, M., W. Yokoyama, Y. Hong, G. Barttley & P. Rupérez (2011) Effect of high-fat diets supplemented with okara soybean by-product on lipid profiles of plasma, liver and faeces in Syrian hamsters. *Food Chemistry*, 124(1), 72–79.

Villares, A., M. A. Rostagno, A. García-Lafuente, E. Guillamón & J. A. Martínez (2011) Content and profile of isoflavones in soy-based foods as a function of the production process. *Food and Bioprocess Technology*, 4(1), 27–38.

Vishwanathan, K., V. Singh & R. Subramanian (2011) Wet grinding characteristics of soybean for soymilk extraction. *Journal of Food Engineering*, 106(1), 28–34.

Vong, W. C., X. Y. Hua & S.-Q. Liu (2018) Solid-state fermentation with Rhizopus oligosporus and Yarrowia lipolytica improved nutritional and flavour properties of okara. *LWT*, 90, 316–322.

Vong, W. C. & S.-Q. Liu (2016) Biovalorisation of okara (soybean residue) for food and nutrition. *Trends in Food Science and Technology*, 52, 139–147.

Vong, W. C. & S. Q. Liu (2017) Changes in volatile profile of soybean residue (okara) upon solid-state fermentation by yeasts. *Journal of the Science of Food and Agriculture*, 97(1), 135–143.

Voss, G., L. Rodríguez-Alcalá, L. Valente & M. Pintado (2018) Impact of different thermal treatments and storage conditions on the stability of soybean byproduct (okara). *Journal of Food Measurement and Characterization*, 12(3), 1981–1996.

Wang, S., Y. Wang, M.-H. Pan & C.-T. Ho (2017) Anti-obesity molecular mechanism of soy isoflavones: Weaving the way to new therapeutic routes. *Food and Function*, 8(11), 3831–3846.

Wen, Y., M. Niu, B. Zhang, S. Zhao & S. Xiong (2017) Structural characteristics and functional properties of rice bran dietary fiber modified by enzymatic and enzyme-micronization treatments. *LWT*, 75, 344–351.

Wu, S. (2003) Preparation of bean dregs cake. *Food Industry*, 24, 23–24.

Xu, H., Y. Wang, H. Liu, J. Zheng & Y. Xin (2000) Influence of soybean fibers on blood sugar and blood lipid metabolism and hepatic-nephritic histomorphology of mich with STZ-induced diabetes. *Acta Nutr Sinica*, 22, 171–174.

Xu, J., D. Mukherjee & S. K. Chang (2018) Physicochemical properties and storage stability of soybean protein nanoemulsions prepared by ultra-high pressure homogenization. *Food Chemistry*, 240, 1005–1013.

Yalcin, S. & A. Basman (2015) Effects of infrared treatment on urease, trypsin inhibitor and lipoxygenase activities of soybean samples. *Food Chemistry*, 169, 203–210.

Yan, X., R. Ye & Y. Chen (2015) Blasting extrusion processing: The increase of soluble dietary fiber content and extraction of soluble-fiber polysaccharides from wheat bran. *Food Chemistry*, 180, 106–115.

Yilmaz, I., I. Demiryilmaz, M. Turan, N. Cetin, M. Gul & H. Suleyman (2015) The effects of thiamine and thiamine pyrophosphate on alcohol-induced hepatic damage biomarkers in rats. *European Review for Medical and Pharmacological Sciences*, 19(4), 664–670.

Yu, H., R. Liu, Y. Hu & B. Xu (2017) Flavor profiles of soymilk processed with four different processing technologies and 26 soybean cultivars grown in China. *International Journal of Food Properties*, 20(sup3), S2887–S2898.

Zhang, G., Y. Zhang, Y. Shu & H. Ma (2015) Screening and identification of three types of soybean lines lacking different seed storage protein subunits. *Soybean Science*, 34, 1–31.

Zhang, Y.-P., F.-F. Jia, X.-M. Zhang, Y.-X. Qiao, K. Shi, Y.-H. Zhou & J.-Q. Yu (2012) Temperature effects on the reactive oxygen species formation and antioxidant defence in roots of two cucurbit species with contrasting root zone temperature optima. *Acta Physiologiae Plantarum*, 34(2), 713–720.

Zhao, G.-l. & J. Kong (2009) Enzymolysis bean dregs biscuit development. *Food Research and Development*, 10, 67–69.

Zhong-Hua, L., G. Hong-Lian, L. Rui-Ling & Z. Jin-Hui (2015) Extraction and antioxidant activity of soybean saponins from Lowtemperature soybean meal by MTEH. *The Open Biotechnology Journal*, 9(1).

Zuo, F., X. Peng, X. Shi & S. Guo (2016) Effects of high-temperature pressure cooking and traditional cooking on soymilk: Protein particles formation and sensory quality. *Food Chemistry*, 209, 50–56.

19 Impact of Drying on Isoflavones in Soybean

Chalida Niamnuy and Sakamon Devahastin

CONTENTS

19.1 INTRODUCTION

Soybean (*Glycine max* [L.] Merrill) is one of the economic crops of several countries. It has been appreciated by consumers as a rich edible oil source and a health-promoting food. There are various important nutrients in soybean, viz. protein, oil, carbohydrate, and several phytonutrients, such as isoflavones, phytic acid, anthocyanin pigments, saponin, lecithin, etc. A large number of researchers have reported the human health benefits of isoflavones (Klein et al., 1995) such as their weak estrogenic property, antioxidant ability, and the ability to reduce the risk of cardiovascular, atherosclerotic, hemolytic, and carcinogenic diseases; additionally, isoflavones assist with the alleviation of osteoporosis, menopausal and blood-cholesterol related symptoms, and with the inhibition of the growth of hormone-related human breast cancer and prostate cancer cell lines (Scambia et al., 2000; Zhang et al., 2003; Kwak et al., 2007; Lee et al., 2008). Furthermore, their ability as an α-glucosidase inhibitor led to the application of isoflavones for type 2 diabetes mellitus treatment, through the lowering of the blood glucose level (Choi et al., 2008).

The distribution of isoflavones in soybean can be altered during growing and various treatment and processing steps after harvest, including storage, fermentation, cooking, frying, roasting, extraction, and drying (Lee & Lee, 2009). The conversion and loss of isoflavones during processing significantly affect the nutraceutical values of soybean. Several works reported the effect of processing on changes of isoflavones and their functional properties (Chen et al., 2007; Wardhani et al.,

2008). It was found that the effect of food processes on the distribution of isoflavones, in processed soybean, depends on various parameters such as soybean derivative, treatment method, process temperature, and processing time. In addition, the number of total isoflavones is found to decrease upon processing in all cases, due to thermal degradation and oxidation reactions during processing (Chien, Hsieh, Kao, & Chen, 2005).

Drying is an important process to extend the shelf-life of or to prepare food, including soybean, for subsequent production. Firstly, the study of soybean drying is based on designing the drying process such that the moisture content of soybean is reduced to a safe level; it is for preventing the spoilage and mycotoxins that occur from the microorganism created during storage. The recommended moisture content for soybeans is 11–13% (wet basis), which corresponds to A_w lower than 0.8 to limit mold growth and deterioration during storage (Atungulu & Olatunde, 2018). However, the soybean is a rich source of nutrients; several researchers, therefore, studied the effect of drying conditions that significantly affect the various qualities of soybeans; for example, Sangkram and Noomhorm (2002) reported a comparative study of soybean drying between one-stage and two-stage (high-low temperature) drying effects on soybean cracking and qualities of extracted crude oil and lecithin. Prachayawarakorn et al. (2006) studied the drying characteristics and inactivation of urease in soybean during hot-air and superheated-steam fluidized bed drying. Agrahar-Murugkar and Jha (2010) reported the effect of drying on protein, trypsin inhibitor, and nitrogen solubility of soy flour from sprouted soybean. In the last few decades, several works reported the health benefits of bioactive compounds in soybean, especially isoflavones. There was also some research regarding the effect of processing conditions on the retention and profile of isoflavones in soy products. In addition, the aglycones form of isoflavones is promoted as per requirement; it is because of their high bioaccessibility and unique property of α-glucosidase inhibition. Drying is among the important processes that are applied to soybean and are known to significantly affect soy isoflavones distribution.

19.2   DRYING PRINCIPLES

Drying is successfully applied to various biomaterials. The objective of drying is to evaporate from a wet surface of solid particles into a surrounding stream of drying medium. Hence, the amount of heat required in the drying process is equal to the latent heat of evaporation, plus the heat necessary to bring the water and materials to the temperature of the drying medium or to the boiling point of water at any pressure of process. Consequently, two transfer phenomena are involved, namely, heat transfer from the drying medium, which is provided in the form of the sensible heat of gas to the solid materials and the mass transfer of water in the solid particles to the drying medium. Transport of heat can be mainly facilitated by convection from drying medium to surface of materials and conduction from surface to inside of materials. Transport of water can be accommodated by liquid diffusion, vapor diffusion, Knudsen diffusion, surface diffusion, capillary action, etc. Removal of water from solids depends on liquid evaporation rate at the material's surface or liquid diffusion from the material's interior to the surface, which depends on differences of materials and drying periods.

A typical drying kinetic was investigated from a batch of wet materials exposed to a hot air/drying medium of fixed temperature, humidity, and flow rate. The drying process starts with a short period of initial or start-up transient period, where a short period of time is needed for the materials to reach the desired temperature. In this period, part of the heat supplied by the heated gas is absorbed by the solid particles, which in turn increases their temperature. After an initial transient period, if the materials have high moisture content, it can be seen that the moisture content of materials decreases rapidly; this is called the constant rate period. The drying rate in this period is controlled by the water evaporation rate at the material's surface into the drying medium. The evaporation rate of moisture can be increased by raising the drying medium temperature or flow rate or reducing the drying medium humidity. The reduction rate of water continues unchanged as long as the film of water covers the particle surface fully. During this period, the temperature of the material surface remains constant at the wet-bulb temperature of the air. Once the particle surfaces become partially dried, the rate of drying is reduced. The drying rate in this stage is controlled by the rate of internal water diffusion to the surface prior to vaporization. The drying characteristic in this stage is called the fall rate period. The moisture content of particles at which the drying changes from the constant rate period to the falling rate period is called the critical moisture content. This critical value is a function of the drying conditions for a material. If the drying is continued for long enough, the surface film disappears, and the surface attains its equilibrium moisture content, corresponding to the drying condition it is exposed to. Drying of some products such as grains often displays a falling rate period which consists of two sections: First and second falling rate periods. The second critical moisture content in the drying curve can be marked at the boundary between the first and second falling rate periods (Mujumdar, 2015). There are several operating parameters that mainly affect the drying rate of materials, which directly relate to the changes and degradation of bioactive compounds as well as other qualities of bio-products. Drying technique is an important factor; it affects the drying kinetics and quality of products as well as the content of bioactive compounds in the products. Some advanced drying techniques used for soybean and soy products, for example, are presented in the next section.

19.3   SOME DRYING TECHNIQUES APPLIED FOR SOYBEAN AND SOY PRODUCTS

19.3.1   BIN DRYING

Bin drying is widely used for agricultural products after harvest. Bin dryers are large insulated containers. They may be several meters in diameter and height. The materials are packed in the dryer bed, and high-temperature air (40–45° C) is then passed through a bed from the bin bottom at low velocities for a long drying time (typically > 36 h). The circulation of drying air is a very important parameter for drying efficiency. These dryers have a high capacity and low capital and operation costs. Therefore, a bin dryer is a conventional dryer used in the large-scale drying of grains (Kudra & Mujumdar, 2009).

19.3.2   VACUUM DRYERS

Vacuum drying occurs with the reduction of the pressure of the system to be lower than the atmospheric pressure. Therefore, the boiling point of water is lowered below 100° C at normal atmospheric pressure. Vacuum drying is suitable for high-value materials and is sensitive to heat and oxygen because of the expensive operating cost. It is because of the reduced pressures that the transfer of heat depends on other modes than convection. Therefore, vacuum drying can be combined with other drying techniques, especially drying with radiation heat transfer (Kudra & Mujumdar, 2009; Fellows, 2009).

19.3.3   DRUM DRYING

Drum drying is one of the methods for the materials in slurry, puree, and liquid forms. Single- and twin-drum dryers are conventionally used. The material is applied on the outer surface of the drum as a thin layer. The drum (made of steel or stainless steel) is heated internally by steam. A uniform film with suitable thickness is required for the success of drum drying. The direct contact with the hot surface leads to a high rate of drying, and the advantages of a drum dryer are low cost and simple operation. Drum drying can be used under atmospheric or vacuum conditions (Kudra & Mujumdar, 2009; Fellows, 2009).

19.3.4   Freeze Drying

Freeze drying is a superior drying method, which can maintain the structure and shape of materials by the rapid removal of water inside the material. Freeze-drying involves a two-stage process of first freezing the water in the materials to be ice, followed by the application of heat to the product, which leads to ice that can be directly sublimed to vapor. Sublimation from ice to water vapor can occur at a temperature and pressure below the triple point of water. As in other forms of drying, freeze-drying represents coupled heat and mass transfer. The advantages of freeze-drying are low shrinkage and high retention of heat-sensitive compounds. However, the limitation of freeze drying is its very high capital and processing costs (van't Land, 1991; Kudra & Mujumdar, 2009).

19.3.5   Microwave Drying

Microwave drying is the technique where heat is generated inside the materials by the dielectric property of materials. The advantages of microwave drying are rapid and uniform heating and selective absorption of the radiation by liquid water. Therefore, the drying time can be shortened, which can preserve the heat-sensitive compounds in the materials. However, microwave drying exhibits a higher initial cost and operation cost compared to conventional hot air drying. Microwave drying can be combined with another drying technique, such as vacuum drying. Microwave vacuum drying can be applied for materials that are sensitive to heat and oxygen (Kudra & Mujumdar, 2009).

19.3.6   Impingement Drying

In the air-jet impingement dryer, hot jets are directed normally onto thin beds of materials. The jets pseudo-fluidize the bed to ensure good gas–solid contact for very high heat and mass transfer. Its advantages are very rapid drying and that it's a small unit. Impingement jet drying is suitable for removing unbound moisture, such as the moisture on the surface of a thin sheet or small pellet materials. Typically, the jet velocity and temperature were recommended to be in the range of 10–100 ms^{-1} and 100–350° C, respectively (Mujumdar, 2015).

19.3.7   Infrared Drying

Infrared drying consists of a radiation mode of heat transfer. Infrared radiation is thermal, which is classified as near-infrared, medium infrared, and far-infrared, with the wavelength of 0.75–3.00, 3.00–25.00, and 25–100 μm, respectively. Solid materials generally absorb infrared radiation in a very narrow layer near the surface. Therefore, infrared drying is suggested to be used for the drying of low-thickness or small-sized materials. The advantages of infrared drying are inexpensive capital and operating costs and high efficiency to convert electrical energy to heat for electrical infrared and uniform heating of products.

19.3.8   Superheated Steam Drying

Superheated steam is produced by heating up the saturated steam to get the temperature higher than the saturation temperature. The typical superheated steam dryer mainly consists of a steam boiler, superheater, and drying chamber. After heat and mass transfer between superheated steam and materials, the temperature of steam should be maintained to be higher than saturation temperature, to avoid the condensation of steam. Superheated steam drying is suggested to be applied for drying oxygen-sensitive materials (Law & Mujumdar, 2007).

19.3.9 FLUIDIZED BED DRYING

A conventional fluidized bed is formed by passing a drying medium stream from the bottom of the column, through a perforated gas distributor, across the bed of particles. It is uniformly distributed across the bed. The particles in the bed, hence, are thoroughly stirred during residence time. At the minimum fluidization velocity, the bed of particles is initially fluidized with the gas stream totally supporting the weight of the whole bed. The fluidized bed is usually operated at a superficial gas velocity that is normally 1–4 times the minimum fluidization velocity. During fluidization, heat from the gas stream transfers to the particles, and the moisture transfers from the particles to the gas stream at a higher rate than the tray drying. After passing through the fluidized bed, the gas stream is moved into gas-cleaning systems to separate the fine particles out from the gas stream before discharging it to the atmosphere or recycling. The advantages of fluidized bed drying are a high rate of moisture removal, high thermal efficiency, suitability for small- and large-scale operations, and low maintenance cost. However, the fluidization quality of fine and irregular-shaped particles is poor (Law & Mujumdar, 2007).

19.3.10 SPRAY DRYING

The spray drying method is essential for drying liquid food products and encapsulation of compounds. Spray drying by definition is the transformation of a feed from a fluid state into a dried form by spraying it into a hot, dry medium. Firstly, fluid is fed and sprayed from the atomizer; fluid is in the droplet form. Then droplets of fluid come into contact with the drying air in the chamber, so heat from the hot air is transferred to the droplets. After that, the saturated vapor, which is in the film form at the surface of droplets, evaporates quickly. Evaporation takes place in two stages. At the first stage, the moisture is suffused within the droplets and they then diffuse towards the droplet surface and replenish the content that was lost by evaporation at the surface during drying. This is called a constant rate period of drying. This stage mostly occurs by the influence of spray drying. When the moisture content in the droplets becomes too low to maintain saturated condition at the surface, it is known as the critical point, and a dried shell is formed at the surface of droplets. The evaporation rate is decreased and depends on the rate of moisture diffusion from inner droplets to the dried surface shell. The thickness of the dried shell increases as the drying time increases, and solid particles form. This is called the falling rate period of drying. When the dried powders have the desired moisture content, they are removed out of the spray dryer, from the bottom, and pneumatically conveyed to a cyclone, where they are separated with the air (Kudra & Mujumdar, 2009; Mujumdar, 2015).

19.4 THE CHANGES OF ISOFLAVONES DURING DRYING

Isoflavones in soybean are mainly found as β-glycosides (genistin, daidzin, glycitin), malonyl-β-glycosides (6''-O-malonyl-genistin, 6''-O-malonyl-daidzin, 6''-O-malonyl-glycitin), acetyl-β-glycosides (6''-O-acetylgenistin, 6''-O-acetyldaidzin, 6''-O-acetylglycitin), and aglycones (genistein, daidzein, glycitein). Malonyl-β-glycosides is the predominant form in conventional raw soybean (Lee & Lee, 2009), and they can be decarboxylated to produce acetyl-β-glycosides. Malonyl- and acetyl-β-glycosides can be transformed to β-glycosides via the de-esterification reaction. Malonyl-, acetyl-, and β-glycosides can be hydrolyzed to produce aglycones. These inter-conversion reactions of isoflavones can be accelerated by heat, acid, alkaline, and enzymes (Riedl et al., 2005; Vaidya et al., 2007) (see Figure 19.1). In addition, the degradation of all derivatives of isoflavones can take place. The degradation of isoflavones during drying is mainly due to thermal and oxidation degradation.

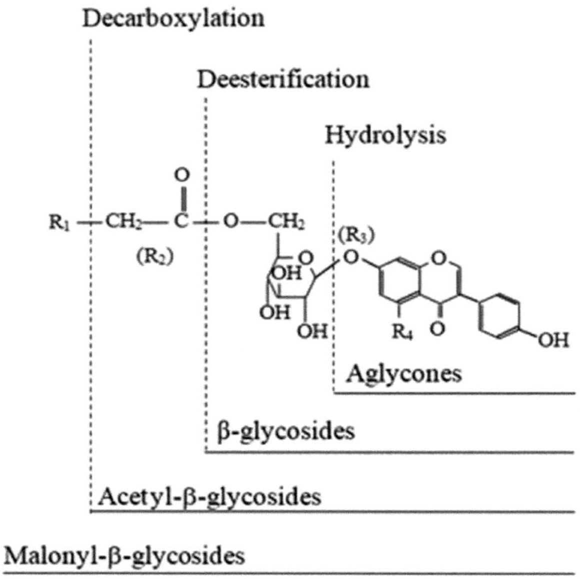

Decarboxylation

Deesterification

Hydrolysis

$$R_1 \!-\! CH_2 \!-\! \overset{\displaystyle O}{\overset{\displaystyle \|}{C}} \!-\! O \!-\! CH_2 \quad (R_3)$$

(R₂)

Aglycones

β-glycosides

Acetyl-β-glycosides

Malonyl-β-glycosides

Compound	R₁	R₂	R₃	R₄
1. Genistin	-	H	-	OH
2. Daidzin	-	H	-	H
3. Glycitin	-	H	-	CH₃
4. 6"-O-malonyl-genistin	COOH	-	-	OH
5. 6"-O-malonyl-daidzin	COOH	-	-	H
6. 6"-O-malonyl-glycitin	COOH	-	-	CH₃
7. 6"-O-acetylgenistin	H	-	-	OH
8. 6"-O-acetyldaidzin	H	-	-	H
9. 6"-O-acetylglycitin	H	-	-	CH₃
10. Genistein	-	-	OH	OH
11. Daidzein	-	-	OH	H
12. Glycitein	-	-	OH	CH₃

FIGURE 19.1 Chemical structures and inter-conversion reactions of isoflavones. Source: modified from Yuan et al. (2009).

The kinetics of inter-conversion and degradation of isoflavones have proved to follow the first-order reaction kinetics (Chein et al., 2005); the first-order kinetic model is then adopted in this work:

$$-\frac{dC_A}{dt} = kC_A \tag{19.1}$$

where C_A is the concentration of various isoflavones (µg/g dry solid) at any time t, k is the reaction rate constant (s⁻¹), t is the drying time (s). The temperature dependency of the reaction rate constant is again represented by the Arrhenius equation.

Figure 19.2 presents the pathways of all inter-conversion and degradation reactions of each isoflavone during drying; k_1 is the reaction rate constant for the decarboxylation of MGL to AGL; k_2 and

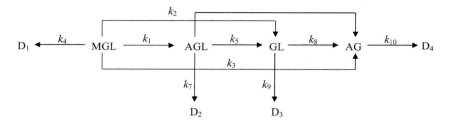

FIGURE 19.2 Typical inter-conversion and degradation reactions of isoflavones during heating. MGL, AGL, GL, and AG are malonyl-β-glycosides, acetyl-β-glycosides, β-glycosides, and aglycones, respectively. Source: modified from Niamnuy et al. (2012).

k_5 are the reaction rate constants for the de-esterification of MGL and AGL to GL, respectively; k_3, k_6, and k_8 are the reaction rate constants for the hydrolysis of MGL, AGL, and GL to AG, respectively; k_4, k_7, k_9, and k_{10} are the reaction rate constants for the degradation of MGL, AGL, GL, and AG, respectively. The reaction rate constants (k_1–k_{10}) at various drying temperatures were determined from the consecutive kinetic models shown below (Chien et al., 2005):

$$\frac{d[MGL]}{dt} = -K_1 [MGL]$$ (19.2)

where $K_1 = k_1 + k_2 + k_3 + k_4$

$$\frac{d[AGL]}{dt} = k_1 [MGL] - K_2 [AGL]$$ (19.3)

where $K_2 = k_5 + k_6 + k_7$

$$\frac{d[GL]}{dt} = k_5 [MGL] + k_2 [AGL] - K_3 [GL]$$ (19.4)

where $K_3 = k_8 + k_9$

$$\frac{d[AG]}{dt} = k_6 [AG] + k_8 [AG] + k_3 [AG] - k_{10} [AG]$$ (19.5)

The temperature dependency of the rate constant is represented by a well-known Arrhenius-type correlation:

$$k = k_0 \exp\left(\frac{-E_a}{RT_{abs}}\right)$$ (19.6)

where k_0 is the pre-exponential factor (m²/s), E_a is the activation energy of degradation/conversion (J/mol), T_{abs} is the absolute particle temperature (K), and R is the universal gas constant (8.314 J/mol•K).

19.5　APPLICATIONS OF DRYING ON ISOFLAVONES IN SOYBEAN AND SOY FOOD PRODUCTS

19.5.1　SOYBEAN AND SOY FOOD PRODUCTS

Concerned about the retention and distribution of isoflavones after drying, several researchers studied the effect of various advanced drying techniques on the isoflavone content in the soybean.

Niamnuy et al. (2011) studied the effects of moving-bed drying methods and temperature on the distribution of isoflavones and their bioactivity in dried soybean. Hot-air fluidized bed drying (HAFBD), superheated-steam fluidized bed drying (SSFBD), and gas-fired infrared combined with hot air vibrating drying (GFIR-HAVD) were carried out at various drying temperatures (50, 70, 130, and 150° C). The results showed that lower drying temperatures (50 and 70° C) did not significantly decrease total isoflavones, whereas higher drying temperatures (130 and 150° C) led to a significant decrease of total isoflavones in soybean. The results indicated that higher drying temperatures led to higher drying rates and higher amounts and levels of β-glycosides but to lower amounts of malonyl-β-glycosides, acetyl-β-glycosides, and total isoflavones. It indicated that dry heating at high temperatures accelerated the de-esterification and degradation more than hydrolysis of isoflavones in soybeans. The drying using GFIR-HAVD at 130° C resulted in the highest levels of aglycones and α-glucosidase inhibitory activity.

Beyond soybeans, there were some works that reported the study of the effect of drying of soy food products and byproducts on their isoflavones content. Muliterno et al. (2017) investigated the effect of drying temperature (50–70° C) of hot-air drying on the retention of isoflavones (6"-*O*-malonyl-glycosides, β-glycosides, and aglycones) in okara. Okara is an insoluble byproduct obtained from soymilk and tofu productions. It contains about 12–40% of raw soybean isoflavones. Therefore, okara can be value-added and applied as a nutritional supplement or as an ingredient of several foods. β-glycoside is the main isoflavone derivative in okara. It indicated that the de-esterification reaction mainly occurred during the soymilk and tofu processes. The most significant reduction in the isoflavones content was observed at 70° C. At 50° C, the conversion of β-glycosides to aglycones occurred during the first 6 h of drying; the loss of isoflavones can be observed in the final period of drying. It indicated that the hydrolysis reaction and the degradation were the main reactions that occurred during the drying of okara. In addition, the degradation of isoflavones in okara directly depended on drying temperature and drying time. Moreover, Weng and Chen (2012) studied the drying of fermented soybean (natto). Three drying methods (hot-air drying [50° C], vacuum drying [40° C, −0.97 kg$_f$/cm^2), and freeze drying (−50° C) were used in this work. Vacuum drying presented the highest isoflavones (daidzein and genistein) content in natto, followed by freeze drying and hot-air drying, respectively. The results suggested that a high rate of hydrolysis of β-glycosides into aglycones occurred at the drying temperatures of 50° C and higher. There are other works that studied the drying of soybean and okara on isoflavone content and related properties. It was concluded and shown in Table 19.1.

In addition, Niamnuy et al. (2012) investigated the kinetic of inter-conversion and degradation of all β-glycoside conjugates and the aglycone form of genistin during the drying of soybeans, using gas-fired infrared combined with hot-air vibrating drying (GFIR-HAVD) at the drying temperatures of 50, 70, 130, and 150° C. The results showed that a first-order kinetic model is able to be used to describe the changes of isoflavones during drying. It was found that the conversion of 6"-*O*-malonyl genistin and 6"-*O*-acetyl genistin to genistin via a de-esterification reaction was the main reaction with the highest rate constant. In addition, 6"-*O*-Malonylgenistin had the highest degradation rate constant, while genistein showed the lowest degradation rate constant. It indicated that the aglycones form had higher stability to heat than conjugated glycosides forms of genistin (see Table 19.1).

19.5.2 Soybean Extract

Recently, soybean extract has been promoted to be a supplement micronutrient and ingredient in cosmetics. However, isoflavones have low solubility because of their hydrophobic structure and are also easily lost during storage. Encapsulation can improve the solubility and protect the degradation of isoflavones in soybean extract. There are several works reporting on the encapsulation of soybean extract using drying. Yatsu et al. (2013) studied the effect of the encapsulation method to improve the solubility of isoflavones (daidzein, genistein, glycitein) in soybean extract. 2-ydroxypropyl-β-cyclodextrin (HPβCD) was used for inclusion encapsulation of soybean extract using several methods,

TABLE 19.1

Effect of Various Drying Conditions on Isoflavones in Soybean and Soy Food Products

Soy Product	Drying Method	Condition	Isoflavones	Results	Reference
Sprouts from dried seeds	DIC process	200° C, 2–10 bar	GE, DE	- Sprouts from DIC seeds showed a significantly higher GE and DE content than that obtained from seeds dried by other methods.	Plaza et al. (2003)
Seeds	Hot air drying	100° C, 0–120 min	GE, GI, AGI, MGI, DE, DI, ADI, MDI, GY, GYI, AGYI, MGYI	- Malonyl derivatives of isoflavones decreased and β-glycosides increased during drying. - MGI had a higher decreasing rate than MDI. - Total isoflavones in 120-min dried soybean were not significantly different compared to those of untreated sample.	Lee & Lee (2009)
Soy germ flours	Drum drying (DD) Freeze drying (FD)	130° C for DD –52° C for FD	GE, GI, DI, DE, GY, GYI	- FD showed a higher aglycones content in samples than DD. - Total isoflavones in samples in cases of FD and DD were comparable but higher than those in case of untreated sample.	Tipkanon et al. (2011)
Seeds	Hot air drying	30–110° C	GE, GI, MGI, DE, DI, MDI, GY, GYI, MGYI	- Drying temperature significantly affected the content of several isoflavones.	Ferreira et al. (2019)
Okara	Air jet impingement drying	50–70° C, air velocity 1.3–2.3 m/s, loading density 3–5 kg/m²	GE	- All independent variables exhibited significant quadratic effect on isoflavone content.	Wang et al. (2016)
Okara	Hot air drying	80, 200° C	GE, GI, DE, DI	- The dried okara at 200° C exhibited higher isoflavones content compared to dried okara at 80° C and fresh okara.	Voss et al. (2018)
Okara	Rotating-pulsed fluidized bed drying	50–90° C, 7.5–24.5 Hz	GE, GI, MGI, DE, DI, MDI, GY, GYI, MGYI	- Malonyl derivatives and β-glycosides of isoflavones decreased but aglycones increased after drying.	Lazarin et al. (2020)
Okara flour	Hot air drying (HAD) Freeze drying (FD) Microwave drying (MD)	70° C for HAD 800 W for MD	MGI, GL, AG	- HAD showed the highest, but FD exhibited the lowest conversion of MG and GL to AG. - The drying technique did not significantly affect the total isoflavones.	Guimaraes et al. (2020)
Fermented okara	Hot air drying (HAD) Freeze drying (FD) Infrared freeze drying (IFD) Microwave vacuum drying (MVD)	60° C for HAD –36° C, 40 Pa vacuum for FD and IFD –60° C, 10 kPa for MVD	GE	- MVD showed the highest but HAD presented the lowest isoflavones content. - Isoflavone in sample dried by FD was not significantly different compared to that dried by IFD.	Shi et al. (2021)

Abbreviations: MGI = 6″-O-malonyl-genistin; AGI = 6″-O-acetyl-genistin; GI = genistin; GE = genidtein; MDI = 6″-O-malonyl-daidzin; ADI = 6″-O-acetyl-daidzin; DI = daidzin; DE = daidzein; MGYI = 6″-O-malonyl-glycitin; AGYI = 6″-O-acetyl-glycitin; GYI = glycitin; GY = glycitein; MGL = malonyl-β-glycosides; GL = β-glycosides; AG = aglycones.

namely, spray drying, freeze drying, kneading/microwave, microwave, and co-evaporation. The different drying methods led to the different sizes and shapes of dried particles. The spray-dried particles (Figure 19.3a) showed a smooth spherical shape with a smaller size compared to the particles treated using other methods. The freeze-dried particles (Figure 19.3b) were in the form of an amorphous solid, while other complexes (Figure 19.3c–e) had a shard shape and broken edges because of crushing. To monitor the interaction between isoflavones and HPβCD as well as crystallinity of complexes, XRD analysis was reported (Figure 19.4 The XRD results illustrated that isoflavones presented diffraction peaks, corresponding to crystalline structure, while HPβCD showed board peaks, suggested as amorphous behavior. The physical mixture exhibited diffraction peaks of isoflavones, while the kneading/microwaved, microwaved, and co-evaporated complexes showed lower diffraction peaks. The results presented that these complexes are less crystalline than the mixture. The freeze-dried and spray-dried complexes exhibited the disappearance of sharp peaks. The results suggested that the inclusion complex of isoflavones and HPβCD is formed. The formation of inclusion complex by internal association (the insertion of B-ring of isoflavones into the cyclodextrin cavity) and also externally association was suggested. The increase of amorphous structure can enhance the solubility of the complex. However, the freeze-drying method showed lower isoflavones content and solubility of the complex than the spray-drying method. This could be due to the use of ethanol as a co-solvent in spray drying. The hydrophobic interaction between isoflavones and HPβCD facilitated by the use of alcoholic solvent can increase the encapsulation efficiency and solubility of the complex. In addition, the larger surface area of spray-dried particles can increase the solubility of the complex.

Spray drying is well known to be suitable for the encapsulation of bioactive compounds. Some researchers subsequently reported more about the study of the effect of co-solvent in spray drying

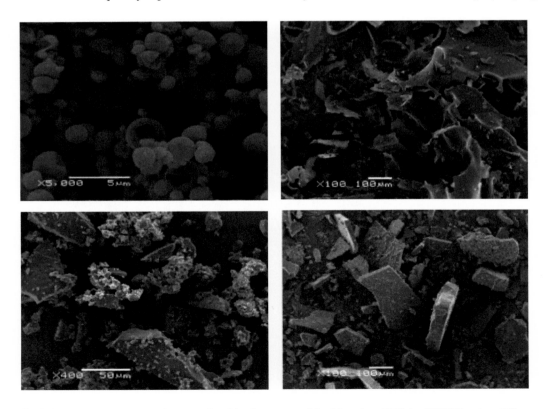

FIGURE 19.3 Morphology of encapsulated isoflavones enriched fraction (IEF) by HPβCD using different methods (a) spray drying, (b) freeze-drying, (c) kneading/microwave, (d) microwave, and (e) co-evaporation. Source: Yatsu et al. (2013).

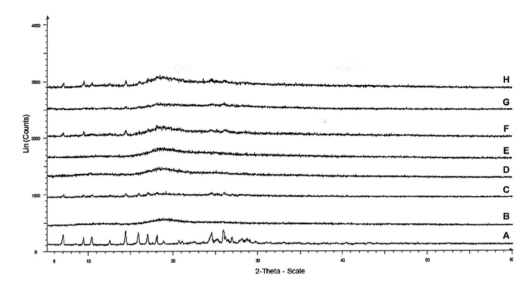

FIGURE 19.4 XRD of (A) isoflavones enriched fraction (IEF), (B) HPβCD, (C) physical mixture of IEF and HPβCD (IEF/ HPβCD), (D) freeze dried IEF/ HPβCD complex, (E) Spray dried IEF/ HPβCD complex, (F) microwaved IEF/ HPβCD complex, (G) kneading/microwaved IEF/ HPβCD complex, and (H) co-evaporated IEF/ HPβCD complex. Source: Yatsu et al. (2013).

on several properties of isoflavones in the soybean extract. Sansone et al. (2013) prepared the microparticles of sodium carboxymethyl cellulose (Na-CMC) loaded with soybean extract, enriched in daidzein and genistein. Different liquid feeds, based on H₂O, ethanol, and acetone, were used as co-solvents. It was found that encapsulation can improve the isoflavone release and dissolution rate. The type and ratio of co-solvent significantly impacted the amorphous physical state, related to the solubility of microparticles. The co-solvent of H₂O and acetone, with the ratio of 50:50, showed the highest solubility and dissolution rate of microparticles. In addition, the presence of Na-CMC enhanced powder wettability led to an increase in the permeation rate of isoflavones through bio-membrane for *in vitro* skin permeability measurement. The solvent of pure H₂O showed the highest loading efficiency of isoflavones, followed by the co-solvent of H₂O, ethanol, and acetone with the ratio of 50:15:35, while the co-solvent of H₂O and ethanol, with the ratio of 50:50, presented the lowest loading efficiency. According to the report of Yatsu et al. (2013) and Sansone et al. (2013), it was indicated that the interaction between type of co-solvent and type of carrier impacted the encapsulation efficiency of isoflavones in soybean extract.

Del Gaudio et al. (2017) studied the encapsulation of soybean extract using nanospray drying. Nanospray drying can provide a nano-sized particle, which can modify the releasing rate of bioactive compounds. In this work, hypromellose acetate succinate (HPMCAS) was used as the carrier to encapsulate the isoflavones (daidzein and genistein) in soybean extract. The ratio of soybean extract and HPMCAS (5:1, 3:1, 2:1, 1:1, 1:2, 1:3) and feed concentration (0.15–0.50%, w/v) in a hydro-alcoholic system were tested. It was found that the encapsulation efficiencies of both isoflavones were in the range of 62–86%, depending on the ratio of soybean extract and HPMCAS and feed concentration. In addition, the presence of HPMCAS improved the amorphization of isoflavones during the drying process and avoided recrystallization during storage. The nanoencapsulation can increase isoflavones penetration through the membrane, related to isoflavones release on the human skin, up to ten-fold compared to raw soybean extract. Niamnuy et al. (2019) reported on the encapsulation of isoflavones in soybean extract (SE) using spray drying. The studied parameters were the type of carrier (gum arabic [GA], maltodextrin [MD], β-cyclodextrin [CD]), inlet air temperature (130–170° C), and storage time (0–6 months). β-glycoside was the major isoflavone in the soybean extract; it indicated that the major conversion of isoflavones during the alcoholic extraction was de-esterification. The hydrolysis reaction for the transformation of all β-glucoside conjugating into aglycone forms was a major inter-conversion of isoflavones for the encapsulation process in all cases. The type of wall material had a more significant impact on the properties of microencapsulated soybean extract than the inlet air temperature in the studied ranges. The degradation of isoflavones during storage increased with an increase of inlet air temperature during encapsulation by spray drying. The use of β-cyclodextrin as a carrier presented the highest preservation of total isoflavones after encapsulation and during storage. To investigate the interaction between isoflavones in soybean extract and carrier, FTIR analysis is used (Figure 19.5). The FTIR spectra showed a peak at 1,606 cm^{-1} for C=C stretching of the aromatic rings. The peak at 1,517 cm^{-1} is for the C=O stretching and CCC and CC=O bending. The peak at 1,138 cm^{-1} corresponds to C–O vibrations of functional groups. These spectra were presented in the case of soybean extract. The results showed that these spectra for all microencapsulated soybean extract decreased, decreased, and increased compared to that of soybean extract itself, in case of the peak at 1,606, 1,517, and 1,138 cm^{-1}, respectively. This means that there are interactions between soybean extract and carriers. The use of β-cyclodextrin as a carrier showed the largest changes of these peaks, indicating the highest interaction during encapsulation compared to other carriers. There are other works that studied the encapsulation of isoflavones in soybean extract using several drying methods; they are shown in Table 19.2.

Beyond the soybean extract, the extracts of other soy products were investigated. Georgetti et al. (2013) studied the encapsulation of genistein in fermented soybean extracts using spray drying. The soybean was ground, fermented, and extracted in 80% aqueous methanol. The extract was spray-dried with different inlet air temperatures (115 and 150° C). The colloidal silicon dioxide (tixosil 333) at the concentration of 45% and 70% was used as the carrier. The results showed that the encapsulated extract at 150° C and 70% concentration of carrier showed higher genistein content than at 115° C and 45% concentration of carrier. It implied that the concentration of carrier had a large

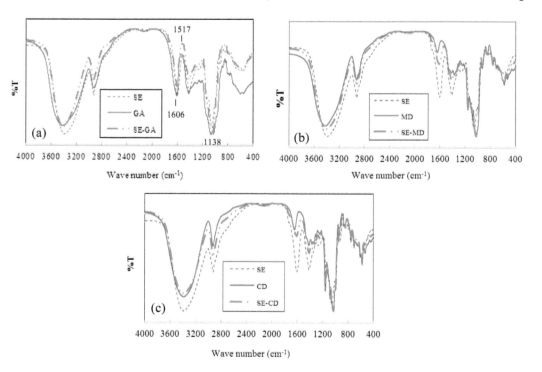

FIGURE 19.5 FTIR of encapsulated soybean extract (a) SE-GA, (b) SE-MD, and (c) SE-βCD. Source: Niamnuy et al. (2019).

TABLE 19.2

Effect of Various Drying Conditions on Isoflavones in Soybean Extract

Drying Method	Carrier	Condition	Isoflavones	Results	Reference
Spray drying (SD) Vacuum drying (VD) Freeze drying (FD)	Sodium alginate Maltodextrin gelatin H-γ-PGA(Na) L-γ-PGA(Na)	Inlet air: 150–180° C, feed rate: 5–25% for SD 60° C for VD –40° C, 60 mtorr	DE, DI, ADI, MDI, GE, GI, AGI, MGI, GY, GYI, AGYI, MGYI	- FD showed the highest content of isoflavones in dried SE, followed by VD and SD, respectively. - Co-carriers of H-γ-PGA(Na) and sodium alginate showed the higher isoflavones content than that of pure sodium alginate.	Kao & Chen (2007)
Spray drying	Colloidal silicon dioxide (tixosil 333)	Inlet air: 80, 115, 150° C	GE	- Inlet air temperature showed significant effect on the GE contents of dried SE.	Georgetti et al. (2008)
Spray drying	-	Inlet air: 180° C Outlet air: 80° C Storage: 12 months	DE, GE	- Dried SE showed stable isoflavones for 12 months, whereas those in extract form were not stable.	Sobharaksha et al. (2011)
Spray drying	Gelatin	Inlet air: 160° C Outlet air: 90° C SE/gelatin: 1:2, 1:3	DE, DI, GE, GI, GY, GYI	- *In vitro* release and solubility of isoflavones in encapsulated SE was higher than those in extract form.	Panizzon et al. (2014)
Spray drying	Milk Gum acacia Maltodextrin (10%)	Inlet air: 175° C Outlet air: 80–85° C Carrier concentration: 4–8% Storage: 30 days	GE	- Combination of co-carrier of maltodextrin and gum acacia showed the highest encapsulation efficiency of isoflavones in SE. - Encapsulation of SE with co-carrier of 10% maltodextrin and 6% gum acacia showed the highest stability of isoflavones during storage.	Mazumder & Ranganathan (2020)

Abbreviations: MGI = 6″-*O*-malonyl-genistin; AGI = 6″-*O*-acetyl-genistin; GI = genistin; GE = genidtein; MDI = 6″-*O*-malonyl-daidzin; ADI = 6″-*O*-acetyl-daidzin; DI = daidzin; DE = daidzein; MGYI = 6″-*O*-malonyl-glycitin; vAGYI = 6″-*O*-acetyl-glycitin; GYI = glycitin; GY = glycitein; SE = soybean extract.

effect on the encapsulation yield of genistein during spray drying. In addition, the result reported that genistein was less sensitive to heat for degradation than total phenolic compounds during spray drying. Batista et al. (2020) study the encapsulation of isoflavones extract from soybean molasses, using spray drying. Soybean molasses was a dark brown waste syrup from the production of soybean protein concentrate. Maltodextrin, modified starch, and their co-mixture were used as carriers at a concentration of 18%, while the inlet air temperature of 120–140° C was tested. The results showed that the type and ratio of carriers as well as inlet air temperature significantly affected the encapsulation efficiency of isoflavones (daidzein, genistein, and total isoflavones) in the extract. The encapsulation with the use of modified starch as a carrier and inlet air temperature of 130° C showed the highest encapsulation efficiency of isoflavones.

Although isoflavones in soybean extract are currently traded in the pharmaceutical market as a natural dietary supplement, the improvement of their water solubility and bioavailability is still required. Among several reported encapsulation by drying methods used to protect the degradation and enhance the dissolution rate of isoflavones, the innovative drying technique as superheated steam spray drying is very interesting. A schematic diagram of the superheated steam spray dryer setup is shown in Figure 19.6. Saturated steam from a boiler was passed to several supported

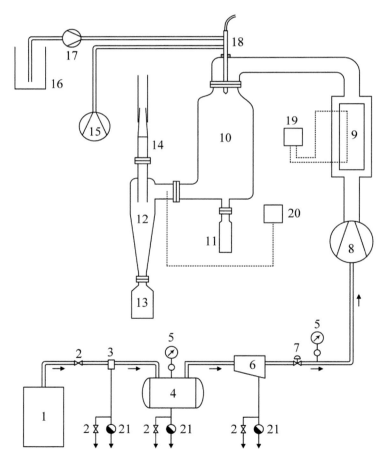

FIGURE 19.6 A schematic diagram of superheated steam spray dryer and associated units. (1) boiler; (2) globe valves; (3) steam pocket; (4) steam header; (5) pressure gauges; (6) steam separator; (7) pressure reducing valve; (8) blower; (9) supplementary electric heater; (10) drying chamber; (11) wide receiving bottle; (12) cyclone; (13) main receiving bottle; (14) exhausted air pipe; (15) air compressor; (16) feed tank; (17) peristaltic pump; (18) two-fluid nozzle atomizer; (19) thermocouple; (20) PID controller; (21) steam trap.

units, namely, steam pocket, steam header, and steam separator for removing excess moisture. An electric heater is used to heat up the steam to be superheated steam. After that, the superheated steam is passed into the drying chamber. A peristaltic pump is used to feed the liquid sample to an atomizer. The droplets of the sample are dried as a powder before falling to the bottom of the chamber. The powder is pneumatically passed to a cyclone for the separation of dried particles from the superheated steam before being collected in the receiving bottle (Fuengfoo et al., 2018). In addition, a venturi nozzle is suggested to be installed between the drying chamber and cyclone for better separation of powder (Linke et al., 2021). Lum et al. (2018) reported that in the first period of spray drying using superheated steam, the steam condensed and presented as the water film on the semi-dried droplet of the sample. The hydrophilicity of the water film promoted the migration of hydrophilic components to the surface and drove hydrophobic components into the core of the droplet. This phenomenon resulted in greater hydrophilicity and wettability over the surface of dried particles. Therefore, the superheated steam spray drying is expected to enhance the encapsulation efficiency, solubility, and bioavailability of isoflavones. In addition, drying with superheated steam can prevent the loss of isoflavones due to oxidation reaction, since the oxygen in the process is limited.

19.6 CONCLUSION

Drying is an important process; it influences the changes of bioactive compounds in plants and plant-based food products. Isoflavones are an important bioactive compound in soybean. Several works confirm that the drying technique and drying conditions impact the inter-conversion/degradation of isoflavones in dried soybean, dried soy products, and soybean extract. In addition, the types of carrier and co-solvent influence the distribution, encapsulation efficiency, solubility, and stability of isoflavones in soybean extract. However, few studies regarding the drying of the isoflavones content in soybean products have been reported in recent years. Therefore, there is very limited information on the correlations among drying conditions, isoflavones distribution, physicochemical properties, and the bioactivities to human health of soybean, for both *in vitro* and *in vivo* studies. In addition, several industrial drying techniques and other kinds of soybean products should be considered for further study.

REFERENCES

Agrahar-Murugkar, D., & Jha, K. (2010). Effect of drying on nutritional and functional quality and electrophoretic pattern of soyflour from sprouted soybean (*Glycine max*). *Journal of Food Science and Technology, 47*(5), 482–487.

Atungulu, G.G., & Olatunde, G.A. (2018). Assessment of new in-bin drying and storage technology for soybean seed. *Drying Technology, 36*(4), 383–399.

Batista, V.S.F., Nunes, G.L., Viegas, G.I., Lucas, B.N., Bochi, V.C., Emanuelli, T., Barin, J.S., de Menezes, C.R., & de Menezes, C.R. (2020). Extraction, characterization and microencapsulation of isoflavones from soybean molasses. *Ciência Rural, 50*(3), e20190341.

Chen, J., Cheng, Y.-Q., Yamaki, K., & Li, L.-T. (2007). Anti-α-glucosidase activity of Chinese traditionally fermented soybean (douchi). *Food Chemistry, 103*(4), 1091–1096.

Chien, J.T., Hsieh, H.C., Kao, T.H., & Chen, B.-H. (2005). Kinetic model for studying the conversion and degradation of isoflavones during heating. *Food Chemistry, 91*(3), 425–434.

Choi, M.S., Jung, U.J., Yeo, J., Kim, M.J., & Lee, M.K. (2008). Genistein and daidzein prevent diabetes on set by elevating insulin level and altering hepatic gluconeogenic and lipogenic enzyme activities in non-obese diabetic (NOD) mice. *Diabetes/Metabolism Research and Reviews, 24*(1), 74–81.

Del Gaudio, P., Sansone, F., Mencherini, T., De Cicco, F., Russo, P., & Aquino, R.P. (2017). Nanospray drying as a novel tool to improve technological properties of soy isoflavone extracts. *Planta Medica, 83*(5), 426–433.

Fellows, P.J. (2009). *Food Processing Technology: Principles and Practice (Woodhead Publishing in Food Science, Technology and Nutrition)*, 3rd edn. CRC Press, New York.

Ferreira, C.D., Ziegler, V., Goebel, J.T.S., Hoffmann, J.F., Carvalho, I.R., Chaves, F.C., & de Oliveira, M. (2019). Changes in phenolic acid and isoflavone contents during soybean drying and storage. *Journal of Agricultural and Food Chemistry*, *67*(4), 1146–1155.

Fuengfoo, M., Devashastin, S., Niamnuy, C., & Soponronnarit, S. (2018). Preliminary study of superheated steam spray drying: A case study with maltodextrin. In *IDS 21st International Drying Symposium*, 1147–1154.

Georgetti, S.R., Casagrande, R., Souza, C.R.F., Oliveira, W.P., & Fonseca, M.J.V. (2008). Spray drying of the soybean: Effects on chemical properties and antioxidant activity. *LWT - Food Science and Technology*, *41*(8), 1521–1527.

Georgetti, S.R., Casagrande, R., Vicentini, F.T.M.C., Baracat, M.M., Verri Jr, W.A., & Fonseca, M.J.V. (2013). Protective effect of fermented soybean dried extracts against TPA-induced oxidative stress in hairless mice skin. *BioMed Research International*, *2013*, 340626.

Guimarães, R.M., Ida, E.I., Falcão, G., de Rezende, T.A.M., Silva, J.S., Alves, C.C.F., da Silva, M.A.P., & Egea, M.B. (2020). Evaluating technological quality of okara flours obtained by different drying processes. *LWT – Food Science and Technology*, *123*, 109062.

Kao, T.H., & Chen, B.H. (2007). Effects of different carriers on the production of isoflavone powder from soybean cake. *Molecules*, *12*(4), 917–931.

Klein, B.P., Perry, A.K., & Adair, N. (1995). Incorporating soy proteins into baked products for use in clinical studies. *American Journal of Clinical Nutrition*, *125* Supplement, 666–674.

Kwak, C.S., Lee, M.S., & Park, S.C. (2007). Higher antioxidant properties of chungkookjang, a fermented soybean paste, may be due to increased aglycone and malonylglycoside isoflavone during fermentation. *Nutrition Research*, *27*(11), 719–727.

Kudra, T., & Mujumdar, A.S. (2009). *Advanced Drying Technologies*, 2nd edn. CRC Press, New York.

Law, C.L., & Mjumdar, A.S. (2007). *Hand Book of Industrial Drying*, 3rd edn. CRC Press, New York.

Lazarin, R.A., Falcão, H.G., Ida, E.I., Berteli, M.N., & Kurozawa, L. (2020). Rotating-pulsed fluidized bed drying of okara: Evaluation of process kinetic and nutritive properties of dried product. *Food and Bioprocess Technology*, *13*(9), 1611–1620.

Lee, S.J., Kim, J.J., Moon, H.I., Ahn, J.K., Chun, S.C., Jung, W.S., Lee, O.K., & Chung, I.-M. (2008). Analysis of isoflavones and phenolic compounds in Korean soybean [*Glycine max* (L.) Merrill] seeds of different seed weights. *Journal of Agricultural and Food Chemistry*, *56*(8), 2751–2758.

Lee, S.W., & Lee, J.H. (2009). Effects of oven-drying, roasting, and explosive puffing process on isoflavone distribution in soybeans. *Food Chemistry*, *112*(2), 316–320.

Linke, T., Happe, J., & Kohlus, R. (2021). Laboratory-scale superheated steam spray drying of food and dairy products. *Drying Technology*, in press.

Lum, A., Mansouri, S., Hapgood, K., & Woo, M.W. (2018). Single droplet drying of milk in air and superheated steam: Particle formation and wettability. *Drying Technology*, *36*(15), 1802–1813.

Mazumder, M.A.R., & Ranganathan, T.V. (2020). Encapsulation of isoflavone with milk, maltodextrin and gum acacia improves its stability. *Current Research in Food Science*, *2*, 77–83.

Mujumdar, A.S. (2015). *Handbook of Industrial Drying*, 4th edn. CRC Press, New York.

Muliterno, M.M., Rodrigues, D., de Lima, F.S., Ida, E.I., & Kurozawa, L.E. (2017). Conversion/degradation of isoflavones and color alterations during the drying of okara. *LWT – Food Science and Technology*, *75*, 512–519.

Niamnuy, C., Nachaisin, M., Laohavanich, J., & Devahastin, S. (2011). Evaluation of bioactive compounds and bioactivities of soybean dried by different methods and conditions. *Food Chemistry*, *129*(3), 899–906.

Niamnuy, C., Nachaisin, M., Poomsa-ad, N., & Devahastin, S. (2012). Kinetic modelling of drying and conversion/degradation of isoflavones during infrared drying of soybean. *Food Chemistry*, *133*(3), 946–952.

Niamnuy, C., Poomkokrak, J., Dittanet, P., & Devahastin, S. (2019). Impacts of spray drying conditions on stability of isoflavones in microencapsulated soybean extract. *Drying Technology*, *37*(14), 1844–1862.

Panizzon, G.P., Bueno, F.G., Nakamura, T.U., Nakamura, C.V., & Filho, B.P.D. (2014). Preparation of spray-dried soy isoflavone-loaded gelatin microspheres for enhancement of dissolution: Formulation, characterization and *in vitro* evaluation. *Pharmaceutics*, *6*(4), 599–615.

Plaza, L., de Ancos, B., & Cano, M.P. (2003). Nutritional and health-related compounds in sprouts and seeds of soybean (*Glycine max*), wheat (*Triticum aestivum L.*) and alfalfa (*Medicago sativa*) treated by a new drying method. *European Food Research and Technology*, *216*(2), 138–144.

Prachayawarakorn, S., Prachayawasin, S., & Soponronnarit, S. (2006). Heating process of soybean using hot-air and superheated-steam fluidized-bed dryers. *LWT - Food Science and Technology*, *39*(7), 770–778.

Riedl, K.M., Zhang, Y.C., Schwartz, S.J., & Vodovotz, Y. (2005). Optimizing dough proofing conditions to enhance isoflavone aglycones in soy bread. *Journal of Agricultural and Food Chemistry*, *53*(21), 8253–5258.

Sangkram, U., & Noomhorm, A. (2002). The effect of drying and storage of soybean on the quality of bean, oil, and lecithin production. *Drying Technology*, *20*(10), 2041–2054.

Sansone, F., Picerno, P., Mencherini, T., Russo, P., Gasparri, F., Giannini, V., Lauro, M.R., Puglisi, G., & Aquino, R.P. (2013). Enhanced technological and permeation properties of a microencapsulated soy isoflavones extract. *Journal of Food Engineering*, *115*(3), 298–305.

Scambia, G., Mango, D., Signorile, P.G., Anselmi Angeli, R.A., Palena, C., Gallo, D., Bombardelli, E., Morazzoni, P., Riva, A., & Mancuso, S. (2000). Clinical effects of a standardized soy extract in post-menopausal women: A pilot study. *Menopause*, *7*(2), 105–111.

Shi, H., Zhang, M., Mujumdar, A.S., Xu, J., & Wang, W. (2021). Influence of drying methods on the drying kinetics, bioactive compounds and flavor of solid-state fermented okara. *Drying Technology*, *39*(5), 644–654.

Sobharaksha, P., Indranupakorn, R., & Luangtana-anan, M. (2011). Engineering of soybean extract powder: Identification, characterisation and stability evaluation. *Advanced in Materials Research*, *194–196*, 507–510.

Tipkanon, S., Chompreeda, P., Haruthaithanasan, V., Prinyawiwatkul, W., No, H.K., & Xu, Z. (2011). Isoflavone content in soy germ flours prepared from two drying methods. *International Journal of Food Science and Technology*, *46*(11), 2240–2247.

Vaidya, N.A., Mathias, K., Ismail, B., Hayes, K.D., & Corvalan, C.M. (2007). Kinetic modeling of malonyl-genistin and malonyldaidzin conversions under alkaline conditions and elevated temperatures. *Journal of Agricultural and Food Chemistry*, *55*(9), 3408–3413.

van't Land, C.M. (1991). *Industrial Drying Equipment: Selection and Application*. Marcel Dekker, New York.

Voss, G.B., Rodríguez-Alcalá, L.M., Valente, L.M.P., & Pintado, M.E. (2018). Impact of different thermal treatments and storage conditions on the stability of soybean byproduct (okara). *Journal of Food Measurement and Characterization*, *12*(3), 1981–1996.

Wang, G., Deng, Y., He, X., Zhao, Y., Zou, Y., Liu, Z., &, Yue, J. (2016). Optimization of air jet impingement drying of okara using response surface methodology. *Food Control*, *59*, 743–749.

Wardhani, D.H., Vazquez, J.A., & Pandiella, S.S. (2008). Kinetics of daidzin and genistin transformations and water absorption during soybean soaking at different temperatures. *Food Chemistry*, *111*(1), 13–19.

Weng, T.M., & Chen, M.T. (2012). Effect of drying methods on Γ-Pga, isoflavone contents and ace inhibitory activity of natto (a fermented soybean food). *Journal of Food Processing and Preservation*, *36*(6), 483–488.

Yatsu, F.K.J., Koester, L.S., Lula, I., Passos, J.J., Sinisterra, R., & Bassani, V.L. (2013). Multiple complexation of cyclodextrin with soy isoflavones present in an enriched fraction. *Carbohydrate Polymers*, *98*(1), 726–735.

Yuan, J.P., Liu, Y.B., Peng, J., Wang, J.H., & Liu, X. (2009). Changes of isoflavone profile in the hypocotyls and cotyledons of soybeans during dry heating and germination. *Journal of Agricultural and Food Chemistry*, *57*(19), 9002–9010.

Zhang, X., Shu, X.O., Gao, Y.T., Yang, G., Li, Q., Li, H., Jin, F., & Zheng, W. (2003). Soy food consumption is associated with lower risk of coronary heart disease in Chinese women. *Journal of Nutrition*, *133*(9), 2874–2878.

20 Soybean-Derived Bioactive Peptides and Their Health Benefits

*Md Minhajul Abedin, Loreni Chiring Phukon,
Rounak Chourasia, Swati Sharma,
Dinabandhu Sahoo, and Amit Kumar Rai*

CONTENTS

20.1 INTRODUCTION

Soybean (*Glycine max*) is one of the most acknowledged sources of plant protein, contributing to a wide range of health benefits (Sanjukta & Rai, 2016). For thousands of years, it has been cultivated in Asia. During the eighteenth century, it was introduced to the European continent and then to the United States in the nineteenth century. Soybean has been an important economic crop in the United States, which produces about 38% of the world's production (Tidke et al., 2015; Chatterjee et al., 2018). On average, an Asian consumes about 20 to 80 g daily of traditional soy foods as a staple diet, whereas an American consumes only about 1 to 3 g daily (Lule et al., 2015). Soybean contains basic nutritive constituents such as vitamins, lipids, free sugars, and minerals, and also contains flavonoids, isoflavones, saponins, and peptides having therapeutic values (Sanjukta & Rai, 2016; Jayachandran & Xu, 2019; Kumar et al., 2020).

Soybeans are exceptionally versatile and can be made into a variety of food products. Soybean is consumed in two forms, unfermented (soy powder, soybean oil, soy butter, roasted and fried soybean, etc.) and fermented (soy cheese, soy sauce, yogurt, pickle, etc.). Since ancient times, fermentation has been used to preserve perishable food materials especially when such foods were scarce. During the process of fermentation, complex organic compounds are broken down into smaller molecules by microbes that exert various physiological functions beyond their nutritional properties. Fermentation of soybean by different microorganisms enhances the biofunctional properties due to the increase in peptides and free isoflavones (Sanjukta et al., 2015; Sanjukta & Rai, 2016; Chourasia et al., 2021). Fermentation can also result in the decrease of antimicrobial components such as phytic acid, oxalic acid, urease, and proteinase-inhibitors (Egounlety & Aworh, 2003; Reddy & Pierson, 1994; Sharma et al., 2021). In many Asian countries, fermented soybean products are quite popular, such as *meju, doenjang, cheonggukjang, kanjang* (Korea), *miso, natto, tofuyo* (Japan), *thua-nao* (Thailand), *sufu, douchi, doubanjiang* (China), *tempeh* (Indonesia), *kinema, tungrymbai*, and *hawaijar* (India).

Microorganisms involved during soybean fermentation produces proteolytic enzymes that hydrolyze soy proteins resulting in the release of peptides. Bioactive peptides consist of protein fragments of amino acids, usually 2–20 residues, and exhibit numerous biological properties, such as antimicrobial, antioxidant, anti-inflammatory, antihypertensive, and immunomodulatory activity. These peptides are inactive within the parent proteins but are released upon fermentation, enzymatic hydrolysis, or during food processing (Liu et al., 2020). Peptides improve the functional properties of fermented foods and can substitute various synthetic drugs. In fermented soybean products, peptide-mediated therapeutic values differ with specific microbes involved in the fermentation process. Bioactive peptides that are present in the fermented soybean have been studied for different therapeutic properties such as antihypertensive, antioxidant, antidiabetic, anti-tumor, and anti-atherosclerosis.

This chapter gives an overview of soybean fermentation by various microbes, their potential bioactive peptides, and their health benefits.

20.2 SOY PROTEINS

Soybean is composed of about 35–40% protein, 20% lipids, 9% dietary fiber, and nearly 8.5% moisture based on the dry weight of mature raw seeds (He & Chen, 2013). Its composition varies with the location and climate of planting. The proteomic studies of soybean protein have shown the basic information about the major storage proteins present in soybean. Soybean proteins consist of two major globulins – 11S glycinin and 7S β-conglycinin – comprising about 40% and 30% of protein content respectively and acting as precursors for most of the isolated bioactive peptides (Sanjukta & Rai, 2016). Glycinin is composed of acidic (A) and basic (B) subunits of ~35,000 and ~20,000 Da molecular weight linked through a disulfide bond: A1aB2, A1bB1b, A2B1a, A3B4, and A5A4B3. Glycinin exhibits molecular heterogenicity due to polymorphism in the subunit, which is due to the differences in the amino acid sequences among the variety of the soybean produced (Tang, 2019). β-conglycinin is a trimer with a molecular weight of 150–200 kDa, belonging to the glycosylated protein class (Hou & Chang, 2004). It is composed of α (MW~68,000 Da), α' (~72,000 Da), β (~52,000 Da), and γ (~52,000 Da) subunits that are associated by hydrophobic interaction with each other (Wu et al., 2017; Shi et al., 2020). The 11S polypeptide has five genetic variants that are divided into two groups, group I and group II, based on the homology of its subunit sequences (Sanjukta & Rai, 2016). It has been reported that heterogeneity in the 11S fraction differs among soybean varieties and also within a single variety of soybean (Wang et al., 2014; Chen et al., 2018). Moreover, genomic studies of these protein globulins revealed that the genes of both 7S and 11S have been derived from a common ancestral gene (Sanjukta & Rai, 2016; Sanjuktaet al., 2017). Soy proteins having a different ratio of glycinin and β-conglycinin are believed to have different physiological and nutritional effects (Singh et al., 2014). The mean ratio of glycinin to β-conglycinin varies between 1.6 and 2.5 among different soybean varieties. This ratio influences the protein quality and functional properties of soybean food products (Chen et al., 2020). Soy proteins having varying subunit compositions have shown different functional properties concerning yield, quality, and texture in tofu production (Chatterjee et al., 2018). Since soybean protein content and composition vary among different varieties, the chances of peptide formation having different functional properties also increase. It has been reported that hydrolyzed soy proteins are absorbed faster and efficiently in humans (Maebuchi et al., 2007). Other minor protein globulins present in soybean include 2S, 9S, and 15S; lectins; lipoxygenases; α- amylases; and trypsin inhibitors (Sanjukta & Rai, 2016). Some potential bioactive peptides have also been reported in raw soybean (Gaspar & Castanho, 2016). Lunasin, a bioactive peptide composed of 43 amino acid residues, which has been isolated from different varieties of soybean, has the potential to exhibit anticancer, antioxidant, and anti-inflammatory properties (Lule et al., 2015; Lee et al., 2019; Ashaolu, 2020a).

20.3 METHODS FOR THE PRODUCTION OF SOYBEAN-DERIVED BIOACTIVE PEPTIDES

Bioactive peptides are encrypted in the amino acid sequences of food proteins, which generally contain 2–20 amino acids residues. These peptides are inactive functionally within the native protein and are released by proteolysis having specific bioactive functions (Figure 20.1). Bioactive peptides derived from soybean are small fragments of amino acids produced by microbial fermentation, gastrointestinal digestion, enzymatic hydrolysis, and food processing of larger soybean proteins which exert various beneficial metabolic effects (De Mejia & De Lumen, 2006; Singh et al., 2014; Lule et al., 2015; Chourasia et al., 2020a, 2020b). The production and composition of bioactive peptides by different methods are affected by the food processing, enzymes, or microorganisms used in fermentation.

20.3.1 MICROBIAL FERMENTATION

Traditionally, people of Asian countries like China, Korea, Japan, and India have been consuming fermented soybean products such as *natto*, *miso*, *tempeh*, and *kinema* for a long time (He & Chen, 2013). Fermentation is an efficient and cost-effective method for the production of bioactive peptides and food-grade hydrolyzed proteins through microbial activity (Singh et al., 2014; Iwaniak et

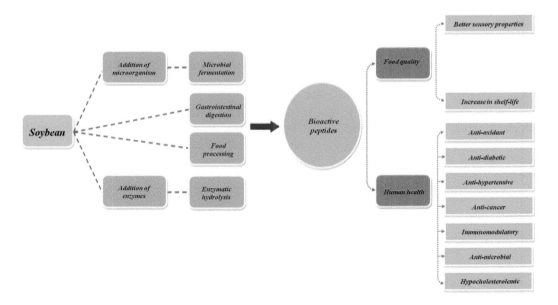

FIGURE 20.1 Production of bioactive peptides from soybean using microbial fermentation and enzymatic hydrolysis, its effect on food quality, and the beneficial effects on human health.

al., 2020). Soybean fermentation affects the texture, aroma, therapeutic, and nutraceutical values of the product. Lactic acid bacteria (LAB), *Bacillus* spp., and some fungi have been reported to be the key microbes in fermented soybean products (Sanjukta et al., 2017). Soybean fermentation using *Bacillus* spp. occurs in an alkaline state, whereas soymilk fermentation with LAB is an acidic fermentation (Sanjukta & Rai, 2016). In the North East region of India, fermented soybean products such as *hawaijar*, *bekang*, *tungrymbai*, and *kinema* are prepared traditionally and consumed. Among all the fermented soybean products, *hawaijar* and *kinema* are most studied. Studies have reported that *kinema* possesses high antioxidant properties due to the enhancement in peptides and polyphenols content during the process of fermentation (Sanjukta et al., 2015, 2021a; 2021b). Soymilk fermentation using LAB is also a source of bioactive peptides and free isoflavones. LAB such as *Lactobacillus acidophilus*, *Lb. delbrueckii*, *Lb. plantarum*, *Streptococcus thermophilus*, *Bifidobacterium longum*, *B. infantis*, and *B. breve* are used for soymilk fermentation, resulting in the production of peptides having health benefits (Sanjukta & Rai, 2016). Similarly, *Enterococcus faecium* strains used for soy milk fermentation have been shown to possess Angiotensin I converting enzyme (ACE) inhibitory and antioxidant properties (Martinez-Villaluenga et al., 2012).

Apart from bacterial starter culture, some filamentous fungi are also used in several fermented soybean products. Some popular fungi fermented soybean products include miso, tempeh, douchi, and tofu using *Aspergillus oryzae*; *Rhizopus* spp., and *Fusarium* spp.; *Mucor* and *Aspergillus*; and *Actinomucor* spp, *Mucor* spp., and *Rhizopus* spp. respectively (Wang et al., 2015; Naresh et al., 2019). These fermented products have been shown to possess ACEinhibitory, antioxidant, α-glucosidase inhibitory, antimicrobial, anticancer, antihypercholesterolaemic, and antithrombotic properties (Sanjukta & Rai, 2016; Rai et al., 2017a). In some fermented soybean products, both bacteria and fungi are used for the development of the final product, such as *Doenjang-meju*, a traditional Korean fermented soybean paste. *B. subtilis* and molds including *Rhizopus* spp., *Geotrichum* spp., *Mucor* spp., and *Aspergillus* spp. are used, and it has been reported that *B. subtilis* isolated from *Doenjang-meju* has shown anticancer surfactin properties (Lee et al., 2012). In another study, it was found that using both *B. subtilis* and *R. Oligosporus* for soybean fermentation has shown a higher degree of protein hydrolysis in comparison to single-starter culture (Weng & Chen, 2011).

20.3.2 GASTROINTESTINAL DIGESTION

Gastrointestinal digestion has an important role in increasing the biological activity of food-derived peptides, allowing the release of new and small active fragments, (González-Montoya et al., 2018). Upon ingestion of soybean food product and its digestion by acid and digestive enzymes from the stomach, such as trypsin and pepsin, and from the pancreas and small intestine, such as trypsin, pepsin, pancreatin, and chymotrypsin, small peptides are released. These are absorbed through the walls of the small intestine into the bloodstream and exert various physiological properties (De Angelis et al., 2017; Chatterjee et al., 2018). The extensive degradation soybean proteins undergo during simulated gastrointestinal digestion generates several peptides having biological properties, though the *in vivo* digestion process is a more complicated one in which several digestive enzymes take part (Capriotti et al., 2015).

20.3.3 ENZYMATIC HYDROLYSIS

Enzymatic hydrolysis is one of the most common and efficient processes for protein modifications. Enzymatic hydrolysis helps in the cleavage of proteins into a low molecular mass of amino acids and peptides and changes the conformational epitopes and linear epitopes of allergens by cleaving the peptides bonds (Huang et al., 2018). Additionally, enzymatic hydrolysis is mild, efficient, and the hydrolysis conditions can be controlled, generating bioactive peptides, which have beneficial effects on human health (Marciniak et al., 2018; Singh et al., 2019). Some parameters such as temperature and pH must be controlled and measured throughout digestion for optimal hydrolysis.

Also, hydrolysis duration is crucial as it is directly related to the degree of hydrolysis, influencing the size and composition of amino acids, and thus the bioactivities of the produced peptides (Sun et al., 2020). *In vitro* enzymatic hydrolysis is usually done for commercial purposes, having better quality control and use of more effective and stable enzymes in obtaining peptides with specific molecular weight and peptide fragments (Singh et al., 2014). *In vitro* enzymatic hydrolysis can utilize both specific and nonspecific proteases such as trypsin, pepsin, papain, chymotrypsin, and peptidase to obtain peptides from soy protein digestion under optimal pH and temperature (Chatterjee et al., 2018), as well as commercial proteases such as Flavourzyme™, Alcalase™, and Protamex™ (Boukil et al., 2018). Various combinations of proteinases including Alcalase™, pancreatin, pepsin, chymotrypsin, and Thermolysin™ as well as enzymes isolated from bacterial and fungal cultures have been used to produce bioactive peptides from various proteins (Zhang & Mu, 2017; Guan et al., 2018). Digestive enzymes are of important interest since they mimic the normal human digestion of proteins to assess the probable bioactive peptides released after the consumption of different matrices (Sun et al., 2020). Moreover, enzymatic hydrolysis is safe and provides promising methods for bioactive peptide production.

20.3.4 FOOD PROCESSING

Food processing can result in the production of bioactive peptides due to alteration in the structural and chemical composition of amino acid residues. Chemical treatments and pH modifications may lead to modification of amino acids composition, changing the functional properties (Singh et al., 2019). The enhancement in functionality can result in digestibility, peptide enrichment, and/or reduction of trypsin inhibitor acidity, arising from acylation, phosphorylation, glycosylation, succinylation, reductive alkylation, or lyophilization (Chatterjee et al., 2018). Moreover, food processing with the help of two or more microbial strains resulting in mixed culture fermentation has a potential synergistic effect to develop the physiological functions and organoleptic properties of the final products (Cui et al., 2020). Some common food processing methods include heat treatment, protein separation, ultra high-pressure processing, pH modification, and most importantly, storage conditions (Korhonen et al., 1998; Coscueta et al., 2019).

20.4 SOYBEAN-DERIVED BIOACTIVE PEPTIDES AND THEIR BIO-FUNCTIONALITY

Over the last decade, numerous studies have been shifted to the identification and characterization of food-based bioactive peptides and their respective physiological functions. Numerous soy-based peptides having multiple physiological properties have been identified, which include antioxidant, anti-inflammatory, antidiabetic, antihypertensive, anticancer, antimicrobial, and immunomodulatory properties (Table 20.1). Bioactive peptides from fermented soybean are either released during soybean protein hydrolysis or with the help of starter cultures. Many studies have reported the presence of bioactive peptides in fermented soybean products from Asian and African countries.

20.4.1 ANTIOXIDANT

Oxidation of biomolecules occurs in the human body due to numerous biochemical processes, leading to tissue damage and cell death. Many disorders such as diabetes, arthritis, arteriosclerosis, cancers, and many neurodegenerative disorders are caused by free radicals such as superoxides, hydroxyl radicals, peroxyl radicals, hydrogen peroxide, and lipid hydroperoxides (Gulcin, 2020). Amino acid residues such as His, Trp, Ala, Phe, Met, Try, Leu, Gly, and Val in fermented soybean have been reported to be components of antioxidant peptides (Sanjukta et al., 2015). Peptides purified from *B. subtilis* SHZ fermented soybean exhibited a quite significant scavenging activity of

TABLE 20.1

List of Potential Bioactive Peptides Derived from Soybean/Soy-Based Products

Source	Condition	Peptide Sequence	Biological Properties	Reference
Douchi (fermented soybean)	Fermentation	Peptide containing Phe, Ile, and Gly in the ratio1:2:5	ACE inhibitory	Zhang et al., (2006)
Tofuyo (fermented soybean)	Fermentation	Trp-Ile and Ile-Phe-Leu	ACE inhibitory	Kuba et al., (2003)
Soy milk	Enzymatic hydrolysis (protease)	Phe-Phe-Tyr-Tyr, Trp-His-Pro, Phe-Val-Pro, Leu-His-Pro-Gly-Asp-Ala-Gln-Arg, Ile-Ala-Val, Val-Asn-Pro, Leu-Glu-Pro-Pro, and Trp-Asn-Pro-Arg	ACE inhibitory	Tomatsu et al., (2013), Yoshikawa (2015)
Fermented soybean	Fermentation	Val-Ala-His-Ile-Asn-Val-Gly-Lys and Tyr-Val-Trp-Lys	ACE inhibitory	Hernandez-Ledesma et al., (2004)
Defatted soybean	Enzymatic hydrolysis (protease)	X-Met-Leu-Pro-Ser-Try-Ser-Pro-Try	Anticancer	Kim et al., (2000)
Soybean seed	Fermentation	Ser-Lys-Trp-Gln-His-Gln-Gln-Asp-Ser-Cys-Arg-Lys-Gln-Lys-Gln-Gly-Val-Asn-Leu-Thr-Pro-Cys-Glu-Lys-His-Ile-Met-Glu-Lys-Ile-Gln-Gly-Arg-Gly-Asp-Asp-Asp-Asp-Asp-Asp-Asp-Asp-Asp	Anticancer	Lule et al., (2015)
Chunghookjang (fermented soybean)	Fermentation	Leu-Glu, Glu-Trp, Ser-Pro, Val-Glu, Val-Leu, Val-Thr, and Glu-Phe	Antidiabetic	Yanget al., (2013a)
Soy glycinin	Enzymatic hydrolysis (trypsin and pepsin)	Ile-Ala-Val-Pro-Thr-Gly-Val-Ala, Ile-Ala-Val-Pro-Gly-Glu-Val-Ala, and Leu-Pro-Tyr-Pro	Antidiabetic	Lammiet al., (2019b)
Fermented soysauce	Fermentation	Ala-Trp, Gly-Trp, Ala-Tyr, Ser-Tyr, Gly-Tyr, Ala-Phe, Val-Pro, Ala-Ile, and Val-Gly	Antihypertensive	Nakahara et al., (2010)
Korean soybean paste	Enzymatic hydrolysis (chymotrypsin)	His-His-Leu	Antihypertensive	Shin et al., (2001)
Japanesesoy sauce	Protease from *B. subtilis*	Pro-Gly-Thr-Ala-Val-Phe-Lys	Antihypertensive	Kitts & Weiler (2003)
Soy milk	Enzymatic hydrolysis (pepsin and pancreatin)	Arg-Gln-Arg-Lys and Val-Ile-Lys	Anti-inflammatory	Dia et al., (2014)
Fermented soy meal	Fermentation	His-Thr-Ser-Lys-Ala-Leu-Leu-Asp-Met-Leu-Lys-Arg-Leu-Gly-Lys	Antimicrobial	Cheng et al., (2017)
Fermented soybean	Fermentation	His, Trp, Ala, Phe, Met, Try, Leu, Gly, and Val residues	Antioxidant	Sanjukta et al., (2015)
Soybean glycinin	Enzymatic hydrolysis	Leu-Pro-Tyr-Pro-Arg	Hypocholesterolemic	Yoshikawa et al., (2000)
Soy protein	Enzymatic hydrolysis (trypsin)	Met-Ile-Thr-Leu-Ala-Ile-Pro-Val-Asn-Lys-Pro-Gly-Arg	Immunomodulatory	Yoshikawa (2015)

hydroxyl (96%) and superoxide (62%) radicals at a concentration of 10 mg/ml (Yu et al., 2008). An increase in antioxidant activity due to an increase in protease activity was observed during the process of douchi fermentation (Fan et al., 2009). Another study on oxygen radical scavenging activity and inhibition of oxidation of low-density lipoprotein of natto suggested that peptides also contribute to the antioxidant effect (Iwai et al., 2002).

Antioxidant peptides isolated from soybean meal fermented by *Lb. plantarum* Lp6 exhibited strong antioxidant activity (Amadou et al., 2010).Fermented soy product increased the antioxidant activity due to the release of bioactive peptides during fermentation and can increase the shelf-life by inhibiting the redox reactions occurring during food spoilage (Tonolo et al., 2019). Mixed culture fermentation using *B. subtilis* GD1, *B. subtilis* N4, *Bacillus velezensia* GZ1, *L. bulgaricus*, and *H. Anomala* CICC 1728 also can be used for the production of soy food products having high peptides, flavonoids, phenolic content, and with high antioxidant capacity (Cui et al., 2020). Similarly, solid-state fermentation of whole soybean flour with *L. casei* can enhance the nutritional value, along with improving antioxidant activity (Li et al., 2020). Fermentation of soybean meal product with *B. amyloliquefaciens* SWJS22 for 54 h increased the antioxidant property (< 3 kDa) and at a dose of > 250 mg/Kg body weight also could inhibit the oxidative damage induced by D-galactose in mice (Yang et al., 2019).

Lunasin also has exhibited potential antioxidant properties and was reported to inhibit 2,20-azino-bis(3-ethylbenzothiazoline-6-sulfonic acid) diammonium salt radical scavenger and reactive oxygen species in mouse RAW 264.7 macrophages (Lule et al., 2015). It acts as a potent superoxide and peroxyl scavenger which can prevent glutathione peroxidase and catalase (Chatterjee et al., 2018).

20.4.2 ANTIDIABETIC

Diabetes is a metabolic disease that is characterized by an increase in blood sugar levels. It is grouped into type I (insulin-dependent), where the pancreas fails to secrete insulin, and type II (non-insulin-dependent), caused by an imbalance in insulin secretion and blood sugar absorption (Hamid et al., 2015). The use of synthetic drugs as therapy for diabetes treatment has adverse effects, increasing the risk of gastrointestinal disorders, obesity, pancreatitis, and other metabolic disorders (Mada et al., 2020). However, bioactive peptides present in soybean food products have been reported to possess antidiabetic properties. Soy peptides such as Ile-Ala-Val-Pro-Thr-Gly-Val-Ala, Ile-Ala-Val-Pro-Gly-Glu-Val-Ala, and Leu-Pro-Tyr-Pro improved glucose metabolism by increasing glucose uptake via glucose transporters (GLUT) 1 and 4 in cultured hepatic cells (Lammiet al., 2019b). Similarly, DPP-IV is also responsible for the hydrolysis of glucagon-like peptides and glucose-dependent insulinotropic polypeptides which are crucial for maintaining glucose levels (Yoshikawa et al., 2000). Insulin-stimulated glucose uptake in 3T3-L1 adipocytes is higher in water extract than in the methanolic extract, which increases the fermentation time, and due to the production of smaller peptides, the antioxidant property is much more (Kwon et al., 2011). Dyslipidemic insulin-resistant Wister rats who were fed a sucrose-rich diet supplemented with soy protein had lower hepatic triglyceride and cholesterol storage, functional muscle GLUT4, and normal glycogen level (Oliva, Chicco, & Lombardo, 2015). Studies have reported that *cheonggukjang* improved insulinotropic activity in islets of type II diabetes rats, due to the presence of isoflavonoid aglycones and small peptides (Yang et al., 2013a).

Tetrapeptides from soy protein hydrolysates (Val-His-Val-Val) having anti-hypertrophic and anti-apoptotic efficiency against high glucose have been reported which could increase insulin secretion in the pancreas of mice and also act as a β-cell protector (Marthandam Asokan et al., 2018). It was also found that it facilitated blood regulation, thus resisting hyperglycemia progression. α-glucosidase inhibitory peptides Glu-Ala-Lys and Gly-Ser-Arg were proficient in minimizing symptoms of hyperglycemia or diabetes in mice models, and may serve as food ingredients for diabetic patients in the future (Jiang et al., 2018; Ashaolu, 2020a). The *in vivo* study of soy hydrolysates peptides showed that soy peptides and soy hydrolysates can possibly reduce fat accumulation and increase fat excretion in tissues (Lecerf et al., 2020). Also, the soy peptide aglycin, having the amino acid sequence of Ala-Ser-Cys-Asn-Gly-Val-Cys-Ser-Pro-Phe-Glu-Met-Pro-Pro-Cys-Gly-Ser-Ser-Arg-Cys-Arg-Cys-Ile-Pro-Val-Gly-Leu-Val-Val-Gly-Tyr-Cys-Arg-His-Pro-Ser-Gly, with a molecular weight of 3,742.3 Da, is resistant to gastrointestinal digestion and can be absorbed in mice. Meanwhile, in *in vivo* studies, it has been reported that soy peptides produced using specific enzymes can inhibit DPP-IV, reduce lipid accumulation and inflammation in adipose tissue, improve insulin sensitivity, and increase glucose intake in muscles and liver (Lecerf et al., 2020).

20.4.3 ANTIHYPERTENSIVE

Hypertension or high blood pressure is a major risk factor for heart disease (Gomes et al., 2020). Antihypertensive peptides are the most widely studied peptides in food. These peptides function by blocking angiotensin-converting peptides (ACE). ACE converts angiotensin I to angiotensin II, which inactivates a potent vasodilator, thereby increasing the blood pressure and risks of cardiovascular diseases. Antihypertensive and ACE inhibitory peptides derived from protein-rich foods can be used for lowering high blood pressure and hypertension. Protein-rich fermented foods are the natural source of antihypertensive peptides and ACE inhibitory peptides. The presence of positively charged amino acids (Lys and Arg) or hydrophobic amino acids (Phe, Try, Ala, Trp, Val, Ile, and Met) along with Pro at the C-terminal position of the ACE inhibitory peptides showed a better affinity with ACE (Haque & Chand, 2008; He et al., 2012; Rai et al., 2017b). ACE inhibitory peptides do not exhibit any side effects and can be preferred above synthetic drugs, such as captopril (Sanjukta & Rai, 2016). Traditional fermented soybean foods such as natto, soy sauce, soybean paste, and tempeh are rich in ACE inhibitory peptides (Yadav et al., 2019; Ashaolu, 2020a).

Korean fermented soybean paste, when treated with chymotrypsin, results in the release of the hypotensive peptides (His-His-Leu), while soybean fermented with *Bacillus subtilis* or *B. natto* is shown to contain two ACE inhibitory peptides (Tyr-Val-Xaa-Lys and Val-Ala-His-Ile-Asn-Val-Gly-Glz-Lys) (Singh et al., 2014). It has been reported that fermented soybean seasoning has higher ACE inhibitory activity compared to soy sauce, due to the presence of peptides Gly-Tyr and Ser-Tyr, which lowered hypertension in salt-sensitive Dahl rats by suppressing the rennin-angiotensin system and decreasing serum aldosterone levels (Martinez-Villaluenga et al., 2017). Soy pulp extract, a by-product of tofu production, has also been shown to have ACE inhibitory activity due to the presence of small antihypertensive peptides (Nishibori et al., 2017). Some other antihypertensive soy peptides include Tyr-Val-Val-Phe-Lys, Pro-Gly-Thr-Ala-Val-Phe-Lys, Ile-Val-Phe, Leu-Asn-Phe, Leu-Glu-Phe, Leu-Leu-Phe, Leu-Ser-Trp, Pro-Asn-Asn-Lys-Pro-Phe-Gln, Ile-Pro-Pro-Gly-Val-Pro-Tyr-Trp-Thr, Asn-Trp-Gly-Pro-Leu-Val, and Thr-Pro-Arg-Val-Phe (Wang et al., 2006; Singh et al., 2014). Treatment of soy milk with industrial proteases, such as PROTIN SD-NY10, produced Trp-His-Pro, Phe-Val-Pro, Trp-Asn-Pro-Arg, Phe-Phe-Tyr-Tyr, Ile-Ala-Val, Val-Asn-Pro, Leu-Glu-Pro-Pro, and Leu-His-Pro-Gly-Asp-Ala-Gln-Arg peptides which enhance ACE inhibitory activity as compared to regular soy milk (Tomatsu et al., 2013). In soy glycinin, it was found to contain the ACE inhibitory peptides (Trp-Leu, Ser-Pro-Tyr-Pro, and Val-Leu-Ile-Val-Pro), especially the A4 and A5 subunits comprising the antihypertensive peptide sequence of Asn-Trp-Gly-Pro-Leu-Val(Onuh & Aluko, 2019). Similarly, in soy β-conglycinin, Leu-Pro-His-Phe and Leu-Ala-Ile-Pro-Val-Asn-Lys-Pro peptides were reported to have ACE inhibitory activity (Tomatsu et al., 2013). Soybean protein isolates hydrolyzed by Prozyme and subsequently fermented with *L. rhamnosus* EBD1 resulted in an increase in *in-vitro* ACE inhibitory peptides, Ile-Ile-Arg-Cys-Thr-Gly-Cys, Gly-Pro-Lys-Ala-Leu-Pro-Ile-Ile, and Pro-Pro-Asn-Asn-Asn-Pro-Ala-Ser-Pro-Ser-Phe-Ser-Ser-Ser-Ser. The oral administration of soybean protein hydrolysates at 200 mg/Kg body weight to spontaneously hypertensive rats for six weeks significantly lowered the systolic blood pressure (Daliri et al., 2019).

20.4.4 ANTICANCER

Cancer is an abnormal cell growth that either occurs at a particular site or keeps spreading within the body. Soybean-derived peptides have drawn much attention over the years in terms of their role in cancer from its promotion and also its prevention stand points. There are many reports on antitumor and anticancer properties of bioactive peptides but few studies have been reported on peptides derived from fermented soybean foods acting against cancer. Anticancer properties of peptides can be due to the presence of surfactins, having cyclic peptides, lipopeptides soybean protein hydrolysis, or production of lipoproteins by the starter culture. *Bacillus stubtilis* isolated from *doenjang* showed that it produces a surfactin-like compound, having anticancer properties. The study also reported that the compound has a survival rate of about 58.3% at pH 3.0 after three hours. The compound was purified and showed three potential isoforms of surfactin, which were identical to the Gly-Leu-Leu-Val-Asp-Leu-Leu amino acid sequence, having the potential to inhibit breast cancer cells (MCF-7) in a dose-dependent matter, with an IC_{50} value of 10 µg/ml at 24 hours (Lee et al., 2012). Moreover, when the same strain CSY191 was used to ferment *cheonggukjang*, the surfactin concentration increased to 48.2 from 0.3 mg/kg and anticancer activity increased to 5.1- from 2.6-fold. Similarly, *Bacillus natto* isolated from natto produced a lipopeptide biosurfactant and was found to exhibit antitumor properties (Zhang et al., 2020).

Lunasin, an anticancer peptide, has been reported in many soybean products, such as tofu, *sujae*, tempeh, soymilk, and soy infant formula (Cavazos et al., 2012; Sanjuktaet al., 2017). Lunasin has been reported for its cancer-preventive ability using cell cultures and animal models against skin, colon, prostate, breast, and liver cancer (Lule et al., 2015). Lunasins (Ser-Lys-Trp-Gln-His-Gln-Gln-Asp-Ser-Cys-Arg-Lys-Gln-Lys-Gln-Gly-Val-Asn-Leu-Thr-Pro-Cys-Glu-Lys-His-Ile-Met-Glu-Lys-Ile-Gln-Gly-Arg-Gly-Asp-Asp-asp-Asp-Asp-Asp-Asp-Asp-Asp) are closely associated with the Bowman-Birk inhibitor (BBI), which is a 43-long residue with C-terminal of nine aspartic acid residues and a cell adhesion motif, RGD, that enables binding to non-acetylated H3 and H4 to prevent acetylation, thus providing its anticancer activity (Wang & de Mejia, 2005; Singh et al., 2014). It has been reported that BBI helps in protecting lunasin from gastrointestinal degradation when soy proteins are consumed orally (Jae, Hyung, & De Lumen, 2007). It was also reported that lunasins decreased skin tumor incidences in the SENCAR mice skin cancer model by about 70%, at a dose of 250 µg, and also induced colony suppression of mammalian cells promoted by carcinogens and viral oncogenes RAS and E1A by 30–43% (Wang et al., 2008; Singh et al., 2019; Ashaolu, 2020a). Moreover, it was also reported that lunasins induced apoptosis in MCF-7 cells by upregulation of tumor suppressor phosphatase and tensin homolog (PTEN), which is similar to that of soy isoflavone genistein (Pabona et al., 2013). Lunasins are also able to deactivate the tumor-suppressing proteins, Rb, pp32, and p53, and compete with histone acetyltransferases for binding to the deacetylated H3 and H4 histones, thus switching off transcription, leading to arrest of the G1/S phase and causing apoptosis (Lule et al., 2015).

20.4.5 IMMUNOMODULATORY

Immunomodulatory peptides are closely associated with antioxidant, anti-inflammatory, and anti-cancer peptides. Immunological impairments are induced by acute or chronic stress, leading to inflammation, allergic and autoimmune diseases, and which can be halted by soy-derived peptides. These immunomodulatory peptides boost immune cells' functions, such as cytokine regulation or natural killer activity (Singh et al., 2014), and have been found in soy protein hydrolysates which are enzymatically digested (Aluko, 2018). The hydrolysates which are prepared from insoluble soy protein with alcalase have phagocytosis capability and murine splenic lymphocyte (Kong et al., 2008; Singh et al., 2014). The peptides Gly-Ala-Pro-Ala and His-Cys-Gly-Ala-Pro-Ala from soy protein's glycinin component stimulate phagocytosis (Yoshikawa, 2015). The sequence Met-Ile-Thr-Leu-Ala-Ile-Pro-Val-Asn-Lys-Pro-Gly-Arg, produced by the digestion of soy protein with trypsin, was able to stimulate phagocytosis in leukocytes. This peptide sequence is derived from α'-subunit of βCG and was named soymetide-13 due to Met at N-terminal. When some of the C-terminal residues of soymetide-13 are removed to form soymetide-9, it has the highest activity. MITL (soymetide-4) is the minimal sequence required to have immunomodulatory activity (Yoshikawa, 2015).

B. subtilis natto and its metabolites have been shown to have a positive effect on other immune measures (Cao et al., 2019). A study has been reported that surfactin (a cyclic lipopeptide) produced by *B. subtilis natto* TK-1 reduced the interferon-gamma (IFN-γ), IL-6, nitric oxide (NO), and inducible nitric oxide synthase (iNOS) expression in LPS-stimulated mouse peritoneal macrophages by downregulating Toll-like receptor (TLR)–induced nuclear factor-kappa B (NF-κB) signaling (Zhang et al., 2015). Similarly, it has also been reported that dietary supplementation of BALB/C mice for eight weeks with *B. subtilis natto* BS02 and BS04 (10^8 CFU/kg feed) increased natural killer cell cytotoxicity and monocyte phagocytic capacity (Gong et al., 2018). Ingestion of BS02 or BS04 strain increased splenic percentages of CD4 T helper (T_H) cells producing the T_H1 signature cytokine IFN-γ, but BS02 strain had higher percentages of CD4 T_H cells than BS04 and influenced innate immune activity by enhancing phagocytic cell respiratory burst activity (Gong et al., 2018).

The immunomodulatory property of fermented soy using LAB strain led to an increase in splenocyte concanavalin A-induced TNFα production in BALB/c mice fed with fermented soy milk (Appukutty et al., 2015). In another study, it was found that soy fermented using *L. helveticus* R0052 and *S. thermophilus* R0083 downregulated IEC expression of several TNFα-induced pro-inflammatory genes regulated by NF-κB, which was compared to acidified soy milk controls (Lin et al., 2016). LAB fermented soy milk also downregulates the production of nitric oxide (NO) which is a key mediator in host defense and a regulator of innate and adaptive immunity (Bogdan, 2015). Soymilk fermented with *L.acidophilus* LA-5, *S. thermophilus*, and *B. bifidum* Bb-12 decreased neutrophil percentages and increase lymphocyte percentages, thus altering immune responses (Niamah, Sahi, & Al-Sharifi, 2017).

20.4.6 ANTIMICROBIAL

The growing problem of resistance to antibiotics and the need for new antibiotics has encouraged an interest in the development of antimicrobial peptides as human therapeutics (Singh et al., 2017). Antimicrobial activity is considered a necessary function in processed foods, having a direct influence on the shelf-life of the product (Görgüç et al., 2020). Antimicrobial peptides can be classified as small, having 20–46 amino acids, rich in arginine or lysine, and have amphipathic properties (Toldrá et al., 2018). They can cause cell membrane disruption and physiological events such as cell division (Yadavalli et al., 2016).

Most of the reports on antimicrobial peptides derived from soybean are produced by starter cultures. *B. subtilis* found in traditional fermented soybeans produces a wide range of antimicrobial compounds. Antimicrobial compounds such as proteins, lipopeptides, bacteriocins, enzymes, etc. are secreted by *Bacillus* spp. and have been reported to act against many pathogens such as *Clostridium botulinum*, *Campylobacter* spp., *Pseudomonas aeruginosa*, *Listeria monocytogenes*, *Staphylococcus aureus*, and *Salmonella typhimurium*. Mucilaginous slime of fermented soybean composed of low molecular peptides and lipids also inhibits the growth of harmful bacteria and yeasts (Sanjukta & Rai, 2016). *Bacillus natto* TK-1 isolated from natto produced lipopeptide has an anti-adhesive agent against several bacterial strains such as *Fusarium moniliforme*, *Botrytis cinerea*, *S. typhimurium*, and *Micrococcus luteus*, and also exhibited antifungal and antibacterial properties (Cao et al., 2009). *B. subtilis* SCK-2, isolated from *kyeopjang*, a traditional Korean fermented soybean paste, produced an antimicrobial peptide, AMP IC-1, having more thermostability than that of peptides isolated from traditionally fermented soybean paste (Yeo et al., 2012). The peptide was composed of 33 residues with 13 different types of amino acids, Cys, Asn or Asp, Gln or Glu, Ser, Ala, Pro, Gly, Arg, Thr, Val, Ile, Leu, and Lys. The antimicrobial activity of bioactive peptides derived from fermented soymilk showed high activity against *E.coli*, followed by *S. dysenteriae*,

L. monocytogenous, and *B. Cereus* (Singh et al., 2015). Growth of *P. aeruginosa* and *L. monocytogenes* decreased by exposure to peptides Pro-Gly-Thr-Ala-Val-Phe-Lys at concentrations above 312.5 µM and Ile-Lys-Ala-Phe-Lys-Glu-Ala-Thr-Lys-Val-Asp-Lys-Val-Val-Val-Leu-Trp-Thr-Ala at a concentration of 37.2 µM, isolated from soybean (Dhayakaran et al., 2016). Similarly, peptides isolated from soybean meal, His-Thr-Ser-Lys-Ala-Leu-Leu-Asp-Met-Leu-Lys-Arg-Leu-Gly-Lys were also reported to contain antimicrobial bioactive peptides (Cheng et al., 2017). Thus, antimicrobial peptides can be used as food preservatives, alternatives to chemical preservatives, and also alternatives to antibiotics.

20.4.7 Hypocholesterolemic

Hypercholesterolemia is one of the major causes of death, being a significant risk factor for the heart. Production of bile, as well as vitamin D and steroid hormones in the body requires around 50 mg/dL levels of moderate low-density lipoprotein (LDL) cholesterol. Excess blood cholesterol may lead plaques to form in arteries causing arteriosclerosis, reducing oxygen levels, and finally causing cardiovascular disease (CVD) (Daliri et al., 2017). Soy peptides reduce LDL cholesterol levels by about an average of 5.7% and a high reduction of 12.4% was achieved in triglyceride levels (Wang et al., 2004). Bioactive peptide Leu-Pro-Tyr-Pro-Arg from soy-peptide glycinin has hypocholesterolemic properties and demonstrated that 11S-globulin had good hypocholesterolemic activity compared to that of native soy peptide (Pak et al., 2005). Daily dietary intake of about 20 g of soy proteins and 80 mg of isoflavones for five weeks would reduce CVDs among high-risk, middle-aged men (Ashaolu, 2020a).

A novel hypocholesterolemic peptide derived from soybean glycinin, Val-Ala-Trp-Trp-Met-Tyr, was reported to have a greater ability to bind bile acid than SPH, and bile binding ability as strong as that of hypocholesterolemic medicine *in vitro* (Nagaoka, 2018). Soy peptides Leu-Pro-Tyr-Pro-Arg and Trp-Gly-Ala-Pro-Ser-Leu, similar to plant sterols and stanols, can significantly increase plasma total cholesterol and LDL cholesterol levels in mice (Zhang et al., 2013). Peptides Tyr-Val-Val-Asn-Pro-Asp-Asn-Asp-Glu-Asn and Tyr-Val-Val-Asn-Pro-Asp-Asn-Asn-Glu-Asn, from soybean β-glycinin, demonstrated that they behave as competitive inhibitors of 3-hydroxy-3-methylglutaryl CoA reductase catalytic site, improving the ability of HepG2 cells to take up LDL from the extracellular environment (Lammi et al., 2015). Soy proteins can also regulate cholesterol homeostasis in HepG2 cell lines due to their potential to reduce LDL production (Lovati et al., 2000).

In the liver, bile acids are synthesized from cholesterol-conjugated glycine or taurine and are secreted into the small intestine, facilitating lipid absorption. Bile acids are deconjugated by gut microbiota with bile salt, hydrolyze in the colon, and are excreted into feces, which might induce more bile acid synthesis from cholesterol, leading to a reduction in serum cholesterol levels (Long et al., 2017). A study reported by Cavallini et al., (2011) found improved lipid profiles in hypercholesterolemic rabbits consuming soy milk fermented with *L. helviticus* 416 and *E. faecium* CRL183. Similarly, a study by Wang et al. (2013) reported that soy fermented with *L. plantarum* had a greater inhibitory effect on serum cholesterol levels than unfermented soy milk in a diet-induced hyperlipidemic rat model. In a recent study, a symbiotic fermented soy product supplemented with okara, a by-product of soybean, was consumed by healthy men with CVD risk markers. Subjects consumed fermented soy-based products with *Bifidobacterium animalis* subsp. *lactis* Bb-12, *Lactobacillus acidophilus* La-5, and *Streptococcus thermophilus* (starter culture), and it was observed that LDL-C means decreased, resulting in improvement of the LDL-C/HDL-C ratio.

20.5 FUTURE PERSPECTIVE AND CONCLUSION

Synthetic peptides possess several challenges including high cost, intensive market competition, and adverse side effects, leading to an increase in demand for food-derived bioactive peptides. In the near future, food-derived bioactive peptides will be abundant and soon sold as nutraceuticals in markets; such peptides could be also regulated as drugs due to their beneficial properties and modes of action (Daliri et al., 2017). Soybean-derived bioactive peptides have high potential in increasing the biofunctionality of food products, having excellent protein sources, and exerting beneficial health effects. Digestion of soy-derived peptides is more effective than that of native proteins, making it a suitable food for infants, patients, and elderly individuals (Ashaolu, 2020b). The presence of emulsifying activity and foam stability, along with high methionine content of about 4.2%, makes soybean fermented foods a good source of protein for baby food, and also due to its hypo-allergenic functions, makes it suitable for fortifying foods especially for medical diets (Barca et al., 2000; Ashaolu & Yupanqui, 2017). Another futuristic use of soybean hydrolysate produced by enzymatic hydrolysis of pepsin and papain for improving viscoelastic and interfacial properties of ice cream has been reported, making it a better option than that made with skimmed milk powder (Chen et al., 2019). It has also been reported that soybean protein hydrolysates and its blend with xanthan gum, produced by enzymatic hydrolysis and heat shearing treatment, can be helpful as a fat replacer in the production of reduced-fat ice cream. Numerous reports also suggest that soybean-derived peptides and their use in futuristic applications will be interesting. Moreover, the use of genetically modified microorganisms is also quite of interest in the near future.

Soybean is a promising source of bioactive peptides, which have a wide range of biofunctional activities such as antioxidant, anti-inflammatory, antidiabetic, antihypertensive, anticancer, and immunomodulatory properties. Further studies are needed to understand the mechanism of action, as well as absorption, metabolism, and tissues being targeted. The bioactive peptides produced during fermentation are either by hydrolysis of protein or by the starter cultures. From the studies to date, it can be concluded that peptides derived from soybean fermentation are potential components promoting health benefits. More studies are required to identify the quantity of the active peptides released during fermentation by different methods, and their impact on gender and age. Moreover, further studies are to be done on the characterization of bioactive peptides and other health benefits, which can lead to their application in pharmaceutical industries and food industries to replace synthetic drugs.

REFERENCES

Aluko, R. E. (2018). Food protein-derived peptides: Production, isolation, and purification. In *Proteins in Food Processing*, 2nd Edition (pp. 389–412). Elsevier Inc. https://doi.org/10.1016/B978-0-08-100722-8.00016-4

Amadou, I., Gbadamosi, O. S., Shi, Y. H., Kamara, M. T., Jin, S., & Le, G. W. (2010). Identification of antioxidative peptides from lactobacillus plantarum Lp6 Fermented Soybean Protein Meal. *Research Journal of Microbiology*, 5(5), 372–380. https://doi.org/10.3923/jm.2010.372.380

Appukutty, M., Ramasamy, K., Rajan, S., Vellasamy, S., Ramasamy, R., & Radhakrishnan, A. K. (2015). Effect of orally administered soy milk fermented with Lactobacillus plantarum LAB12 and physical exercise on murine immune responses. *Beneficial Microbes*, 6(4), 491–496. https://doi.org/10.3920/BM2014.0129

Ashaolu, T. J. (2020a). Health applications of soy protein hydrolysates. In *International Journal of Peptide Research and Therapeutics*. Springer. https://doi.org/10.1007/s10989-020-10018-6

Ashaolu, T. J. (2020b). Applications of soy protein hydrolysates in the emerging functional foods: A review. *International Journal of Food Science and Technology*, 55(2), 421–428. Blackwell Publishing Ltd. https://doi.org/10.1111/ijfs.14380

Ashaolu, T. J., & Yupanqui, C. T. (2017). Suppressive activity of enzymatically-educed soy protein hydrolysates on degranulation in IgE-antigen complex-stimulated RBL-2H3 cells. *Functional Foods in Health and Disease*, 7(7), 545. https://doi.org/10.31989/ffhd.v7i7.356

Barca, A. M. C., Ruiz-Salazar, R. A., & Jara-Marini, M. E. (2000). Enzymatic hydrolysis and synthesis of soy protein to improve its amino acid composition and functional properties. *Journal of Food Science*, 65(2), 246–253. https://doi.org/10.1111/j.1365-2621.2000.tb15988.x

Bogdan, C. (2015). Nitric oxide synthase in innate and adaptive immunity: An update. *Trends in Immunology*, 36(3), 161–178. Elsevier Ltd. https://doi.org/10.1016/j.it.2015.01.003

Boukil, A., Suwal, S., Chamberland, J., Pouliot, Y., & Doyen, A. (2018). Ultrafiltration performance and recovery of bioactive peptides after fractionation of tryptic hydrolysate generated from pressure-treated B-lactoglobulin. *Journal of Membrane Science, 556*, 42–53. https://doi.org/10.1016/j.memsci.2018.03.079

Cao, X.-H., Liao, Z.-Y., Wang, C.-L., Yang, W.-Y., & Lu, M.-F. (2009). Evaluation of a lipopeptide biosurfactant from Bacillus natto TK-1 as a potential source of anti-adhesive, antimicrobial and antitumor activities. *Brazilian Journal of Microbiology, 40*(2), 373–379. https://doi.org/10.1590/s1517-83822009000200030

Cao, Z.H., Green-Johnson, J.M., Buckley, N.D., & Lin, Q Y. (2019). Bioactivity of soy-based fermented foods: A review. *Biotechnology Advances, 37*(1), 223–238. https://doi.org/10.1016/j.biotechadv.2018.12.001

Capriotti, A.L., Caruso, G., Cavaliere, C., Samperi, R., Ventura, S., Zenezini Chiozzi, R., & Laganà, A. (2015). Identification of potential bioactive peptides generated by simulated gastrointestinal digestion of soybean seeds and soy milk proteins. *Journal of Food Composition and Analysis, 44*, 205–213. https://doi.org/10.1016/j.jfca.2015.08.007

Cavallini, D.C., Suzuki, J.Y., Abdalla, D.S., Vendramini, R.C., Pauly-Silveira, N.D., Roselino, M.N., Pinto, R.A., & Rossi, E.A. (2011). Influence of a probiotic soy product on fecal microbiota and its association with cardiovascular risk factors in an animal model. *Lipids in Health and Disease, 10*, 126. https://doi.org/10.1186/1476-511X-10-126

Cavazos, A., Morales, E., Dia, V.P., & De Mejia, E.G. (2012). Analysis of lunasin in commercial and pilot plant produced soybean products and an improved method of lunasin purification. *Journal of Food Science, 77*(5). https://doi.org/10.1111/j.1750-3841.2012.02676.x

Chatterjee, C., Gleddie, S., & Xiao, C.W. (2018). Soybean bioactive peptides and their functional properties. *Nutrients, 10*(9), 8–11. https://doi.org/10.3390/nu10091211

Chen, J., Liu, G., Pantalone, V., & Zhong, Q. (2020). Physicochemical properties of proteins extracted from four new Tennessee soybean lines. *Journal of Agriculture and Food Research, 2*, 100022. https://doi.org/10.1016/j.jafr.2020.100022

Chen, N., Zhang, J., Mei, L., & Wang, Q. (2018). Ionic strength and pH responsive permeability of soy glycinin microcapsules. *Langmuir, 34*(33), 9711–9718. https://doi.org/10.1021/acs.langmuir.8b01559

Chen, W., Liang, G., Li, X., He, Z., Zeng, M., Gao, D., Qin, F., Goff, H. D., & Chen, J. (2019). Effects of soy proteins and hydrolysates on fat globule coalescence and meltdown properties of ice cream. *Food Hydrocolloids, 94*, 279–286. https://doi.org/10.1016/j.foodhyd.2019.02.045

Cheng, A.C., Lin, H.L., Shiu, Y.L., Tyan, Y.C., & Liu, C.H. (2017). Isolation and characterization of antimicrobial peptides derived from Bacillus subtilis E20-fermented soybean meal and its use for preventing Vibrio infection in shrimp aquaculture. *Fish and Shellfish Immunology, 67*, 270–279. https://doi.org/10.1016/j.fsi.2017.06.006

Chourasia, R., Abedin, M. M., Chiring Phukon, L., Sahoo, D., Singh, S. P., & Rai, A. K. (2020a). Biotechnological approaches for the production of designer cheese with improved functionality. *Comprehensive Reviews in Food Science and Food Safety*. https://doi.org/10.1111/1541-4337.12680

Chourasia, R., Padhi, S., Phukon, C. P., Abedin, M., Singh, S. P., & Rai, A. K. (2020b). A potential peptide from soy cheese produced using *Lactobacillus delbrueckii* WS4 for effective inhibition of SARS-CoV-2 main protease and S1 glycoprotein. *Frontiers in Molecular Biosciences*. https://doi.org/10.3389/fmolb.2020.601753

Chourasia, R., Phukon, C. L., Abedin, M. M., Sahoo, D., & Rai, A. K. (2021). Microbial Transformation during Gut Fermentation. In *Bioactive Compounds in Fermented Foods*, (pp. 365–402). New York: CRC Press. https://doi.org/10.1201/9780429027413-18

Coscueta, E. R., Campos, D. A., Osório, H., Nerli, B. B., & Pintado, M. (2019). Enzymatic soy protein hydrolysis: A tool for biofunctional food ingredient production. *Food Chemistry: X, 1*, 100006. https://doi.org/10.1016/j.fochx.2019.100006

Cui, J., Xia, P., Zhang, L., Hu, Y., Xie, Q., & Xiang, H. (2020). A novel fermented soybean, inoculated with selected Bacillus, Lactobacillus and Hansenula strains, showed strong antioxidant and anti-fatigue potential activity. *Food Chemistry, 333*, 127527. https://doi.org/10.1016/j.foodchem.2020.127527

Daliri, E. B. M., Ofosu, F. K., Chelliah, R., Park, M. H., Kim, J. H., & Oh, D. H. (2019). Development of a soy protein hydrolysate with an antihypertensive effect. *International Journal of Molecular Sciences, 20*(6), 1496. https://doi.org/10.3390/ijms20061496

Daliri, E., Oh, D., & Lee, B. (2017). Bioactive peptides. *Foods, 6*(5), 32. https://doi.org/10.3390/foods6050032

De Angelis, E., Pilolli, R., Bavaro, S. L., & Monaci, L. (2017). Insight into the gastro-duodenal digestion resistance of soybean proteins and potential implications for residual immunogenicity. *Food and Function, 8*(4), 1599–1610. https://doi.org/10.1039/c6fo01788f

De Mejia, E., & De Lumen, B. O. (2006). Soybean bioactive peptides: A new horizon in preventing chronic diseases. *Sexuality, Reproduction and Menopause*, 4(2), 91–95. No longer published by Elsevier. https://doi.org/10.1016/j.sram.2006.08.012

Dhayakaran, R., Neethirajan, S., & Weng, X. (2016). Investigation of the antimicrobial activity of soy peptides by developing a high throughput drug screening assay. *Biochemistry and Biophysics Reports*, 6, 149–157. https://doi.org/10.1016/j.bbrep.2016.04.001

Dia, V. P., Bringe, N. A., & De Mejia, E. G. (2014). Peptides in pepsin-pancreatin hydrolysates from commercially available soy products that inhibit lipopolysaccharide-induced inflammation in macrophages. *Food Chemistry*, 152, 423–431. https://doi.org/10.1016/j.foodchem.2013.11.155

Egounlety, M., & Aworh, O. C. (2003). Effect of soaking, dehulling, cooking and fermentation with Rhizopus oligosporus on the oligosaccharides, trypsin inhibitor, phytic acid and tannins of soybean (Glycine max Merr.), cowpea (Vigna unguiculata L. Walp) and groundbean (Macrotyloma geocarpa Harms). *Journal of Food Engineering*, 56(2–3), 249–254. https://doi.org/10.1016/S0260-8774(02)00262-5

Fan, J., Zhang, Y., Chang, X., Saito, M., & Li, Z. (2009). Changes in the radical scavenging activity of bacterial-type douchi, a traditional fermented soybean product, during the primary fermentation process. *Bioscience, Biotechnology and Biochemistry*, 73(12), 2749–2753. https://doi.org/10.1271/bbb.90361

Gaspar, D., & Castanho, M. A. R. B. (2016). Anticancer peptides: Prospective innovation in cancer therapy. In *Host Defense Peptides and Their Potential as Therapeutic Agents* (pp. 95–109). Springer International Publishing. https://doi.org/10.1007/978-3-319-32949-9_4

Gomes, C., Ferreira, D., Carvalho, J. P. F., Barreto, C. A. V., Fernandes, J., Gouveia, M., Ribeiro, F., Duque, A. S., & Vieira, S. I. (2020). Current genetic engineering strategies for the production of anti-hypertensive acei peptides. *Biotechnology and Bioengineering* 117(8), 2610–2628. https://doi.org/10.1002/bit.27373

Gong, L., Huang, Q., Fu, A., Wu, Y., Li, Y., Xu, X., Huang, Y., Yu, D., & Li, W. (2018). Spores of two probiotic *Bacillus* species enhance cellular immunity in BALB/C mice. *Canadian Journal of Microbiology*, 64(1), 41–48. https://doi.org/10.1139/cjm-2017-0373

González-Montoya, M., Hernández-Ledesma, B., Silván, J. M., Mora-Escobedo, R., & Martínez-Villaluenga, C. (2018). Peptides derived from in vitro gastrointestinal digestion of germinated soybean proteins inhibit human colon cancer cells proliferation and inflammation. *Food Chemistry*, 242, 75–82. https://doi.org/10.1016/j.foodchem.2017.09.035

Görgüç, A., Gençdağ, E., & Yılmaz, F. M. (2020). Bioactive peptides derived from plant origin by-products: Biological activities and techno-functional utilizations in food developments – A review. *Food Research International*, 136, 109504. https://doi.org/10.1016/j.foodres.2020.109504

Guan, H., Diao, X., Jiang, F., Han, J., & Kong, B. (2018). The enzymatic hydrolysis of soy protein isolate by Corolase PP under high hydrostatic pressure and its effect on bioactivity and characteristics of hydrolysates. *Food Chemistry*, 245, 89–96. https://doi.org/10.1016/j.foodchem.2017.08.081

Gulcin, İ. (2020). Antioxidants and antioxidant methods: An updated overview. *Archives of Toxicology*, 94(3), 651–715. Springer. https://doi.org/10.1007/s00204-020-02689-3

Hamid, H. A., Yusoff, M. M., Liu, M., & Karim, M. R. (2015). α-glucosidase and α-amylase inhibitory constituents of Tinospora crispa: Isolation and chemical profile confirmation by ultra-high performance liquid chromatography-quadrupole time-of-flight/mass spectrometry. *Journal of Functional Foods*, 16, 74–80. https://doi.org/10.1016/j.jff.2015.04.011

Haque, E., & Chand, R. (2008). Antihypertensive and antimicrobial bioactive peptides from milk proteins. *European Food Research and Technology*, 227(1), 7–15. https://doi.org/10.1007/s00217-007-0689-6

He, F.-J., & Chen, J.-Q. (2013). Consumption of soybean, soy foods, soy isoflavones and breast cancer incidence: Differences between Chinese women and women in Western countries and possible mechanisms. *Food Science and Human Wellness*, 2(3–4), 146–161. https://doi.org/10.1016/j.fshw.2013.08.002

He, R., Ma, H., Zhao, W., Qu, W., Zhao, J., Luo, L., & Zhu, W. (2012). Modeling the QSAR of ACE-inhibitory peptides with ANN and its applied illustration. *International Journal of Peptides*, 2012. https://doi.org/10.1155/2012/620609

Hernandez-Ledesma, B., Amigo, L., Ramos, M., & Recio, I. (2004). Angiotensin converting enzyme inhibitory activity in commercial fermented products. Formation of peptides under simulated gastrointestinal digestion. *Journal of Agricultural and Food Chemistry*, 52(6), 1504–1510. https://doi.org/10.1021/jf034997b

Hou, D. H. J., & Chang, S. K. C. (2004). Structural characteristics of purified glycinin from soybeans stored under various conditions. *Journal of Agricultural and Food Chemistry*, 52(12), 3792–3800. https://doi.org/10.1021/jf035072z

Huang, T., Bu, G., & Chen, F. (2018). The influence of composite enzymatic hydrolysis on the antigenicity of β-conglycinin in soy protein hydrolysates. *Journal of Food Biochemistry*, 42(5), 1–8. https://doi.org/10.1111/jfbc.12544

Iwai, K., Nakaya, N., Kawasaki, Y., & Matsue, H. (2002). Antioxidative functions of natto, a kind of fermented soybeans: Effect on LDL oxidation and lipid metabolism in cholesterol-fed rats. *Journal of Agricultural and Food Chemistry*, 50(12), 3597–3601. https://doi.org/10.1021/jf0117199

Iwaniak, A., Hrynkiewicz, M., Minkiewicz, P., Bucholska, J., & Darewicz, M. (2020). Soybean (Glycine max) protein hydrolysates as sources of peptide bitter-tasting indicators: An analysis based on hybrid and Fragmentomic approaches. *Applied Sciences*, 10(7), 2514. https://doi.org/10.3390/app10072514

Jae, H. P., Hyung, J. J., & De Lumen, B. O. (2007). In vitro digestibility of the cancer-preventive soy peptides lunasin and BBI. *Journal of Agricultural and Food Chemistry*, 55(26), 10703–10706. https://doi.org/10.1021/jf072107c

Jayachandran, M., & Xu, B. (2019). An insight into the health benefits of fermented soy products. *Food Chemistry*, 271(July), 362–371. https://doi.org/10.1016/j.foodchem.2018.07.158

Jiang, M., Yan, H., He, R., & Ma, Y. (2018). Purification and a molecular docking study of α-glucosidase-inhibitory peptides from a soybean protein hydrolysate with ultrasonic pretreatment. *European Food Research and Technology*, 244(11), 1995–2005. https://doi.org/10.1007/s00217-018-3111-7

Kim, S. E., Kim, H. H., Kim, J. Y., Kang, Y. I., Woo, H. J., & Lee, H. J. (2000). Anticancer activity of hydrophobic peptides from soy proteins. *BioFactors*, 12(1–4), 151–155. https://doi.org/10.1002/biof.5520120124

Kitts, D., & Weiler, K. (2003). Bioactive proteins and peptides from food sources. Applications of bioprocesses used in isolation and recovery. *Current Pharmaceutical Design*, 9(16), 1309–1323. https://doi.org/10.2174/1381612033454883

Kong, X. Z., Guo, M. M., Hua, Y. F., Cao, D., & Zhang, C. M. (2008). Enzymatic preparation of immuno-modulating hydrolysates from soy proteins. *Bioresource Technology*, 99(18), 8873–8879. https://doi.org/10.1016/j.biortech.2008.04.056

Korhonen, H., Pihlanto-Leppälä, A., Rantamäki, P., & Tupasela, T. (1998). Impact of processing on bioactive proteins and peptides. *Trends in Food Science and Technology*, 9(8–9), 307–319. Elsevier. https://doi.org/10.1016/S0924-2244(98)00054-5

Kuba, M., Tanaka, K., Tawata, S., Takeda, Y., & Yasuda, M. (2003). Angiotensin I-converting enzyme inhibitory peptides isolated from tofuyo fermented soybean food. *Bioscience, Biotechnology and Biochemistry*, 67(6), 1278–1283. https://doi.org/10.1271/bbb.67.1278

Kumar, S., Abedin, M. M., Singh, A. K., & Das, S. (2020). Role of Phenolic Compounds in Plant-Defensive Mechanisms. In *Plant Phenolics in Sustainable Agriculture* (pp. 517–532). Singapore: Springer. https://doi.org/10.1007/978-981-15-4890-1_22

Kwon, D. Y., Hong, S. M., Ahn, I. S., Kim, M. J., Yang, H. J., & Park, S. (2011). Isoflavonoids and peptides from meju, long-term fermented soybeans, increase insulin sensitivity and exert insulinotropic effects in vitro. *Nutrition*, 27(2), 244–252. https://doi.org/10.1016/j.nut.2010.02.004

Lammi, C., Aiello, G., Boschin, G., & Arnoldi, A. (2019a). Multifunctional peptides for the prevention of cardiovascular disease : A new concept in the area of bioactive food-derived peptides. *Journal of Functional Foods*, 55(November 2018), 135–145. https://doi.org/10.1016/j.jff.2019.02.016

Lammi, C., Arnoldi, A., & Aiello, G. (2019b). Soybean peptides exert multifunctional bioactivity modulating 3-hydroxy-3-methylglutaryl-CoA reductase and dipeptidyl peptidase-IV targets in vitro [research-article]. *Journal of Agricultural and Food Chemistry*, 67(17), 4824–4830. https://doi.org/10.1021/acs.jafc.9b01199

Lammi, C., Zanoni, C., Arnoldi, A., & Vistoli, G. (2015). Two peptides from soy β-conglycinin induce a hypo-cholesterolemic effect in HepG2 cells by a statin-like mechanism: Comparative in vitro and in silico modeling studies. *Journal of Agricultural and Food Chemistry*, 63(36), 7945–7951. https://doi.org/10.1021/acs.jafc.5b03497

Lecerf, J. M., Arnoldi, A., Rowland, I., Trabal, J., Widhalm, K., Aiking, H., & Messina, M. (2020). Soyfoods, glycemic control and diabetes. *Nutrition Clinique et Métabolisme*, 34(2), 141–148. Elsevier Masson SAS. https://doi.org/10.1016/j.nupar.2020.02.437

Lee, J. H., Nam, S. H., Seo, W. T., Yun, H. D., Hong, S. Y., Kim, M. K., & Cho, K. M. (2012). The production of surfactin during the fermentation of cheonggukjang by potential probiotic Bacillus subtilis CSY191 and the resultant growth suppression of MCF-7 human breast cancer cells. *Food Chemistry*, 131(4), 1347–1354. https://doi.org/10.1016/j.foodchem.2011.09.133

Lee, S. H., Lee, H., & Kim, J. C. (2019). Anti-inflammatory effect of water extracts obtained from doenjang in lps-stimulated raw 264.7 cells. *Food Science and Technology*, 39(4), 947–954. https://doi.org/10.1590/fst.15918

Li, S., Jin, Z., Hu, D., Yang, W., Yan, Y., Nie, X., Lin, J., Zhang, Q., Gai, D., Ji, Y., & Chen, X. (2020). Effect of solid-state fermentation with Lactobacillus casei on the nutritional value, isoflavones, phenolic acids and antioxidant activity of whole soybean flour. *LWT, 125*, 109264. https://doi.org/10.1016/j.lwt.2020 .109264.

Lin, Q., Mathieu, O., Tompkins, T. A., Buckley, N. D., & Green-Johnson, J. M. (2016). Modulation of the TNFα-induced gene expression profile of intestinal epithelial cells by soy fermented with lactic acid bacteria. *Journal of Functional Foods, 23*, 400–411. https://doi.org/10.1016/j.jff.2016.02.047

Liu, L., Li, S., Zheng, J., Bu, T., He, G., & Wu, J. (2020). Safety considerations on food protein-derived bioactive peptides. *Trends in Food Science and Technology, 96*, 199–207. Elsevier Ltd. https://doi.org/10.1016 /j.tifs.2019.12.022

Long, S. L., Gahan, C. G. M., & Joyce, S. A. (2017). Interactions between gut bacteria and bile in health and disease. *Molecular Aspects of Medicine, 56*, 54–65. Elsevier Ltd. https://doi.org/10.1016/j.mam.2017.06.002

Lovati, M. R., Manzoni, C., Gianazza, E., Arnoldi, A., Kurowska, E., Carroll, K. K., & Sirtori, C. R. (2000). Soy protein peptides regulate cholesterol homeostasis in Hep G2 cells. *Journal of Nutrition, 130*(10). https://academic.oup.com/jn/article-abstract/130/10/2543/4686098

Lule, V. K., Garg, S., Pophaly, S. D., Hitesh, & Tomar, S. K. (2015). Potential health benefits of lunasin: A multifaceted soy-derived bioactive peptide. *Journal of Food Science, 80*(3), R485–R494. https://doi.org /10.1111/1750-3841.12786

Mada, S. B., Ugwu, C. P., & Abarshi, M. M. (2020). Health promoting effects of food-derived bioactive peptides: A review. *International Journal of Peptide Research and Therapeutics, 26*(2), 831–848. Springer. https://doi.org/10.1007/s10989-019-09890-8

Maebuchi, M., Samoto, M., Kohno, M., Ito, R., Koikeda, T., Hirotsuka, M., & Nakabou, Y. (2007). Improvement in the intestinal absorption of soy protein by enzymatic digestion to oligopeptide in healthy adult men. *Food Science and Technology Research, 13*(1), 45–53. https://doi.org/10.3136/fstr.13.45

Marciniak, A., Suwal, S., Naderi, N., Pouliot, Y., & Doyen, A. (2018). Enhancing enzymatic hydrolysis of food proteins and production of bioactive peptides using high hydrostatic pressure technology. *Trends in Food Science and Technology, 80*, 187–198. https://doi.org/10.1016/j.tifs.2018.08.013

Marthandam Asokan, S., Wang, T., Su, W., & Lin, W. (2018). Short tetra-peptide from soy-protein hydrolysate attenuates hyperglycemia associated damages in H9c2 cells and ICR mice. *Journal of Food Biochemistry, 42*(6), e12638. https://doi.org/10.1111/jfbc.12638

Martinez-Villaluenga, C., Peñas, E., & Frias, J. (2017). Bioactive peptides in fermented foods: Production and evidence for health effects. In *Fermented Foods in Health and Disease Prevention* (pp. 23–47). Elsevier Inc.. https://doi.org/10.1016/B978-0-12-802309-9.00002-9

Martinez-Villaluenga, C., Torino, M. I., Martín, V., Arroyo, R., Garcia-Mora, P., Estrella Pedrola, I., Vidal-Valverde, C., Rodriguez, J. M., & Frias, J. (2012). Multifunctional properties of soy milk fermented by Enterococcus faecium strains isolated from raw soy milk. *Journal of Agricultural and Food Chemistry, 60*(41), 10235–10244. https://doi.org/10.1021/jf302751m

Nagaoka, S. (2018). Structure – Function properties of hypolipidemic peptides. October 2017, 1–8. https://doi .org/10.1111/jfbc.12539

Nakahara, T., Sano, A., Yamaguchi, H., Sugimoto, K., Chikata, H., Kinoshita, E., & Uchida, R. (2010). Antihypertensive effect of peptide-enriched soy sauce-like seasoning and identification of its angiotensin I-converting enzyme inhibitory substances. *Journal of Agricultural and Food Chemistry, 58*(2), 821–827. https://doi.org/10.1021/jf903261h

Naresh, S., Ong, M. K., Thiagarajah, K., Muttiah, N. B. S. J., Kunasundari, B., & Lye, H. S. (2019). Engineered soybean-based beverages and their impact on human health. In *Non-Alcoholic Beverages* (pp. 329–361). Elsevier.https://doi.org/10.1016/b978-0-12-815270-6.00011-6

Niamah, A. K., Sahi, A. A., & Al-Sharifi, A. S. N. (2017). Effect of feeding soy milk fermented by probiotic bacteria on some blood criteria and weight of experimental animals.*Probiotics and Antimicrobial Proteins, 9*(3), 284–291. https://doi.org/10.1007/s12602-017-9265-y

Nishibori, N., Kishibuchi, R., & Morita, K. (2017). Soy pulp extract inhibits angiotensin I-converting enzyme (ACE) activity in vitro: Evidence for its potential hypertension-improving action. *Journal of Dietary Supplements, 14*(3), 241–251. https://doi.org/10.1080/19390211.2016.1207744

Oliva, M. E., Chicco, A., & Lombardo, Y. B. (2015). Mechanisms underlying the beneficial effect of soy protein in improving the metabolic abnormalities in the liver and skeletal muscle of dyslipemic insulin resistant rats. *European Journal of Nutrition, 54*(3), 407–419. https://doi.org/10.1007/s00394-014-0721-0

Onuh, J. O., & Aluko, R. E. (2019). Metabolomics as a tool to study the mechanism of action of bioactive protein hydrolysates and peptides: A review of current literature. *Trends in Food Science and Technology, 91*, 625–633. https://doi.org/10.1016/j.tifs.2019.08.002

Pabona, J. M. P., Dave, B., Su, Y., Montales, M. T. E., De Lumen, B. O., De Mejia, E. G., Rahal, O. M., & Simmen, R. C. M. (2013). The soybean peptide lunasin promotes apoptosis of mammary epithelial cells via induction of tumor suppressor PTEN: Similarities and distinct actions from soy isoflavone genistein. *Genes and Nutrition*, 8(1), 79–90. https://doi.org/10.1007/s12263-012-0307-5

Pak, V. V., Koo, M. S., Kasymova, T. D., & Kwon, D. Y. (2005). Isolation and identification of peptides from soy 11S-globulin with hypocholesterolemic activity. *Chemistry of Natural Compounds*, 41(6), 710–714. https://doi.org/10.1007/s10600-006-0017-6

Rai, A. K., Sanjukta, S., Chourasia, R., Bhat, I., Bhardwaj, P. K., & Sahoo, D. (2017). Production of bioactive hydrolysate using protease, B-glucosidase and A-amylase of Bacillus spp. isolated from kinema. *Bioresource Technology*, 235, 358–365. https://doi.org/10.1016/j.biortech.2017.03.139

Rai, A. K., Sanjukta, S., & Jeyaram, K. (2017). Production of angiotensin I converting enzyme inhibitory (ACE-I) peptides during milk fermentation and their role in reducing hypertension. *Critical Reviews in Food Science and Nutrition*, 57(13), 2789–2800. https://doi.org/10.1080/10408398.2015.1068736

Reddy, N. R., & Pierson, M. D. (1994). Reduction in antinutritional and toxic components in plant foods by fermentation. *Food Research International*, 27(3), 281–290. https://doi.org/10.1016/0963-9969(94)90096-5

Sanjukta, S., Rai, A. K., & Sahoo, D. (2017) *Bioactive Molecules in Fermented Soybean Products and Their Potential Health Benefits* (pp. 97–121). https://doi.org/10.1201/9781315205359-5

Sanjukta, S., Padhi, S., Sarkar, P., Singh, S. P., Sahoo, D., & Rai, A. K. (2021a). Production, characterization and molecular docking of antioxidant peptides from peptidome of kinema fermented with proteolytic Bacillus spp. *Food Research International*, 141, 110161. https://doi.org/10.1016/j.foodres.2021.110161

Sanjukta, S., Sahoo, D., & Rai, A. K. (2021b). Fermentation of black soybean with Bacillus spp. for the production of kinema: changes in antioxidant potential on fermentation and gastrointestinal digestion. *Journal of Food Science and Technology*, 1–9. https://doi.org/10.1007/S13197-021-05144-Y

Sanjukta, S., & Rai, A. K. (2016). Production of bioactive peptides during soybean fermentation and their potential health benefits. *Trends in Food Science and Technology*, 50, 1–10. https://doi.org/10.1016/j.tifs.2016.01.010

Sanjukta, S., Rai, A. K., Muhammed, A., Jeyaram, K., & Talukdar, N. C. (2015). Enhancement of antioxidant properties of two soybean varieties of Sikkim Himalayan region by proteolytic Bacillus subtilis fermentation. *Journal of Functional Foods*, 14, 650–658. https://doi.org/10.1016/j.jff.2015.02.033

Sharma, S., Padhi, S., Kumari, M., Kumar Rai, A., & Sahoo, D. (2021). Bioactive Compounds in Fermented Foods. In *Bioactive Compounds in Fermented Foods*, (pp. 48–69). https://doi.org/10.1201/9780429027413-3

Shi, Y. G., Yang, Y., Piekoszewski, W., Zeng, J. H., Guan, H. N., Wang, B., Liu, L.L., Zhu, X.Q., Chen, F.L., & Zhang, N. (2020). Influence of four different coagulants on the physicochemical properties, textural characteristics and flavour of tofu. *International Journal of Food Science and Technology*, 55(3), 1218–1229. https://doi.org/10.1111/ijfs.14357

Shin, Z. I., Yu, R., Park, S. A., Chung, D. K., Ahn, C. W., Nam, H. S., Kim, K. S., & Lee, H. J. (2001). His-His-Leu, an angiotensin I converting enzyme inhibitory peptide derived from Korean soybean paste, exerts antihypertensive activity in vivo. *Journal of Agricultural and Food Chemistry*, 49(6), 3004–3009. https://doi.org/10.1021/jf001135r

Singh, B. P., Vij, S., & Hati, S. (2014). Functional significance of bioactive peptides derived from soybean. In *Peptides*, 54, 171–179. Elsevier. https://doi.org/10.1016/j.peptides.2014.01.022

Singh, B. P., Vij, S., Hati, S., Singh, D., Kumari, P., & Minj, J. (2015). Antimicrobial activity of bioactive peptides derived from fermentation of soy milk by Lactobacillus plantarum C 2 against common foodborne pathogens. *International Journal of Fermented Foods*, 4(1and2), 91. https://doi.org/10.5958/2321-712x.2015.00008.3

Singh, B. P., Yadav, D., & Vij, S. (2017). Soybean bioactive molecules: Current trend and future prospective. In *Reference Series in Phytochemistry* (pp. 1–29). Cham: Springer. https://doi.org/10.1007/978-3-319-54528-8_4-1

Singh, B. P., Yadav, D., & Vij, S. (2019). *Soybean Bioactive Molecules: Current Trend and Future Prospective* (pp. 267–294). Springer. https://doi.org/10.1007/978-3-319-78030-6_4

Sun, X., Acquah, C., Aluko, R. E., & Udenigwe, C. C. (2020). Considering food matrix and gastrointestinal effects in enhancing bioactive peptide absorption and bioavailability. *Journal of Functional Foods*, 64, 103680. https://doi.org/10.1016/j.jff.2019.103680.

Tang, C. H. (2019). Nanostructured soy proteins: Fabrication and applications as delivery systems for bioactives (a review). *Food Hydrocolloids*, 91, 92–116. Elsevier B.V. https://doi.org/10.1016/j.foodhyd.2019.01.012

Tidke, S. A., Ramakrishna, D., Kiran, S., Kosturkova, G., & Ravishankar, G. A. (2015). Nutraceutical potential of soybean: Review. *Asian Journal of Clinical Nutrition*, 7(2), 22–32. https://doi.org/10.3923/ajcn.2015.22.32

Toldrá, F., Reig, M., Aristoy, M. C., & Mora, L. (2018). Generation of bioactive peptides during food processing. *Food Chemistry*, 267, 395–404. https://doi.org/10.1016/j.foodchem.2017.06.119

Tomatsu, M., Shimakage, A., Shinbo, M., Yamada, S., & Takahashi, S. (2013). Novel angiotensin I-converting enzyme inhibitory peptides derived from soya milk. *Food Chemistry*, 136(2), 612–616. https://doi.org/10.1016/j.foodchem.2012.08.080

Tonolo, F., Moretto, L., Folda, A., Scalcon, V., Bindoli, A., Bellamio, M., Feller, E., & Rigobello, M. P. (2019). Antioxidant properties of fermented soy during shelf life. *Plant Foods for Human Nutrition*, 74(3), 287–292. https://doi.org/10.1007/s11130-019-00738-6

Wang, T., Qin, G. X., Sun, Z. W., & Zhao, Y. (2014). Advances of research on glycinin and β-conglycinin: A review of two major soybean allergenic proteins. *Critical Reviews in Food Science and Nutrition*, 54(7), 850–862. https://doi.org/10.1080/10408398.2011.613534

Wang, W., & de Mejia, E. G. (2005). A new frontier in soy bioactive peptides that may prevent age-related chronic diseases. *Comprehensive Reviews in Food Science and Food Safety*, 4(4), 63–78. https://doi.org/10.1111/j.1541-4337.2005.tb00075.x

Wang, W., Dia, V. P., Vasconez, M., de Mejia, E. G., & Nelson, R. L. (2008). Analysis of soybean protein-derived peptides and the effect of cultivar, environmental conditions, and processing on lunasin concentration in soybean and soy products. *Journal of AOAC International*, 91(4), 936–946

Wang, Y. C., Yu, R. C., & Chou, C. C. (2006). Antioxidative activities of soymilk fermented with lactic acid bacteria and bifidobacteria. *Food Microbiology*, 23(2), 128–135. https://doi.org/10.1016/j.fm.2005.01.020

Wang, Yanwen, Jones, P. J. H., Ausman, L. M., & Lichtenstein, A. H. (2004). Soy protein reduces triglyceride levels and triglyceride fatty acid fractional synthesis rate in hypercholesterolemic subjects. *Atherosclerosis*, 173(2), 269–275. https://doi.org/10.1016/j.atherosclerosis.2003.12.015

Wang, Yurong, Li, F., Chen, M., Li, Z., Liu, W., & Wang, C. (2015). Angiotensin I-converting enzyme inhibitory activities of Chinese traditional soy-fermented Douchi and Soypaste: Effects of processing and simulated gastrointestinal digestion. *International Journal of Food Properties*, 18(4), 934–944. https://doi.org/10.1080/10942912.2014.913180

Wang, Z., Bao, Y., Zhang, Y., Zhang, J., Yao, G., Wang, S., & Zhang, H. (2013). Effect of soymilk fermented with Lactobacillus plantarum P-8 on lipid metabolism and fecal microbiota in experimental hyperlipidemic rats. *Food Biophysics*, 8(1), 43–49. https://doi.org/10.1007/s11483-012-9282-z

Weng, T. M., & Chen, M. T. (2011). Effect of two-step fermentation by Rhizopus oligosporus and bacillus subtilis on protein of fermented soybean. *Food Science and Technology Research*, 17(5), 393–400. https://doi.org/10.3136/fstr.17.393

Wu, H., Zhang, Z., Huang, H., & Li, Z. (2017). Health benefits of soy and soy phytochemicals. *American Medical Journal*, 2, 162–162. https://doi.org/10.21037/amj.2017.10.04

Yadav, A. N., Yadav, N., Kour, D., Kumar, A., Yadav, K., Kumar, A., Rastegari, A. A., Sachan, S. G., Singh, B., Chauhan, V. S., & Saxena, A. K. (2019). Bacterial community composition in lakes. In *Freshwater Microbiology*. Elsevier Inc.. https://doi.org/10.1016/b978-0-12-817495-1.00001-3

Yadavalli, S. S., Carey, J. N., Leibman, R. S., Chen, A. I., Stern, A. M., Roggiani, M., Lippa, A. M., & Goulian, M. (2016). Antimicrobial peptides trigger a division block in Escherichia coli through stimulation of a signalling system. *Nature Communications*, 7(1), 12340. https://doi.org/10.1038/ncomms12340

Yang, H. J., Kim, H. J., Kim, M. J., Kang, S., Kim, D. S., Daily, J. W., Jeong, D. Y., Kwon, D. Y., & Park, S. (2013a). Standardized chungkookjang, short-term fermented soybeans with Bacillus lichemiformis, improves glucose homeostasis as much as traditionally made chungkookjang in diabetic rats. *Journal of Clinical Biochemistry and Nutrition*, 52(1), 49–57. https://doi.org/10.3164/jcbn.12-54

Yang, H. J., Kwon, D. Y., Moon, N. R., Kim, M. J., Kang, H. J., Jung, D. Y., & Park, S. (2013b). Soybean fermentation with Bacillus licheniformis increases insulin sensitizing and insulinotropic activity. *Food and Function*, 4(11), 1675–1684. https://doi.org/10.1039/c3fo60198f

Yang, J., Wu, X.B., Chen, H.L., Sun-waterhouse, D., Zhong, H.B., & Cui, C. (2019). A value-added approach to improve the nutritional quality of soybean meal byproduct: Enhancing its antioxidant activity through fermentation by Bacillus amyloliquefaciens SWJS22. *Food Chemistry*, 272, 396–403. https://doi.org/10.1016/j.foodchem.2018.08.037

Yeo, I. C., Lee, N. K., & Hahm, Y. T. (2012). Genome sequencing of Bacillus subtilis SC-8, antagonistic to the Bacillus cereus group, isolated from traditional Korean fermented- soybean food. *Journal of Bacteriology*, 194(2), 536–537. https://doi.org/10.1128/JB.06442-11

Yoshikawa, M. (2015). Bioactive peptides derived from natural proteins with respect to diversity of their receptors and physiological effects. *Peptides*, *72*, 208–225. Elsevier Inc. https://doi.org/10.1016/j.peptides.2015.07.013

Yoshikawa, M., Fujita, H., Matoba, N., Takenaka, Y., Yamamoto, T., Yamauchi, R., Tsuruki, H., & Takahata, K. (2000). Bioactive peptides derived from food proteins preventing lifestyle-related diseases. *BioFactors*, *12*(1–4), 143–146. https://doi.org/10.1002/biof.5520120122

Yu, B., Lu, Z. X., Bie, X. M., Lu, F. X., & Huang, X. Q. (2008). Scavenging and anti-fatigue activity of fermented defatted soybean peptides. *European Food Research and Technology*, *226*(3), 415–421. https://doi.org/10.1007/s00217-006-0552-1

Zhang, H., Bartley, G. E., Zhang, H., Jing, W., Fagerquist, C. K., Zhong, F., & Yokoyama, W. (2013). Peptides identified in soybean protein increase plasma cholesterol in mice on hypercholesterolemic diets. *Journal of Agricultural and Food Chemistry*, *61*(35), 8389–8395. https://doi.org/10.1021/jf4022288

Zhang, J., Bilal, M., Liu, S., Zhang, J., Lu, H., Luo, H., Luo, C., Shi, H., Iqbal, H. M. N., & Zhao, Y. (2020). Isolation, identification and antimicrobial evaluation of bactericides secreting Bacillus subtilis Natto as a biocontrol agent. *Processes*, *8*(3), 259. https://doi.org/10.3390/pr8030259

Zhang, J. H., Tatsumi, E., Ding, C. H., & Li, L. Te (2006). Angiotensin I-converting enzyme inhibitory peptides in douchi, a Chinese traditional fermented soybean product. *Food Chemistry*, *98*(3), 551–557. https://doi.org/10.1016/j.foodchem.2005.06.024

Zhang, M., & Mu, T. H. (2017). Identification and characterization of antioxidant peptides from sweet potato protein hydrolysates by alcalase under high hydrostatic pressure. *Innovative Food Science and Emerging Technologies*, *43*, 92–101.

Zhang, Y., Liu, C., Dong, B., Ma, X., Hou, L., Cao, X., & Wang, C. (2015). Anti-inflammatory activity and mechanism of surfactin in lipopolysaccharide-activated macrophages. *Inflammation*, *38*(2), 756–764. https://doi.org/10.1007/s10753-014-9986-y

21 Fatty Acid Composition in the Soybean Sprout

*Hyun Jo, Syada Nizer Sultana,
Jong Tae Song, and Jeong-Dong Lee*

CONTENTS

21.1 INTRODUCTION

Fat is an essential dietary source derived from fatty acids with various carbon chain lengths and degrees of saturation. The composition of foodstuff containing fatty acids is essential to regulate how the body responds to energy intake. Fatty acids are powerful signaling molecules that regulate gene transcription (Afman & Müller, 2012; Petrus & Arner, 2019). In general, oils are triacylglycerols, composed of a glycerol backbone attached to three fatty acids, whose composition influences the application of the oil (Ensminger & Ensminger, 1993). Five major fatty acids are found in oils derived from crops, including palmitic acid, stearic acid, oleic acid, linoleic acid, and α-linolenic acid. The composition of fatty acids differs for each oil crop, as shown in Table 21.1. Based on the number of double bond(s) found in the carbon chains of fatty acids, oils can be defined as either saturated or unsaturated fatty acids. The saturated fatty acids are palmitic and stearic acids, while the unsaturated fatty acids in crop oil are divided into two categories, monounsaturated fatty acids (oleic acid) and polyunsaturated fatty acids (linoleic and α-linolenic acid). Linoleic (ω-6) and α-linolenic acid (ω-3) are essential fatty acids and are the precursors of eicosapentaenoic acid (EPA) and docosahexaenoic acid (DHA). α-linolenic acid is found mainly in plant oils, such as flaxseed, perilla, soybean, and canola oils.

Soybean (*Glycine max* [L.] Merrill) seeds consist of 40% protein and 20% oil. Because of its seed composition, it is one of the most economically important oil crops globally. Fifty-nine percent

TABLE 21.1

Fatty Acid Composition of Different Oil Crops

Oil Crop Seed	Palmitic Acid	Stearic Acid	Oleic Acid	Linoleic Acid	α-Linolenic Acid	References
			Fatty Acid (%)			
Alfalfa	15.9	3.2	10.4	34.3	24.9	Márton et al.,
Lentil	26.2	1.6	14.0	19.4	3.3	2010b
Wheat	31.2	1.9	10.7	25.6	2.0	
Radish	8.2	3.2	35.1	15.5	0.1	
Sunflower	5.8	5.4	21.7	65.0	0.1	
Almond	6.8	2.3	67.2	22.8	–	Orsavova et al.,
Coconut oil	–	2.7	6.2	1.6	–	2015
Grape	6.6	3.5	14.3	74.7	0.2	
Olive	16.5	2.3	66.4	16.4	1.6	
Peanut	7.5	2.1	71.1	18.2	–	
Rapeseed	4.6	1.7	63.3	19.6	1.2	
Safflower	6.7	2.4	11.5	79.0	0.2	
Sesame	9.7	6.5	41.5	40.9	0.2	
Soybean	11.0	4.0	23.0	55.0	8.0	Dhakal et al., 2009

FIGURE 21.1 Soybean sprouts (left) and soybean sprout products (right).

of the world's oilseed production derives from soybean seeds followed by rapeseed, sunflower, peanut, cotton, and palm kernel (Soystats, 2020). Fatty acids present in soybean oil consist of 11% palmitic acid, 4% stearic acid, 23% oleic acid, 55% linoleic acid, and 8% α-linolenic acid (Fehr, 2007). We will focus on the variation in fatty acid content in soybean sprouts. In addition, factors affecting the fatty acid composition when grown from seeds to soybean sprouts are presented.

21.2 SOYBEAN SPROUTS

For centuries, soybean sprouts have been a popular vegetable, particularly in regions where seasonal vegetables are not readily available during the winter season (Figure 21.1) (Ghani et al., 2016; Shi et al., 2010; Yang et al., 2015). Soybean sprouts are simple to cultivate, which makes them desirable vegetables throughout the year (Ghani et al., 2016; Lee et al., 2007a). Soybean sprouts (the Korean name is "Kongnamul") have been chosen as food in Korea, Japan, and China. Different dishes use soybean sprouts as an ingredient, such as mixed with rice, soybean sprout soup, and mixed in salads.

Generally, small soybean seeds, which weigh less than 120 mg/seed, are usually preferred for sprout production. Small soybean seeds are known to have better water absorption, longer hypocotyl length, higher percent germination, and higher sprout yield. The growth period for soybean sprouts is 5–7 days after germination. Delaying harvest may have a negative impact on soybean sprout quality due to the growth of lateral roots and leaves (Ghani et al., 2016). Soybean sprouts consist of a cotyledon, hypocotyl, and root. The hypocotyl is the main part of the sprout. A study indicated that a hypocotyl length ranging from 8–12 cm and a hypocotyl thickness in the range of 2.0–2.2 mm, after five days of sprouting, are desirable for the sprout's utilization in food products (Lee et al., 2007a).

21.3 COMPARISON OF THE NUTRITIONAL COMPOSITION OF SOYBEAN SEEDS AND SPROUTS

Sprouts are good sources of protein, vitamins, minerals, phytochemicals, enzymes, and amino acids, which are the most useful for human health (Finley, 2005; Webb, 2006). Sprouting from seeds increases the nutritional and functional properties of cowpea (Devi et al., 2015; Giami, 1993), sorghum (Abbas & Musharaf, 2008), and wheat (Hussain & Uddin, 2012; Islam et al., 2017). In soybean, differences in nutritional components have been observed during sprouting, such as altered protein and oil levels. Specifically, crude protein content increases during soybean sprouting, whereas oil content gradually decreases from 15% to 10% (Kim, 1981; Lee & Chung, 1982; Shi et al., 2010).

Free amino acid content also varies during the sprouting process and across soybean genotypes (Ghani et al., 2016). Generally, levels of most free amino acids are high in soybean sprouts and increase during the sprouting process (Mizuno & Yamada, 2006; Yang, 1981). Significant increases have been observed in the concentrations of asparagine, glutamate, histidine, alanine, proline, lysine, valine, and isoleucine during sprouting (Friedman & Brandon, 2001; Villaluenga et al., 2006). Of these, asparagine content in the soybean sprout (7,423.3 mg/100 g) is dramatically increased from that found in the mature soybean seed (42.9 mg/100 g) (Byun et al., 1977). Soybean sprouts are a well-known antidote for hangovers because asparagine has been reported to detoxify the acetaldehyde produced during alcohol metabolism in humans (Lee & Hwang, 1996).

Total isoflavone content ranges from 0.05% to 0.5% of the soybean seed based on dry weight (Lee et al., 2004). Generally, the isoflavone content in soybean sprouts is higher than in the seed (Kim et al., 2003, 2004, 2006). Isoflavone is present in the roots, cotyledons, and hypocotyls of soybean plants and sprouts. The accumulation of isoflavone content in the soybean seed and sprout is determined by the genotype and environmental factors including the year of seed production, temperature, light, and soybean field conditions (Hoeck et al., 2000; Kim et al., 2003, 2006; Seguin et al., 2004; Xu et al., 2003; Zhu et al., 2005).

Soybean seeds contain vitamins A, B1, E, and C (Collin & Sanders, 1976). Studies have revealed that the soybean sprout contains higher vitamin content than the soybean seed (Plaza et al., 2003; Youn et al., 2011). A two-fold increase of vitamin B1 levels has been reported to occur during sprouting (Collin & Sanders, 1976). After five days, vitamin C content in the sprout has been reported to increase up to 11 mg/100 g (Liu, 1999). Further, Plaza et al. (2003) revealed that the sprouting soybean seed resulted in marked increases in vitamin A, C, E, and group B of vitamins.

Saponins are secondary plant metabolites of considerable interest because of their health benefits. Kang et al. (2010) evaluated 79 Korean soybean cultivars to determine soyasapogenol content. Soybean sprouts had higher levels of total soyasapogenol (4.88%) and soyasapogenol B (8.31%), whereas soyasapogenol A content in soybean sprouts was approximately 1.6% lower compared to the soybean seed. In addition, the crude saponin content increased from 4.59 mg/g in seeds to 5.33 mg/g in six-day-old sprouts (Oh et al., 2003).

Soybean seeds contain approximately 16.6% soluble carbohydrates (Hymowitz & Collins, 1974). Generally, mature soybean seeds contain 41–68% sucrose, 5–16% raffinose, and 12–35% stachyose soluble carbohydrates (Verma & Shoemaker, 1996). The sugar content in the soybean seed decreases during the sprouting process. Shi et al. (2010) found that total sugar content decreases from 19.9%

in the soybean seed to 14% in seven-day-old sprouts. Similarly, Silva et al. (1990) reported a greater than 90% decrease in the stachyose and raffinose content in Brazilian soybean cultivars after four days of germination. However, sucrose content decreases up to four days after germination but then increases six days after germination.

Soybean sprouts contain different minerals important for human nutrition. The changes in mineral content from the seed to sprout stages of the soybean vary from 23.93 µg/g to 85.88 µg/g for zinc, from 810 µg/g to 2,770 µg/g for calcium, from 2,610 to 8,60 µg/g for sodium, from 6.72 to 19.6 µg/g for manganese, from 2,070 to 10,170 µg/g for potassium, from 11.43 to 19.61 µg/g for copper, from 1,330 to 1,510 µg/g for magnesium, and from 48.87 to 35.29 µg/g for iron (Plaza et al., 2003).

21.4 VARIATION IN FATTY ACID COMPOSITIONS ACROSS DIFFERENT CROPS BETWEEN SEEDS AND SPROUTS

Differences in fatty acid compositions in different crops between seeds and sprouts are shown in Table 21.2. The alfalfa sprout contains approximately 22.4% palmitic acid, 4.4% stearic acid, 9.8% oleic acid, 29.1% linoleic acid, and 15.9% α-linolenic acid in the seven-day-old alfafa sprout, whereas these levels are approximately 15.9% palmitic acid, 3.2% stearic acid, 10.4% oleic acid, 34.3% linoleic acid, and 24.9% α-linolenic acid in the seed (Márton et al., 2010). For the lentil, palmitic acid was one of the highest fatty acid components in the seed (26.2%) and in the three-day-old sprout (27.0%). Linoleic acid content increases to 27.4% in the lentil sprout, while oleic acid decreases to 9.3% in the sprout compared to 14.0% in the seed. The radish presents the most abundant fatty acid component, oleic acid, which is present at 35.1% in the seed and 34.6% in the sprout. The fatty acid composition in the sunflower consists of linoleic acid, which has the highest concentration in the seed (65.0%) followed by the sprout (64.8%). The highest fatty acid in wheat is palmitic acid, which is present at 31.2% in the

TABLE 21.2
Comparison of Fatty Acid Composition of Seeds and Sprouts in Oil Crops

Crop Sprouts	Sprouting Period	Fatty Acid (%)					Reference
		Palmitic Acid	Stearic Acid	Oleic Acid	Linoleic Acid	α-Linolenic Acid	
Alfalfa	Seed	15.9	3.2	10.4	34.3	24.9	Márton et al., 2010b
	7 days	22.4	4.4	9.8	29.1	15.8	
Lentil	Seed	26.2	1.6	14	19.4	3.3	
	3 days	27.0	2.2	9.3	27.4	4.7	
Radish	Seed	8.2	3.2	35.1	15.5	13.6	
	6 days	8.0	3.0	34.6	15.9	14.0	
Sunflower	Seed	5.8	5.4	21.7	65.0	0.1	
	5 days	5.8	5.6	20.9	64.8	1.0	
Wheat	Seed	31.2	1.9	10.7	25.6	2.0	
	3 days	33.5	1.2	7.8	27.3	2.5	
Barley	Seed	19.5	1.4	14.8	53.2	5.7	Chung et al., 1989
	5 days	19.9	1.4	14.2	54.3	5.5	
Canola	Seed	4.4	1.8	60.1	20.5	10.9	
	5 days	4.3	1.8	55.0	20.7	11.1	
Mung bean	Seed	16.2	6.3	15.8	33.1	19.5	Abdel-Rahman et al., 2007
	3 days	17.5	7.7	14.4	30.6	18.3	
Soybean	Seed	11.0	2.3	18.5	59.8	8.4	Lee et al., 2002
	5 days	10.1	2.7	18.0	61.7	7.6	

seed and 33.5% in the sprout. Chung et al. (1989) reported the fatty acid composition in the seed and sprouts of barley and canola. After five days of sprouting of these crops, the oleic acid levels declined to 4.0% and 9.8% in barley and canola, respectively. The study assumed that oleic acid was involved in lipid metabolism in germinating barley and canola (Chung et al., 1989). On the third day of sprouting, the mung bean contains approximately 17.5% palmitic acid, 7.7% stearic acid, 14.4% oleic acid, 30.6% linoleic acid, and 18.3% α-linolenic acid (Abdel-Rahman et al., 2007).

Tokiko and Koji (2006) investigated functional components including oil and fatty acid compositions of sprouts. The authors found that the oil content ranged from 0.4% to 1.6% in various sprouts. With regard to the fatty acid content, α-linolenic acid was present in the highest concentration, with 23% in buckwheat, 48% in soybeans, 47.7% in clover, and 40.6% in peas, respectively (Márton et al., 2010; Tokiko & Koji, 2006). In comparison with other sprouted legumes, soybean sprouts appear to be nutritionally superior (Hedges & Lister, 2006). The soybean sprout contains approximately 16.2% palmitic acid, 4.9% stearic acid, 28.8% oleic acid, 42.9% linoleic acid, and 7.2% α-linolenic acid (Dhakal et al., 2009). Soybean sprouts as well as seeds are a good alternative source of essential fatty acids (Narina et al., 2013; Orhan et al., 2007).

21.5   CHANGES IN FATTY ACID LEVELS FROM SEED TO SPROUT IN SOYBEAN

Soybean is one of the most important oil crops for food and other applications. Studies have evaluated changes in oil content and the fatty acid compositions from seed to sprout in the soybean. The oil content generally decreases from the seed to sprout; it is 18.3% in seeds based on an average of five cultivated soybeans and becomes 13.6% in soybean sprouts, but is 10.7% in seeds based on an average of five wild soybeans (*Glycine soja* Sieb. and Zucc.) and decreases to 5.9% in wild soybean sprouts (Lee et al., 2002). Shi et al. (2010) also reported that oil content decreased from 15% in the seed to 10% in the seven-day-old soybean sprout.

The quality of soybean oil is determined by the respective fatty acid profiles. Fatty acid compositions vary slightly during soybean sprouting. For example, in the soybean seed, palmitic acid levels range from 8.0–15.1%, stearic acid from 2.6–6.2%, oleic acid from 21.4–82.4%, linoleic acid from 2.1–54.3%, and α-linolenic acid from 3.7–10.3%, whereas in soybean sprouts, palmitic acid levels range from 7.9–18.0%, stearic acid from 2.6–6.4%, oleic acid from 17.7–77.6%, linoleic acid from 6.6–57.2%, and α-linolenic acid from 4.5–12.3% (Table 21.3). By evaluating different sprouting days of six soybean cultivars, it was determined that the genotype had a significant effect on the accumulation of fatty acid composition in the soybean sprout, and sprouting days were an important factor determining the accumulation of palmitic, linoleic, and α-linolenic acids (Dhakal et al., 2009). Dhakal et al. (2009) also reported that the trend of fatty acid composition in soybean sprouts grown for five, six, and seven days was similar to those in soybean seeds, except palmitic and α-linolenic acids. Twelve soybean accessions with six high oleic acid lines, three parents of high oleic acid lines, and three checks with normal and elevated oleic acid concentrations are showed in Table 21.3 (Dhakal et al., 2014). The study concluded that oleic acid, linoleic acid, and α-linolenic acid in prouts from each accession were similar to those found in the soybean seed. The oleic acid concentration in the sprouts of high-containing oleic acid species was still high. Thus, the authors reported that high oleic soybean lines could provide positive health benefits via their sprouts to individuals who consume soybean sprouts as a vegetable.

Changes in fatty acid compositions from seeds to sprouts of the wild soybean are shown in Table 21.4 (Lee et al., 2002). On day five of sprouting, *Glycine soja* (n = 5) contained 11.3% palmitic acid in seeds and 10.2% in the sprout, 2.8% stearic acid in the seed and 2.6% in the sprout, 14.5% oleic acid in the seed and 13.4% in the sprout, 57.9% linoleic acid in the seed and 61.8% in the sprout, and 13.5% α-linolenic acid in the seed and 12.1% in the sprout (Lee et al., 2002). Although there were no significant differences in hypocotyl length and sprout yield between sprouts of cultivated and wild soybeans, the rate of good quality of sprouts was higher in the cultivated soybean (Lee et al., 2002).

TABLE 21.3

Changes in Fatty Acid Compositions During Sprouting Periods in Cultivated Soybeans

Crop Sprouts	Sprouting Period	Palmitic Acid	Stearic Acid	Oleic Acid	Linoleic Acid (ω-6)	α-Linolenic Acid (ω-3)	ω-6/ω-3	Note	Reference
Pungsannamul	Seed	15.1	4.2	30.9	43.8	6.0	7.3	Soybean sprout cultivar	Dhakal et al., 2009
	5 days	16.2	4.7	28.2	43.9	7.1	6.2		
	7 days	16.2	4.9	28.8	42.9	7.2	6.0		
Cheongja	Seed	14.5	3.7	25.4	47.7	8.7	5.5	Black soybean cultivar	
	5 days	16.6	5.1	17.7	52.5	8.1	6.5		
	7 days	18.0	5.1	20.9	49.2	6.9	7.2		
Eunha	Seed	15.1	5.0	27.0	45.5	7.4	6.2	Soybean sprout cultivar	
	5 days	14.3	4.0	22.6	51.4	7.7	6.7		
	7 days	15.8	5.4	20.4	50.9	7.5	6.8		
Hwangkeum	Seed	13.7	3.9	37.3	38.9	6.2	6.3	Yellow soybean cultivar	
	5 days	15.2	4.3	31.9	41.5	7.1	5.8		
	7 days	16.1	4.4	30.2	41.3	8.1	5.1		
KLG12072	Seed	11.8	4.7	51.7	23.6	8.2	2.9	Elevated oleic acid	
	5 days	12.4	4.4	41.7	33.3	8.3	4.0		
	7 days	14.4	3.1	43.1	30.8	8.6	3.6		
KLG12228	Seed	14.1	6.2	23.1	52.9	3.7	14.4	Elevated α-linolenic acid	
	5 days	14.3	6.0	18.0	57.2	4.5	12.7		
	7 days	14.8	6.4	18.2	56.0	4.6	12.2		

(Continued)

TABLE 21.3 (CONTINUED)

Changes in Fatty Acid Compositions During Sprouting Periods in Cultivated Soybeans

Crop Sprouts	Sprouting Period	Palmitic Acid	Stearic Acid	Oleic Acid	Linoleic Acid (ω-6)	α-Linolenic Acid (ω-3)	ω-6/ω-3	Note	Reference
S08-14719	Seed	8.3	2.7	82.4	2.1	4.5	0.5	High oleic acid	Dhakal et al, 2014
	5 days	8.3	2.6	77.6	6.6	4.8	1.4	(FAD2-1A and FAD2-1B)*	
	6 days	8.3	2.6	75.2	8.9	5.1	1.7		
S08-14692	Seed	8.4	3.4	76.2	5.4	6.6	0.8	High oleic acid	
	5 days	8.5	3.3	74.5	8.2	5.6	1.5	(FAD2-1A and FAD2-1B)	
	6 days	8.3	3.3	73.0	9.8	5.6	1.8		
S08-14622	Seed	8.9	3.6	59.0	20.7	7.8	2.7	High oleic acid	
	5 days	9.5	3.8	55.1	24.5	7.3	3.4	(FAD2-1A and FAD2-1B)	
S08-14709	Seed	8.4	2.6	69.2	11.7	8.3	1.4	High oleic acid	
	5 days	7.9	2.8	70.4	11.4	7.7	1.5	(FAD2-1A and FAD2-1B)	
S08-14610	Seed	8.3	2.6	62.9	16.3	10.1	1.6	High oleic acid	
	5 days	8.4	2.8	61.2	18.8	9.0	2.1	(FAD2-1A and FAD2-1B)	
S08-14700	Seed	8.0	2.9	74.3	7.0	8.0	0.9	High oleic acid	
	5 days	7.9	2.9	71.4	10.4	7.6	1.4	(FAD2-1A and FAD2-1B)	
S08-14717	Seed	8.4	3.1	75.7	6.0	6.9	0.9	High oleic acid	
	5 days	8.1	3.3	74.0	8.2	6.5	1.3	(FAD2-1A and FAD2-1B)	
5002T	Seed	11.4	4.3	21.4	54.3	8.7	6.3	Soybean cultivar	
	5 days	11.9	4.8	22.4	53.1	7.9	6.8		
Kwangan	Seed	13.4	4.0	29.1	44.6	9.0	5.0	Soybean cultivar	
	5 days	13.8	4.0	27.6	45.5	9.2	4.9		
PI567189A	Seed	12.1	3.4	26.8	48.3	9.5	5.1	FAD2-1B	
	5 days	12.1	3.6	20.9	53.0	10.5	5.1		
PI283327	Seed	9.7	3.9	42.3	33.9	10.3	3.3	FAD2-1B	
	5 days	10.3	4.0	20.2	53.3	12.3	4.3		
M23	Seed	10.1	3.9	44.1	33.7	8.2	4.1	FAD2-1A	
	5 days	9.9	3.5	44.3	34.2	8.1	4.2		
	6 days	9.9	3.6	44.0	34.4	8.2	4.2		

*The *FAD2-1A* and *FAD2-1B* indicates a mutation in the microsomal delta-12 fatty acid desaturase 2 (*FAD2*) genes that encode enzymes involved in the conversion of oleic acid to linoleic acid in fatty acid biosynthesis, which are similar to the Williams 82 reference genome for *Glyma.10g278000* and *Glyma.20g111000*, respectively.

TABLE 21.4

Comparison of Fatty Acid Compositions in Seed and Sprouts of Wild Soybeans

Soybean Accessions	Species	Seed or Sprout	Fatty Acid (%)						Reference
			Palmitic Acid	Stearic Acid	Oleic Acid	Linoleic Acid (ω-6)	α-Linolenic Acid (ω-3)	ω-6/ω-3	
IT183049	*Glycine soja*	Seed	10.8	2.5	16.8	55.9	14.1	4.0	Lee et al., 2002
		5 days	9.9	2.0	15.4	61.0	11.6	5.3	
IT184172		Seed	11.6	3.3	13.2	57.2	14.8	3.9	
		5 days	10.2	2.9	11.3	64.1	11.6	5.5	
IT184178		Seed	10.7	2.6	14.6	59.9	12.1	5.0	
		5 days	9.5	2.1	12.8	63.2	12.5	5.1	
IT184256		Seed	10.2	2.8	11.9	59.7	15.5	3.9	
		5 days	9.6	3.2	12.9	61.0	13.4	4.6	
KLG10079		Seed	13.4	2.6	16.2	56.8	11.0	5.2	
		5 days	11.7	2.7	14.6	59.6	11.4	5.2	

TABLE 21.5

Differences in Fatty Acid Composition Across Soybean Cultivars for Fresh and Boiled Sprouts

Soybean Cultivar	Status of Sprout	Fatty Acid (%)					Reference
		Palmitic Acid	Stearic Acid	Oleic Acid	Linoleic Acid	α-Linolenic Acid	
Nokchae	Fresh sprout	13.2	4.1	22.2	50.4	10.0	Kim et al., 2011
	Boiled sprout	13.1	4.1	22.3	50.8	9.9	
Dawon	Fresh sprout	13.7	5.1	14.1	56.3	10.8	
	Boiled sprout	13.5	5.3	15.4	56.0	9.9	
Seonam	Fresh sprout	13.9	4.1	19.4	53.7	8.3	
	Boiled sprout	13.3	4.1	20.0	53.1	9.5	
Orialtae	Fresh sprout	13.1	4.4	13.8	54.3	14.3	
	Boiled sprout	13.4	4.5	13.6	54.1	14.4	
Pungsannamul	Fresh sprout	13.2	4.2	22.8	51.1	8.5	
	Boiled sprout	13.4	4.2	23.7	50.6	8.1	

In addition, fatty acid compositions of fresh soybean sprouts and boiled soybean sprouts are reported in Table 21.5 (Kim et al., 2011). Upon evaluation of five soybean cultivars of soybean sprouts, the fatty acid compositions achieved were not significantly different between fresh and boiled soybean sprouts, and boiling did not affect the fatty acid composition in soybean (Kim et al., 2011).

21.6   DIFFERENCES OF FATTY ACID CONCENTRATION ACROSS SOYBEAN GENOTYPES

Genotype is an important factor that determines the fatty acid composition in seeds and sprouts. The fatty acid profiles in soybean are primarily determined by genetic variants (Abdelghany et al.,

2020; Chae et al., 2015; Ginting et al., 2018; Goffman et al., 2003; Kulkarni et al., 2017; Lee et al., 2007b; Mourtzinis et al., 2017). The variations in fatty acid content based on species differences are described in Table 21.3. Soybean cultivars (Pungsannamul and Eunha) developed for the cultivation of soybean sprouts are shown to be slightly changed in their fatty acid composition on sprouting (Table 21.3). The Kwangan cultivar contains approximately 42% oleic acid (Lee, 2012) with 39.7% oleic acid (Dhakal et al., 2006) in the sprout. The high oleic acid content variety S08-14719 containing 82.4% oleic acid, 2.1% linoleic acid, and 4.5% linolenic acid in the seed contains 77.6%, 6.6%, and 4.8%, respectively in the five-day-old sprout (Table 21.3). Such variations offer possibilities for the development of new cultivars with improved fatty acid content in the soybean sprout. Variations in fatty acid composition due to varietal differences have also been reported for canola and flaxseed (Bhardwaj & Hamama, 2009; Kanmaz & Ova, 2015).

21.7 CHANGES IN FATTY ACID COMPOSITIONS ON DIFFERENT SPROUTING DAYS

The harvesting time of the soybean sprouts may vary based on the germination rate and water temperature (Ghani et al., 2016). Sprouts can be ready for harvest after five to seven days after germination. Delaying harvest may have negative impacts on sprout quality due to the undesired growth of lateral roots and leaves (Liu, 1999; Silva et al., 2013). Brown et al. (1962) measured the fatty acid compositions in soybean cotyledons and the seedling axis of the soybean cultivar Chippewa following germination in the dark for up to 12 days. The proportion of saturated fatty acids increased from 15.5% on the first day after germination to 17.4%, 12 days after germination. Oleic acid increased up to the fourth day after germination and then decreased to 10.9% oleic acid after 12 days of germination. The linoleic acid content was constant up to four days after germination and then increased to 57.5% at 12 days after germination. Levels of α-linolenic acid concentrations showed no significant changes during 12 days of germination (Brown et al., 1962). Joshi et al. (1973) reported a decrease in palmitic and oleic acids in soybean cotyledons of the soybean cultivar Lee during germination in the dark for 6–12 days. Dhakal et al. (2009) also reported that the trend in fatty acid composition in soybean sprouts and cotyledons grown in conditions for five, six, and seven days was similar to that in the soybean seed, except for palmitic and α-linolenic acids. In addition, Kim et al. (2011) reported that fatty acid levels in the whole and cotyledon portions of the soybean sprout did not change during sprouting, whereas the content in the hypocotyl increased by approximately 13.2% of the total unsaturated fatty acid composition compared to whole soybean sprout. Taken together, this evidence suggests that since five-day soybean sprouts are recommended for soybean sprout production and quality, fatty acid compositions remain relatively unchanged during the recommended soybean sprouting period.

21.8 CHANGES IN FATTY ACID COMPOSITIONS DUE TO TEMPERATURE AND LIGHT CONDITIONS DURING SPROUTING

Temperature is an environmental factor that affects the composition of fatty acid as it strongly influences their biosynthesis (Abdelghany et al., 2020; Rennie & Tanner, 1989; Tsukamoto et al., 1995). Water and air temperature exert a significant influence on soybean germination, sprout quality, and yield. Low temperatures have adverse effects on the time of initial germination to sprout harvest, hypocotyl length, thickness, and sprout quality. Generally, soybean sprouts are grown at temperatures between 20° C and 23° C (Ghani et al., 2016). A water temperature above 20° C is generally ideal during seed imbibition (Lee et al., 2007c), whereas Koo et al. (2015) reported that a germination temperature of 25° C was more effective for sprouting. The protein content and antioxidant activity (DPPH and ADH activity) of soybean sprouts was increased at 25° C, whereas the total fatty acid concentration was reduced at 25° C. Kumar et al. (2006) examined the effects of temperature

(25° C or 35° C) on sprouting and reported there were limited changes in fatty acid composition in six-day-old seedlings. For both temperature conditions, there was no significant difference in fatty acid composition until after five days of germination. In the six-day-old seedling, lower levels of oleic acid content were observed, although α-linolenic acid levels were not significantly altered (Kumar et al., 2006).

Light has a negative effect on soybean sprout quality during germination. Exposure to light during soybean sprouting results in root elongation (short root preferred) and initiates photosynthesis, giving the cotyledons a green color, which is not preferred by consumers (Liu, 1999; Shi et al., 2010). However, Lee et al. (2007c) reported that there were no significant differences in total isoflavone concentrations between green sprouts exposed to light and yellow sprouts grown in the dark, but altered its root, cotyledon, and hypocotyl structure. Yang et al. (1982) investigated the effects of exposure to light for different times on fatty acid compositions in the root, cotyledon, and hypocotyl of soybean sprouts. The authors reported that the proportion of linoleic and α-linolenic acid in the hypocotyl of the soybean sprout increased with increasing exposure to light, while levels of palmitic and stearic acids decreased in the hypocotyls. In the root, palmitic and α-linolenic acids decreased with increasing exposure to light.

21.9 SOYBEAN SEED COAT COLOR AFFECTS FATTY ACID COMPOSITIONS IN SOYBEAN SPROUTS

Soybean seed coats exist in a range of colors from black to brown, green, and yellow. Commercially grown soybean cultivars (*Glycine max*) have yellow seeds and some accumulate anthocyanins within the epidermal layer of the seed coat leading to a colored seed that is black and brown (Todd & Vodkin, 1993; O'Bryan et al., 2014). Dhakal et al. (2006) reported a significant variation in fatty acid composition in soybean sprouts of different seed coat colors (Table 21.6). The oleic acid content in the soybean sprouts with green and yellow seed coat colors was higher than those having black or brown seed coat colors (Table 21.6). However, the linoleic acid content in soybean seeds that are green and yellow in color was lower than in seeds having black and brown seed coat color. Yellow soybeans presented the highest levels of α-linolenic acid content and the lowest ratio of ω-6/ω-3 content (Table 21.6). Seed coat colors of the soybean may be associated with the fatty acid compositions in soybean sprouts (Dhakal et al., 2006).

21.10 GENETIC VARIATION OF SOYBEAN GERMPLASM FOR SOYBEAN SPROUTS

Great genetic diversity and marked phenotypic variation have been reported for *Glycine soja*, *Glycine max* germplasm (Lee et al., 2008), and Korean soybean landraces (Cho et al., 2008).

TABLE 21.6
Fatty Acid Compositions in Soybean Sprouts with Different Seed Coat Colors

Seed Coat Color	Number of Accessions	Fatty Acid (%)			ω-6/ω-3	Reference
		Oleic Acid	Linoleic Acid (ω-6)	α-linolenic Acid (ω-3)		
Black	198	24.7	54.8	10.1	5.6	Dhakal et al., 2006
Green	114	27.4	52.4	9.8	5.5	
Brown	47	26.9	53.7	9.5	5.7	
Yellow	300	21.0	56.7	11.1	5.2	

Specific alleles from these diverse species could be transferred to elite soybean cultivars to develop new soybean cultivars for sprout production. Kwon et al. (1981) assessed 164 soybean accessions, which produced smaller than 15 g/100 seeds and exhibited marked genetic variation in terms of sprout characteristics and yield. In addition, Jeong et al. (2007) reported sprout characteristics and yield for 783 soybean accessions distributed by the National Agrobiodiversity Center of Korea. Among the 783 accessions, 18 lines showed better performance in terms of soybean sprout quality and production, and were used for the breeding and development of soybean sprout cultivars.

21.11   ENVIRONMENTAL FACTORS INFLUENCING FATTY ACID COMPOSITIONS IN SOYBEAN SEED

There is concern about the instability of fatty acid composition across cultivars; thus, many studies have explored the effects of exposure to different environments on genotype and on fatty acid composition in soybean seeds. Studies have determined that temperature plays an important role in the synthesis of oil and fatty acids. Generally, higher temperature increases oil content in the soybean seed (Wilson, 2004). Soybeans grown under higher average temperatures present reduced linoleic acid and linolenic acid, and increased oleic acid content. However, the contents of saturated fatty acids showed limited variability under different environments (Dornbos & Mullen, 1992; Hou et al., 2006; Howell & Collins, 1957; Rennie & Tanner, 1989; Wilson, 2004; Wolf et al., 1982). Primomo et al. (2002) concluded that year effects had the largest impact on all fatty acid levels and location effects were significant only for oleic and linolenic acids.

Soybean lines with modified fatty acid compositions were evaluated to determine the stability of fatty acid profiles in different environments (Lee et al., 2007b, 2009). The highly concentrated oleic acid soybeans carrying two mutant alleles of *FAD2-1A* and *FAD2-1B* genes showed relatively stable oleic acid levels (Kim et al., 2015; Lee et al., 2012). In addition, in the variant *M23*, the elevated oleic acid composition remained stable, whereas soybean genotypes with higher α-linolenic acid concentration were less stable than genotypes having lower α-linolenic acid content in different environments (Oliva et al., 2006; Primomo et al., 2002). However, soybeans with elevated linolenic acid obtained by an interspecies cross between *Glycine max* and *Glycine soja* had stable levels of α-linolenic fatty acid concentrations under different environmental conditions (Asekova et al., 2014; Jo et al., 2020).

21.12   BREEDING SOYBEAN CULTIVARS WITH IMPROVED FATTY ACID COMPOSITIONS IN SOYBEAN SPROUTS

Modification of fatty acid profiles in soybean sprouts has mainly been achieved by altering fatty acid compositions in soybean seeds. The genetic basis for altered fatty acid profiles in soybean seed oil is encoded in the genome. Microsomal delta-12 fatty acid desaturase 2 (*FAD2*) genes encode enzymes involved in the conversion of oleic acid to linoleic acid (ω-6) in the fatty acid biosynthesis pathway (Schlueter et al., 2007). The microsomal ω-3 fatty acid desaturases (*FAD3*) catalyze the transformation of linoleic acid into α-linolenic acid (Yadav et al., 1993). Soybean oil with more than 80% oleic acid (high oleic soybean) has been developed from the combination of mutant alleles from both *FAD2-1A* on chromosome 10 (*Glyma.10g278000* for Wm82.a2. v1 assembly) and *FAD2-1B* on chromosome 20 (*Glyma.20g111000*) (Pham et al., 2010, 2012). In soybeans, the levels of α-linolenic acid are controlled by three *FAD3* genes, and combining the mutant alleles of these *FAD3* genes results in soybean oil having ~1% α-linolenic acid (Bilyeu et al., 2005, 2006, 2011; Reinprecht et al., 2009). Recently, a potential gene (*Glyma.05g221500, HD*) was identified as responsible for elevated α-linolenic acid concentrations in soybean seeds (Jo et al., 2021). *HD* is a homeodomain-like transcriptional regulator that may regulate the expression of *FAD3* genes and increases the α-linolenic acid concentrations in soybean oil. The utilization of these genes can improve fatty acid compositions in soybean sprouts with benefits to human health.

Six high oleic acid genotypes are shown in Table 21.3 (*S08-14719*, *S08-14692*, *S08-14622*, *S08-14709*, *S08-14610*, and *S08-14717*), which are homozygous recessive for *FAD2-1A* and *FAD2-1B* genes, were compared to determine differences between soybean seed and sprout fatty acid compositions (Dhakal et al., 2014). The oleic acid concentration in the sprouts remained still high. To evaluate α-linolenic acid content, Ha (2018) examined five soybean lines with elevated α-linolenic acid levels in seeds and sprouts and concluded that the α-linolenic acid levels did not change during a seven-day sprouting period. Those could represent elite accessions for high-quality fatty acid profiles to be used in sprout breeding programs.

21.13 SUMMARY

Soybean sprouts contain approximately 16.2% palmitic acid, 4.9% stearic acid, 28.8% oleic acid, 42.9% linoleic acid, and 7.2% α-linolenic acid. The trends in fatty acid composition in soybean sprouts grown for five, six, and seven days are similar to those in soybean seed, except palmitic and α-linolenic acids. In addition, since five-day soybean sprouts are recommended for soybean sprout production and quality, the fatty acid compositions should remain relatively constant during the recommend soybean sprouting period. Fatty acid compositions in soybean sprouts do not differ significantly between fresh and boiled soybean sprouts, and boiling does not affect the fatty acid composition in soybean. However, the seed coat colors of the soybean may be associated with differences in fatty acid composition in soybean sprouts. To obtain better oil quality in soybean sprouts, it is important to develop soybean seeds with the modification of fatty acid profiles that ensure similar patterns of fatty acid compositions in soybean sprouts. Thus, genotypic effects are an important factor in the determination of fatty acid composition in seeds and sprouts. Those may represent elite accessions of high-quality fatty acid soybean profiles to be used in the implementation of sprout breeding programs.

ACKNOWLEDGMENTS

This work was supported by the National Research Foundation of Korea (NRF) grant funded by the Korean government (MSIT) (No. NRF-2020R1C1C1008759).

REFERENCES

Abbas, T.E., Musharaf, N.A. (2008). The effects of germination of low–tannin sorghum grains on its nutrient contents and broiler chicks performance. *Pakistan Journal of Nutrition*, 7(3): 470–474.
Abdelghany, A.M., Zhang, S., Azam, M., Shaibu, A.S., Feng, Y., Li, Y., Tian, Y., Hong, H., Li, B., Sun, J. (2020). Profiling of seed fatty acid composition in 1025 Chinese soybean accessions from diverse ecoregions. *The Crop Journal*, 8(4): 635–644.
Abdel-Rahman, E.S.A., El-Fishawy, F.A., El-Geddawy, M.A., Kurz, T., El-Rify, M.N. (2007). The changes in the lipid composition of mung bean seeds as affected by processing methods. *International Journal of Food Engineering*, 3(5): 1–10.
Afman, L.A., Müller, M. (2012). Human nutrigenomics of gene regulation by dietary fatty acids. *Progress in Lipid Research*, 51(1): 63–70.
Asekova, S., Chae, J.H., Ha, B.K., Dhakal, K.H., Chung, G., Shannon, J.G., Lee, J.D. (2014). Stability of elevated α-linolenic acid derived from wild soybean (*Glycine soja* Sieb. & Zucc.) across environments. *Euphytica*, 195(3): 409–418.
Bhardwaj, H.L., Hamama, A.A. (2009). Cultivar and growing location effects on oil content and fatty acids in canola sprouts. *Hortscience*, 44(6): 1628–1631.
Bilyeu, K., Palavalli, L., Sleper, D., Beuselinck, P. (2005). Mutations in soybean microsomal omega-3 fatty acid desaturase genes reduce linolenic acid concentration in soybean seeds. *Crop Science*, 45(5): 1830–1836.
Bilyeu, K., Palavalli, L., Sleper, D.A., Beuselinck, P. (2006). Molecular genetic resources for development of 1% linolenic acid soybeans. *Crop Science*, 46(5): 1913–1918.

Bilyeu, K., Gillman, J.D., LeRoy, A.R. (2011). Novel FAD3 mutant allele combinations produce soybeans containing 1% linolenic acid in the seed oil. *Crop Science*, 51(1): 259–264.

Brown, B.E., Meade, E.M., Butterfield, J.R. (1962). The effect of germination upon the fat of the soybean. *Journal of the American Oil Chemists' Society*, 39(7): 327–330.

Byun, S.M., Huh, N.E., Lee, C.Y. (1977). Asparagine biosynthesis in soybean sprouts. *Applied Biological Chemistry*, 20(1): 33–42.

Chae, J.H., Ha, B.K., Chung, G., Park, J.E., Park, E., Ko, J.M., Shannon, J.G., Song, J.T., Lee, J.D. (2015). Identification of environmentally stable wild soybean genotypes with high alpha-linolenic acid concentration. *Crop Science*, 55(4): 1629–1636.

Cho, G.T., Lee, J., Moon, J.K., Yoon, M.S., Baek, H.J., Kang, J.H., Kim, T.S., Paek, N.C. (2008). Genetic diversity and population structure of Korean soybean landrace [*Glycine max* (L.) Merr.]. *Journal of Crop Science and Biotechnology*, 11(2): 83–90.

Chung, T.Y., Nwokolo, E.N., Sim, J.S. (1989). Compositional and digestibility changes in sprouted barley and canola seeds. *Plant Foods for Human Nutrition*, 39(3): 267–278.

Collins, J.L., Sanders, G.G. (1976). Changes in trypsin inhibitory activity in some soybean varieties during maturation and germination. *Journal of Food Science*, 41(1): 168–172.

Devi, C.B., Kushwaha, A., Kumar, A. (2015). Sprouting characteristics and associated changes in nutritional composition of cowpea (*Vigna unguiculata*). *Journal of Food Science and Technology*, 52(10): 6821–6827.

Dhakal, K.H., Jeong, Y., Baek, I., Kang, N., Yeo, Y., Hwang, Y. (2006). Composition of oil and fatty acid in sprout and wild soybeans for specific food and industrial applications. *Korean Journal of Breeding*, 38(4): 236–241.

Dhakal, K.H., Jeong, Y.S., Lee, J.D., Baek, I.Y., Ha, T.J., Hwang, Y.H. (2009). Fatty acid composition in each structural part of soybean seed and sprout. *Journal of Crop Science and Biotechnology*, 12(2): 97–101.

Dhakal, K.H., Jung, K.H., Chae, J.H., Shannon, J.G., Lee, J.D. (2014). Variation of unsaturated fatty acids in soybean sprout of high oleic acid accessions. *Food Chemistry*, 164: 70–73.

Dornbos, D.L., Mullen, R.E. (1992). Soybean seed protein and oil contents and fatty acid composition adjustments by drought and temperature. *Journal of the American Oil Chemists Society*, 69(3): 228–231.

Ensminger, M.E., Ensminger, A.H. (1993). *Foods & Nutrition Encyclopedia, Two Volume Set*. CRC Press, Boca Raton, FL.

Fehr, W.R. (2007). Breeding for modified fatty acid composition in soybean. *Crop Science*, 47: S-72.

Finley, J.W. (2005). Proposed criteria for assessing the efficacy of cancer reduction by plant foods enriched in carotenoids, glucosinolates, polyphenols and selenocompounds. *Annals of Botany*, 95(7): 1075–1096.

Friedman, M., Brandon, D.L. (2001). Nutritional and health benefits of soy proteins. *Journal of Agricultural and Food Chemistry*, 49(3): 1069–1086.

Ghani, M., Kulkarni, K.P., Song, J.T., Shannon, J.G., Lee, J.D. (2016). Soybean sprouts: A review of nutrient composition, health benefits and genetic variation. *Plant Breeding and Biotechnology*, 4(4): 398–412.

Giami, S.Y. (1993). Effect of processing on the proximate composition and functional properties of cowpea (*Vigna unguiculata*) flour. *Food Chemistry*, 47(2): 153–158.

Ginting, E., Yulifianti, R., Kuswantoro, H., Lee, B.W., Baek, I.Y. (2018). Protein, fatty acid, and isoflavone contents of soybean lines tolerant to acid soil. *Journal of the Korean Society of International Agricultue*, 30(3): 167–176.

Goffman, F.D., Pinson, S., Bergman, C. (2003). Genetic diversity for lipid content and fatty acid profile in rice bran. *Journal of the American Oil Chemists' Society*, 80(5): 485–490.

Ha, N.T. (2018). *Evaluation of Agronomic and Sprout Traits for Soybeans with High Linolenic Acid* (Master's thesis). Kyungpook National University, South Korea.

Hedges, L.J., Lister, C.E. (2006). The nutritional attributes of legumes. *Crop & Food Research Confidential Report*, 1745: 50.

Hoeck, J.A., Fehr, W.R., Murphy, P.A., Welke, G.A. (2000). Influence of genotype and environment on isoflavone contents of soybean. *Crop Science*, 40(1): 48–51.

Hou, G., Ablett, G.R., Pauls, K.P., Rajcan, I. (2006). Environmental effects on fatty acid levels in soybean seed oil. *Journal of the American Oil Chemists' Society*, 83(9): 759–763.

Howell, R.W., Collins, F.I. (1957). Factors affecting linolenic and linoleic acid content of soybean oil 1. *Agronomy Journal*, 49(11): 593–597.

Hussain, I., Uddin, M.B. (2012). Optimization effect of germination on functional properties of wheat flour by response surface methodology. *International Research Journal of Plant Science*, 3(3): 031–037.

Hymowitz, T., Collins, F.I. (1974). Variability of sugar content in seed of *Glycine max* (L.) Merrill and *G. soja* Sieb. and Zucc.1. *Agronomy Journal*, 66(2): 239–240.

Islam, M.J., Hassan, M.K., Sarker, S.R., Rahman, A.B., Fakir, M.S.A. (2017). Light and temperature effects on sprout yield and its proximate composition and vitamin C content in lignosus and mung beans. *Journal of the Bangladesh Agricultural University*, 15(2): 248–254.

Jeong, Y., Dhakal, K.H., Lee, J., Hwang, Y. (2007). Selection of superior lines based on the practical and useful characteristics in Korean indigenous soy-sprout germplasm. *Korean Journal of Breeding Science*, 39(1): 20–26.

Jo, H., Kim, M., Ali, L., Tayade, R., Jo, D., Le, D.T., Phommalth, S., Ha, B.K., Kang, S., Song, J.T., Lee, J.D. (2020). Environmental stability of elevated α-linolenic acid derived from a wild soybean in three Asian countries. *Agriculture*, 10(3): 70.

Jo, H., Kim, M., Cho, H., Ha, B.K., Kang, S., Song, J.T., Lee, J.D. (2021). Identification of a potential gene for elevating ω-3 concentration and its efficiency for improving the ω-6/ω-3 ratio in soybean. *Journal of Agricultural and Food Chemistry*, 69(13): 3836–3847.

Joshi, A.C., Chopra, B.K., Collins, L.C., Doctor, V.M. (1973). Distribution of fatty acids during germination of soybean seeds. *Journal of the American Oil Chemists' Society*, 50(8): 282–283.

Kang, E.Y., Kim, S.H., Kim, S.L., Seo, S.H., Kim, E.H., Song, H.K., Ahn, J.K., Chung, I.M. (2010). Comparison of soyasapogenol A, B concentrations in soybean seeds and sprouts. *Korean Journal of Crop Science*, 55: 165–176.

Kanmaz, E.Ö., Ova, G. (2015). The effect of germination time on moisture, total fat content and fatty acid composition of flaxseed (*Linum usitatissimum* L.) sprouts. GIDA. *The Journal of Food*, 40(5): 249–254.

Kim, H.J., Ha, B.K., Ha, K.S., Chae, J.H., Park, J.H., Kim, M.S., Asekova, S., Shannon, J.G., Son, C.K., Lee, J.D. (2015). Comparison of a high oleic acid soybean line to cultivated cultivars for seed yield, protein and oil concentrations. *Euphytica*, 201(2): 285–292.

Kim, J.S., Kim, J.G., Kim, W.J. (2004). Changes in isoflavone and oligosaccharides of soybeans during germination. *Korean Journal of Food Science and Technology*, 36(2): 294–298.

Kim, K.H. (1981). Studies on the growing characteristics of soybean sprout. *Korean Journal of Food Science and Technology*, 13(3): 247–252.

Kim, S.Y., Lee, K.A., Yun, H.T., Kim, J.T., Kim, U.H., Kim, Y.H. (2011). Analyses of fatty acids and dietary fiber in soy sprouts. *Korean Journal of Crop Science*, 56(1): 29–34.

Kim, Y.H., Hwang, Y.H., Lee, H.S. (2003). Analysis of isoflavones for 66 varieties of sprout beans and bean sprouts. *Korean Journal of Food Science and Technology*, 35(4): 568–575.

Kim, Y.J., Oh, Y.J., Cho, S.K., Kim, J.G., Park, M.R., Yun, S.J. (2006). Variations of isoflavone contents in seeds and sprouts of sprout soybean cultivars. *Korean Journal of Crop Science*, 51: 160–165.

Koo, S.C., Kim, S.G., Bae, D.W., Kim, H.Y., Kim, H.T., Lee, Y.H., Kang, B.K., Baek, S.B., Baek, I.Y., Yun, H.T., Choi, M.S. (2015). Biochemical and proteomic analysis of soybean sprouts at different germination temperatures. *Journal of the Korean Society for Applied Biological Chemistry*, 58(3): 397–407.

Kulkarni, K.P., Kim, M., Song, J.T., Bilyeu, K.D., Lee, J.D. (2017). Genetic improvement of the fatty acid biosynthesis system to alter the ω-6/ω-3 ratio in the soybean seed. *Journal of the American Oil Chemists' Society*, 94(11): 1403–1410.

Kumar, V., Rani, A., Chauhan, G.S. (2006). Influence of germination temperature on oil content and fatty acid composition of soy sprouts. *Journal of Food Science and Technology-Mysore*, 43(3): 325–326.

Kwon, S.H., Lee, Y.I., Kim, J.R. (1981). Evaluation of important sprouting characteristics of edible soybean sprout cultivars. *Korean Journal of Breeding*, 13(3): 202–206.

Lee, J., Renita, M., Fioritto, R.J., St. Martin, S.K., Schwartz, S.J., Vodovotz, Y. (2004). Isoflavone characterization and antioxidant activity of Ohio soybeans. *Journal of Agricultural and Food Chemistry*, 52(9): 2647–2651.

Lee, J.C., Hwang, Y.H. (1996). Variation of asparagine and aspartic acid contents in beansprout soybeans. *Korean Journal of Crop Science*, 41: 592–599.

Lee, J.D., Hwang, Y.H., Cho, H.Y., Kim, D.U., Choung, M.G. (2002). Comparison of characteristics related with soybean sprouts between *Glycine max* and *G. soja. Korean Journal of Crop Science*, 47(3): 189–195.

Lee, J.D., Shannon, J.G., Jeong, Y.S., Lee, J.M., Hwang, Y.H. (2007a). A simple method for evaluation of sprout characters in soybean. *Euphytica*, 153(1–2): 171–180.

Lee, J.D., Bilyeu, K.D., Shannon, J.G. (2007b). Genetics and breeding for modified fatty acid profile in soybean seed oil. *Journal of Crop Science and Biotechnology*, 10: 201–210.

Lee, J.D., Yu, J.K., Hwang, Y.H., Blake, S., So, Y.S., Lee, G.J., Nguyen, H.T., Shannon, J.G. (2008). Genetic diversity of wild soybean (*Glycine soja* Sieb. and Zucc.) accessions from South Korea and other countries. *Crop Science*, 48(2): 606–616.

Lee, J.D., Woolard, M., Sleper, D.A., Smith, J.R., Pantalone, V.R., Nyinyi, C.N., Cardinal, A., Shannon, J.G. (2009). Environmental effects on oleic acid in soybean seed oil of plant introductions with elevated oleic concentration. *Crop Science*, 49(5): 1762–1768.

Lee, J.D. (2012). Environmental effects on oleic acid in soybean seed oil of cultivar Kwangan. *Korean Journal of Breeding Science*, 44(1): 29–44.

Lee, J.D., Bilyeu, K.D., Pantalone, V.R., Gillen, A.M., So, Y.S., Shannon, J.G. (2012). Environmental stability of oleic acid concentration in seed oil for soybean lines with FAD2-1A and FAD2-1B mutant genes. *Crop Science*, 52(3): 1290–1297.

Lee, S.H., Chung, D.H. (1982). Studies on the effects of plant growth regulator on growth and nutrient compositions in soybean sprout. *Applied Biological Chemistry*, 25(2): 75–82.

Lee, S.J., Ahn, J.K., Khanh, T.D., Chun, S.C., Kim, S.L., Ro, H.M., Song, H.K., Chung, I.M. (2007c). Comparison of isoflavone concentrations in soybean (*Glycine max* (L.) Merrill) sprouts grown under two different light conditions. *Journal of Agricultural and Food Chemistry*, 55(23): 9415–9421.

Liu, K. (1999). *Soybeans: Chemistry, Technology, and Utilization.* Aspen Publishers, Inc., Gaithersburg, MD.

Márton, M., Mándoki, Z.S., Csapo, J. (2010). Evaluation of biological value of sprouts-I. Fat content, fatty acid composition. *Acta Universitatis Sapientiae: Alimentaria*, 3: 53–65.

Mizuno, T., Yamada, K. (2006). Proximate composition, fatty acid composition and free amino acid composition of sprouts. *Journal for the Integrated Study of Dietary Habits*, 16(4): 369–375.

Mourtzinis, S., Marburger, D., Gaska, J., Diallo, T., Lauer, J., Conley, S. (2017). Corn and soybean yield response to tillage, rotation, and nematicide seed treatment. *Crop Science*, 57(3): 1704–1712.

Narina, S.S., Hamama, A.A., Bhardwaj, H.L. (2013). Fatty acid composition of flax sprouts. *Journal of Agricultural Science*, 5(4): 75–79.

O'Bryan, C.A., Kushwaha, K., Babu, D., Crandall, P.G., Davis, M.L., Chen, P., Lee, S.O., Ricke, S.C. (2014). Soybean seed coats: A source of ingredients for potential human health benefits-a review of the literature. *Journal of Food Research*, 3(6): 188.

Oh, B.Y., Park, B.H., Ham, K.S. (2003). Changes of saponin during the cultivation of soybean sprout. *Korean Journal of Food Science and Technology*, 35(6): 1039–1044.

Oliva, M.L., Shannon, J.G., Sleper, D.A., Ellersieck, M.R., Cardinal, A.J., Paris, R.L., Lee, J.D. (2006). Stability of fatty acid profile in soybean genotypes with modified seed oil composition. *Crop Science*, 46(5): 2069–2075.

Orsavova, J., Misurcova, L., Ambrozova, J.V., Vicha, R., Mlcek, J. (2015). Fatty acids composition of vegetable oils and its contribution to dietary energy intake and dependence of cardiovascular mortality on dietary intake of fatty acids. *International Journal of Molecular Sciences*, 16(6): 12871–12890.

Orhan, I., Ozcelik, B., Kartal, M., Aslan, S., Sener, B., Ozguven, M. (2007). Quantification of daidzein, genistein and fatty acids in soybeans and soy sprouts, and some bioactivity studies. *Acta Biol Cracov*, 49(2): 61–68.

Petrus, P., Arner, P. (2019). The impact of dietary fatty acids on human adipose tissue. *Proceedings of the Nutrition Society*, 79(1): 42–46.

Pham, A.T., Lee, J.D., Shannon, J.G., Bilyeu, K.D. (2010). Mutant alleles of FAD2-1A and FAD2-1B combine to produce soybeans with the high oleic acid seed oil trait. *BMC Plant Biology*, 10(1): 195.

Pham, A.T., Shannon, J.G., Bilyeu, K.D. (2012). Combinations of mutant FAD2 and FAD3 genes to produce high oleic acid and low linolenic acid soybean oil. *Theoretical and Applied Genetics*, 125(3): 503–515.

Plaza, L., de Ancos, B., Cano, P.M. (2003). Nutritional and health-related compounds in sprouts and seeds of soybean (Glycine max), wheat (Triticum aestivum.L) and alfalfa (Medicago sativa) treated by a new drying method. *European Food Research and Technology*, 216(2): 138–144.

Primomo, V.S., Falk, D.E., Ablett, G.R., Tanner, J.W., Rajcan, I. (2002). Genotype x environment interactions, stability, and agronomic performance of soybean with altered fatty acid profiles. *Crop Science*, 42(1): 37–44.

Reinprecht, Y., Luk-Labey, S.Y., Larsen, J., Poysa, V.W., Yu, K., Rajcan, I., Ablett, G.R., Pauls, K.P. (2009). Molecular basis of the low linolenic acid trait in soybean EMS mutant line RG10. *Plant Breeding*, 128(3): 253–258.

Rennie, B.D., Tanner, J.W. (1989). Fatty acid composition of oil from soybean seeds grown at extreme temperatures. *Journal of the American Oil Chemists' Society*, 66(11): 1622–1624.

Schlueter, J.A., Vasylenko-Sanders, I.F., Deshpande, S., Yi, J., Siegfried, M., Roe, B.A., Schlueter, S.D., Scheffler, B.E., Shoemaker, R.C. (2007). The FAD2 gene family of soybean: Insights into the structural and functional divergence of a paleopolyploid genome. *Crop Science*, 47(S1): S-14.

Seguin, P., Zheng, W., Smith, D.L., Deng, W. (2004). Isoflavone content of soybean cultivars grown in eastern Canada. *Journal of the Science of Food and Agriculture*, 84(11): 1327–1332.

Shi, H., Nam, P.K., Ma, Y. (2010). Comprehensive profiling of isoflavones, phytosterols, tocopherols, minerals, crude protein, lipid, and sugar during soybean (*Glycine max*) germination. *Journal of Agricultural and Food Chemistry*, 58(8): 4970–4976.

Silva, H.C., Braga, G.L., Bianchi, M.L.P., Rossi, E.A. (1990). Effect of germination on oligosaccharide and reducing sugar contents of Brazilian soybean cultivars. *Alimentos e Nutrição Sao Paulo*, 2: 13–19.

Silva, L.R., Pereira, M.J., Azevedo, J., Gonçalves, R.F., Valentão, P., de Pinho, P.G., Andrade, P.B. (2013). *Glycine max* (L.) Merr., *Vigna radiata* L. and *Medicago sativa* L. sprouts: A natural source of bioactive compounds. *Food Research International*, 50(1): 167–175.

SoyStats. (2020). Available online: http://www.soystats.com (accessed on 1 May 2021).

Todd, J.J., Vodkin, L.O. (1993). Pigmented soybean (*Glycine max*) seed coats accumulate proanthocyanidins during development. *Plant Physiology*, 102(2): 663–670.

Tokiko, M., Koji, Y. (2006). Proximate composition, fatty acid composition and free amino acid composition of sprouts. *Journal for the Integrated Study of Dietary Habits*, 16(4): 369–375.

Tsukamoto, C., Shimada, S., Igita, K., Kudou, S., Kokubun, M., Okubo, K., Kitamura, K. (1995). Factors affecting isoflavone content in soybean seeds: Changes in isoflavones, saponins, and composition of fatty acids at different temperatures during seed development. *Journal of Agricultural and Food Chemistry*, 43(5): 1184–1192.

Verma, D.P.S., Shoemaker, R.C. (1996). *Soybean: Genetics, Molecular Biology and Biotechnology*. CABI. Wallingford, United Kingdom.

Villaluenga, C.M., Kuo, Y.H., Lambein, F., Frias, J., Valverde, C.V. (2006). Kinetics of free protein amino acids, free non protein amino acids and trigonelline in soybean (*Glycine max* L.) and lupin (*Lupinus angustifolius* L.) sprouts. *European Food Research and Technology*, 224(2): 177–186.

Webb, G.P. (2006). *Dietary Supplements and Functional Foods*. Blackwell Publishing Ltd., Oxford, 1–120.

Wilson, R.F. (2004). Seed composition. 3rd ed., p.621–795. In H.R. Boerma, J.E. Specht (eds.), *Soybeans: Improvement, Production, and Uses*. American Society of Agronomy Crop Science Society of America-Soil Science Society of America, Madison, WI.

Wolf, R.B., Cavins, J.F., Kleiman, R., Black, L.T. (1982). Effect of temperature on soybean seed constituents: Oil, protein, moisture, fatty acids, amino acids and sugars. *Journal of the American Oil Chemists' Society*, 59(5): 230–232.

Xu, M., Zhu, M., Gu, Q. (2003). Light-induced accumulation of isoflavone in soybean sprouts. *Zhongguo Liangyou Xuebao*, 18: 74–77.

Yadav, N.S., Wierzbicki, A., Aegerter, M., Caster, C.S., Pérez-Grau, L., Kinney, A.J., Hitz, W.D., Booth Jr, J.R., Schweiger, B., Stecca, K.L., Allen, S.M. (1993). Cloning of higher plant [omega]-3 fatty acid desaturases. *Plant Physiology*, 103(2): 467–476.

Yang, C.B. (1981). Changes of nitrogen compounds and nutritional evaluation of soybean sprout-Part III. Changes of free amino acid composition. *Applied Biological Chemistry*, 24(2): 101–104.

Yang, H., Gao, J., Yang, A., Chen, H. (2015). The ultrasound-treated soybean seeds improve edibility and nutritional quality of soybean sprouts. *Food Research International*, 77: 704–710.

Yang, M.S., Kim, K.S., Ha, H.S. (1982). Effect of light on fatty acid and sterol composition in soybean seeding. *Korean Journal of Soil Science and Fertilizer*, 15(4): 251–257.

Youn, J.E., Kim, H.S., Lee, K.A., Kim, Y.H. (2011). Contents of minerals and vitamins in soybean sprouts. *Korean Journal of Crop Science*, 56(3): 226–232.

Zhu, D., Hettiarachchy, N.S., Horax, R., Chen, P. (2005). Isoflavone contents in germinated soybean seeds. *Plant Foods for Human Nutrition*, 60(3): 147–151.

22 Peptides Derived from High Oleic Acid Soybean and Their Health Benefits

Navam Hettiarachchy, Soma Mukherjee, Darry L. Holliday, Srinivasan J. Rayaprolu, and Sriloy Dey

CONTENTS

22.1 INTRODUCTION

Cultivation of soybean (*Glycine Max*) started about 5,000 years ago in China and then in Japan (He & Chen, 2013; Valliyodan et al., 2016). Since 1940, soybean has been cultivated as a major economic crop and currently, the United States is the second leading soy producer with over 35% of the total world's production (He & Chen, 2013; Soybean Production, 2021). In 1999, the FDA approved the health claim associated with soy protein for decreasing the risk of cardiovascular disease, and later in 2017, the FDA recommended additional changes to follow up on the health claim of cardioprotective ability in soy product labeling (FDA, 2017). Soybean is rich in high-quality protein containing most essential amino acids except the two sulfur-containing amino acids, methionine and cysteine, in inadequate quantities (Glodfus et al., 2006). However, modifications in soybean breeding and cultivation to enhance the two sulfur-containing amino acids have been reported (Imsande, 2001; Panthee et al., 2006; Singer et al., 2019).

DOI: 10.1201/9781003030294-22

Epidemiological evidence suggests that consumption of soybean products is linked to a reduction in disease conditions such as obesity, coronary heart disease, diabetes, cancer, and immunological conditions (Omoni & Aluko, 2005; De Mejia & Ben, 2006; Velasquez & Bhathena, 2007; Xiao, 2008; Erdmann et al., 2008; Kwon et al., 2010; He & Chen, 2013). The health benefits are believed to be associated with the isoflavones present in soybean. However, the specific functional property or mechanism of action of the bioactive component of soybean is not well understood. Current research on soybean-derived peptide functionality is focused on the biological activity of the peptides and their behavior while passing through the gastrointestinal system. This chapter summarizes current knowledge on the functional activity of the peptides and their specific role in the alteration of physiological activity, as a preventive agent in combating chronic diseases, as well as the bioactivities of high oleic acid peptides toward health benefits.

22.2 SOYBEAN SEED COMPOSITION

Mature raw seeds of soybean are generally composed of about 30–40% protein. Dried soybean seeds contain important constituents such as lipid (20%) and carbohydrates or dietary fiber (30%) (He & Chen, 2013). The main components of high oleic acid soybean are protein, carbohydrate, crude fiber, moisture, and vitamins (Table 22.1). Peptides and proteins are the important components that contribute to the bioactivity of soybeans (Omoni & Aluko, 2005; Wang et al., 2008).

The major storage proteins that constitute 80–90% of the whole protein in raw soybeans are β-conglycinin (βCG, 7S) and glycinin (11S) (Zarkadas et al., 2007; Velasquez & Bhathena, 2007; Xiao, 2008; Cam & Mejia, 2012). The β-conglycinin protein consists of three subunits of α, α', and β configuration, and glycinin contains acidic (A) and alkaline or basic (B) subunits designated as A1aB2, A1bB1b, A2B1a, A3B4, and A5A4B3 (Xiao, 2008; Wang et al., 2008). Minor proteins include three smaller storage proteins of 2S, 9S, and 15S; lectin; and a protease inhibitor (also named Kunitz and Bowman-Birk [BBI]) (Zarkadas et al., 2007). The ratio of beta-conglycinin in soy proteins might contribute to varied nutritional benefits and also impart diverse physiological effects (Clarke & Wiseman, 2000; Singh et al., 2014). Various subunit compositions of soy proteins are attributed to different functionality in terms of product quality and protein yield and affect the texture of prepared soy products such as tofu (Poysa et al., 2006). Bioactive peptides are functional only when they are not a part of the parent protein sequence and are cleaved or released by enzymatic and gastrointestinal digestion. Various food processing techniques such as fermentation can also separate peptides from the parent molecule and can activate the peptides for nutraceutical functions (De Mejia & Ben, 2006; Erdmann et al., 2008; Singh et al., 2014). Administration of 11S peptides via beverage showed a significant increase in amino acid concentration in blood upon intestinal absorption, compared to 11S globulin or other amino acids in human subjects (Maebuchi

TABLE 22.1
Composition and Nutrient Content per 100 g Soy Flour

Type	Moisture (g)	Protein (Nx5.71)	Fat (g)	Carbohydrate (g)	Crude Fiber (g)	Calcium (mg)	Iron (mg)
Defatted	7.3	47.0	1.2	38.4	4.3	241	9.24
Full fat	5.2	34.5	20.6	35.2	4.7	206	6.37
Type	Zinc (mg)	Thiamin (mg)	Riboflavin (mg)	Niacin (mg)	Vitamin B6 (mg)	Folacin (mcg)	Sodium (mg)
Defatted	2.46	0.70	0.25	2.61	0.57	305.4	20
Full fat	3.92	0.58	1.16	4.32	0.46	345.0	13

Source: USDA (1986).

et al., 2007). In the same study, it was demonstrated that aromatic, branched-chain amino acids and hydrolyzed soy proteins are absorbed by humans in a short period of time with greater efficiency of absorption through the gut. The soy-based products (*cheonggukjang*) made from high oleic acid varieties of soybean showed differences in functional property in comparison to other soy varieties and it has been reported that soybeans with high oleic acid content can be ingredients for food products with better functional quality (Lee et al., 2017).

22.3 METHODS OF PEPTIDE PRODUCTION FROM HIGH OLEIC ACID SOYBEAN SEED

Peptides are protein fragments produced from a large parent protein and they could be occurring naturally in the soybean seeds. Preparation of peptides is also possible by enzymatic hydrolysis, solid and liquid state fermentation, and various other food processing techniques and procedures using simulated gastrointestinal digestion (Figure 22.1). The peptides derived after fragmentation exhibit a large number of beneficial physiological functional properties (Rayaprolu et al., 2013; Singh et al., 2014; Rayaprolu et al., 2015). The production of functional soy peptides is largely affected by the type of enzymes used in hydrolysis and the type of bacteria used in fermentation.

22.3.1 *IN VITRO* ENZYMATIC HYDROLYSIS

This is the main approach to producing bioactive peptides using *in vitro* enzymatic hydrolysis. The first step in peptide generation from the parent protein is to isolate the protein from the ground flour of seeds. Conditions are optimized for enzymatic digestion of the protein isolate with a selected enzyme, an endopeptidase, which has the unique ability to cleave within the protein at certain amino acid sites. The conditions enable the production of peptides that may be short or long chains and prevent the complete digestion of the protein into individual amino acid residues. Common enzymes used in this process are pepsin, trypsin, chymotrypsin, papain, and endopeptidases. Using centrifugation, the peptides in the supernatant are separated from the precipitate and the undigested materials. The peptides can be fractionated based on molecular cut-off sizes using ultrafiltration membranes (Rayaprolu, 2015) and freeze drying. A modified procedure can also be performed,

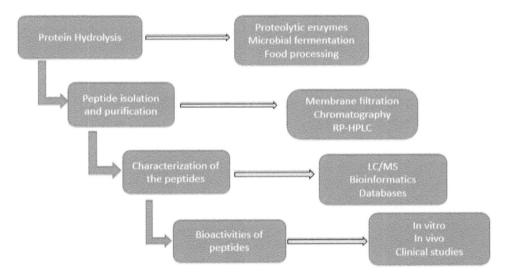

FIGURE 22.1 Isolation, purification, and characterization of bioactive peptides. Source: Recreated from Singh et al., 2021.

where freeze-dried supernatant is separated into molecular cut-off sizes using ultrafiltration. This approach of peptide production, however, is less expensive and therefore convenient (Aluko, 2012). Smaller size peptides increase the bioavailability, and the solubility of these peptides can enable absorption by the intestinal wall in the human digestive system. *In vitro* gastrointestinal digestion of high oleic acid peptides produced potent bioactive peptides (Rayaprolu, 2015).

22.3.2 FOOD PROCESSING

Structural modification in proteins takes place during food processing primarily because of heat, followed by pressure and pH changes. These factors modify the amino acid profile and the functional properties. Protein modification during food processing produces peptides that have good digestibility. The major modification occurs via acylation, glycosylation, phosphorylation, oxidation, succinylation, and lyophilization. Thermal treatment is the most frequently used food processing methodology (Korhonen et al., 1998; Wang & Meija, 2005). It is important to note that food processing has a significant effect on the bioactivity of peptides in high oleic acid soybean (Table 22.2, Table 22.3, Table 22.4) (Kim, 2000).

22.3.3 FERMENTATION

Korean, Chinese, and Japanese have traditionally consumed fermented soy food products such as soy sauce and paste, natto, tempeh, and miso. These fermented foods are a rich source of bioactive soy peptides since fermentation is an effective method to generate them. Generally, fermentation alone is not always sufficient to hydrolyze the parent soy protein structure. Other potent enzymes including pronase, trypsin, and plasma proteases can also cleave large polypeptides into smaller bioactive peptides (Korhonen et al., 1998; Wang & Meija, 2005; Singh et al., 2014). High oleic acid content in soybean seed affects the growth of fermentative bacteria and also the fermentation process. Cleavage of the polypeptide chain is affected by the growth rate of the bacteria during the fermentation of proteins (Lee et al., 2017). The fermentation process also produces new flavors and modifies the structure of soy protein (Singh et al., 2014; Lee et al., 2017). There is limited research on high oleic acid soybean fermentation and the formation of peptides. In addition, many researchers reported that fermentation causes a significant effect on flavor development.

An alternative process of peptide production involves microbial fermentation where microbial enzymes hydrolyze the parent protein into soluble peptides. Bioactive peptides are recovered from the supernatant after centrifugation (De Angelis et al., 2017). This process is employed to produce peptides in large quantities. This process is pH-specific and thermally controlled and reported to be more efficient in recovering low-molecular-weight-specific peptides (Singh et al., 2014).

22.4 PREPARATION AND PROPERTIES OF SOY PEPTIDES

22.4.1 PREPARATION OF SOYMEAL AND PROTEIN ISOLATE

The seeds from two high oleic acid soybean lines (N98-4445A and S03-543CR) and one high protein line (R95-1705) were ground and sieved (250 microns) to obtain a uniform particle size. The lipid extraction was performed using the solvent n-hexane (1:3 w/v ratio of flour to solvent). The meals were suspended in de-ionized (DI) water and soy protein isolates (SPI) were prepared by alkali extraction. The supernatants were separated by centrifugation followed by isoelectric precipitation to isolate the proteins which were then freeze dried (Rayaprolu, 2015).

22.4.2 OPTIMIZATION OF CONDITIONS FOR ENZYMATIC HYDROLYSIS WITH ALCALASE

The conditions for optimization of enzymatic hydrolysis using alcalase were determined by the Response Surface Methodology (RSM) (Ma & Ooraikul, 1986). The SPI was mixed with DI water

TABLE 22.2

Antioxidant Activity of Bioactive Peptides from Soybean

Source	Enzymes/Treatment/Method	MW of Peptides/Name of the Fractions	Antioxidant Assay	Reference
Lunasin	Natural	SKWQHQQSCRKQLQ, GVNLTPC, DDDDDDDD, EKHIMEKIQGRGDDD, DDDDDD, EKHIMEKIQ	ABTS, inhibited ROS generation in LPS stimulated RAW	Hernández-Ledesma et al., 2004
SPI	Alcalase, SEC, IEC	Pentapeptide (FDPAL), 561 Da	HRSA, SRSA, inhibition of ROS	Ma et al., 2016
High oleic acid-rich soybean flakes	Alcalase, pepsin, pancreatin, SEC	823.38–1,216.58 Da	DPPS, ABTS, Inhibition of ROS in caco-2 cells	Zhang et al., 2018
Fractionated soybeans	Flavorzyme, UF	Less than 10 kDa	HRSA, ABTS, FRAP	Moure et al., 2006
Black soy sauce	SEC	40–4000 Da	ABTS	Wang et al., 2006
Soy meal protein	Extrution, alcalase, UF	Fraction I (> 3 kDa), II (3–1 kDa), III (< 1 KDa)	HRSA	Sami, 2017
Germinated soybean	Pepsin, pancreatin	Fraction (42.1, 31.4, 19.9, 15.5, 13.6, 12 kDa)	Reducing power, –Cu2+ and Fe2+–chelating, HRSA chelation, HRSA	Marcela et al., 2016
Soy milk	*Lactobacillus rhamnosus* strains	Fermentates	ABTS, DPPH, HRSA	Singh et al., 2020
Soy milk	*Lactobaccilus planterum*	Fermentates	DPPH, HRSA, inhibition of ROS generation and lipid peroxidation in Caco-2 cells, enhanced levels of CAT, SOD, GSH-Px.	Li et al., 2018
Soy milk with 2% whey protein concentrate	*Lactobacillus rhamnosus*	Fermentates	ABTS, DPPH, FRAP	Subrota et al., 2013
SPI	*Chryseobacterium*	Fermentates	ABTS, DPPH, Fe2+ chelation	De Oliveira et al., 2015
SPI	Germination	Germinates	DPPH, ABTS, HRSA	Guo et al., 2020

TABLE 22.3

Soy Bioactive Peptides and Their Health Beneficial Activities

Soy Protein	Functional Peptide	Functionality	Experimental Model	Reference
βCG	YVVNPDNDEN YVVNPDNNEN	Reduces cholesterol	Liver (human) cell line HepG2	Lammi et al., 2015; Lammi et al., 2015
βCG	LAIPVNKP; LPHF	ACE I inhibitor	ACE I inhibitory activity assay	Wang et al., 2008; Kuba et al., 2005
βCG (α′ sub unit)	Soymetide-13: MITLAIPVNKPGR Soymetide-9: MITLAIPVN	Immunity booster	ICR mice, phagocytosis assay Anti-alopecia in neonatal rat model	Wang & Mejia, 2005; Yoshikawa et al., 2015 Tsuruki et al., 2003; Tsuruki and Yoshikawa, 2004
βCG (α′ sub unit)	KNPQLR; EITPEKNPQLR; RKQEEDEDEQQORE	Reduces FAS activity	FAS inhibition studies; 3T3-L1 mouse adipocyte	Martinez-Villaluenga et al., 2010; Singh et al., 2014
βCG (β sub unit)	Soymorphin-5: YPFVV	Antidiabetic; triglyceride-lowering; immunostimulating; suppress feeding and intestinal transit	Guinea pig ileum assay opioid activity; diabetic KKAy mice Elevated Plus Maze Test in male ddY mice Male BALB/c and ddY mice	Nishi et al., 2003; Nishi et al., 2003; Ohinata et al., 2007; Kaneko et al., 2010; Nakajima et al., 2010; Pak et al., 2012; Yamada et al., 2012, Yoshikawa et al., 2015
	Soymorphin-6: YPFVVN Soymorphin-7: YPFVVNA VRIRLLQRFNKRS	Suppressed appetite	Male BALB/c and ddY mice; male Sprague-Dawley rat; mouse intestinal STC-1 cells	Nishi et al., 2003; Nishi et al., 2003; Erdmann et al., 2008; Nakajima et al., 2010
Glycinin	IAVPGEVA; IAVPTGVA; LPYP	Reduces cholesterol; antidiabetic	HMGR activity assay; HepG2 human liver cells; DPP-IV activity assay	Yoshikawa et al., 2000; Pak et al., 2005; Erdmann et al., 2008; Pak et al., 2012; Lammi et al., 2015; Yoshikawa et al., 2015
Glycinin	VLIVP; SPYP; WL	Antihypertensive	ACE1 inhibition assay	Wang et al., 2006
	SFGVAE	Reduces cholesterol	HMGR activity	Pak et al., 2012
	HCQRPR QRPR	Phagocytosis stimulatory peptide	Macrophages; human polymorphonuclear leukocytes; C3H/He mouse	Yoshikawa et al., 1993
Defatted soy protein	X-MLPSYSPY	Anticancer	Arrest P38BD1 mouse monocyte macrophages at G2/M phase to block cell cycle progression	Kim et al., 2000; Wang et al., 2008; Singh et al., 2014
Soy protein	YVVFK; IPPGVPYWT; PNNKPFQ; NWGPLV; TPRVF	Hypotensive	Spontaneously hypertensive rats	
Soy protein	WGAPSL; VAWWMY; FVVNATSN	Hypocholesterolemic	Rats; HepG2 cells	Wang et al., 2008, Singh et al., 2014; Yoshikawa et al., 2015
Protease (PROTIN SD-NY10) treated soy milk	FFFY; WHP; FVP; LHPGDAQR; IAV; VNP; LEPP; WNPR	ACE I inhibitor	ACE inhibitory activity assay	Tomatsu et al., 2013; Yoshikawa et al., 2015
Fermented soybean, *Bacillus natto* or *subtilis*	VAHINVGK; YVWK		ACE inhibitory activity assay; *in vitro* gastrointestinal digestion	Hernandez-Ledesma et al., 2004
Fermented soybean seasoning	SY, GY		Spontaneously hypertensive rats	Nakahara et al. 2010; Nakahara, et al., 2011

TABLE 22.4

High Oleic Acid Soy Peptides and Functional Properties

Peptide Size/Sequence	Functionality	Cell Line/Assay	Reference
1,157 Da/X-Met-Leu-Pro-Ser-Tye-Ser-Pro-Tyr	Inhibited cancer	Monocyte macrophage cell line P388D1; *in vitro* cytotoxicity test	Kim et al., 2000
< 5 kDa, 5–10 kDa, and 10–50 kDa	Inhibited colon, liver, and lung cancer	(HCT-116, Caco-2); liver HepG2; NCL-H1299	Rayaprolu et al., 2013
Not determined	Antihypertensive	Assay	Rayaprolu et al., 2015
< 5, 5–10, and 10–50 kDa	Inhibited blood, breast, prostate cancer	Blood (CCRF-CEM and Kasumi-3), breast (MCF-7), and prostate (PC-3)	Rayaprolu et al., 2017

and homogenized. The suspensions were subjected to treatments involving four factors at three levels in the design viz., pH (6.5, 7.5, and 8.5), time (30, 60, and 90 min), temperature (55, 62.5, and 70° C), and enzyme concentration (1, 3.5, and 5 Anson units). The enzyme was inactivated at 85° C for 3 min; the hydrolysates were cooled to ambient temperature and centrifuged and the degree of hydrolysis (DH) was determined using the OPA method in the supernatants (Rayaprolu et al., 2015).

22.4.3 PREPARATION OF GASTROINTESTINAL ENVIRONMENT RESISTANT PEPTIDE HYDROLYSATES

The simulated gastrointestinal (GI) juice was prepared mimicking the human GI tract secretions for treating the protein hydrolysates resulting from the alcalase treatment (Kannan et al., 2008). Resistance to gastric juice was determined by treating the protein hydrolysates with pepsin. The enzyme was inactivated by adjusting the pH to 7.2 and the contents were centrifuged to separate the supernatant. The pH of the supernatant was adjusted to 8.0 using 3N sodium hydroxide solution and pancreatin was added to test their resistance against the intestinal juice. The GI treated peptide hydrolysates were centrifuged at 3,000 g and the supernatants were passed through the membrane filters with molecular cut-off sizes of 5, 10, and 50 kDa. The specific peptide fractions obtained, < 5, 5–10, and 10–50 kDa, were freeze dried and stored at 5° C. *In vitro* gastrointestinal digestion of high oleic acid peptides fractions produced highly bioactive smaller peptides (Rayaprolu et al., 2015).

22.5 PROPERTIES OF SOY PEPTIDES IN GENERAL AND HEALTH BENEFITS

In the last ten years, the major emphasis on soy protein and peptide research has been concentrated on the identification and sequencing of bioactive peptides and their specific physiological benefits. Various beneficial soy peptides with a multitude of beneficial functions have been reported (Table 22.2). The major physiological functions of soy peptides studied include antihypertensive, antidiabetic, and lipid-lowering effects in various model organisms. Table 22.2, Table 22.3, and Table 22.4 list beneficial properties of high oleic acid peptides such as anticancer and antihypertensive effects.

22.5.1 CHOLESTEROL LOWERING PEPTIDES

The most studied bioactivity of soy protein–derived peptides is their hypolipidemic function. A number of soy peptides produced by various methodologies have been reported to decrease cholesterol and triglycerides, and to inhibit fat synthesis and absorption in different experimental model systems. The first hypoglycemic peptide LPYPR (leucine-proline-tyrosine-proline-arginine) from soybean was derived from the glycinin subunit of soy protein (Yoshikawa et al., 2000). Administration of LPYPR peptide in rats at 50 mg/kg of body weight for two days decreased the total and LDL cholesterol in serum by 25% (Yoshikawa et al., 2000). Further studies showed that peptide LPYPR has the potential to act as a competitive inhibitor of 3-hydroxy-3-methyl-glutaryl-CoA reductase (HMGR) which is the major enzyme in the biosynthetic pathway of cholesterol synthesis (Pak et al., 2012). LPYPR activates one receptor protein (LDLR) and one sterol regulator binding protein 2 (SREBP2) and this activity has been confirmed in cultured liver cells (Lammi et al., 2015). Three other peptides have also been tested in HepG2 cells and reported to inhibit HMGR following the same biosynthetic pathway (Yoshikawa et al., 2000; Pak et al., 2012; Lammi et al., 2015).

Several other peptides have been derived from glycinin, βCG, and lunasin (2S) of soy protein (Choi et al., 2004; Wang et al., 2008; Cho et al., 2008; Nagaoka et al., 2010; Singh et al., 2014; Lammi et al., 2015; Lule et al., 2015). These peptides have the potential to inhibit several disease-causing enzymes, decrease fatty acid synthesis in the hepatic tissue, and reduce serum triglyceride in a rat model. Researchers have found that fatty acid synthase (FAS) activity was suppressed by three other peptides (KNPQLR, EITPEKNPQLR, RKQEEDEDEEQQRE) (Martinez-Villaluenga et al., 2010; Singh et al., 2014). The βCG subunits of soy protein have demonstrated better ability to lower lipid profile in mouse (3T3-L1) fatty tissues and embed more bioactive peptide fractions compared to glycinin soy protein subunit (Martinez-Villaluenga et al., 2008). The β-conglycinin subunit has been shown to suppress lipid synthesis by down-regulating lipoprotein lipase and fatty acid synthase (FAS). No significant studies have been conducted on the cholesterol-lowering activity of high oleic acid soybean peptides and other low oleic acid soybean varieties.

22.5.2 Anti-Diabetic Soy Peptides

Three metabolic diseases, diabetes, hyperlipidemia, and heart disease, are associated with metabolic disease syndrome. It is significant that many soy protein–derived peptides (LPYP, IAVPGEVA, IAVPTGVA) possess both hypolipidemic and antidiabetic activity as shown by studies with several experimental organisms. For example, it has been shown that hypolipidemic soy peptides also increase glucose metabolism and uptake in hepatic cells through glucose-transporting proteins (GLUT4) (Lammi et al., 2015; Lammi et al., 2015). Studies conducted by *in vitro* and *in silico* experiments showed that IAVPTGVA peptide was a potent inhibitor of DPP-IV (dipeptidyl peptidase) and can also inhibit serine exopeptidases. DPP-IV is an important enzyme that plays a crucial role in the hydrolysis of glucagon and other insulinotropic enzymes in sustaining glucose homeostasis (Lammi et al., 2016). It is also evidenced that similar hypocholesterolemic function of the peptide might not confer by inhibiting the vital enzymes of the glucose metabolic pathway due to the difference in the peptide sequence, specifically the absence of proline at the specific site of the N-terminal (Lammi et al., 2016).

Administration of soy protein and isoflavones together have been associated with a positive impact in diabetic mouse models in reducing glucose, enhancing insulin level, and decreasing plasma glucose level at fasting. However, peptide-rich fermented soybean foods like natto have been shown to prevent the initiation of type 2 diabetes in human and rodent models (Kwon et al., 2010). Consumption of mixed protein diet (animal, soy, and other plant proteins) for six weeks by female participants aged 18–40 with gestational diabetes was linked with a notable enhancement of plasma glucose, serum insulin, and two homeostasis assessments such as insulin sensitivity and resistance index compared with a control group of the diet without soy protein (Martinez-Villaluenga et al., 2009).

Feeding of the βCG subunit of soy protein enhanced the muscle glucose absorption of plasma adipokinin, enhanced GLUT4 translocation, and increased the levels of AMPK (adenosine monophosphate protein kinase). Another similar peptide, soymorphin-5, which is a meu-opioid peptide obtained from the beta-subunit of βCG, decreased sugar and lipid profiles in diseased mice by triggering the activity of adipokinin and another receptor protein, PPARα. Application of low molecular peptides isolated by ultrafiltration and electrodialysis from mixed soy proteins enhanced glucose uptake in cultured rat muscle cell-based *in vitro* tests by Roblet et al. (2014). Furthermore, activation of AMPK has also been reported in the above study. Digestion of the resistant peptide of soybean, Vglycin (composed of 37 amino acids with six half-cysteines that are part of three pairs of disulfide bonds) has been shown to function as antidiabetic when administered in diseased Wister rats for four weeks and lowered fasting glucose levels to normal, enhanced insulin, reestablished insulin signaling, and improved pancreas function (Jiang et al., 2014). However, oleic acid–rich soybeans with high nutritional value for specific comparative studies have not been reported in the literature.

22.5.3 Antioxidative Peptides from Soybeans

Modification and optimization of soy peptides by various methods introduced several opportunities to the peptide industry to their multiple antioxidative benefits. For example, the free radical scavenging activity of proteins can be enhanced by approximately five times through enzyme treatment/digestion. Nwachukwu and Aluko reported that low molecular weight peptides exhibit significantly higher hydroxy radical scavenging activity after a higher degree of hydrolysis (Nwachukwu & Aluko, 2019). Other literature reports have also demonstrated evidence of increased antioxidant activity of fermented soy products such as soybean meal, soy food, and soy milk (Chatterjee et al., 2018; Shahidi & Zhong, 2010). The biofunctionality and health implications of soy peptides in reducing the risk of non-communicable diseases such as heart disease and cancer are linked with their action against oxidation in cell components, like genetic material, and reduction of inflammation by the suppression of specific triggering factors.

Most of the scientific evidence of antioxidant activity is obtained from soy protein hydrolysates that are not well characterized. Peptide functionality research needs to target specific molecular weight, size, and completely sequenced peptides. Table 22.2 lists the various production methods employed in the production and characterization of functionality – specifically, the antioxidant properties of hydrolysates and peptides from soy protein. Limited research on the bioactivity of high oleic acid peptides demonstrated significant results (Rayaprolu, 2015). However, the significant contribution has not been compared with lower oleic acid and other varieties of soybean seeds.

22.5.4 Anti-Hypertensive Soybean Peptides

Hypertension is a major risk factor for causing cardiovascular disease. Among all other functional peptides, antihypertensive peptides are most abundant in various common food sources (De Mejia & Ben, 2006; Singh et al., 2014). Severe acute respiratory syndrome (SARS) and coronavirus disease 2019 (COVID-19, also known as SARS-CoV-2) have a significant correlation with the angiotensin-converting enzyme (ACE), which is the key regulator of hypertension. Inactive angiotensinogen is converted to angiotensin I (AT I) by the enzyme renin, followed by the conversion of AT I to angiotensin II (AT II) by the ACE-1 converting enzyme. ACE I inhibitors are used to block the conversion of AT I to AT II to treat high blood pressure since AT II is a vasoconstrictor that causes high blood pressure. When the concentration of AT II is low, the angiotensin receptor 1 (ATR 1) can bind with ACE 2 and metabolize AT II to ACE 1, which is a vasodilator and thereby lowers the blood pressure. ACE 2 is also the key receptor of SARS-CoV-2. When the concentration of AT II is high, the ACE 2 receptor can bind to SARS-CoV-2, enabling it to enter the cells, replicate its genetic material, and multiply. Therefore, AT II cannot be metabolized to a vasodilator (ACE 1, 7) resulting in vasoconstriction, leading to an increase in blood pressure, pulmonary edema, and presumably acute respiratory disease syndrome (ARDS). A few studies with synthetic peptides have proven the effectiveness by coupling liposome and SARS coronavirus-specific cytotoxic T lymphocytes to confer viral clearance (Ohno et al., 2009). Furthermore, oxidative stress regulators play an important role in protecting from SARS-CoV-2 infection (Sharma et al., 2020).

The functional pathway of hypertensive peptides starts by blocking the angiotensin-converting enzyme that is associated with the renin-angiotensin system, hence controlling blood pressure (De

Mejia & Ben, 2006; Wang & De Mejia, 2005). The enzyme dipeptidyl carboxypeptidase of ACE alters the ACE1 into vasoconstrictor peptide angiotensin II (octapeptide), triggering blood pressure (De Mejia & Ben, 2006; Chappell, 2019; Yang et al., 2011). Fermented soy foods (soybean paste, natto, soy sauce, tempeh) rich in soy peptides are good sources of antihypertensive/ACE inhibitory peptides (Okamoto et al., 1995a; Okamoto et al., 1995b; Shin et al., 2001; Gibbs et al., 2004; Hernandez-Ledesma et al., 2004). Antihypertensive tripeptide (HHL) has been isolated from fermented soybean paste treated with the enzyme chymotrypsin. Soybean products fermented with the *Bacillus* species (*B. natto, B. subtilis*) were shown to produce two very potent antihypertensive peptides (VAHINVGZK and YVWK). The quantity and sources of the peptides from soybean products vary widely. It has been found that fermented seasoning from soybean has a greater ACE suppressing ability compared to sauce prepared from fermented soybean (Nakahara et al., 2010; Chatterjee et al., 2018). The greater ability was confirmed by administering peptides obtained from both soy seasoning and sauce in the same research. Dipeptides SY and GY isolated from soy seasoning showed ACE inhibition and decreased aldosterone level in blood serum in a salt-sensitive rat model (Dahl rat) (Nakahara et al., 2010). ACE inhibitory small antihypertensive peptides were also isolated from a by-product of tofu known as okara (Nishibori et al., 2017). Various antihypertensive peptides have been provided in Table 22.2. Soy BCG and two other peptides (LAIPVNKP and LPHF) were reported to contain antihypertensive activity while glycinin was a potent source of ACE inhibiting peptides, mainly the subunits of A4 and A5, which contained a characterized anti-hypertensive peptide with a sequence of NWGPLV. More peptide sequencing alignment and characterization revealed that some structural homology occurs among these ACE inhibiting peptides. The occurrence of Pro or OH-Pro at the C-terminal of the peptide imparts resistance to the digestive enzymes. On the other hand, the occurrence of Pro, Lys, or Arg at the C-terminal was attributed to their antihypertensive potential (Wang et al., 2008; Kuba et al., 2005). Two dipeptides with different C-terminal structures exhibited a different level of functionality. As an example, Tyr at the C-terminal of a peptide reportedly showed a greater antihypertensive potential compared to the one with phenylalanine (Erdmann et al., 2008).

The ultrafiltration of GI-resistant protein hydrolysates obtained from high oleic acid soybean lines (protein content range: 89–92%) using 5, 10, and 50 kDa molecular cut-offs columns provided fractions < 5, 5–10, 10–50, and > 50 kDa (yield: 1.8–2.1% d. b. by mass balance, derived from the isolate) for each soybean line (Rayaprolu, 2015). Initial screening showed comparatively poor ACE-I inhibition by the > 50 kDa fractions which were eliminated during further testing. The non-GI fractions showed overall low ACE-I inhibition when compared to the GI fractions, which was consistent with previous research on rice bran–derived peptides by Kannan (2009).

High oleic acid–rich soybean line peptides with a concentration of 500 µg/mL were used in a study, which showed an overall low activity (highest being: 48.9% by the 5–10 kDa fraction obtained from the R95-1705 soybean line) in comparison to the positive control, captopril (approximately 75% inhibition) (Rayaprolu, 2015). Peptides of both large and small sizes have been shown to be bioactive in previous studies and have exhibited useful functionalities for food products (Wu et al., 1998; Wu & Ding, 2002). The GI-resistant protein hydrolysates with anti-hypertensive activity could potentially be available for absorption through the intestine when consumed as food. Previous research has shown that peptides of various molecular sizes are absorbed through the intestinal wall but the ability of absorption decreases with an increase in molecular size (Roberts et al., 1998).

22.5.5 ANTICANCER SOYBEAN PEPTIDES

Carcinogenesis depends on many associated closely linked factors such as pro-inflammatory and pro-oxidant factors and immune suppression which led to abnormal proliferation of cancer cells. Proteins/peptides with anticancer activity often possess antioxidant and anti-inflammatory properties also (Lule et al., 2015; Roberts et al., 1998). Soy protein has long been known as an oxidative stress reducer and has the potential to inhibit nuclear factor kappa B (NF-kB) which promotes inflammation. It has also been shown to block cytokinins that promote pro-inflammation in a stress-induced rat model, hyperlipidemic rats, human subjects with terminal stage renal failure, and healthy women over 70 years of age (Draganidis et al., 2016). Smaller peptides obtained from soy milk after digestion with pepsin and pancreatin have been shown to reduce inflammation in murine macrophages. Protein hydrolysates derived from soy milk inhibited multiple oxidative stress-inducing enzymes such as interleukin and cyclooxygenase-2 (Dia et al., 2014). In this same study, protein hydrolysates were also shown to inhibit the production of nitric acid and nitric oxide synthase.

Research has shown that the anticancer property of soybean peptides is associated with a small 2S fraction of soybean-derived protein fractions: Lunasin and Bowman-Birk inhibitor (BBI). BBI can inhibit the enzymatic activity of digestive enzymes trypsin and chymotrypsin and has long been known as an anti-nutrient (Losso, 2008). Lunasin is a potent chemo-preventive peptide with 43 amino acid residues and a C-terminal with nine Asp residues. It has been associated with BBI and a cell adhesion motif called arginylglycylaspartic acid (RGD) that inhibits binding to non-acetylated H3 and H4 histones, which are known to impart anticarcinogenic functionality (Singh et al., 2014; Wang & De Meija, 2005). The BBI can protect lunasin from digestive gastrointestinal enzymes when consumed along with soy protein (Park et al., 2007). Topical administration of lunasin, as a treatment, in mice skin cancer cell lines (SENCAR) decreased tumor incidence by 70%. Lunasin is also known to inhibit colony suppression of mammalian cells induced with carcinogen by 30–40%. However, inconsistent results were reported on the efficacy of Lunasin in human breast cancer cell lines (MCF-7). A recent study reported that the bioactivity pathway of lunasin takes place by upregulating the tumor suppressor PTEN similar to the isoflavone genistein (Galvez et al., 2001; Lam et al., 2003; Jeong et al., 2003; De Mejia & Ben, 2006, Jeong et al., 2007; Wang et al., 2008). Another study also confirmed the ability of lunasin in inactivating different tumor suppressor proteins and can stop the transcription resulting in the arrest of the cell cycle G1/S phase, causing cellular apoptosis (Lule et al., 2015).

Peptide lunasin possesses anticancer activity that stems from the dual beneficial role as an anti-oxidative and anti-inflammatory reagent. Lunasin was reported to inhibit free radicals induced by 2,2'-Azino-bis sulfonic acid, ROS production, and also can suppress the release of cytokinins which are pro-oxidants in mouse RAW-254.7 macrophages (Hernandez-Ledesma et al., 2004; Lule et al., 2015). Lunasin can efficiently scavenge peroxyl and superoxide radicals, which can inhibit the activity of two important metabolic enzymes (glutathione peroxidase and catalase) (Lam et al., 2003). The terminal end of lunasin contains an RGD motif, which can efficiently block inflammation by interacting with integrin through the protein kinase B (Akt) mediated NF-kB pathway (Cam & De Mejia 2012). Oral ingestion of 50 g of soy-derived protein in healthy male individuals resulted in increased absorption of lunasin (about 4.5%) (Dia et al., 2009).

Many soy peptides have been isolated and are proven to have anticancer properties in multiple experimental model systems (De Mejia & Ben, 2006; Wang et al., 2008; Pabona et al., 2013; Singh et al., 2014; Lule et al., 2015). An important discovery in this context started with the purified hydrophobic peptide from high oleic acid, which stopped the cell cycle at the G2/M phase in murine macrophage cancer cells (P388D1) (Kim et al., 2000). Further studies conferred the mechanism of action involving apoptosis via ROS-induced damage in mitochondria because of suppression of proteasomal activity and angiogenesis (Tsai et al., 2020). The high oleic acid peptides derived from soybean lines (S03-543CR, N98-4445A) showed anticancer activity in several cancer cell lines. Gastrointestinal resistant fractionated peptide hydrolysates of molecular sizes of < 5 kDa, 5–10 kDa, and 10–50 kDa were tested against human colon (HCT-116, Caco-2), liver (HepG-2), and lung (NCL-H1299) cancer cell lines (Rayaprolu, 2015). A cytotoxicity assay using MTS, 3-(4,5-dimethyl thiazol-2-yl)-5-(3-carboxymethoxyphenyl)-2-(4-sulfophenyl)-2 H-tetrazolium was performed to test *in vitro* cancer cell viability upon treatment with peptide fractions. The peptide fractions from high oleic acid soybean lines showed cell growth inhibitions of 73% for colon cancer (HCT-116), 70% for liver cancer cells, and 68% for lung cancer cells. Dose-response analysis showed that the peptides had a significantly higher inhibitory effect at higher concentrations (600 µg/mL to 1,000 µg/mL). Reverse-phase HPLC identified three peptides from the 10–50 kDa fractions of a high oleic acid soy line, N98-4445A, that had the potential for enhanced activity (Rayaprolu, 2015). Research has shown that soybean peptide fractions can be an excellent source of bioactivity against colon, liver, and lung cancer cell proliferation (Rayaprolu et al., 2013).

22.5.6 IMMUNOMODULATORY PEPTIDES FROM SOYBEANS

The immunomodulatory function of peptides is closely linked with other beneficial health-promoting peptides (anticancer, antioxidant, and anti-inflammatory). These peptides can boost the function of immune cells and the activity of natural killer cells, and also regulates the activity of cytokinin (Singh et al., 2014). These peptides are also derived following the preparation of protein hydrolysate and then digested by enzymes. Hydrophobic insoluble soy protein digested with alcalase had greater efficacy in engulfing peritoneal macrophages (Kong et al., 2008; Singh et al., 2014). Two peptides derived from glycinin of soy protein hydrolysate had the capability to stimulate phagocytosis. Another peptide (MITLAIPVNKPGR) was also reported to be able to perform a similar function in leucocytes (Wang & De Mejia, 2005; Yoshikawa, 2015). This peptide was obtained from the alpha-subunit of beta-conglycinin, called soymetide, and later renamed as soymetide-13 with methionine at its N-terminal and responsible for its functionality (Wang & De Mejia, 2005; Yoshikawa, 2015).

22.5.7 NEUROMODULATORY PEPTIDES FROM SOYBEANS

Peptide sequencing has provided us with the information to identify similar peptide sequences from other sources for investigation purposes. An opioid-derived peptide that has morphine-like functionality has been discovered in the beta subunit of BCG. This has led to the discovery of three neuromodulatory peptide sequences (soymorphin-5, -6, -7) with anxiolytic functionality. These peptides were able to suppress food ingestion through a meu opioid receptor and inhibit intestinal food transit via coupling with neurotransmitters in a mice model (Kaneko et al., 2010). Ingestion of soymorphin-5 in diabetic mice model exhibited an increased level of sugar and triglyceride (Yamada et al., 2012). The action of these peptides does not require them to be absorbed into the bloodstream. Extracellular calcium receptors can mediate the functionality of these peptides and inhibit food intake or intestinal transit by mediating satiety and plasma cholecystokinin (Nishi et al., 2003a; Nishi et al., 2003b; Nakajima et al., 2010). There have been no studies reported with high oleic acid soybean peptides and neuromodulatory effects.

22.6 CHALLENGES

Processing and production procedures of peptides create special challenges and consequences. Protein hydrolysates and the individual peptides often fail to attribute palatable taste. A significant number of these products are bitter in taste, which limits their acceptability (Bumberger & Belitz, 1993; Maehashi & Huang, 2009). It has been reported that peptides with higher molecular weight can impart significantly higher bitterness compared to the low molecular weight analogs (Kim & Li-Chan, 2006). The molecular mechanism of bitterness and the regulatory mechanism are not fully understood. Hence, alteration rather than a preventive approach to reduce bitterness may be an option. Another important challenge in the preparation of bioactive peptides is their stability. Most consumable peptides are easily degraded in the gut and therefore do not show any activity when tested *in vivo*. Insertion of the structure at the end of the inducing probe and splicing the peptide sequence of the purified peptide with good activity can be stabilized to prevent digestion and solve this problem. This stabilization strategy can also improve the bioavailability of the consumed peptides. Absorption of intact peptides, either individually or as part of the protein hydrolysate is an active area of current research and is critical for successful oral delivery of these soy peptides in food products. Examination of the specific mechanism of the therapeutic approach is needed to move forward. Research and development should target palatability, digestion, location of action, and oral use of bioactive high oleic acid peptides and protein hydrolysates. Fermentation and processing can offer a number of unique challenges that require further effort. Safety and toxicity are the two major questions that require regulatory approvals. Efficient methods for synthesis, isolation, and characterization of these peptides at a larger scale and with a reliable profile and amino acid sequence using the latest technology including 2-D and 3-D NMR spectroscopy will add value to high oleic acid–derived peptides from soybeans (Martin et al., 2004; Haney & Vogel, 2009; Porcelli et al., 2013).

22.7 CONCLUSION

Soybean seeds are one of the most protein-rich food sources for human consumption. Hydrolysis of soybean proteins by enzymes occurring in digestion or fermentation can release an enormous amount of bioactive soy peptides whose activities span a broad spectrum of health benefits. Bioactive peptides derived from high oleic acid soybeans have the potential to be used in superfoods as ingredients for health and wellness. Purified bioactive soy peptides are reported to have antimicrobial, antioxidant, antidiabetic, antihypertensive, anticancer, and immunomodulatory properties. Hypertension is reported to be closely linked with the progression of COVID-19. Despite a significant amount of research on the isolation, purification, and assessment of the bioactivities of bioactive high oleic acid soy peptides, there are significant challenges to overcome: Enhancing the shelf-life stability, sensory acceptability, and mechanisms of bioactivities. Technological advancement to manufacture shelf-stable soybean high oleic acid peptides on a large scale without losing bioactivities is essential for application in "food for health" products, especially when currently there is a trend and demand for plant-based proteins.

REFERENCES

Aluko, R. E. (2012). Bioactive peptides. In: *Functional Foods and Nutraceuticals* (pp. 37–61). New York: Springer.

Bumberger, E., & Belitz, H. D. (1993). Bitter taste of enzymic hydrolysates of casein. Z. *Lebensm. Unters. Forsch*, 197(1), 14–19.

Cam, A., & de Mejia, E. G. (2012). Role of dietary proteins and peptides in cardiovascular disease. *Mol. Nutr. Food Res.*, 56(1), 53–66.

Chappell, M. C. (2019). The angiotensin-(1–7) axis: Formation and metabolism pathways. A comprehensive review. *Angiotensin* (1–7), 1–26. https://doi.org/10.1007/978-3-030-22696-1_1.

Chatterjee, C., Gleddie, S., & Xiao, C. W. (2018). Soybean bioactive peptides and their functional properties. *Nutrients*, 10(9), 1211.

Cho, S. J., Juillerat, M. A., & Lee, C. H. (2008). Identification of LDL-receptor transcription stimulating peptides from soybean hydrolysate in human hepatocytes. *J. Agric. Food Chem.*, 56(12), 4372–4376.

Choi, S. K., Adachi, M., & Utsumi, S. (2004). Improved bile acid-binding ability of soybean glycinin a1A polypeptide by the introduction of a bile acid-binding peptide (VAWWMY). *Biosci. Biotechnol. Biochem.*, 68(9), 1980–1983.

Clarke, E. J., & Wiseman, J. (2000). Developments in plant breeding for improved nutritional quality of soya beans I. Protein and amino acid content. *J. Agric. Sci.*, 134(2), 111–124.

De Angelis, E., Pilolli, R., Bavaro, S. L., & Monaci, L. (2017). Insight into the gastro-duodenal digestion resistance of soybean proteins and potential implications for residual immunogenicity. *Food Funct.*, 8(4), 1599–1610.

De Mejia, E., & Ben, O. (2006). Soybean bioactive peptides: A new horizon in preventing chronic diseases. *Sex. Reprod. Menopause*, 4(2), 91–95. ISSN 1546-2501.

De Oliveira, C. F., Corrêa, A. P. F., Coletto, D., Daroit, D. J., Cladera-Olivera, F., & Brandelli, A. (2015). Soy protein hydrolysis with microbial protease to improve antioxidant and functional properties. *J. Food Sci. Technol.*, 52(5), 2668–2678.

Dia, V. P., Bringe, N. A., & De Mejia, E. G. (2014). Peptides in pepsin–pancreatin hydrolysates from commercially available soy products that inhibit lipopolysaccharide-induced inflammation in macrophages. *Food Chem.*, 152, 423–431.

Dia, V. P., Torres, S., De Lumen, B. O., Erdman Jr, J. W., & Gonzalez De Mejia, E. (2009). Presence of lunasin in plasma of men after soy protein consumption. *J. Agric. Food Chem.*, 57(4), 1260–1266.

Draganidis, D., Karagounis, L. G., Athanailidis, I., Chatzinikolaou, A., Jamurtas, A. Z., & Fatouros, I. G. (2016). Inflammaging and skeletal muscle: Can protein intake make a difference? *J. Nutr.*, 146(10), 1940–1952.

Erdmann, K., Cheung, B. W., & Schröder, H. (2008). The possible roles of food-derived bioactive peptides in reducing the risk of cardiovascular disease. *J. Nutr. Biochem.*, 19(10), 643–654.

FDA. (2017). Food labeling: Health claims; soy protein and coronary heart disease https://www.fda.gov/media/108701/download (accessed 05/19/2021).

Galvez, A. F., Chen, N., Macasieb, J., & de Lumen, B. O. (2001). Chemopreventive property of a soybean peptide (lunasin) that binds to deacetylated histones and inhibits acetylation. *Cancer Res.*, 61(20), 7473–7478.

Gibbs, B. F., Zougman, A., Masse, R., & Mulligan, C. (2004). Production and characterization of bioactive peptides from soy hydrolysate and soy-fermented food. *Food Res. Int.*, 37(2), 123–131.

Goldflus, F., Ceccantini, M., & Santos, W. (2006). Amino acid content of soybean samples collected in different Brazilian states: Harvest 2003/2004. *Rev. Bras. Cienc. Avic.*, 8(2), 105–111.

Guo, J., Du, M., Tian, H., & Wang, B. (2020). Exposure to high salinity during seed development markedly enhances seedling emergence and fitness of the progeny of the extreme halophyte Suaeda salsa. *Front. Plant Sci.*, 11, 1291. https://doi.org/10.3389/fpls.2020.01291.

Haney, E. F., & Vogel, H. J. (2009). NMR of antimicrobial peptides. *Annu. Rep. NMR Spectrosc.*, 65, 1–51.

He, F. J., & Chen, J. Q. (2013). Consumption of soybean, soy foods, soy isoflavones and breast cancer incidence: Differences between Chinese women and women in Western countries and possible mechanisms. *Food Sci. Hum. Well*, 2(3–4), 146–161.

Hernández-Ledesma, B., Amigo, L., Ramos, M., & Recio, I. (2004). Angiotensin converting enzyme inhibitory activity in commercial fermented products. Formation of peptides under simulated gastrointestinal digestion. *J. Agric. Food Chem.*, 52(6), 1504–1510.

Imsande, J. (2001). Selection of soybean mutants with increased concentrations of seed methionine and cysteine. *Crop Sci.*, 41(2), 510–515.

Jeong, H. J., Jeong, J. B., Kim, D. S., & de Lumen, B. O. (2007). Inhibition of core histone acetylation by the cancer preventive peptide lunasin. *J. Agric. Food Chem.*, 55(3), 632–637.

Jeong, H. J., Park, J. H., Lam, Y., & de Lumen, B. O. (2003). Characterization of lunasin isolated from soybean. *J. Agric. Food Chem.*, 51(27), 7901–7906.

Jiang, H., Feng, J., Du, Z., Zhen, H., Lin, M., Jia, S., Li, T., Huang, X., Ostenson, C. G., & Chen, Z. (2014). Oral administration of soybean peptide V glycin normalizes fasting glucose and restores impaired pancreatic function in Type 2 diabetic Wistar rats. *J. Nutr. Biochem.*, 25(9), 954–963.

Kaneko, K., Iwasaki, M., Yoshikawa, M., & Ohinata, K. (2010). Orally administered soymorphins, soy-derived opioid peptides, suppress feeding and intestinal transit via gut μ1-receptor coupled to 5-HT1A, D2, and GABAB systems. *Am. J. Physiol. Gastrointest. Liver Physiol.*, 299(3), G799–G805.

Kannan, A. (2009). Preparation, separation, purification, characterization and human cell line anti-cancer evaluation of rice bran peptides. Theses and Dissertations. 181.

Kannan, A., Hettiarachchy, N., Johnson, M. G., & Nannapaneni, R. (2008). Human colon and liver cancer cell proliferation inhibition by peptide hydrolysates derived from heat-stabilized defatted rice bran. *J. Agric. Food Chem.*, 56(24), 11643–11647.

Kim, H. O., & Li-Chan, E. C. (2006). Quantitative structure– activity relationship study of bitter peptides. *J. Agric. Food Chem.*, 54(26), 10102–10111.

Kim, S. E., Kim, H. H., Kim, J. Y., Kang, Y. I., Woo, H. J., & Lee, H. J. (2000). Anticancer activity of hydrophobic peptides from soy proteins. *BioFactors*, 12(1–4), 151–155. https://doi.org/10.1002/biof.5520120124.

Kong, X., Guo, M., Hua, Y., Cao, D., & Zhang, C. (2008). Enzymatic preparation of immunomodulating hydrolysates from soy proteins. *Bioresour. Technol.*, 99(18), 8873–8879.

Korhonen, H., Pihlanto-Leppälä, A., Rantamäki, P., & Tupasela, T. (1998). Impact of processing on bioactive proteins and peptides. *Trends Food Sci. Technol.*, 9(8–9), 307–319.

Kuba, M., Tana, C., Tawata, S., & Yasuda, M. (2005). Production of angiotensin I-converting enzyme inhibitory peptides from soybean protein with Monascus purpureus acid proteinase. *Process Biochem.*, 40(6), 2191–2196.

Kwon, D. Y., Daily III, J. W., Kim, H. J., & Park, S. (2010). Antidiabetic effects of fermented soybean products on type 2 diabetes. *Nutr. Res.*, 30(1), 1–13. https://doi.org/10.1016/j.nutres.2009.11.004.

Lam, Y., Galvez, A., & de Lumen, B. O. (2003). Lunasin suppresses E1A-mediated transformation of mammalian cells but does not inhibit growth of immortalized and established cancer cell lines. *Nutr. Cancer*, 47(1), 88–94.

Lammi, C., Zanoni, C., & Arnoldi, A. (2015). IAVPGEVA, IAVPTGVA, and LPYP, three peptides from soy glycinin, modulate cholesterol metabolism in HepG2 cells through the activation of the LDLR-SREBP2 pathway. *J. Funct. Foods*, 14, 469–478.

Lammi, C., Zanoni, C., & Arnoldi, A. (2015). Three peptides from soy glycinin modulate glucose metabolism in human hepatic HepG2 cells. *Int. J. Mol. Sci.*, 16(11), 27362–27370.

Lammi, C., Zanoni, C., Arnoldi, A., & Vistoli, G. (2016). Peptides derived from soy and lupin protein as dipeptidyl-peptidase IV inhibitors: *In vitro* screening and in silico. *Molecular Modelling* 2. Study 3. In *Congresso Italiano di Chimica degli Alimenti* (pp. 105–105). Società chimica italiana.

Lee, S., Lee, S., Singh, D., Oh, J. Y., Jeon, E. J., Ryu, H. S., Lee, D. W., Kim, B. S., & Lee, C. H. (2017). Comparative evaluation of microbial diversity and metabolite profiles in doenjang, a fermented soybean paste, during the two different industrial manufacturing processes. *Food Chem.*, 221, 1578–1586.

Li, G., Long, X., Pan, Y., Zhao, X., & Song, J. L. (2018). Study on soybean milk fermented by *Lactobacillus plantarum* YS-1 reduced the H2O2-induced oxidative damage in Caco-2 cells. *Biomed. Res.*, 29(2), 357–364.

Losso, J. N. (2008). The biochemical and functional food properties of the Bowman-Birk inhibitor. *Crit. Rev. Food Sci. Nutr.*, 48(1), 94–118.

Lule, V. K., Garg, S., Pophaly, S. D., & Tomar, S. K. (2015). Potential health benefits of lunasin: A multifaceted soy-derived bioactive peptide. *J. Food Sci.*, 80(3), R485–R494.

Ma, A., & Ooraikul, B. (1986). Optimization of enzymatic hydrolysis of canola meal with response surface methodology. *J. Food Process. Pres.*, 10(2), 99–113. https://doi.org/10.1111/j.1745-4549.1986.tb00010.x.

Ma, H., Liu, R., Zhao, Z., Zhang, Z., Cao, Y., Ma, Y., & Xu, L. (2016). A novel peptide from soybean protein isolate significantly enhances resistance of the organism under oxidative stress. *PLOS ONE*, 11(7), e0159938.

Maebuchi, M., Samoto, M., Kohno, M., Ito, R., Koikeda, T., Hirotsuka, M., & Nakabou, Y. (2007). Improvement in the intestinal absorption of soy protein by enzymatic digestion to oligopeptide in healthy adult men. *Food Sci.*, 13(1), 45–53.

Maehashi, K., & Huang, L. (2009). Bitter peptides and bitter taste receptors. *Cell. Mol. Life Sci.*, 66(10), 1661–1671. https://doi.org/10.1007/s00018-009-8755-9..

Marcela, G. M., Eva, R. G., Del Carmen, R. R. M., & Rosalva, M. E. (2016). Evaluation of the antioxidant and antiproliferative effects of three peptide fractions of germinated soybeans on breast and cervical cancer cell lines. *Plant Foods Hum. Nutr.*, 71(4), 368–374.

Martin, N. I., Sprules, T., Carpenter, M. R., Cotter, P. D., Hill, C., Ross, R. P., & Vederas, J. C. (2004). Structural characterization of lacticin 3147, a two-peptide lantibiotic with synergistic activity. *Biochemistry*, 43(11), 3049–3056.

Martinez-Villaluenga, C., Bringe, N. A., Berhow, M. A., & Gonzalez de Mejia, E. (2008). β-Conglycinin embeds active peptides that inhibit lipid accumulation in 3T3-L1 adipocytes *in vitro*. *J. Agric. Food Chem.*, 56(22), 10533–10543.

Martinez-Villaluenga, C., Dia, V. P., Berhow, M., Bringe, N. A., & Gonzalez de Mejia, E. (2009). Protein hydrolysates from β-conglycinin enriched soybean genotypes inhibit lipid accumulation and inflammation *in vitro*. *Mol. Nutr. Food Res.*, 53(8), 1007–1018.

Martinez-Villaluenga, C., Rupasinghe, S. G., Schuler, M. A., & Gonzalez de Mejia, E. (2010). Peptides from purified soybean β-conglycinin inhibit fatty acid synthase by interaction with the thioesterase catalytic domain. *FEBS J.*, 277(6), 1481–1493.

Moure, A., Domínguez, H., & Parajó, J. C. (2006). Antioxidant properties of ultrafiltration-recovered soy protein fractions from industrial effluents and their hydrolysates. *Process Biochem.*, 41(2), 447–456.

Nagaoka, S., Atsushi, N., Haruhiko, S., & Yoshihiro, K. (2010). Soystatin (VAWWMY), a novel bile acid-binding peptide, decreased micellar solubility and inhibited cholesterol absorption in rats. *Biosci. Biotechnol. Biochem.*, 74(8), 1738–1741.

Nakahara, T., Sano, A., Yamaguchi, H., Sugimoto, K., Chikata, H., Kinoshita, E., & Uchida, R. (2010). Antihypertensive effect of peptide-enriched soy sauce-like seasoning and identification of its angiotensin I-converting enzyme inhibitory substances. *J. Agric. Food Chem.*, 58(2), 821–827.

Nakahara, T., Sugimoto, K., Sano, A., Yamaguchi, H., Katayama, H., & Uchida, R. (2011). Antihypertensive mechanism of a peptide-enriched soy sauce-like seasoning: The active constituents and its suppressive effect on renin–angiotensin–aldosterone system. *J. Food Sci.*, 76(8), H201–H206.

Nakajima, S., Hira, T., Eto, Y., Asano, K., & Hara, H. (2010). Soybean β51–63 peptide stimulates cholecystokinin secretion via a calcium-sensing receptor in enteroendocrine STC-1 cells. *Regul. Pept.*, 159(1–3), 148–155.

Nishi, T., Hara, H., Asano, K., & Tomita, F. (2003a). The soybean β-conglycinin β 51–63 fragment suppresses appetite by stimulating cholecystokinin release in rats. *J. Nutr.*, 133(8), 2537–2542.

Nishi, T., Hara, H., & Tomita, F. (2003b). Soybean β-conglycinin peptone suppresses food intake and gastric emptying by increasing plasma cholecystokinin levels in rats. *J. Nutr.*, 133(2), 352–357.

Nishibori, N., Kishibuchi, R., & Morita, K. (2017). Soy pulp extract inhibits angiotensin I-converting enzyme (ACE) activity *in vitro*: Evidence for its potential hypertension-improving action. *J. Diet. Suppl.*, 14(3), 241–251.

Nwachukwu, I. D., & Aluko, R. E. (2019). Structural and functional properties of food protein-derived antioxidant peptides. *J. Food Biochem.*, 43(1), e12761.

Ohinata, K., Agui, S., & Yoshikawa, M. (2007). Soymorphins, novel μ opioid peptides derived from soy β-conglycinin β-subunit, have anxiolytic activities. *Biosci. Biotechnol. Biochem.*, 71(10), 2618–2621.

Ohno, S., Kohyama, S., Taneichi, M., Moriya, O., Hayashi, H., Oda, H., Mori, M., Kobayashi, A., Akatsuka, T., Uchida, T., & Matsui, M. (2009). Synthetic peptides coupled to the surface of liposomes effectively induce SARS coronavirus-specific cytotoxic T lymphocytes and viral clearance in HLA-A* 0201 transgenic mice. *Vaccine*, 27(29), 3912–3920.

Okamoto, A., Hanagata, H., Kawamura, Y., & Yanagida, F. (1995a). Anti-hypertensive substances in fermented soybean, natto. *Plant Foods Hum. Nutr.*, 47(1), 39–47.

Okamoto, A., Hanagata, H., Matsumoto, E., Kawamura, Y., Koizumi, Y., & Yanagida, F. (1995b). Angiotensin I converting enzyme inhibitory activities of various fermented foods. *Biosci. Biotechnol. Biochem.*, 59(6), 1147–1149.

Omoni, A. O., & Aluko, R. E. (2005). Soybean foods and their benefits: Potential mechanisms of action. *Nutr. Rev.*, 63(8), 272–283.

Pabona, J. M. P., Dave, B., Su, Y., Montales, M. T. E., Ben, O., De Mejia, E. G., Rahal, O. M., & Simmen, R. C. (2013). The soybean peptide lunasin promotes apoptosis of mammary epithelial cells via induction of tumor suppressor PTEN: Similarities and distinct actions from soy isoflavone genistein. *Genes Nutr.*, 8(1), 79–90.

Pak, V. V., Koo, M. S., Kasymova, T. D., & Kwon, D. Y. (2005). Isolation and identification of peptides from soy 11S-globulin with hypocholesterolemic activity. *Chem. Nat. Comp.*, 41(6), 710–714.

Pak, V. V., Koo, M., Kwon, D. Y., & Yun, L. (2012). Design of a highly potent inhibitory peptide acting as a competitive inhibitor of HMG-CoA reductase. *Amino Acids*, 43(5), 2015–2025.

Panthee, D. R., Pantalone, V. R., Sams, C. E., Saxton, A. M., West, D. R., Orf, J. H., & Killam, A. S. (2006). Quantitative trait loci controlling sulfur containing amino acids, methionine and cysteine, in soybean seeds. *Theor. Appl. Genet.*, 112(3), 546–553.

Park, J. H., Jeong, H. J., & Lumen, B. O. D. (2007). *In vitro* digestibility of the cancer-preventive soy peptides lunasin and BBI. *J. Agric. Food Chem.*, 55(26), 10703–10706.

Porcelli, F., Ramamoorthy, A., Barany, G., & Veglia, G. (2013). On the role of NMR spectroscopy for characterization of antimicrobial peptides. *Membr. Proteins*, 159–180.

Poysa, V., Woodrow, L., & Yu, K. (2006). Effect of soy protein subunit composition on tofu quality. *Food Res. Int.*, 39(3), 309–317.

Rayaprolu, S. J. (2015). Extraction, purification and characterization of a pure peptide from soybean to demonstrate anti-proliferation activity on human cancer cells and test the ability of soy peptide fractions in reducing the activity of angiotensin-I converting enzyme. Theses and dissertations Retrieved from https://scholarworks.uark.edu/etd/1343, p. 164.

Rayaprolu, S. J., Hettiarachchy, N. S., Chen, P., Kannan, A., & Mauromostakos, A. (2013). Peptides derived from high oleic acid soybean meals inhibit colon, liver and lung cancer cell growth. *Food Res. Int.*, 50(1), 282–288.

Rayaprolu, S. J., Hettiarachchy, N. S., Horax, R., Kumar-Phillips, G., Liyanage, R., Lay, J., & Chen, P. (2017). Purification and characterization of a peptide from soybean with cancer cell proliferation inhibition. *J. Food Biochem.*, 41(4), e12374. https://doi.org/10.1111/jfbc.12374.

Roberts, P. R., Latg, K. W. B., Santamauro, J. T., & Zaloga, G. P. (1998). Dietary peptides improve wound healing following surgery. *Nutrition*, 14(3), 266–269.

Roblet, C., Doyen, A., Amiot, J., Pilon, G., Marette, A., & Bazinet, L. (2014). Enhancement of glucose uptake in muscular cell by soybean charged peptides isolated by electrodialysis with ultrafiltration membranes (EDUF): Activation of the AMPK pathway. *Food Chem.*, 147, 124–130.

Sami, R. (2017). Antioxidant properties of peptides from soybean meal protein hydrolysates evaluated by electron spin resonance spectrometry. *Adv. Environ. Biol.*, 11(4), 12–18.

Shahidi, F., & Zhong, Y. (2010). Novel antioxidants in food quality preservation and health promotion. *Eur. J. Lipid Sci. Technol.*, 112(9), 930–940.

Sharma, A., Garcia Jr, G., Wang, Y., Plummer, J. T., Morizono, K., Arumugaswami, V., & Svendsen, C. N. (2020). Human iPSC-derived cardiomyocytes are susceptible to SARS-CoV-2 infection. *Cell Rep.*, 1(4), 100052.

Shin, Z. I., Yu, R., Park, S. A., Chung, D. K., Ahn, C. W., Nam, H. S., Kim, K. S., & Lee, H. J. (2001). His-His-Leu, an angiotensin I converting enzyme inhibitory peptide derived from Korean soybean paste, exerts antihypertensive activity *in vivo*. *J. Agric. Food Chem.*, 49(6), 3004–3009.

Singer, W. M., Zhang, B., Mian, M. A., & Huan, H. (2019). Soybean amino acids in health, genetics, and evaluation. In: *Soybean for Human Consumption and Animal Feed*. IntechOpen.

Singh, B., Aluko, R. E., Hati, S., & Solanki, D. (2021). Bioactive peptides in the management of lifestyle-related diseases: Current trends and future perspectives. *Crit. Rev. Food Sci. Nutr.*, 28(1–14). https://doi .org/10.1080/10408398.2021.1877109.

Singh, B. P., Bhushan, B., & Vij, S. (2020). Antioxidative, ACE inhibitory and antibacterial activities of soy milk fermented by indigenous strains of lactobacilli. *Legum*, 2(4), e54. https://doi.org/10.1002/leg3.54.

Singh, B. P., Vij, S., & Hati, S. (2014). Functional significance of bioactive peptides derived from soybean. *Peptides*, 54, 171–179.

Soybean production. (2021). https://www.statista.com/statistics/263926/soybean-production-in-selected -countries-since-1980/ (accessed 05/19/2021).

Subrota, H., Shilpa, V., Brij, S., Vandna, K., & Surajit, M. (2013). Antioxidative activity and polyphenol content in fermented soy milk supplemented with WPC-70 by probiotic *Lactobacilli*. *Int. J. Food Res.*, 20(5), 2125.

Tomatsu, M., Shimakage, A., Shinbo, M., Yamada, S., & Takahashi, S. (2013). Novel angiotensin I-converting enzyme inhibitory peptides derived from soya milk. *Food Chem.*, 136(2), 612–616.

Tsai, C.-K. B., Kuo, W., Day, C. H., Hsieh, D. J., Kuo, C., Daddam, J., Chen, R., Vijaya Padma, V., Wang, G., & Huang, C. (2020). The soybean bioactive peptide VHVV alleviates hypertension-induced renal damage in hypertensive rats via the SIRT1-PGC1α/Nrf2 pathway. *J. Funct. Foods*, 75(104255). ISSN 1756-4646.

Tsuruki, T., & Yoshikawa, M. (2004). Design of soymetide-4 derivatives to potentiate the anti-alopecia effect. *Biosci. Biotechnol. Biochem.*, 68(5), 1139–1141.

Tsuruki, T., Kishi, K., Takahashi, M., Tanaka, M., Matsukawa, T., & Yoshikawa, M. (2003). Soymetide, an immunostimulating peptide derived from soybean β-conglycinin, is an fMLP agonist. *FEBS Lett.*, 540(1–3), 206–210.

USDA. (1986). Soybeans, mature seeds, raw. Retrieved from https://fdc.nal.usda.gov/fdc-app.html#/food -details/174270/nutrients (accessed 05/19/2021).

Valliyodan, B., Qiu, D., Patil, G., Zeng, P., Huang, J., Dai, L., & Nguyen, H. T. (2016). Landscape of genomic diversity and trait discovery in soybean. *Sci. Rep.*, 6(1), 1–10.

Velasquez, M. T., & Bhathena, S. J. (2007). Role of dietary soy protein in obesity. *Int. J. Med. Sci.*, 4(2), 72.

Wang, W., & De Mejia, E. G. (2005). A new frontier in soy bioactive peptides that may prevent age-related chronic diseases. *CRFSFS*, 4(4), 63–78.

Wang, W., Dia, V. P., Vasconez, M., De Mejia, E. G., & Nelson, R. L. (2008). Analysis of soybean protein-derived peptides and the effect of cultivar, environmental conditions, and processing on lunasin concentration in soybean and soy products. *J. AOAC Int.*, 91(4), 936–946.

Wang, Y. C., Yu, R. C., & Chou, C. C. (2006). Antioxidative activities of soymilk fermented with lactic acid bacteria and bifidobacteria. *Food Microbiol.*, 23(2), 128–135.

Wu, J., & Ding, X. (2002). Characterization of inhibition and stability of soy-protein-derived angiotensin I-converting enzyme inhibitory peptides. *Food Res. Int.*, 35(4), 367–375.

Wu, W. U., Hettiarachchy, N. S., & Qi, M. (1998). Hydrophobicity, solubility, and emulsifying properties of soy protein peptides prepared by papain modification and ultrafiltration. *J. Am. Oil Chem. Soc.*, 75(7), 845–850.

Xiao, C. W. (2008). Health effects of soy protein and isoflavones in humans. *J. Nutr.*, 138(6), 1244S–1249S.

Yamada, Y., Muraki, A., Oie, M., Kanegawa, N., Oda, A., Sawashi, Y., & Ohinata, K. (2012). Soymorphin-5, a soy-derived μ-opioid peptide, decreases glucose and triglyceride levels through activating adiponectin and PPARα systems in diabetic KKAy mice. *Am. J. Physiol. Endocrinol. Metab.*, 302(4), E433–E440.

Yang, R., Smolders, I., & Dupont, A. G. (2011). Blood pressure and renal hemodynamic effects of angiotensin fragments. *Hypertens. Res.*, 34(6), 674–683. https://doi.org/10.1038/hr.2011.24.

Yoshikawa, M. (2015). Bioactive peptides derived from natural proteins with respect to diversity of their receptors and physiological effects. *Peptides*, 72, 208–225.

Yoshikawa, M., Fujita, H., Matoba, N., Takenaka, Y., Yamamoto, T., Yamauchi, R., & Takahata, K. (2000). Bioactive peptides derived from food proteins preventing lifestyle-related diseases. *Biofactors*, 12(1–4), 143–146.

Yoshikawa, M. K., Takahashi, M., Watanabe, A., Miyamura, T., Yamazaki, M., & Chiba, H. (1993). Immunostimulating peptide derived from soybean protein. *Ann. N. Y. Acad. Sci.*, 685, 375–376.

Zarkadas, C. G., Gagnon, C., Poysa, V., Khanizadeh, S., Cober, E. R., Chang, V., & Gleddie, S. (2007). Protein quality and identification of the storage protein subunits of tofu and null soybean genotypes, using amino acid analysis, one-and two-dimensional gel electrophoresis, and tandem mass spectrometry. *Food Res. Int.*, 40(1), 111–128.

Zhang, Q., Tong, X., Qi, B., Wang, Z., Li, Y., Sui, X., & Jiang, L. (2018). Changes in antioxidant activity of alcalase-hydrolyzed soybean hydrolysate under simulated gastrointestinal digestion and transepithelial transport. *J. Funct. Foods*, 42, 298–305.

Index